Lignin

This book covers the current know-how on lignin, its sustainable isolation from lignocellulose, characterization of the isolated lignin, and transforming it to add value. The book is divided into three sections. A section on lignin isolation elaborates the well-established commercial delignification methods, including recent developments. The lignin characterization section includes traditional methods based on wet chemistry and advanced characterization techniques. It covers lignin value addition, with the possibility of transforming lignin into functional materials, backed by figures and graphical presentations.

Features:

- Provides exclusive information on lignin isolation, including conventional processes as well as recent advances,
- Covers reaction mechanisms, kinetics, mass transfer, and unit operations, including all aspects of laboratory-based lignin characterization,
- Focuses on lignin as a starting material for obtaining functional materials,
- Discusses the current state of advancement of lignin value addition.

This book is aimed at researchers, professionals, and graduate students in chemical engineering and industrial chemistry with particular interest in lignocelluloses, pulp, paper, biorefineries, and sustainable materials.

AF728912

Lignin
Isolation, Characterization, and Value Addition

Himadri Roy Ghatak

CRC Press
Taylor & Francis Group
Boca Raton London New York

CRC Press is an imprint of the
Taylor & Francis Group, an **informa** business

Designed cover image: Himadri Roy Ghatak

First edition published 2025
by CRC Press
2385 NW Executive Center Drive, Suite 320, Boca Raton FL 33431

and by CRC Press
4 Park Square, Milton Park, Abingdon, Oxon, OX14 4RN

CRC Press is an imprint of Taylor & Francis Group, LLC

© 2025 Himadri Roy Ghatak

Reasonable efforts have been made to publish reliable data and information, but the author and publisher cannot assume responsibility for the validity of all materials or the consequences of their use. The authors and publishers have attempted to trace the copyright holders of all material reproduced in this publication and apologize to copyright holders if permission to publish in this form has not been obtained. If any copyright material has not been acknowledged please write and let us know so we may rectify in any future reprint.

Except as permitted under U.S. Copyright Law, no part of this book may be reprinted, reproduced, transmitted, or utilized in any form by any electronic, mechanical, or other means, now known or hereafter invented, including photocopying, microfilming, and recording, or in any information storage or retrieval system, without written permission from the publishers.

For permission to photocopy or use material electronically from this work, access www.copyright.com or contact the Copyright Clearance Center, Inc. (CCC), 222 Rosewood Drive, Danvers, MA 01923, 978-750-8400. For works that are not available on CCC please contact mpkbookspermissions@tandf.co.uk

Trademark notice: Product or corporate names may be trademarks or registered trademarks and are used only for identification and explanation without intent to infringe.

ISBN: 978-1-032-86954-4 (hbk)
ISBN: 978-1-032-98442-1 (pbk)
ISBN: 978-1-003-59866-4 (ebk)

DOI: 10.1201/9781003598664

Typeset in Times
by SPi Technologies India Pvt Ltd (Straive)

Contents

Preface.....ix
About the Author.....xi

Chapter 1 Introduction.....1

1.1 Need to Find an Alternative to Fossils.....1
1.2 Comprehensive Utilization of Lignocellulose.....2
1.3 Historical Perspective of Lignin.....3
1.4 Lignin Availability, Potential, Market, and Value-Added Applications.....6
1.4.1 Kraft Lignin.....6
1.4.2 Soda Lignin.....7
1.4.3 Lignosulphonate.....7
1.4.4 Organosolv Lignin.....8
1.4.5 Hydrolysis Lignin.....8
1.4.6 Some Commercialized Value-Added Lignin Applications.....8
References.....10

SECTION 1 Lignin Isolation

Chapter 2 Lignin Isolation through Dissolution from Lignocelluloses.....13

2.1 Dissolution in Alkaline Media.....13
2.1.1 Chemistry of Alkaline Delignification.....15
2.1.2 Process Parameters in Lignin Dissolution in Alkaline Media.....18
2.2 Dissolution in Acidic Media.....27
2.2.1 Chemistry of Acidic Delignification.....27
2.2.2 Process Parameters in Lignin Dissolution in Acidic Media.....31
2.3 Dissolution in Organic Media.....38
2.3.1 Chemistry of Organosolv Delignification.....38
2.3.2 Process Parameters in Organosolv Delignification.....39
References.....60

Chapter 3 Separation of Dissolved Lignin from Solution.....64

3.1 Separation of Dissolved Lignin from Alkaline Spent Liquids.....64
3.1.1 Black Liquor Characteristics from Alkaline Delignification Processes.....65
3.1.2 Lignin Separation by Acid Precipitation.....68
3.1.3 Lignin Separation by Membrane Filtration.....77
3.2 Separation of Dissolved Lignin from Acidic Spent Liquors.....83
3.2.1 Spent Liquor Characteristics from Acidic Delignification Processes.....83
3.2.2 Monosaccharides Management of Spent Liquors.....83
3.2.3 Lignosulphonates through Membrane Separation.....86
3.2.4 Lignosulphonates through Solvent Extraction.....88

3.2.5 Lignosulphonates through Ion Exchange 90
3.2.6 Drying of Lignosulphonates 91
3.3 Separation of Dissolved Lignin from Organosolv Spent Liquors 91
3.3.1 Spent Liquor Characteristics from Organosolv Delignification Processes 92
3.3.2 Lignin Separation by Solvent Evaporation 92
3.3.3 Lignin Precipitation by Antisolvent Addition 93
3.3.4 Lignin Separation by Membrane Filtration 97
References 97

Chapter 4 Other Methods of Lignin Isolation 101

4.1 Isolation through Carbohydrate Hydrolysis from Lignocelluloses 101
4.1.1 Dilute Acid Hydrolysis of Lignocelluloses 102
4.1.2 Concentrated Acid Hydrolysis of Lignocelluloses 102
4.1.3 Enzymatic Hydrolysis of Lignocelluloses 104
4.1.4 Subcritical and Supercritical Fluid Hydrolysis of Lignocelluloses 106
4.2 Milled wood Lignin 108
References 110

SECTION 2 Lignin Characterization

Chapter 5 Physicochemical Characterization of Lignin 115

5.1 Elemental Analysis of Lignin 115
5.2 Molecular Structure of Lignin 115
5.2.1 Functional Groups in Lignin 117
5.2.2 Intermonomer Linkages in Lignin 124
5.2.3 Lignin Carbohydrate Complexes 129
5.3 Molecular Weight and Molecular Weight Distribution in Lignin 130
5.3.1 Gel Permeation Chromatography 131
5.3.2 Vapour Pressure Osmometry 137
5.3.3 LASER Light Scattering 138
5.4 Solution Properties of Lignin 143
5.4.1 Viscosity of Lignin Solutions 143
References 145

Chapter 6 Spectroscopic Characterization of Lignin 148

6.1 Lignin Characterization by UV Spectroscopy 148
6.1.1 UV Spectroscopy for Determination of Concentration of Lignin Solution 148
6.1.2 Lignin Structural Elucidation from UV Spectrum 150
6.2 Lignin Characterization by Fourier Transform Infrared Spectroscopy 153
6.2.1 Lignin Structural Elucidation from FTIR Spectrum 154
6.3 Lignin Characterization by Raman Spectroscopy 166
6.4 Lignin Characterization by Nuclear Magnetic Resonance Spectroscopy 169
6.4.1 Lignin Characterization by ^{1}H NMR Spectroscopy 170

6.4.2 Lignin Characterization by ^{13}C NMR Spectroscopy 178
6.4.3 Lignin Characterization by ^{31}P NMR Spectroscopy 186
6.4.4 Lignin Characterization by Heteronuclear 2D NMR Spectroscopy 191
References 198

Chapter 7 Thermochemical Characterization of Lignin 201

7.1 Lignin Characterization by Thermogravimetry 201
7.1.1 Soda Lignin 202
7.1.2 Kraft Lignin 204
7.1.3 Organosolv Lignin 204
7.1.4 Milled Wood Lignin 207
7.1.5 Hydrolysis Lignin 207
7.1.6 Lignin Characterization by TG-FTIR 209
7.1.7 Kinetic Interpretation from Thermogravimetry 212
7.2 Lignin Characterization by Differential Thermal Analysis 215
7.2.1 Alkali Lignin 215
7.2.2 Milled Wood Lignin 217
7.3 Lignin Characterization by Differential Scanning Calorimetry 217
7.3.1 Lignin from Alkaline Delignification Processes 217
7.3.2 Organosolv Lignin 218
7.3.3 Hydrolysis Lignin 219
7.3.4 Milled Wood Lignin 219
7.4 Lignin Characterization by Pyrolysis and Product Identification and Quantification 219
7.4.1 Lignin Characterization by Py-GC-MS 221
7.5 Comparison of Characterization Methods, Lignin Source, and Isolation Methods 228
References 241

SECTION 3 Lignin Value Addition

Chapter 8 Value-Added Carbons from Lignin 247

8.1 Biochar and Activated Carbon from Lignin 247
8.1.1 Biochar from Lignin 248
8.1.2 Activated Carbon from Lignin 250
8.2 Carbon Fibres from Lignin 256
8.2.1 Spinning of Lignin-Based Carbon Fibres 257
8.2.2 Carbonization of Spun Fibres and Fibre Characteristics 260
8.3 Carbon Nanotubes Modified with Lignin 263
8.3.1 Lignin as Dispersing Agent for Carbon Nanotubes 265
8.3.2 Composites from Lignin and Carbon Nanotubes 265
8.3.3 Lignin-Based Carbon Fibres Incorporated with Carbon Nanotubes 266

8.3.4 Lignin as Carbon Precursor for Carbon Nanotubes 268
8.4 Other Lignin-Based Carbon Products 269
8.5 Lignin-Based Carbon Materials in Energy Storage and Electrochemical Applications 270
8.5.1 Anode in Rechargeable Cells 270
8.5.2 Capacitive Energy Storage 273
References 275

Chapter 9 Value-Added Polymer Applications of Lignin 277

9.1 Lignin in Adhesives and Binder Resins 277
9.1.1 Lignin-Based Phenolic Resins 277
9.1.2 Lignin-Based Epoxy Resins 283
9.2 Lignin-Based Polyurethanes 292
9.2.1 Soda and Kraft Lignin 294
9.3 Lignin Blended with Other Polymers 302
9.3.1 Soda and Kraft Lignin 303
9.3.2 Organosolv Lignin 306
9.3.3 Hydrolysis Lignin 307
9.3.4 Lignosulphonate 309
References 312

Chapter 10 Value-Added Low Molecular Weight Organics from Lignin 314

10.1 Catalytic Oxidation of Lignin with Molecular Oxygen 314
10.1.1 Reaction Mechanism and Kinetics 315
10.1.2 Yield and Product Profile 317
10.1.3 Catalysts in Lignin Oxidation with Molecular Oxygen 323
10.2 Lignin Pyrolysis to Obtain Low Molecular Weight Organics 327
10.2.1 Non-Catalytic Pyrolysis of Lignin 328
10.2.2 Reactor Configurations and Operational Issues 333
10.2.3 Catalytic Pyrolysis of Lignin 336
10.2.4 Catalysts in Lignin Pyrolysis 342
10.3 Advanced Oxidation of Lignin to Obtain Low Molecular Weight Organics 346
10.3.1 Electrochemical Oxidation of Lignin 346
10.4 Separation and Purification of Low Molecular Weight Organic Compounds 355
10.4.1 Separation of Unreacted Lignin from the Reaction Mixture 356
10.4.2 Separation of Low Molecular Weight Compounds from the Reaction Mixture 356
References 362

Index 365

Preface

The journey of human development over the last 150 years has been one with acute reliance on petroleum and other fossil resources for energy and functional material needs. In the context of the current emphasis on sustainable development, there is ever-increasing interest in developing processes and technologies for producing energy and functional organic materials from renewable sources. For this to happen, sustainable and comprehensive use of biomass in a biorefinery framework is needed.

Lignin, the second most abundant terrestrial organic substance and the most abundant aromatic material, is yet to be utilized to its full potential. Naturally occurring as a constituent of lignocellulose biomass, it is often discarded as an undesirable liability or a low-value byproduct. Barring some very limited commercial applications, most of the lignin ends up as a low-value fuel, which is hardly its optimal utilization. The major roadblock to lignin value addition and its large-scale application as a possible petroleum substitute is our limited and incomplete understanding of its physical and chemical nature. We must appreciate the fact that 'lignin' is a generic term denoting a class of substance derived from vegetative biomass, rather than a well-defined compound. Thus, there is heterogeneity in the characteristics of lignin based on its lignocellulose source and the method of its isolation.

Despite all the recent technological advancements, the isolation of lignin from lignocellulose biomass still remains a challenge. The isolation method must conform to the requirements of comprehensive and sustainable biomass utilization. In the context of lignocellulose, it implies the utilization of all its main chemical constituents, cellulose and hemicellulose, besides lignin. Considering its physical and chemical heterogeneity, it is imperative to characterize the isolated lignin if we intend to put it to a value-added application. Finally, we must find a means to transform the lignin, again in a sustainable framework, so that it finds an application as a functional material in modern human lives. The proposed book is an attempt to bring forth the current state of know-how on lignin – its sustainable isolation from lignocellulose, characterization of the isolated lignin, and transforming it further to add value to this important renewable material.

This book aims to act as an advanced text and a reference volume for students, researchers, scientists, engineers, technologists, and others who have a stake in the sustainable utilization of lignin. It begins with an introduction that sheds light on the historical developments in lignin science, engineering, and technology graduating into the contemporary relevance of lignin. The chapter also provides insight into the current state of lignin utilization and the major commercial players operating today for lignin isolation and valorization. The main part of the book is arranged into three sections dealing, respectively, with lignin isolation from lignocellulose, lignin characterization, and lignin value addition. All the chapters in the book are enriched with figures and graphical presentations to augment the text. Wherever necessary, information is provided in tabular form for better interpretation and contextual relevance. Mathematical enumerations are included to strengthen the scientific concepts. There are extensive citations from technical literature and relevant patents to bring forth the state of the art in lignin know-how. Overall, the book strives to serve as a scientific and technological knowledge pool in the quest for sustainable utilization of lignin.

The section on lignin isolation elaborates the well-established commercial delignification methods, including recent developments in the field. Further, it also discusses new and emerging isolation methods that have evoked interest in the scientific community but still await widespread commercialization. This section is further divided into three chapters. The first chapter in this section (Chapter 2 of the book) is dedicated to the reactive dissolution of lignin from different lignocelluloses using alkaline, acidic, or organic liquids. Well-established commercial processes of soda, kraft, sulphite, and organosolv delignification are deliberated along with the recent developments.

Process chemistry, major process parameters, and general engineering aspects are discussed. The second chapter in this section (Chapter 3) deals with the isolation of lignin from the reacting system after it is dissolved from the lignocellulose. Processes based on acidification and membrane filtration are discussed with associated unit operations and parameters controlling them. In the third chapter of the section (Chapter 4), the process of lignin isolation from lignocellulose by hydrolyzing the carbohydrates and the isolation of milled wood lignin is discussed.

The lignin characterization section includes not only the traditional methods based on wet chemistry but also the sophisticated advanced characterization techniques. Before the characteristics of different lignins as determined by a particular characterization technique are elaborated, the fundamentals of the characterization technique are explained. This section is subdivided into also three chapters. The first chapter in the section (Chapter 5) is devoted to such physicochemical characteristics of lignin such as elemental analysis, functional groups, intermonomer linkages, molecular weight, and solution properties. The second chapter in the section (Chapter 6) deals with spectroscopic methods of UV spectroscopy, FTIR spectroscopy, Raman spectroscopy, and NMR spectroscopy as applied to lignin characterization. In the third chapter of this section (Chapter 7), lignin characteristics based on thermogravimetry, differential thermal analysis, differential scanning calorimetry, and pyrolysis coupled with gas chromatography mass-spectroscopy are described. A comparative account of different characterization methods with their merits and limitations is also included towards the end of this chapter, which also portrays a picture of the heterogeneity and diversity in lignin characteristics obtained from different lignocellulose sources using different isolation methods.

The third section of the book is devoted to lignin value addition with a detailed presentation on the possibility of transforming lignin into functional materials necessary for modern human life. Besides process and product details, the section emphasizes the critical importance of lignin source, its isolation method, and the resulting lignin characteristics on its possible value addition. The section provides a vivid account of the lignin valorization possibilities with the current state of scientific and technological development and the road that lies ahead. This section comprises three chapters: Chapters 8, 9, and 10. The utilization of lignin for obtaining value-added carbon products of activated carbon, biochar, carbon fibres, carbon nanotubes, and carbon materials for energy storage is deliberated in Chapter 8. Chapter 9 is devoted to lignin-based polymer formulations, blends, and composites. Chapter 10 enumerates lignin valorization possibilities through its depolymerization and associated derivatization to obtain low molecular weight organic chemicals.

Himadri Roy Ghatak

About the Author

Himadri Roy Ghatak received his BE and ME degrees from the erstwhile University of Roorkee (now Indian Institute of Technology Roorkee), Roorkee, India, and his PhD from Sant Longowal Institute of Engineering and Technology, Longowal, India. He worked in Nepa Ltd., Nepanagar, India, from 1991 to 1995 before joining the faculty of Sant Longowal Institute of Engineering and Technology, where he rose to the position of Professor in Chemical Engineering and presently serves as the Head of the Department. He also served there as Associate Dean (Research and Consultancy). He has taught courses in Waste to Energy, Chemical Recovery Processes in Pulp and Paper Industry, Pulping Processes, and Biorefineries. Prof. Ghatak's research focuses on advanced oxidation treatment of nonwood lignins, with special emphasis on electrochemical oxidation. As a related domain, he also works in the area of electrolytic hydrogen production, electrochemical wastewater treatment, and integrating these processes.

1 Introduction

1.1 NEED TO FIND AN ALTERNATIVE TO FOSSILS

For almost the last 150 years, human development has been largely driven by petroleum and other fossil-based economies. These substances fulfil around 85% of our primary energy requirement. The transportation sector, especially automotive transport, is heavily dependent on liquid fuels sourced from petroleum and gaseous fuels in the form of natural gas. Similarly, coal and natural gas have been the mainstay of electricity generation, which is a ubiquitous energy carrier or secondary source of energy. Our overwhelming dependence on fossils for energy needs arises from the following:

- Large proven resources and their possible need-based extraction.
- Proven technology as a result of scientific developments of the last century.
- Competitive economics.
- Ever-growing human population and per capita energy consumption.

Fossils, especially petroleum and natural gas, not only act as a valuable primary source of energy but also provide most of the functional organic materials synonymous with modern human life. Plastics and polymers sourced from petroleum and natural gas find numerous applications in day-to-day common articles as well as in high-end niche products. Raw materials for structural plastic items that are routinely used in households, offices, and other commercial establishments are all derived from petroleum or natural gas. Synthetic fibres used in textiles, furnishings, and packaging are all made from organic materials obtained from petroleum and natural gas. Petrochemicals find their way into the manufacturing of substances essential for the synthesis and formulation of paints, coatings, membranes, personal care products, and pharmaceuticals. Many of the automotive parts and aircraft and spacecraft components are produced from materials that originate from petroleum and natural gas.

Despite all the advantages and positive contributions of fossils – particularly petroleum – they come with several serious drawbacks. These can be summarized as follows:

- *Fossils are non-renewable natural resources.* Their reserves are finite and bound to finish in the not-so-distant future. From the most optimistic conservative estimates, petroleum is likely to last another half a century, natural gas for a century, and coal for the next 2–3 centuries. For sustainable development, therefore, it is strongly mandated to find an alternative.
- *Highly non-uniform distribution of reserves across the globe.* For many countries, it is a cause of geopolitical stress and supply uncertainty.
- *Very unfavourable environmental impact.* In energy applications, fossils are the main contributors to anthropogenic carbon dioxide emissions. The use of fossils in the energy sector results in the emission of significant amounts of oxides of nitrogen and sulphur, and suspended particulate matter. Most of the petroleum and natural gas-derived synthetic materials – plastics, for example – are non-biodegradable and fail to integrate into the natural biogeochemical cycles.

With the debate on human progress finally settling in favour of sustainable development, it is imperative to find an alternative to fossils as the main natural resource for energy and materials. Among the renewable and alternative sources of energy, the available choices are solar, wind, hydropower,

DOI: 10.1201/9781003598664-1

biomass, tidal, geothermal, and nuclear. Of these, biomass is the only natural resource that can provide functional organic materials along with energy. From this standpoint, it can be the only true substitute for fossils. However, for this to happen, mindset stereotypes need to break and all stakeholders – policymakers, administrators, scientists, and technologists – must join hands to develop the necessary technological and economic framework.

1.2 COMPREHENSIVE UTILIZATION OF LIGNOCELLULOSE

Plants account for the largest stock of standing terrestrial biomass and the largest sink of atmospheric carbon on land. Naturally, these are the most abundant of all biomass resources and, therefore, the most sought-after. The main edible chemical constituents of plants are starch, sucrose, other minor mono and disaccharides, proteins, and oils. Besides their utility of being consumed as food, many of these chemical components have important other applications. Starch and sucrose are used in large quantities to produce ethanol, a part of which may be used as an energy source. Similarly, edible vegetable oils may also be converted into biodiesel for targeted energy applications. Such biofuels, obtained from edible plant components, are termed first-generation biofuels. With the increasing human population, it is a continuous challenge to keep pace with the increasing demand for food items. Given this, any non-food application of an edible plant component may lead to strife and conflict.

The major part of the body of a plant is made up of structural entities like stems, trunks, and branches. These parts provide the structural integrity, strength, and rigidity to the plant body while providing the transport channels for the transfer of material between different parts of the plant. Once dead, these residual plant body parts – wood, straws, grasses, for example – are available as important biomass resources. In the context of sustainable use of biomass, these resources have the following important advantages:

- *Large abundance*. In terms of availability, vegetative biomass resources such as wood and straws far outscore starch, sucrose, and edible oils. The estimated annual availability of lignocelluloses is reported to be 181.5 billion tonnes with less than 5% put to meaningful use (Mujtaba et al. 2023).
- *No food availability conflict*. Being unfit for consumption as food for humankind, their utilization in energy and other industrial applications does not raise any ethical debate.

The use of wood and other non-edible vegetative biomass resources as sources of energy is as old as human civilization. However, for the most part, these biomass resources have been used in an unorganized and inefficient manner. This notion has started to change only very recently. For these resources to be a true substitute for fossils, we must come out of the old dictum of looking at them only as low-value fuel fit for marginal use. Rather, it is desirable to use them as raw materials in the biorefinery framework and transform them into value-added energy and material products. Herein comes the idea of second-generation biorefineries, industrial establishments for the comprehensive utilization of non-edible vegetative biomass resources. These biorefineries strive to utilize these raw materials in their entirety and put all the raw material components to their optimal value-added applications.

The major chemical constituents of the non-edible residual plant body parts are cellulose, hemicellulose, and lignin. That is why these materials – wood, straws, grasses, bagasse – are known as lignocelluloses. Other organic and inorganic substances are also present in small quantities. Cellulose, a polysaccharide of repeating β, D-glucopyranose units, is the most abundant terrestrial organic matter. It makes up nearly half of the mass of a typical lignocellulose. Hemicellulose is a generic term used for the polysaccharides, other than cellulose, present in the lignocelluloses. Together, they make up about 20–25% of the mass of a typical lignocellulose. Some of the important hemicelluloses are xylan, glucomannan, and galactoglucomannan. Lignin, an aromatic macromolecule, is the second most abundant terrestrial organic matter. It is present in lignocelluloses to the extent of about 15–30%. The techno-economic viability of a second-generation biorefinery based on lignocelluloses is closely linked to the sustainable and comprehensive utilization of all three major chemical constituents of lignocelluloses – cellulose, hemicellulose, and lignin.

Apart from the utilization of lignocelluloses as a renewable energy source, timber, or construction material, its major utilization is in the manufacturing of paper and paperboard. Here, cellulose is the desirable lignocellulose component. Once isolated and purified, cellulose has several other important applications. It can be derivatized and converted to fibres, films, and membranes with advanced scientific and industrial applications. During the manufacturing of paper and paperboard, especially the bleached varieties, lignin and most of the hemicelluloses are discarded. They end up being used in industrial boilers as fuel – not a very high-value utilization. Of late, interest is growing in the conversion of lignocelluloses in second-generation biorefineries to bioethanol – the so-called second-generation bioethanol. Even in these conversion processes, the main emphasis is on cellulose. Some proprietary process configurations do convert the hemicelluloses to bioethanol. However, the lignin still remains unutilized or only marginally utilized. This is, certainly, not the optimal and sustainable use of an otherwise abundant bioresource. We need to add value to lignin in order to realize its full potential and also to lift the economic bottom line of lignocellulose biorefineries.

1.3 HISTORICAL PERSPECTIVE OF LIGNIN

Experiments during the early 19th century on wood digestion with strong acids led to the discovery of cellulose (Payen 1838). This treatment also left an acid-insoluble brownish residue that was termed lignin (Schulze 1857). The word itself is derived from 'lignum', Latin for wood. Scientists realized that it was the glue that held the fibres together in the plant stem providing its structural rigidity. As wood pulping picked up and acquired the prime position in paper manufacturing industries, there was a renewed interest in the study of lignin. This was primarily to devise better ways to remove lignin as it was considered an unwanted wood component to get rid of. It was realized that the presence of lignin prevented fibre separation and conversion of wood into pulp. It was also responsible for the generally unwanted brown colour of the pulp. Nevertheless, these initial scientific studies provided crucial insights and helped in building a strong foundation for scientists and technologists to come.

Besides the pulp as the desired product, wood pulping produced a lignin-rich spent liquid that also contained the inorganic chemicals used in pulping. Economics necessitated the recovery and recycling of these chemicals. Organic components in the spent liquid, mainly lignin, were burnt for this purpose. It was soon realized that the energy generated in this combustion process exceeded the energy required for the recovery of chemicals. This excess energy could be used elsewhere in the paper manufacturing process, leading to further cost benefits. This is how lignin found its application as a low-value fuel, a situation that would prevail for the next 100 years and beyond. Technological advancements over the years improved the process of chemical recovery, including the energy efficiency of the process, but the role of lignin as a low-value fuel did not change. Other than processing wood to produce papermaking pulp, another interesting way of industrial utilization of wood was its saccharification. This line of thinking gained ground during the first half of the 20th century to meet the increasing demand for ethanol at the time of the two world wars. At this time, most of the wood saccharification processes were based on acid hydrolysis of the polysaccharides. Wood saccharification left a lignin-rich solid residue. Again, its primary use was a fuel to meet part of the mill energy demand.

Scientists, however, kept on looking deeper at the physicochemical nature of lignin. Not very long after the initial understanding of lignin as a distinct chemical constituent of wood, there were indications of the aromatic nature of lignin. Among others, the work of Klason during the late 19th and early 20th centuries had a profound influence on the shaping of lignin chemistry. He was able to chemically relate lignin to coniferyl aldehyde, thereby establishing the phenylpropane building blocks of lignin (Klason 1897). Subsequently, he confirmed the methoxyl groups in lignin and also the ether linkages between the phenylpropane units, findings that withstood the test of time and hold true even today (Klason 1920a; Klason 1920b). It was only a quarter century later that the main ether linkages in lignin were identified to be β-aryl type (Gardner and Hibbert 1944). It was much later, during the late 1950s and 1960s, that carbon–carbon linkages were observed in lignin. Almost at the same time, aryl–aryl ether linkages were determined in lignin structures (Freudenberg and Chen 1967). During this time, mechanistic studies on the cleavage of alkyl–aryl ether linkages provided crucial insight into the process of alkaline delignification. It was observed that phenolic hydroxyl and aliphatic hydroxyl groups played an important role in these reactions (Gierer and Kunze 1961). The mechanism for the action of hydrosulphide ions in kraft pulping, the predominant process to produce pulp, was also proposed (Gierer and Smedman 1964). Much later, in the mid-1990s, dibenzodioxin structural moieties were determined in lignin. These structures had three phenylpropane units joined by 5–5′, and two alkyl–aryl ether linkages (Karhunen et al. 1995). It is now firmly believed that the occurrence and abundance of intermonomer bonds are altered to a considerable extent during delignification for the conversion of wood and other lignocelluloses to pulp. The presence of phenolic hydroxyl groups in lignin was already conclusively proven in early lignin research. Erdtman, and also Freudenberg, proposed that plants synthesized lignin from coniferyl alcohol through dehydrogenation (Erdtman 1933; Freudenberg 1933). This further strengthened the understanding of the phenylpropane skeleton of lignin building blocks. Conclusive evidence that plants synthesized lignin from coniferyl alcohol came another 20-plus years later (Freudenberg and Eisenhut 1955). The mechanism of biosynthesis of lignin in plants was fully understood only by the end of the 20th century (Sarkanen 1998).

The early studies almost exclusively dealt with softwood – the prevalent raw material for pulp production – lignin and lignosulphonates – sulphite delignification being the prevalent pulping method. It was recognized that lignin was sulphonated during sulphite delignification. Alkaline nitrobenzene oxidation and analysis of reaction products marked an important milestone in lignin structural elucidation. Experiments with softwood lignin yielded vanillin in good yield while hardwood lignin produced both vanillin and syringaldehyde (Creighton et al. 1941; Freudenberg et al. 1940). Thioacidolysis as a method for controlled degradation of lignin macromolecule was developed in the late 1960s and contributed a lot to the estimation of ether linkages and functional groups (Nimz 1969).

Scientists had a feeling since the early stages of lignin research that it was a macromolecule. It was believed to have a crosslinked structure which resisted its dissolution in many solvents. Concrete evidences were put forth in the 1950s and 1960s regarding the polymeric nature of lignin, particularly lignosulphonates, and their behaviour in solutions. Later, it was observed that lignin in solution behaved as a colloidal system (Lindströmn 1979). This knowledge was of help in better understanding the lignin separation processes from spent pulping liquids. Accordingly, lignin characterization techniques included the measurement of properties like zeta potential and hydraulic radius. Of late, new experimental evidence has suggested phenylpropane units are joined in a linear fashion in lignin macromolecules (Funaoka 2013).

Advances in spectroscopy and chromatography facilitated lignin research. Lange's work on UV spectroscopy of lignin put to rest any doubt about its aromatic nature, and more specifically, its structural proximity to coniferyl aldehyde (Lange 1944). Methods were developed for the determination of lignin concentration in solution based on UV spectroscopy (Patterson et al. 1951). Around this time, UV spectroscopy also assisted in the quantitative determination of phenolic hydroxyl groups in lignin (Aulin-Erdtman 1954). Infrared spectroscopy also started contributing to lignin molecular structure determination (Hergert 1960). ^{1}H NMR spectroscopy was first applied to lignin characterization in the mid-1960s and started an important phase in the scientific understanding of lignin (Ludwig et al. 1964). NMR spectroscopy gave a huge boost to the understanding of different functional groups and intermonomer linkages in lignin. Lignin characterization by ^{13}C NMR spectroscopy was first reported about a decade after the use of the ^{1}H NMR technique (Lüdemann and

Nimz 1974). Quantitative ^{13}C NMR characterization was applied to lignin research, and it was soon to be found of immense utility for quantitative estimation of functional groups, linkages, and structural moieties in lignin macromolecule (Robert and Brunow 1984). Structural analysis of lignin macromolecules was vastly facilitated by the application of different techniques of quantitative ^{13}C NMR spectroscopy, and it has become one of the main analytical methods of choice for lignin characterization at present. Similarly, ^{19}F and ^{31}P NMR spectroscopy techniques broadened the scope of lignin characterization. Gel permeation chromatography was effectively introduced in lignin research to fractionate and estimate the size of lignin macromolecules (Jensen et al. 1962). Initial procedures detected the different molar mass fractions by the difference in their refractive indices. Later, the phenomenon of LASER beam scattering was utilized in particle size determination and classification. Detectors working on this principle were combined with gel permeation chromatography, which improved the capability of this technique in lignin molecular mass and polydispersity characterization. Combining gas chromatographic techniques for product identification – with controlled lignin degradation procedures such as nitrobenzene oxidation or thioacidolysis – formed a powerful analytical tool in lignin characterization. The method was further expanded and strengthened with mass spectroscopic detection of components after gas chromatographic separation. Modern computing capabilities also started to contribute to the lignin structural analysis around the mid-1970s with simulation-based computational chemistry assisting in arriving at the molecular structure of lignin (Glasser and Glasser 1974).

There was unanimity among scientists in the fact that lignin was substantially altered from its native state during the digestion of wood with acid or alkali. The method of Klason, as well as the sulphite and alkaline delignification processes, introduced significant changes in the chemical nature of lignin. As a result, lignin isolated from the same lignocellulose with different processes would have significantly different characteristics. The quest for isolating lignin from wood without causing its modification, or with minimal modification, has been an integral and important part of lignin research. Brauns was the first to isolate lignin from wood by extracting it with ethanol with the belief that the resulting material represented lignin in its native state (Brauns 1939). It was presumed that ethanol did not react with lignin and only interacted physically during lignin dissolution. Though this presumption was found to be incorrect in subsequent studies, it opened an important avenue in lignin research through its isolation with organic solvents. The concept of solubility parameter, first by Hildebrand and later by Hansen, was put to use in the selection of solvents for lignin dissolution. To date, the discovery of a solvent that removes lignin from lignocelluloses only by physical interaction, without any chemical reaction, remains elusive. Lignin in its substantially unaltered form could be isolated by Bjorkman by prolonged milling of wood meal in toluene and extracting the resulting material with dioxane. He termed this milled wood lignin (Bjorkman 1956).

Despite most of the lignin removed from wood during pulp production being burnt, there were some early efforts to isolate lignin from spent pulping liquids and put them to non-fuel use. The use of lignosulphonates as dispersing agents and tanning agents was introduced in the 1930s. At around the same time, the commercial production of vanillin from softwood lignosulphonates started using the process of oxidation in alkaline conditions. An important application of lignosulphonate derivatives was found in the form of an additive in the petroleum drilling process. It became a common industrial practice to isolate lignosulphonates from sulphite-spent pulping liquids and sell them. Fermented lignosulphonates were also produced and sold in the market. The practice declined only with the worldwide shifting of pulping technology to the kraft process. The synthesis of vanillin from lignosulphonates also met a similar fate due to decreasing availability of lignosulphonates. Moreover, the process of synthesizing vanillin from guaiacol, a petrochemical, proved to be more economical than synthesizing it from lignosulphonates. It was only after the idea of sustainable development gained ground that there was a renewed interest in lignin. The isolation of kraft lignin from pulping spent liquid was first practised in North America in the early 1940s and it was sold by the name Indulin, which still continues. Ethanol organosolv pulping was commercialized in a limited way with the lignin separated and sold by the name Alcell. The commercial production of dimethyl sulphide and dimethyl sulphoxide from kraft black liquor was practised for a brief period. Attempts to make lignin-based adhesive dates back to the mid-20th century but they did not find much commercial success. Efforts to develop lignin-based binder resins were again at the forefront in the 1970s. It achieved reasonable success in producing thermosets, like phenol formaldehyde resins, with the use of lignin as a phenol substitute. These efforts continued and studies were also conducted on blending lignin with thermoplastics such as polypropylene and polyethylene. These studies provided an important understanding of the compatibility of lignin with different polymers. The development of lignin-containing epoxy resins, and their successful application in fabricating printed circuit boards during the last decade of the 20th century, marked an important milestone in the commercial utilization of lignin-based polymers. Attempts to use lignin in polyurethane synthesis started around this time. With the current understanding, it is generally accepted that lignosulphonates and organosolv lignins are more suited to such applications compared to kraft lignin. Kraft lignin, on the other hand, is the most widely available industrial lignin. Therefore, consistent research and development efforts have gone into finding ways for physical or chemical modification of kraft lignin to make it more compatible with polymer synthesis and blending. The medicinal properties of lignin and its use in animal feed formulations were also recognized during the 1980s and 1990s. A relatively high carbon content in lignin has always been recognized and there have been limited uses of lignin to produce chars and activated carbons. Very recently, we have found efforts to produce high-end carbon products like fibres, films, and nanotubes from lignin. However, the commercial applications are extremely limited.

1.4 LIGNIN AVAILABILITY, POTENTIAL, MARKET, AND VALUE-ADDED APPLICATIONS

The status of lignin as an abundant renewable resource, only after cellulose in terms of occurrence, is beyond doubt. It is all the more important given its aromatic character as no other renewable aromatic material with such vast natural occurrence is available. Therefore, lignin valorization is vital to have a sustainable replacement of aromatic chemicals derived from petroleum. It is believed that the biosphere has more than 300 billion tons of lignin present in it (Mastrolitti et al. 2021). The global lignin market was reported to be worth close to 1 billion USD in 2022 and is expected to increase to 1.178 billion USD by 2029 (Global Lignin Market 2023). The production of pulp from lignocellulosic raw materials is the main industrial enterprise that leaves lignin as a by-product. The other industrial process of importance is second-generation bioethanol production. In these industries, the carbohydrate fraction of the lignocellulose is hydrolysed to fermentable sugars, leaving behind a lignin-rich residue. However, the second-generation bioethanol production is yet to be commercialized on a large scale. In terms of lignocellulose processing capacity, the pulp and paper industries far outweigh the second-generation bioethanol plants. From a lignocellulose raw material perspective, hardwood lignin has the largest market share, followed by softwood lignin and straws. Of the various lignin types, kraft lignin has the largest market share followed by lignosulphonate. Most of the lignin is utilized and expected to be utilized as a source of aromatic platform chemicals, speciality chemicals, and polymer ingredients. North America has the largest lignin market, followed by Europe. The Asia Pacific region also has a reasonable lignin market.

1.4.1 Kraft Lignin

The kraft process is the predominant process for the delignification of lignocellulose and its conversion to pulp. Therefore, it is no surprise that kraft lignin accounts for the major share of industrial lignin. Its worldwide annual production is reported to be nearly 70 million tons, of which 28 million tons could be available as a surplus after meeting the internal energy demand of pulp mills (Argyropoulos et al. 2023). This enormous potential is a strong incentive for finding viable value-added applications for this important renewable resource. However, at present, only 6.3×10^5 tons are separated for non-fuel applications, which is still about 85% of all lignin production (Mastrolitti et al. 2021). This lignin, usually, is much altered from the native lignin for functional groups and intermonomer linkages. Particularly, new carbon–carbon linkages form between adjacent aromatic nuclei. It is also a sulphur-containing lignin. Kraft lignin is distinctly hydrophobic and dissolves only in strong alkali or other highly polar liquids. That is why acidification is the most common method of isolating it from pulping spent liquid. Despite the kraft process being the most prevalent delignification process, the separation of kraft lignin at the commercial level is scantly practised. Worldwide annual separation for non-fuel use is about 1.7×10^5 tons (Mastrolitti et al. 2021). Some of the commercial products available in the market, and some major commercial players in the kraft lignin market are as follows:

- Ingevity, a North American paper industry, sells softwood kraft lignin by the name Indulin.
- Domtar, a North American paper industry, used to sell softwood kraft lignin by the name Bio-choice lignin. Their lignin business is now taken over by UPM Biochemicals and the lignin product line goes under the trade name of BioPiva.
- West Fraser, a North American forestry industry, sells softwood kraft lignin.
- Stora Enso, a European paper industry, sells different grades of softwood kraft lignin by the name Lineo. They also produce and market lignin-based binder, NeoLigno, and lignin-based energy storage material Lignode.

1.4.2 Soda Lignin

This is a lignin resulting from non-sulphur alkaline delignification. The lignin is similar to kraft lignin in many respects but has no or negligible amount of sulphur. At the commercial scale, soda delignification is mostly restricted to agricultural residues and grasses. Greencone Environs is an Indian venture that produces and markets soda lignin from nonwood raw materials such as wheat straw and sarkanda that sell by the name Protobind.

1.4.3 Lignosulphonate

Of all the lignins separated from the industrial use of lignocellulose raw materials, lignosulphonates have the most industrial application and level of commercialization. Even some of the kraft lignin separated is subsequently sulphonated to different degrees for better industrial acceptability. Lignosulphonates are the only lignin type that is hydrophilic and water soluble. This sulphur-containing lignin is obtained by drying and purification of spent liquid from the sulphite delignification process. Depending on the base used in the process, lignosulphonates can be in calcium, magnesium, ammonium, and sodium form. Worldwide annual separation for non-fuel use is about 1.5–1.8 million tons (Ekielski and Mishra 2021; Mastrolitti et al. 2021). Unless extensively purified, lignosulphonates have appreciable carbohydrate impurities. Some of the commercial products available in the market, and some major commercial players in the lignosulphonate market are as follows:

- Borregaard, a Scandinavian biorefinery, produces and sells softwood lignosulphonate-derived products.
- Domsjo, a Scandinavian pulp and paper industry and a part of Aditya Birla Group, produces and sells softwood lignosulphonate.
- Nippon Paper Chemical, a Japanese company, produces and sells lignosulphonates by the names SAN X, VANILLEX, and PEARLLEX.
- Rayonier Advanced Materials produces and sells lignosulphonate by the name ARBO from its Canadian facility.

- Azelis, a European company, sells sodium lignosulphonate-based dispersant by the name Bretax S.
- LENZING, a European company, produces and sells hardwood magnesium lignosulphonate.

1.4.4 Organosolv Lignin

Though a few pilot-scale facilities are in operation, there is hardly any commercial venture that uses organosolv delignification. Among the different organic solvents, ethanol, formic acid, and acetic acid find the most application in delignification. The produced lignin is distinctly hydrophobic and dissolves only in strong alkali or some organic liquids. There can be more variability in organosolv lignin characteristics compared to kraft, soda, or lignosulphonate. This is due to the choice of solvents and the range of process parameters. Noteworthy patented technologies in the area of organosolv delignification are Alcell (aqueous ethanol-based), Formasolv (formic acid-based), and Acetosolv (acetic acid-based). Worldwide annual production of organosolv lignin, at 3000–4000 tons, is hardly worth mentioning (Mastrolitti et al. 2021). However, this scenario can change if the research being currently undertaken finds an economically viable industrial application for organosolv lignin.

1.4.5 Hydrolysis Lignin

Though commercial production and availability are still very low, this lignin has gained prominence in recent years due to the developments in the field of second-generation bioethanol production from lignocelluloses. Hydrolysis and saccharification of the polysaccharide fraction leave the hydrolysis lignin as a solid residue. Hydrolysis may be carried out with mineral acids or enzymes, the latter being preferred in the overall process configuration. Hydrolysis lignin may have considerable carbohydrate impurities requiring purification. With enzymatic hydrolysis preceded by physical pretreatment, the lignin separated is chemically much less modified compared to the lignins obtained from delignification processes. Reaction conditions during acid hydrolysis can induce chemical modifications in lignin. Some important commercial second-generation bioethanol facilities are run by Raizen, GranBio, Clariant, and Indian Oil Corporation Ltd. At present, hydrolysis lignin produced during second-generation bioethanol production has hardly any market penetration for added value application. Most of this lignin is burnt as fuel to meet the in-house energy demand of the plant. However, estimates suggest that 60–65% of the hydrolysis lignin produced in second-generation bioethanol production could be in excess of that required to meet the in-house energy demand (Obydenkova et al. 2019).

1.4.6 Some Commercialized Value-Added Lignin Applications

There is no denying the fact that lignin, an abundant renewable resource, is grossly underutilized. Yet there are commercial ventures that put lignin to value-added applications outside its more common use as a low-value fuel. There are enterprises that produce lignin-based polymer formulations and composites – binder resins, thermoplastics, elastomers, boards, panels, and laminates. There is reasonable use of lignin in construction activities – superplasticizers in concrete admixtures, bitumen replacement in asphalt formulations, and dust suppressant in gravel roads. Lignin-based formulations are being used in skin care products and nutraceuticals. Here are some of the commercial ventures in the lignin-based value-added products business.

- Borregard, the Scandinavian company engaged in the biorefining of wood, has a diverse portfolio of lignin products. These include the thermoplastic LignoPol, soil conditioner BoreGro, and vanillin.

- Zero Furniture Board, produced by a Scandinavian company Koskisen, uses the kraft lignin-based binder NeoLigno from Stora Enso. This application completely replaces fossil-based ingredients in the furniture board.
- Latvijas Finieris, a European plywood company, uses the kraft lignin Lineo from Stora Enso to replace phenol in phenol formaldehyde resin.
- Tecnaro, a European company, produces a fibre-reinforced injection mouldable composite material ARBOFORM, where the binding material is lignin along with plant fibres as the reinforcement.
- Arpa Industries, a European company, produces laminates – ARPA Bloom – and panels – FENIX NTM Bloom – for interior design applications. The resin used in these materials has 50% phenol substitution with lignin.
- Lignin Industries AB, a Scandinavian company, produces a thermoplastic RENOL that is made from 100% lignin. The company claims it to be carbon-negative and can be blended with common thermoplastics like polyethylene and polypropylene.
- RenFuel, a Scandinavian company, produces the proprietary thermoplastic LIGNISOL, thermosetting resin LIGNISET, and elastomer LIGNIFLEX using lignin.
- PRISMA Renewable Composites, a North American company, produces the thermoplastic BioLAN from kraft lignin which the company claims is a renewable replacement for acrylonitrile butadiene styrene (ABS). The resin can be injection or compression moulded with existing machinery. Domtar now holds the majority stake in PRISMA and Yanfeng Automotive Interiors uses this resin to replace petroleum-derived ABS for the production of automobile interior parts.
- The European construction company, Holding de Vier Ambachten (H4A), replaces 50% of petroleum bitumen with Lineo, the softwood kraft lignin from Stora Enso, in its asphalt formulation for road construction.
- The Scandinavian company, MetGen, fractionates lignin with the use of enzymes to fractions tailored for specific applications. The commercial product line includes SHIELD (additive to improve paperboard water repellence and strength), METNIN Resin (phenol replacement in phenol formaldehyde resin), and METNIN NANOpolyol (replacement for conventional petroleum-based polyols used in polyurethane synthesis).
- Lignopure, a European company, has a product line of lignin-based proprietary functional materials. LignoBase is sold to cosmetics manufacturers for use as an additive in skin care products for antioxidant and UV protection. LignoLite is a nutraceutical product that imparts antioxidant, anti-inflammatory, and disinfectant properties.

The dictum of sustainable development requires comprehensive utilization of biomass resources. This is all the more important for the countries and regions with scant petroleum reserves. When selecting a biomass resource, abundance and availability are paramount, as they affect economic viability. Lignin certainly fits into this parameter, being the aromatic substance with the highest natural occurrence. This is an opportunity to be tapped. There are numerous methods to separate lignin from the lignocellulose material, each producing a distinct type with its own set of characteristics. Given the pre-eminence of cellulose among the chemical constituents of lignocellulose, we may also note that all the current lignin isolation methods are primarily designed to discard it rather than claim it. Therefore, a thorough characterization of any lignin entity is indispensable if it is to be put to its optimal use. This can also help to step back and fine-tune the lignin isolation and purification process. With all the scientific advancements and resulting sophistication in the physicochemical analysis of materials, characterizing lignin is still a challenge. Finally, scientific, technological, and entrepreneurial efforts are needed to add value to this wonderful substance and realize its full potential. Let us isolate lignin, characterize it to evaluate its optimal use, and add value to this abundant renewable bioresource.

REFERENCES

Argyropoulos, DDS, Crestini, C., Dahlstrand, C, Furusjö, E, Gioia, C, Jedvert, K, Henriksson, G, Hulteberg, C, Lawoko, M, Pierrou, C, Samec, JSM, Subbotina, E, Wallmo, H, Wimby, M. *ChemSusChem* 16:e202300492, 2023.

Aulin-Erdtman, G. *Svensk. Papperstidn.* 57:745–760, 1954.

Bjorkman, A. *Svensk. Papperstidn.* 59:477–485, 1956.

Brauns, FE. *J. Am. Chem. Soc.* 61:2120–2127, 1939.

Creighton, RHJ, McCarthy, JL, Hibbert, H. *J. Am. Chem. Soc.* 63:3049–3052, 1941.

Ekielski, A, Mishra, PK. *Int. J. Mol. Sci.* 22:63, 2021.

Erdtman, H. *Ann. Chem.* 503:283–294, 1933.

Freudenberg, K. Tannin Cellulose Lignin. Springer, Berlin, 114–146, 1933.

Freudenberg, K, Lautsch, W, Engler, K. *Chem. Ber.* 73:167–171, 1940.

Freudenberg, K, Chen, CL. *Chem. Ber.* 100:3683–3688, 1967.

Freudenberg, K, Eisenhut, W. *Chem. Ber.* 88:626–633, 1955.

Funaoka, M. *React. Funct. Poly.* 73:396–404, 2013.

Gardner, JAF, Hibbert, H. *J. Am. Chem. Soc.* 66:607–610, 1944.

Gierer, J, Kunze, I. *Acta Chem. Scand.* 15:803–807, 1961.

Gierer, J, Smedman, LA. *Acta Chem. Scand.* 18:1244–1248, 1964.

Glasser, WG, Glasser, HR. *Holzforschung* 28:5–11, 1974.

Global Lignin Market 2023. Global Lignin Market, Industry Analysis and Forecast 2023. Available online at www.maximizemarketresearch.com

Hergert, HL. *J. Org. Chem.* 25:405–413, 1960.

Jensen, W, Fremer, K, Forss, K. *Tappi J.* 45:122–127, 1962.

Karhunen, P, Rummakko, P, Sipilä, J, Brunow, G, Kilpeläinen, I. *Tetrahedron Lett.* 36:169–170, 1995.

Klason, P. Svensk. *Kem. Tidskr.* 9:133–140, 1897.

Klason, P. *Chem. Ber.* 53:706–711, 1920a.

Klason, P. *Chem. Ber.* 53:1864–1873, 1920b.

Lange, PW. *Svensk Papperstidn.* 47:262–265, 1944.

Lindströmn, T. *Colloid Polymer Sci.* 257:277–285, 1979.

Lüdemann, HD, Nimz, H. *Makromol. Chem.* 175:2409–2422, 1974.

Ludwig, CH, Nist, BJ, McCarthy, JL. *J. Am. Chem. Soc.* 86:1196–1202, 1964.

Mastrolitti, S, Borsella, E, Giuliano, A, Petrone, MT, De Bari, I, Gosselink, R, van Erven, G, Annevelink, E, Triantafyllidis, KS, Stichnothe, H. IEA Bioenergy: Task 42 Report, pp. 5, 2021. Available online at https://task42.ieabioenergy.com/wp-content/uploads/sites/10/2021/11/Sustainable-Lignin-Valorization_rev26-01-2022.pdf

Mujtaba, M, Fraceto, LF, Fazeli, M, Mukherjee, S, Savassa, M, de Medeiros, GA, Pereira, AES, Mancini, SD, Lipponen, J. *J. Cleaner Prod.* 402:136815, 2023.

Nimz, H. *Chem. Ber.* 102:799–810, 1969.

Obydenkova, SV, Kouris, PD, Hensen, EJM, Smeulders, DMJ, van der Meer, Y, Boot, MD. *Bioresource Technol.* 291:121805, 2019.

Patterson, RF, Keays, JL, Hart, JS, Strapp, RK, Luner, P. *Pulp Paper Mag. Can.* 52(12):105–111, 1951.

Payen, A. *Comptes Rendus* 7:1052–1056, 1838.

Robert, DR, Brunow, G. *Holzforschung* 38:85–90, 1984.

Sarkanen, S. Lignin and Lignan Biosynthesis. Lewis, N.S., Sarkanen, S. (Eds.), ACS Symposium Series. 194–209, 1998.

Schulze, F. *Chem. Zentr.* 21:321–325, 1857.

Section 1

Lignin Isolation

2 Lignin Isolation through Dissolution from Lignocelluloses

In lignocelluloses, the two principal components – cellulose and lignin – are intricately linked to each other through physical as well as chemical interactions. For plants, this association is vital for the structural conformity and rigidity that provide for the transport of materials and other important metabolic activities. The same, however, poses the primary challenge in the isolation of lignin from lignocelluloses. One of the approaches is to use a chemical agent that would selectively depolymerize the lignin macromolecule to make it soluble, leaving the carbohydrate fraction more or less intact. A representative depiction of different repeating monomer units and different inter-unit linkages in the lignin macromolecule are shown in Figure 2.1. Inter-unit linkages in lignin macromolecules are of two types viz. ether linkages – two carbon atoms joined together through an oxygen bridge, and C–C type – two carbon atoms joined by a direct covalent bond (Capanema et al. 2005). Of these, β-O-4′ linkages account for almost half of all the intermonomer unit linkages in the lignin macromolecule (Ralph et al. 2004). Lignin isolation through dissolution from lignocelluloses is essentially a reactive extraction process. The solid matrix of lignocellulose interacts with a reactive solvent. The lignin macromolecule is depolymerized into smaller fragments. These reaction products then leach out of the solid matrix into the liquid reaction medium which acts both as the depolymerizing agent and the solvent. Cleavage of intermonomer linkages, therefore, constitutes the most important part of the process. Different lignocelluloses differ in their abundance and distribution of monomer unit types as well as the intermonomer linkages. This diversity along with the choice of the reactive solvent is an important consideration. A brief description of lignin structural features from different lignocelluloses is presented in Table 2.1. This variability in molecular structure affects the susceptibility of different lignocelluloses towards the dissolution of lignin.

Of the different intermonomer linkages, the ether linkages have lower bond energies in the range of 50–80 kcal/mol. In comparison, the carbon-to-carbon linkages have higher bond energies, varying between 90 and 125 kcal/mol (Lou et al., 2018). Consequently, the ether linkages are easier to cleave compared to carbon-to-carbon linkages.

Therefore, lignocelluloses with a higher proportion of ether linkages – hardwoods and agroresidues – are easier to delignify than those with a lower proportion of these linkages – softwoods.

2.1 DISSOLUTION IN ALKALINE MEDIA

Ether bonds, especially the β-O-4′ linkages, are susceptible to cleavage under alkaline conditions where the hydroxyl ions catalyse the reaction. This is a commercially established process, which is widely utilized in the production of pulp for papermaking as well as for producing dissolving grade pulp for rayon and other high-end cellulose products. In its original form, the process utilizes a hot solution of sodium hydroxide to delignify wood and other lignocelluloses to convert them into pulp, hence the name soda pulping. Credit for inventing soda pulping goes to Burgess of USA and Watts of England who patented the process in 1851. The first commercial pulp mill based on soda delignification, commissioned in the USA, traces its origin to 1866. Since then, the process has undergone

DOI: 10.1201/9781003598664-3

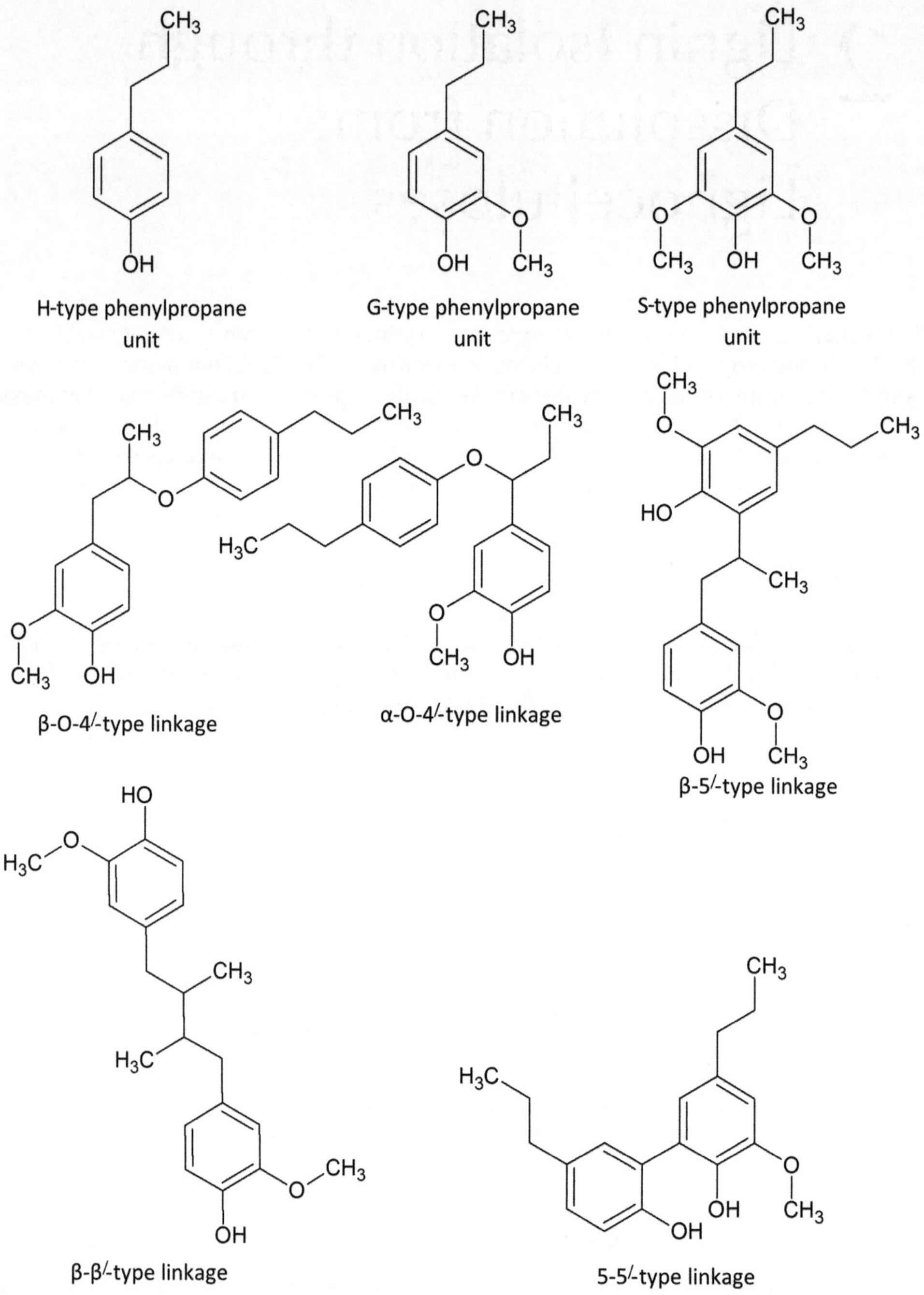

FIGURE 2.1 Monomeric repeating units and intermonomer linkages in lignin macromolecule.

several modifications, upgradations, and improvements. Most notable of these are the kraft process – where sodium sulphide is used in addition to sodium hydroxide – and the soda anthraquinone process – where anthraquinone is used as an additive to sodium hydroxide. All these processes – soda, kraft, and soda anthraquinone – are premised on base-catalysed depolymerization of lignin macromolecule and subsequent dissolution of the resulting fragments in hot alkali. The alkaline solution thus serves the dual purpose of depolymerizing agent and the solvent. In this way, the

TABLE 2.1
Lignin Characteristics from Different Lignocelluloses[a]

	Softwood	Hardwood	Agroresidues and Bamboo
β-O-4′ linkages	40–50%	60–62%	74–84%
α-O-4′ linkage	6–8%	3–11%	5–11%
β-5′ linkage	9–12%		
β-β′ linkage	2–3%	2–14%	1–7%
5-5′ linkage	18–25%	<1%	1–2%
H-type moieties	5–15%	2–4%	5–15%
G-type moieties	85–95%	15–55%	40–65%
S-type moieties	0–2%	40–85%	30–40%

[a] *Source:* Chakar and Ragauskas 2004; Guadix-Montero and Sankar 2018; del Río et al. 2012.

process yields a liquid spent stream, known as the black liquor because of its colour, which contains mainly the dissolved lignin and some carbohydrates degraded and solubilized through inevitable side reactions. Traditionally, the lignin in black liquor has only been considered as a liability at best being used as a low-value fuel. This notion is, however, changing due to more and more attention being paid to sustainable development and the need for renewable feedstock for organic chemicals.

2.1.1 Chemistry of Alkaline Delignification

Base-catalysed cleavage of ether linkages in lignin – β-O-4′ linkages in particular, owing to their greater abundance – has been extensively studied. The ease of breaking the ether linkages depends on the presence or otherwise of a free phenolic moiety at position 4 of the benzene ring. When a free phenolic hydroxyl is absent and the carbon at position 4′ is etherified, the reaction is initiated by the hydroxyl ion through the abstraction of a proton from the α (or γ) position. The presence of the hydroxyl group at the α or γ position facilitates this step. The resulting alkoxide anion rearranges by charge dislocation forming an oxirane moiety and breaking of the β-O bond in the β-O-4′ linkage. The net result is the defragmentation of the lignin macromolecule. This is represented in reaction scheme 1 of Figure 2.2. In the presence of hydroxyl ions, the phenolic hydroxyl is largely present as the corresponding phenolate anion through loss of proton. This intermediate ion rearranges to the quinone form by charge dislocation, which is further stabilized by the formation of a quinone methide structure if a suitable leaving group is present at the α position (Santos et al. 2013). For α-O-4′ linkages, this amounts to the severing of the α-O bond leading to lignin depolymerization. For β-O-4′ linkages, the methide structure may convert to an enol ether entity by expulsion of formaldehyde. This manifests as the relative difficulty of delignification in soda pulping. This is shown by reaction scheme 2 of Figure 2.2.

The limitation is surmounted by the presence of hydrosulphide ions, a stronger nucleophile, in the reacting system. The addition of hydrosulphide ion to the quinone methide moiety and the induced charge dislocation forms an intermediate with a mercaptide anion at the α position and the original negative charge is restored to the oxygen at position 4. What follows next is the intramolecular nucleophilic attack of the mercaptide anion at the α position on the β carbon. The result is the formation of a sulphur bridge between the α and β carbons – giving rise to a thiirane structure – and the cleavage of the β-O bond. The original lignin macromolecule is thus defragmented. The thiirane structure stabilizes itself by the expulsion of elemental sulphur and the formation of coniferyl alcohol. The reaction is schematically shown in Figure 2.3. Under the conditions of the reaction the elemental sulphur would ultimately convert to hydrosulphide and thiosulphate ions.

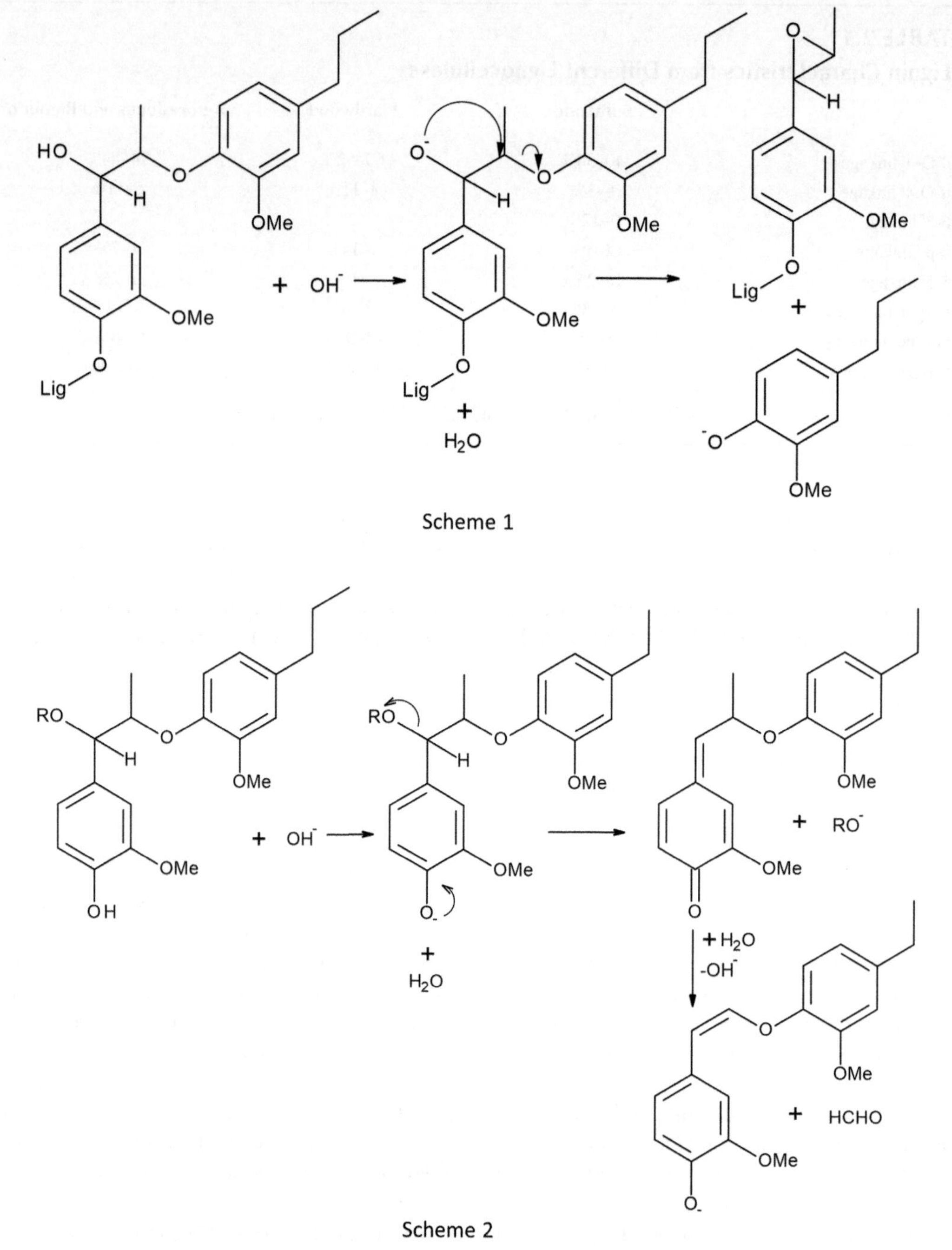

FIGURE 2.2 Breaking of ether linkages in the presence of hydroxyl ions.

Another additive that assists in delignification under alkaline conditions is anthraquinone (AQ). It first reacts with the reducing end of the carbohydrate chain to render it unreactive to a peeling reaction. In the process, AQ is converted to anthrahydroquinone (AHQ). AHQ attacks the α carbon of the quinone methide structure, forming an addict, so that the phenolate ion is restored. The adduct dismembers by the formation of a double bond between the α and β carbons and breaking of the bond between β carbon and oxygen. AQ is regenerated and is available for the next cycle of reaction.

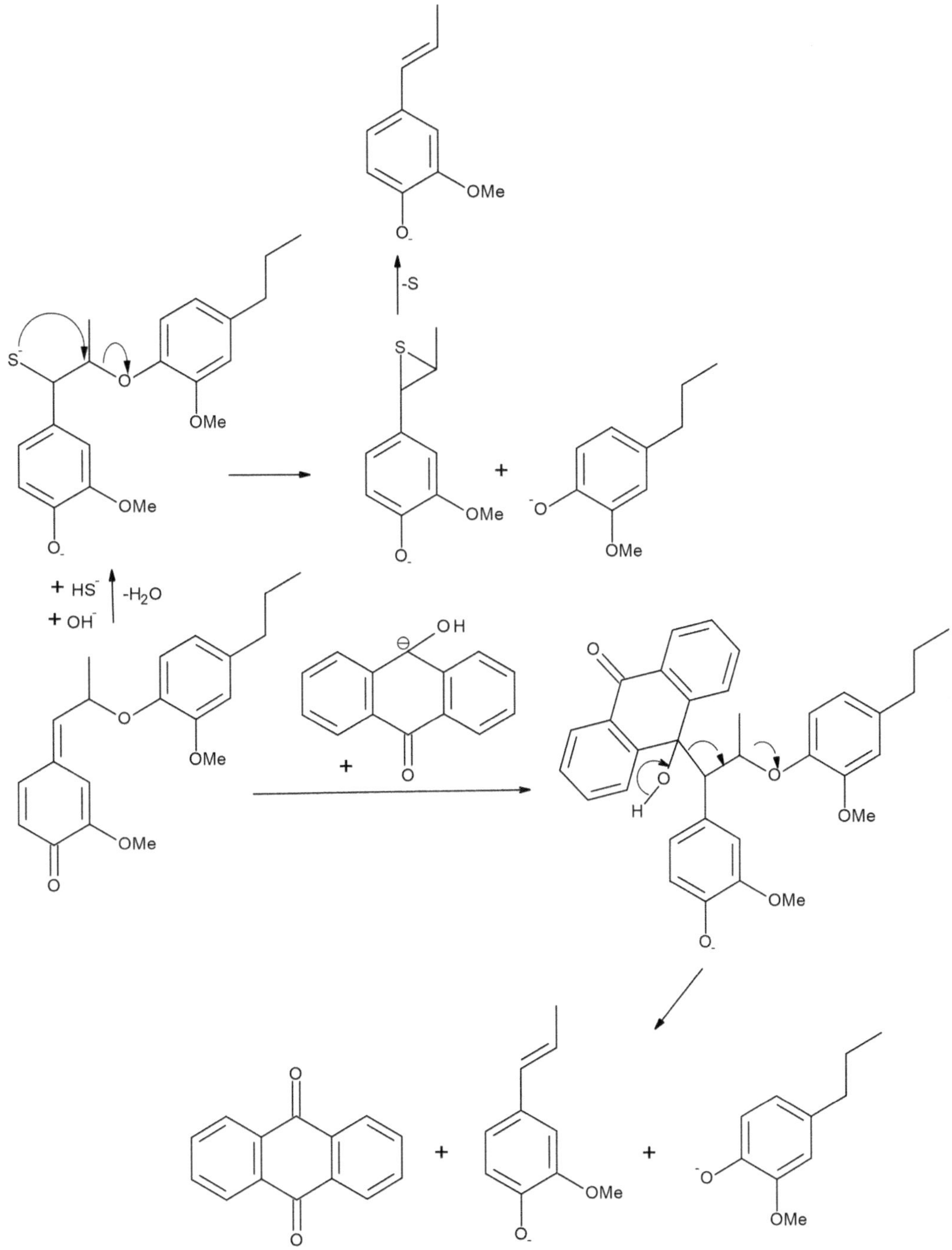

FIGURE 2.3 Breaking of ether linkages in the presence of hydrosulphide ions, and anthraquinone, respectively.

The net result is the depolymerization of the lignin macromolecule. Such depolymerization by the cleavage of α-O-4′ or β-O-4′ linkages generates a free phenolic hydroxyl at the 4′ position thereby increasing the hydrophilicity of the entity and its subsequent dissolution, when the molecular size is sufficiently reduced. The reaction is schematically shown in Figure 2.3.

As mentioned earlier, intermonomer carbon-to-carbon linkages in lignin have higher bond energies compared to the ether linkages. Under conditions of alkaline delignification, these linkages are not cleaved. However, structural moieties involving these linkages may participate in secondary reactions. This recalcitrance does not, in fact, affect the lignin dissolution as the breaking of ether linkages sufficiently defragment the lignin macromolecule.

While breaking of ether linkages successively defragment the lignin, the fragments so formed may react with each other forming bonds which are stable to degradation under the prevailing conditions. These so-called condensation reactions have a counterbalancing effect on the overall process and limit the extent of alkaline delignification. These reactions arise from the fact that carbanions are generated from phenolic lignin structures. The carbanions so formed are sufficiently strong nucleophiles so as to compete with hydroxyl ions, hydrosulphide ions, and AHQ in the reacting system (Gierer and Pettersson 1977).

2.1.2 Process Parameters in Lignin Dissolution in Alkaline Media

Lignocellulose substrates are porous solids with complex physical and morphological structures. Alkaline delignification involves the treatment of lignocelluloses with aqueous alkaline solution at elevated temperatures. Obviously, the process involves heat and mass transfer besides chemical reaction. For optimal results, it is imperative that the entire reacting system is quickly and uniformly heated to the desired reaction temperature and maintained at that temperature during the course of delignification. Equally important is to ensure that the liquid reactants, from the bulk of liquid, uniformly reach the innermost locations in the microstructure of lignocellulose at a rate matching the rate of chemical reaction. Similarly, lignin depolymerized and dissolved at the reaction sites within the lignocellulose microstructure must move out to the bulk of the liquid phase. On the reaction front, we strive for a high rate of delignification while suppressing the reactions involving carbohydrates. These multiple, often contradicting, objectives require the right combination of process parameters.

Process parameters that affect the outcome of alkaline delignification can be listed as

1. Characteristics of lignocellulose raw material
2. Size of lignocellulose raw material
3. Amount of delignification chemicals used
4. Composition and concentration of delignification chemicals used
5. Liquid-to-solid ratio
6. Reaction temperature
7. Time required to heat the reacting system to the desired reaction temperature
8. Reaction time at the desired reaction temperature

These parameters are interactive rather than mutually exclusive.

2.1.2.1 Characteristics of Lignocellulose Raw Material

Lignocelluloses from different sources are highly variable in terms of their chemical and morphological characteristics. It strongly affects the process of alkaline delignification. Substrates with a higher amount of lignin require a higher amount of delignification chemicals for complete dissolution of lignin. At the same time, raw materials with higher amounts of hemicelluloses – carbohydrates with a lower degree of polymerization and amorphous nature – react readily during alkaline delignification and constitute an important but undesirable side reaction. Lignocelluloses having more open morphological structures facilitate mass transfer. Table 2.2 highlights the chemical and morphological characteristics of common lignocelluloses.

TABLE 2.2
Chemical and Morphological Characteristics of Some Lignocelluloses

	Lignin (%)	Hemicellulose (%)	Ash (%)	Morphology	Reference
Pine	24–30	18–23	<1	Long thick and robust fibres	Gonultas and Ucar 2013; Shemfe et al. 2015
Spruce	25–28	17–30	<1	Long thick and robust fibres	Ban et al. 2004; Molteberg and Høibø 2006; Tutus et al. 2010
Bamboo	21–31	25–35	1.5–5	Long sturdy fibres	Ghatak 2002; Khantayanuwong et al. 2023
Poplar	21–24	27–30	1	Thin, flexible fibres with moderate length	Ghatak 2002; Sun et al. 2002
Eucalyptus	25–29	18–20	<1	Thin, flexible fibres with moderate length	Ghatak 2002; Chen et al. 2020; Xu et al. 2018
Wheat straw	16–21	25–35	4.5–11	Short and thin fibres	del Río et al. 2012; Ghatak 2002; Sun et al. 2002
Rice straw	12–24	26–34	10–15	Short and thin fibres	Lim et al. 2019; Sun et al. 2002
Bagasse	18–23	17–25	2–5	Thin, flexible fibres with moderate length	Ghatak 2002; Hamzeh et al. 2009; Khristova et al. 2006; Kininge and Gogate 2022; Rocha et al. 2012

2.1.2.2 Size of Lignocellulose Raw Material

The process being a heterogeneous one, mass transfer plays a critical role in the overall outcome. Schematically, it can be visualized as consisting of the following steps:

1. Transfer of active pulping chemicals from the bulk of the liquid to the external surface of the lignocellulose particle.
2. Movement of active pulping chemicals through the microstructure of the lignocellulose to the reaction sites.
3. Reaction of active pulping chemicals with the constituents of lignocellulose – lignin and carbohydrates.
4. Movement of reaction products through the microstructure of the lignocellulose to the external surface of the lignocellulose particle.
5. Transfer of products from the external surface of the lignocellulose particle to the bulk of the liquid.

Of these, steps 2, 3, and 4 are not essentially sequential and proceed simultaneously.

The more the external exposed surface area of the lignocellulose in contact with the liquid delignification chemicals, the faster would be the external mass transfer step. The size of the lignocellulose raw material, therefore, is an important process parameter. The smaller the size, the larger the external surface area per unit mass. Even more crucial is the movement of active pulping chemicals through the microstructure of the lignocellulose to the reaction sites. Here also, small particle size is beneficial as the ions have to diffuse through smaller distances. At the beginning of the process, the diffusion rates would be different in different directions. The movement of mass in the fibre direction is faster than in the radial and tangential directions. This disparity, however, evens out as the reactions proceed with material removed from the fibre wall making it more porous and amenable to mass transfer across it. However, the size reduction of lignocellulose consumes energy, especially for the denser and harder woody raw materials. One must strike a balance between the benefits of a smaller size of the lignocellulose raw material and the increase in energy consumption to obtain ever-decreasing particle size. The size reduction operation is also associated with the production of fines – size so small that it hinders the operation of the delignification reactor. Taking all this into account, the optimal size for a given alkaline delignification process is determined by experience. For processes where the cellulose fraction is to be converted to paper or paperboard, the size of the lignocellulose raw material is a few mm. This is to avoid excessive fines generation and associated damage to the cellulose fibres. In one laboratory experiment holm oak wood was milled into pieces to pass 10–20 mm mesh (Alaejos et al. 2006). Several hardwood species were chipped and screened into size fractions of 27–23 mm, 23–17 mm, and 17–11 mm prior to alkaline delignification (Santos et al. 2011). In a Swedish study reed canary grass was hammer milled to pass through a 20 mm screen and studied for kraft and soda anthraquinone delignification (Finell and Nilsson 2004). Congress grass was chopped to 10–20 mm pieces prior to delignification (Ghatak 2002). One Scandinavian research group used 2–4-mm-sized spruce chips for their pulping studies (Iakovlev et al. 2010). Norway spruce chips measuring 5 mm in the grain direction and 6–7 mm in the tangential direction were used in a kraft pulping study (Molteberg and Høibø 2006). For alkaline delignification of rapeseed stalk, the raw material was cut into 3–5 cm lengths (Tofanica and Puitel 2018). In contrast, where the primary motive behind delignification is to facilitate carbohydrate hydrolysis for ethanol production, the raw material is often ground to an even sub-mm particle size. A mixture of spruce and pine was ground to 150 μm size to delignify it by kraft or soda anthraquinone method to obtain cellulose (Reguant et al. 1997; Angles et al. 2003). Researchers report grinding of bagasse to 0.5 mm size before subjecting it to alkaline delignification (Kininge and Gogate 2022).

2.1.2.3 Amount of Delignification Chemicals Used

Being a chemical reaction, the amount of delignification chemicals used plays a key role in the process. For near-complete delignification, the amounts of chemicals used must exceed a minimum threshold value. This depends on the lignin content of the lignocellulose. Softwoods, like spruce and pine, and some species of bamboo have higher lignin content and require more delignification chemicals to be used. On the other hand, straws and bagasse contain lower amounts of lignin and require lesser amount of delignification chemicals. Chemicals requirement for hardwoods, like eucalyptus and poplar, would be intermediate between softwoods and straws. Besides the extent of delignification, the amount of chemicals used also affects delignification kinetics. The concentration of reacting species in the liquid phase depends on the amount of chemicals used. Along with chemical reactions, mass transfer steps are strongly influenced by the concentration of migrating species in the bulk of the liquid phase. It is not only the rate and extent of delignification that increases with the increased amount of chemicals used but also the rate and extent of carbohydrate degradation and loss. A balance, therefore, must be maintained. In alkaline delignification processes, sodium hydroxide is the primary source of hydroxyl ions and alkalinity in the system. However, other delignification chemicals, like sodium sulphide in the kraft process, also contribute to the system's alkalinity. In determining the amount of chemicals to be used all the sources of hydroxyl ions must be accounted for. In the kraft process, the amount of delignification chemicals used is normally expressed either as active alkali – sum of the amounts of sodium hydroxide and sodium sulphide – or as effective alkali – sum of sodium hydroxide and half the amount of sodium sulphide. Moreover, the amounts of different chemicals are expressed in terms of the equivalent amount of sodium oxide. Kraft delignification of black spruce – a softwood species – with 18.5% active alkali on the dry mass of raw material is reported (Ban et al. 2004). Similarly, Norway spruce was delignified with 21% active alkali as NaOH by the kraft process (Iakovlev et al. 2010). Kraft delignification of oriental spruce was performed with two active alkali charges of 20% and 30% (Tutus et al. 2010). Soda anthraquinone delignification of a mixture of spruce and pine was carried out with 20–40% sodium hydroxide – 15.5% to 31% as Na_2O – addition (Martinez et al. 1997; Reguant et al. 1997). Kraft delignification of holm oak – a hardwood species – was studied using 10%, 15%, and 20% active alkali on dry wood (Alaejos et al. 2006). In another study, softwood pine was delignified with 19% active alkali and hardwoods acacia, sweetgum, and eucalyptus were delignified with 16% active alkali (Jardim et al. 2022). Reed canary grass was delignified by the kraft process using 14–21% effective alkali to study the effect of the amount of added chemicals (Finell and Nilsson 2004). Congress grass was delignified by soda and soda anthraquinone process by addition of 14.3% and 16.7% sodium hydroxide (Ghatak 2002). Kraft delignification of rapeseed stalk using 12.4–17% active alkali charge was performed to optimize the process parameters (Tofanica and Puitel 2018). Bagasse delignification by soda process with 16% sodium hydroxide on the mass of raw material is reported (Hamzeh et al. 2009; Khristova et al. 2006). Another study reports using 17% sodium hydroxide for the soda delignification of bagasse. Sodium hydroxide addition was reduced to 15.5% when the bagasse was hot water extracted prior to delignification by the soda anthraquinone process (Lei et al. 2010). 19% sodium hydroxide based on dry raw material was added for soda delignification of wheat straw and bagasse (Singh and Ghatak 2017). Wheat straw was delignified by soda anthraquinone process using 16% sodium hydroxide on raw material (Tripathi et al. 2019).

2.1.2.4 Composition and Concentration of Delignification Chemicals Used

As already mentioned, the alkaline delignification process has several variations. In all the processes, the principal source of alkali is sodium hydroxide. In the kraft process, sodium sulphide is used in addition to sodium hydroxide. Sodium sulphide contributes both hydroxyl ions and hydrosulphide ions to the reacting system. While hydroxyl ions are reactive to lignin as well as carbohydrates, hydrosulphide ions react exclusively with lignin. The addition of sodium sulphide, therefore, improves the selectivity of delignification. Keeping this in view, relative amounts of sodium hydroxide and sodium sulphide used in kraft delignification have an important bearing on the process outcome. The parameter sulphidity, expressed as a percentage, is defined as

$$\text{Sulphidity} = \frac{\text{Amount of } Na_2S}{\text{Amount of NaOH} + \text{Amount of } Na_2S} \times 100$$

To an extent, increased sulphidity improves delignification while reducing carbohydrate degradation and loss. Excessive sulphidity, however, is not warranted as the increased possibility of sulphur emissions overshadows the marginal benefits of delignification. Kraft delignification of black spruce employed 30% sulphidity (Ban et al. 2004). Norway spruce kraft delignification at 35% sulphidity is reported (Iakovlev et al. 2010). Oriental spruce delignification at 20 and 30% sulphidity is reported (Tutus et al. 2010). Holm oak was delignified by kraft process at 25% sulphidity (Alaejos et al. 2006). For the kraft delignification of pine, acacia, sweetgum, and eucalyptus, researchers used 25% sulphidity (Jardim et al. 2022). Finell and Nilsson delignified reed canary grass with a relatively high sulphidity of 38% (Finell and Nilsson 2004). Kraft delignification of rapeseed stalk used 22.1% sulphidity (Tofanica and Puitel 2018). In another important process variation of alkaline delignification, anthraquinone is added along with sodium hydroxide. In its catalytic role, anthraquinone accelerates delignification while simultaneously protecting carbohydrates against attack by alkali. A small amount of anthraquinone is generally sufficient to get the desired effect. Soda anthraquinone delignification of holm oak was carried out with anthraquinone addition of 0.05% of the dry mass of raw material (Alaejos et al. 2006). Steam prehydrolysed saw dust mixture of spruce and pine was delignified by soda anthraquinone process using 40% sodium hydroxide and 0.1% anthraquinone based on dry raw material mass (Angles et al. 2003). A softwood mixture of spruce and pine was delignified with 0.1% anthraquinone in the presence of different amounts of sodium hydroxide (Martinez et al. 1997). 0.1% anthraquinone was added for soda anthraquinone delignification of reed canary grass (Finell and Nilsson 2004). Anthraquinone addition varied from 0.06% to 0.12% during soda anthraquinone delignification of congress grass (Ghatak 2002). Bagasse delignification is reported with 0.1% anthraquinone at 14% sodium hydroxide charge (Khristova et al. 2006). In another study, bagasse was extracted with hot water prior to soda anthraquinone delignification with 0.1% anthraquinone addition on raw material mass (Lei et al. 2010). Soda anthraquinone delignification of wheat straw was performed with an addition of 0.05% anthraquinone (Tripathi et al. 2019). One research group reports the addition of 0.5% anthraquinone on raw material mass during the kraft delignification of oriental spruce (Tutus et al. 2010).

2.1.2.5 Liquid-to-Solid Ratio

Delignification processes are heterogeneous processes where chemicals present in aqueous solution react with lignin and carbohydrates present in the solid matrix of the lignocellulose. In this context, the amount of liquid phase present, measured as the liquid-to-solid ratio, plays an important role. It should be sufficient to immerse the entire solid mass with free liquid available for circulation. The circulation of free liquid provides for the necessary turbulence in the bulk of the liquid phase and lowers the resistance to external mass transfer steps. This is important for process kinetics and uniformity of delignification. Efficient liquid circulation also ensures effective heat transfer and temperature uniformity essential to maintain uniform delignification. The liquid-to-solid ratio, however, should not be very high. For the same amount of delignification chemicals used, a high liquid-to-solid ratio requires more water addition. This lowers the chemical concentration and can have adverse effects on both reaction rate and rate of mass transfer. A high liquid-to-solid ratio will also be demanding on the size of the delignification reactor and auxiliary equipment like heat exchangers and pumps. Energy consumption will also increase as the additional water present in the system has to be heated to the reaction temperature. Finally, the concentration of dissolved lignin and other constituents is reduced in the spent liquid if the liquid-to-solid ratio is low. This poses problems in the recovery of chemicals. Kraft delignification of black spruce was performed at a liquid-to-solid ratio of 4 (Ban et al. 2004). Similarly, holm oak was delignified at a liquid-to-solid ratio of 4 by kraft, and soda anthraquinone process (Alaejos et al. 2006). Recently, another research group working on alkaline delignification of different softwoods and hardwoods used a liquid-to-solid ratio of 4 (Jardim et al. 2022). Kraft delignification of Norway spruce utilized a liquid-to-solid ratio of 6 (Iakovlev et al. 2010). On the other hand, steam exploded softwood mixture of spruce and pine was delignified by soda anthraquinone process at a liquid-to-solid ratio of 10 (Angles et al. 2003). Kraft and soda anthraquinone delignification of reed canary grass at a liquid-to-solid ratio of 5 is reported (Finell and Nilsson 2004). Similarly, rapeseed stalk was delignified at a liquid-to-solid ratio of 5 by kraft, and soda anthraquinone processes (Tofanica and Puitel 2018). Soda and soda anthraquinone delignification of bagasse was carried out at a liquid-to-solid ratio of 4 (Khristova et al. 2006). In another study, bagasse was hot water extracted prior to soda and soda anthraquinone delignification at a liquid-to-solid ratio of 5.5 (Lei et al. 2010). The liquid-to-solid ratio was 5.2 during the soda delignification of wheat straw and bagasse (Singh and Ghatak 2017). Soda anthraquinone delignification of wheat straw was carried out at a liquid-to-solid ratio of 4 (Tripathi et al. 2019). Soda and soda anthraquinone delignification of congress grass was performed at a liquid-to-solid ratio of 6 (Ghatak 2002).

2.1.2.6 Reaction Temperature

The temperature at which the delignification reaction is carried out has a profound influence on the process. An increase in temperature increases the reaction rate, both for delignification as well as for carbohydrate degradation. Delignification temperature, therefore, affects the process selectivity. The activation energy for kraft delignification of softwoods is reported to be 140 kJ/mol (Santos et al. 2011; Sjöström 1981) and for several hardwoods, it was found to vary between 131 and 139 kJ/mol (Santos et al. 2011). For soda delignification of Western Hemlock, researchers have determined an activation energy of 131 kJ/mol (Dolk et al. 1989). In contrast, the activation energy for cellulose degradation by alkaline hydrolysis is estimated to be 150 kJ/mol (Fearon et al. 2020). For xylan, an important hemicellulose, the activation energy for alkaline degradation is reported to be 167 kJ/mol (Christensen et al. 1983). The higher activation energy for carbohydrate degradation compared to delignification suggests that increased temperature would accelerate the carbohydrate degradation more than the delignification reaction. Thus, increasing temperature increases the reaction rate but decreases the selectivity for delignification. The choice of temperature is governed by a trade-off between these two contrasting trends. If the temperature is too low, the delignification may be so sluggish that abnormally large reactors would be required. Also, at very low temperatures, carbohydrate degradation through a peeling reaction may be significant though alkaline hydrolysis is suppressed. If the temperature is too high, the selectivity of delignification would be compromised. Higher temperature also increases the pressure inside the reactor necessitating thicker shells. Finally, processes at higher temperatures consume more energy. Kraft delignification of black spruce was carried out at 172 °C (Ban et al. 2004). Kraft and soda anthraquinone delignification of a sawdust mixture of pine and spruce was performed at 170 °C (Angles et al. 2003). Kraft delignification temperature was set at 170 °C for Norway spruce (Iakovlev et al. 2010). The reaction temperature for kraft and kraft anthraquinone delignification of oriental spruce was 170 °C (Tutus et al. 2010). One research group applied 170 °C temperature for softwood and 160 °C temperature for hardwoods during kraft delignification (Jardim et al. 2022). Kraft and soda anthraquinone delignification of holm oak was studied at three temperature levels of 135, 165, and 195 °C. More delignification was observed at higher temperatures, but carbohydrate degradation was most pronounced at 135 °C (Alaejos et al. 2006). Kraft delignification kinetics of hardwood sawdust of different species at 150 °C is reported (Santos et al. 2011). Reed canary grass was delignified by kraft, and soda anthraquinone process at 165 °C (Finell and Nilsson 2004). Soda delignification of bagasse is reported at 165 °C (Hamzeh et al. 2009). Delignification temperature was 165 and 155 °C, respectively, for soda and soda anthraquinone process for bagasse (Khristova et al. 2006). Bagasse was delignified by soda process at 160 °C and the process was compared with soda anthraquinone delignification at 155 °C with the bagasse pre-extracted with hot water (Lei et al. 2010). The reaction temperature was 160 °C for the soda delignification of wheat straw and bagasse (Singh and Ghatak 2017). A delignification temperature of 166 °C was used in the soda anthraquinone process for wheat straw (Tripathi et al. 2019). To study the effect of delignification temperature, kraft and soda anthraquinone delignification of rapeseed stalks was performed at 150, 160, and 170 °C. More delignification as well as carbohydrate degradation was observed at higher temperatures (Tofanica and Puitel 2018). The reaction temperature was 175 °C for the delignification of congress grass by soda and soda anthraquinone process (Ghatak 2002).

2.1.2.7 Time Required to Heat the Reacting System to the Desired Reaction Temperature

The time required to heat the reacting system to the desired reaction temperature is important for several reasons. For batch delignification processes it increases the time required to complete a batch, thereby reducing the plant output. For a continuous process, it increases the equipment size for the same throughput or reduces the throughput for the same equipment size. On the reaction front, during the heat-up period before the system attains 130–140 °C temperature, the rate of delignification is substantially low and commercially insignificant. Yet, even at these low temperatures, considerable carbohydrate degradation may occur due to peeling reaction. From these considerations, one should strive for rapid heating. However, too rapid a heating is detrimental. Being a heterogeneous process mass transfer steps are as important as the chemical reactions. Uniformity of delignification requires that all the reaction sites within the solid matrix of the lignocellulose are uniformly exposed to the delignification chemicals. For this to happen, mass transfer steps must keep pace with the chemical reactions. With an increase in temperature, the rates of chemical reactions exponentially increase according to Arrhenius law. The increase in the rate of mass transfer with temperature, on the other hand, is not so rapid. It is highly desirable to ensure adequate impregnation of delignification chemicals throughout the lignocellulose mass before the temperature is high enough for the chemical reactions to be rapid. This sets a limit on the rate of heating to the final delignification temperature. Alternatively, the reacting mass is first heated to an intermediate temperature of 100–120 °C and allowed impregnation time before heating to the final temperature. In plant operation, the size of heat exchange equipment and the availability of steam also restrict the rate of heating. Kraft delignification of black spruce utilized a ramp time of 100 min to 172 °C (Ban et al. 2004). The heating time to the final temperature of 165 °C was 60 min for kraft and soda anthraquinone delignification of canary reed grass (Finell and Nilsson 2004). Bagasse was delignified by soda anthraquinone process after heating to 155 °C in 57 min (Khristova et al. 2006). Heating to 160 °C required 60 min during soda delignification of wheat straw and bagasse (Singh and Ghatak 2017). The reacting system was heated to the final temperature of 175 °C in 90 min during soda and soda anthraquinone delignification of congress grass (Ghatak 2002). In one kraft delignification protocol, Norway spruce chips were first impregnated with chemicals for 70 min at 120 °C, followed by heating to 170 °C in 20 min (Iakovlev et al. 2010). Similarly, hot water extracted bagasse was delignified by soda anthraquinone process in which the reacting system was first heated to 110 °C in 30 min, held at this temperature for 15 min to allow impregnation, and finally heated to 155 °C in another 45 min (Lei et al. 2010). Rapeseed stalk was delignified by kraft and soda anthraquinone process at 150, 160, and 170 °C. In each case, the heating time from 100 °C to the final temperature was 60 min (Tofanica and Puitel 2018). During the kraft delignification of pine and several hardwoods, the final delignification temperature of 170 and 160 °C, respectively, were attained within 30–40 min (Jardim et al. 2022). One study reports rapid heating during kraft, and soda anthraquinone delignification of holm oak with the reacting mass reaching the final delignification temperature of 165 and 195 °C within 17 min (Alaejos et al. 2006).

2.1.2.8 Reaction Time at the Desired Reaction Temperature

Major part of delignification and dissolution of depolymerized lignin fragments take place once the reacting mass is heated to the final delignification temperature. Therefore, the reacting system must be maintained at this temperature for adequate time. Not only delignification but also the reactions degrading carbohydrates are at their fastest during this period. A balance between extent of delignification and the extent of carbohydrate loss is to be maintained. Accordingly, prolonged reaction time at final delignification temperature is best avoided. Reaction time also affects plant operation in terms of productivity. Extended reaction time increases the time required to complete a batch in batch operation. For continuous operations larger reactor would be required if the reaction time is more. Hold time was 90 min at 172 °C for kraft delignification of black spruce (Ban et al. 2004). Steam exploded sawdust mixture of spruce and pine was delignified by soda anthraquinone process. Reaction time at 170 °C was 135, 150, and 180 min in different experiments to study its effect (Angles et al. 2003). Reaction time at 170 °C was 125 min for Kraft delignification of Norway spruce (Iakovlev et al. 2010). Effect of reaction time at final delignification temperature was studied for kraft, and soda anthraquinone delignification of holm oak by keeping the reaction time at 30, 60, and 90 min (Alaejos et al. 2006). Kraft delignification of pine, and several hardwoods was performed at 170 and 160 °C, respectively. In different experiments the extent of delignification was varied as the time at final temperature changed from 30 to 150 min (Jardim et al. 2022). For kraft delignification of reed canary grass, the time at 165 °C delignification temperature was varied from 5 to 75 min. In soda anthraquinone delignification the time varied from 5 to 85 min (Finell and Nilsson 2004). Bagasse was delignified by soda process with a reaction time of 60 min at 165 °C (Hamzeh et al. 2009). Soda delignification of bagasse was performed with 90 min reaction time at 165 °C. For soda anthraquinone delignification these parameters were 60 min and 155 °C (Khristova et al. 2006). Reaction time was 30 min at 155 °C for soda anthraquinone delignification of hot water extracted bagasse (Lei et al. 2010). Soda anthraquinone delignification of wheat straw required 20 min at 166 °C (Tripathi et al. 2019). For soda delignification of wheat straw and bagasse, the reaction time at 160 °C was 150 min (Singh and Ghatak 2017). Rapeseed stalk was subjected to kraft, and soda anthraquinone delignification by providing a reaction time of 60 min at different delignification temperatures (Tofanica and Puitel 2018). Congress grass delignification by soda and soda anthraquinone process required 120 min at 175 °C (Ghatak 2002).

2.1.2.9 Interaction among Process Parameters

The outcome of any alkaline delignification process depends not only on the individual process parameters but also on the interaction among them. Mass transfer effects during delignification depend on the size of the lignocellulose raw material, the concentration of the delignification chemicals, the liquid-to-solid ratio, and the time-temperature protocol during delignification. Mass transfer restriction due to lowering of the concentration of delignification chemicals, to some extent, can be compensated by the use of lignocellulose particles of smaller size. With mass transfer being facilitated by smaller particle sizes, it may be possible to delignify with a somewhat lower liquid-to-solid ratio. Ease of impregnation with smaller particle size also allows faster heating to final delignification temperature without compromising on uniformity. The concentration of chemicals during delignification depends on the amount of chemicals used and the liquid-to-solid ratio. Provided they remain above the threshold values, both these parameters can be lowered by using lignocellulose particles of smaller size. Above their respective threshold values, the parameters amount of chemicals used, the liquid-to-solid ratio, and the final delignification temperature are closely interlinked. The combination of the amount of chemicals used and the liquid-to-solid ratio determines the concentration of chemicals. Increased chemical concentration accelerates the reactions, making it possible to achieve the same extent of delignification at lower temperatures. The interactive influence of alkali concentration, delignification temperature, and delignification time at the final temperature,

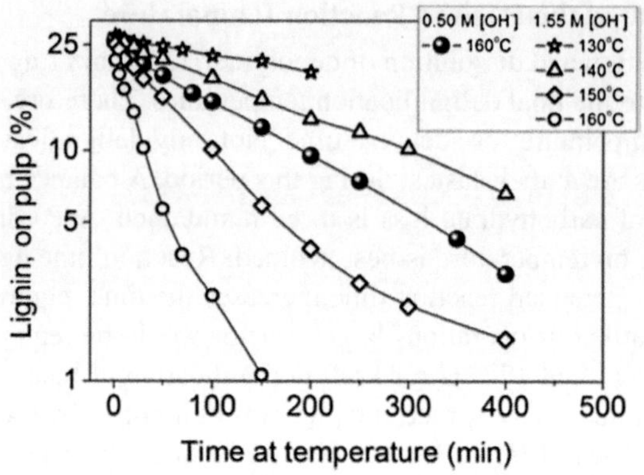

FIGURE 2.4 Interactive effects of chemicals concentration, final delignification temperature, and reaction time during kraft delignification. (Reprinted from Paananen, M, Liitia, T, Sixta H, *Ind. Eng. Chem. Res.*, 52, 12777–12784, 2013, with permission from American Chemical Society.)

for kraft delignification of scot pine wood meal, is depicted in Figure 2.4. Delignification is carried out at a sulphidity of 33% and a high liquid-to-solid ratio of 200 to ensure uniform concentration of chemicals during the entire course of the reaction. At the same temperature, increasing the alkali concentration substantially reduces the time required for the same extent of delignification. Conversely, at the same alkali concentration, increasing the temperature reduces the time required for the same extent of delignification (Paananen et al. 2013).

The most significant interaction is observed among the parameters of reaction temperature, time required to heat the reacting system to the desired reaction temperature, and reaction time at the desired reaction temperature. These three parameters are combined together into a single parameter known as the H-factor (Vroom 1957). H-factor for delignification can be represented as

$$H = \int_0^t k_r dt \tag{2.1}$$

Where k_r, the relative rate constant, is the ratio of the reaction rate at a given temperature to the reaction rate at 100 °C. Time, t, is in hours. For kraft delignification, with an activation energy of 134 kJ/mol assumed by Vroom, k_r is

$$k_r = \exp\left[43.2 - \left(\frac{16113}{T}\right)\right] \tag{2.2}$$

It is believed that the extent of delignification is a direct function of the H-factor and increasing the H-factor results in more delignification irrespective of the exact time temperature protocol. Kraft delignification of pine resulted in 39.4% lignin removal at an H-factor of 503. Delignification increased to 89.6% when the H-factor increased to 1400 (Jardim et al. 2022). Figure 2.5 illustrates the dependence of the extent of delignification on the H-factor and its interaction with the concentration and composition of delignification chemicals for soda anthraquinone delignification of scot pine wood meal. For all concentrations and compositions, the extent of delignification increases with an increase in H-factor.

Increasing the alkali concentration increases the extent of delignification at any given H-factor. Significantly higher delignification can be achieved even at a much lower H-factor if the alkali

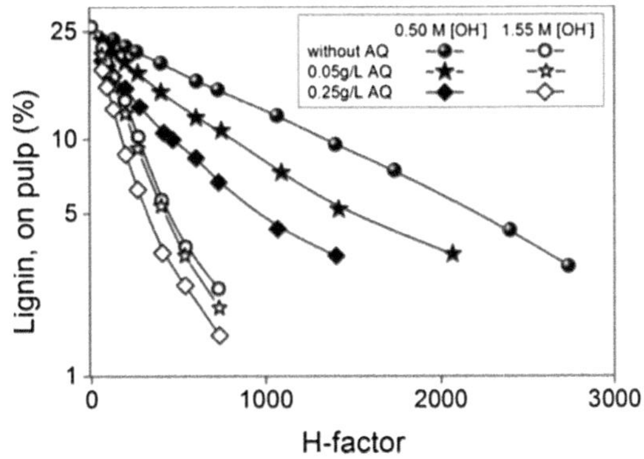

FIGURE 2.5 Interactive effects of chemicals concentration, H-factor, and anthraquinone addition during soda anthraquinone delignification. (Reprinted from Paananen, M, Liitia, T, Sixta H, *Ind. Eng. Chem. Res.*, 52, 12777–12784, 2013, with permission from American Chemical Society.)

concentration is higher. Anthraquinone addition has a profound influence on delignification, especially with low alkali concentrations.

2.2 DISSOLUTION IN ACIDIC MEDIA

Lignin can be depolymerized and dissolved by reacting it with bisulphite- and sulphite-based delignification chemicals in aqueous acidic media. The sulphite delignification process, as it is known, predates the kraft process – the major alkaline delignification process. Credit for inventing the sulphite process goes to Tilghman who obtained a patent in 1867 for the delignification of wood with the help of calcium bisulphite. Before this, Julius Roth in 1857 had used sulphurous acid to produce paper pulp from wood. In Tilghman's process calcium ions were the counterions – commonly known as base – to bisulphite ions. Magnesium-based sulphite delignification process was developed by Ekman and commercialized in 1879. Both bisulphite and sulphite ions, as well as an aqueous solution of sulphur dioxide, are capable of depolymerizing lignin at elevated temperatures. The lignin fragments – the lignosulphonates – remain dissolved in the liquid under the prevailing conditions. The spent liquid from the sulphite process, therefore, constitutes an important lignin-rich liquid byproduct stream. Compared to kraft delignification, acid sulphite processes are less selective to lignin and all the hemicelluloses are also degraded and dissolved in the spent liquor. As a result, this process is preferred over the kraft process to produce dissolving grade pulps. These pulps are the starting materials for industrially important cellulose derivatives like rayon, and cellophane.

2.2.1 Chemistry of Acidic Delignification

Ether linkages in lignin can be broken to a moderate extent by acid-catalysed reactions in processes known as acidolysis or autohydrolysis. Mildly acidic conditions are maintained at elevated temperatures leading to partial depolymerization and softening of lignin. These processes, however, are not meant for dissolution and removal of lignin from the solid matrix of lignocellulose. Rather, their importance lies in reducing the recalcitrance of lignocellulose and facilitating cellulose hydrolysis to release fermentable sugars. Near-complete depolymerization and dissolution of lignin fragments can be achieved by reaction with aqueous sulphur dioxide and/or bisulphite ions. In the process, lignin is extensively sulphonated to impart hydrophilic character and water solubility. Sulphonated lignins have several important industrial applications.

2.2.1.1 Equilibria of Sulphite-Based Delignification Chemicals

Acidic delignification processes are premised on the dissolution of sulphur dioxide in water to form hydrated sulphur dioxide and its subsequent dissociation and resulting ionic equilibria. These can be schematically represented as

$$SO_2 + H_2O \leftrightarrow SO_2{\bullet}H_2O \qquad \{2.1\}$$

$$SO_2{\bullet}H_2O \leftrightarrow H^+ + HSO_3^- \qquad \{2.2\}$$

$$HSO_3^- \leftrightarrow H^+ + SO_3^{2-} \qquad \{2.3\}$$

The first reversible process represents a dissolution equilibrium dependent on the temperature and pressure of the system. The other two processes represent ionic equilibria governed by temperature and hydrogen ion concentration. However, being in sequence, these equilibria are interdependent. For these sequential reversible processes, increasing solution pH drives the equilibria to the right and so does an increasing partial pressure of sulphur dioxide gas. Increasing sulphur dioxide partial pressure has practical limits in industrial operations due to the strength of material consideration. The solution pH can be increased by the addition of a base, thereby allowing more sulphur dioxide to dissolve and get hydrated to subsequently dissociate and form bisulphite and sulphite ions. Hydroxides of calcium, magnesium, ammonium, and sodium, have all been used as bases for this purpose. For the equilibrium process {2.2} the equilibrium constant at 25 °C is reported between 0.008 and 0.02. For the equilibrium {2.3} it is estimated between 2 and 10 at 25 °C (Goldberg and Parker 1985). Temperature plays an important role in the equilibria as well, especially the dissolution equilibrium {2.1}. Increased temperature decreases the amount of sulphur dioxide dissolved which must be compensated by increasing the pressure, which has its practical limits. In a system containing only sulphur dioxide and water the amount of $SO_2{\bullet}H_2O$ can be empirically estimated as (Evtuguin 2016)

$$C_{SO_2{\bullet}H_2O} = 0.03 p_{SO_2} 0.0363^{-T} \qquad (2.3)$$

Here, $C_{SO_2{\bullet}H_2O}$ is the molar concentration of $SO_2{\bullet}H_2O$, p_{SO_2} is the sulphur dioxide partial pressure measured in mm of Hg, and T is the temperature in °C. In an aqueous medium, $SO_2{\bullet}H_2O$, HSO_3^-, and SO_3^{2-}, can all react with lignin and cause its depolymerization and dissolution. At strongly acidic conditions – typical pH 1 to 2 – $SO_2{\bullet}H_2O$ is the predominant moiety responsible for delignification. These processes are commonly known as acid sulphite processes (Deshpande et al. 2018; Fiskari and Kilpeläinen 2021). Besides delignification, the harsh acidic conditions lead to near-complete hydrolysis and dissolution of hemicelluloses. The resulting residue is almost pure cellulose and useful for further derivatization. Under relatively mild acidic conditions – typical pH 4 to 5 – HSO_3^- is the predominant ion in the solution and acts as the major delignification agent. These delignification processes are referred to as bisulphite processes (Deshpande et al. 2018). In sulphite-based acidic delignification terminology, the amounts and concentrations of different sulphur species are referred to as free SO_2, combined SO_2, and total SO_2. While free SO_2 refers to dissolved and hydrated SO_2, combined SO_2 stands for sulphur present as bisulphite and sulphite ions. The sum of free and combined SO_2 is the total SO_2.

2.2.1.2 Lignin Reactions

Sulphonation and acid-catalysed cleavage of ether linkages are the main lignin reactions in acidic media. The reaction is invariably initiated by a hydronium ion attack on lignin moieties producing carbonium or oxonium ions. These ions are susceptible to further attack by nucleophiles such as $SO_2{\bullet}H_2O$, and HSO_3^- producing the sulphonated derivative. Acid-catalysed cleavage of ether linkages is schematically presented in Figure 2.6. Initial attack by hydrogen ion abstracts the moiety

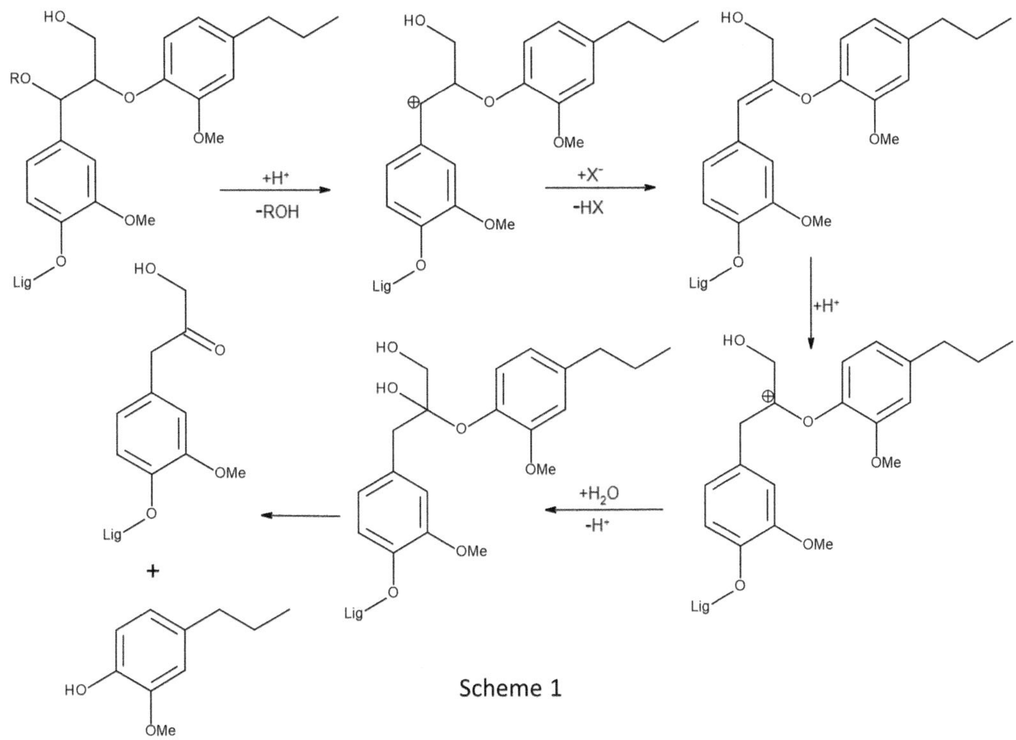

Scheme 1

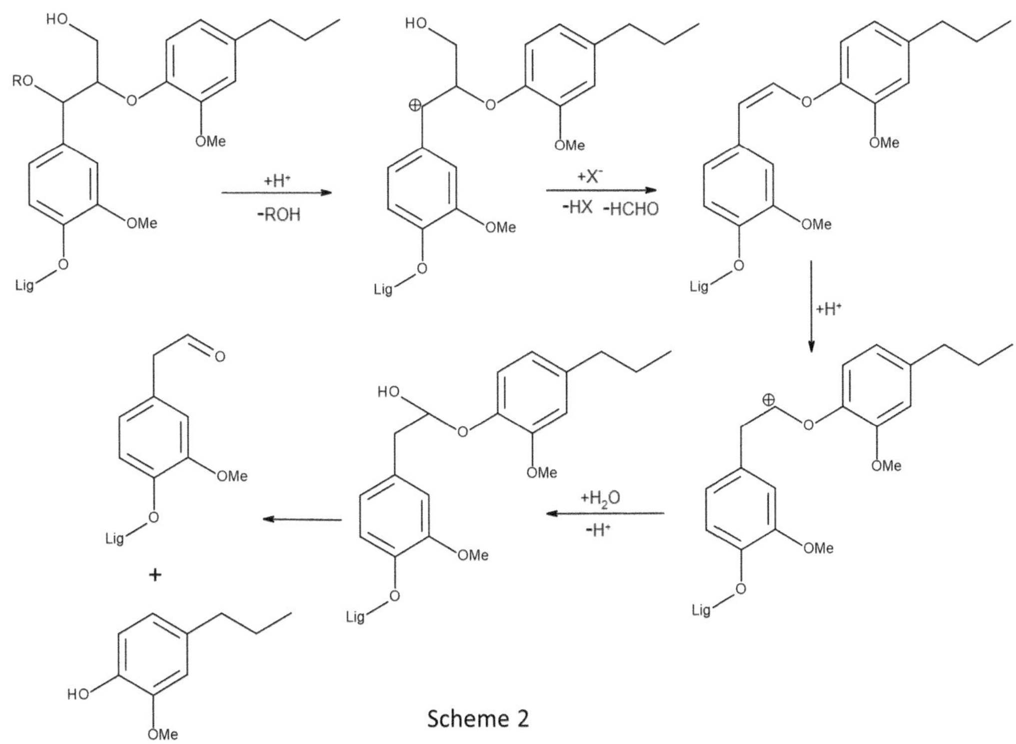

Scheme 2

FIGURE 2.6 Acid-catalysed breaking of ether linkages.

ROH from the lignin structure leaving behind a carbonium ion with a positive centre at the α carbon. In one possible transformation (Scheme 1), the carbonium ion intermediate undergoes a nucleophilic attack by the conjugate base on the α carbon, elimination of hydrogen ion from β carbon, and forming of a double bond between the α, and the β carbons. The resulting intermediate undergoes a second attack by hydrogen ion forming another carbonium ion, this time with the positive centre at the β carbon. A nucleophilic addition follows, resulting in an intermediate with hydroxyl at the β carbon. This unstable intermediate quickly dissociates with the breaking of the β-O-4/ ether linkage releasing two phenylpropane subunits with one having a ketone functional group (Lundquist and Lundgren 1972; Santos et al. 2013; Sturgeon et al. 2014). In another possible transformation (Scheme 2), the carbonium ion intermediate undergoes a nucleophilic attack by the conjugate base on the α carbon, elimination of a hydrogen ion, and a molecule of formaldehyde from β carbon, and forming of a double bond between the α, and the β carbons. The rest of the transformations are similar to those in Scheme 1, finally breaking the β-O-4/ ether linkage. One of the subunits in this case has a phenyl ethane skeleton with the β carbon having an aldehyde functional group (Lundquist and Ericsson 1970; Santos et al. 2013; Sturgeon et al. 2014). If an α-O-4/ linkage is present instead of β-O-4/ linkage, it is cleaved during the initial attack by hydrogen ion. The ROH leaving group in this case would be a 4-hydroxy phenylpropane unit.

When nucleophiles such as $SO_2{\bullet}H_2O$, and HSO_3^- are present in the reacting system, the overall reaction is a nucleophilic substitution reaction. $SO_2{\bullet}H_2O$, and HSO_3^- readily attack and add to the positive centre at the α carbon in the carbonium ion produced as a result of the initial attack by hydrogen ion (Deshpande et al. 2018; Evtuguin 2016). The product is a sulphonated structure with SO_3^- functional group at the α carbon. This functional group imparts sufficient polar nature and hydrophilic character to the lignin derivative making it water-soluble. The reaction is schematically shown in Figure 2.7.

An undesirable side reaction that limits the extent of delignification is the condensation reaction forming new carbon–carbon bonds between fragmented lignin moieties. Intramolecular rearrangements involving charge dislocation cause negatively polarized nucleophilic centres in the benzene ring. These can interact with the positively charged centre formed by the attack of hydrogen ions on

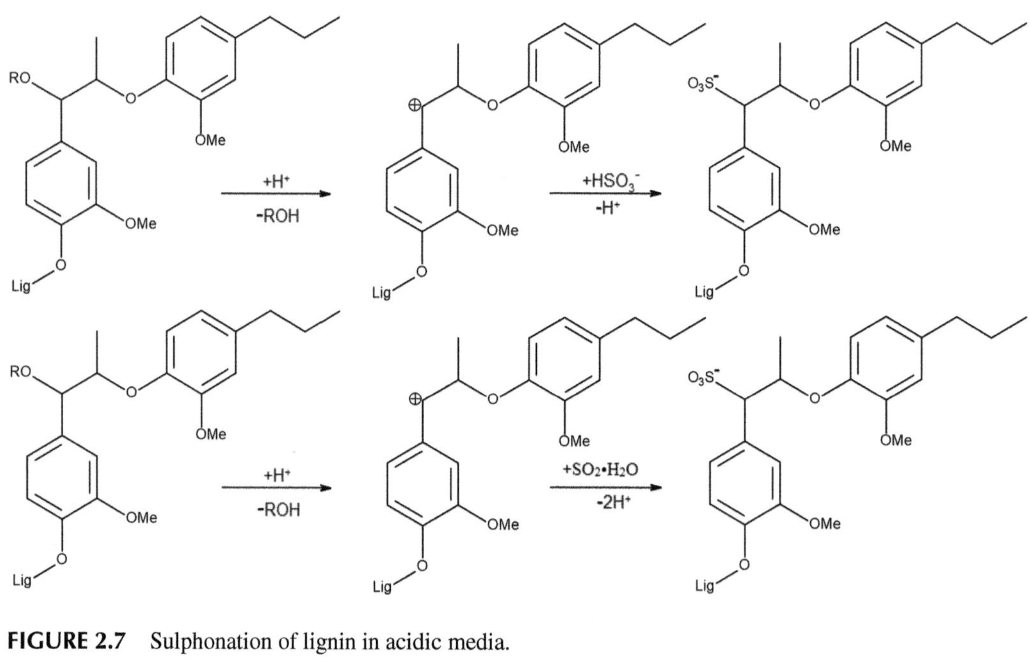

FIGURE 2.7 Sulphonation of lignin in acidic media.

the α carbon forming a carbon–carbon bond. Compared to alkaline media, acidic conditions favour the hydrolysis of carbohydrates. At the high acidity of the acid sulphite process, the breaking of glycosidic bonds is near-complete to produce the corresponding monosaccharides. In the bisulphite process, oligosaccharides are also formed in addition to the monosaccharides.

2.2.2 Process Parameters in Lignin Dissolution in Acidic Media

Acid sulphite and bisulphite delignification processes have complex heterogeneous reacting systems comprising solid, liquid, and gaseous phases. Besides chemical reactions, physical processes of heat and mass transfer are involved. Preparation of the delignification chemicals involves the dissolution of sulphur dioxide in a solution of magnesium hydroxide. This is a reactive absorption process involving the following steps.

1. Transfer of sulphur dioxide from the bulk of gas to the gas side of gas–liquid interface.
2. Transfer of sulphur dioxide across the interface to the liquid side.
3. Transfer of sulphur dioxide from the liquid side of gas–liquid interface to the bulk of liquid phase.
4. Reaction of sulphur dioxide and other sulphur species according to reactions {2.1}–{2.3}.

Steps 3 and 4 are not essentially sequential and proceed simultaneously. A high partial pressure of sulphur dioxide would increase the rate of step 1. Vigorous and intimate contact between the gas and the liquid phases accelerates step 2. The rate of step 3 increases by increasing the liquid phase turbulence. It is also affected by the rates of reactions and positions of reaction equilibria {2.1}–{2.3}, i.e., step 3. Reaction rates and equilibria in turn are governed by the composition of different sulphur species, system pH, and temperature. The liquid delignification chemicals, so formed, come in contact with the solid matrix of the lignocellulose. The process can be thought of as consisting of the following steps:

1. Transfer of active pulping chemicals from the bulk of the liquid to the external surface of the lignocellulose particle.
2. Movement of active pulping chemicals through the microstructure of the lignocellulose to the reaction sites
3. Reaction of active pulping chemicals with the constituents of lignocellulose – lignin and carbohydrates.
4. Movement of reaction products through the microstructure of the lignocellulose to the external surface of the lignocellulose particle.
5. Transfer of products from the external surface of the lignocellulose particle to the bulk of the liquid.

Based on these, the important process parameters are

1. Characteristics of lignocellulose raw material
2. Size of lignocellulose raw material
3. Amount of delignification chemicals used
4. Composition and concentration of delignification chemicals used
5. Liquid-to-solid ratio
6. Reaction temperature
7. Time required to heat the reacting system to the desired reaction temperature
8. Reaction time at the desired reaction temperature

2.2.2.1 Characteristics of Lignocellulose Raw Material

As in alkaline delignification, lignocelluloses with a higher amount of lignin require a higher amount of delignification chemicals. Acid sulphite and bisulphite delignification processes are less selective for lignin compared to alkaline processes. Though lignosulphonates are hydrophilic and water soluble, the breaking of β-O-4/ ether bonds is difficult. This, along with the possibility of condensation reactions, limits the extent of delignification compared to alkaline delignification. Hardwoods have a higher percentage of β-O-4/ ether bonds relative to softwoods and, therefore difficult to delignify by acid sulphite and bisulphite delignification processes (Santos et al. 2013). Another important raw material characteristic is the syringyl-to-guaiacyl ratio in the lignin structure. There is little variation in this ratio among different species of softwoods. Different hardwood species, however, have largely different syringyl-to-guaiacyl ratios (Pinto et al. 2005). The presence of polyphenolic extractives in the raw material adversely affects delignification as these moieties facilitate condensation reactions. These are even more noticeable for acid sulphite delignification (Fiskari and Kilpeläinen 2021). Softwood species having higher extractive content, like pine, are difficult to delignify compared to species like spruce and fir with low extractive contents. Extractives profiles of some selected lignocelluloses are presented in Table 2.3.

Acid sulphite delignification of eucalyptus removed 85% of extractives (Rodrigues et al. 2018). In another study, a limited extent of delignification of eucalyptus was attributed to dense wood structures (Fiskari and Kilpeläinen 2021). Carbohydrate degradation and formation of monosaccharides were reported during acid sulphite delignification of eucalyptus (Rueda et al. 2015). Acid sulphite and bisulphite delignification of pine was performed to assess the recalcitrance to delignification due to lignin carbohydrate linkages (Deshpande et al. 2018).

2.2.2.2 Size of Lignocellulose Raw Material

Mass transfer considerations in the overall process outcome are equally valid for acid sulphite, and bisulphite processes as they are for alkaline delignification. However, the anisotropy of lignocellulose to diffusion rate is not as irrelevant in acidic delignification as for alkaline processes. This is because of the propensity of the alkaline solutions to swell lignocellulose, especially when the

TABLE 2.3
Extractives Profiles of Select Lignocelluloses

	Alcohol Benzene Solubles (%)	Alcohol Solubles (%)	Acetone Solubles (%)	Reference
Pine	13–18		4	Kebbi-Benkeder et al. 2015; Park et al. 2021
Spruce	2		1–5	Caron et al. 2013; Füchtner et al. 2020; Kebbi-Benkeder et al. 2015; Kohlbrenner and Schuerch 1959
Bamboo	3–11	8.5		Ghatak 2002; Sugesty et al. 2015; Zhang et al. 2013.
Poplar	2–6		2.5–3.3	Ghatak 2002; Krutul et al. 2019; Myeong et al. 2010;
Eucalyptus	2.5–3.5		1.5	Bin et al. 2012; Ghatak 2002; Gutiérrez et al. 1999; Rodrigues et al. 2018
Wheat straw	0.5–2		2.3	Ghatak 2002; Tripathi et al. 2019
Rice straw	3–7	1.6		Inglesby et al. 2005; Tajmirriahi et al. 2021; Xiao et al. 2001
Bagasse	2–10	2.9		Ghatak 2002; Neves et al. 2016

temperature is raised. While in alkaline delignification the chip thickness is the controlling dimension, it is the chip length which is critical for the acidic processes (Hartler 1996). Acidic solutions must enter the lignocellulose structure through the lumen of the fibre as the diffusion across the cell wall is practically irrelevant. Reduced size of the lignocellulose raw material along the fibre direction and the associated increase in specific surface area, therefore, facilitates impregnation of chemicals and their uniform distribution throughout the solid matrix. The mass transfer rate should match the rate of chemical reaction. Pine and spruce chips of size 13–45 mm were used for acid sulphite and bisulphite delignification (Deshpande et al. 2016a; Deshpande et al. 2018). Nitric acid delignification of birch was carried out on raw material pieces of 30 × 30 × 2 mm size (Yang et al. 2019). Bisulphite delignification of fungal-treated pine and spruce chips of 18 mm nominal size is reported (Fischer et al. 1994). In similar experiments, pine chips of 16 mm average size were fungal-treated prior to delignification by acid sulphite and bisulphite processes (Scott et al. 1995). Eucalyptus chips of 15–30 mm length were used for acid bisulphite delignification (Rueda et al. 2015). The chips were 2–4 mm thick. In another study, the eucalyptus chips were 21.9 ± 5.5 mm long and 3.4 ± 5.5 mm thick for acid sulphite delignification (Sixta et al. 2004).

2.2.2.3 Amount of Delignification Chemicals Used

β-O-4/ ether linkages are not readily cleaved during acid sulphite and bisulphite processes. As a result, the lignin defragmentation is not as extensive as in alkaline processes. Therefore, the larger molecular fragments from partial lignin depolymerization must be sufficiently hydrophilic to be soluble in the reaction medium. This can be achieved only by extensive sulphonation. To ensure this, an adequate amount of desired sulphur species is charged at the beginning of the reaction. Since $SO_2•H_2O$ is a weaker nucleophile than HSO_3^-, more chemicals are required in acid sulphite delignification than in the bisulphite process. Compared to alkaline processes, acidic environments are capable of hydrolysing the carbohydrates to their corresponding monosaccharides. The amount of chemicals should be commensurate with this if carbohydrate hydrolysis is one of the process objectives. Along with total SO_2, the type and amount of base is important in maintaining the system pH. It follows from the equilibria {2.1}–{2.3} that higher sulphur dioxide partial pressure is required to maintain the same total SO_2 if the pH is low. Bisulphite processes fare better on this count. However, acid sulphite processes are effective for complete hemicellulose saccharification. In acid sulphite delignification of fungal-treated pine, 32% total SO_2 based on dry raw material is reported. This amount was 20% in the bisulphite process (Scott et al. 1995). Bisulphite delignification of spruce was studied with 12.4% total SO_2 charge on dry raw material (Deshpande et al. 2015). In another study, pine was delignified by both the acid sulphite and bisulphite processes. For the acid sulphite process, the total SO_2 charge was 17.5%, and for the bisulphite process, it was 12.4% (Deshpande et al. 2018). Other researchers applied 26% total SO_2 on raw material mass for acid sulphite delignification of pine (Shen and Chen 2009). 17% total SO_2 on raw material was charged for bisulphite delignification of pine and spruce which were subjected to fungal treatment beforehand (Fischer et al. 1994). Acid sulphite delignification of eucalyptus was carried out with 17.9–18.4% total SO_2 on raw material (Sixta et al. 2004). Hardwood species, acacia, and eucalyptus were delignified by acid sulphite process with 25–34% total SO_2 added to raw material mass (Fiskari and Kilpeläinen 2021). The total SO_2 charge was 21% for acid sulphite delignification of beech (Fischer and Schmidt 1991). In one particular study on acid sulphite delignification of eucalyptus, where the production of monosaccharides was the objective, a high total SO_2 charge of 150–170% was used (Rueda et al. 2015).

2.2.2.4 Composition and Concentration of Delignification Chemicals Used

In sulphite-based acidic delignification, the predominant delignifying moiety may be $SO_2•H_2O$ or HSO_3^- depending on the system pH. This in turn is controlled by the amounts and concentrations of the employed base, total SO_2, free SO_2, and combined SO_2. Acid sulphite processes work at lower pH and require less base concentration compared to bisulphite processes that work at relatively higher pH. The resulting ratio of combined SO_2 to free SO_2 is higher in bisulphite processes. Combined SO_2 is more effective in sulphonating lignin moieties than free SO_2. The concentration of chemicals not only affect the reaction rates but also the rate of mass transfer. For acid sulphite delignification of fungal-treated pine, combined SO_2 was between 30% and 34% of the total SO_2. For bisulphite delignification, this ratio was 50% (Scott et al. 1995). In bisulphite delignification of spruce, 56% of total SO_2 was present as combined SO_2 at a pH of 4.5. For this, a 5% base (sodium) charge as Na_2O on raw material mass was required (Deshpande et al. 2015). Similar chemical composition was employed for bisulphite delignification of pine. However, for acid sulphite delignification combined SO_2 was 16% of total SO_2 at 1.5 pH requiring 2.5% base charge as Na_2O (Deshpande et al. 2018). These researchers found the rate of bisulphite delignification to be first order with respect to total SO_2 concentration (Deshpande et al. 2016a). However, for acid sulphite delignification of birch, the delignification rate is reported to be second-order in SO_2 concentration, especially in the early stage of the reaction (Fischer and Schmidt 1991). Yet others estimated the reaction order to be 0.75 with respect to digester SO_2 pressure during acid sulphite delignification of pine (Shen and Chen 2009). In one study, calcium base reaction medium for acid sulphite delignification of pine had 15.5–19.5% of total SO_2 as combined SO_2 (Shen and Chen 2009). Bisulphite delignification of pine and spruce was carried out at 3.8 pH with 1.7% MgO and 5.1% total SO_2 concentration (Fischer et al. 1994). Of the total delignification chemicals charged, 65% was present as free SO_2 for acid sulphite delignification of eucalyptus (Sixta et al. 2004). Acid sulphite delignification of acacia and eucalyptus used 20% of total SO_2 as combined SO_2 with sodium base (Fiskari and Kilpeläinen 2021). One study reports the bisulphite delignification of spruce, birch, and aspen in a reaction medium at pH 3.5 with 4.2% total SO_2 concentration of which 52% was combined SO_2 (Khakimova et al. 2017). Calcium base acid sulphite treatment of eucalyptus to have delignification along with conversion of carbohydrates to monosaccharides used 20–21% of total SO_2 as combined SO_2 with pH between 1.1 and 1.5 (Rueda et al. 2015).

2.2.2.5 Liquid-to-Solid Ratio

Adequate mixing and agitation in the bulk liquid phase is a prerequisite for the macro-level uniformity of delignification. Turbulence in the liquid phase improves the mass transfer coefficient, thereby increasing the rate at which the delignification chemicals reach the external surface of the lignocellulose particle. Turbulence in the liquid phase also improves the rate at which the gaseous sulphur dioxide is transferred to the liquid phase. Mixing and agitation in the liquid phase is also essential for efficient convective heat transfer and temperature uniformity throughout the reacting mass. Therefore, sufficient free liquid must be present inside the reactor. The amount of liquid phase also affects the concentration of chemicals. The more the liquid phase, the less the concentration of chemicals for a given amount of delignification chemicals charged. From these considerations, the ratio of liquid to solid is an important process parameter. Too much liquid is also undesirable as it occupies reactor space at the cost of lignocellulose raw material. It also consumes additional thermal energy during the heat-up period of the reaction. Acid sulphite delignification of pine was carried out at a liquid-to-solid ratio of 6–7. The ratio was 4 for bisulphite pulping (Scott et al. 1995). Liquid-to-solid ratio was 4.6 during acid sulphite, and bisulphite delignification of spruce and pine (Deshpande et al. 2015; Deshpande et al. 2018). Pine and spruce were delignified by the bisulphite process at a liquid-to-solid ratio of 3.3 (Fischer et al. 1994). Industrial acid sulphite delignification of pine at a liquid-to-solid ratio of 4 is reported (Shen and Chen 2009). Beech was delignified by the acid sulphite process at a liquid-to-solid ratio of 2.7 and 3 (Fischer and Schmidt 1991). Mixed eucalyptus chips were delignified by the acid sulphite process at a liquid-to-solid ratio of 2.7 (Sixta et al. 2004). In one study, a high liquid-to-solid ratio of 20 was used during acid sulphite delignification of eucalyptus (Rueda et al. 2015). However, the industrial practice of acid sulphite delignification of eucalyptus at a liquid-to-solid ratio of 3 is reported (Marques et al. 2009).

2.2.2.6 Reaction Temperature

Temperature affects the chemical reactions; the rate increases with temperature according to the Arrhenius equation. For bisulphite delignification of spruce the activation energy was estimated to be 130 kJ/mol (Deshpande et al. 2015). For pine, the calculated activation energy was 121 kJ/mol (Deshpande et al. 2016b). For acid sulphite delignification of beech, the reported activation energy was 91 kJ/mol. The activation energy for cellulose degradation for the same process was found to be 176 kJ/mol (Fischer and Schmidt 1991). For hemicellulose degradation, the reported activation energy was 87 kJ/mol for the bisulphite process (Deshpande et al. 2015). It is evident that cellulose degradation is far less sensitive to increasing temperature than hemicellulose degradation. Hydrolysis of hemicelluloses to corresponding monosaccharides is near-complete at the industrially prevalent delignification temperatures. Temperature also affects mass transfer rates but not as intensely as it affects the reaction rates. The reacting system in acid sulphite and bisulphite processes has all three phases – gas, liquid, and solid. Besides reaction and mass transfer rates, the system pressure is strongly related to temperature. The sharp rise in system pressure with increasing temperature sets the limit for reaction temperature due to structural constraints on the material of construction. Bisulphite delignification of spruce was studied in the temperature range of 142–165 °C (Deshpande et al. 2015; Deshpande et al. 2016a). The reaction temperature was 142 °C for the acid sulphite process and 154 °C for the bisulphite process during the delignification of pine (Deshpande et al. 2018). Pine was delignified at a temperature of 138–139 °C using the acid sulphite process. Bisulphite delignification of fungal-treated spruce and pine was carried out at a temperature of 155 °C (Fischer et al. 1994). Acid sulphite delignification of beech was performed at temperatures ranging from 132 to 145 °C (Fischer and Schmidt 1991). Temperatures ranging from 130 to 150 °C were employed during acid sulphite delignification of eucalyptus with simultaneous hydrolysis of hemicellulose to monosaccharides (Rueda et al. 2015). Researchers maintained a delignification temperature of 138 °C for acid sulphite delignification of eucalyptus (Sixta et al. 2004). Industrial acid sulphite delignification of eucalyptus carried out at 130 °C is reported (Marques et al. 2009). Another study reports acid sulphite delignification of eucalyptus at 142 °C (Rodrigues et al. 2018). From the foregone discussion, it is evident that acid sulphite delignification is generally practised at a lower temperature than bisulphite delignification. The reason behind this is the system pH and the relative positions of the equilibria {2.1}–{2.3}. Under the low pH conditions of the acid sulphite process, there is very little dissociation of $SO_2{\bullet}H_2O$. As a result, higher sulphur dioxide partial pressure is required to maintain the desired total SO_2 concentration. This poses a restriction on the working temperature so that the design limit of the reactor is not exceeded.

2.2.2.7 Time Required to Heat The Reacting System to the Desired Reaction Temperature

A heating period is inevitable to bring the reacting system from its initial temperature to the final reaction temperature. This time is an important process parameter due to several reasons. Rapid heating to the final temperature with a reduced heating period can increase the production rate for the same equipment. However, impregnation of lignocellulose mass with the delignification chemicals before the temperature is sufficiently high is equally important. It is necessary to maintain uniformity of delignification and hemicellulose hydrolysis. This is due to the fact that mass transfer rates lag behind chemical reaction rates with increasing temperature. Acidic delignification liquor has to enter the lignocellulose mass almost entirely along the fibre direction. This necessitates a longer impregnation time than alkaline systems. Heat transfer infrastructure and availability of steam are other considerations that decide the rate of heating. A study reports a heating time of 5 h from 70 to 140 °C during acid sulphite delignification of pine. This duration was 1.5 h for bisulphite delignification (Scott et al. 1995). Heating time to 140 °C for acid sulphite delignification of fungal-treated eucalyptus was 5 h 20 min (Christov et al. 1998). Other researchers provided a 2 h heating time during acid sulphite delignification of eucalyptus irrespective of final delignification temperature (Rueda et al. 2015). In another acid sulphite delignification study of pine, reacting mass was heated to 142 °C in two stages. First, the material was heated from 100 to 120 °C in 45 min and held at this temperature for 15 min to allow for liquor impregnation. Thereafter, it was further heated to 142 °C in another 1 h (Deshpande et al. 2018). A similar heating protocol was also used for bisulphite delignification of pine and spruce with different maximum temperatures (Deshpande et al. 2016b; Deshpande et al. 2016c). In a Chinese pulp mill using pine as raw material, the reacting mass was first heated to 116 °C in 115 min, held at this temperature for 30 min, and finally heated to 138 °C in another 110 min (Shen and Chen 2009). Another heating protocol for bisulphite delignification of fungal pretreated pine and spruce reports initial heating from 20 to 90 °C in 30 min followed by heating to 155 °C in the next 90 min (Fischer et al. 1994). In one study, the temperature protocol for acid sulphite delignification of eucalyptus was: heating from 85 to 110 °C in 80 min to ensure impregnation followed by heating to 138 °C in the next 120 min (Sixta et al. 2004).

2.2.2.8 Reaction Time at the Desired Reaction Temperature

Most of the delignification takes place during the high-temperature phase of the process. Adequate reaction time at final delignification temperature is, therefore paramount to achieve proper lignin isolation. Extended reaction time at the final delignification temperature leads to a higher degree of delignification but also puts cellulose at an increased risk of degradation. This is important for mills producing paper-grade pulp. The time of reaction also affects the productivity of the mill. Acid sulphite delignification of pine required 4–5 h at 140 °C while bisulphite delignification was completed in 3.5–4.5 h at 161 °C (Scott et al. 1995). The pulp kappa number – a measure of the residual lignin – was 31.27 when the delignification time was 3.75 h at maximum temperature during bisulphite delignification. Kappa number dropped to 22.6 – a significant increase in the extent of delignification – when the time at maximum temperature was 4.25 h. In acid sulphite delignification, the pulp kappa number was 26.75 with a 4.5 h reaction time at maximum temperature but dropped to 14 when the reaction time was increased to 5 h. The reaction time at a final reaction temperature of 154 °C was 2 h for bisulphite delignification of spruce (Deshpande et al. 2016a). The same researchers delignified pine by acid sulphite, and bisulphite processes with 2 h reation time at 142 °C (Deshpande et al. 2018). Bisulphite delignification of pine and spruce was performed with a 4.5 h reaction time at 155 °C (Fischer et al. 1994). 85 min delignification time at 140 °C was provided for acid sulphite delignification of fungal pretreated eucalyptus (Christov et al. 1998).

2.2.2.9 Interaction among Process Parameters

Individual process parameters affect acidic delignification as much in their own right as they interact with other parameters in affecting the final outcome. The size of the lignocellulose raw material and the composition and concentration of delignification chemicals together influence the mass transfer steps in the heterogeneous reacting environment. Small raw material pieces, to some extent, offset the adverse effects of diminished concentration on mass transfer and vice versa. Improper mass transfer manifests itself in the form of inadequate delignification in parts of the lignocellulose chip which remains short of defibration and comes out as screening rejects. For the same initial chip size, the rejects percentage on initial raw material mass dropped from 1.9 to 1.1, and then further to 0.7 as the total SO_2 charge was increased from 21% to 24%, and then to 28% during acid sulphite delignification of eucalyptus (Fiskari and Kilpeläinen 2021). The most significant interaction between process parameters is that between reaction temperature and reaction time. Increased delignification temperature and delignification time both result in higher lignin removal. A decrease in one parameter can, therefore, be compensated to some extent by an increase in the other parameter. These interactive influences are well documented in technical literature and provide the mills with some flexibility in operation. Such interactive effects are illustrated in Figure 2.8 for bisulphite delignification of pine. It is evident that a decrease in delignification temperature cannot be completely compensated by an increase in reaction time. The extent of delignification achieved after a 3 h reaction at 165 °C cannot be achieved even after 6 h reaction time at 142 °C. But the extent of

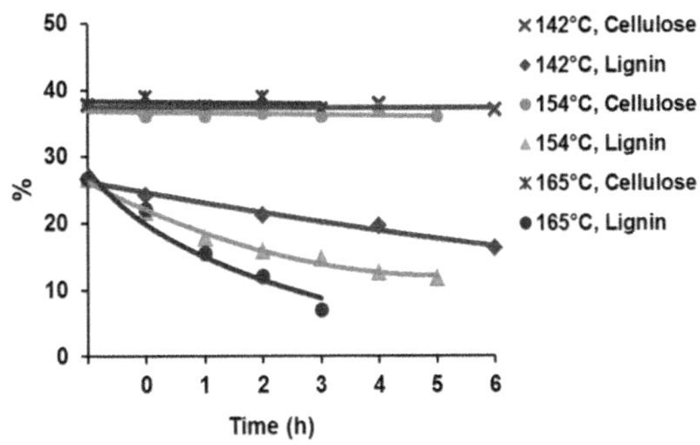

FIGURE 2.8 Interactive effects of delignification temperature and reaction time during bisulphite delignification of pine. (Reprinted from Deshpande, R, Sundvall, L, Grundberg H, Germgård U, *Nordic. Pulp. Pap. Res. J.*, 31(3), 379–385, 2016, with permission from De Gruyter.)

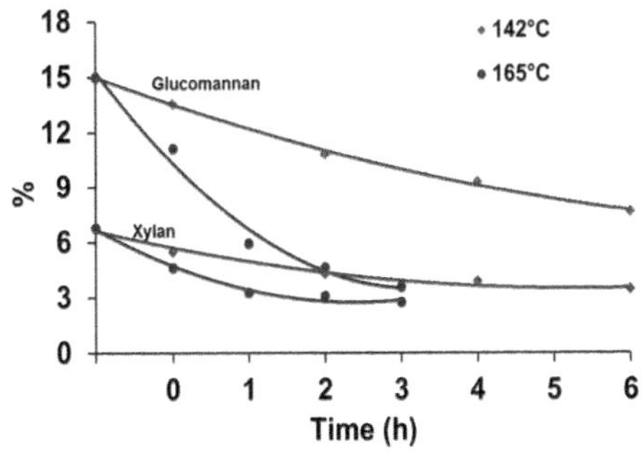

FIGURE 2.9 Interactive effects of delignification temperature and reaction time on hemicellulose removal during bisulphite delignification of pine. (Reprinted from Deshpande, R, Sundvall, L, Grundberg H, Germgård U, *Nordic. Pulp. Pap. Res. J.*, 31(3), 379–385, 2016, with permission from De Gruyter.)

delignification after 2 h at 165 °C is the same as that achieved after 5 h at 154 °C. Along with delignification, carbohydrate degradation reactions also take place simultaneously. Cellulose degradation is minimal during bisulphite delignification as depicted in Figure 2.8, but here also higher temperature results in marginally higher cellulose loss. However, the interactive effects of delignification time and temperature are significant for hemicelluloses as shown in Figure 2.9. A sharp increase in hemicellulose removal rate is observed with an increase in reaction temperature from 142 to 165 °C which cannot be compensated by reducing reaction time. This is important in situations where the target product is dissolving grade pulp or where hemicelluloses are to be hydrolysed to corresponding monosaccharides. Among the hemicelluloses, the removal rate of glucomannan is faster than that of xylan in bisulphite delignification of softwood species spruce. One detrimental aspect of using a higher delignification temperature is the increased occurrence of side reactions producing thiosulphate (Deshpande et al. 2016b). Researchers have proposed a delignification factor in terms of delignification temperature, SO_2 partial pressure, and reaction time to estimate the amount of lignin dissolved during the acid sulphite delignification of pine. It can be mathematically represented as (Shen and Chen 2009)

$$D = \int_0^t k_0 e^{\frac{-E_a}{RT}} \left(P_{SO_2}\right)^{0.75} dt \qquad (2.4)$$

2.3 DISSOLUTION IN ORGANIC MEDIA

The delignification of lignocelluloses through reaction and dissolution in alkaline and acidic media is well-established as a commercial process. Historically, the objective of these processes was to obtain cellulose, and the removed lignin was only meant for marginal use as a low-grade fuel. However, this notion is changing as the development debate is decisively tilting in favour of sustainable development. Today, lignocelluloses are considered an important renewable resource to be comprehensively utilized in a biorefinery framework. For this purpose, the fractionation of lignocellulose into its major components and their subsequent utilization is gaining in importance. With proper choice of reaction conditions delignification with organic chemicals is much more selective towards lignin removal. The characteristics of the isolated lignin substantially differ from the lignin present in the native lignocellulose and depend on the method of isolation. Over the years, researchers have attempted different reaction and dissolution media for isolating lignin from lignocelluloses, other than the conventional alkaline and acidic delignification processes. In this context, several organic liquids have been used as delignification agents, either as standalone chemicals or in combination with conventional alkaline or acidic delignification chemicals, with varying success. Principal among these are carboxylic acids, alcohols, and ketones and the processes are commonly referred to as organosolv delignification processes. Other used organic delignification agents include phenols, ethers, esters, aldehydes, amines, and ionic liquids. One distinct feature of these processes is the relatively mild process conditions and, therefore, less energy intensive. Some organic delignification chemicals are also easier to recover than conventional alkaline, and acidic agents.

2.3.1 Chemistry of Organosolv Delignification

Solvolysis and autohydrolysis are the principal reactions responsible for the depolymerization of lignin into fragments small enough to be dissolved into the reaction medium. Cleavage of intermonomer ether linkages in the lignin macromolecule takes place. α-O-4/ linkages are easier to break than β-O-4/ bonds, but even the latter can be broken with stronger reaction conditions. One of the reasons for this disparity is the higher bond dissociation energy of 290 kJ/mol for the β-O-4/ bond compared to only 215 kJ/mol for the α-O-4 bond. Also, the higher activation energy of 150 kJ/mol for β-O-4/ bond breaking compared to only 80–118 kJ/mol for α-O-4 bond breaking contributes to the ease of cleavage for the later (Jasiukaitytė-Grojzdek et al. 2020). Breaking of β-O-4/ ether bonds is considered to be the rate-limiting step in delignification. In addition, the lignin carbohydrate linkages, that join the α carbon in lignin to the 4-O-methylglucuronic ester moiety in the hemicellulose through an oxygen bridge, are also attacked and contribute significantly to overall lignin removal. Condensation reactions involving reaction intermediates also take place as an undesired side reaction forming high molecular weight moieties and inhibiting delignification. Besides reactions of lignin, hydrolysis of hemicellulose polysaccharides and degradation of the resulting monosaccharides to carbonyl products like hydroxymethyl furfural and furfural also take place under the prevailing acidic environment. A schematic of reaction pathways for α-O-4 bond breaking in organosolv delignification is presented in Figure 2.10. It is believed that the reaction follows S_N1 mechanism, owing to the formation of a stable benzylic carbocation intermediate (El Hage et al. 2010; Jasiukaitytė-Grojzdek et al. 2020). The formation of the carbonium ion is facilitated by the acidity of the reaction system and the nucleophilic and polar nature of the solvent. Owing to this reason, organosolv delignification is often catalysed by acids. The reaction is completed in the second, non-rate determining, step of nucleophilic attack by either OH^- (hydrolysis) or OR^- (solvolysis).

A schematic of the reaction pathway for β-O-4/ bond breaking is shown in Figure 2.11. The reaction is initiated by the formation of a benzylic carbocation intermediate which subsequently leads to the breaking of the β-O-4/ bond with the formation of a phenolic hydroxyl group and a Hibbert's ketone structure. Alternatively, the reaction can also proceed through the loss of γ carbon as formaldehyde and the formation of another aldehyde (El Hage et al. 2010; Meyer et al. 2020). However,

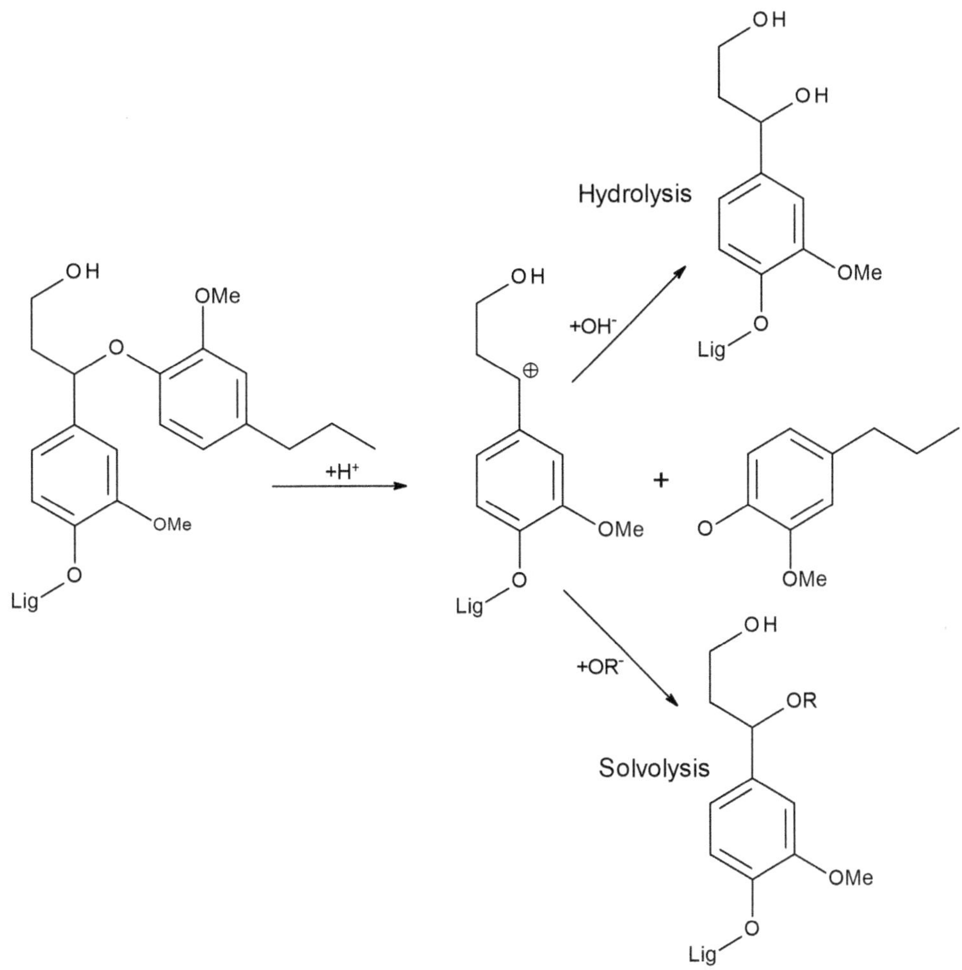

FIGURE 2.10 α-O-4/ bond cleavage during organosolv delignification.

the breaking of the β-O-4/ bond is only partial owing to the competing reaction of nucleophilic addition of OR^- to the formed carbonium ion. The benzylic carbonium ion can also react with carbon centres on nearby lignin monomer units having negative charge densities. These condensation reactions form new carbon–carbon linkages forming larger lignin fragments which are difficult to dissolve and remove from the solid matrix.

2.3.2 Process Parameters in Organosolv Delignification

Lignin fragmentation through the breaking of ether bonds, particularly the β-O-4/ bond, and the condensation reactions depend on the severity of reaction conditions which in turn depend on the process parameters. Along with fragmentation, dissolution of the resulting fragments is also affected by conditions. We can list the important process parameters for organosolv delignification as

1. Solvent characteristics
2. Use of catalyst
3. Liquid-to-solid ratio
4. Temperature of treatment
5. Time of treatment

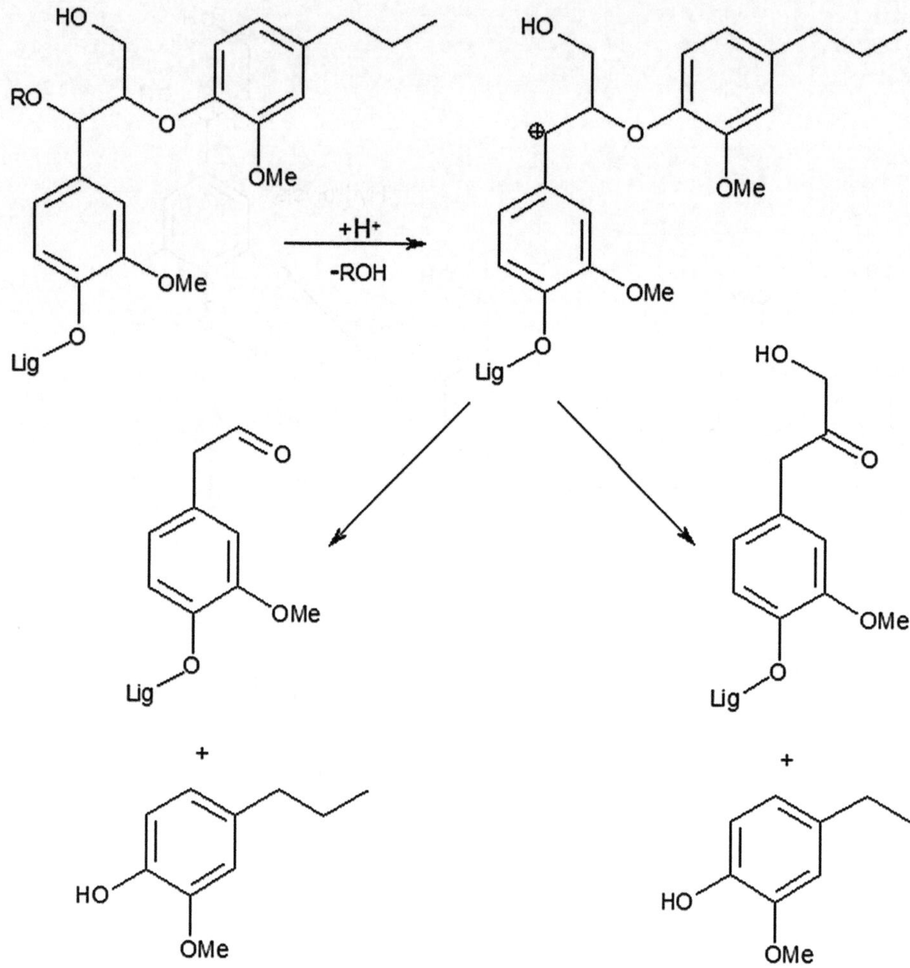

FIGURE 2.11 β-O-4/ bond cleavage during organosolv delignification.

2.3.2.1 Solvent Characteristics

One of the reasons for the versatility of organosolv delignification is the possibility to select from many organic compounds and their combinations. The selection of solvent is dictated by several considerations, both technical and economic. Adequate depolymerization of the lignin macromolecule and dissolution of the depolymerized lignin remains the primary consideration. The possibility to recover and reuse the solvent adds to economic and environmental sustainability. Important solvent characteristics to look for are polar nature, acidic character, volatility, viscosity, and toxicity.

In technical literature reporting organosolv delignification, it is customary to assess the effectiveness of an organic solvent for lignin dissolution in terms of the Hildebrand solubility parameter, expressed empirically as (Quesada-Medina et al. 2010; Zhang et al. 2016)

$$\delta = \left[\frac{\Delta H - RT}{V_m}\right]^{1/2} \tag{2.5}$$

where δ is the Hildebrand solubility parameter, measured in the units of $(\text{MPa})^{1/2}$. ΔH is the latent heat of vaporization, J/mol; R is the universal gas constant, J/(K.mol); T is the normal boiling temperature; and V_m is the molar volume, cm^3/mol. Hildebrand solubility parameter is an empirical measure of the energy density of the material. It is proposed that the closer the value of δ of a solvent to that of lignin to be dissolved, the better the solvent is in dissolving lignin. It is argued that for the best dissolution capability, a solvent should have a δ value within ± 2 $(MPa)^{1/2}$ of that of lignin (Venkatram et al. 2019). Mixtures of organic solvents or their aqueous solutions are often preferred over pure compounds for better delignification performance. The solubility parameter of any solvent is a function of pressure. Considering the complexity and structural heterogeneity of lignin fragments produced by the depolymerization of the native lignin macromolecule, the estimation of δ for lignin requires the estimation of $\left(\Delta H - RT\right)$ and V_m, respectively, for the different structural moieties. The values for $\left(\Delta H - RT\right)$ and V_m for the composite lignin sample are then determined by adding the weighted contributions from individual structural moieties. Using this approach, δ for lignin in hydrolysed almond shells was estimated to be 29.1 $(MPa)^{1/2}$ (Quesada-Medina et al. 2010). For several lignin samples from wheat straw, bagasse, and softwood, the estimated δ values were between 27 and 31 $(MPa)^{1/2}$ (Ma et al. 2021). Another study reports the value of δ for lignin between 24.3 and 25.5 $(MPa)^{1/2}$ (Mohan et al. 2022). The delignification performance of some common organic solvents for the organosolv process is presented below.

Monohydroxy Alcohols

Alcohols, especially the lower homologues, are among the first to be used in organic delignification chemicals. Kleinert and Tayenthal were granted a patent for fractionating lignocellulose into its components by using alcohol as the reagent (Kleinert and Tayenthal 1932). The process involved the treatment of the lignocellulosic material with ethanol–water mixture at an elevated temperature for lignin removal. δ value for methanol is 29.6 $(MPa)^{1/2}$ which is close to that of lignin (Zhang et al. 2016). Ethanol has a δ value of 26.35 $(MPa)^{1/2}$ at 0.1 Mpa pressure which increases to 29.5 $(MPa)^{1/2}$ at 300 MPa pressure, (Rai et al. 2007). This value is again very close to the value for lignin. Owing to this, both methanol and ethanol are excellent as organosolv delignification solvents. However, the use of methanol is discouraged of late because of its higher toxicity. Ethanol–water mixture in equal volumetric ratio is the most frequently used organic solvent for lignocellulose delignification. Aqueous ethanol concentration was 50% for lignin removal from eucalyptus (Oliet et al. 2000). Organosolv delignification of jute was carried out with 50% v/v aqueous solution of ethanol to obtain partial delignification (Sahin 2003). Pine delignification with 50% aqueous solution of ethanol was practised to obtain a cellulose-rich material for subsequent saccharification (Park et al. 2010). Aqueous ethanol concentration was 50% for delignifying poplar with 50–60% lignin removal (Koo et al. 2011). Similarly, organosolv delignification of eucalyptus with 50% v/v aqueous solution of ethanol resulted in 63% lignin recovery (Yanez-S et al. 2014). Sugarcane bagasse was delignified with aqueous ethanol at 50% concentration. Delignification to the extent of 92% could be achieved (Vallejos et al. 2014). 50% concentration aqueous ethanol served as the organosolv solvent for the delignification of *Liriodendron tulipifera* (tulip tree), and the lignin removal varied between 60% and 80% (Jang et al. 2016). Similarly, the aqueous ethanol concentration was 50% for the delignification of Broussonetia papyrifera, usually known as paper mulberry, and about 71% lignin removal was reported (Yao et al. 2018). Alkali ethanol organosolv delignification of Arundo donax used 40% concentration aqueous ethanol (Shatalov and Pereira 2005). Similarly, previously autohydrolysed wheat straw was delignified with aqueous ethanol having 40% concentration (Ruiz et al. 2011). Aqueous ethanol concentration was 65% for lignin isolation from pine (Sannigrahi et al. 2010).

In a study to find the best ethanol concentration for the delignification of hydrolysed almond shells, the investigators reported 72.8% concentration to be the best for lignin removal (Quesada-Medina et al. 2010). The studied variation of lignin removal, selectivity of removal, and extractability, with ethanol concentration is illustrated in Figure 2.12. The figure also depicts the behaviour of two other organic solvents, acetone and dioxane. Organosolv delignification of poplar reported the use of 70% aqueous ethanol as the solvent (Wang et al. 2012). Aqueous ethanol with 75%

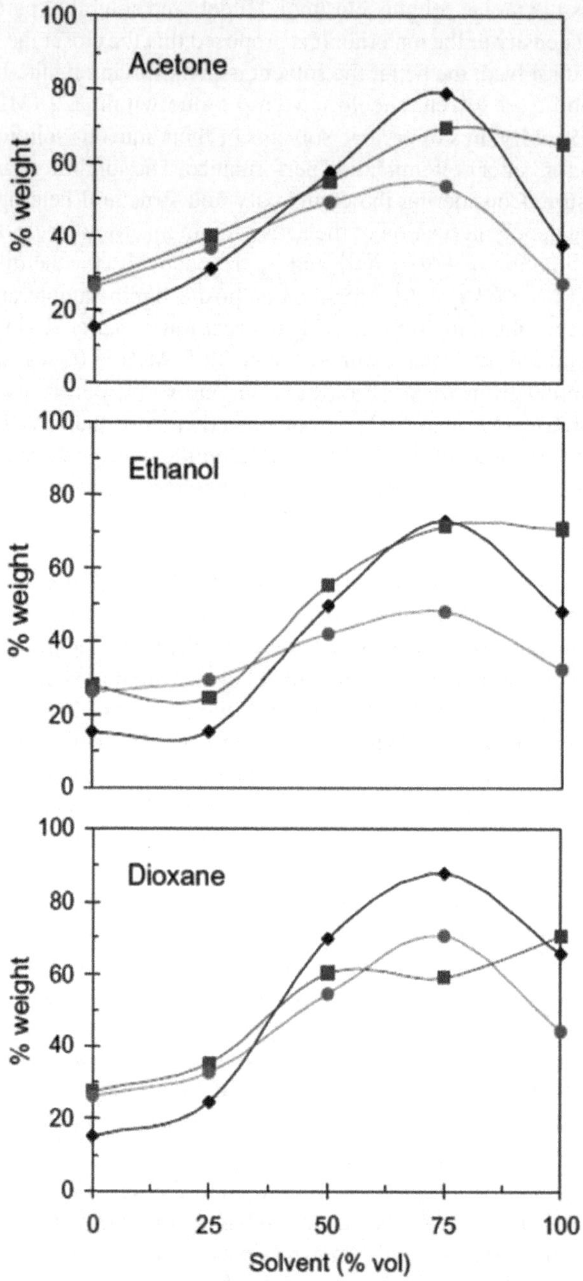

FIGURE 2.12 Effect of solvent composition on organosolv delignification. ◆ marker represents delignification, ■ marker represents selectivity, and ● marker represents extractability. (Reprinted from Quesada-Medina, J, López-Cremades, FJ, Olivares-Carrillo P, *Bioresource Technol.*, 101, 8252–8260, 2010, with permission from Elsevier.)

volumetric concentration was used for the delignification of larch sawdust and lignin was isolated in good yield (Hochegger et al. 2019). Wheat straw delignification used a 60–65% aqueous solution of ethanol (Chen et al. 2015). Another research reports the use of 50–80% aqueous ethanol for wheat straw delignification (de la Torre et al. 2013). Aqueous ethanol in 50–65% concentration was used for the organosolv delignification of miscanthus, and 15–50% lignin recovery was reported

(El Hage et al. 2010). Similar ethanol concentrations were also used in the delignification of Buddleja davidii wood chips and a maximum of 64% lignin removal was achieved with an ethanol concentration of 65% (Hallac et al. 2010). A study on poplar delignification with aqueous ethanol of 65% concentration is reported (Meyer et al. 2020). Ethanol concentration in the water mixture varied from 60% to 80% for organosolv delignification of date palm fronds (Cybulska et al. 2017). 80% concentration aqueous ethanol could partially delignify walnut shells (Zijlstra et al. 2020). One particular research result reports the use of alkaline ethanol in 10–30% concentration for the delignification of previously acid-pretreated sugarcane bagasse (Mesa et al. 2011). Pilot scale batch delignification of beech is reported using 50 weight% aqueous ethanol (Schulze et al. 2016).

Methanol, though sparingly used compared to ethanol, is an effective delignification agent. In one study, both methanol and ethanol as aqueous solutions in 38–62% concentration were used as the organosolv solvent to delignify eucalyptus. However, no clear choice for the alcohol type emerged (Oliet et al. 2002). A relatively high methanol concentration of 80% is reported for the delignification of spruce (Yawalata and Paszner 2004a). The use of 50% aqueous solution of methanol to delignify wheat straw is reported (Xu et al. 2006). In one study, beech sawdust was delignified using pure methanol, pure ethanol, as well as alcohols in 75% aqueous solution. The highest lignin removal was achieved with aqueous ethanol followed by aqueous methanol (Jasiukaitytė-Grojzdek et al. 2020).

δ values for 1-butanol and 2-propanol are 23.2, and 23.6 $(MPa)^{1/2}$, respectively (Dastpak et al. 2020). For higher alcohols, 1-heptanol, 1-octanol, 1-nonanol, and 1-decanol, the values of δ are reported to be 20.39, 20.88, 20.3, and 20.2 $(MPa)^{1/2}$, respectively, at 0.1 MPa pressure and 25 °C temperature (Sohani et al. 2020). Lower alcohols like methanol and ethanol have low boiling points that require pressurized operations at elevated temperatures on the one hand but also allow for their easy recovery through distillation from the spent liquid after delignification. For higher alcohols with higher boiling points, delignification is possible at normal pressure. The use of higher alcohols as delignification solvents is far less common compared to ethanol or methanol. Hydrophobic higher alcohols 1-butanol and 1-pentanol were used as delignification agents for sorghum bagasse at 12.5% concentration. 1-pentanol resulted in higher lignin removal of 75% compared to 65% with 1-pentanol. Interestingly, these two solvents resulted in two liquid fractions, a pale yellow aqueous layer containing monosaccharides from hemicellulose, and a black organic layer containing the dissolved lignin. This was attributed to the stronger hydrophobic character of the higher alcohols (Teramura et al. 2016). 1-butanol and 1-pentanol in 80% aqueous concentration were investigated for their effectiveness in removing lignin from walnut shells. Both the higher alcohols, particularly 1-pentanol, were found to be less effective than ethanol in removing lignin. For 1-pentanol the miscibility with water at such a high concentration was reported to be an issue as separation of the organic layer was noticed (Zijlstra et al. 2020).

Polyhydroxy Alcohols

Among the polyhydroxy alcohols, ethylene glycol, and glycerol have been used as delignification agents in organosolv processes. Ethylene glycol has a δ value of 32.9 $(MPa)^{1/2}$ while it is 36.1 $(MPa)^{1/2}$ for glycerol (Zhang et al. 2016). The use of aqueous ethylene glycol with 70–80% glycol concentration is reported for the limited delignification of oil palm empty fruit bunches (Rodriguez et al. 2008). Another Spanish research group used an aqueous solution of ethylene glycol and sodium hydroxide for the delignification of olive tree trimmings. Pulping liquids had 5–15% glycol concentration and 2.5–7.5% sodium hydroxide concentration. Higher glycol and alkali concentration was found to increase delignification, but the effect of glycol concentration was not very pronounced compared to the effect of alkali concentration (Jimenez et al. 2004). Sugarcane bagasse was delignified with aqueous glycerol with 20–80% concentration (Novo et al. 2011). Aqueous glycerol concentration was 80% for wheat straw organosolv delignification (Saberikhah et al. 2011). Glycerol–water mixture in 40–80% concentration was studied for lignin removal from sugarcane bagasse and increased glycerol concentration was found to favour delignification (Martin et al.

2011). Researchers used aqueous glycerol with 40–80% glycerol concentration to delignify eucalyptus and obtained 80% delignification with high glycerol concentration (Romani et al. 2013). 97% concentration glycerol was the solvent during organosolv delignification of beech to achieve close to 90% delignification (Hundt et al. 2013). One research group examined glycerol, ethylene glycol, and propylene glycol in 90% aqueous concentration for their delignification effectiveness of sugarcane bagasse. The use of glycols resulted in 79% lignin removal while glycerol could remove only 29% lignin (Zhang et al. 2013). In a recent study, 99% pure glycerol was the organic solvent for the delignification of sugarcane bagasse (Pascal et al. 2019).

Carboxylic Acids

Both formic acid and acetic acid have been used as organosolv delignification agents with reasonable success. δ value for formic acid is 24.9 $(MPa)^{1/2}$ and that for acetic acid is 21.4 $(MPa)^{1/2}$ (Zhang et al. 2016). The use of acids in organosolv delignification processes serves the dual role of acting as a solvent as well as a catalyst for breaking ether bonds in lignin macromolecules. On this account, formic acid is particularly attractive due to its solubility parameter being very near to that of lignin. The use of peroxyacids can give even better delignification as oxidative cleavage of intermonomer lignin bonds is far more effective than acid-catalysed reaction alone. In most cases, the peroxyacid is prepared in situ by the reaction of the corresponding acid or anhydride with hydrogen peroxide. The disadvantage of using acidic solvents is their inherent corrosive nature which takes a toll on the material of construction. 90% concentration aqueous acetic acid was used as the delignification solvent for lignin removal from eucalyptus (Villa et al. 2003). Organosolv delignification of eucalyptus was performed with aqueous formic acid of 80–95% concentration as well as with aqueous acetic acid of 60–90% concentration. The extent of delignification was more with formic acid (Ligero et al. 2008). Olive tree prunings were delignified by both formic acid and acetic acid as the organosolv solvent. Aqueous acids had 80% and 90% concentration, respectively. In addition, the raw material was also delignified with a mixture containing both acids in water with a 30:60:10 volumetric concentration. Of all the solvents, formic acid provided the best delignification performance (Erdocia et al. 2014). Mixed acid organosolv delignification of wheat straw was performed with a mixture of acetic acid, formic acid, and water in a volumetric ratio of 60:20:20 (Kham et al. 2005). Similarly, acetic acid, formic acid, and water mixed in a 50:30:20 ratio were used for the delignification of olive kernels, peach kernels, and grape pomace (Manara et al. 2014). Wheat straw delignification with 66–88% concentration aqueous formic acid resulted in 72–78% lignin removal. In comparison, delignification achieved was only 36–54% with aqueous acetic acid of 70–90% concentration as the delignification solvent (Chen et al. 2015). Aqueous acetic acid of 80% concentration could remove 58–65% lignin from birch during organosolv delignification (Kangas et al. 2015). In one study, cedar chips were soaked in a peracetic acid solution of 0.1–2% concentration. The soaked chips were then steam exploded (Ando et al. 1988). In a comparative study, 88% aqueous formic acid, and a mixture of 88% aqueous formic acid with 3% hydrogen peroxide (peroxyacid) on raw material mass, were used as solvents for delignification of bamboo. Lignin removal was 85% with peroxyacid and 75% with formic acid (Li et al. 2012). Another study reports the use of mixed solvents for the delignification of the residue from the medicinal herb *Acanthopanax senticosus* after the extraction of pharmaceutical ingredients. The solvents used were acetone–acetic acid–water in an 80:10:10 ratio, ethanol–acetic acid–water in an 80:10:10 ratio, 1,4-butanediol–acetic acid–water in an 80:10:10 ratio, or acetic acid–water in an 80:20 ratio. The highest delignification efficiency was achieved with the solvent combination of 1,4-butanediol–acetic acid–water followed by acetic acid–water (Lu et al. 2012).

Acetone

With a δ value of 20 $(MPa)^{1/2}$ (Zhang et al. 2016), it appears at first glance that acetone should not be a solvent of choice for organosolv delignification. However, a lot of research work has gone into developing delignification processes based on acetone. One of the reasons for this is the low boiling point of acetone which makes for easy recovery post lignin removal. Also, the difference between the solubility parameters of acetone and lignin can, to some extent, be reduced by using acetone–water mixtures in different proportions. There are several research reports of acetone organosolv delignification of wheat straw. An acetone–water mixture with 65% acetone was found to achieve partial delignification of wheat straw (Jimenez et al. 1998). Similarly, acetone–water mixture in 25–60% concentration was studied for wheat straw delignification and the highest lignin removal of 68.6% was obtained when the acetone concentration was 40% (Huijgen et al. 2010). 79% acetone concentration in an aqueous mixture provided the best delignification from hydrolysed almond shells (Quesada-Medina et al. 2010). The studied variation of lignin removal, selectivity of removal, and extractability, with acetone concentration is illustrated in Figure 2.12. Wheat straw was delignified with 50–80% concentration aqueous acetone (de la Torre et al. 2013). Aqueous acetone with 50% concentration was used as the delignifying solvent for wheat straw, corn stover, birch, poplar, spruce, and pine. Good delignification was achieved for hardwoods, and nonwoods, ranging from 80% to 87% but the softwoods were not adequately delignified (Smit and Huijgen 2017).

Ionic Liquids

The use of ionic liquids for organosolv delignification is a recent development compared to other solvents. These are molten salts at room temperature with organic ions. Their solvent properties are superior to many conventional solvents. The greatest advantage of using ionic liquids is their ability to effectively fractionate the lignocellulosic biomass into its principal components which can then be subjected to value addition. With the right choice of solvent and process conditions, any one or all the components of lignocellulose can be dissolved. The addition of an antisolvent, like water, precipitates the components from the solution. Further, the negligible vapour pressure exerted by these liquids prevents the possibility of fugitive air emissions. The recovery of the ionic liquids is possible, making the process promising. However, the high cost of these solvents presently prevents their commercial use. Several nitrile-based ionic liquids were used to extract lignin from bamboo and 53% delignification was reported (Muhammad et al. 2013). One recent advancement is the development of choline-cation-based ionic liquids that can be synthesized from renewable materials. These room-temperature ionic liquids are highly selective in dissolving lignin. Organosolv delignification of rice straw was carried out with ionic liquids with cholinium cation and different amino acids as the anion. 43–60% lignin removal was reported with lysine anion as the best performer (Hou et al. 2012). In another study, different cholinium ionic liquids were used for organosolv fractionation of rice straw, wheat straw, bagasse, eucalyptus, and pine in pulverized form to obtain a lignin-rich fragment with 50–70% lignin (An et al. 2015). Ionic liquid 1-ethyl-3-methyl imidazolium acetate was able to remove 65% lignin from hinoki cypress wood (Moniruzzaman and Goto 2019).

Other Organic Solvents

Some early studies have reported on the use of phenol as an organic solvent for delignification. Phenol as 50–80% aqueous solution was studied as an organosolv delignification agent for wheat straw and 60–80% lignin removal was obtained (Jimenez et al. 1997). In another study, the phenol concentration varied between 42% and 58% for the organosolv delignification of miscanthus sinensis and the lignin removal was 30–80%. As reported by the authors, one of the important properties of phenol is its immiscibility with water at room temperature but miscibility at a higher temperature of around 100 °C, which allows for easy phase separation and component separation (Vega et al. 1997). In the studied concentration range, higher phenol concentration caused more delignification. Bagasse organosolv delignification was carried out with a dimethyl formamide–water mixture with 40–60% weight percentage of the organic solvent to achieve moderate lignin removal (Rezayati-Charani et al. 2006). The use of aqueous monoethanolamine or diethanolamine in 70–80% concentration resulted in efficient delignification of oil palm empty fruit bunches with the cellulosic pulp having 17–18 kappa number (Rodriguez et al. 2008). Aqueous dioxane with 87.6% concentration was the best for the delignification of hydrolysed almond shells. With pure dioxane, however, lignin was removed with the highest selectivity (Quesada-Medina et al. 2010).

Use of Catalysts

Though the common organic solvents, described above, are capable of dissolving lignin from a solid lignocellulose matrix, the effectiveness of lignin removal is improved by the use of catalysts. Bond cleavage reactions in lignin are catalysed by acids, bases, as well as inorganic salts and oxides. The cleavage of lignin carbohydrate ether bonds is also facilitated by the use of catalysts. Besides lignin removal, carbohydrate reactions and their removal are also affected by the use of catalysts. It is believed that catalysts accelerate the solvolysis reactions while condensation reactions are unaffected. The net result is improved lignin defragmentation and its dissolution in the solvent. Under comparable conditions, acids fare better than alkali in catalysing organosolv delignification. Other than the catalytic activity, downstream processing requirements like possible recovery or neutralization of catalyst, and economics influence the choice of the catalyst.

Acid-Catalysed Organosolv Delignification

Mineral acids are the most effective and hence the most widely used catalysts for organosolv delignification. Near-complete hydrolysis of hemicellulose carbohydrates to monomeric sugars is the added advantage of carrying out the reactions under acidic conditions. The monosaccharides so obtained serve as the starting material for bioethanol or furfural production improving the overall viability of the process. Hydrochloric acid and sulphuric acid find the largest applications primarily because of economic considerations. The use of other mineral acids is less frequent.

Numerous studies report the catalytic use of sulphuric acid in organosolv delignification using various solvents. The amount of catalyst to be used is either fixed as a percentage of lignocellulose mass or as a concentration in a delignifying solvent. In a kinetic study ethanol organosolv delignification was performed in the presence of sulphuric acid to attain a starting pH of 2.48 (Oliet et al. 2000). Sulphuric acid was the catalyst for organosolv delignification of miscanthus with aqueous ethanol. Acid concentrations ranged between 0.5 and 1.6% (El Hage et al. 2010). Sulfuric-acid-catalysed delignification of Buddleja davidii with aqueous ethanol was studied at acid concentrations of 1.25%, 1.5%, and 1.75%. Delignification was more when using higher acid concentrations (Hallac et al. 2010). Aqueous ethanol was fortified with a 1% concentration of sulphuric acid for the delignification of pine (Park et al. 2010). Sulphuric acid catalyst addition was 1% on lignocellulose mass for the delignification of poplar with aqueous ethanol as the organic solvent (Koo et al. 2011). Sulphuric-acid-catalysed organosolv delignification of pine was performed with aqueous ethanol with a catalyst addition of 1.1% on lignocellulose mass (Sannigrahi et al. 2010). In one study on sulphuric-acid-assisted delignification of sugarcane bagasse with aqueous glycerol as the solvent, the acid loading was 9.4% on lignocellulose mass which was quite high from catalytic standards (Martin et al. 2011). Similarly, sulphuric acid dosing was about 11% on raw material mass for the delignification of sorghum bagasse using different alcohols in aqueous medium as the delignifying agent (Teramura et al. 2016). The volumetric concentration of sulphuric acid catalyst was between 0.001 N during organosolv delignification of wheat straw using aqueous ethanol solvent and the lignin removal was close to 61% (de la Torre et al. 2013). For organosolv lignin removal using ethylene glycol, and propylene glycol, sulphuric acid was the catalyst. The solvent had 0.2–1.2% acid concentration and the highest lignin removal of 89% was reported at 1.2% catalyst concentration (Zhang et al. 2013). 40 mM sulphuric acid was used as a catalyst alongside aqueous ethanol for the delignification of wheat straw and the lignin removal was 58% with acid catalysis compared to 46% without catalyst. Similarly, the use of sulphuric acid as a catalyst to the extent of 0.1–0.3 weight% improved delignification during the organosolv process with acetic acid. Sulphuric acid addition was 0.05–0.1 wight % for improved lignin removal when formic acid was the delignification solvent (Chen et al. 2015). Sulphuric acid catalyst in 5–95 mM concentration was used for the delignification of tulip tree with aqueous ethanol. The resulting solution pH varied between 0.72 and 2.01 (Jang et al. 2016). Sulphuric acid between 40 and 60 mM concentrations was used to catalyse the organosolv delignification of several lignocelluloses with aqueous acetone as the solvent (Smit and Huijgen 2017). Organosolv delignification of date palm fronds was performed with aqueous ethanol and 1.5 weight% sulphuric acid as catalyst. Enhanced lignin removal was obtained with catalyst addition compared to the case when no catalyst was used (Cybulska et al. 2017). Sulphuric acid was the catalyst during the ethanol organosolv delignification of larch. The amount of acid used was 0.75–1.65% based on raw material mass. The highest delignification was achieved with the highest catalyst dose. Even with the lowest catalyst dose, the lignin removal was almost threefold that obtained without a catalyst (Hochegger et al. 2019). Sugarcane bagasse was delignified with aqueous glycerol catalysed with sulphuric acid in 0.02–0.1% concentration. Increased acid concentration had a marginal positive influence on lignin removal but a profound effect on hemicellulose dissolution (Pascal et al. 2019). Lignin extraction from beech was carried out with aqueous ethanol or aqueous methanol organosolv process using 1% sulphuric acid on raw material mass and more than 80% lignin recovery was achieved (Jasiukaitytė-Grojzdek et al. 2020). During a pilot scale batch delignification of beech with aqueous ethanol, 0.5% sulphuric acid on raw material mass was used as the catalyst (Schulze et al. 2016).

Organosolv delignification of miscanthus was carried out with phenol in the presence of hydrochloric acid as a catalyst added in the range of 0.1–6% on the raw material mass. Lignin removal was considerably improved with increased acid addition (Vega et al. 1997). Hydrochloric acid was the catalyst added, 2% on the lignocellulose mass, when delignifying eucalyptus with acetic acid as the organic solvent (Villa et al. 2003). Eucalyptus was delignified with aqueous formic acid in the presence of hydrochloric acid catalyst. Catalyst doses in the range of 0.1–0.3 weight% on lignocellulose mass provided the optimum delignification results (Ligero et al. 2008). Hydrochloric acid-catalysed delignification of Siam weed was performed with acetic acid as the organic solvent. Hydrochloric acid concentration in the liquid phase was 0.1% but the reported delignification was only 13% (Zhao and Liu 2010). Compared to organosolv delignification of bamboo with aqueous formic acid alone, a marginal improvement in lignin removal was observed when 1% hydrochloric acid on lignocellulose mass was added as a catalyst (Li et al. 2012). Organosolv delignification of wheat straw was performed with aqueous ethanol or aqueous acetone as the solvent with 0.001–0.01 N hydrochloric acid as the catalyst. Higher catalyst concentration always provided higher lignin removal in all cases. The study also established that hydrochloric acid was the best catalyst among the different mineral acids, namely, sulphuric acid, nitric acid, and phosphoric acid (de la Torre et al. 2013). Hydrochloric acid in 0.2% concentration acted as the catalyst for the organosolv delignification of olive tree pruning with acetic acid or formic acid as the solvent (Erdocia et al. 2014). Walnut shells were delignified with several aqueous alcohols in the presence of a hydrochloric acid catalyst. Acid concentration in the solvent mixture was 0.3 M. Notably, the research group found hydrochloric acid to be more efficient in lignin extraction than sulphuric or hydrobromic acid (Zijlstra et al. 2020).

While sulphuric acid and hydrochloric acid are the catalysts of choice in acid-catalysed organosolv processes, there are a few reports on other acidic ingredients acting as catalysts. In one study, 3.5% H_3PO_2 on lignocellulose mass was the catalyst for birch organosolv delignification in aqueous acetic acid solvent (Kangas et al. 2015). There have been attempts to use dissolved carbon dioxide as a substitute for mineral acid for catalysing organosolv delignification. In such cases, the reaction mixture is pressurized with carbon dioxide to maintain the desired concentration of carbon dioxide. Organosolv degradation of lignin model compound was studied using aqueous ethanol and the reaction mixture was acidified by fortifying with carbon dioxide to a pressure of 16 MPa. It was found that the reaction was significantly faster in the presence of carbon dioxide than in its absence. The activation energy for the carbon-dioxide-assisted reaction was determined to be 42 kJ/mol against a much higher value of 68.9 kJ/mol in its absence (Schrems et al. 2012). A recent study reports the use of carbon dioxide at 5.0×10^5 Pa pressure to acidify the reaction mixture in aqueous ethanol for delignifying miscanthus. 73% lignin removal was obtained (Kubota et al. 2018).

Base-Catalysed Organosolv Delignification

Base catalysis for organosolv delignification is less common compared to acid catalysis. However, alkaline reacting systems are less corrosive and do not require high-end materials for construction. The use of NaOH in combination with ethylene glycol was found to give more delignification with better lignin removal selectivity compared to delignification with either of the two reagents alone (Jimenez et al. 2004). Base-catalysed organosolv delignification of pine used aqueous ethanol with 1–2% sodium hydroxide concentration in the solvent (Park et al. 2010). Poplar was delignified using aqueous ethanol as a solvent with the addition of 1% sodium hydroxide on lignocellulose mass as a catalyst (Koo et al. 2011). 2% sodium hydroxide on the mass of wheat straw was the catalyst used along with aqueous glycerol as the solvent. Delignification was distinctly better than the non-catalytic runs (Saberikhah et al. 2011). Sugarcane bagasse was delignified in aqueous glycerol with 3% sodium hydroxide on lignocellulose mass added as a catalyst. It was observed that the catalyst not only improved the extent of delignification but also its selectivity (Martin et al. 2011). Catalyst dosing of sodium hydroxide was 3% on lignocellulose mass for the delignification of acid-pretreated sugarcane bagasse with aqueous ethanol solvent (Mesa et al. 2011). Delignification of autohydrolysed wheat straw was performed with aqueous ethanol in the presence of 0.1% concentration of sodium hydroxide as a catalyst (Ruiz et al. 2011). Poplar was delignified with different aqueous alcohols in the presence of 1% sodium hydroxide on lignocellulose mass acting as a catalyst (Wang et al. 2012). Base-catalysed organosolv delignification of beech, with glycerol as the solvent, used 8% potassium hydroxide as a catalyst, based on the solvent mass (Hundt et al. 2013).

Inorganic Salts and Oxides Catalysed Organosolv Delignification

Compared to acids and bases, the possible use of metal salts as delignification catalysts can bring the reaction pH closer to neutral with obvious advantages in operating the process. Researchers studied the use of the chlorides of sodium, potassium, calcium, magnesium, manganese, zinc, aluminium, and trivalent iron as possible catalysts in aqueous methanol organosolv delignification. Experiments were conducted with salt concentrations of 0.025, 0.05, and 0.075 M. Among the salts studied, magnesium chloride, calcium chloride, manganese chloride, and zinc chloride were found to be effective delignification catalysts, in decreasing order of effectiveness, for several lignocellulose biomass. Higher salt concentrations resulted in more lignin removal while still maintaining the reaction pH only mildly acidic (Yawalata and Paszner 2004a). Among the different magnesium salts, chlorides fared the best as delignification catalyst followed by nitrate, sulphate, and acetate. For all the salts except acetate the reaction pH was between 6 and 7. For acetate, the pH was between 8.5 and 8.8 (Yawalata and Paszner 2004b). Delignification of pine with aqueous ethanol was carried out with 1% concentration of magnesium chloride in the solvent acting as a catalyst (Park et al. 2010). Wheat straw was delignified with aqueous ethanol in the presence of calcium chloride, aluminium chloride, or ferric chloride as catalyst. The catalyst concentration used was 0.001 N. Calcium chloride, and aluminium chloride fared marginally better than ferric chloride in lignin removal efficacy (de la Torre et al. 2013). Mixed metal oxides of Cu-Ni-Fe and Cu-Ni-Al were studied for their effectiveness as catalysts for methanol organosolv degradation of lignin model compounds having α-O-4 bonds. The ether bond cleavage was negligible when no catalyst was used. In comparison, 20–40% bond breaking was observed in the presence of a catalyst with Cu-Ni-Fe mixed metal oxide performing better (Awan et al. 2021).

2.3.2.2 Liquid-to-Solid Ratio

For any heterogeneous reaction or transfer process the relative amount of the phases is an important parameter and organosolv delignification is no exception. A higher amount of liquid relative to the solid raw material has several ramifications. It amounts to more solvent available for component dissolution during delignification but also results in spent liquor of lower concentration. A very low liquid-to-solid ratio is equally undesirable as it severely restricts the uniform distribution of solvent through the lignocellulose mass. Process engineering is affected by the liquid-to-solid ratio. More liquid relative to the solid raw material demands a bigger reactor for the same lignocellulose mass. Mixing, and the resultant turbulence, in the liquid phase is also affected by the liquid-to-solid ratio. The size of the lignocellulose material along with the liquid phase turbulence influences the mass transfer effectiveness. This ratio has a profound influence on the energy consumption of the process. More liquid relative to the solid raw material would require more energy to heat the reacting system to the desired temperature.

Laboratory investigations of organosolv delignification processes have used liquid-to-solid ratios in a wide range. Aqueous phenol was used at a liquid-to-solid ratio of 15 to delignify miscanthus that was milled to 0.1–1 mm particle size (Vega et al. 1997). Wheat straw was delignified using aqueous acetone at a very high liquid-to-solid ratio of 40 (Jimenez et al. 1998). Aqueous ethanol or methanol at a liquid-to-solid ratio of 7 acted as the solvent for the delignification of eucalyptus chips (Oliet et al. 2002). The liquid-to-solid ratio was 10 for aqueous acetic acid delignification of eucalyptus chips of thickness 2 mm (Villa et al. 2003). Eucalyptus chips of thickness less than 2 mm were delignified with an aqueous mixture of formic, and acetic acid with a liquid-to-solid ratio of 5 (Sixta et al. 2004). Spruce was cut into chips of size 20 mm × 13 mm × 3 mm and delignified with aqueous methanol at a liquid-to-solid ratio of 10 (Yawalata and Paszner 2004a). For wheat straw pieces of about 3 cm in length, an agricultural waste, the liquid-to-solid ratio was 15 during organosolv delignification with carboxylic acids or peroxy-carboxylic acids (Kham et al. 2005). Aqueous formamide organosolv delignification of sugarcane bagasse was performed at a liquid-to-solid ratio of 10 (Rezayati-Charani et al. 2006). Delignification of oil palm empty fruit bunches with ethylene glycol or ethanolamine was performed at a liquid-to-solid ratio of 6 (Rodriguez et al. 2008). Pine sawdust was milled to pass 40 mesh screens and delignified with 10 L of aqueous ethanol for 1 kg of lignocellulose mass (Park et al. 2010). Milled pine particles of size less than 6 mm were delignified by the organosolv process in aqueous ethanol at a liquid-to-solid ratio of 7 (Sannigrahi et al. 2010). The liquid-to-solid ratio was 8 for the organosolv delignification of miscanthus, milled to a particle size of 1–3 mm, in aqueous ethanol (El Hage et al. 2010). For the organosolv delignification of Siam weed, particles of size between 20 and 40 mesh, aqueous acetic acid was the solvent at a liquid-to-solid ratio of 10 (Zhao and Liu 2010). Aqueous acetone at a liquid-to-solid ratio of 14.2 was the organic solvent used to delignify wheat straw cut to less than 2 cm size (Huijgen et al. 2010). For its delignification, hydrolysed almond shells were ground to 0.2–0.5 mm size. The material was treated with different organic solvents, i.e., aqueous acetone, aqueous ethanol, or aqueous 1,4-dioxane. In each case, the liquid-to-solid ratio was 18.75 (Quesada-Medina et al. 2010). Wheat straw, cut and sorted into pieces of about 3 cm in length, was delignified with aqueous glycerol. The liquid-to-solid ratio was 10 (Saberikhah et al. 2011). Wheat straw was milled to less than 1 mm particle size and autohydrolysed with steam. Subsequently, the material was delignified with aqueous ethanol using a liquid-to-solid ratio of 10 (Ruiz et al. 2011). Organosolv delignification of sugarcane bagasse was performed with aqueous glycerol keeping a liquid-to-solid ratio of 10 (Martin et al. 2011). Sugarcane bagasse was cut into pieces of less than 1 cm size and subjected to dilute acid pretreatment. The pretreated material was subsequently delignified with aqueous ethanol with a liquid-to-solid ratio of 7 (Mesa et al. 2011). Depithed sugarcane bagasse of size bigger than 1.19 mm was subjected to organosolv delignification using aqueous glycerol at a liquid-to-solid ratio of 15 (Novo et al. 2011). Ground bamboo of 40–60 mesh size was delignified with 20 L aqueous formic acid as a solvent for every kg of lignocellulose (Li et al. 2012). Poplar sawdust of size less than 40 mesh was delignified with aqueous ethanol at a liquid-to-solid ratio of 10. Aqueous glycerol organosolv delignification of milled eucalyptus particles of less than 2 mm size used a liquid-to-solid ratio of 10 (Romani et al. 2013). Sugarcane bagasse of particle size 0.25–0.5 mm was delignified with aqueous ethylene glycol or propylene glycol and the liquid-to-solid ratio was 10 (Zhang et al. 2013). Beech delignification with glycerol used a liquid-to-solid ratio of about 13.5. Three size fractions of wood were studied: 0.1–1 mm, 2–4 mm, and 4–6.3 mm (Hundt et al. 2013). Chips of olive tree pruning of size 4–6 cm were subjected to organosolv delignification. The solvent was aqueous acetic acid or formic acid at a liquid-to-solid ratio of 10 (Erdocia et al. 2014). Rice straw delignification was carried out at a liquid-to-solid ratio of 20 with cholinium-based ionic liquids (Hou et al. 2012). Lignin was extracted from olive kernels, grape pomace and seeds, and peach kernels, ground to less than 1 mm size, using the organosolv procedure. Aqueous mixture of formic acid, and acetic acid was the solvent used at a high liquid-to-solid ratio of 25 (Manara et al. 2014). Several lignocellulose materials were milled to a particle size of 0.25–0.42 mm and delignified with cholinium-based ionic liquids at a liquid-to-solid ratio of 15 (An et al. 2015). The liquid-to-solid ratio was 5 for the delignification of birch in acetic acid (Kangas et al. 2015). However, organosolv delignification of about 2 cm long wheat straw pieces with formic acid or acetic acid used a liquid-to-solid ratio of 10 (Chen et al. 2015). Bamboo chips were delignified with aqueous ethanol with 10 L of solvent for every kg of lignocellulose mass (Correia et al. 2016). Tulip tree, milled to a particle size of 0.5 mm, was delignified with aqueous ethanol and the liquid-to-solid ratio was 10 (Jang et al. 2016). For organosolv lignin removal, sorghum bagasse was ground to pass a 2 mm screen. The powder so formed was delignified with different aqueous alcohol solvents using a liquid-to-solid ratio of 13.3 (Teramura et al. 2016). The liquid-to-solid ratio was 9 and the size of biomass was less than 1 mm for the organosolv delignification of date palm fronds using aqueous ethanol as solvent (Cybulska et al. 2017). Broussonetia papyrifera was milled to pass a 2 mm screen for its organosolv delignification. Aqueous ethanol was used as the solvent at a liquid-to-solid ratio of 9 (Yao et al. 2018). Larch sawdust with a particle size range of 0.1–1.5 mm was delignified with aqueous ethanol with a liquid-to-solid ratio of 7 (Hochegger et al. 2019). The ratio of aqueous glycerol to sugarcane bagasse, approximately 1 cm scrap size, was 10 during its organosolv delignification (Pascal et al. 2019). Milled poplar of size 3.5–4.5 mm was delignified with aqueous ethanol at a liquid-to-solid ratio of 10 (Meyer et al. 2020). Delignification of beech sawdust of size less than 24 mesh was performed at a liquid-to-solid ratio of 7 using different aqueous alcohols as solvents (Jasiukaitytė-Grojzdek et al. 2020). For lignin extraction from walnut shells, the material was ground to a fine powder. Organosolv delignification was carried out at a liquid-to-solid ratio of 8 with several aqueous alcohols as solvents (Zijlstra et al. 2020). Aqueous ethanol pilot scale batch delignification of beech is reported at a liquid-to-solid ratio of 3.2 (Schulze et al. 2016).

There are few studies which explicitly ponder on the effect of liquid-to-solid ratio on organosolv delignification. Base-catalysed delignification of olive tree trimmings, chipped to 5–10 mm size, with ethylene glycol as the organic solvent was carried out at three different liquid-to-solid ratios of 4, 6, and 8. Improvement in lignin removal was reported with an increase in the liquid-to-solid ratio (Jimenez et al. 2004). Eucalyptus chips of about 5 cm in size were delignified with aqueous formic acid with a liquid-to-solid ratio ranging from 10 to 30. No clear advantage could be established for increasing the liquid-to-solid ratio beyond 10 (Ligero et al. 2008). The liquid-to-solid ratio was varied between 3 and 9 to study its effect on aqueous ethanol delignification of sugarcane bagasse, with particle size between 1 and 5 cm. The researchers contended that increasing the amount of liquid with respect to the solid mass improved lignin solubilization due to efficient mass transfer (Vallejos et al. 2014). In a particular study, several woody and nonwoody biomasses were delignified with aqueous acetone. For woody biomass, the chip size was 2 mm and the liquid-to-solid ratio was 5 while the nonwoody biomass was cut to 10 mm size and delignified with a liquid-to-solid ratio of 10 (Smit and Huijgen 2017).

2.3.2.3 Temperature of Treatment

One of the most important process parameters in lignin removal is the temperature at which the process is carried out. For organosolv delignification, it is even more crucial given that both reaction rate and solubility are affected by the temperature. For low-boiling organic solvents, the reacting system must be pressurized if the temperatures are high. This adds to engineering complexities and enhanced cost. For deciding the right temperature for organosolv delignification, a trade-off between a high reaction rate and high system pressure at high temperature is often required. Temperature affects the viscosity of liquids which in turn affects the mass transfer rate. This can be a controlling factor for solvents like ethylene glycol or glycerol. Above all these, the reaction temperature strongly affects energy consumption and process economics. Given the variety of solvents available for organosolv delignification, these processes are performed at a wide range of temperature, from ambient to well over 200 °C.

Organosolv delignification processes that work at temperatures near or below 100 °C are not uncommon. These relatively low-temperature processes have several advantages in terms of simplicity of equipment design and less energy consumption. Phenol–water mixture was used at its normal boiling temperature of 100 °C for the delignification of miscanthus (Vega et al. 1997). The delignification temperature was 107 °C for lignin extraction from wheat straw using an aqueous mixture of acetic and formic acids (Kham et al. 2005). Milled eucalyptus was delignified with aqueous acetic acid or formic acid at their respective boiling temperatures at atmospheric pressure (Ligero et al. 2008). Acetic acid–water mixture was used at its normal boiling temperature of 107 °C to delignify Siam weed stem but the results were not very encouraging (Zhao and Liu 2010). Ground bamboo was delignified with aqueous formic acid, with or without an acid catalyst, at a temperature of 101 °C (Li et al. 2012). Rice straw delignification with cholinium-based ionic liquids was performed at temperatures of 50–130 °C. Lignin removal was 33% at 50 °C but increased to 53% at 70 °C and further to 60% at 90 °C. At 110, and 130 °C the lignin removal was 69 and 71%, respectively, but the selectivity of delignification was compromised (Hou et al. 2012). Another research group used cholinium-based ionic liquids at 90 °C as a solvent for lignin extraction from several lignocellulose biomass (An et al. 2015). The delignification temperature was 107 °C and the organic solvent was an aqueous mixture of formic and acetic acids for removing lignin from olive kernels, grape pomace and seeds, and peach kernels. However, the lignin removal was not encouraging (Manara et al. 2014). Aqueous acetic acid and formic acid were used as delignification solvents at their respective boiling points, well below 100 °C, for lignin removal from wheat straw (Chen et al. 2015). Lignin was partially extracted from hydrolysed almond shells at a relatively low temperature of 80 °C using different aqueous alcohols as solvents. 20–40% lignin could be extracted with n-butanol being the most efficient (Zijlstra et al. 2020).

There are organosolv processes that work in the intermediate temperature range of 100–150 °C. The delignification rate was observed to be consistently increasing during aqueous ethanol treatment of eucalyptus in the temperature range of 110–150 °C (Oliet et al. 2000). Eucalyptus delignification was carried out at 130 °C using aqueous acetic acid as the solvent (Villa et al. 2003). The reaction temperature was 110 °C for acetic acid delignification of eucalyptus (Sixta et al. 2004). Tulip tree powder was delignified with aqueous ethanol at different reaction temperatures. For the acid-catalysed process, the temperatures were 120, 130, and 140 °C and the lignin removal consistently increased with increasing temperature; the highest lignin removal being 43% at 140 °C. Higher temperatures also allowed for more complete hemicellulose dissolution. For base catalysis, higher reaction temperatures of 140, 150, and 160 °C were employed. Lignin removal improved substantially when increasing the temperature from 140 to 150 °C, but beyond this temperature appreciable benefit was not achieved (Koo et al. 2011). Lignin removal from sugarcane bagasse was practised at 130 °C using ethylene glycol or propylene glycol mixed with water and delignification close to 80% was reported (Zhang et al. 2013). Delignification of olive tree pruning was performed at 130 °C temperature using acetic acid or formic acid (Erdocia et al. 2014). Birch was delignified in aqueous acetic acid maintained at 150 °C temperature (Kangas et al. 2015). Aqueous ethanol delignification of tulip tree was carried out at a temperature of 130 °C and the lignin removal was 58%. For comparison, experiments were performed at 150, and 170 °C and the lignin removal was 73, and 79%, respectively (Jang et al. 2016). Several lignocellulose materials were delignified with aqueous acetone at the intermediate temperature of 140 °C (Smit and Huijgen 2017).

In most instances, organosolv delignification is practised at elevated temperatures of well above 150 °C to take advantage of increased reaction rate and solubility. Organosolv delignification of wheat straw was performed with aqueous phenol at three different temperatures of 150, 175, and 200 °C. Delignification consistently increased with increased temperature (Jimenez et al. 1997).

Similar trends were obtained when the delignification solvent was aqueous acetone (Jimenez et al. 1998). Effect of reaction temperature on aqueous alcohol, ethanol, and methanol, organosolv delignification in the temperature range 176–194 °C revealed significantly higher lignin removal at higher temperatures (Oliet et al. 2002). Delignification of eucalyptus chips with an aqueous mixture of acetic acid, and formic acid at 170 °C is reported (Sixta et al. 2004). For spruce delignification with aqueous methanol, the temperature was maintained at 205 °C (Yawalata and Paszner 2004a). The reaction temperature was varied between 165, 180, and 195 °C for the delignification of olive tree trimmings using ethylene glycol. The highest lignin removal, as measured by the kappa index of the solid residue, was obtained at the highest temperature of 195 °C (Jimenez et al. 2004). Three different reaction temperatures of 190, 200, and 210 °C were examined for removing lignin from sugarcane bagasse using dimethyl formamide as solvent. The extent of delignification, as measured by the kappa number of the solid residue, was substantially improved with an increase in reaction temperature from 190 to 210 °C (Rezayati-Charani et al. 2006). Oil palm empty fruit bunches were delignified with aqueous ethylene glycol or ethanolamine at two reaction temperatures of 170 or 180 °C (Rodriguez et al. 2008). Aqueous ethanol organosolv delignification of miscanthus was performed at temperatures of 170, and 190 °C (El Hage et al. 2010). The effect of reaction temperature on lignin extraction from hydrolysed almond shells was studied for three different solvents, aqueous ethanol, aqueous acetone, and aqueous dioxane. For this purpose, the temperature was varied from 150 to 230 °C. For all the solvents, delignification increased up to 210 °C. Beyond this temperature, lignin removal marginally decreased for ethanol and acetone but a marked reduction in lignin removal was noticed for dioxane (Quesada-Medina et al. 2010). The reaction temperature was between 180, and 195 °C for aqueous ethanol organosolv delignification of Buddleja davidii (Hallac et al. 2010). Pine was delignified at 170 °C with aqueous ethanol as the solvent (Sannigrahi et al. 2010). The effect of reaction temperature was studied between 160 and 220 °C for aqueous acetone organosolv delignification of wheat straw. These effects are shown in Figure 2.13. Delignification increased from 11% to 79% as the reaction temperature increased from 160 to 205 °C. The lignin

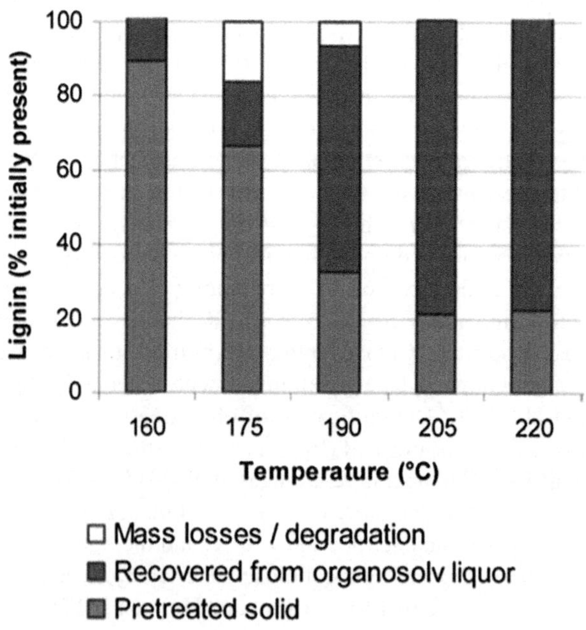

FIGURE 2.13 Temperature effects in aqueous acetone organosolv delignification. (Adapted and reprinted from Huijgen, WJJ, Reith, JH, den Uil H, *Ind. Eng. Chem. Res.*, 49, 10132–10140, 2010, with permission from American Chemical Society.)

removal, however, dropped due to condensation reactions as the temperature was further increased to 220 °C. It is also important to note that the reactor pressure was around 23 bar at a temperature of 190 °C (Huijgen et al. 2010).

The organic solvent was aqueous ethanol for the delignification of acid-pretreated sugarcane bagasse at three temperature levels of 175, 185, and 195 °C (Mesa et al. 2011). Raw sugarcane bagasse was delignified with aqueous glycerol with temperatures varying from 150 to 190 °C. Almost three times more delignification was obtained at 190 °C, compared to at 150 °C (Novo et al. 2011). Aqueous glycerol organosolv delignification of wheat straw was carried out under complete reflux with the temperature varying between 195 and 205 °C (Saberikhah et al. 2011). Similarly, reacting mixture was maintained at 190 °C for delignifying sugarcane bagasse with aqueous glycerol (Martin et al. 2011). Aqueous glycerol was used at reaction temperatures of 180, 190, and 200 °C for delignifying eucalyptus. Significantly higher delignification was noticed at 200 °C compared to 180 °C. The benefit was the most pronounced when a high solvent concentration was used along with a high reaction temperature. In addition to higher delignification more complete hemicellulose dissolution was also achieved at higher temperatures (Romani et al. 2013). Beech chips and sawdust were delignified with alkaline glycerol at temperatures varying from 190 to 230 °C. Three reaction temperatures of 150, 170, and 190 °C were studied for organosolv delignification of wheat straw with either aqueous ethanol or aqueous acetone as the solvent. Lignin removal was substantially improved when the temperature increased from 150 to 170 °C. In comparison, the benefit of temperature increase was only marginal as the reaction temperature was further increased to 190 °C (de la Torre et al. 2013). Ethanol–water delignification of sugarcane bagasse was performed at three different temperatures of 160, 175, and 190 °C. More than 90% delignification was achieved at the highest reaction temperature of 190 °C. Lignin removal was around 60% at 175 °C and varied between 20% and 50% when the temperature was 160 °C (Vallejos et al. 2014). Aqueous ethanol delignification of wheat straw at 190 °C in the presence of a catalyst and 220 °C without catalysts is reported (Chen et al. 2015). The reaction temperature was 190 °C for organosolv delignification of bamboo using aqueous ethanol (Correia et al. 2016). Several alcohols were used for delignifying sorghum bagasse at a temperature of 180 °C (Teramura et al. 2016). Aqueous ethanol was the solvent for delignifying date palm fronds. Two reaction temperatures of 140 and 200 °C were studied. Lignin extraction increased from 13% to 43% when the reaction temperature was increased from 140 to 200 °C (Cybulska et al. 2017). The reaction temperature was 200 °C when miscanthus was delignified with aqueous ethanol in the presence of pressurized carbon dioxide. 73% of the lignin present in the original lignocellulose material was removed (Kubota et al. 2018). Aqueous ethanol was used at 160 °C for delignifying Broussonetia papyrifera and 71% delignification was reported (Yao et al. 2018). Larch sawdust was delignified at 167, 177, 182, and 187 °C temperatures with aqueous ethanol as solvent. The highest lignin extraction was achieved at the highest temperature of 187 °C. However, the investigators also pointed out that beyond 182 °C the rate of carbohydrate degradation exceeded the rate of delignification (Hochegger et al. 2019). Sugarcane bagasse was treated with aqueous glycerol for organosolv delignification. Reaction temperature was maintained at four different levels of 140, 160, 180, and 200 °C. Delignification increased from 27% to 31% as the temperature increased from 140 to 200 °C while hemicellulose removal increased from 81% to 90% (Pascal et al. 2019). Aqueous ethanol delignification of poplar was performed at temperatures of 150, 180, and 210 °C. Lignin removal kinetics was found to be first order with an activation energy of 30 kJ/mol and the highest rate constant of 1.1 h^{-1} at 210 °C (Meyer et al. 2020). Beech sawdust was delignified at 180 °C using methanol or ethanol either in pure form or as aqueous solutions (Jasiukaitytė-Grojzdek et al. 2020). Pilot scale batch delignification of beech is reported using 50 weight% aqueous ethanol at a delignification temperature of 170 °C (Schulze et al. 2016).

2.3.2.4 Time of Treatment

The time of organosolv delignification treatment is an equally important process parameter. It affects the extent of bond breaking as well as the dissolution of the depolymerized fractions. Lignin removal and carbohydrate, primarily hemicellulose, removal are both affected. Therefore, treatment time has an influence on product yield and product purity. The heat-up time must also be accounted for in the total treatment time. Treatment time will certainly have a bearing on scale-up. Longer treatment time reduces the production rate for batch and continuous operations. Given the versatility of organosolv processes, we find treatment time to vary from a low of a few minutes to a high of a few hours.

Moderate-to-high delignification can be achieved with a reaction time of less than an hour to around 90 min. This is very important from techno-economic consideration as a rapid delignification procedure would allow for increased production with smaller reactors. Aqueous phenol delignification of miscanthus was carried out for treatment times varying in the range of 22–140 min (Vega et al. 1997). Reaction time was varied between 30 and 120 min for aqueous acetone delignification of wheat straw and reaction time between 75 and 90 min was reported to be the most effective for lignin removal. The study also reports the delignification to reduce if the treatment time is prolonged beyond this value (Jimenez et al. 1998). Wheat straw was delignified under similar conditions with aqueous phenol as the solvent. The highest lignin removal of 85% was obtained after 120 min treatment time. At 30 min treatment time, only 35% of lignin could be removed (Jimenez et al. 1997). For aqueous ethanol or methanol delignification of eucalyptus, treatment time was varied between 56 and 104 min (Oliet et al. 2002). 90 min treatment time was provided for the delignification of eucalyptus for producing viscose-grade pulps with either aqueous acetic acid or aqueous formic acid (Sixta et al. 2004). Aqueous methanol organosolv delignification was practised on spruce. The reacting mixture was heated from 25 to 205 °C in 40 min and treated at this temperature for 20 min (Yawalata and Paszner 2004a). Wheat straw was treated with a mixture of formic acid, acetic acid, and water. Lignocellulose mass was impregnated with solvent at 50 °C for 30 min followed by heating to a final reaction temperature of 107 °C in another 28 min. Thereafter, a treatment time of 90 min was provided for delignification (Kham et al. 2005). Lignin removal was between 15% and 50% within 60–80 min of treatment time during organosolv delignification of miscanthus in aqueous ethanol (El Hage et al. 2010). 40–60 min treatment of Buddleja davidii with aqueous ethanol resulted in 20–60% lignin removal (Hallac et al. 2010). Organosolv delignification of wheat straw in acetone–water mixture was carried out for 15, 30, 60, and 120 min and the effect on lignin removal is shown in Figure 2.14. 40–50% delignification is achieved in 15–30 min reaction time which increased to about 65% at 60 min reaction time (Huijgen et al. 2010).

For the aqueous ethanol organosolv treatment of pine, the reacting system was preheated to the final temperature in 50 min and allowed to react for another 20 min (Park et al. 2010). The reaction time was 1 h for 27% pine delignification with aqueous ethanol (Sannigrahi et al. 2010). Several organic solvents were examined for their suitability to delignify hydrolysed almond shells with a 10–55 min reaction time. For acetone and ethanol, the highest lignin removal of 79, and 73%, respectively, was obtained after 40 min of reaction. Delignification did not increase beyond 40 min treatment time. For dioxane, 87% delignification was achieved within 25 min and no further lignin removal was noticed with extended treatment time (Quesada-Medina et al. 2010). 40 min heating time was followed by a 10 min reaction at the final temperature for delignifying tulip tree with aqueous ethanol. Lignin removed was between 23 and 43% (Koo et al. 2011). Aqueous ethanol organosolv delignification of acid-pretreated sugarcane bagasse was carried out for treatment times of 20, 40, and 60 min (Mesa et al. 2011). Aqueous glycerol delignification of wheat straw for 30, 45, 90, and 180 min resulted in pulps of kappa numbers 64, 62, 55, and 51, respectively (Saberikhah et al. 2011). Reaction time was between 40 and 90 min for lignin removal from eucalyptus using aqueous glycerol. 30–60% lignin removal could be achieved even with 40 min reaction time and the delignification exceeded 80% with 90 min of reaction (Romani et al. 2013). Delignification varied between 54 and 94% for aqueous ethanol treatment of wheat straw for 30 and 75 min. It was 35–89% when the solvent used was aqueous acetone. No appreciable benefit was observed when the treatment time increased to 120 min (de la Torre et al. 2013). 30 min treatment of sugarcane bagasse with aqueous solutions of glycerol, ethylene glycol, and propylene glycol was carried out. Delignification was 89% with the two glycols, and 57% with glycerol (Zhang et al. 2013). Aqueous glycerol treatment

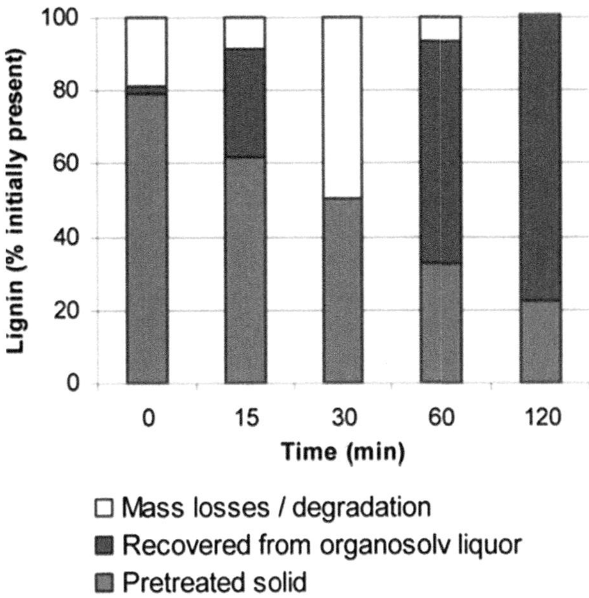

FIGURE 2.14 Treatment time effects in aqueous acetone organosolv delignification. (Adapted and reprinted from Huijgen, WJJ, Reith, JH, den Uil H, *Ind. Eng. Chem. Res.*, 49, 10132–10140, 2010, with permission from American Chemical Society.)

of beech reportedly removed 87–96% lignin with treatment times varying between 15 and 25 min (Hundt et al. 2013). Reaction time was 90 min for aqueous acetic acid or formic acid or mixed acid treatment of olive tree pruning (Erdocia et al. 2014). For aqueous acetic acid or formic acid treatment of wheat straw, the treatment time was varied between 30 and 90 min. Delignification varied between 70% and 80% for formic acid and 35–55% for acetic acid. For aqueous ethanol treatment, the reaction time was 60 min and the measured delignification was 58% (Chen et al. 2015). 60 min heating to final temperature was followed by treatment time of 30, 60, 90, and 120 min for aqueous acetic acid delignification of birch. Lignin removed was 39, 56, 64, and 88%, respectively, with increasing treatment time (Kangas et al. 2015). Prehydrolysed sugarcane bagasse was delignified with aqueous ethanol reacting the material between 30 and 150 min. The achieved delignification was 63–87%, with more treatment time giving more lignin removal. For untreated bagasse, less lignin was removed for 30 and 90 min treatment, compared to treated bagasse (Vallejos et al. 2014). Treatment time was 10, 30, and 50 min for aqueous ethanol organosolv delignification of tulip tree. Lignin removal between 60 and 80% was reported (Jang et al. 2016). 45 min treatment removed 65% lignin from sorghum bagasse using 1-butanol as solvent. Lignin removal was 74% with 1-pentanol (Teramura et al. 2016). A treatment time of 60 min with 15–20 min of heating was required for 13–43% lignin extraction during aqueous ethanol delignification of date palm fronds (Cybulska et al. 2017). Miscanthus was soaked for 20 min in aqueous ethanol and ground. The mixture was pressurized with carbon dioxide while heating to 200 °C and maintained at this temperature for 60 min. 73% lignin removal was reported (Kubota et al. 2018). Aqueous ethanol treatment of Broussonetia papyrifera was conducted for 5, 10, and 60 min. Reported lignin solubilization was 40, 61, and 72%, respectively (Yao et al. 2018). 30 min reaction time could remove 60–80% lignin from larch sawdust during aqueous ethanol organosolv treatment (Hochegger et al. 2019). A rapid organosolv treatment of sugarcane bagasse with aqueous glycerol was reported wherein 27–38% delignification was obtained in a 5–40 min reaction (Pascal et al. 2019). Lignin was extracted from beech sawdust in 76% yield by reacting with aqueous methanol or ethanol for 60 min (Jasiukaitytė-Grojzdek et al. 2020).

Treatment time of more than 90 min to several hours may lead to more complete lignin removal but also increases the possibility of cellulose damage. Reaction time was 180 min for aqueous acetic acid delignification of eucalyptus and 90% lignin was solubilized (Villa et al. 2003). Sugarcane bagasse was treated with dimethyl formamide between 180 and 240 min. After 240 min reaction, the resulting pulp had a 27–65 kappa number. At the lower end of reaction time, the kappa number ranged from 35 to 83 (Rezayati-Charani et al. 2006). Delignification of oil palm empty fruit bunches was performed for two treatment times of 90 min and 150 min. The solvent was an aqueous solution of ethylene glycol or ethanolamine (Rodriguez et al. 2008). Eucalyptus was treated with aqueous acetic acid, or formic acid with treatment times of 60, 120, and 180 min. Formic acid was more effective in lignin removal. Lignin remaining in pulp, as indicated by kappa number, was between 30 and 34 for 60 min reaction time. Kappa number ranged between 22 and 30 when the material was treated for 120 min. For 180 min treatment time, the highest lignin removal was achieved with pulp kappa number varying between 23 and 29. For the acetic acid process, pulp kappa number ranged from 32 to 48 at 60 min treatment time. The kappa number range dropped to 32–40 as the reaction time increased to 120 min. The upper limit of the kappa number range further dropped to 38 with 180 min treatment (Ligero et al. 2008). Siam weed was delignified with aqueous acetic acid for 3 h but the lignin removed was only 13% (Zhao and Liu 2010). Three reaction times of 60, 150, and 240 min were employed for delignifying sugarcane bagasse with aqueous glycerol. After 60 min, lignin removal was between 25 and 70%. Obtained lignin removal was 52–54% for 150 min treatment. 33–80% lignin removal was achieved after 240 min (Novo et al. 2011). Sugarcane bagasse was delignified with aqueous glycerol for 60, 120, 180, and 240 min. The results reported were interesting as increasing treatment time did not always improve lignin removal. For base-catalysed and autocatalysed reactions, delignification improved up to 180 min but decreased slightly at 240 min reaction time. Though the authors did not give any explanation, this could be due to condensation reactions and lignin reprecipitation predominating when the organosolv treatment was prolonged. For acid-catalysed reaction, on the contrary, lignin removal was the highest after 60 min and sharply dropped when the reaction time was increased to 120 min. Thereafter, lignin removal consistently increased with treatment time up to 240 min. This is curious and even the authors did not put forward an explanation for such a behaviour (Martin et al. 2011). Several lignocellulose materials were treated with a mixture of formic acid, acetic acid, and water. The raw material was impregnated with solvent at 50 °C for 30 min followed by heating to a final reaction temperature of 107 °C in an unspecified time. Thereafter, a treatment time of 180 min was provided for delignification. Lignin removal was 31% for the olive kernel, and 39% for the peach kernel, but it was abysmally low for grape pomace and seeds (Manara et al. 2014). A prolonged treatment time of 12 h was required for about 65% lignin removal from different lignocellulose materials using cholinium-based ionic liquid as solvent (An et al. 2015). Similarly, rice straw was treated with cholinium-based ionic liquids for 24 h to extract 40–60% lignin from biomass (Hou et al. 2012). Delignification time was 120 min for the aqueous formic acid organosolv treatment of bamboo. Lignin yield was between 19 and 22% (Li et al. 2012). Bamboo was delignified for 120 min with aqueous ethanol with an additional 20 min of preheating time, but the resulting pulp still had 14% lignin (Correia et al. 2016). 120 min treatment time was employed for lignin removal from several agricultural residues, hardwoods, and softwoods. The organic solvent was aqueous acetone, and the delignification was close to 80% for agricultural residues, and hardwoods. The softwoods, however, proved to be resistant to delignification with only about 30% lignin removed (Smit and Huijgen 2017). Kinetics of poplar delignification using aqueous ethanol was studied over a period of 25 h. Rapid delignification was achieved in the first 2 h. It was observed that a treatment time of more than 5 h was detrimental to lignin removal, especially at high reaction temperatures (Meyer et al. 2020). The effect of extraction time on lignin recovery was studied between 1 and 72 h for organosolv treatment of walnut shells with different alcohols mixed with water. With n-pentanol and n-butanol, lignin recovery was 10–12% after 5h, around 20% after 16 h, and 40–45% after 72 h (Zijlstra et al. 2020). Reaction time was 100 min at 170 °C for pilot scale batch delignification of beech using aqueous ethanol as solvent (Schulze et al. 2016).

2.3.2.5 Interaction among Process Parameters

Individual parameters affect organosolv delignification but not entirely in their own right. Rather they work in tandem to influence the outcome of the process. Solvent composition and liquid-to-solid ratio together make up for the total amount of organic solvent added to the reacting system. Temperature and solvent composition interact as temperature affects the lignin solubility through the δ value. Miscibility of many organic liquids, like higher alcohols and phenols, with water depends on temperature. One of the approaches to evaluate parameter interactions on delignification outcomes is through statistical modelling. In an early study on aqueous phenol delignification of wheat straw, the authors determined an appreciable interaction between treatment time and reaction temperature on lignin removal (Jimenez et al. 1997). Similarly, for aqueous acetone delignification of wheat straw, experiments suggested an appreciable interaction between acetone concentration and treatment time, and between treatment time and reaction temperature on lignin removal (Jimenez et al. 1998). Statistical analysis revealed appreciable interaction between the process parameters – catalyst charge, reaction time, and solvent concentration – on lignin removal during hydrochloric-acid-catalysed aqueous phenol delignification of miscanthus. Parameters were varied in the range; catalyst charge 1–6%, reaction time 22–141 min, and phenol concentration 41.6–58.4 weight% (Vega et al. 1997). Aqueous alcohol, ethanol, and methanol delignification of eucalyptus was studied to statistically ascertain the effects of treatment temperature (176–194 °C), treatment time (56–104 min), and alcohol concentration (38–62%), and their mutual interactions. Reaction temperature, reaction time, and alcohol concentration, in that order of significance, were found to individually affect lignin removal. Increasing temperature and time had a positive influence on delignification while increasing alcohol concentration had a negative effect. Interaction between temperature and alcohol concentration had the most positive influence on lignin removal followed by time and alcohol concentration. Interaction between temperature and time, on the other hand, had a negative effect on lignin removal (Oliet et al. 2002). Parameter effects and their interactions were evaluated for aqueous dimethyl formamide delignification of sugarcane bagasse. The investigated parameters and their ranges were, the temperature varying from 190 to 210 °C, treatment time varying from 180 to 210 °C, and solvent concentration from 40% to 60% v/v. Reaction temperature had the strongest effect on delignification followed by reaction time and solvent concentration. Further, solvent concentration did not interact with temperature or treatment time in affecting lignin removal. However, a moderate interaction between treatment temperature and treatment time was observed (Rezayati-Charani et al. 2006). During the delignification of eucalyptus with aqueous formic or acetic acid, hydrochloric acid was the catalyst. For the formic acid process at unspecified treatment temperature, catalyst concentration was varied from 0.05 to 1.0 weight% along with formic acid concentration (60–95 wight%), liquid-to-solid ratio (10–30), and reaction time (60–150 min). The optimal combination of low pulp kappa number (20.5) and high pulp yield (41.5%) would be attained with the parameter combination of 0.2% catalyst concentration, 80% acid concentration, liquid-to-solid ratio of 10, and 180 min reaction. Similar interactions were observed for the acetic acid process in which the liquid-to-solid ratio was fixed at 10; the minimum attainable kappa number was 31 at 0.5% catalyst concentration, 90% acid concentration, and 180 min reaction (Ligero et al. 2008). Sulphuric-acid-catalysed aqueous ethanol treatment of Buddleja davidii was carried out with three different combinations of process parameters catalyst concentration, ethanol concentration, treatment temperature, and treatment time. The conditions were (1) 1.25%, 50%, 180 °C, 60 min, (2) 1.75%, 50%, 180 °C, 40 min, (3) 1.5%, 65%, 195 °C, 60 min. The resulting delignifications were 15, 30, and 64%, respectively (Hallac et al. 2010). Among the three process variables, reaction temperature, reaction time, and solvent concentration, only temperature was found to be statistically significant in influencing lignin removal during aqueous glycerol delignification of sugarcane bagasse. In the studied range of temperature (150–190 °C). time (60–240 min), and glycerol concentration (20–80% v/v), the best lignin removal of 80% was achieved when all the parameters were at their highest values. When the temperature was reduced to 150 °C with 240 min reaction and 80% glycerol concentration, the lignin removal dropped to 33%. On the contrary, the lignin removal was still a healthy

70% when bagasse was treated for 60 min at 190 °C with 80% concentration glycerol (Novo et al. 2011). Wheat straw was delignified with aqueous formic acid or acetic acid in the presence of sulphuric acid catalyst. The studied process variables for formic acid treatment were acid concentration (68–88%), catalyst concentration (0–0.1%), and treatment time (0.5–1.5 h). The highest delignification of 78% was achieved with 88% acid concentration, 0.1% catalyst concentration, and 1h reaction time. For acetic acid treatment, the studied ranges of parameters were acid concentration (70–90%), catalyst concentration (0.1–0.3%), and treatment time (1–3 h). The combination of 90% acid concentration, 0.3% catalyst concentration, and 2 h reaction provided the highest delignification of 54% (Chen et al. 2015).

Hydrochloric-acid-catalysed delignification of wheat straw was performed with two aqueous organic solvents, ethanol, and acetone. Treatment conditions were varied as temperature (150–190 °C), time (30–120 min), solvent concentration (50–80%), and catalyst concentration (0.001–0.01 N). Optimal lignin removal of 94% with aqueous ethanol resulted with a reaction temperature of 179 °C, a treatment time of 30 min, and 0.01N catalyst concentration. Ethanol concentration in the studied range did not affect delignification. For acetone, the optimal concentration was 80% to remove 98% lignin, with the conditions being 170 °C temperature, 120 min treatment, and 0.01 N catalyst concentration (de la Torre et al. 2013). Delignification of hot water pretreated sugarcane bagasse with aqueous ethanol revealed that treatment time (30–150 min) was the most significant parameter followed by treatment temperature (160–190 °C), while liquid-to-solid ratio (3–9) had the least effect (Vallejos et al. 2014). Aqueous ethanol delignification of date palm frond in the presence of sulphuric acid catalyst was carried out by varying the three process parameters, treatment temperature (140 and 200 °C), catalyst concentration (0 and 1.5%), and ethanol concentration (60 and 80%). The highest lignin recovery of 43% occurred in the absence of a catalyst at 200 °C and 80% ethanol concentration (Cybulska et al. 2017). Sulphuric-acid-catalysed sugarcane bagasse delignification with aqueous glycerol exhibited significant interaction between the parameters, treatment temperature (140–200 °C) and catalyst dose (0.02–0.1%), and treatment temperature and treatment time (5–40 min), in affecting lignin removal. Individually, temperature had the strongest influence on lignin removal and selectivity (Pascal et al. 2019). Poplar was delignified in aqueous ethanol and the interaction between reaction temperature (150–210 °C), and reaction time (0.25–25 h) was evaluated. The interactions are shown in Figure 2.15. We can see that at higher temperatures (180 and 210 °C), maximum lignin removal is achieved at some intermediate treatment time as prolonged reaction leads to excessive condensation and reprecipitation (Meyer et al. 2020).

2.3.2.6 Severity Factor

Parameters that strongly affect the lignin removal are catalyst dose, treatment temperature, and treatment time and these are the parameters that strongly interact in influencing organosolv delignification. The combination effects of time and temperature on organosolv lignin removal are clubbed into a single parameter, referred to as Reaction Ordinate, R_0. This concept was first proposed by Overend and Chornet (Overend and Chornet 1987). Its instantaneous value at a given time can be written as

$$R_0 = t \times e^{\frac{(T-373)}{14.75}} \tag{2.6}$$

Here, t is the reaction time in min, and T is the reaction temperature in Kelvin. The temperature of the reacting system changes during the initial heating and post-reaction cooling. When these changes are taken into account, the weighted average value of the reaction ordinate can be written as

$$R_0 = \int_0^t e^{\frac{(T-373)}{14.75}} \mathrm{d}t \tag{2.7}$$

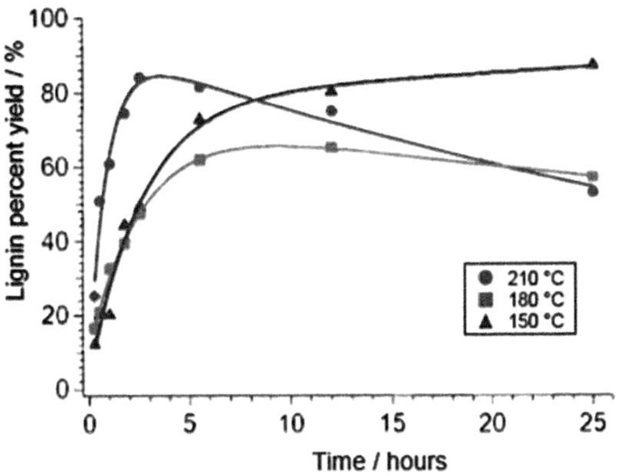

FIGURE 2.15 Parameter interaction between treatment temperature, and treatment time effects in aqueous ethanol organosolv delignification. (Reprinted from Meyer, JR, Li, H, Zhang J, Foston, MB, *ChemSusChem*, 13(17), 4557–4566, 2020, with permission from Wiley.)

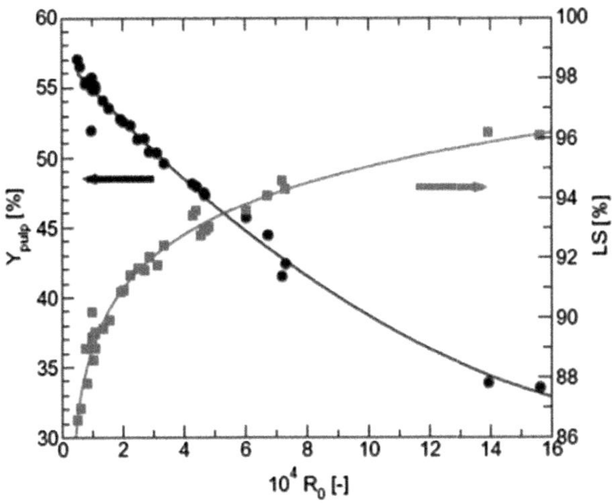

FIGURE 2.16 Effect of reaction ordinate in aqueous glycerol organosolv delignification. (Reprinted and adapted from Hundt, M, Schnitzlein, K, Schnitzlein MG, *Bioresource Technol.*, 136, 672–679, 2013, with permission from Elsevier.)

A parameter, severity factor (SF), is defined as (Galbe and Zacchi 2007)

$$SF = \log_{10}\left(R_0\right) = \log_{10}\left[t \times e^{\frac{(T-373)}{14.75}}\right] \tag{2.8}$$

For base-catalysed delignification of beech with glycerol, R_0 was varied from 0.53 × 10^4 to 15.64 × 10^4 by varying the temperature between 190 and 230 °C, and the treatment time between 15 and 25 min. A strong correlation between lignin solubilized (LS) and R_0 was observed as illustrated in Figure 2.16. Similarly, the pulp yield correlated with R_0 (Hundt et al. 2013).

In acid-catalysed organosolv delignification, when the effect of catalyst dose is also factored into, the reaction ordinate is modified as (Abatzoglou et al. 1992)

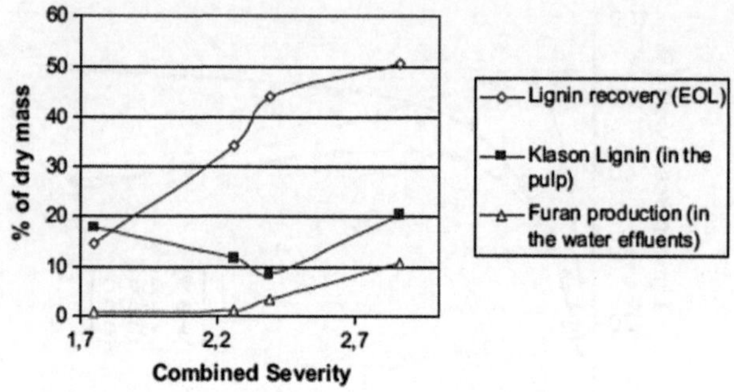

FIGURE 2.17 Effect of CSF in aqueous ethanol organosolv delignification. (Reprinted from El Hage, R, Brosse, N, Sannigrahi P, Ragauskas, A, *Polymer Deg. Stab.*, 95, 997–1003, 2010, with permission from Elsevier.)

$$R_0^/ = R_0\left[\mathrm{H}^+\right] \tag{2.9}$$

where $\left[\mathrm{H}^+\right]$ is the concentration of hydrogen ions in the reacting system. From this, the Combined Severity Factor (CSF) is defined as

$$\mathrm{CSF} = \log_{10}\left[t \times e^{\frac{(T-373)}{14.75}}\right] - \mathrm{pH} \tag{2.10}$$

Miscanthus was delignified with aqueous ethanol in the presence of a sulphuric acid catalyst. In the range of reaction temperature 170–190 °C, treatment time 60–80 min, ethanol concentration 50–65%, and catalyst concentration 0.5–1.6%, the CSF varied between 1.75 and 2.93 (El Hage et al. 2010). The increase in CSF clearly favours organosolv lignin recovery, almost linearly, as shown in Figure 2.17. *L. tulipifera* was delignified with aqueous ethanol under sulphuric acid catalysis. The catalyst concentration was varied between 5 and 95 mM, temperature between 130 and 170 °C, and treatment time between 10 and 50 min, to have a constant CSF of 1.16 for each experiment. However, the lignin removal was not the same always. Treatment temperature was found to have a disproportionate effect with the low temperature of 130 °C not conducive to sufficient lignin removal. For other combinations of process parameters, with the same CSF of 1.16, lignin removal varied in a narrow range of 73–80%. (Jang et al. 2016). Acid-catalysed delignification of larch sawdust with aqueous ethanol resulted in increased lignin removal with increased CSF. The studied CSF range was 1.65–2.42. The study also revealed that higher process severity decreases the selectivity of lignin removal over that of carbohydrate removal (Hochegger et al. 2019).

REFERENCES

Abatzoglou, N, Chornet, E, Belkacemi, K, Overend, RP. *Chem. Eng. Sci.* 47(5):1109–1122, 1992.
Alaejos, J, López, F, Eugenio, ME, Tapias, R. *Bioresour. Technol.* 97:2110–2116, 2006.
An, YX, Zong, MH, Wu, H, Li, N. *Bioresour. Technol.* 192:165–171, 2015.
Ando, S, Kakimoto, T, Itoh, K, Arai, I, Kiyoto, K, Hanai, S. *Biotechnol. Bioeng.* 31:802–804, 1988.
Angles, MN, Reguant, J, Garcia-Valls, R, Salvado, J. *Wood Sci. Technol.* 37:309–320, 2003.
Awan, IZ, Beltrami, G, Bonincontro, D, Gimello, O, Cacciaguerra, T, Tanchoux, N, Martucci, A, Albonetti, S, Cavani, F, Di Renzo, F. *Appl. Catal. A. Gen.* 609:117929, 2021.
Ban, W, Mancosky, D, Lucia, LA. Cellulose *Chem. Technol.* 38(1-2):79–85, 2004.
Bin, Z, Bin, WS, Liang, GD. *For. Stud. China* 14(3): 235–239, 2012.
Capanema, EA, Balakshin, MY, Kadla, JF. *J. Agric. Food Chem.* 53:9639–9649, 2005.

Caron, A, Altaner, CM, Gardiner, B, Jarvis, MC. *Eur. J. Wood Prod.* 71:697–704, 2013.

Chakar, FS, Ragauskas, AJ. *Ind. Crop Prod.* 20:131–141, 2004.

Chen, H, Zhao, J, Hu, T, Zhao, X, Liu, D. *Appl. Energy* 150:224–232, 2015.

Chen, X, Zhang, K, Xiao, LP, Sun, RC, Song, G. *Biotechnol. Biofuels* 13:2, 2020.

Christensen, T, Albright, LF, Williams, TJ. *Proc. Tappi Annual Meeting*, 239–246, 1983.

Christov, LP, Akhtar, M, Prior, BA. *Enzyme Microb. Technol.* 23:70–74, 1998.

Correia, VD, dos Santos, V, Sain, M, Santos, SF, Leao, AL, Junior, HS. *Cellul.* 23:2971–2987, 2016.

Cybulska, I, Brudecki, G, Rosentrater, K, Julson, JL, Lei, H. *Appl. Energy* 185:1040–1050, 2017.

Dastpak, A, Lourençon, TV, Balakshin, M, Hashmi, SF, Lundström, M, Wilson, BP. *Ind. Crop Prod.* 148:112310, 2020.

de la Torre, MJ, Moral, A, Hernández, MD, Cabeza, E, Tijero, A. *Ind. Crop Prod.* 45:58–63, 2013.

del Río, JC, Rencoret, J, Prinsen, P, Martínez, AT, Ralph, J, Gutiérrez, A. *J. Agric. Food Chem.* 60: 5922–5935, 2012.

Deshpande, R, Sundvall, L, Grundberg, H, Germgard, U. *O Papel* 76(4):56–61, 2015.

Deshpande, R, Sundvall, L, Grundberg, H, Germgard, U. *Cellulose Chem. Technol.* 50(2):293–300, 2016a.

Deshpande, R, Sundvall, L, Grundberg, H, Germgard, U. *Nordic Pulp Paper Res. J.* 31(3):379–385, 2016b.

Deshpande, R, Sundvall, L, Grundberg, H, Germgard, U. *BioResources* 11:5961–5973, 2016c.

Deshpande, R, Giummarella, N, Henriksson, G, Germgard, U, Sundvall, L, Grundberg, H, Lawoko, M. *Ind. Crop Prod.* 115:315–322, 2018.

Dolk, M, Yan, JF, McCarthy, JL. Lignin 25. *Holzforschung* 43:91–98, 1989.

El Hage, R, Brosse, N, Sannigrahi, P, Ragauskas, A. *Pol. Deg. Stabil.* 95:997–1003, 2010.

Erdocia, X, Prado, R, Corcuera, MA, Labidi, J. *J. Ind. Eng. Chem.* 20:1103–1108, 2014.

Evtuguin, DV, Sulphite Pulping. Belgacem, N, Pizzi, A (eds.) Lignocellulosic Fibers and Wood Handbook, 225–244. Scrivener Publishing LLC. 2016.

Fearon, O, Nykänen, V, Kuitunen, S, Ruuttunen, K, Alén, R, Alopaeus, V, Vuorinen, T. *AIChE J.* 66(8): 16252, 2020.

Finell, M, Nilsson, C. *Ind. Crop Prod.* 19:155–165, 2004.

Fischer, K, Schmidt, I. *Tappi J.* 74(1):181–185, 1991.

Fischer, K, Akhtar, M, Blanchette, RA, Burnes, TA, Messner, K, Kirk, TK. *Holzforschung* 48:285–290, 1994.

Fiskari, J, Kilpeläinen, P. Asia Pacific J. *Chem. Eng.* 16(6):e2707, 2021.

Füchtner, S, Brock-Nannestad, T, Smeds, A, Fredriksson, M, Pilgård, A, Thygesen, LG. *Front. Plant Sci.* 11:855, 2020.

Galbe, M, Zacchi, G. *Adv. Biochem. Eng. Biotechnol.* 108:41–65, 2007.

Ghatak, HR. *Tappi J.* 1(1):24–27, 2002.

Gierer, J, Pettersson, I. *Can. J. Chem.* 55, 593–599, 1977.

Goldberg, RN, Parker, VB. *J. Res. Natl. Bur. Stand.* 90(5):341–358, 1985.

Gonultas, O, Ucar, MB. *Lignocellulose* 2(1):262–269, 2013.

Guadix-Montero, S, Sankar, M. *Topics in Catalysis.* 61:183–198, 2018.

Gutiérrez, del Rio, JC, González-Vila, FJ, Martín, F. *Holzforschung* 53:481–486, 1999.

Hallac, BB, Pu, Y, Ragauskas, AJ. *Energy Fuel* 24:2723–2732, 2010.

Hamzeh, Y, Abyaz, A, Niaraki, MOM, Abdulkhani, A. *BioResources* 4:1267–1275, 2009.

Hartler, N. *Tappi J.* 79(2):259–264, 1996.

Hochegger, M, Cottyn-Boitte, B, Cezard, L, Schober, S, Mittelbach, M. *Int. J. Chem. Eng.* 2019:1734507, 2019.

Hou, XD, Smith, TJ, Li, N, Zong, MH. *Biotechnol. Bioeng.* 109(10):2484–2493, 2012.

Huijgen, WJJ, Reith, JH, Uil, HD. *Ind. Eng. Chem. Res.* 49:10132–10140, 2010.

Hundt, M, Schnitzlein, K, Schnitzlein, MG. *Bioresour. Technol.* 136:672–679, 2013.

Iakovlev, M, Hiltunen, E, van Heiningen, A. *Nordic Pulp Paper Res. J.* 25(4):428–443, 2010.

Inglesby, MK, Gray, GM, Wood, DF, Gregorski, KS, Robertson, RG, Sabellano, GP. *Colloids Surf. B. Biointerfaces* 43:83–94, 2005.

Jang, SK, Kim, HY, Jeong, HS, Kim, JY, Yeo, H, Choi, IG. *Renew. Energy* 87:599–606, 2016.

Jardim, JM, Hart, PW, Lucia, LA, Jameel, H, Chang, HM. *Fibers* 10:16, 2022.

Jasiukaitytė-Grojzdek, E, Huš, M, Grilc, M, Likozar, B. *Sci. Rep.* 10:11037, 2020.

Jimenez, L, de la Torre, MJ, Maestre, F, Ferrer, JL, Perez, I. *Bioresour. Technol.* 60:199–205, 1997.

Jimenez, L, de la Torre, MJ, Bonilla, JL, Ferrer, JL. *Process Biochem.* 33(4):401–408, 1998.

Jimenez, L, Rodriguez, A, Diaz, MJ, Lopez, F, Ariza, J. *Holzforschung* 58:122–128, 2004.

Kangas, H, Liitiä, T, Rovio, S, Ohra-aho, T, Heikkinen, H, Tamminen, T, Poppius-Levlin, K. *Holzforschung* 69:247–256, 2015.

Kebbi-Benkeder, Z, Colin, F, Dumarçay, S, Gérardin, P. *Annal. Forest Sci.* 72:277–284, 2015.

Khakimova, FK, Khakimov, RR, Noskova, OA. Russian *J. Appl. Chem.* 90(3): 440–445, 2017.
Kham, L, Bigot, YL, Delmas, M, Avignon, G. *Ind. Crop Prod.* 21:9–15, 2005.
Khantayanuwong, S, Yimlamai, P, Chitbanyong, K, Wanitpinyo, K, Pisutpiched, S, Sungkaew, S, Sukyai, P, Puangsin, B. *J. Natural Fibres* 20: 2150924, 2023.
Khristova, P, Kordsachia, O, Patt, R, Karar, I, Khider, T. *Ind. Crop Prod.* 23:131–139, 2006.
Kininge, MM, Gogate, PR. *Ultrason. Sonochem.* 82:105870, 2022.
Kleinert, T, Tayenthal, K. US Patent 1856567, 1932.
Kohlbrenner, PJ, Schuerch, C. *J. Org. Chem.* 24:166–172, 1959.
Koo, BW, Park, N, Jeong, HS, Choi, JW, Yeo, H, Choi, IG. *J. Ind. Eng. Chem.* 17:18–24. 2011.
Krutul, D, Antczak, A, Radomski, A, Drozdzek, M, Kłosińska, T, Zawadzki, J. *Ann. Warsaw Univ. Life Sci. – SGGW, For. and Wood Technol.* 107:131–138, 2019.
Kubota, AM, Kalnins, R, Overton, TW. *Biomass Bioenergy* 111:52–59, 2018.
Lei, Y, Liu, S, Li, J, Sun, R. *Biotechnol. Adv.* 28:609–612, 2010.
Li, MF, Sun, SN, Xu, F, Sun, RC. *Chem. Eng. J.* 179:80–89, 2012.
Ligero, P, Villaverde, JJ, de Vega, A, Bao, M. *Ind. Crop Prod.* 27:110–117, 2008.
Lim, WL, Gunny, AAN, Kasim, FH, AlNashef, IM, Arbain, D. *Cellul.* 26:4085–4098, 2019.
Lou, R, Lyu, G, Wu, S, Zhang, B, Zhao, H, Lucia, LA. *ACS Sustainable Chem. Eng.* 6:430–437, 2018.
Lu, Q, Liu, W, Yang, L, Zu, Y, Zu, B, Zhu, M, Zhang, Y, Zhang, X, Zhang, R, Sun, Z, Huang, J, Zhang, X, Li, W. *Food Chem.* 131:313–317, 2012.
Lundquist, K, Ericsson, L. *Acta Chem. Scand.* 24:3681–3686, 1970.
Lundquist, K, Lundgren, R. *Acta Chem. Scand.* 26:2005–2023, 1972.
Ma, Q, Wang, L, Zhai, H, Ren, H. *Int. J. Biol. Macromol.* 182:51–58, 2021.
Manara, P, Zabaniotou, A, Vanderghem, C, Richel, A. *Catal. Today* 223:25–34, 2014.
Marques, AP, Evtuguin, DV, Magina, S, Amado, FML, Prates, A. *J. Wood Chem. Technol.* 29:322–336, 2009.
Martin, C, Puls, J, Saake, B, Schreiber, A. *Cellulose Chem. Technol.* 45(7-8):487–494, 2011.
Martinez, JM, Reguant, J, Salvado, J, Farriol, X. *Bioresour. Technol.* 60:161–167, 1997.
Mesa, L, González, E, Cara, C, González, M, Castro, E, Mussatto, SI. *Chem. Eng. J.* 168:1157–1162, 2011.
Meyer, JR, Li, H, Zhang, J, Foston, MB. *ChemSusChem* 13(17):4557–4566, 2020.
Mohan, M, Huang, K, Pidatala, VR, Simmons, BA, Singh, S, Sale, KL, Gladden, JM. *Green Chem.* 24(3): 1165–1176, 2022.
Molteberg, D, Høibø, O. *Wood Sci. Technol.* 40:173–189, 2006.
Moniruzzaman, M, Goto, M. *Adv. Biochem. Eng. Biotechnol.* 168:61–77, 2019.
Muhammad, N, Man, Z, Bustam, MA, Mutalib, MIA, Rafiq, S. *J. Ind. Eng. Chem.* 19(1):207–214, 2013.
Myeong, S, Han, SH, Shin, SJ. *J. Korean For. Soc.* 99(5):706–710, 2010.
Neves, PV, Pitarelo, AP, Ramos, LP. *Bioresour. Technol.* 208:184–194, 2016.
Novo, LP, Gurgel, LVA, Marabezi, K, Curvelo, APD. *Bioresour. Technol.* 102:10040–10046, 2011.
Oliet, M, Rodriguez, F, Santos, A, Gilarranz, MA, Garcia-Ochoa, F, Tijero, J. *Ind. Eng. Chem. Res.* 39: 34–39, 2000.
Oliet, M, Garcia, J, Rodriguez, F, Gilarranz, MA. *Chem. Eng. J.* 87:157–162, 2002.
Overend, RP, Chornet, E. *Phil. Trans. R. Soc. Lond. A* 321:523–536, 1987.
Paananen, M, Liitiä, T, Sixta, H. *Ind. Eng. Chem. Res.* 52:12777–12784, 2013.
Park, N, Kim, HY, Koo, BW, Yeo, H, Choi, IG. *Bioresour. Technol.* 101:7046–7053, 2010.
Park, SY, Kim, JH, Kim, JC, Yang, SY, Yeo, H, Choi, IG. *BioResources* 16(3):5301–5312, 2021.
Pascal, K, Ren, H, Sun, FF, Guo, S, Hu, J, He, J. *ACS Omega* 4(22):20015–20023, 2019.
Pinto, PC, Evtuguin, DV, Neto, CP. *Ind. Eng. Chem. Res.* 44(26):9777–9784, 2005.
Quesada-Medina, J, López-Cremades, FJ, Olivares-Carrillo, P. *Bioresour. Technol.* 101:8252–8260, 2010.
Rai, N, Siepmann, JI, Schultz, NE, Ross, RB. *J. Phys. Chem.* 111:15634–15641, 2007.
Ralph, J, Lundquist, K, Brunow, G, Lu, F, Kim, H, Schatz, PF, Marita, JM, Hatfield, RD, Ralph, SA, Christensen, JH, Boerjan, W. *Phytochem. Rev.* 3:29–60, 2004.
Reguant, J, Martinez, JM, Montane, D, Salvado, J, Farriol, X. *J. Wood Chem. Technol.* 17(1–2):91–110, 1997.
Rezayati-Charani, P, Mohammadi-Rovshandeh, J, Hashemi, SJ, Kazemi-Najafi, S. *Bioresour. Technol.* 97:2435–2442, 2006.
Rocha, GJM, Goncalves, AR, Oliveira, BR, Olivares, EG, Rossell, CEV. *Ind. Crop Prod.* 35:274–279, 2012.
Rodriguez, A, Serrano, L, Moral, A, Perez, A, Jimenez, L. *Bioresour. Technol.* 99:1743–1749, 2008.
Rodrigues, PF, Evtyugin, DD, Evtuguin, DV, Prates, A. *J. Wood Chem. Technol.* 38(5):397–408, 2018.
Romani, A, Ruiz, HA, Pereira, FB, Domingues, L, Teixeira, JA. *Ind. Eng. Chem. Res.* 52:14342–14352, 2013.
Rueda, C, Fernández-Rodríguez, J, Ruiz, G, Llano, T, Coz, A. *Carbohydr. Polym.* 116:18–25, 2015.

Ruiz, HA, Ruzene, DS, Silva, DP, da Silva, FFM, Vicente, AA, Teixeira, JA. *Appl. Biochem. Biotechnol.* 164:629–641, 2011.
Saberikhah, E, Rovshandeh, JM, Rezayati-Charani, P. Cellulose *Chem. Technol.* 45(1-2):67–75, 2011.
Sahin, HT. *J. Chem. Technol. Biotechnol.* 78:1267–1273, 2003.
Santos, RB, Capanema, EA, Balakshin, MU, Chang, HM, Jameel, H. *BioResources* 6(4):3623–3637, 2011.
Santos, RB, Hart, PW, Jameel, H, Chang, HM. *BioResources* 8(1):1456–1477, 2013.
Sannigrahi, P, Ragauskas, AJ, Miller, SJ. *Energy Fuel* 24:683–689, 2010.
Schrems, M, Liebner, F, Betz, M, Zeilinger, M, Bohmdorfer, S, Rosenau, T, Potthast, A. *J. Wood Chem. Technol.* 32:225–237, 2012.
Schulze, P, Seidel-Morgenstern, A, Lorenz, H, Leschinsky, M, Unkelbach, G. *Bioresour. Technol.* 199: 128–134, 2016.
Scott, GM, Akhtar, M, Lentz, M. *Proc. Tappi Pulping Conference*, 355–361, 1995.
Shatalov, AA, Pereira, H. *Ind. Crop Prod.* 21:203–210, 2005.
Shemfe, MB, Gu, S, Ranganathan, P. *Fuel* 143:361–372, 2015.
Shen, W, Chen, X. *Ind. Eng. Chem. Res.* 48:8980–8984, 2009.
Singh, S, Ghatak, HR. *J. Wood Chem. Technol.* 37:407–422, 2017.
Sixta, H, Harms, H, Dapia, S, Parajo, JC, Puls, J, Saake, B, Fink, HP, Order, T. *Cellul.* 11:73–83, 2004.
Sjöström Wood Chemistry: *Fundamentals and Applications*, Academic Press, New York. 1981.
Smit, A, Huijgen, W. *Green Chem.* 19:5505–5514. 2017.
Sohani, A, Pedram, MZ, Hoseinzadeh, S. *J. Mol. Liq.* 297:111847, 2020.
Sugesty, S, Kardiansyah, T, Hardiani, H. *Procedia Chem.* 17:194–199, 2015.
Sturgeon, MR, Kim, S, Lawrence, K, Paton, RS, Chmely, SC, Nimlos, M, Foust, TD, Beckham, GT. *ACS Sustain. Chem. Eng.* 2:472–485, 2014.
Sun, R, Sun, XF, Wang, SQ, Zhu, W, Wang, XY. *Ind. Crop Prod.* 15:179–188, 2002.
Tajmirriahi, M, Momayez, F, Karimi, K. *Bioresour. Technol.* 319:124167, 2021.
Teramura, H, Sasaki, K, Oshima, T, Matsuda, F, Okamoto, M, Shirai, T, Kawaguchi, H, Ogino, C, Hirano, K, Sazuka, T, Kitano, H, Kikuchi, J, Kondo, A. *Biotechnol. Biofuels* 9:27, 2016.
Tofanica, BM, Puitel, AC. *Chem. Eng. Communic.* 206(3):378–386, 2018.
Tripathi, SK, Bhardwaj, NK, Ghatak, HR. Cellulose *Chem. Technol.* 53(1-2):105–112, 2019.
Tutus, A, Ates, S, Deniz, I. African *J. Biotechnol.* 9:1648–1654, 2010.
Vallejos, ME, Zambon, MD, Area, MC, Curvelo, AAD. *Ind. Crop Prod.* 65:349–353, 2014.
Vega, A, Bao, M, Lamas, J. *Bioresour. Technol.* 61:1–7, 1997.
Venkatram, S, Kim, C, Chandrasekaran, A, Ramprasad, R. *J. Chem. Inf. Model.* 59(10):4188–4194, 2019.
Villa, C, Santos, V, Parajo, JC. *Ind. Eng. Chem. Res.* 42:349–356, 2003.
Vroom, KE. *Pulp Pap. Mag. Canada* 58:228–231, 1957.
Wang, K, Yang, H, Guo, S, Tang, Y, Jiang, J, Xu, F, Sun, RC. *Process Biochem.* 47:1503–1509, 2012.
Xiao, B, Sun, XF, Sun, RC. *J. Wood Chem. Technol.* 21:397–411, 2001.
Xu, F, Subramanian, HMSR, Latha, R, Rao, CS. *Ind. Crop Prod.* 23:180–193, 2006.
Xu, JY, Yuan, TQ, Xiao, L, Sun, RC. *Carbohydr. Polym.* 195:114–119, 2018.
Yanez-S, M, Matsuhiro, B, Nuñez, C, Pan, S, Hubbell, CA, Sannigrahi, P, Ragauskas, AJ. *Polym. Degrad. Stab.* 110:184–194, 2014.
Yang, D, Gao T, Mao, Z. *BioResources* 14:8892–8903, 2019.
Yao, L, Chen, C, Yoo, CG, Meng, X, Li, M, Pu, Y, Ragauskas, AJ, Dong, C, Yang, H. *ACS Sustainable Chem. Eng.* 6:14767–14773, 2018.
Yawalata, D, Paszner, L. *Holzforschung* 58:7–13, 2004a.
Yawalata, D, Paszner, L. *Holzforschung* 58:1–6, 2004b.
Zhang, Z, O'Hara, IM, Doherty, WOS. *Cellul.* 20:3179–3190, 2013a.
Zhang, Z, Xue, Q, Huang, K, Ma, Q, Guo, Y. *Przegląd Elektrotechniczny* 2013(3b): 212–215, 2013b.
Zhang, Z, Harrison, MD, Rackemann, DW, Doherty, WOS, O'Hara, IM. *Green Chem.* 18:360–381, 2016.
Zhao, X, Liu, D. *Ind. Crop Prod.* 32:284–291, 2010.
Zijlstra, DS, Lahive, CW, Analbers, CA, Figueirêdo, MB, Wang, Z, Lancefield, CS, Deuss, PJ. *ACS Sustainable Chem. Eng.* 8:5119–5131, 2020.

3 Separation of Dissolved Lignin from Solution

Delignification causes the native lignin in the lignocellulose to depolymerize and break into fragments. Some changes in the functional groups in these lignin fragments also occur, making them soluble in the reacting liquid phase under prevalent conditions. However, the solubilized lignin, still having a distinct polymeric character, is liable to separation using suitable means. After the delignification of lignocellulose, the separation of lignin from the spent liquid is an important step for its possible derivatization or value addition. This is mandated for the comprehensive utilization of lignocellulose biomass in a biorefinery framework. Native lignin is hydrophobic and, in most cases, its degraded fragments are insoluble in water. A change in the pH, polarity, or ionic character of the spent liquid can result in the separation of lignin from the solution as a separate solid phase. This is one of the approaches for lignin separation from the spent liquid after delignification. Owing to its polymeric character, lignin fragments in the spent liquid vastly differ in molecular size from the solvent molecules. This makes them liable to separation using an appropriate membrane. Depending on the method of delignification, the spent liquid and the dissolved lignin fragments differ in their characteristics. Alkaline delignification processes – soda, kraft, or anthraquinone – produce a spent liquid that is strongly alkaline. The lignin fragments dissolved in it are grossly hydrophobic. Sulphite-based delignification processes – acid sulphite or bisulphite – on the other hand, result in a spent liquid that is strong to moderately acidic. During these processes, the lignin is extensively sulphonated, making them water-soluble. Organosolv delignification processes are far too diverse in the choice of solvent and catalyst. The resulting spent liquid, therefore, can also significantly differ in its physicochemical character. Besides solution pH, the concentration of dissolved solids may differ significantly in all delignification processes. In addition to the lignin fragments, monomeric sugars and their derivatives from carbohydrate degradation may also be present. Extractives present in the lignocellulose often find their way into the spent liquid. The nature of the lignin fragments dissolved also varies quite a bit. All these make the task of separating lignin from the spent liquid after the delignification of lignocellulose a challenging one.

3.1 SEPARATION OF DISSOLVED LIGNIN FROM ALKALINE SPENT LIQUIDS

Shortcomings notwithstanding, alkaline delignification processes find the widest application in commercial ventures in the pulp and paper industry. Most of the mills prefer the kraft process, though. The spent liquid generated in these processes, called the black liquor, contains not only the dissolved components from the lignocellulose material but also the used delignification chemicals in modified form. Traditionally, economic and environmental concerns demanded that the black liquor be processed to recover the delignification chemicals for reuse while utilizing the dissolved organic components, including lignin, for energy generation. Separating lignin from black liquor, at least partially, is now being considered as a serious alternative to traditional black liquor processing due to reasons (Ghatak 2012):

1. Sustainability concerns: Need to comprehensively utilize and add value to lignocellulose components.
2. Increasing throughput in existing mills without overloading the recovery boiler.
3. Additional revenue stream for mills by capitalizing on the expanding lignin market.

DOI: 10.1201/9781003598664-4

The commercial production capacity of kraft lignin was reported to be 265,000 t, and that of soda lignin was 75,000 t, in the year 2018 (Dessbesell et al. 2020). There have been several studies, mostly in Scandinavian countries, to ascertain the optimal utilization of black liquor from the standpoint of the energy dynamics of mills. These studies recommend a part of lignin present in black liquor be separated for either energy use outside the chemical recovery system or for possible derivatization into industrially important chemicals. One case study suggested the separation of around 10% lignin present in the black liquor for external utilization (Loutfi et al. 1991). Computer simulations estimated the viability of lignin separation from kraft black liquor to the extent of 0.15 t of lignin for 1 t pulp production. This could allow for a production increase of 25% of the existing throughput (Axelsson et al. 2006; Olsson et al. 2006). Yet another simulation established the possibility of removing 37–50% lignin from kraft black liquor (Wising et al. 2006).

3.1.1 Black Liquor Characteristics from Alkaline Delignification Processes

As deliberated in Chapter 2, alkaline delignification processes are subdivided into soda, kraft, and soda anthraquinone. A strong solution of sodium hydroxide is the principal constituent of the delignification chemicals. Black liquor, the spent liquid after delignification, is characterized by its distinctly alkaline nature. Its composition is important, the main constituents being the inorganics that come from the delignification chemicals used and the organics that originate from the lignocellulose raw material. The amount of water present is another crucial aspect of black liquor characteristics. The physical characteristics of black liquor are additionally important. The main characteristics of black liquor that deserve mention are:

1. Dissolved solids concentration
2. Organics-to-inorganics ratio of dissolved solids
3. Residual active alkali
4. Major organic constituents present
5. Density
6. Viscosity and rheological characteristics
7. Surface tension

3.1.1.1 Dissolved Solids Concentration

One of the important characteristics of black liquor is its dissolved solids concentration. Almost all the other characteristics of black liquor are dependent on its dissolved solids concentration. It primarily depends on the amount of delignification chemicals used, the extent of delignification and dissolution of other components of the lignocellulose material, and the water added during delignification and subsequent washing of the solid residue. The dissolved solids concentration of black liquor strongly influences its processing, including the possible separation of dissolved lignin. This has serious techno-economic ramifications as lignin separation processes work best in a given range of black liquor concentration. Very dilute black liquors require evaporation to remove excess water. This is an energy-intensive process involving complex engineering. Highly, concentrated black liquors, on the other hand, are difficult to handle and not very amenable to lignin separation. Dissolved solids concentrations normally found in black liquor samples are presented in Table 3.1.

3.1.1.2 Organics-to-Inorganics Ratio of Dissolved Solids

Organic components in the black liquor come from the lignocellulose raw material while the inorganic components mainly originate from the delignification chemicals used. Some inorganic constituents present in the lignocellulose may also dissolve during the delignification process and pass into the black liquor. This ratio, therefore, is closely linked to the prevalent process parameters during delignification. Conditions that favour the degradation and dissolution of lignocellulose

TABLE 3.1
Typical Range of Black Liquor Characteristics from Alkaline Delignification Processes

Lignocellulose and Delignification Process	Dissolved Solids Concentration (Weight%)	Organics-to-Inorganics Ratio	Residual Active Alkali (g/L) or pH	Viscosity (mPa.s) and Rheological Characteristics	Reference
Softwood kraft	13–17	Organics 47–54%	pH: 13.3–13.6	Viscosity: 0.98–1.13 Shear thinning, mildly non-Newtonian at high concentration	Arkell et al. 2014; Manttari et al. 2015; Velez and Thies 2016
Hardwood kraft	16–18	Organics 51–69	pH: 12.7–13.5	Pseudoplastic, strongly non-Newtonian at high concentration	Cardoso et al. 2009; Costa et al. 2011; Costa et al. 2018; Velez and Thies 2016
Agroresidue soda/kraft	7.7–10.6	Organics 54–56	pH: 11.5–12.4	Shear thinning, strongly non-Newtonian at high concentration	Cao et al. 2019; Singh and Ghatak 2017

components increase the proportion of organics. These are higher temperatures and delignification times. Paradoxically, a higher amount of delignification chemicals used also causes more dissolution of organic material from the lignocellulose. The ratio of organic components to inorganic components is an important characteristic of the black liquor that affects its subsequent processing during lignin separation. Many black liquor characteristics are also dependent on this ratio. Especially, the rheological complexity of black liquor is dependent on its organic content. Typical values for the organic-to-inorganic ratio in black liquor are shown in Table 3.1.

3.1.1.3 Residual Active Alkali

The white liquor used for alkaline delignification processes is strongly alkaline and caustic in nature with a pH well above 13. Delignification and carbohydrate degradation reactions consume alkali, and the pH drops. Some acidic reaction products are also formed which further reduce the alkalinity. Soda and kraft lignins are soluble in alkali. If the pH of the reacting system drops excessively during the reaction, it may cause lignin fragments to reprecipitate on the fibres. This is an unwanted situation. To avoid this, the initial delignification chemicals charged should be such that it is not entirely consumed during the reaction and sufficient alkali should remain unreacted at the end of the process. This unreacted alkali becomes part of the black liquor and is referred to as the residual active alkali (RAA). It is an important black liquor characteristic. Besides ensuring that lignin remains in solution, it affects other black liquor characteristics like the organic-to-inorganic ratio, viscosity, rheological behaviour, and surface tension. RAA concentrations commonly encountered in black liquor are given in Table 3.1.

3.1.1.4 Major Organic Constituents Present

Organics in black liquor originate from the reaction products generated during delignification. Depolymerized and degraded lignin fragments are the main constituents present in black liquor. Carbohydrates, especially the hemicelluloses, are also degraded and/or dissolved during delignification and constitute an important component of black liquor organics. Degraded carbohydrate components are different monomeric, dimeric, or oligomeric sugars, or acidic derivatives of these sugars. Besides these two major constituents, black liquor contains several low molecular organic compounds either directly dissolved from the starting lignocellulose material or produced as reaction products. The organic constituents and their relative proportion in black liquor affect many of its characteristics like viscosity, rheological behaviour, and surface tension. Most importantly, the presence of organics other than lignin complicates the lignin separation process and the purity of the final lignin product. Black liquor composition with respect to major organic constituents present is shown in Table 3.2.

TABLE 3.2
Major Organic Constituents in Black Liquor

Major Constituents	Proportion (% of dry Solids)	Reference
Lignin	33–44	Arkell et al. 2014; Ela et al. 2020; Jönsson et al. 2008
Xylose	0.6–4.6	Arkell et al. 2014; Ela et al. 2020; Jönsson et al. 2008
Glucose	0.18–0.3	Arkell et al. 2014; Ela et al. 2020; Jönsson et al. 2008
Mannose	0.12–0.2	Arkell et al. 2014; Ela et al. 2020; Jönsson et al. 2008
Galactose	0.09–0.6	Arkell et al. 2014; Ela et al. 2020; Jönsson et al. 2008
Arabinose	0.4–0.5	Arkell et al. 2014; Jönsson et al. 2008

3.1.1.5 Density

Density is a physical property of black liquor that has a bearing on its handling and transportation through pumping. It is primarily dependent on the dissolved solids concentration of black liquor with the density increasing as the solids concentration increases. Through this interconnect, density also affects the viscosity and rheological behaviour of black liquor. It is also temperature-dependent. It is customary to express black liquor density as a correlation of the form

$$\rho = a_1 + a_2(\mathrm{DS}) - a_3 T \tag{3.1}$$

where ρ is black liquor density, DS is the dissolved solids concentration in black liquor, and T is the temperature. The correlation coefficients, a_1, a_2, and a_3, are dependent on the type of black liquor.

3.1.1.6 Viscosity and Rheological Characteristics

Viscosity and rheological characteristics of black liquor govern its flow behaviour. Process equipment used in the separation of lignin from black liquor works best if the black liquor viscosity falls within a certain range. The viscosity and rheological behaviour of black liquor are important parameters in the design and operation of pumps and piping systems. The viscosity of black liquor is very strongly dependent on its dissolved solids concentration and so is its rheological characteristics. From a low-viscosity Newtonian fluid at low concentrations, black liquor turns into a highly viscous and rheologically complex material at high concentrations. The rheological behaviour of high-concentration black liquor is generally expressed as a power law (Yue et al. 2017)

$$\mu \propto \gamma^{\mathrm{n}} \tag{3.2}$$

where μ is the viscosity, γ is the shear rate, and n is the experimentally determined exponent. Viscosity is also affected by the organic content and its nature in black liquor. Organics of higher molecular weight polymeric character tend to increase black liquor viscosity. Higher RAA, on the other hand, has a lowering effect on viscosity. Finally, increasing temperature has a moderating effect on viscosity. Typical values and their range for black liquor viscosities are shown in Table 3.1.

3.1.1.7 Surface Tension

Black liquor usually has a low surface tension when its dissolved solids concentration is low. Resinous materials dissolved from lignocellulose during the delignification process are the main reason for its low surface tension. Surface tension is also lowered due to the presence of RAA. Low surface tension causes foaming in black liquor which is detrimental to the functioning of many process equipment used in the separation of lignin from black liquor.

3.1.2 Lignin Separation by Acid Precipitation

Black liquor from alkaline delignification processes contains lignin that is polydisperse in nature. The molecular weight of the lignin fragments varies in a wide range. The system borders a colloidal state where the negatively charged molecular clusters of lignin, stabilized by the positive counter-ions, are prevented from aggregating into bigger particles and separating out of the solution. The negative charge on the lignin cluster mainly arises from the phenolic hydroxyl present in the lignin molecule, and to a lesser extent, due to the presence of carboxylate (COOR) groups. Phenolic hydroxyls are acidic in character and dissociate into phenolate ions in the presence of hydroxyl ions through the abstraction of protons. A sufficiently high concentration of hydroxyl ions is, therefore, necessary to stabilize the colloidal system and prevent the lignin from coagulating. The addition of an acid to the black liquor neutralizes the hydroxyl ions causing the pH to drop. With the decrease in the concentration of hydroxyl ions, more and more phenolic hydroxyls remain in undissociated form. For softwood kraft lignin Indulin, the degree of dissociation varied between 0 at pH 6.5 to 1 at pH 13.2. The degree of dissociation generally decreased with increasing temperature (Norgren and Lindstrom 2000). Other investigators have reported the degree of dissociation of phenolic hydroxyl in softwood kraft LignoBoost lignin between 0.027 at 45 °C, reducing to only 0.001 at 77 °C (Sewring et al. 2019). This reduces the negative charge on the lignin moiety and opens the possibility of lignin fragments growing to sizes big enough to coagulate. A measure of the negative charge on lignin moieties is its zeta potential: the more negative the value of zeta potential, the larger the charge on the lignin fragment and the lesser the possibility of agglomeration. The zeta potential of lignin fragments in mixed hardwood kraft black liquor of intermediate concentration was found to be pH-dependent. However, the change in zeta potential with pH was irregular and temperature-dependent as well. It was argued that besides pH, carbohydrate and inorganics composition of black liquor affected zeta potential. Another important observation that transpired from this study is the absence of a direct correlation between zeta potential and the hydrodynamic size of the agglomerates (Ela et al. 2020). The mean hydrodynamic radii of lignin molecular clusters can increase by as much as ten times when the pH drops from 13 to 6.5 (Garver and Callaghan 1991). This is the primary mechanism of lignin separation from alkaline black liquors by acid precipitation. At sufficiently low pH, even the COOR groups are protonated paving the way for more lignin fragments to coagulate. Acid precipitation is the oldest method of separating lignin from alkaline black liquors and has been in practice for more than half a century. Besides pH, the colloidal stability depends on the temperature. There are contradicting claims on the effect of temperature on lignin precipitation. Some investigators report enhanced lignin precipitation at lower temperatures (Theliander 2010). Others contend that increasing temperature facilitates lignin separation in the higher pH range (8–14), however, in the lower pH range the trend is reversed (Ela et al. 2020). High-temperature heat treatment prior to acidification is also found to affect lignin precipitation and filtration by degrading the carbohydrate and lignin moieties present in black liquor. One of the effects of this heat treatment is to decrease the viscosity of the black liquor. When softwood black liquor of 41% dissolved solids concentration and 13 initial pH was heat treated at 150 °C for 24 h, its viscosity decreased from 33.9 mPa.s to 21.1 mPa.s. Viscosity dropped to 19.27 mPa.s when treated for 48 h at 150 °C. In contrast, viscosity reduced to 17.92 mPa.s within 24 h when heat treated at 200 °C. These heat-treated black liquors, however, yielded lesser lignin upon acid precipitation. When each of these black liquors was acidified to pH 9 with carbon dioxide, the lignin yield was 66% for the untreated black liquor. The yield dropped to 47% when the liquor was heat treated for 24 h at 150 °C. Yield was only 43% when the treatment temperature was 200 °C (Pichler et al. 2019).

3.1.2.1 Precipitation of Lignin

Lignin can be precipitated from alkaline black liquor of all types, be it soda, kraft or soda anthraquinone prepared from any type of lignocellulose material. The black liquor characteristics and the acidification process, however, have a profound influence on the final outcome. It affects the subsequent step of solid–liquid separation as well as the yield and purity of the lignin product. Sequential acidification to progressively lower pH can fractionate lignin into fractions of different purity and structural features. Mineral acids can be used for acidification. Black liquor can also be acidified with carbon dioxide which can be sourced from the flue gases, thereby reducing net carbon emissions. Studies indicate that lignin precipitation from softwood and hardwood kraft black liquors by carbon dioxide is best performed at intermediate solids concentrations of 27–35 weight% (Alén et al. 1985). For softwood kraft black liquor, lignin yield from acid precipitation, as affected by temperature, pH, and ionic concentration of sodium and potassium ions, can be empirically correlated as (Theliander 2010)

$$Y = ae^{\frac{b}{T}}\left[H^{+}\right]^{c}\left[Na^{+} + K^{+}\right]^{d} \tag{3.3}$$

Here, Y is the lignin yield, $[H^+]$ is the hydrogen ion concentration, $[Na^+ + K^+]$ is the combined concentration of sodium and potassium ions, and T is the temperature. a, b, c, and d are experimentally determined constants.

The presence of certain anions, especially sulphate, increases the lignin yield from acid precipitation. A similar yield increase was also observed for the addition of cationic surfactants (Norgren and Mackin 2009). For mixed hardwood kraft black liquor, lignin yield from acid precipitation was statistically correlated to pH, temperature, and residence time as (Ela et al. 2020)

$$Y = 65.95 - 3.86\text{pH} + 0.11T + 4.55t \tag{3.4}$$

Here, t is the residence time at the end pH, which varied between 30 and 90 min. pH is, by far, the most significant parameter affecting lignin yield. Its effect on acidification of soda black liquor is depicted in Figure 3.1.

The reaction mixture should be well agitated during acidification. This is all the more important when carbon dioxide is the acidification agent as a gas–liquid mass transfer step is involved. Agitation is also a parameter affecting the size and shape of lignin agglomerates formed during acidification. Primary lignin agglomerates are spheroidal in shape but decrease in size and become more compact with increasing agitation (Kannangara et al. 2016). Morphological characteristics – size, size distribution, sphericity, and density – of the lignin particles formed during acidification will be an influencing factor when the formed slurry is subjected to solid–liquid separation under different force fields or filtered. Formation of lignin agglomerates can be either diffusion-controlled or reaction-controlled. Experiments with kraft LignoBoost lignin suggested that both the number of lignin agglomerates as well as their size increased as the pH decreased. The size distribution of lignin agglomerates as they evolve during acidification is shown in Figure 3.2.

Black liquor was obtained from mild soda delignification of wheat and barley straw. Wheat straw black liquor had an initial pH of 10.72 and the initial pH of barley straw black liquor was 10.24. The dissolved solids concentrations were 3.26% and 4.57%, respectively. Dissolved solids in wheat straw black liquor had 51% organics of which 70% was lignin. For barley straw black liquor, organics content in dissolved solids was 64% with only 43% being lignin. Both the black liquors were acidified to pH 2 with concentrated sulphuric acid and allowed to sediment for 24 h. The yield of precipitated solids after centrifugation, based on dissolved solid content in black liquor, was 56% for

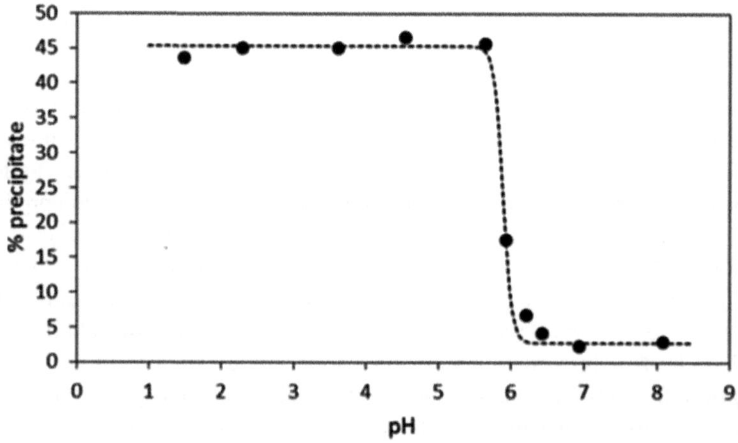

FIGURE 3.1 Amount of precipitate obtained as a percentage of dissolved solids at different end pH during acidification of miscanthus soda black liquor. (Reprinted from Lauwaert, J, Stals, I, Lancefield, CS, Deschaumes, W, Depuydt D, Vanlerberghe B, Devlamynck, T, Bruijnincx, PCA, Verberckmoes, A, *Sep. Purific. Technol.*, 221, 226–235, 2019, with permission from Elsevier.)

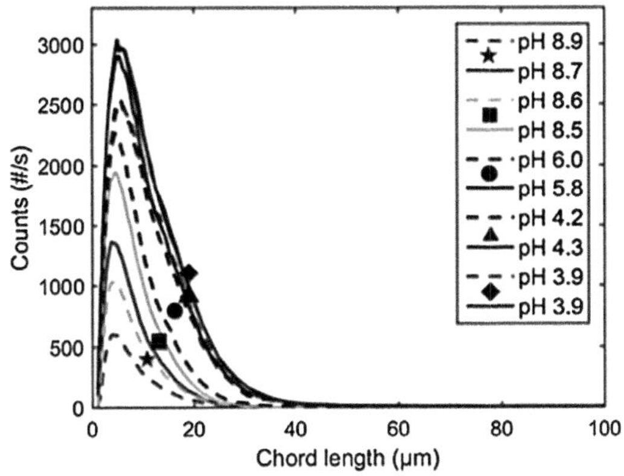

FIGURE 3.2 Size distribution of lignin agglomerates during acidification of kraft LignoBoost lignin. Temperature 65 °C, Na^+ concentration 1M. Dashed lines show experiments in the presence of 5 g xylan per 95 g lignin. (Reprinted from Sewring, T, Durruty, J, Schneider, L, Schneider, H, Mattsson T, Theliander H, *J. Wood Chem. Technol.*, 39(1), 1–13, 2019, with permission from Taylor and Francis.)

wheat straw, and 41% for barley straw. Lignin purities were 69% and 54%, respectively (Domínguez-Robles et al. 2017). In a pilot scale study, 700 kg soda black liquor from miscanthus, with a dissolved solids concentration of 4.1% and initial pH of 12.5, was acidified to pH 4.8 by the addition of 20% sulphuric acid under stirring. After acidification, the mixture was heated to 50 °C. While maintaining this temperature, 1.7 kg Accellerase, an enzyme mixture to hydrolyse cellulose and hemicellulose, was added to the mixture and allowed to incubate for 15 h under stirring (Lauwaert et al. 2019).

Soda anthraquinone delignification of birch resulted in a black liquor of initial pH 13.5. Lignin was separated from this black liquor in multistep acidification. In the first step, the liquor was acidified to pH 8.5 with pressurized carbon dioxide for 40 min at 1.5 bar and 80 °C. Approximately 50% of the original lignin in black liquor precipitated out in this acidification step, which was separated by centrifugation. The remaining black liquor was further acidified to pH 2 by the addition of 2M sulphuric acid. This resulted in additional lignin precipitation. Total lignin precipitation in the combined process was approximately 90% of the original lignin (Kumar et al. 2016). Soda anthraquinone black liquor was obtained from the delignification of leguminous plants *Leucaena leucocephala*, and *Chamaecytisus proliferus*. Wheat straw black liquor had an initial pH of *L. leucocephala* black liquor was 9.85, and that of *C. proliferus* black liquor was 10.47. The dissolved solids concentrations were 5.6% and 5.2%, respectively. The organic content in *L. leucocephala* black liquor dissolved solids was 58%, with 75% lignin in it. These values were 60% and 61%, respectively, for *C. proliferus* black liquor. Both the black liquors were acidified to pH 2 with concentrated sulphuric acid and allowed to sediment for 24 h. Precipitated solids were separated by centrifugation. 100 g dissolved solids in *L. leucocephala* black liquor yielded 48 g lignin of 55% purity. Similarly, 100 g dissolved solids in *C. proliferus* black liquor yielded 44 g lignin of 61% purity (Domínguez-Robles et al. 2017).

Softwood kraft black liquor samples from different industries were subjected to acid precipitation. Dissolved solids concentration in black liquor varied from 25% to 34% and pH varied from 13.3 to 13.8. Each black liquor was filled in an agitated vessel fitted with a Rushton turbine-type impeller. Carbon dioxide was the acidification medium, sparged into the black liquor as small bubbles from below the impeller. Acidification was done at 80 °C for 60 min to a pH of about 9. Post-acidification agitation was continued for another 60 min under the same temperature to improve precipitation. Lignin yield varied between 58% and 74% (Ohman et al. 2007). Kraft black liquor from a Brazilian mill was acidified with mineral acids, sulphuric, and hydrochloric to different end pHs 2, 4, and 6. For sulphuric acid, the highest lignin precipitation occurred at pH 4. The best end pH with hydrochloric acid was 2 (dos Santos et al. 2014). Kraft black liquor samples of eucalyptus and pine were sequentially acidified with dropwise addition of hydrochloric acid of 12M and 1M concentrations. In the first step, the liquor was acidified to pH 9. Centrifugation of precipitated solids resulted in 52% lignin yield from eucalyptus black liquor and 50% lignin yield from pine black liquor. The remaining black liquor samples were further acidified with hydrochloric acid to pH 7. Additional 34% and 37% lignin precipitated from eucalyptus black liquor and pine black liquor, respectively. Subsequent acidification to pH 5 of the remaining black liquor precipitated some more lignin; 10% from eucalyptus black liquor, and 11% from pine black liquor. Additional lignin yields were nominal when the remaining black liquors were further acidified to pH 3 and then to pH 1 (Lourençon et al. 2015). Acid precipitation of kraft black liquor from softwood, and mixed softwood and hardwood was carried out with 32 weight% sulphuric acid to pH 2. An agitated vessel with a propeller-type impeller, operating under ambient temperature, was used for this purpose. The system was kept under agitation for 30 min after the final pH was reached. More than 90% of lignin precipitated out of the black liquor (Manttari et al. 2015). Mixed softwood kraft black liquor of 32 weight% dissolved solids concentration, and 2.87 mol/kg Na^+ and K^+ concentration, was acidified with 6 M sulphuric acid to different end pH in the range 11–9.5 at temperatures varying between 45 and 75 °C. The highest lignin yield was 80–87% when the liquor was acidified to pH 9.5 at 45 °C. The yield dropped to around 30–43% when the end pH was 11 and the acidification temperature was 65 °C. At a given end pH or temperature, lignin yield increased with increasing ionic strength of Na^+ and K^+ (Zhu and Theliander 2015). The effect of the presence of xylan and temperature on lignin precipitation behaviour was examined by acidifying alkaline solutions of LignoBoost lignin in 1 M sodium hydroxide solution with 8.8–8.9 weight% lignin concentration. The solution was acidified to a different target pH from an initial pH of 13.4 by the addition of 6M sulphuric acid under 250 rpm agitation with a pitch blade agitator. The onset of lignin precipitation, defined as the condition when stable lignin agglomerates of size ≥ 1μm were formed, was observed in the pH range 9.1–9.4 irrespective of the presence of xylan or acidification temperature. However, when the Na^+ concentration was increased to 4M, the onset of lignin precipitation occurred at a pH of 11.3. As the system was acidified to pH below the onset of precipitation, more and more lignin was precipitated (Sewring et al. 2019). A two-step acidification of mixed hardwood kraft black liquor of 38.5% dissolved solids concentration, and pH in excess of 13, was carried out for lignin precipitation. In the first step, the liquor, placed in a stirred open vessel, was bubbled for 1 h with carbon dioxide stored under 80 kPa pressure. This was followed by the addition of 72% sulphuric acid to achieve the desired end pH (Ela et al. 2020). Kraft black liquor from several lignocellulose species, softwood and hardwood, of different compositions but with constant dissolved solids concentration of 20%, was acidified in two steps. The first step consisted of bubbling with carbon dioxide to pH 9.5 and allowing it to stand for 30 min. In the second step, the filtrate from the first step was further acidified with 6N hydrochloric acid to pH 2.5. Lignin yield from pine black liquor after acidification to 9.5 pH was 68% which was 92% pure. Another 22.5% lignin, of 90% purity, precipitated after the second step of acidification to pH 2.5. Lignin yield from sweetgum was 38% at 9.5 pH and 31.5% at pH 2.5. Their purities were 76%, and 93%, respectively. Eucalyptus black liquor yielded 56% lignin (purity 85%) at pH 9.5 and another 26% (purity 91.5%) at pH 2.5 (Jardim et al. 2022).

3.1.2.2 Separation of Precipitated Lignin

Being solid, the lignin that precipitates out of the black liquor during acidification constitutes a separate phase. It can be separated from the remaining liquid phase by an appropriate solid–liquid separation technique. The unit operations that are suitable for this purpose are centrifugation, sedimentation, and filtration. Among these, filtration is the most used in large-scale industrial installations followed by centrifugation. These unit operations are profoundly influenced by the hydrodynamic characteristics of the lignin agglomerates. To facilitate, the unit operations are often assisted by flocculation, a physicochemical change that alters the size, size distribution, and morphology of lignin agglomerates, thereby changing their hydrodynamic characteristics.

In centrifugation, the solid particulates are acted upon by a centrifugal force field. The magnitude of the centrifugal force depends on the mass of the particle, its rotational speed, and the radial distance from the centre, and can be represented as

$$F_C = \frac{mv_T^2}{r} \tag{3.5}$$

where F_C is the centrifugal force, m is the mass of the particle, v_T is the tangential component of the particle velocity, and r is the radial distance from the centre. The mass of the particle is dependent on its density, and size. The bigger and denser the particle, the more its mass and the larger the centrifugal force acting on it. The density of the particle is strongly affected by its morphological characteristics. Highly porous particles tend to be less dense with occluded liquid phase trapped inside. In addition to the centrifugal force, a drag force also acts on the particle in a direction opposite to the centrifugal force. It can be written as

$$F_D = \frac{1}{2} C_D A \rho_p v_r^2 \tag{3.6}$$

where F_D is the drag force, C_D is the drag coefficient, A is the projected cross-sectional area of the particle perpendicular to the direction of motion, ρ_p is the density of the particle, and v_r is the radial component of the particle velocity. The quantity A depends on the size, shape, and orientation of the particle. Drag coefficient, C_D, is a function of the Reynolds number, which besides particle size and shape, depends on v_r, and the density and viscosity of the liquid. The resulting force on the particle is

$$F = F_C - F_D \tag{3.7}$$

in the outward radial direction. In equation (3.7), gravitational force in the downward direction and buoyant force in the upward direction are neglected in comparison to F_C and F_D. The resultant acceleration of the particle, in the outward radial direction, is given by

$$\frac{dv}{dt} = \frac{v_T^2}{r} - \frac{C_D A \rho_p v_r^2}{2m} \tag{3.8}$$

where v, the resultant velocity of the particle, is the vector sum of v_T and v_r. The tangential velocity component, v_T, remains constant in magnitude but constantly changes in direction. In contrast, the radial velocity component, v_r, changes in magnitude but is always directed radially outwards. As v_r increases in magnitude, the drag force on the particle, F_D, also increases. When it becomes equal to the centrifugal force F_C, the net force on the particle, from equation (3.7), is zero. Beyond this point, the particle stops accelerating and acquires a terminal velocity in the radial direction. The magnitude of this velocity, from equation (3.8) is

$$\left(v_{tm}\right)_C = v_T \sqrt{\frac{2m}{rC_D A \rho_p}} \tag{3.9}$$

where $\left(v_{tm}\right)_C$ is the terminal velocity under the centrifugal force field.

In filtration, a porous filter medium is placed in the flow path of the slurry containing the solid particulates. The filter medium allows the liquid to pass through its pores while retaining the particles. As the operation progresses, a layer of solid particles gradually builds up on the filter medium. This layer of solid particles, referred to as the filter cake, acts as an additional filter medium and assists in the overall process of filtration. The filter medium along with the filter cake offers resistance to the flow of the liquid and causes a pressure drop along the flow path. In filtration, the liquid flows through numerous tortuous paths formed by the congregation of particles. Every particle exerts a drag force on the flowing liquid and the total resistive force experienced by the liquid is the sum of all these drag forces. The total pressure drop is the sum of pressure drops across the filter medium and the filter cake. If a substantially thick filter cake is formed, it causes a pressure drop that far exceeds that offered by the filter medium. The pressure gradient across the filter cake can be expressed by the Kozeny–Carman equation as (Carman 1937)

$$\frac{dp}{dL} = \frac{K\mu v(1-\epsilon)^2}{(\Phi_S d_P)^2 \epsilon^3} \tag{3.10}$$

Here, p is local pressure, L is the position in the filter cake, μ is liquid viscosity, v is superficial liquid velocity, ϵ is the porosity of the filter cake, Φ_S is the sphericity of the particles, and d_P is the diameter of the equivalent sphere. The value of the coefficient, K, in the equation, depends on the flow regime. Its value in the original equation, valid for low particle Reynolds numbers, is 150. The superficial velocity, v, is calculated as the ratio of the volumetric flow rate of the filtrate to the filter cross-sectional area. Equation (3.10), in its integrated form over the thickness of the filter cake, can be rearranged in terms of the mass of the particles in the filter cake, m_{CK}, as (McCabe et al. 1993).

$$\Delta p_{CK} = \frac{K\mu v(1-\epsilon)m_{CK}}{\rho_P A_F (\Phi_S d_P)^2 \epsilon^3} \tag{3.11}$$

During integration, while getting equation (3.11), the quantity $\frac{K(1-\epsilon)}{\rho_P (\Phi_S d_P)^2 \epsilon^3}$ is taken to be constant and defined as the specific filter cake resistance. For this quantity to be constant, the cake porosity, ϵ, the particle size $\Phi_S d_P$, and the particle density, ρ_P, should be constant across the thickness of the filter cake for the entire duration of filtration irrespective of the pressure drop. Such filter cakes are known as incompressible. In practice, however, these quantities are seldom constant. The specific filter cake resistance normally increases over time as the filtration progresses. It also increases as we move closer to the filter medium. Such filter cakes are termed compressible. Total pressure drop during filtration, Δp, is the sum of Δp_{CK} and Δp_{FM}, where Δp_{FM} is the pressure drop due to the filter medium. This can be expressed as (McCabe et al. 1993)

$$\Delta p_{FM} = R_{FM}\mu v \tag{3.12}$$

R_{FM}, the filter medium resistance, is a characteristic property of the filter medium. Combining equations (3.11), and (3.12),

$$\Delta p = \left[\frac{K(1-\epsilon)m_{CK}}{\rho_P A_F (\Phi_S d_P)^2 \epsilon^3} + R_{FM}\right]\mu v \tag{3.13}$$

As the mass of the filter cake, m_{CK}, increases during filtration, one must apply more and more pressure difference to maintain constant liquid velocity, v. This mode of filtration is known as constant rate filtration. Alternatively, filtration can be performed while keeping the pressure drop, Δp, constant. This is known as constant pressure filtration. In such case, liquid velocity, v, steadily decreases as filtration continues to maintain the validity of equation (3.13). Filtration operations can be further classified as pressure filtration or vacuum filtration. In pressure filtration, downstream pressure is one atmosphere while upstream pressure exceeds atmospheric pressure. In vacuum filtration, the upstream pressure is one atmosphere, and the downstream side is under vacuum.

Lignin agglomerates can form during acidification under diffusion-controlled or reaction-controlled regimes. The former mechanism is rapid leading to the formation of disordered and loose agglomerates with a higher mass fractal dimension. Reaction control leads to the slow formation of agglomerates which are closely packed and dense having a lower mass fractal dimension. A stability ratio is defined as (Norgren et al. 2002)

$$W = \frac{\text{Rate constant of diffusion} - \text{controlled aggregation}}{\text{Rate constant of reaction} - \text{controlled aggregation}} \tag{3.14}$$

This ratio drops as the total ionic strength of the system increases. The size distribution of agglomerates is strongly pH dependent and also to some extent on the presence of carbohydrates. When an alkali lignin solution of 0.3 g/L concentration was acidified to pH 6, agglomerates were found to have a bimodal size distribution with two particle size ranges of 80–142 nm and 190–615 nm. Larger size particles were predominant in the slurry. In the presence of 0.5 g/L xylan, most of the particles were in the size range of 220–531 nm (Hu et al. 2021). In a laboratory study with several intermediate concentration softwood or hardwood kraft black liquors, constant pressure filtration of acidified lignin slurries at pH 9 at a pressure difference of 1.5 bar resulted in filter cakes with 40–45% dry solids content. 95–98% of this dry matter was lignin and about 2–5% carbohydrates. The specific filter cake resistance varied between 1×10^{10} and 1×10^{13} m/kg. In general, the black liquors concentrated through evaporation had lower cake resistance than black liquors concentrated by membrane separation. Similarly, black liquors heat-treated prior to concentration had lower cake resistance than non-heat-treated black liquors (Ohman et al. 2007). Heat treatment of black liquor, however, should be practised with caution as it lowers the lignin yield from acid precipitation (Pichler et al. 2019).

In most laboratory studies, precipitated lignin is separated from the slurry either by filtration or centrifugation. Lignin was separated from acid precipitated slurry of mixed hardwood black liquor by two-stage vacuum filtration through filter paper of porosity 20–25 μm. After the first stage of filtration, the filter cake was redispersed in an aqueous solution of the same pH as the acid-precipitated slurry. The slurry so obtained was again vacuum filtered and the formed filter cake was rinsed with ultrapure water (Ela et al. 2020). In a pilot scale study, black liquor acidified to pH 4.8 was treated with the enzyme mix Accelerase to hydrolyse the carbohydrates. The final slurry was filtered in a chamber filter press to get 13.4 weight% of black liquor dissolved solids as the lignin product with 77% purity. Filtration was stopped after 4 h when the rate of filtration reduced considerably due to filter clogging. The remaining mixture was flocculated by adding epichlorohydrin-dimethylamine and again filtered through the chamber filter press to get 22.6 weight% of black liquor dissolved solids as the second lignin product with 83% purity. The chamber filter press consisted of several plates of 400 × 400 mm size covered with polypropylene filter cloths of 20 μm porosity. Air permeability through a clean filter cloth was 8–10 L/(min.dm^2) at a pressure differential of 200 Pa. The chamber press could handle a maximum pressure drop of 15 bar (Lauwaert et al. 2019).

Lignin separation, after acid precipitation of kraft black liquor to different pH, was achieved by centrifugation for 15 min at 3500 rpm. Separated solids were dispersed in aqueous solutions at the pH of acidified black liquor and centrifuged again and the supernatant was discarded. The precipitate was filtered, and the retained solids were washed with acid at pH 2. The resultant lignin was then dried (Lourençon et al. 2015). Lignin particles were separated from black liquor acidified to different pH with carbon dioxide or mineral acid. For this, the slurry was centrifuged at 3000 rpm for 30 min (Kumar et al. 2016). Slurries from acid-precipitated black liquor at pH 2 were centrifuged for 20 min at 8000 rpm to separate the lignin particles. Solids were washed two times with distilled water using the procedure of dispersion followed by centrifugation (Domínguez-Robles et al. 2017). Lignin was separated and purified from carbon dioxide-precipitated slurry at pH 9.5 using three centrifugation stages. Solids obtained after the first stage centrifugation were dispersed in hydrochloric acid at pH 2.5 to remove inorganic impurities. The resulting slurry was again centrifuged to separate the solids. These solids were dispersed in acetic acid at pH 3.5 to remove the residual hydrochloric acid and centrifuged again. The resulting solids were washed with deionized water till acid-free with successive dispersion and centrifugation (Jardim et al. 2022).

3.1.2.3 Commercial Processes

To date, lignin separation from alkaline black liquors has limited commercial application. The lignoBoost process was developed jointly by Innventia, presently a subsidiary of the Rise Research Institutes of Sweden, Chalmers University of Technology, and ÅF Process, under the Future Resource-Adapted pulp Mill (FRAM) R&D program. The intellectual properties and commercial rights are currently vested in Valmet which demerged from the parent company Metso. Intermediate concentration black liquor is acidified with carbon dioxide to precipitate lignin. The slurry is filtered in a chamber filter press but not washed. Instead, the filter cake is redispersed and acidified further. The resulting slurry is then filtered in a second chamber filter press and the cake is washed. The key inventive step of the process is the redispersion of the filter cake from the first press filter. During the redispersion, the slurry pH is adjusted to equal the pH of the final wash water – around 2–3. This pH adjustment is claimed to improve the filterability of the slurry during cake formation and substantially improve lignin cake permeability facilitating easy washing. The overall result is less filter area requirement, less acid consumption in the lignin washing stage, higher dry solids content of the lignin cake, and higher lignin purity (Ohman et al. 2013; Ohman et al. 2014; Tomani 2010). Stora Enso operates its commercial LignoBoost-based 50,000 tonnes per annum lignin production facility in Finland. In the USA, Domtar runs a commercial LignoBoost facility of 25,000 tonnes per annum capacity. Recently, a small 500 tonnes per annum capacity plant has been commissioned in Brazil.

Another patented and commercialized acidification-based lignin separation process from alkaline black liquors is the LignoForce process. It was developed and patented by FPInnovations (Quebec, Canada) and NORAM (NORAM Engineering and Constructors Ltd., British Columbia, Canada). The process consists of evaporating the black liquor to an intermediate concentration of 30–40% dissolved solids concentration followed by oxidation with molecular oxygen. The oxidized black liquor is acidified with carbon dioxide at 65–75 °C to pH 9.5 to coagulate the lignin. The coagulated lignin is allowed to age at 60–65 °C to obtain an easily filterable lignin slurry which is separated by filtration in a filter press. The filter cake is washed in two stages with sulphuric acid and water and has a solid content of 45–65%. The washed filter cake is air dried to give the final lignin product at 88–98% total solids. The process of oxidation, carried out with controlled oxygen amount, reaction temperature of 80 °C, and reaction time, is claimed to facilitate the filtration of precipitated lignin and washing of filter cake in a single stage. It also claims a 20–40% reduction in carbon dioxide required for acidification. The process also eliminates the emission of malodorous reduced sulphur compounds by oxidizing them to odourless non-volatile substances (Kouisni and Paleologou 2014; Kouisni et al. 2016). FPInnovations runs a semi-continuous demo pilot plant at Thunder Bay, Ontario, Canada, with a capacity to produce 50 kg of lignin per day. The process works with kraft black liquor from softwoods or hardwoods. Commercial application of the technology was started in 2016 by West Fraser's Hinton pulp mill located in Alberta, Canada, to produce 30 tonnes of lignin per day.

A-Recovery^{+} is a lignin separation technology developed and commercialized by Andritz. Intermediate concentration black liquor at 35–45% dissolved solids concentration is acidified with carbon dioxide. Lignin precipitate is separated in a membrane filter press generating a lignin cake with 60–62% dry solids. It can be further washed by dispersing in sulphuric acid and filtered through another filter press. The washed and filtered lignin cake is dried in a flash dryer with flue gas and/or steam to a final dryness of 95% (Spectrum Magazine 2017).

A recently developed patented technology is the Sequential Liquid Lignin Recovery and Purification (SLRP) process which acidifies the black liquor with carbon dioxide at an elevated temperature. The initial acidification is carried out in a pressurized bubble column reactor operating at about 120–150 °C. Intermediate concentration black liquor is fed from the top and reacts countercurrently with the carbon dioxide injected from the bottom. The target pH is close to 9. The reaction produces small liquid lignin droplets that coalesce as they descend through the column. This causes the lignin to separate out from the black liquor as a hydrated liquid phase rather than as agglomerated particulates. The liquid phase being denser than the residual black liquor can be separated by settling and does not require filtration or centrifugation. A settler at the bottom of the column decants the residual black liquor. The settled liquid lignin is transferred to an acidification tank where the pH is reduced to between 2 and 3 by the addition of sulphuric acid. The resulting lignin slurry is filtered in a filter press, dewatered, and dried. A key feature of the process is the absorption of vent gases from the bubble column in sodium hydroxide solution to prevent the escape of malodorous gases. Vent from the acidification tank is recycled to the top of the bubble column along with the black liquor feed. The advantages claimed by the process are lower carbon dioxide consumption, and elimination of the first filter press after carbonation (Lake and Blackburn 2014; Lake and Blackburn 2016).

3.1.3 Lignin Separation by Membrane Filtration

Membrane filtration is a technique that separates molecules from a solution based on their size, shape, ionic character, and other physicochemical characteristics. Lignin present in alkaline black liquor is macromolecular and polydisperse in nature with a distinct negative charge on the moiety. It can be separated from water, the solvent, and other low molecular weight components by forcing the black liquor through a semipermeable membrane of suitable pore structure and polar character. The membrane allows the smaller low molecular weight entities to cross over to the other side while preventing the larger entities from doing so. This splits the original feed into two product streams, a permeate containing most of the water and inorganic contents of the black liquor that passes across the membrane, and a retentate containing lignin and some other polymeric entities like carbohydrate fractions. Since the process is essentially a segregation based on molecular size, lignin can be fractionated into different molecular weight fractions by successive membrane filtration. Thus, black liquor can be subjected to both ultrafiltration and nanofiltration, the former preceding the latter in a sequential operation. One advantage of membrane filtration over acid precipitation is that no chemicals are to be added from outside which can adversely affect the mill mass balance of components. Practically, however, lignin-rich streams from membrane separation require an acidification step to finally get the desired product in solid form. Membranes, however, are often costly and require delicate handling.

3.1.3.1 Fundamentals of Membrane Filtration

The driving force for membrane separation is the transmembrane pressure (TMP), the pressure difference between the retentate side and the permeate side of the membrane. It can be expressed as (Ghatak 2012)

$$\Delta p = p_{\mathrm{ret}} - p_{\mathrm{perm}} \tag{3.15}$$

where Δp is the TMP, p_{ret} is the pressure on the retentate side, and p_{perm} is the pressure on the permeate side. On the retentate side, the liquid flows across the membrane and suffers a frictional pressure drop. As a result, p_{ret} at the inlet end of the membrane module is higher than that on the exit end. Consequently, Δp, the TMP, also varies over the surface of the membrane. For normal calculations, the arithmetic mean of the TMP at the two ends provides reasonably good results. For more precise calculations, a log mean TMP is recommended, which can be written as

$$(\Delta p)_{LM} = \frac{(\Delta p)_{in} - (\Delta p)_{out}}{\ln\left(\frac{(\Delta p)_{in}}{(\Delta p)_{out}}\right)} \tag{3.16}$$

Here, $(\Delta p)_{LM}$ is the log mean transmembrane pressure. $(\Delta p)_{in}$, and $(\Delta p)_{out}$ are the TMP at the two ends. Therefore, the crossflow velocity of the liquid over the membrane surface is an important operating parameter. The throughput in a membrane filtration process is expressed in terms of the permeate flux. This quantity, having the dimensions of velocity, is the volume of permeate flowing across a unit surface area of the membrane in unit time. Mathematically, it can be expressed as (Ghatak 2012)

$$J = \frac{1}{(A)_{mem}} \frac{dV}{dt} \tag{3.17}$$

Here, J is the permeate flux, $(A)_{mem}$ is the surface area of the membrane, and $\frac{dV}{dt}$ is the volumetric flow rate of the permeate. The permeate flux is directly proportional to the TMP, and inversely proportional to the viscosity of the permeate and can be represented as (Costa et al. 2018)

$$J = \kappa_M \frac{\Delta p}{\mu_{perm}} \tag{3.18}$$

κ_M is the permeability of the membrane, having dimensions of length, and μ_{perm} is the viscosity of the permeate.

3.1.3.2 Membranes Used in Lignin Separation

Membranes used for lignin separation from black liquor are either ceramic type or polymeric type. Ceramic membranes may have a relatively robust backbone with a thin layer of another active material to top it for controlled pore size distribution. Such membranes are often used for ultrafiltration. Alternatively, they can be of a single layer of active material. Such membranes are preferred for nanofiltration. One characterizing feature of a membrane is its molecular weight cut-off (MWCO) which specifies the biggest molecular size that can permeate through it. Frequently, the membranes are formed into single- or multichannel tubular forms, or spirals wound to accommodate a large surface area in a compact volume. Typical characteristics of membranes used in the treatment of black liquors are described in Table 3.3.

3.1.3.3 Ultrafiltration of Black Liquor

Ultrafiltration with ceramic membranes in itself may not be a very suitable method for separating lignin from alkaline black liquors. However, it can serve as a pre-filtration step before other more specific membrane separation stages. Hardwood kraft black liquor at 17% dissolved solids concentration was fractionated with a 15,000 Da MWCO ceramic ultrafiltration membrane. The conditions maintained were 100 kPa TMP, 5 m/s crossflow velocity, and 90 °C temperature with 90% of feed permeating through the membrane. The permeate recovered 82.4% lignin, 32.6% carbohydrates, and 90% inorganics. The operation therefore increased the lignin-to-carbohydrate ratio in the permeate and decreased it in the retentate. Therefore, the process was not considered suitable for lignin separation as a standalone operation (Jönsson et al. 2008). Ultrafiltration of softwood kraft black liquor of 16.7% dissolved solids concentration and 13.4 pH was performed with 20,000 Da MWCO

TABLE 3.3
Membrane Types and Characteristics Used for Lignin Separation from Black Liquor

Type	Use type	Material	MWCO (Da)	Trade Name	Supplier	Reference
Ceramic	Ultra	Al_2O_3 supported TiO_2	20,000	Atech Type 7/6	Atech Innovations GmbH, Germany	Arkell et al. 2014; Atech Data Sheet, n.d.
	Ultra	TiO_2-Al_2O_3 supported ZrO_2	50,000		Céramiques Techniques Industrielles, France	Costa et al. 2018
	Ultra	TiO_2-Al_2O_3 supported ZrO_2	15,000		Céramiques Techniques Industrielles, France	Costa et al. 2018
	Ultra	Al_2O_3-TiO_2-ZrO_2 supported TiO_2	5000	Filtanium	Tami Industries, France	Costa et al. 2018
	Nano	TiO_2	1000	INSIDE Ceram	Tami Industries, France	Arkell et al. 2014
Polymer	Ultra	PES	4000	ES404	PCI Membranes, UK	Jönsson et al. 2008; PCI Product Info, n.d.
	Ultra	Modified PES	6000	EM006	PCI Membranes, UK	Jönsson et al. 2008; PCI Product Info, n.d.
	Ultra	PS	20,000	PU120	PCI Membranes, UK	Jönsson et al. 2008; PCI Product Info, n.d.
	Ultra	PVDF	100,000	FP100	PCI Membranes, UK	Jönsson et al. 2008; PCI Product Info, n.d.
	Ultra	PES	5000		MANN + HUMMEL, Germany	Mendes et al. 2021
	Nano	Proprietary	1000	MPF-36	Koch Separation Solutions, Inc., USA	Arkell et al. 2014
	Nano	Proprietary	600	MPF-31	Koch Separation Solutions, Inc., USA	Arkell et al. 2014
	Nano	Proprietary	200	MPF-34	Koch Separation Solutions, Inc., USA	Arkell et al. 2014
	Electro-dialysis	PFSA and PTFE copolymer		Nafion	Chemours Company, USA	Jin et al. 2013

PES: Polyethersulphone; PS: Polysulphone; PVDF: Polyvinyldifluoride; PFSA: Perfluorosulfonic acid; PTFE: Polytetrafluoroethylene.

ceramic membrane at 2 bar TMP, 5 m/s crossflow velocity, and 90 °C operating temperature. When 90% of the feed volume permeated through the membrane, the permeate dissolved solids concentration dropped to 16%. The operation resulted in 18.7% lignin, 46.4% hemicellulose, and 14.7% inorganics being retained in the retentate. Thus, the ultrafiltration only marginally reduced the lignin-to-inorganics ratio from 0.76 to 0.72. The lignin to hemicellulose ratio in the permeate stream increased to 27.2 from its value of 17.9 in the black liquor. It further brings forth the fact that ultrafiltration with ceramic membrane as a standalone operation is not very suitable for separating lignin from alkaline black liquor (Arkell et al. 2014). Eucalyptus kraft black liquor of 17.6% dissolved solids concentration was fractionated by ultrafiltration in three stages using ceramic membranes. The black liquor was first passed through the membrane with 50,000 Da MWCO, and 63% volume was allowed to permeate. The permeate was passed through the second membrane with 15,000 Da MWCO with 72% volume of the feed to the second stage allowed to permeate. Permeate from the second stage was fractionated in the third and final stage with 5000 Da MWCO membrane. Permeate volume in the third stage was 44% of the feed to this stage. Initial black liquor had a ratio of lignin:inorganics:carbohydrate of 46.9:49.1:4. This ratio changed to 49.1:42.5:8.3 in the retentate from first stage ultrafiltration with the dissolved solids concentration increasing to 20.9%. It is apparent that ultrafiltration with 50,000 Da MWCO membrane retained more carbohydrates than lignin. Interestingly, this ratio was 46.9:51.8:1.3 (dissolved solids concentration 16.5%) in the second stage retentate. It amounts to a significant increase in the lignin-to-carbohydrate ratio but at the cost of more inorganics being present. The ratio further changed to 43.6:55.2:1.2 (dissolved solids concentration 15.7%) in the third stage retentate. The ratio of inorganics further increased without a change in the lignin-to-carbohydrate ratio (Costa et al. 2018).

3.1.3.4 Nanofiltration of Black Liquor

In many instances, the black liquor is treated in a sequence of ultrafiltration followed by nanofiltration. Occasionally, nanofiltration of black liquor itself is also reported. Generally, retentates from nanofiltration have significantly enhanced lignin concentration, and much lower ionic strength and water content. These retentates are more amenable to lignin separation by mild acidification compared to untreated black liquor, or the permeate after ultrafiltration. Ultrafiltration permeate after processing of hardwood kraft black liquor with a ceramic membrane was further treated by nanofiltration using a polymeric membrane of 1000 MWCO. TMP of 2.5 MPa, crossflow velocity of 4 m/s, and a temperature of 60 °C was maintained. Dissolved solids content in nanofiltration retentate increased to 29% from 16% in the feed, but the lignin-to-inorganics ratio had a marked increase to 1.85 from its value of 0.72 in the feed. Overall, nanofiltration preferentially allowed to permeate water and inorganics while retaining the lignin and carbohydrates (Jönsson et al. 2008). Permeate from the ultrafiltration of softwood kraft black liquor of 16.7% dissolved solids concentration was further fractionated by nanofiltration at 70 °C by a polymeric membrane or a ceramic membrane, both with 1000 Da MWCO. Lignin retention by the polymeric membrane was about 90% and so was the hemicellulose retention when 82% volume was allowed to permeate. In contrast, the ceramic membrane allowed 95% volume to permeate with 75% lignin retention and a little more carbohydrate retention. Inorganic retention was not reported for both the membranes. Lignin retention was 90% when the black liquor was directly treated with the polymeric nanofiltration membrane with 50% volume permeating through the membrane. Black liquor directly treated with ceramic nanofiltration membrane resulted in 80–85% lignin retention with 80% feed volume permeating (Arkell et al. 2014).

3.1.3.5 Electrodialysis of Black Liquor

A variation of the membrane filtration process applied for the treatment of alkaline black liquor is electrodialysis. In this process, an anode and a cathode are placed on either side of the intervening ion perm-selective membrane, thereby forming a two-chamber electrolysis cell. One of the chambers – mostly anode – is filled with black liquor, while the other is filled with a suitable electrolyte – mostly an aqueous solution of sodium hydroxide. A DC potential is applied across the electrodes alongside the TMP. The applied potential causes a preferential migration of cations or anions across the membrane, depending on the perm-selectivity of the membrane. The process, in addition to the usual fractionation of components by membrane separation also reduces the black liquor pH to a point suitable for lignin to precipitate. Electrodialysis of 10% dissolved solids concentration black liquor was carried out at a current density of 150 mA/cm^2 in a two-compartment cell separated by a Nafion membrane. The anode chamber was filled with black liquor while the cathode chamber was filled with aqueous sodium hydroxide. pH of black liquor dropped from 13 to 4.6 with a passage of 22.5 Q electrical charge. This resulted in about 20% of dissolved solids precipitating out of the black liquor. In comparison, acidification with mineral acid to the same end pH caused the precipitation of about 30% dissolved solids (Jin et al. 2013). Softwood kraft black liquor with 20–30% dissolved solids concentration was subjected to electrodialysis using an array of bipolar membranes, and cation exchange membranes, arranged in an alternate manner. In the two-compartment cell, black liquor formed the catholyte in the cathode chamber, and an aqueous solution of sodium hydroxide formed the anolyte in the anode chamber. Under the applied electric potential, electrodialysis caused a preferential migration of Na^+ (and other cations) towards the cathode compartment along with electro-oxidation of hydroxyl ions to oxygen. An equivalent OH^- ions would be generated in the anode chamber through hydrogen production by water splitting. The net result was a decrease in OH^- ions in the catholyte leading to acidification of black liquor. In different experiments, the black liquor pH decreased by 1.5–2 units after 40–55 min of electro dialysis (Haddad et al. 2016).

3.1.3.6 Permeate Flux and Its Variation

The pore structure of the membrane may progressively get clogged during operation by the entrapment of large molecules. In addition, unfavourable concentration gradients may also arise due to adsorption, and concentration polarization (Humpert et al. 2016). This causes the permeate flux to decrease over time. This can be compensated by progressively increasing the TMP. However, an excessive increase in TMP is not mandated to prevent structural damage to the membrane and excessive pumping costs. At this stage, the membrane requires cleaning to unclog its pore structure. The effect of crossflow velocity and TMP on permeate flux is presented in Figure 3.3. At a given crossflow velocity and TMP, the flux is almost similar for the polymeric membranes of MWCO 200, and 600 Da. However, there is a substantial increase in flux for 1000 Da MWCO polymeric membrane. At the same MWCO, TMP, and crossflow velocity, the ceramic membrane had a flux higher than the polymeric membrane. For each membrane, the permeate flux increased with an increase in TMP at a given crossflow velocity. Similarly, flux increased when the crossflow velocity was increased at a given TMP. Similar results were reported for ultrafiltration of hardwood kraft black liquor of 31% dissolved solids concentration with a 15,000 Da MWCO ceramic membrane.

At a given crossflow velocity, permeate flux increased up to a certain maximum with the increase in TMP and attained a plateau thereafter. This plateau occurred between 100 and 200 kPa TMP and was attributed to cake formation on the membrane. At any TMP, higher crossflow velocity resulted in higher flux. In addition, higher crossflow velocity required higher frictional pressure drop. Moreover, for any operating condition (crossflow velocity and TMP), flux continuously declined with time during ultrafiltration. A drop in permeate flux to almost 30% of the initial value was noticed after 65% of the feed permeated through the membrane (Jönsson et al. 2008). This behaviour, typical of membrane separation processes, is depicted in Figure 3.4. Here, volume reduction refers to the fraction of feed volume permeating through the membrane.

Permeate flux was found to decline with time for two ceramic ultrafiltration membranes during the treatment of eucalyptus kraft black liquor. For the 50,000 Da MWCO membrane, the flux dropped from 15.5 to 6.36 L/(m^2.h) over a treatment period of 36 h, with the initial 3 h causing the sharpest flux decline. For the same treatment duration, permeate flux decreased from 40.2 to 6.46 L/(m^2.h) for the 5000 Da MWCO membrane. In this case, the first 2 h of operation had a sharp decline in flux (Costa et al. 2018). However, the authors did not explain the reason for the higher

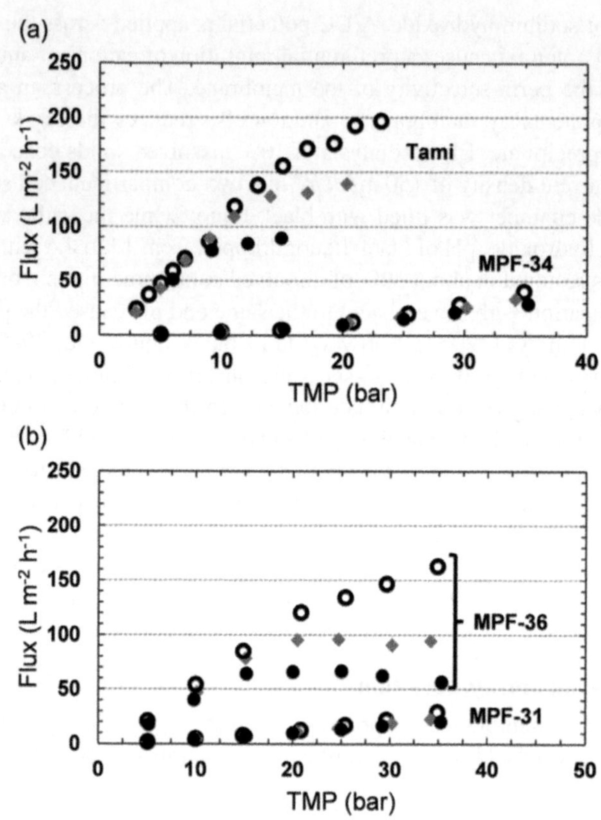

FIGURE 3.3 Variation in permeate flux with change in TMP and crossflow velocity for nanofiltration of softwood kraft black liquor of 16.7% dissolved solids concentration. Tami: ceramic membrane of MWCO 1000 Da; MPF 34: polymeric membrane of MWCO 200 Da; MPF 36: polymeric membrane with MWCO 1000 Da; MPF 31: polymeric membrane with MWCO 600 Da. Crossflow velocities are: 2 m/s (●), 3 m/s(◊), and 4 m/s(○). (Reprinted from Arkell, A, Olsson, J, Wallberg, O, *Chem. Eng. Res. Des.*, 92, 1792–1800, 2014, with permission from Elsevier.)

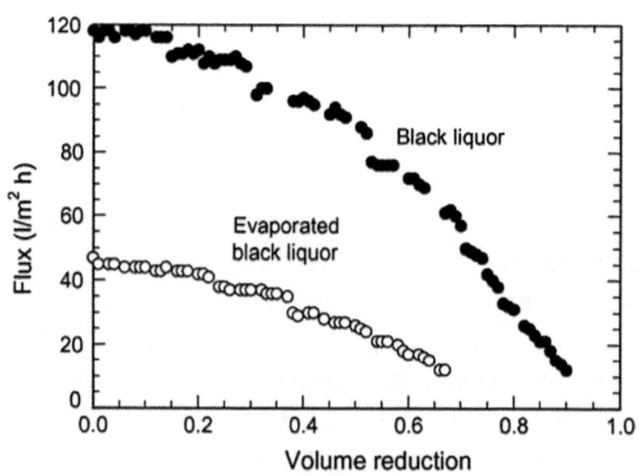

FIGURE 3.4 Decline in permeate flux with the progress of ultrafiltration of hardwood kraft black liquor of 17% dissolved solids concentration, and the same black liquor evaporated to 30% dissolved solids concentration. Process conditions: Polymer membrane of MWCO 15,000 Da; TMP 100 kPa; crossflow velocity 5 m/s; temperature 90 °C. (Reprinted from Jönsson, A, Nordin, A, Wallberg, O, *Chem. Eng. Res. Des.*, 86, 1271–1280, 2008, with permission from Elsevier.)

initial flux for lower MWCO membranes, though they report higher water permeability for the membrane having higher MWCO. The two ceramic membranes reported in the study had different ceramic supports and active layers. It highlights the fact that the structural features of membranes other than its MWCO have a significant effect on permeate flux.

3.2 SEPARATION OF DISSOLVED LIGNIN FROM ACIDIC SPENT LIQUORS

Acidic delignification processes – bisulphite or acid sulphite – sulphonates the lignin in the lignocellulose material. The resulting lignosulphonate, being hydrophilic, goes into solution under the prevailing conditions. Besides the chemical nature of lignin, acidic spent liquors have compositions vastly different from that of spent liquors from other delignification processes. Under acidic reaction conditions, the low molecular weight and largely amorphous hemicellulose polysaccharides are completely hydrolysed to monomer sugars and some oligomer fragments. These spent liquors have appreciable amounts of monomeric sugars from the hydrolysis of carbohydrates, mainly hemicelluloses. They constitute an important organic byproduct, other than lignin, suitable for further derivatization. Therefore, the conversion of monosaccharides to alcohols, and other value-added products, often forms an integral part of the processing of spent liquors from acidic delignification processes. The inorganic content of the spent liquor is made up of sulphite and/or bisulphite anions and base cations. Sulphate and thiosulphate ions are also formed as a result of side reactions occurring during delignification. Unlike alkali lignins – soda or kraft – lignosulphonates cannot be easily precipitated from the spent liquid, owing to their hydrophilic character. In the old Howard process, excess calcium hydroxide is added to the concentrated spent liquor to precipitate calcium lignosulphonate. Nowadays, the general strategy in the separation of lignosulphonates from acidic spent liquors involves fractionation and purification of the liquor to obtain a solution of relatively pure lignosulphonate. The solution is then dried to evaporate the water and the solid product is obtained as a powder.

3.2.1 Spent Liquor Characteristics from Acidic Delignification Processes

Physical and chemical characteristics of the spent liquors from acidic delignification processes are important in their processing. Physical characteristics, like density and viscosity, affect the storage, and pumping of the liquor. Liquor viscosity is an important consideration in its evaporation as it affects the heat transfer behaviour in the evaporators. Viscosity also affects the flux in membrane-based fractionation and purification techniques. It governs the energy consumption during agitation of the liquor in different processing steps. One important characteristic that affects almost all the other characteristics and also the processing is the dissolved solids concentration. Among the chemical characteristics, one important aspect is the liquor pH. The organics composition of the spent liquor is very important, especially the respective amounts of lignosulphonates and different monosaccharides present. Xylan is the principal hemicellulose in hardwoods while glucomannan is predominant in softwoods. This is also reflected in the sugar compositions of the spent liquors from different lignocellulose sources. Xylose and mannose are present in appreciable quantities in hardwood and softwood spent liquors, respectively. Low molecular weight carboxylic acids, different aldonic acids, and furfural are present in minor quantities. Typical organics composition from acidic delignification processes is presented in Table 3.4.

3.2.2 Monosaccharides Management of Spent Liquors

Acid sulphite and bisulphite delignification processes cause the almost complete acid-catalysed hydrolysis of hemicellulose, and also to some extent that of cellulose. The spent liquor from these processes, besides lignosulphonates, is rich in pentoses and hexoses as exemplified in Table 3.4. From the standpoint of lignosulphonates, these monosaccharides are impurities that downgrade the

TABLE 3.4
Major Organic Constituents in Spent Liquor from Acidic Delignification

Major Constituents	Proportion (% of Dry Solids)	Reference
	From hardwoods	
Lignin	40–56	Restolho et al. 2009; Ringena et al. 2005; Rueda et al. 2014; Xavier et al. 2010
Xylose	17–22	Rueda et al. 2014; Xavier et al. 2010
Glucose	1.5–2.5	Rueda et al. 2014; Xavier et al. 2010
Mannose	1.5–6	Rueda et al. 2014; Xavier et al. 2010
Galactose	2–3	Rueda et al. 2014; Xavier et al. 2010
Arabinose	1.9–5.3	Rueda et al. 2014; Xavier et al. 2010
Furfural	0.1–1.4	Rueda et al. 2014; Xavier et al. 2010
Acetic acid	5.5–7	Rueda et al. 2014; Xavier et al. 2010
	From softwoods	
Lignin	43–63	Al-Rudainy et al. 2020; Schlackl et al. 2020
Xylose	1.8	Al-Rudainy et al. 2020; Fatehi and Ni 2011
Glucose	1.7–2	Al-Rudainy et al. 2020; Fatehi and Ni 2011
Mannose	4.6–6	Al-Rudainy et al. 2020; Fatehi and Ni 2011
Galactose	1.9	Al-Rudainy et al. 2020
Arabinose	0.6–0.7	Al-Rudainy et al. 2020; Fatehi and Ni 2011
Furfural	0.1–0.2	Fatehi and Ni 2011; Schlackl et al. 2020
Acetic acid	2.7	Fatehi and Ni 2011
Xylonic acid	1.6	Fatehi and Ni 2011
Mannonic acid	2	Fatehi and Ni 2011

product quality. At the same time, the monomer sugars constitute an important starting point for biochemical transformation to several industrially important chemicals. Bioethanol production from the monosaccharides present in the spent sulphite liquors is a well-established commercial process. Traditionally, hexoses were converted to ethanol. With the progress in enzyme technology, it is now possible to convert both hexoses and pentoses to bioethanol, thereby increasing the yield. Monosaccharides present in the spent sulphite liquors can also convert to biobutanol and acetone, which are important industrial chemicals. Similarly, it is possible to produce succinic acid through the biochemical transformation of sugars present in spent liquor. These biochemical processes can be schematically represented as

$$\text{Monosaccharides} + \text{enzyme} \rightarrow C_2H_5OH + CO_2 \quad \{3.1\}$$

$$\text{Monosaccharides} + \text{enzyme} \rightarrow C_2H_5OH + C_4H_9OH + CH_3COCH_3 + CO_2 \quad \{3.2\}$$

$$\text{Monosaccharides} + CO_2 + \text{enzyme} \rightarrow C_2H_4(COOH)_2 + CH_3COOH + HCOOH \quad \{3.3\}$$

Some common microorganisms used in the fermentation of monosaccharides present in spent sulphite liquors are presented in Table 3.5.

The yeast species widely used in commercial ethanol fermentation is *Saccharomyces cerevisiae* with robust capability to convert hexoses to ethanol. It is reasonably tolerant to the presence of fermentation inhibitors like lignosulphonates, carboxylic acids, and furfural. However, its inability to ferment pentoses restricted its use in fermenting the sugars in spent sulphite liquors, as these liquors have considerable amounts of pentoses. Genetically engineered *S. cerevisiae* with gene insertion

TABLE 3.5
Common Microorganisms, their Principal Substrates, and Main Products in the Microbial Treatment of Spent Sulphite Liquors

Microorganism	Type	Primary Substrate	Product	References
Actinobacillus succinogenes	Bacteria	Hexose	Succinic acid, acetic acid, formic acid	Alexandri et al. 2016
Basfia succiniciproducens	Bacteria	Hexose	Succinic acid, acetic acid, formic acid	Alexandri et al. 2016
A. succinogenes	Bacteria	Hexose	Succinic acid, acetic acid, formic acid	Ladakis et al. 2018
Scheffersomyces stipitis	Yeast	Pentose	Ethanol	Pereira et al. 2015
Pachysolen tannophilus	Yeast	Pentose	Ethanol	Harner et al. 2014
Saccharomyces cerevisiae	Yeast	Hexose	Ethanol	Helle et al. 2004
Clostridium acetobutylicum	Bacteria	Hexose	Ethanol, butanol, acetone	He and Chen 2020

from *Scheffersomyces stipitis* is found to be reasonably successful in co-fermenting hexoses and pentoses to ethanol (Helle et al. 2004). Pretreatment of the spent liquor, lime treatment, or air stripping, prior to fermentation was found to improve ethanol yield (Helle et al. 2008).

Spent liquor from eucalyptus acid sulphite delignification was fermented with the *S. stipitis* strain adapted to high concentrations of lignosulphonates and acetic acid. During 62 h fermentation, 0.26 g ethanol was produced per g of substrate amounting to 51% conversion efficiency. 71% xylose was consumed while glucose consumption was close to 100%. The final ethanol concentration in the broth was 6.93 g/L (Pereira et al. 2015). Acid sulphite spent liquor from hardwood delignification was fermented with inhibitor-tolerant mutant strains of *Pachysolen tannophilus* (Harner et al. 2014).

Eucalyptus spent sulphite liquor from acid sulphite delignification, after pre-extraction with isopropanol, was fermented with either *Basfia succiniciproducens* or *Actinobacillus succinogenes*, both bacterial strains. For each fed-batch fermentation run, 50 mL bacterial inoculum was added to 500 mL of pre-extracted and detoxified liquor, with an initial total sugar concentration of 25 g/L, and fermented at 37 °C and 6.6 pH under 300 rpm agitation and sparging with carbon dioxide. With *A. succinogenes*, after 29 h of fermentation, the total sugar concentration dropped to 3 g/L and the broth had a succinic acid concentration of about 10 g/L. At this point, the fermentation broth was pulse-fed with concentrated pre-extracted and detoxified liquor, having 170 g/L total sugars and 10% inoculum, to raise the total sugar concentration to 13 g/L. After 48 h, the broth had a succinic acid concentration of 19.1 g/L amounting to a yield of 0.55 g per g total sugars. Productivity of succinic acid in the fermentation broth was 0.4 g/(L.h). Formic and acetic acids were produced as undesired byproducts. For the bacterial strain *B. succiniciproducens*, the initial total sugar concentration was 15 g/L in the diluted pre-extracted and detoxified liquor. Two feed pulses were added, one after 11 h to raise the total sugar concentration from 3 g/L to 11.5 g/L, and the second after 22 h to raise the total sugar concentration from 2 g/L to 11 g/L. 29 h fermentation yielded 0.58 g succinic acid per g total sugar with an overall productivity of 0.67 g/(L.h). The final succinic acid concentration in the broth was 19.6 g/L. Formic and acetic acids were produced as undesired byproducts for both the bacterial strains. *B. succiniciproducens* fared better than *A. succinogenes* in terms of yield and productivity but was less tolerant to the presence of lignosulphonates and phenolics (Alexandri et al. 2016). Nanofiltered spent liquor from acid sulphite delignification of eucalyptus – 44.7 g/L sugars, 5.7 g/L lignosulphonate, and 0.2 g/L phenolics – was fermented in continuous bioreactors using *A. succinogenes*, and *B. succiniciproducens*. For *A. succinogenes*, the succinic acid yield was 0.26–0.48 g/g sugar, and productivity was 0.38–0.68 g/(L.h). For *B. succiniciproducens*, these quantities were 0.4–0.55 g/g sugar, and 0.4–1.6 g/(L.h), respectively (Ladakis et al. 2018).

3.2.3 Lignosulphonates through Membrane Separation

Lignosulphonates can be fractionated from spent sulphite liquors by membrane separation. Both ultrafiltration and nanofiltration have been studied for their efficacy in retaining lignosulphonate. Nanofiltration is commonly employed that allows the monomer sugars to permeate while retaining the lignosulphonates. The monosaccharide-rich rich permeate is usually fermented for its value addition. The retentate is further processed for the lignosulphonate product.

Spent acid sulphite liquor from eucalyptus delignification was treated with different polymeric ultrafiltration and nanofiltration membranes. During ultrafiltration at TMP of 2–3 bar and 935 Reynolds number, both 10,000 and 20,000 Da MWCO membranes retained 41–50% lignosulphonates after 60 min treatment, though the authors did not report the volume fraction permeating. Sugar retention was between 4% and 27%. Permeation flux was also similar for the two membranes between 10 and 36 kg/(h.m^2). The 10,000 Da MWCO membrane was made up of hydrophilic polysulphone while the 20,000 Da MWCO membrane was fluoro polymer type. Since the two membranes with vastly different MWCO gave similar performance in terms of lignocellulose and sugar retention, the interaction of membrane material with the components in the spent liquor was crucial. This is further exemplified by the ultrafiltration performance of another 10,000 Da MWCO PES membrane which retained 68% lignosulphonates, but only 3% sugars, at 4 bar TMP with a much reduced permeate flux of 6 kg/(h.m^2). PES membrane of 5000 Da MWCO could retain 80% lignosulphonates, but it came at a cost of 49% sugar retention and a permeate flux of only 2 kg/(h.m^2). Nanofiltration of the same spent liquor with different membranes made up of polyamide thin film composite allowed for 95–99% lignosulphonates retention along with similarly high sugar retention. PES nanofiltration membranes retained 92% lignosulphonates but 77% sugars (Restolho et al. 2009). Hardwood spent acid sulphite liquor of 7.2% dissolved solids concentration, 1.02 mm^2/s viscosity, and 3.3 pH was ultrafiltered with different multichannel ceramic membranes made of TiO_2. TMP ranged from 180 to 200 kPa and the feed flow rate was 650–690 L/h. Though the authors did not specify the volume fraction permeating, 65% lignosulphonate retention along with 12.9% sugar retention was reported with 5000 Da MWCO membrane. In addition to retention, the membrane also increased the lignosulphonate concentration in the retentate from 39.9 to 63.7 g/L. When treating in a series arrangement of membranes, the permeate from 15,000 Da MWCO membrane was fed into 5000 Da MWCO membrane, and the permeate from 5000 Da MWCO membrane was fed into 1000 Da MWCO membrane. The final permeate from the 1000 Da MWCO membrane recovered 73% sugars from the initial spent liquor with a lignosulphonate-to-sugar ratio of 0.64 compared to 1.72 in the spent liquor. The three membranes combined recovered 72.6% lignosulphonates in their retentates with 44.3% being retained in the 15,000 Da MWCO membrane. This retentate had a lignosulphonate-to-sugar ratio of 3.32 with a concentration increasing to 96.9 g/L (Fernández-Rodríguez et al. 2015). There is a report of nanofiltration of eucalyptus acid sulphite spent liquor by the Belgian Firm AVECOM NV. 18% dissolved solids concentration liquor was nanofiltered in a vibratory shear-enhanced module employing PES membranes of 800 Da MWCO, and 1.5 m^2 area. 96% lignosulphonates along with 16% sugars were retained (Ladakis et al. 2018). Beech acid sulphite spent liquor with 12.5% dissolved solids concentration was fractionated with different ceramic membranes, one tubular membrane of 20,000 Da MWCO, and three hollow fibre membranes of 30, 8, and 3 nm cut off, respectively. At a TMP of 3 bar, 12,720 Reynolds number, and 80% feed volume permeating, lignosulphonate retention was 68%, 66%, 48%, and 65%, respectively, for 3 nm, 8 nm, 30 nm, and 20,000 Da cut-off membranes. In addition to the MWCO, membrane filtration performance – permeation flux and lignosulphonate retention – strongly depends on TMP, Reynolds number, and the fraction of feed volume permeating (Humpert et al. 2019). The effects are shown in Figs. 3.5–3.7. Ultrafiltration of softwood bisulphite spent liquor with a flat-sheet-regenerated cellulose membrane of 10,000 Da MWCO was performed. After 90% volume had permeated through the membrane, 60% acid-soluble lignin, and more than 95% acid-insoluble

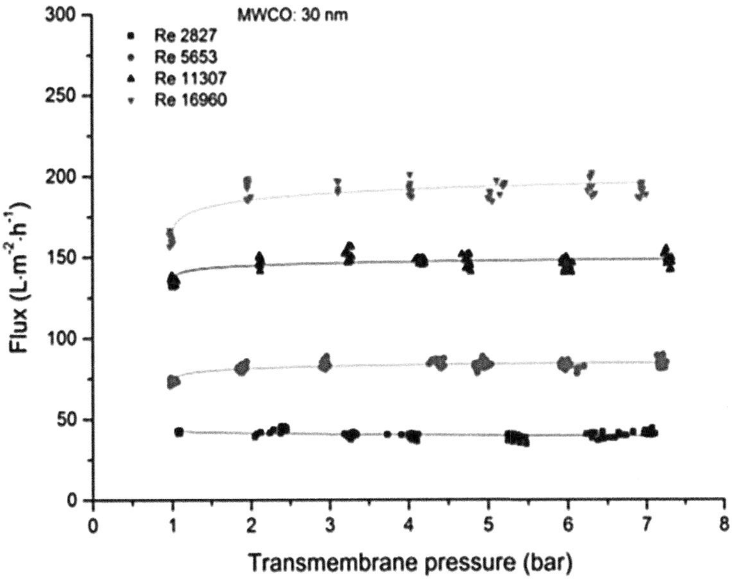

FIGURE 3.5 Dependence of permeation flux on TMP and Reynolds number during membrane filtration of hardwood acid sulphite spent liquor with ceramic hollow fibre membrane. (Reprinted from Humpert, D, Ebrahimi, M, Stroh, A, Czermak, P, *Membranes*, 9(4), 45, 2019, originally published by MDPI under Creative Commons Attribution License.)

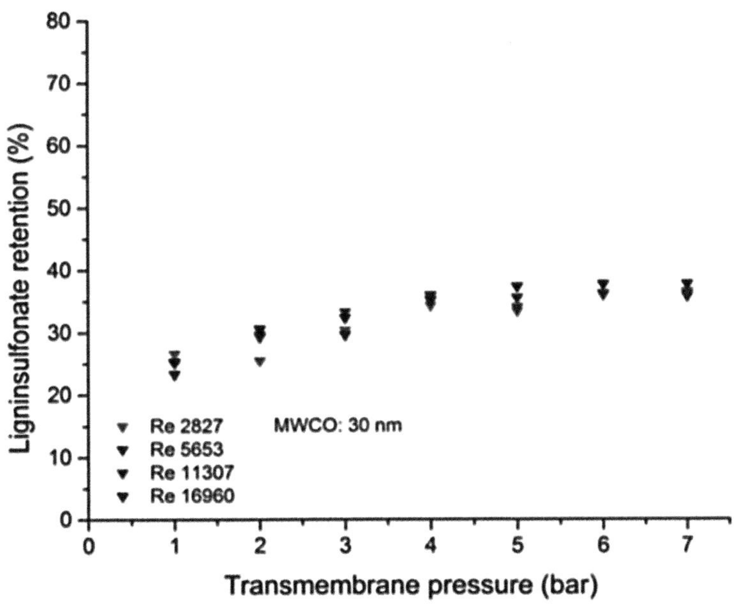

FIGURE 3.6 Dependence of lignosulphonate retention on TMP and Reynolds number during membrane filtration of hardwood acid sulphite spent liquor with ceramic hollow fibre membrane. (Reprinted from Humpert, D, Ebrahimi, M, Stroh, A, Czermak, P, *Membranes*, 9(4), 45, 2019, originally published by MDPI under Creative Commons Attribution License.)

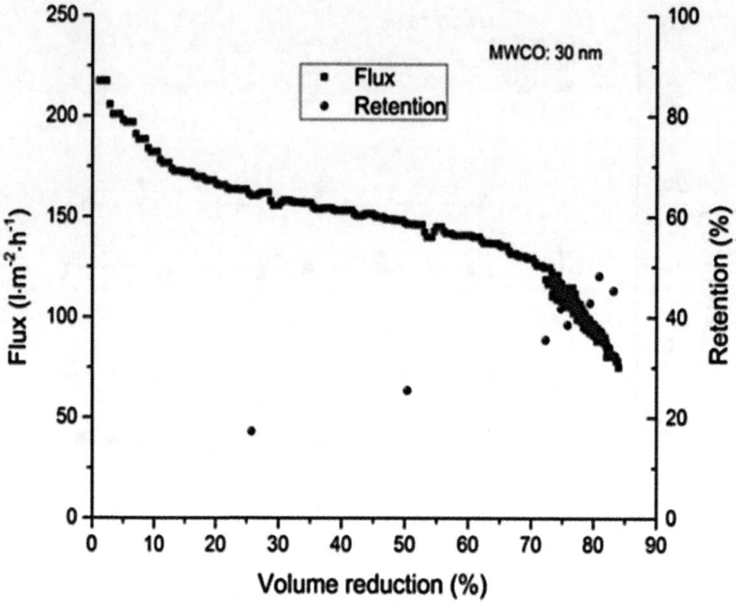

FIGURE 3.7 Dependence of permeation flux and lignosulphonate retention on the volume fraction of feed permeating during membrane filtration of hardwood acid sulphite spent liquor with ceramic hollow fibre membrane. (Reprinted from Humpert, D, Ebrahimi, M, Stroh, A, Czermak, P, *Membranes*, 9(4), 45, 2019, originally published by MDPI under Creative Commons Attribution License.)

lignin were recovered in the retentate. Permeation flux varied between 80 and 20 L/(m^2.h) (Al-Rudainy et al. 2020).

Another recent study reports the ultrafiltration and nanofiltration of spruce sulphite spent liquor. Ultrafiltration with a PES membrane of 5000 Da MWCO, at 10 bar TMP, and 0.8 m/s crossflow velocity, resulted in a concentration increase of liquor from 18.1% to 32% dissolved solids with the proportion of lignosulphonates increasing from 63% to 83% (Schlackl et al. 2020).

3.2.4 Lignosulphonates through Solvent Extraction

Lignosulphonates can be separated from the acidic spent liquors by liquid–liquid extraction using a suitable solvent. A liquid suitable as a solvent should have a high selectivity and distribution coefficient for lignosulphonate. It should have very low miscibility with water at all concentrations to form a separate liquid phase. Additionally, the physical characteristics of the solvent are important. In the first step of liquid–liquid extraction, the solvent and the spent liquor are intimately mixed under strong agitation to form an oil-in-water-type emulsion where the solvent forms the dispersed phase, and the spent liquor constitutes the continuous phase. A high interfacial area is desirable in the formed emulsion for efficient transfer of lignosulphonate from the aqueous phase – termed the raffinate – to the solvent phase – termed the extract. A low interfacial tension will facilitate this with the formation of smaller globules of the dispersed phase in the emulsion. In the second step of the operation, agitation is stopped to allow the globules of the dispersed phase to coalesce and the emulsion to break into two distinct phases: extract and raffinate. This will be facilitated by a high interfacial tension. In essence, the interfacial tension should neither be too high nor too low. An appreciable difference between the densities of the solvent and the spent liquor is also helpful for the ease of separation of the two phases after the emulsion breaks. Another important consideration in liquid–liquid extraction is the recoverability of the solvent from the extract for reuse in the process. This is desired from the standpoint of both economic and environmental sustainability. The quantities, selectivity and distribution coefficient, measured at equilibrium condition, are defined as (Treybal 1981)

$$\text{Selectivity} = \frac{\text{Weight fraction lignosulphonate in extract / Weight fraction water in extract}}{\text{Weight fraction lignosulphonate in raffinate / Weight fraction water in raffinate}} \quad (3.19)$$

$$\text{Distribution coefficient} = \frac{\text{Weight fraction lignosulphonate in extract}}{\text{Weight fraction lignosulphonate in raffinate}} \quad (3.20)$$

One of the problems associated with solvent extraction of spent liquors is crud formation, the forming of a stable emulsion with solids embedded.

Eucalyptus acid sulphite spent liquor at 9% dissolved solids concentration (65.5 g/L lignosulphonate, and 25 g/L total sugars) was extracted with isopropanol as solvent. 7.33 g solvent was added for every g of water in the spent liquor. After thorough mixing followed by phase separation, 73% lignosulphonate was transferred to the extract phase at 5.5 pH. Lignosulphonate extraction was 75% at pH 6.5 (Alexandri et al. 2016).

A variation of liquid–liquid extraction is reactive extraction where the lignosulphonate, present as the acid form in acidic spent sulphite liquors, reacts with the reactive component of the solvent in the extract phase. This reaction increases the overall mass transfer coefficient and accelerates the transfer of lignosulphonate across the phase boundary. Generally, the reaction is one of complex formation between the solvent and the lignosulphonic acid. Lignosulphonate is recovered, in base form, from the extract phase by reacting with a suitable base, known as the stripping phase, having the desired counter ion. It can be schematically represented as

$$\text{RC (org)} + \text{HLS (org)} \rightarrow \text{RC·HLS (org)} \quad \{3.4\}$$

$$\text{RC·HLS (org)} + \text{MOH} \rightarrow \text{RC (org)} + \text{MLS} + H_2O \quad \{3.5\}$$

Here, RC (org) is the reactive component in the organic solvent, HLS (org) is the acid form of lignosulphonate (lignosulphonic acid) in the extract phase, RC·HLS (org) is the formed complex, MOH is the base with a monovalent counterion, and MLS is the recovered base form of lignosulphonate. The most common counter ion is the sodium ion. Another counter ion frequently used is the magnesium ion, a bivalent cation. Different amines, dodecylamine, dioctylamine, and trioctylamine, have been found to be suitable reactive components for the reactive extraction of lignosulphonates, the effectiveness decreasing in that order (Kontturi et al. 1986). In a recent work, spent liquor from a magnesium base bisulphite process with 137 g/L lignosulphonate concentration was extracted with several amines used as reactive components. The solvent for reactive extraction consisted of a solution of the respective amine in 1-octanol in 20 weight% concentration. The volumetric ratio of feed to solvent and the extract to stripping solution was 1. The stripping phase was 0.3 M aqueous solution of sodium hydroxide. The distribution coefficient for reactive extraction varied between 0.28 and 7.8. Dodecylamine exhibited the highest distribution coefficient of 7.8 and 89% extraction efficiency followed by octylamine and decylamine having a distribution coefficient of 3.6 and extraction efficiency of 78%. Distribution coefficients during stripping were 0.75, 0.57, and 0.51, respectively, for dodecylamine, octylamine, and decylamine and the stripping efficiencies were 57, 64, and 66%. The study also highlighted that the distribution coefficients and extraction efficiencies are influenced by the presence of other components, like monosaccharides, in the system as these quantities had different values when model lignosulphonate solutions were extracted instead of spent liquor (Demmelmayer et al. 2020).

In another variation of the liquid–liquid extraction process, the feed and the solvent are separated by a liquid membrane. Components are transferred across the membrane due to differences in their chemical potential on the two sides. The liquid membrane can be a supported membrane, a bulk membrane, or an emulsion membrane. In the supported liquid membrane process, a solid support is preferentially impregnated with the solvent phase owing to its hydrophobic nature. Various support materials have been investigated for their efficacy to embed the solvent and form a liquid membrane. Their morphological characteristics, such as thickness, porosity, pore size, and tortuosity are important. During studies with model solutions of sodium lignosulphonate, the liquid membrane was prepared by immersing the support material in the solvent, a solution of 4 vol% trioctylamine – reactive component – in dichloroethane, for 24 h. The support material impregnated with the solvent constituted the liquid membrane and separated the model lignosulphonate solution from the 0.5 M sodium hydroxide solution, the stripping phase. Nylon 6,6 was found to be the best support material recovering 40% lignosulphonate. The membrane was found to be sufficiently robust and maintained its integrity over a period of 45 days (Chakrabarty et al. 2009a). In one study, cellulose films of thickness varying between 53 and 70 μm, were used as the support for liquid membrane. The solvent used was 20 weight% n-trioctylamine dissolved in 1-octanol, and the stripping phase was an aqueous solution of sodium hydroxide. For extracting lignosulphonates, the spent liquor and the amine solvent were separated by the cellulose film (Kienberger et al. 2020). In the bulk liquid membrane process, the feed and the stripping solution are physically separated by a barrier but are simultaneously in contact with a layer of the solution phase. The transferring component, lignocellulose in the present case, goes into the solution and forms the complex with the reactive component. Since the solution phase is also in contact with the stripping solution, aqueous sodium hydroxide in the present case, the formed complex reacts with the base forming sodium lignosulphonate and passes into the stripping solution. In a laboratory experiment with model lignosulphonate solution of 108 mg/L concentration as feed and 1.25 M sodium hydroxide as stripping solution, 60 mL each of feed and stripping solution were simultaneously in contact with 50 mL of solvent consisting of 4 vol% trioctylamine dissolved in dichloroethane. The assembly was so constructed that the interface area between the solvent and the two aqueous phases was 18.2 cm^2 each. 70% lignosulphonate separation was achieved after 4 h with the aqueous phases stirred at 500 rpm (Chakrabarty et al. 2009b). In another laboratory study with model lignosulphonate solution of 200 mg/L, the solvent phase was 4 vol% trioctylamine dissolved in sunflower oil, and the stripping solution was 0.2 M sodium sulphate solution. The volumetric ratio of solvent:feed:stripping solution was 1:2:2.75% lignosulphonate recovery was reported at a stirring speed of 300 rpm after 3 h (Kumar et al. 2018). However, the results of model solution studies should be read with caution when relating to actual spent liquors owing to their vastly different composition and lignosulphonate concentration. Lignosulphonate concentration in actual spent liquors is several orders of magnitude higher than those used for the model solutions. In addition, the actual liquors contain appreciable amounts of sugars.

3.2.5 Lignosulphonates through Ion Exchange

Ion exchange is an important unit operation capable of fractionating dissolved substances from solution based on their ionic character. Anion exchange resins, mostly with exchangeable hydroxyl ions in the resin backbone, adsorb the anions from the solutions and release the hydroxyl ions into the solution in return. Lignosulphonate anions can be separated in this way. On the other hand, cation exchange resin, mostly with exchangeable hydrogen ions in the resin backbone, adsorbs the cations from the solution exchanging them with hydrogen ions that go into the solution. This comes in handy in the separation of Mg^{++}, or Ca^{++}, the base ions, from the spent liquor. Contacting the liquor sequentially with both types of resins effectively fractionates the liquor. Monosaccharides and oligosaccharides being nonpolar entities do not interact with the ion exchange resins and elute through the process.

Ion exchange fractionation of eucalyptus acid sulphite spent liquor is reported by a two-stage treatment. In the first stage, the liquor was treated with Dowex 50Wx2 cation exchange resin in H^+ form. Treated liquor from the first stage was then contacted with Amberlite IRA-93 anion exchange resin in OH^- form. Almost complete removal of lignosulphonates was achieved, and the liquor obtained after the two-stage ion exchange treatment was essentially a pure sugar solution (Xavier et al. 2010). Acid sulphite spent liquor – dissolved solids concentration 15.8% – from eucalyptus delignification was fractionated with sequential ion exchange. First, the liquor was treated with cation exchange resin, H^+ form of Dowex 50WX2, to remove the base cations, Mg^{++} and Ca^{++}. Eluted liquor from cation exchange was then treated with anion exchange resin, OH^- form of Amberlite IRA-96, to remove the lignosulphonate and other anions. Cation exchange removed 95% Ca^{++}, and 90% Mg^{++}. About 12.5% of lignosulphonates were also retained during cation exchange due to physical adsorption of low molecular weight fractions. Subsequent anion exchange treatment removed more than 99% of the lignosulphonates along with complete removal of acetate anions (Fernandes et al. 2012). Spent liquor from sodium base bisulphite delignification of softwood, with a dissolved solids concentration of 84.7 g/L, was treated with strong basic anion exchange resin, Amberlite IRA958, in chloride form. 80% lignosulphonate adsorption was achieved in 3 h with 0.1 g resin per g spent liquor, with minimal adsorption of sugars. Of the adsorbed lignosulphonates, 97.6% could be recovered by washing three times with 10 g/L sodium chloride solution, regenerating the resin in the process (Al-Rudainy et al. 2020).

3.2.6 Drying of Lignosulphonates

Though liquid formulations of lignosulphonates are available, the preferred physical form of lignosulphonates is dry powder. This facilitates storage, transportation, as well as its subsequent end use. Careful control of the drying conditions affects the product characteristics, mainly the particle size, size distribution, and morphology. The process also affects a mill's economics as drying is an energy-intensive operation and requires capital investment as well. Spray-drying is the most common technique for getting lignosulphonate powder from acidic spent liquors. Spray-drying experiments with Aquasolv lignin determined that the process is strongly affected by the atomization pressure. It is also dependent on the nozzle flow rate and the drying temperature (Gil-Chávez et al. 2020). Laboratory experiments with 1200 mL calcium lignosulphonate solution of 17% dissolved solids concentration were performed to ascertain the effects of drying temperature (150–170 °C), throughput (30–40% of the maximum), and nozzle flow rate (30–40 mL/min) on product characteristics. Product yield was found to decrease with increasing throughput but increased with increasing temperature. With the increase in nozzle flow rate, there was an initial increase in product yield which reached a maximum and then started declining. Very high nozzle flow rates are, therefore, detrimental to product yield. This is most likely due to the formation of a large proportion of very fine liquid droplets at a high nozzle flow rate which get entrained in the airflow causing a drop in yield. The obtained lignosulphonate powder had globular particles with double-concave-shaped geometry (Padhi et al. 2022).

3.3 SEPARATION OF DISSOLVED LIGNIN FROM ORGANOSOLV SPENT LIQUORS

Organosolv delignification is commercially much less utilized compared to alkaline or sulphite processes. As deliberated in Chapter 2, organosolv delignification processes are much varied in the choice of solvent and process conditions. The resulting spent liquor obtained is equally diverse in characteristics. There are a variety of methods, and their combinations, that can be used for the separation of lignin from organosolv spent liquors. Lignins obtained from organosolv spent liquors have aroused much research interest in recent years due to the relatively pure nature of these lignins. The selection of the separation method is dictated by several considerations. The method should separate lignin in high yield and purity. Minimal alterations in the chemical structure of lignin sets apart organosolv delignification from other delignification processes. The purity of the lignin product is, therefore, all the more important. The separation method should not be too demanding in terms of energy. It should integrate with the other operations of the biorefinery. Overall, the method should be technically, economically, and environmentally sustainable. Accordingly, solvent evaporation, precipitation by the addition of antisolvent, filtration, membrane filtration, and liquid–liquid extraction have all been used for the separation of lignin from organosolv spent liquors.

3.3.1 Spent Liquor Characteristics from Organosolv Delignification Processes

3.3.1.1 Spent liquor from alcohol Organosolv Processes

Spent liquors from aqueous alcohol organosolv delignification typically contain – besides the alcohol – macromolecular lignin fragments, monomer and oligomer sugars, lignin carbohydrate complexes, extractives, and other low molecular weight organics like carboxylic acids. In catalytic processes, the liquor would be acidic or basic depending on the nature of the catalyst. Typical compositions of spent liquors from alcohol organosolv processes are presented in Table 3.6.

3.3.2 Lignin Separation by Solvent Evaporation

If the solvent used for delignification is a low boiling liquid, lignin can be separated from the solution by distilling off the solvent, leaving behind an aqueous dispersion of lignin. This is also essential for the recovery and reuse of the solvent, without which the process would be economically and environmentally unsustainable. Lignin in solid form can be obtained by filtering the dispersion and drying the filter cake.

Organosolv delignification processes with low boiling liquids such as aqueous alcohols or aqueous acetone are normally carried out at elevated temperatures and associated high pressure. One useful way of evaporating the solvent is flashing the reactor content to atmospheric pressure after delignification. This not only increases the dissolved solids concentration in the spent liquor but also increases the water content in the liquor due to the higher volatility of the solvent. Flashing of reactor content after aqueous ethanol organosolv delignification reportedly decreased the solvent ethanol content from 50% to 42.6%. For methanol organosolv delignification, a decrease from 50% to 38.6% is reported (Botello et al. 1999). Rotary vacuum evaporation of aqueous ethanol organosolv spent liquor could precipitate the lignin which was separated by centrifugation at reduced temperature (Matsakas et al. 2019). Poplar aqueous ethanol organosolv spent liquor was first extracted with hexane. The non-hexane phase after extraction was vacuum evaporated to dryness to obtain solid lignin which was further purified by water washing (Meyer et al. 2020). Crude lignin was extracted from walnut shells through different aqueous alcohol-based organosolv delignification followed by evaporation of spent liquors to dryness. Lignin was purified by redissolving the solids in acetone and precipitating by dilution with excess water (Zijlstra et al. 2020). One particular study reports

TABLE 3.6
Typical Compositions of Spent Liquors from Organosolv Delignification Processes

Process	Lignin Concentration (%)	Solvent Concentration (%)	Sugar Concentration (%)	References
Ethanol organosolv	1.35–3.35	43–55	3.37	Alriols et al. 2010; Schulze et al. 2019
Methanol organosolv	2.9	50	1.34	Botello et al. 1999
Acetone organosolv		25–60	0.11–0.69	Huijgen et al. 2010
Acetic acid organosolv	1.8–4	80	0.15–0.16	Kangas et al. 2015; Vila et al. 2003

acid-catalysed organosolv delignification of different lignocelluloses by an aqueous mixture of ethyl acetate and ethanol. After delignification, the spent liquor separated into an organic phase, and an aqueous phase. Lignin was obtained in solid form by evaporating the solvent from the organic phase (Cybulska et al. 2012). Spent liquor from aqueous ethylene glycol organosolv delignification was evaporated to dryness at 100–105 °C and less than 1 mm Hg pressure. Lignin was precipitated while the sugars went into solution when the residual mass was redispersed by sonication in cold water. However, the evaporation process was very slow, and the residue was difficult to recover due to stickiness (Thring and Chornet 1990). Lignin was precipitated from spent liquor obtained from aqueous tetrahydrofuran organosolv delignification of poplar by boiling off the organic solvent. The residual water from the resulting resinous mass was removed by vacuum drying to yield the solid lignin (Meng et al. 2019). Aqueous 2-methyl tetrahydrofuran organosolv delignification of beech resulted in a spent liquid that separated into an organic and an aqueous phase. Lignin could be obtained in solid form by evaporating the solvent, 2-methyl tetrahydrofuran, from the organic phase of the spent liquor (Weidener et al. 2020). Spent liquor from oxygen-assisted aqueous acetone organosolv delignification of hardwoods was vacuum evaporated to remove the acetone. Lignin could be precipitated by acidifying the concentrated liquor to pH 2 (Gong and Bujanovic 2014).

3.3.3 Lignin Precipitation by Antisolvent Addition

One common method of lignin separation from organosolv spent liquors is its precipitation by antisolvent addition. The addition of a suitable third liquid into the organosolv spent liquor destabilizes the phase equilibrium, causing the lignin to agglomerate and precipitate out. In many instances, water itself acts as an antisolvent. As the water content in the spent liquor is increased, with an associated drop in the solvent concentration, lignin starts agglomerating.

3.3.3.1 Spent Liquor from Alcohol Organosolv Processes

Lignin solubility in aqueous ethanol is strongly dependent on ethanol concentration. Accordingly, diluting the spent liquor with water or acidified water is frequently used to precipitate lignin. It is useful to understand the phase behaviour of the ethanol–water–lignin system. Under certain conditions, the lignin can form a separate liquid phase with a sticky nature. The process of agglomeration is one of the formation of solid bridges between lignin aggregates, which themselves are formed when macromolecular lignin fragments overcome charge repulsion and cluster together. The zeta potential of the system is critical in controlling the process. Accordingly, temperature, pH, and presence of ions play an important role in the agglomeration and precipitation process. For the success of the process, it is critical to have a precipitate that is easily filterable. Lignin macromolecular structures are softened with an increase in temperature and a decrease in alcohol concentration. Too little softening can be detrimental to aggregation and too much softening can lead to the formation of a sticky mass. Generally, there is a temperature window and a range of solvent concentrations where we get a dispersion that is filterable (Schulze et al. 2016; Schulze et al. 2019). The size of the lignin agglomerates formed correlates strongly to the severity during the delignification step. One study suggests that a delignification temperature higher than 200 °C results in significantly larger lignin particles during precipitation (Adamcyk et al. 2021). This is depicted in Figure 3.8.

Lignin precipitation was carried out by dilution with acidified water from eucalyptus aqueous ethanol, and aqueous methanol organosolv spent liquor. For ethanol liquor, 2.68 times dilution and 3 pH were found to be optimum precipitating 67% lignin. These values were 1.63 times dilution, and 2 pH for methanol liquor, precipitating 70% lignin. One possible reason for this disparity is the lower methanol concentration in the liquor – 38.6% – compared to ethanol concentration in the liquor – 42.6% – (Botello et al. 1999). Lignin was precipitated, from different liquor fractions after ultrafiltration, by using aqueous acidic solution at pH 3 as the antisolvent. After initial precipitation, the particles were allowed to settle slowly for 24 h during which time agglomeration occurred. Separation was carried out by centrifuging at 3000 rpm for 12 min. Residual sugars were separated

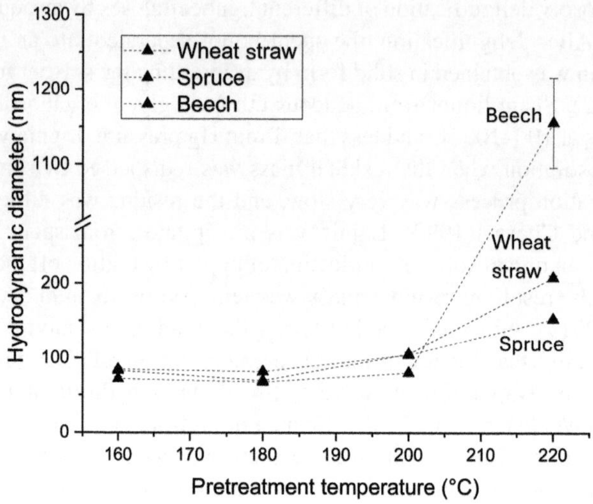

FIGURE 3.8 Dependence of mean particle size in lignin precipitate on process severity during aqueous ethanol organosolv delignification. (Reprinted from Adamcyk, J, Beisl, S, Amini, S, Jung, T, Zikeli, F, Labidi, J, Friedl, A, *Polymers*, 13(3), 384, 2021, originally published by MDPI under Creative Commons Attribution License.)

by additional water wash (Alriols et al. 2010). In a modification of the separation process, spent liquor from aqueous ethanol delignification of *Salix schwerinii* was diluted with water or aqueous ethanol at different pH. The diluting solution was saturated with nitrogen under pressure. Spent liquor was added to this at a controlled rate. As pressure was released, lignin was separated from the dispersion formed by air floatation. The authors found that nitrogen pressure during saturation, rate of spent liquor addition, and temperature were the factors controlling lignin yield (Macfarlane et al. 2009).

Lignin precipitation was achieved from sugarcane bagasse aqueous ethanol organosolv spent liquor by dilution with water at 20 °C. Dilution with four volumes of water for 1 volume of spent liquor precipitated 68% lignin – 55% of that present in the original bagasse. Lignin precipitation was 60% – 52% of that present in the original bagasse – when the dilution factor was 2. Centrifugation for 15 min at 3500 rpm separated the precipitate, and the lignin obtained was 80–90% pure (Fernando et al. 2010). The results of lignin precipitation and recovery, however, were strongly dependent on the characteristics of the spent liquor. Spent liquor obtained at a higher severity factor during delignification resulted in less lignin precipitation and recovery. Lignin from spent liquor of acid-catalysed aqueous ethanol organosolv delignification of miscanthus was precipitated by diluting one volume of liquor with three volumes of water (El Hage et al. 2010). Dilution of spent liquor from acid-catalysed spruce aqueous ethanol delignification with water in a 1:4 ratio precipitated the lignin (Parot et al. 2022). Lignin was precipitated from aqueous ethanol spent liquor of tulip tree delignification by diluting with water in a 1:2 ratio followed by 10 min centrifugation at 12,000 rpm. Lignin precipitation was 40–63% depending on the severity of the delignification process that produced the spent liquor (Jang et al. 2016). Aqueous ethanol organosolv spent liquor from wheat straw, spruce, and beech, 3.52 g/L lignin, was treated with water as an antisolvent to precipitate lignin. Five volumes of water were added to one volume of spent liquor at room temperature (Adamcyk et al. 2021). In a batch study, beech spent liquor from aqueous ethanol organosolv delignification was diluted with water to decrease the ethanol concentration from about 50% to 10% to form the initial lignin dispersion. The resulting slurry was then vacuum evaporated at 100 mbar and less than 50 °C temperature to half the initial volume to selectively evaporate ethanol. Finally, the lignin was filtered in a filter press (Schulze et al. 2016). Lignin was precipitated from acid-catalysed larch aqueous ethanol organosolv spent liquor by adding one volume of spent liquor into three volumes of water. 78% lignin was precipitated from the solution (Hochegger et al. 2019). Spent liquor from acid-catalysed aqueous ethanol or methanol organosolv delignification of beech was diluted with three volumes of water for one volume of liquor to precipitate lignin. The solids could be separated by 10 min centrifugation at 4500 rpm. The authors contended that lignin isolated by aqueous methanol was comparatively difficult to precipitate (Jasiukaitytė-Grojzdek et al. 2020). In the pilot plant study, spent liquor from aqueous ethanol organosolv delignification was evaporated in a falling film evaporator to preferentially distil off ethanol. The process resulted in a liquor with an increased dissolved solids content, increased temperature, and lower ethanol concentration. This liquor was fed into a stirred tank where lignin dispersion was formed. Water was added to the tank to further lower the ethanol concentration as required. Lignin dispersion product was withdrawn from the tank outlet with a part being recycled into the evaporator with the spent liquor feed. Vapours generated in the evaporator were condensed and reused in delignification (Schulze et al. 2016; Schulze et al. 2019).

3.3.3.2 Spent Liquor from Organic Acid Organosolv Processes

Water acts as an effective antisolvent for separating lignin from the spent liquors of carboxylic acid-based organosolv processes. Experiments suggest that concentrating the liquor by vacuum evaporation prior to dilution with water facilitates precipitation of lignin. It was found that the higher the liquor concentration, the lower the amount of water required to precipitate lignin. This makes for an interesting optimization as energy is required to concentrate the spent liquor while less addition of dilution water would result in a filtrate that is relatively concentrated. The temperature of the dilution water is another factor that affects precipitation; normally, a slightly elevated temperature is found to be beneficial. Most researchers recommend adding the concentrated spent liquor to water and not the other way around. A mild agitation is also prescribed in most cases followed by allowing the suspension to stand for 18–24 h to complete precipitation. For eucalyptus aqueous acetic acid spent liquor it was reported that if the liquor was concentrated to 10–20% of the initial mass, about 80% lignin could be precipitated by adding two volumes of water per volume of concentrated liquor (Vila et al. 2003). The effect of the ratio of dilution water to spent liquor on lignin precipitation is presented in Figure 3.9 for aqueous acetic acid organosolv spent liquor.

Marabou wood was delignified with aqueous acetic acid and the spent liquor was vacuum evaporated to one-third of the initial volume. Concentrated liquor was added to ten volumes of water at

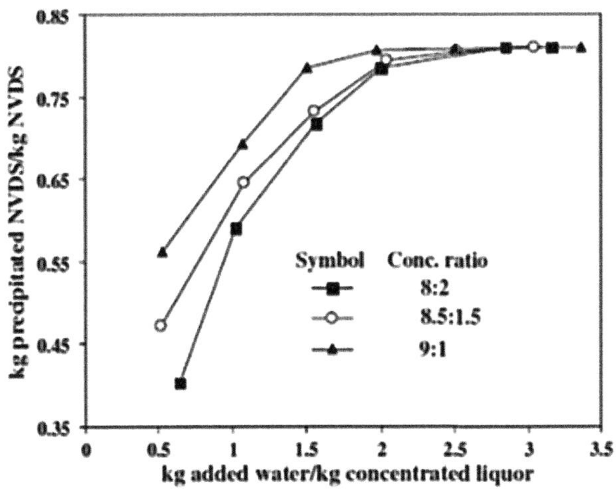

FIGURE 3.9 Variation in lignin precipitation with solvent evaporation and addition of dilution water to concentrated spent liquor. (Reprinted from Vila, C, Santos V, Parajo, JC, *Bioresource Technol.*, 90, 339–344, 2003, with permission from Elsevier.)

80 °C and then left standstill for 24 h (Soudham et al. 2011). Bamboo spent liquor from acid-catalysed aqueous acetic acid delignification was vacuum evaporated at 60 °C to less than one-third of the initial volume before adding to ten volumes of water to precipitate lignin (Li et al. 2012). Spent liquor from aqueous acetic acid delignification of spruce and corncob was concentrated to 10% of the initial amount. This was added under agitation to warm water at a 1:10 ratio over a period of 30 min to initiate precipitation. Precipitation was completed after letting the suspension stand for 18 h (Ditzel et al. 2017). Lignin was isolated from the spent liquors of aqueous acetic acid, and aqueous formic acid organosolv delignification of wheat straw by a combination of solvent evaporation and antisolvent (water) addition. First, the carboxylic acid was preferentially evaporated from the spent liquor at 320 mbar pressure and 60 °C temperature to obtain a predominantly aqueous concentrated liquor of 60% dissolved solids. Lignin precipitation was achieved by diluting the liquor with water in a 1:1 ratio (Snelders et al. 2014). Lignin precipitation was carried out from the spent liquors obtained from acid-catalysed delignification of olive trees by aqueous acetic or formic acid. Five volumes of water were added to one volume of spent liquor for lignin precipitation which was centrifuged at 5500 rpm for 15 min (Erdocia et al. 2014). Spent liquor obtained from the delignification of several agro-industrial wastes with an aqueous mixture of formic and acetic acids was diluted with water in a 1:9 ratio to precipitate the lignin (Manara et al. 2014). One volume of birch aqueous acetic acid organosolv spent liquor was diluted with five volumes of water to precipitate the dissolved lignin which was centrifuged for 7–27 min at 9000 rpm. 66–78% lignin precipitation was reported depending on the severity of the delignification reaction (Kangas et al. 2015). Similar results were reported for lignin isolation from spent liquor from delignification with an aqueous mixture of acetic and formic acid. Precipitation was carried out at 105 °C with five volumes of water per volume of liquor (Watkins et al. 2015). Aqueous acetic acid spent liquor from the treatment of sugarcane bagasse was first concentrated under vacuum. The concentrated liquor was then diluted with ten volumes of water at 60 °C to precipitate lignin (Pinheiro et al. 2017).

3.3.3.3 Spent Liquor from Other Organosolv Processes

Dilute hydrochloric acid of 0.005–0.1 weight% concentration was used as the antisolvent for lignin precipitation from spent liquor from aqueous ethylene glycol organosolv delignification. Spent liquor was slowly added to the hot acid under agitation. The best precipitation was observed at a spent-liquor-to-acid ratio of 3, an acid concentration of 0.05%, and a temperature between 50 and 60 °C. Precipitated lignin could be separated by filtration (Thring and Chornet 1990). Spent liquor from the delignification of palm oil empty fruit bunches with alkali-catalysed ethylene glycol was acidified to different pH with sulphuric acid to precipitate lignin. 97% pure lignin could be precipitated with 75% yield by acidifying to pH 2 (Chin et al. 2020). Efficient lignin precipitation and fractionation are reported from the organic phase of the spent liquor from aqueous 2-methyl tetrahydrofuran organosolv delignification of beech using n-pentane or n-hexane as the antisolvent. The antisolvent-to-tetrahydrofuran ratio was varied between 0.5 and 2.2. n-pentane caused better precipitation than n-hexane owing to lower lignin solubility in the former. Lignin fractionation can be effected by gradually increasing the antisolvent-to-tetrahydrofuran ratio (Weidener et al. 2020). Lignin was precipitated from wheat straw aqueous acetone and dioxane spent liquors by acidifying with HCl to pH 1.5–1.8. Solids were separated by filtration (Sun et al. 1998). In one protocol, wheat straw aqueous acetone organosolv spent liquor was slowly sprayed into cold water. Three volumes of water were used for one volume of liquor. The procedure not only precipitated lignin but also allowed the formed particles to float to the surface for easy separation afterwards. 60% lignin recovery is reported (Huijgen et al. 2010). Lignin could be precipitated from the spent liquor of aqueous acetone organosolv delignification of different lignocelluloses by using water as the antisolvent. Dilution to five times the initial volume at 4 °C was found to be effective (Smit and Huijgen 2017). Acidic water was the antisolvent for lignin recovery from spent liquor of cholinium ionic liquid-based organosolv delignification of several lignocelluloses. Lignin was precipitated when concentrated spent liquor was acidified to pH 0.9–2. Hemicelluloses were also coprecipitated. To obtain purified lignin, the concentrated spent liquor was first acidified to 6.8 pH and diluted with three volumes of 96% aqueous ethanol. This precipitated the hemicellulose, which was filtered, leaving behind the lignin in solution. Ethanol was evaporated from this solution under vacuum evaporation followed by acidification to pH 2. This precipitated the lignin which was separated by filtration (An et al. 2015). There are reports of organosolv delignification using a mixture of γ-valerolactone, benzene, and water as solvent at elevated temperatures. At 20 weight% water, the spent liquor remains as a single phase at high temperature but splits into two phases – organic and aqueous – at normal temperature (Motagamwala et al. 2016). The aqueous phase contains the sugars. Lignin can be precipitated from the organic phase by the addition of a suitable antisolvent like diethyl ether, n-butyl acetate, or toluene. 89% lignin precipitation was achieved from eucalyptus spent liquor with toluene as the antisolvent (Sun et al. 2018).

3.3.4 Lignin Separation by Membrane Filtration

Spent liquors from organosolv processes can be concentrated and fractionated by membrane filtration techniques. Ultrafiltration does not preferentially retain lignin. The obtained retentate still contains significant amounts of sugars and solvent and requires precipitation by the addition of antisolvent. However, the technique comes in handy in segregating the lignin into different molecular mass fractions by using a series of membranes.

3.3.4.1 Spent Liquor from Alcohol Organosolv Processes

Miscanthus aqueous ethanol organosolv spent liquor of 6.3% dissolved solids concentration was treated by sequential ultrafiltration with tubular ceramic membranes at 300 kPa TMP, and 5.6 m/s crossflow velocity. Spent liquor was fed into the 15,000 Da MWCO membrane and 63% lignin was retained – dissolved solids concentration 4.7% and lignin to sugars ratio 0.8 in the retentate. Permeate was fed into the second membrane of 10,000 Da MWCO and 20% lignin was retained – dissolved solids concentration of 5.9% and a lignin-to-sugar ratio of 1.33 in the retentate. Permeate from the second membrane was further fractionated with a membrane of 5000 Da MWCO retaining 11% lignin – dissolved solids concentration of 6.7% and a lignin-to-sugar ratio of 1.01 in the retentate (Alriols et al. 2010). Wheat straw aqueous ethanol organosolv spent liquor, with 60% ethanol and 1.01% lignin concentration, was concentrated and fractionated in a series of microfiltration, ultrafiltration, and nanofiltration. Starting with the microfiltration with 0.2 μm nylon filter, the permeate was fed into the next membrane of decreasing MWCO. Six ultrafiltration and two nanofiltration membranes were used. All the ultrafiltration membranes had active layers of PES with MWCO of 15,0000, 50,000, 30,000, 5000, 4000, and 1000 Da. Nanofiltration membranes had polyamide active layer and MWCO of 150–300, and 100–200 Da. The first ultrafiltration stage, operating at 5.5 bar TMP, retained 52% lignin at a volume reduction of 18.5%. Lignin retention was negligible during ultrafiltration with 50,000, 30,000, and 4000 Da MWCO membranes, operating at TMP of 1, 1, and 3 bar, respectively. 5000, and 1000 Da MWCO membranes retained about 6% lignin each with volume reductions of 91.6%, and 91.1%. Nanofiltration membrane of 150–300 Da MWCO, operating at 6 bar TMP and 88.9% volume reduction, retained 60% of lignin fed. The last nanofiltration membrane of 100–200 Da MWCO operated at 6 bar TMP and 87.5% volume reduction, retaining 41% of input lignin. All the retentates combined separated 90% lignin from the microfiltered liquor (Weinwurm et al. 2016).

REFERENCES

Adamcyk, J, Beisl, S, Amini, S, Jung, T, Zikeli, F, Labidi, J, Friedl, A. *Polymers* 13:384, 2021.
Alexandri, M, Papapostolou, H, Komaitis, M, Stragier, L, Verstraete, W, Danezis, GP, Georgiou, CA, Papanikolaou, S, Koutinas, AA. *Bioresource Technol.* 214:504–513, 2016.
Alén, R, Sjöström, E, Vaskikari, P. Cellulose *Chem. Technol.* 19(5): 537–541, 1985.

Alriols, MG, Garcia, A, Llano-ponte, R, Labidi, J. *Chem. Eng. J.* 157:113–120, 2010.
Al-Rudainy, B, Galbe, M, Wallberg, O. *Molecules* 25:3435, 2020.
An, YX, Zong, MH, Wu, H, Li, N. *Bioresource Technol.* 192:165–171, 2015.
Arkell, A, Olsson, J, Wallberg, O. *Chem. Eng. Res. Des.* 92:1792–1800, 2014.
Atech Data Sheet, n.d. https://www.atech-innovations.com/fileadmin/downloads/atech_innovations_product_data_en.pdf
Axelsson, E, Olsson, MR, Berntsson, T. *Nord. Pulp Pap. Res. J.* 21:485–492, 2006.
Botello, JI, Gilarranz, MA, Rodriguez, F, Oliet, M. *Sep. Sci. Technol.* 34:2431–2445, 1999.
Cao, C, Zhang, Y, Li, L, Wei, W, Wang, G, Bian, C. *Int. J. Hyd. En.* 44:15737–15745, 2019.
Cardoso, M, de Oliveira, ED, Passos, ML. *Fuel* 88:756–763, 2009.
Carman, PC. *Transact. Inst. Chem. Eng.* 15:150–166, 1937.
Chakrabarty, K, Saha, P, Ghoshal, AK. *J. Membrane Sci.* 340:84–91, 2009a.
Chakrabarty, K, Krishna, KV, Saha, P, Ghoshal, AK. *J. Membrane Sci.* 330:135–144, 2009b.
Chin, DWK, Lim, S, Pang, YL, Lim, CH, Shuit, SH. *IOP Conf. Ser.: Earth Environ. Sci.* 463:012003, 2020.
Costa, HF, Egas, APV, Ferreira, AGM, Lobo, LQ. *Appl. Rheol.* 21:42533, 2011.
Costa, CAE, Pinto, PCR, Rodrigues, AE. *Sep. Purific. Technol.* 192:140–151, 2018.
Cybulska, I, Brudecki, G, Rosentrater, K, Julson, JL, Lei, H. *Bioresource Technol.* 118:30–36, 2012.
Demmelmayer, P, Hilgert, J, Wijaya, R, Kienberger, M. *Ind. Eng. Chem. Res.* 59:16420–16426, 2020.
Dessbesell, L, Paleologou, M, Leitch, M, Pulkki, R, Xu, C. *Ren. Sustain. En. Rev.* 123:109768, 2020.
Ditzel, FI, Prestes, E, Carvalho, BM, Demiate, IM, Pinheiro, LA. *Carbohydrate Polym.* 157:1577–1585, 2017.
Domínguez-Robles, J, Sánchez, R, Espinosa, E, Savy, D, Mazzei, P, Piccolo, A, Rodríguez, A. *Int. J. Mol. Sci.* 18:327–343, 2017.
dos Santos, PSB, Erdocia, X, Gatto, DA, Labidi, J. *Ind. Crops Prod.* 55:149–154, 2014.
Ela, RCA, Spahn, L, Safaie, N, Ferrier Jr, RC, Ong, RG. *ACS Sust. Chem. Eng.* 8:13997–14005, 2020.
El Hage, R, Brosse, N, Sannigrahi, P, Ragauskas, A. *Pol. Deg. Stabil.* 95:997–1003, 2010.
Erdocia, X, Prado, R, Corcuera, MA, Labidi, J. *J. Ind. Eng. Chem.* 20:1103–1108, 2014.
Fatehi, P, Ni, Y. Integrated Forest Biorefinery – Sulfite Process. Zhu, J, Zhang, X, Pan, X, (Eds.) Sustainable Production of Fuels, Chemicals, and Fibers from Forest Biomass. *ACS Symposium Series, American Chemical Society, Washington DC*, 409–441, 2011.
Fernandes, DLA, Silva, CM, Xavier, AMRB, Evtuguin, DV. *J. Biotechnol.* 162:415–421, 2012.
Fernández-Rodríguez, J, Garcia, A, Coz, A, Labidi, J. *Chem. Eng. Transact.* 45:553–558, 2015.
Fernando, EF, Vallejos, EM, Area, MC. Cellulose *Chem. Technol.* 44(9):311–318, 2010.
Garver, TM, Callaghan, PT. *Macromolecules* 24:420–430, 1991.
Ghatak, HR. Isolation of lignin from alkaline pulping liquors. Paterson, RJ, (Ed.) *Lignin: properties and applications in biotechnology and bioenergy*. Nova Science Publishing Inc., New York, USA, 257–277, 2012.
Gil-Chávez, J, Padhi, SSP, Hartage, U, Heinrich, S, Smirnova, I. *Adv. Powder Technol.* 31:2348–2356, 2020.
Gong, C, Bujanovic, BM. *Energies* 7:857–873, 2014.
Haddad, M, Labrecque, R, Bazinet, L, Savadogo, O, Paris, J. *Chem Eng. J.* 304:977–985, 2016.
Harner, NK, Bajwa, PK, Habash, MB, Trevors, JT, Austin, GD, Lee, H. *Antonie van Leeuwenhoek* 105: 29–43, 2014.
He, Q, Chen, H. *J. Cleaner Prod.* 263:121356, 2020.
Helle, SS, Murray, A, Lam, J, Cameron, DR, Duff, SJB. *Bioresource Technol.* 92:163–171, 2004.
Helle, SS, Lin, T, Duff, SJB. *Enzyme Microbial Technol.* 42:259–264, 2008.
Hochegger, M, Cottyn-Boitte, B, Cezard, L, Schober, S, Mittelbach, M. *Int. J. Chem. Eng.* 2019:1734507, 2019.
Hu, G, Hu, J, Chen, H, Song, S, Chu, F. *BioResources* 16:7608–7622, 2021.
Huijgen, WJJ, Reith, JH, Uil, HD. *Ind. Eng. Chem. Res.* 49:10132–10140, 2010.
Humpert, D, Ebrahimi, M, Czermak, P. *Membranes* 6:42, 2016.
Humpert, D, Ebrahimi, M, Stroh, A, Czermak, P. *Membranes* 9:45, 2019.
Jang, SK, Kim, HY, Jeong, HS, Kim, JY, Yeo, H, Choi, IG. *Ren. Energy* 87:599–606, 2016.
Jardim, JM, Hart, PW, Lucia, LA, Jameel, H, Chang, HM. *Fibers* 10:16, 2022.
Jasiukaitytė-Grojzdek, E, Huš, M, Grilc, M, Likozar, B. *Sci. Rep.* 10:11037, 2020.
Jin, W, Tolba, R, Wen J, Li, K, Chen, A. *Electrochim. Acta* 107:611–618, 2013.
Jönsson, AS, Nordin, AK, Wallberg, O. *Chem. Eng. Res. Des.* 86:1271–1280, 2008.
Kangas, H, Liitiä, T, Rovio, S, Ohra-aho, T, Heikkinen, H, Tamminen, T, Poppius-Levlin, K. *Holzforschung* 69(3):247–256, 2015.
Kannangara, M, Marinova, M, Fradette, L, Paris, J. *Chem. Eng. Res. Des.* 105:94–106, 2016.
Kienberger, M, Demmelmayer, P, Weißl, M, Zankl, A, Spirk, S. *Solv. Extract. Ion Exchange* 38(1):132–141. 2020.

Kontturi, AK, Sundholm, G. *Acta Chem. Scand.* 40a:121–125, 1986.
Kouisni, L, Paleologou, M. US Patent No. 8,771,464, 2014.
Kouisni, L, Gagne, A, Maki, K, Holt-Hindle, P, Paleologou, M. *ACS Sust. Chem. Eng.* 4:5152–5159, 2016.
Kumar, H, Alen, R, Sahoo, G. *BioResources* 11:9869–9879, 2016.
Kumar, V, Singh, RK, Chowdhury, P. *J. Ind. Eng. Chem.* 67:109–122, 2018.
Ladakis, D, Michailidi, K, Vlysidis, A, Koutinas, A, Kookos, IK. *Biochem. Eng. J.* 137:262–272, 2018.
Lake, MA, Blackburn, JC. *Cellulose Chem. Technol.* 48:799–804, 2014.
Lake, MA, Blackburn, JC. US Patent No. 9260464, 2016.
Lauwaert, J, Stals, I, Lancefield, CS, Deschaumes, W, Depuydt, D, Vanlerberghe, B, Devlamynck, T, Bruijnincx, PCA, Verberckmoes, A. *Sep. Purific. Technol.* 221:226–235, 2019.
Li, MF, Sun, SN, Xu, F, Sun, RC. *J. Agri. Food Chem.* 60:1703–1712, 2012.
Lourençon, TV, Hansel, FA, da Silva, TA, Ramos, LP, de Muniz, GIB, Magalhães, WLE. *Sep. Purific. Technol.* 154:82–88, 2015.
Loutfi, H, Blackwell, B, Uloth, V. *Tappi J.* 74(1):203–210, 1991.
Macfarlane, AL, Prestidge, R, Farid, MM, Chen, JJJ. *Chem. Eng. J.* 148:15–19, 2009.
Manara, P, Zabaniotou, A, Vanderghem, C, Richel, A. *Catal. Today* 223:25–34, 2014.
Manttari, M, Lahti, J, Hatakka, H, Louhi-Kultanen, M, Kallioinen, M. *J. Membrane Sci.* 490:84–91, 2015.
Matsakas, L, Raghavendran, V, Yakimenko, O, Persson, G, Olsson, E, Rova, U, Olsson, L, Christakopoulos, P. *Bioresource Technol.* 273:521–528, 2019.
McCabe, WL, Smith, JC, Harriot, P. *Unit Operations of Chemical Engineering*. McGraw Hill, Singapore, 1016–1021, 1993.
Mendes, SF, Rodrigues, JS, de Lima, VH, Botaro, VR, Cardoso, VL, Reis, MHM. *Appl. Biochem. Biotechnol.* 193:3079–3097, 2021.
Meng, X, Parikh, A, Seemala, B, Kumar, R, Pu, Y, Wyman, CE, Cai, CM, Ragauskas, AJ. *Bioresource Technol.* 272:202–208, 2019.
Meyer, JR, Li, H, Zhang, J, Foston, MB. *ChemSusChem* 13:4557–4566, 2020.
Motagamwala, AH, Won, W, Maravellias, CT, Dumesic, JA. *Green Chem.* 18:5756–5763, 2016.
Norgren, M, Lindstrom, B. *Holzforschung* 54:519–527, 2000.
Norgren, M, Edlund, H, Wågberg, L. *Langmuir* 18:2859–2865, 2002.
Norgren, M, Mackin , S. *Ind. Eng. Chem. Res.* 48:5098–5104, 2009.
Öhman, F, Wallmo, H, Theliander, H. *Nord. Pulp Pap. Res. J.* 22:188–193, 2007.
Öhman, F, Theliander, H, Tomani, P, Axegard, P. US Patent No. 8486224, 2013.
Öhman, F, Theliander, H, Norgren, M, Tomani, P, Axegard, P. US Patent No. 8815052, 2014.
Olsson, MR, Axelsson, E, Berntsson, T. *Nord. Pulp Pap. Res. J.* 21:476–484, 2006.
Parot, M, Rodrigue, D, Stevanovic, T. *Bioresource Technol. Reports* 17:100880, 2022.
Padhi, SSP, Bartolome, MJ, Nyanhongo, Schwaiger, N, Pellis, A, van Herwijnen, HWG, Guebitz, GM. *ACS Omega* 7:23749–23758, 2022.
Pereira, SR, Nogué, VS, Frazão, CJR, Serafim, LS, Gorwa-Grauslund, MF, Xavier, AMRB. *Biotechnol. Biofuels* 8:50, 2015.
PCI Product Information, n.d. https://www.pcimembranes.com/wp-content/uploads/2022/02/PCI_Tubular_Membranes.pdf
Pichler, T, Maitz, S, Kienberger, M. *Holzforschung* 74:286–292, 2019.
Pinheiro, FGC, Soares, AKL, Santaella, ST, e Silva, LMA, Canuto, KM, Cáceres, CA, Rosa, MD, Feitosa, JPD, Leitão, RC. *Ind. Crops Prod.* 96:80–90, 2017.
Restolho, JA, Prates, A, de Pinho, MN, Afonso, MD. *Biomass Bioen.* 33:1558–1566, 2009.
Ringena, O, Saake, B, Lehnen, R. *Holzforschung* 59:604–611, 2005.
Rueda, C, Calvo, PA, Moncalián, G, Ruiz, G, Coz, A. *J. Chem. Technol. Biotechnol.* 90:2218–2226, 2014.
Schlackl, K, Bischof, RH, Fackler, K, Samhaber, W. *Sep. Purific. Technol.* 250:117177, 2020.
Schulze, P, Seidel-Morgenstern, A, Lorenz, H, Leschinsky, M, Unkelbach, G. *Bioresource Technol.* 199: 128–134, 2016.
Schulze, P, Leschinsky, M, Seidel-Morgenstern, A, Lorenz, H. *Ind. Eng. Chem. Res.* 58:3797–3810, 2019.
Sewring, T, Durruty, J, Schneider, L, Mattsson, T, Theliander, H. *J. Wood Chem. Technol.* 39:1–13, 2019.
Singh, S, Ghatak, HR. *J. Wood Chem. Technol.* 37:407–422, 2017.
Smit, A, Huijgen, W. *Green Chem.* 19:5505–5514. 2017.
Snelders, J, Dornez, E, Benjelloun-Mlayah, B, Huijgen, WJJ, de Wild, PJ, Gosselink, RJA, Gerritsma, J, Courtin, CM. *Bioresource Technol.* 156:275–282, 2014.
Soudham, VP, Rodriguez, D, Rocha, GJM, Taherzadeh, MJ, Martin, C. *Forest. Stud. China* 13(1):64–70, 2011.

Spectrum Magazine 2017. 35(1):48. Available online https://www.andritz.com/resource/blob/228380/72f6d8956fde1bbcf0c9ddad33da758e/spectrum-2017-issue-35-en-data.pdf
Sun, RC, Lawther, JM, Banks, WB. *Wood Fiber Sci.* 30(1):56–63, 1998.
Sun, W, Trevorah, R, Othman, MZ. *Bioresource Technol.* 269:255–261, 2018.
Theliander, H. *Proc. International Chemical Recovery Conference, Williamsburg, VA*. 2: 33–42, 2010.
Thring, RW, Chornet, E. *Biomass* 23:289–305, 1990.
Tomani, P. *Cellulose Chem. Technol.* 44:53–58, 2010.
Treybal, RE. *Mass Transfer Operations*. McGraw Hill, Singapore, 477–520, 1981.
Velez, J, Thies, MC. *J. Wood Chem. Technol.* 36:27–41, 2016.
Vila, C, Santos, V, Parajo, JC. *Bioresource Technol.* 90:339–344, 2003.
Watkins, D, Nuruddin, M, Hosur, M, Tcherbi-Narteh, A, Jeelani, S. *J. Mater. Res. Technol.* 4(1):26–32, 2015.
Weidener, D, Holtz, A, Klose, H, Jupke, A, Leitner, W, Grande, PM. *Molecules* 25:3330, 2020.
Weinwurm, F, Drljo, A, Waldmüller, W, Fiala, B, Niedermayer, J, Friedl, A. *J. Cleaner Prod.* 136:62–71, 2016.
Wising, U, Algehed, J, Berntsson, T, Delin, L. *Tappi J.* 5(1):3–8, 2006.
Xavier, AMRB, Correia, MF, Pereira, SR, Evtuguin, DV. *Bioresource Technol.* 101:2755–2761, 2010.
Yue, X, Du, X, Xu, Y. *BioResources* 12:9357–9365, 2017.
Zhu, W, Theliander, H. *BioResources* 10:1696–1714, 2015.
Zijlstra, DS, Lahive, CW, Analbers, CA, Figueirêdo, MB, Wang, Z, Lancefield, CS, Deuss, PJ. *ACS Sustainable Chem. Eng.* 8:5119–5131, 2020.

4 Other Methods of Lignin Isolation

Lignin isolation methods discussed in Chapters 2 and 3 have a common pattern in them. The processes are based on reacting the lignocellulose with a chemical that preferentially reacts with lignin, causing it to pass into a solution. Lignin is then isolated from this solution, the spent liquor from the delignification process, by suitable physicochemical means. As of now, the commercial and pilot scale lignin isolation facilities are based on these methods only. However, during delignification, lignin undergoes mild-to-extensive chemical modification so that the substance finally obtained is much different from the original material present in the lignocellulose – the native lignin. The quest to isolate the lignin in its pristine unaltered state from the lignocellulose has always motivated the scientific community. This not only enables us to have a more comprehensive understanding of lignin chemistry but also opens new avenues for its utilization and value addition. In this context, two methods are worth mentioning that isolate lignin from lignocellulosic materials with minimal alteration to its chemical nature. The first method is based on the hydrolysis of carbohydrates present in the lignocellulose – cellulose and hemicellulose – that leaves a predominantly lignin-rich residue. The second method is based on extracting lignin from finely ground lignocellulose material with a suitable aprotic non-swelling liquid.

4.1 ISOLATION THROUGH CARBOHYDRATE HYDROLYSIS FROM LIGNOCELLULOSES

Conceptually, hydrolysing the carbohydrate fraction of the lignocellulose will leave the lignin as a solid residue. Practically, though, it is not an easy proposition given the intricate association between lignin and the carbohydrates, mainly the hemicelluloses. The most celebrated lignin determination method, attributed to Klason, is based on dissolving the carbohydrate in lignocellulose with concentrated sulphuric acid (Klason 1908). In this method, powdered extractive-free lignocellulose is digested with concentrated sulphuric acid of 72% concentration to degrade and dissolve the carbohydrates, both cellulose and hemicellulose. After filtration and washing, the gravimetrically assayed solid residue is taken as the quantitative estimation of lignin in the original lignocellulose. In the Tappi Standard Test Method, T 222 om-02, powdered lignocellulose is digested in 72% sulphuric acid, 15 mL/g lignocellulose, for 2 h at 20 °C. The digested material is added to distilled water to dilute to an acid concentration of 3% and boiled under reflux for 4 h. The lignin, obtained as a solid residue, is washed free of acid and assayed gravimetrically (Tappi Test Method 2006). Among the processes capable of hydrolysing the carbohydrates in lignocellulose, acid hydrolysis and enzymatic hydrolysis are the most important. Acid hydrolysis is the oldest, with attractive monosaccharide yields. Based on the acid concentration used, we can have dilute acid hydrolysis or concentrated acid hydrolysis. Generally, the physical and chemical recalcitrance of carbohydrates present in lignocellulose to hydrolysis can be overcome with sufficiently strong reaction conditions. The corrosive nature of the acids can act as an impediment to the process requiring specific construction materials. Downstream processing also requires additional care, owing to the acidic nature of the spent liquid and solid residues. Enzymatic hydrolysis processes are more benign in terms of process conditions used. However, lignocellulose recalcitrance is to be overcome. In the past, enzymatic hydrolysis was not as complete as acid hydrolysis. With biotechnological advancements, these constraints are largely removed.

DOI: 10.1201/9781003598664-5

4.1.1 Dilute Acid Hydrolysis of Lignocelluloses

In dilute acid hydrolysis, the lignocellulose is treated with a mineral acid at low concentration. To compensate for low acid concentration, the process often operates at elevated temperatures and may require a longer reaction time. The primary target of dilute acid hydrolysis is the hemicelluloses due to their higher reactivity and ease of hydrolysis. Cellulose is comparatively difficult to hydrolyse and complete conversion to monosaccharides is rare. One of the commercially used dilute acid hydrolysis processes is the Scholler–Tornesch process. This was invented in Germany in 1926 by Heinrich Scholler and the first commercial plant was started in Tornesch in 1931. Originally developed for wood, this process is based on treating the lignocellulose material with dilute sulphuric acid of 0.2–1.2% concentration at 170–180 °C temperature. The heart of the process was a pressurized acid-resistant brick-lined percolator to contact the wood sawdust with dilute sulphuric acid. Initially, the raw material mass was preheated to 130 °C with live low-pressure steam. Dilute sulphuric acid of 0.5% concentration was added in successive batches while increasing the temperature to 175 °C. Each batch of acid was allowed to contact the lignocellulose for 45 min and the hydrolysate was extracted thereafter, replacing it with a fresh batch of dilute acid (Faith 1945). Some of the modifications to the process, suggested by the American investigators, included a shorter preheating pressure, with the use of steam at higher temperatures, and pressing the resulting lignin residue to squeeze the liquid out of it (Faith and Hall 1944). In the modification suggested by Harris and Begliuger, known as the Madison wood-sugar process, less heating requirement and better sugar yield in the hydrolysate were claimed (Harris and Begliuger 1946). The solid residue left after the process is predominantly lignin with inorganic and organic impurities. Organic impurities are mainly the non-hydrolysed or partially hydrolysed polysaccharides. Some of the polysaccharide impurities have sulphur in them as the hydroxyl groups present in the polysaccharide molecules are also sulphated during the process, schematically represented as (Evdokimov et al. 2019)

$$ROH + H_2SO_4 \rightarrow ROSO_3H + H_2O \quad \{4.1\}$$

Research interest in dilute acid hydrolysis of lignocelluloses has been revived in recent years with an emphasis on second-generation biofuel production. Most of the studies focus on the production of monomer sugars and the lignin isolation is not investigated. Nevertheless, there are a few studies which explore the aspect of lignin separation. Dried potato peels milled to 1 mm particle size were subjected to dilute phosphoric acid hydrolysis. The material was hydrolysed at a liquid-to-solid ratio of 19 with acid concentration varied between 2.5 and 10 weight% and temperature studied in the range of 135–200 °C. About 40–50% of the starting lignocellulose was left as lignin-rich solid residue (Lenihan et al. 2010).

4.1.2 Concentrated Acid Hydrolysis of Lignocelluloses

The development of concentrated acid hydrolysis of lignocelluloses is attributed to Bergius, another German scientist of the early 19th century. In this process, the lignocellulose is treated at ordinary temperature with 40% concentration fuming hydrochloric acid. Several reactors, 8–10, are operated in series. Fresh acid comes in contact with the lignocellulose in one reactor. Liquid from this reactor, consisting of residual acid and hydrolysate, is fed into the next reactor where it comes in contact with a fresh batch of lignocellulose. After each reactor, the sugar concentration in the hydrolysate increases. The process goes on until the concentrated hydrolysate is obtained from the last reactor. After hydrolysis, the solid residue from each reactor, which is mainly lignin, is washed with hydrochloric acid to remove any associated sugars and the washings are mixed with the hydrolysate from the last reactor. The techno-economic success of the process lies in the possibility of recovering and recycling the acid by vacuum distillation. Compared to sulphuric acid-based dilute acid hydrolysis, the process has faster and more complete hydrolysis due to high acid concentration and less resistance to mass transfer for penetration of hydrochloric acid into the lignocellulose matrix. The first commercial facility was built in Rheinau in Germany and the process came to be known as the Bergius Rheinau process (Bergius 1937). The process distilled the hydrochloric acid from the sugar solution under vacuum at 36 °C using evaporators having heat conducting ceramic tubes.

The initial success of the Bergius Rheinau process notwithstanding, recovering and recycling the acid always posed a bottleneck for the wide application of the process. Hydrochloric acid forms an azeotrope with water at around 25% concentration, which sets the upper limit in ordinary distillation. One way to overcome this problem is to use reactive liquid extraction where a predominantly water-insoluble liquid is used as a solvent to extract the hydrochloric acid from the aqueous phase. This can be accomplished by using a solution of tris-2-ethylhexyl amine and octanoic acid in dodecane. The amine, carboxylic acid, and hydrochloric acid form a complex through polar interactions involving nitrogen, hydrogen, oxygen, and chlorine atoms. The extract phase is stripped of hydrochloric acid by passing xylene vapours through it. This is the basis of the proprietary Virdia process in which steam-exploded lignocellulose material is first extracted with a suitable solvent to remove the extractives. The extractive-free material is then subjected to concentrated acid hydrolysis. The possibility of efficient acid recovery without energy-intensive distillation allows for very thorough washing of the lignin residue and the resulting product is substantially free from acids (Zviely 2013).

Another variation of the Bergius Rheinau process is the Udic Rheinau process. Here, hydrolysis is split into three steps. In the first step, the lignocellulose is hydrolysed by treating it with 35% hydrochloric acid at normal temperature. In the second stage, the hydrolysed solids from the first stage are treated with 40% hydrochloric acid at normal temperature. In the third and final stage, the solids from the second stage are treated with dilute acid at elevated temperature (Radillo et al. 2011). In the improved variation of the Udic Rheinau process, there are only two hydrolysis stages. The first stage is a prehydrolysis stage with 1% hydrochloric acid at 130 °C, targeting mainly the hemicellulose. The solid residue from the prehydrolysis stage is then hydrolysed with 40% hydrochloric acid at normal temperature (Kucuk and Demirbas 1997). In the Prodor process, the reactor consists of a vertical multistage tray tower where the lignocellulose to be hydrolysed is contacted with hydrogen chloride gas in a countercurrent fashion (Motoyoshi et al. 1959). Lignocellulose material, ground into small particle size, is fed to the top tray and moves downwards while hydrogen chloride gas enters the column at the bottom and moves upwards. During its descent through each tray, the lignocellulose successively contacts the hydrogen chloride gas coming from the tray below. Completely hydrolysed solids are discharged from the bottom tray. The use of hydrogen chloride gas requires prodorite cement as the construction material to withstand the severely corrosive environment. The Hering process is a variation of the Prodor process where the lignocellulose material is first impregnated with concentrated hydrochloric acid and fed to the top tray. It moves downwards through successive trays. Hydrogen chloride gas is fed to the reactor from the bottom. As it moves upwards, it contacts the solids in each tray. Completely hydrolysed solids are discharged from the bottom tray. Typical residence time is 8 h (Kucuk and Demirbas 1997; Oshima et al. 1959).

There are processes that use concentrated sulphuric acid in place of hydrochloric acid for the hydrolysis of polysaccharides in lignocelluloses. The Hokkaido process is a multistep process that employs a steam prehydrolysis step at 180 °C to remove the hemicellulose (Clausen et al. 1993). The hydrolysate, mainly consisting of furfural, is extracted by pressing and the prehydrolysed solid material is treated with 80% sulphuric acid at normal temperature. The lignin residue is washed, and the washings are mixed with the hydrolysate. If the prehydrolysis step is carried out with 1.2–1.5% sulphuric acid at 140 °C, the prehydrolysate contains xylose instead of furfural.

In recent years, concerns for sustainable development have revived the research interest in concentrated acid hydrolysis of lignocellulose. *Lupinus rotundiflorus*, a leguminous plant native to Central America, was hydrolysed by concentrated hydrochloric acid using the Udic Rheinau protocol. The first stage of two-stage hydrolysis consisted of the treatment of lignocellulose for 24 h at 20 °C with a 32% concentration acid. 5 mL of acid was used for each gram of lignocellulose. Solid residue from the first stage was treated with 42.6% acid for 24 h at 20 °C. The volume of acid was the same as in the first stage. The solid residue after two-stage hydrolysis was 22.5% of the original lignocellulose mass with 64% lignin purity, which amounted to 92% lignin recovery (Radillo et al. 2011). Ground lignocellulose of 1 mm particle size from softwood, hardwood, and oil palm empty fruit bunch were hydrolysed with two-stage concentrated acid hydrolysis. In the first stage, acid concentration was 65–80%, the liquid-to-solid ratio was 2, the reaction temperature was 30 °C, and the reaction time was 30 min. After the first stage, the reactor content was diluted with water to reduce the acid concentration to 30% and hydrolysis continued at 80–100 °C. The best results in terms of polysaccharide hydrolysis were obtained at 70% first-stage acid concentration and second-stage hydrolysis at 80 °C (Wijaya et al. 2014).

4.1.3 Enzymatic Hydrolysis of Lignocelluloses

Enzyme systems present in various microorganisms are capable of hydrolysing the polysaccharides in lignocelluloses to the corresponding monomer sugars. These processes are much sought after nowadays for the development of second-generation lignocellulosic biorefineries. Compared to acid hydrolysis, enzyme-driven processes require mild operating conditions making them environmentally more benign. On the downside, enzymatic hydrolysis is slow compared to acid hydrolysis and often requires pretreatment to overcome the recalcitrance of lignocellulose to enzymatic action. Moreover, the lignin carbohydrate complexes are not effectively hydrolysed. Cellulolytic enzyme systems comprise a mixture of individual enzymes that together convert the cellulose to glucose. These are endoglucanase, exoglucanase, and β-glucosidase (Bhardwaj et al. 2021). Commercially available cellulase enzyme systems are described in Table 4.1.

There are several early reports of enzymatic hydrolysis of lignocellulose to isolate lignin. According to one study, 48 h ball-milled spruce or sweetgum sawdust was added to 40 g/L Onozuka SS 1500 enzyme solution buffered to 4.8 pH with a mixture of sodium acetate and acetic acid. The ratio of enzyme solution to lignocellulose was 20. Incubation was done at 50 °C for 72 h. After the first stage of incubation, the solid residue was filtered and added to another serving of enzyme solution half the amount of the first stage. It was then incubated for another 72 h. At the end of the second

TABLE 4.1
Commercial Cellulolytic Enzyme Systems for Hydrolysis of Lignocelluloses

Enzyme System	Producer	Source Microorganism	Main Enzymes	Reference
Accellerase® 1500	Genencor	*Trichoderma reesei*	Exoglucanase, endoglucanase, hemi-cellulase, and β-glucosidase	Genencor technical document 2009
Cellic® CTec2	Novozymes	*T. reesei, Aspergillus niger*	Cellulases, β-glucosidases, hemi-cellulases	Novozymes, n.d.-a; Rodrigues et al. 2015
Cellic® CTec3 HS	Novozymes	*T. reesei, A. niger*	Cellulases, β-glucosidases, hemi-cellulases	Novozymes, n.d.-b; Tian et al. 2017
Multicel	American Biosystems	*Trichoderma viridae*	Cellulase, hemi-cellulase	American Biosystems Product Specification Sheet 2015
Cellulysin	Calbiochem	*T. viridae*	Cellulase, hemi-cellulase	Merc data sheet, n.d.; Rencoret et al. 2015.
FibreZyme® G4	Dyadic	*Myceliophthora thermophila*	Cellulase	Dyadic product information 2014; Hu et al. 2006.
Novozyme 188	Novozymes	*A. niger*	β-glucosidases	Jiang et al. 2013

stage of enzymatic treatment, lignin was recovered as a solid residue by centrifugation for further purification (Chang et al. 1975). Extractive-free pine sawdust was extensively milled by different milling techniques and subjected to enzymatic hydrolysis with Dyadic Fibrezyme cellulase enzyme. 100 g/L suspension of milled wood was prepared in acetate buffer to pH 4.9, added with enzyme, and incubated for 24 h at 50 °C. Solids were centrifuged and hydrolysed a second time using the same procedure. Finally, the solid residue was water-washed. The lignin was purified by two-stage extraction with 96% aqueous dioxane. Purified lignin was precipitated from the combined extracts by adding dropwise into excess deionized water. The highest lignin yield of 86% was obtained by planetary ball milling the wood for 15 h in an argon environment (Hu et al. 2006). Spruce wood meal after extractives removal was subjected to vibratory ball milling for 2–96 h. The resulting material was hydrolysed with a cellulase enzyme dose of 40 Filter Paper Units (FPU) per gram wood buffered to a pH of 4.5 with citrate buffer. Incubation time was 48 h at 40 °C. Crude lignin, centrifuged from the broth, was washed with dilute acid of pH 2. It was further purified with 85% v/v aqueous dioxane acidified with 0.01 M HCl under argon. The supernatant was separated by centrifugation, neutralized with sodium bicarbonate, and added dropwise to excess of dilute acid of pH 2, to precipitate lignin. Lignin yields varied between 15% and 75% with increased milling but the corresponding purities remained relatively non-variant at about 80% (Guerra et al. 2006). 1 mm size miscanthus powder was ball-milled for 2 h and made into suspension of 10 g/L in a citric acid buffer of pH 5. Enzyme doses were 20 FPU cellulase and 40 IU β-glucosidase per g lignocellulose. Hydrolysis was performed for 48 h at 200 rpm and 50 °C. The residue was water-washed, dried, and extracted with 96% aqueous dioxane for 48 h. The extract was evaporated to dryness. 26% lignin at 87% purity was isolated (Wang et al. 2012). Dilute acid pretreated or ionic liquid pretreated sugarcane bagasse was subjected to enzymatic hydrolysis with a cellulase enzyme mixture of Celluclast 1.5L, and Novozyme188. The objective of the pretreatment was to hydrolyse the hemicellulose. Enzyme loading was 60 FPU/g pretreated lignocellulose. Reaction broth consisted of 98 mL of 50 mM citrate buffer at pH 4.8, and 1 mL of 2% sodium azide solution for each gram of biomass. The reaction was performed at 50 °C and 90 rpm agitation for 72 h (Jiang et al. 2013). Ground corncob residue was prepared into a suspension of 50 g/L in sodium acetate solution buffered to pH 4.8 with the addition of 1.8 g/L tea saponin as a natural surfactant. Hydrolysis was carried out with a cellulase mixture of Celluclast 1.5L, and Novozyme188 at an enzyme dose of 7 FPU/g cellulose. Hydrolysis was completed in 120 h at 40 °C under agitation of 180 rpm. The residue was separated by centrifugation. It was observed that the hydrolysis was better in the presence of surfactant than in its absence (Feng et al. 2013). Milled brewers spent grain was enzymatically hydrolysed in three stages with Cellulysin, a mixture of cellulase and hemicellulase enzymes. Enzyme dose was 4% in each stage based on raw material mass. The reaction broth was buffered to pH 5 with sodium acetate buffer. For each stage, the incubation time was 48 h at 30 °C. Solids were centrifuged for 20 min at 8000 rpm from the reaction broth after each stage and mixed with an additional dose of enzyme and buffer. Complete lignin isolation with 95% purity was reported (Rencoret et al. 2014). For isolating enzymatic hydrolytic lignin, dewaxed poplar sawdust of 20–40 mesh was milled at 500 rpm in a planetary ball mill for 5 h in nitrogen atmosphere. The milled lignocellulose was hydrolysed with cellulase enzyme CTec2 and buffered to pH 4.8 with acetate buffer. The enzyme dose was 50 FPU/g of milled wood in a broth containing 5% milled wood. After incubating for 72 h at 50 °C and 180 rpm, the lignin residue was separated from the hydrolysate by centrifugation. Lignin was extracted from the residue with aqueous dioxane of 80 or 96% concentration and lignin yield was 31% and 41%, respectively (Chen et al. 2017). Steam-pretreated poplar, pine, and corn stover were hydrolysed with the Cellic CTec3 cellulase enzyme system. The enzymatic broth had 2% lignocellulose loading in an acetate buffer of pH 4.8. Enzyme loading was 40 mg for every g of glucan present. 72 h incubation at 150 rpm and 50 °C yielded a lignin-rich solid residue; 42% yield and 70% purity from poplar, 52% yield and 75% purity from pine, and 43% yield and 51% purity from corn stover (Tian et al. 2017). In a recent study, lignin-rich solid residue from industrial bioethanol production from corn wstalks was used as the starting material for lignin isolation. The material was extracted in a nitrogen atmosphere for 6 h under reflux with 90% v/v aqueous dioxane acidified with 0.01 M hydrochloric acid. The extract was concentrated and added dropwise into 0.01 M hydrochloric acid to precipitate lignin. Though the yield was not reported, the obtained lignin had a purity in excess of 99% (Liu et al. 2018). 80 mesh size poplar powder was planetary ball-milled for 6 h at 500 rpm. Enzymatic hydrolysis of the milled mass was carried out in two stages with a CTec2 enzyme dose of 100 FPU/g lignocellulose and the addition of 0.02% sodium azide in each stage. The incubation time was 48 h for each stage. The residue was thoroughly washed, first with water, and then with saturated sodium chloride acidified to pH 2.5 with hydrochloric acid. Obtained lignin had 70% purity but yield was not reported (Wen et al. 2021).

One way of increasing the extent of carbohydrate hydrolysis and improving the lignin yield is to treat the milled lignocellulose with a solvent to make it more accessible to enzymes. In one protocol, ball-milled wood was swollen in *N,N*-dimethylacetamide containing 70 g/L lithium chloride for 48 h. The swollen wood meal at 5% concentration was hydrolysed with Onozuka R-10 cellulase enzyme buffered to pH 5 with disodium hydrogen phosphate and citric acid buffer. Incubation time was 24 h at 50 °C. An almost-quantitative isolation of lignin with high purity was reported indicating a near-complete hydrolysis (Chen et al. 1995). Extractive-free basswood and pine powders of 0.5 mm size were milled in a planetary ball mill at 600 rpm. Basswood was milled for 8 h and pine for 10 h. The milled material was dissolved in a mixture of dimethyl sulfoxide (10 mL/g lignocellulose) and N-methylimidazole (5 mL/g lignocellulose). Lignocellulose was regenerated by mixing with excess ethyl acetate and allowing it to precipitate for 2 h. Regenerated material was formed into a suspension in an acetate buffer of pH 5 at a liquid-to-solid ratio of 30. To this, 4% cellulase enzyme was added based on lignocellulose mass and incubated at 35 °C for 48 h. The residue after hydrolysis was washed with water, dried, and extracted twice for 24 h with 96% aqueous dioxane. Lignin was isolated by evaporating the solvent from the extract. The lignin yield for basswood improved from 36.5% (89% purity) for non-regenerated material to 45.8% (85% purity) with regeneration. Yield improvement was 30.5% (86% purity) to 35.3% (85% purity) for pine (Zhang et al. 2010). In a recent work, ball-milled poplar wood meal was dissolved in a 2% solution of the ionic liquid 1-ethyl-3-methylimidazolium acetate in dimethyl sulfoxide. The dissolution was completed in 2 h at 80–100 °C. Lignocellulose was precipitated from the solution by adding three volumes of water. The precipitate was centrifuged to get the regenerated material in 82–91% yield. The regenerated material was hydrolysed with cellulase enzyme CTec2 and buffered to pH 4.8 with acetate buffer. The enzyme dose was 50 FPU/g regenerated lignocellulose in a broth containing 5% substrate. After incubating for 72 h at 50 °C and 180 rpm, the lignin residue was separated from the hydrolysate by centrifugation. Lignin was extracted from the residue with aqueous dioxane of 80% concentration and lignin yield was 70–77% of the original wood against only 31–41% without regeneration (Chen et al. 2017).

4.1.4 Subcritical and Supercritical Fluid Hydrolysis of Lignocelluloses

The problems associated with the acid hydrolysis of lignocelluloses are (1) equipment corrosion and the need for exotic materials of construction, (2) problems with acid recovery and recycling, (3) stream pollution, and (4) not-so-pure lignin. To address some of these problems, new hydrolysis methodologies are being developed. These include supercritical and/or subcritical fluid hydrolysis.

Compared to water at ordinary temperature, supercritical water – with temperature exceeding 374 °C and pressure exceeding 22.1 MPa – shows the characteristics of a weak acid with excellent capability of diffusion into the lignocellulose matrix. Similarly, subcritical water extraction presents good results for the hydrolysis of the carbohydrate fraction in lignocelluloses. The physical properties of subcritical and supercritical water are depicted in Figure 4.1.

Sugi wood sawdust – initial lignin 32% – of particle size 0.105–0.15 mm was hydrolysed at a liquid-to-solid ratio of 10–25 with subcritical water at 260–360 °C and 15–25 MPa. The water flow rate through the reactor varied between 20 and 130 g/min. The lignin-rich solid residue left after hydrolysis was 11–33% of the initial mass of wood; higher reaction temperature resulted in less

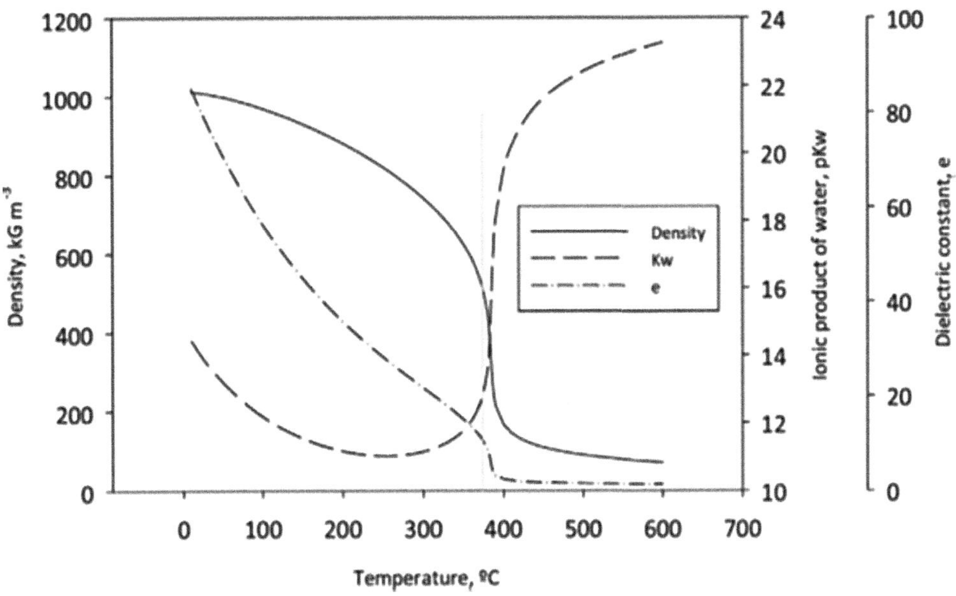

FIGURE 4.1 Physical properties of water near its critical point. (Reprinted from Cocero, MJ, Cabeza A, Abad, N, Adamovic, T, Vaquerizo, L, Martínez, CM, Pazo-Cepeda, MV, *J. Supercritical Fluids*, 133, 550–565, 2018, with permission from Elsevier.)

amount of residue (Matsunaga et al. 2008). Subcritical water hydrolysis of rye straw is reported at a liquid-to-solid ratio of 10 with a 4 g/min flow rate at 280 °C for 10 min reactor space time. 50% of initial lignin in straw was recovered as solid residue with 99% purity (Ingram et al. 2009). Subcritical water hydrolysis of 2 g sugarcane bagasse was carried out in a 50 mL reactor – with a liquid-to-solid ratio 25 – at 200 °C and 20 MPa with a water flow rate of 11–55 mL/min, and a reaction time of 30–48 min. In another set of experiments, 11 g of bagasse – with a liquid-to-solid ratio of 4.55 – was hydrolysed at 213–290 °C and 20 MPa with a water flow rate of 33 mL/min and 30 min reaction time. Lignin-rich solid residue obtained varied between 27% and 32% of the bagasse mass for reaction temperatures up to 213 °C (Prado et al. 2014). Wheat straw chopped to 2–10 cm in size was hydrolysed in subcritical water in the temperature range of 130–270 °C. After 10 min hydrolysis, with a water-to-solid ratio of 5 mL/g, the lignin-rich solid residue varied from 82% at 130 °C to 30% at 270 °C. The amount of solid residue decreased with increasing temperature and increasing reaction time (Abdelmoez et al. 2014). Subcritical water hydrolysis of sugarcane bagasse of 1 mm size was carried out at a water flow rate of 9 or 12.5 mL/min for 22 or 20 min. At 150 °C and 10 MPa, the solid residue after hydrolysis was 39% for 9 mL/min water flow and 22 min hydrolysis time compared to 50% for 12.5 mL/min water flow and 20 min hydrolysis time. Solid residue decreased to about 35% at 200 °C, and further to about 20% at 250 °C for 9 mL/min water flow and 22 min hydrolysis time at 5 MPa (Lachos-Perez et al. 2016). In a pilot scale study, ground Mongolian oak of particle size less than 0.5 mm was converted into a slurry with water with a liquid-to-solid ratio of 50. The sulphuric acid catalyst dose varied between 0.01% and 0.1%. The slurry was injected at 50 kg/h into a stream of supercritical water at 380 °C and 230 bar flowing at 44 kg/h. Post hydrolysis, the reaction mixture was cooled to normal temperature and the solids were filtered from the liquid hydrolysate (Jeong et al. 2017). Dried and milled grape pomace was hydrolysed for 60 min in subcritical water at 15 MPa pressure at different pH. The water-to-solid ratio in the reactor was 30 g/g with water flowing at a rate of 5 mL/min. At 210 °C, about 30% solid residue was obtained both at pH 6 and 10, but the residue at pH 10 had about 50% lignin compared to only about 40% at pH 6 (Castro et al. 2023).

The hydrolysis of lignocelluloses can also be carried out with subcritical or supercritical carbon dioxide. Sugarcane bagasse of size less than 1 mm was pretreated with subcritical carbon dioxide for

100 min at 180 °C and 5 MPa at 15% solids concentration slurry. The pretreated material was further hydrolysed with cellulase enzyme (20 FPU/g lignocellulose) for 72 h at 4.8 pH acetic acid sodium acetate buffer. The solids concentration in the broth was 2%, at a temperature of 50 °C and shaking at 150 rpm. 95.6% lignin recovery was reported with 96.3% purity (Zhang and Wu 2014).

4.2 MILLED WOOD LIGNIN

Lignin isolation methods based on its dissolution and subsequent precipitation result in the alteration of its chemical nature. For a scientific understanding of lignin chemistry, it is vital to isolate lignin from the lignocellulose with no or minimal chemical change from its native state. Milled wood lignin is one such attempt, pioneered by Bjorkman (Bjorkman 1956). The essence of the method lies in milling the extractive-free lignocellulose powder to very fine particle size and extracting the resulting material with a neutral aprotic solvent. The standard solvent used for extraction is aqueous dioxane of 96% v/v concentration, but the use of 80% concentration dioxane is not uncommon. The method is not primarily for a high lignin yield but for obtaining lignin in a form as it was present in the parent lignocellulose. The milling condition and the choice of solvent concentration affect the outcome as lignin degradation cannot be entirely prevented. In a study, the researchers varied the milling condition – environment, temperature, and duration – and observed the effects on the nature of lignin isolated from extractive-free sweetgum wood meal of less than 20 mesh size. The different milling environments used were air, toluene, toluene, tetralin, argon, and hydrogen. Similarly, planetary, rotary, and vibratory ball milling have all been used. The authors concluded that there was always degradation of the lignin structure during milling through the breaking of β-O-4/ linkages. The more the extent of milling to increase the lignin yield, the more the degradation of the lignin (Fujimoto et al. 2005; Guerra et al. 2006).

20 mesh extractive-free pine wood powder was rotary ball-milled in air with glass balls for 7 days, then vibratory ball-milled in toluene for 48 h. The milled mass was extracted twice with 96% aqueous dioxane at a ratio of 2.5 mL liquid to 1 g of starting wood. Each extraction of 24 h duration was performed under nitrogen. Combined extracts from dioxane extractions were vacuum evaporated and added dropwise to deionized water to precipitate the crude lignin. Lignin was purified by dissolving in 90% aqueous acetic acid and reprecipitating it by dilution with 10 volumes of water. The precipitated material was again dissolved in 1,2-dichloroethane and ethanol (2:1 ratio) and reprecipitated in petroleum ether. Purified lignin of 91% purity was obtained by evaporating the ether (Holtman et al. 2004). For isolating milled wood lignin from pine, 20 mesh extractive-free wood powder was milled in different ways to study the effect on the resulting lignin structure. In the first method, the material was rotary ball-milled in air with glass balls for 7 days and then vibratory ball-milled in toluene for 48 h. In the second method, the material was rotary ball-milled in air with glass balls for 7 days and then vibratory ball-milled in nitrogen atmosphere for 48 h. In the third method, the material was rotary ball-milled in air with porcelain balls for 6 weeks. In each case, the milled mass was extracted twice with 96% aqueous dioxane at a ratio of 2.5 mL liquid to 1 g of starting wood. Each extraction of 24 h duration was performed under nitrogen. Combined extracts from dioxane extractions were vacuum evaporated and added dropwise to deionized water to precipitate the crude lignin. Lignin yields were comparable in the first and the third methods of milling, but prolonged rotary ball milling caused more structural changes in lignin compared to short-duration vibratory milling (Holtman et al. 2006). 10–15% of the lignin present in eucalyptus could be isolated as milled wood lignin. For this, extractive-free sawdust was centrifugal ball-milled for 150 h in toluene. Milled lignocellulose was extracted four times with 90% v/v aqueous dioxane. Each extraction was performed with a solvent-to-solid ratio of 10 and lasted 24 h. All the extracts were combined and evaporated to dryness at 40 °C under vacuum. Crude lignin was further purified by redissolving in 20 volumes of 90% aqueous acetic acid and reprecipitating in water and centrifuged (Rencoret et al. 2008). Spruce wood meal after extractives removal was subjected to vibratory ball milling for 2–96 h. Extractive-free sawdust from *Paulownia fortunei* was ball-milled in the

presence of toluene for 50 h in a centrifugal mill. The milled lignocellulose was extracted three times with aqueous dioxane of 90% v/v concentration with a solvent-to-solid ratio of 25 mL/g. Each extraction stage lasted for 8 h. After third-stage extraction, the extract was centrifuged, and vacuum evaporated to dryness at 40 °C to obtain the crude lignin. The dried mass was dissolved in 90% v/v aqueous acetic acid with a solvent-to-solid ratio of 20 mL/g. Milled wood lignin was precipitated from the solution by dilution with water and centrifuged (Rencoret et al. 2009). Ground miscanthus straw, free of extractives, was ball-milled in the presence of toluene for 130 h and dried free of toluene. The milled material was extracted twice, each time for 24 h in the dark, with 96% v/v aqueous dioxane at 190 rpm. Combined extracts were vacuum evaporated to half the volume and added dropwise to two volumes of deionized water to precipitate the crude lignin. The purification of the lignin was done by redissolving the material in aqueous acetic acid and precipitation in excess water (El Hage et al. 2009). Extractive-free ground corncob of 40 mesh size was milled at 600 rpm for 72 h in a vibratory ball mill using toluene as the non-swelling liquid. The milled material was then extracted twice with 96 % v/v aqueous dioxane in an inert nitrogen atmosphere. Each extraction stage lasted 24 h with a solvent-to-solid ratio of 10 after which the liquid was centrifuged. The combined extract from the two stages was freeze-dried to give crude lignin. To obtain the milled wood lignin, the crude lignin was dissolved in 90% v/v aqueous acetic acid. The produced solution was added dropwise to deionized water to precipitate milled wood lignin (Bu et al. 2011). Extractives and tannins-free pine sawdust of less than 2 mm in size were ball-milled for 7 days in nitrogen atmosphere at 96 rpm. The milled material was extracted with 96% aqueous dioxane at a liquid-to-solid ratio of 10. The extraction was carried out in the dark for 24 h and the extracts were collected by centrifugation. The extract was evaporated to dryness under vacuum and redissolved in 20 volumes of 90% aqueous acetic acid. The solution was added dropwise to excess water to precipitate the lignin. The lignin yield varied between 11.4% and 12.2% of the lignin present in the starting material; bark provided a higher yield (Huang et al. 2011). Milled wood lignin was extracted from sweetgum and pine sawdust after removing the extractives. Extractive-free sawdust was subjected to planetary ball milling at 600 rpm for 2–12 h. Milled lignocellulose was extracted thrice with 96% v/v aqueous dioxane. Each extraction was performed for 24 h at a solvent-to-biomass ratio of 15 and a temperature of 20 or 45 °C. The combined extract was dried under vacuum at 35 °C, washed with dioxane, and redried to obtain the lignin product. The effect of milling time and extraction temperature on lignin yield was as expected; the more the milling time and/or higher temperature, the higher the yield (Min et al. 2013). These effects are shown in Figure 4.2.

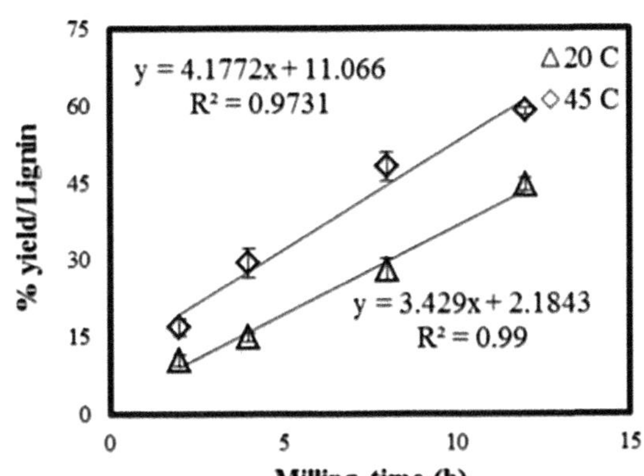

FIGURE 4.2 Variation in milled wood lignin yield with milling time and extraction temperature. (Reprinted from Min, D, Smith SW, Chang, H, Jameel, H, *BioResources*, 8(2), 1790–1800, 2013, originally published by BioResources under their open access policy.)

In a similar study, different lignocelluloses – bamboo, poplar, and larch sawdust – were planetary ball-milled at 600 rpm for 3, 5, or 7 h. In each case, the milled material was extracted thrice for 24 h with 96% aqueous dioxane at a solvent-to-solid ratio of 20. Both lignin yield and purity increased with milling time, albeit with associated structural damage to the lignin. For bamboo, yield varied between 39% and 54% (91.5–94.6% purity). For poplar, yield was 16–36% (purity 90.5–92.8%). Larch yielded 23–26% lignin with 90.1–92.9% purity (Yang et al. 2021). Wheat straw stems and wheat straw leaves were ground to a fine powder and treated for extractives removal. The material was then planetary ball-milled at 600 rpm for 2–6 h. Milled lignocellulose was enzymatically hydrolysed in acetate buffer at 4.8 pH at a liquid-to-solid ratio of 20. The enzyme used was a mixture of cellulase (NS 50013, 0.2 FPU/g lignocellulose), β-glucosidase (NS 50010, 0.2 CBU/g lignocellulose), and xylanase (NS 50014, 0.24 FXU/g lignocellulose). The hydrolysis time was 72 h at 50 °C and 180 rpm. Post hydrolysis, the solid residue was separated, washed with acetate buffer and water, and freeze-dried. Crude preparation was extracted twice for 24 h under nitrogen with 96% aqueous dioxane. Lignin was obtained from the combined extracts by vacuum evaporating the solvent. Lignin recovery was 63% from the stem and 42% from the leaves. Lignin purities were 90% and 92%, respectively (Jiang et al. 2017). Ground spruce was milled in a planetary ball mill for 18 h at 300 rpm either in ambient atmosphere or nitrogen atmosphere. The milled material was extracted for 72 h at room temperature with 96% aqueous dioxane with a solvent-to-solid ratio of 25. The extract was dried by vacuum evaporation, redissolved in 90% acetic acid at a solvent-to-solid ratio of 20, and added dropwise in excess water to precipitate lignin. The lignin yield was 17% for milling in ambient conditions and 24% for milling under nitrogen (Sapouna and Lawoko 2021).

REFERENCES

Abdelmoez, W, Nage, SM, Bastawess, A, Ihab, A, Yoshida, H. *J. Cleaner Prod.* 70:68–77, 2014.

American Biosystems Product Specification Sheet 2015. https://dev-american-biosystems.pantheonsite.io/wp-content/uploads/2015/09/Multicel-Product-Series-Spec-Sheet-1.pdf

Bjorkman, A. *Sven. Papperstidn.* 59:477–485, 1956.

Bergius, F. *Current Sci.* 5:632–637, 1937.

Bhardwaj, N, Kumar, B, Agrawal, K, Verma, P. *Bioresource Bioprocess.* 8:95, 2021.

Bu, L, Tang, Y, Gao, Y, Jian, H, Jiang, J. *Chem. Eng. J.* 175:176–184, 2011.

Castro, LEN, Sganzerla, WG, Barroso, TLCT, Maciel-Silva, FW, Colpini, LMS, Bittencourt, PRS, Rostagno, MA, Forster-Carneiro, T. *J. Supercritical Fluids* 196:105894, 2023.

Chang, H, Cowling, EB, Brown, W, Adler, E, Miksche, G. *Holzforschung* 29:153–159, 1975.

Chen, JY, Shimizu, Y, Takai, M, Hayashi, J. *Wood Sci. Technol.* 29:295–306, 1995.

Chen, TY, Wang, B, Shen, XJ, Li, HY, Wu, YY, Wen, JL, Liu, QY, Sun, RC. *RSC Adv.* 7:3376–3387, 2017.

Clausen, EC, Gaddy, JL. US Patent No. 5188673, 1993.

Dyadic product information 2014. https://www.dyadic.com/wp-content/uploads/2014/08/FibreZyme-G4P.pdf

Genencor technical document 2009. http://www.genencor.com/fileadmin/user_upload/genencor/documents/ACCELLERASE_R__1500_0020704.pdf

El Hage, R, Brosse, N, Chrusciel, L, Sanchez, C, Sannigrahi, P, Ragauskas, A. *Polymer Degrad. Stabil.* 94:1632–1638, 2009.

Evdokimov, AN, Kurzin, AV, Fedorova, OV, Trifonova, A. *Eur. J. Wood Prod.* 77:491–493, 2019.

Faith, WL, Hall, JA. *Chem. Eng. News* 22:525–526, 1944.

Faith, WL. *Ind. Eng. Chem.* 37(1):9–11, 1945.

Feng, Y, Jiang, J, Zhu, L, Yue, L, Zhang, J, Han, S. *Biotechnol. Biofuels* 6:161, 2013.

Fujimoto, A, Matsumoto, Y, Chang, HM, Meshitsuka, G. *J. Wood. Sci.* 51:89–91, 2005.

Guerra, A, Filpponen, I, Lucia, LA, Saquing, C, Baumberger, S, Argyropoulos, DS. *J. Agric. Food Chem.* 54:5939–5947, 2006.

Harris, EE, Begliuger, E. Report No. R1617. US Department of Agriculture, 1946.

Holtman, KM, Chang, HM, Kadla, JF. *J. Agric. Food Chem.* 52:720–726, 2004.

Holtman, KM, Chang, HM, Jameel, H, Kadla, JF. *J. Wood Chem. Technol.* 26:21–34, 2006.

Hu, Z, Yeh, TF, Chang, HM, Matsumoto, Y, Kadla, JF. *Holzforschung* 60:389–397, 2006.

Huang, F, Singh, PM, Ragauskas, AJ. *J. Agric. Food Chem.* 59:12910–12916, 2011.

Ingram, T, Rogalinski, T, Bockemühl, V, Antranikian, G, Brunner, G. *J. Supercritical Fluids* 48:238–246, 2009.
Jeong, H, Park, YC, Seong, YJ, Lee, SM. *Bioresource Technol.* 245:351–357, 2017.
Jiang, LQ, Fang, Z, Li, XK, Luo, J, Fan, SP. *Process Biochem.* 48:1942–1946, 2013.
Jiang, B, Cao, T, Gu, F, Wu, W, Jin, Y. *ACS Sustainable Chem. Eng.* 5:342–349, 2017.
Klason, P. *Ark. Kem. Mineral. Geol.* 3(5):9, 1908.
Kucuk, MM, Demirbas, A. *Ener. Conv. Manage.* 38(2):151–165, 1997.
Lachos-Perez, D, Martinez-Jimenez, F, Rezende, CA, Tompsett, G, Timko, M, Forster-Carneiro, T. *J. Supercritical Fluids* 108:69–78, 2016.
Lenihan, P, Orozco, A, O'Neill, E, Ahmad, MNM, Rooney, DW, Walker, GM. *Chem. Eng. J.* 156:395–403, 2010.
Liu, C, Wang, X, Lin, F, Zhang, H, Xiao, R. *Fuel Process. Technol.* 169:50–57, 2018.
Matsunaga, M, Matsui, H, Otsuka, Y, Yamamoto, S. *J. Supercritical Fluids* 44:364–369, 2008.
Merc data sheet, n.d. https://www.merckmillipore.com/IN/en/product/CELLULYSIN-Cellulase-Trichoderma-viride,EMD_BIO-219466#anchor_PDS
Min, DY, Smith, SW, Chang, HM, Jameel, H. *BioResources* 8:1790–1800, 2013.
Motoyoshi, O, Kusama, J, Tadao, I. US Patent No. 2900284, 1959.
Novozymes, n.d.-a https://biosolutions.novozymes.com/en/bioenergy/products/biomass-conversion/cellic-ctec2
Novozymes, n.d.-b https://biosolutions.novozymes.com/en/bioenergy/products/biomass-conversion/cellic-ctec3-hs
Oshima, M, Shibuya-ku, Tokyo-to, Kusama, J, Ota-ku, Tokyo-to, Ishii, T, Tsurumi-ku, Yokohama-shi. US Patent No. 2900284, 1959.
Prado, JM, Follegatti-Romero, LA, Forster-Carneiro, T, Rostagno, MA, Filho, FM, Meireles, MAA. *J. Supercritical Fluids* 86:15–22, 2014.
Radillo, JJV, Ruiz-Lopez, MA, Macias, RR, Ramirez, LB, Garcia-Lopez, PM, Toral, FALD. *BioResources* 6:344–355, 2011.
Rencoret, J, Marques, G, Gutierrez, A, Ibarra, D, Li, J, Gellerstedt, G, Santos, JI, Jimenez-Barbero, J, Martinez, AT, del Rio, JC. *Holzforschung* 62:514–526, 2008.
Rencoret, J, Marques, G, Gutierrez, A, Nieto, L, Jimenez-Barbero, J, Martinez, AT, del Rio, JC. *Ind. Crops Prod.* 30:137–143, 2009.
Rencoret, J, Prinsen, P, Gutierrez, A, Martinez, AT, del Rio, JC. *J. Agric. Food Chem.* 63:603–613, 2015.
Rodrigues, AC, Haven, MØ, Lindedam, J, Felby, C, Gama, M. *Enzyme Microbial Technol.* 79–80:70–77, 2015.
Sapouna, I, Lawoko, M. *Green Chem.* 23:3348–3364, 2021.
Tappi Test Method, T 222 om-02, 2006.
Tian, D, Hu, J, Chandra, RP, Saddler, JN, Lu, C. *ACS Sustainable Chem. Eng.* 5:2702–2710, 2017.
Wijaya, YP, Putra, RDD, Widyaya, VT, Ha, JM, Suh, DJ, Kim, CS. *Bioresource Technol.* 164:221–231, 2014.
Wang, K, Bauer, S, Sun, RC. *J. Agric. Food Chem.* 60:144–152, 2012.
Wen, P, Zhang, T, Wei, L, Wang, J, Ragauskas, AJ, Zhang, Y, Xu, Y, Zhang, J. *ACS Sustainable Chem. Eng.* 9:11361–11371, 2021.
Yang, G, An, X, Yang, S. Frontiers Bioeng. *Biotechnol.* 9:807625, 2021.
Zhang, A, Lu, F, Sun, RC, Ralph, J. *J. Agric. Food Chem.* 58:3446–3450, 2010.
Zhang, H, Wu, S. *Bioresource Technol.* 158:161–165, 2014.
Zviely, M. Converting lignocellulosic biomass to low cost fermentable sugars. Fang, Z (Ed.) *Pretreatment techniques for biofuels and biorefineries.* Springer, Berlin, 133–150, 2013.

Section 2

Lignin Characterization

5 Physicochemical Characterization of Lignin

Native lignin, as it exists in the plant biomass, is physically and chemically heterogeneous. Lignin isolated by different isolation methods carries this heterogeneity. This may be in terms of molecular weight and its distribution, occurrence and abundance of different functional groups, and the way the phenylpropane units are joined together. Differences in these attributes in lignin may affect its utility for a particular application. In recent years, the idea of comprehensive utilization of lignocellulose biomass in the biorefinery framework has gained ground due to the emphasis on sustainability. Any lignocellulose biorefinery would be incomplete without adequate attention being paid to the gainful utilization of lignin. The first step to this end is to ascertain the nature and character of the lignin to find an application best suited to it. Lignin characterization for its physicochemical nature is, therefore, important for its subsequent value addition.

5.1 ELEMENTAL ANALYSIS OF LIGNIN

Carbon, hydrogen, and oxygen are the major atomic contributions in the lignin macromolecule. Depending on the lignocellulose source, and the isolation method, nitrogen, and sulphur may also be present. The basic principle of standard Carbon, Hydrogen, Nitrogen, and Sulphur (CHNS) analysis lies in the combustion of the sample in excess air at elevated temperature to fully oxidize carbon to carbon dioxide, hydrogen to water vapour, nitrogen to oxides of nitrogen, and sulphur to sulphur dioxide. These product gases are flown out of the combustion chamber, normally with an inert carrier gas like helium. Next, the gaseous stream is made to react with copper at elevated temperature which consumes the unreacted oxygen and also reduces the oxides of nitrogen to nitrogen. The product gases are detected and quantified using infrared absorption and/or thermal conductivity measurement. Additionally, the inorganic impurities in the lignin may introduce elements such as sodium, potassium, and silicon. Atomic absorption spectroscopy is one analytical tool to quantitatively estimate the inorganic elements. Another method includes inductively coupled plasma with mass spectroscopic identification and quantification.

An elemental analysis of lignin is desirable as the first step towards the comprehensive chemical characterization of lignin. It is also important for lignins meant to be used for energy applications. The calorific value of lignin is dependent on its elemental analysis. The air requirement for its combustion or gasification depends on the elemental analysis. Similarly, the flue gas and syngas compositions are related to the elemental analysis. In many applications, sulphur-free lignins are sought after. Lignins with no or very little inorganic impurities are highly valued. Typical elemental analyses of different lignins are shown in Table 5.1.

5.2 MOLECULAR STRUCTURE OF LIGNIN

Native lignin is a three-dimensional macromolecule with aromatic phenylpropane fundamental units joined together by intermonomer linkages. Each phenylpropane unit may, in turn, have different functional groups attached to the benzene ring as well as to the side chain.

Though chemically modified during the isolation process, technical lignins resemble their native counterparts, and the molecular structures of different lignins can differ in these two aspects. Therefore, one important part of lignin characterization is having proper insight into its molecular structure. This includes the identification and quantitative estimation of the different functional

DOI: 10.1201/9781003598664-7

TABLE 5.1
Elemental Analyses of Different Lignins

Lignin Type	C	H	N	S	O	Ash	Reference
Softwood soda/soda anthraquinone	65.4–67.6	5.8–6.1			26.4–28.7	0.01–1.98	Angles et al. 2003
Softwood kraft	62.7–65.8	5.7–6.09	0.7	1.8	28.1–29.1	2.84	Angles et al. 2003; Faix et al. 1994
Softwood MWL	62.2–63.2	5.4–6.2	0.2		28.7–31.2		Holtman et al. 2004; Yang and Goring 1980
Softwood organosolv	68	5.96	0.16		25.9	0.1	Rossberg et al. 2019
Softwood Klason	59.3	5.26			35.5		Zhao et al. 2014
Hardwood MWL	60.8	6.4			31.32		Yanez-S et al. 2014
Hardwood ethanol organosolv	62–67	5-8.16	0.22–0.56		27.5–31		Hepditch and Thring 1997; Yanez-S et al. 2014
Hardwood kraft	59.5–61.3	5.56–5.79	0.13–0.19	2.14–2.54	30.8–32	1.54	Chen et al. 2020; Sameni et al. 2016
Hardwood soda	62.7–65.7	5.95–6.12	0.08–0.15		27.8–30.8		Li and Wu 2014
Hardwood enzyme hydrolysis	58.9–62.4	5.89–6.08	0.49	0.03	31.2–35	0.45	Faix et al. 1994; Sameni et al. 2016
Hardwood Klason	55.7	4.02			39.7		Zhao et al. 2014
Nonwood acid hydrolysis	66.1	5.82			27.8		Faix et al. 1994
Nonwood enzymatic/ acidolysis	50.4	6.24	0.35	0.09	40.1		Lou et al. 2018
Nonwood MWL	58.8–63.5	5.7–5.96	0.2–0.28		29.2–34.97	0.1	El Hage 2009; Li et al. 2012
Nonwood soda	52.8–65.4	5.7–8.48	0.4–0.59	0.38	22.9–38.9	1.38-–2	Ghatak et al. 2010; Mousavioun and Doherty 2010; Sameni et al. 2016
Nonwood organosolv	58.3–61.7	5.57–5.6			25.6–35.8	0.44–2.93	Cybulska et al. 2012; Sheng et al. 2022

groups present in the lignin macromolecule. Similarly, it is important to identify and quantify the linkages that join the phenylpropane units in the lignin macromolecule. There are several excellent wet chemistry methods available for these characterizations. In addition, advanced spectroscopic and chromatography techniques assist in the characterizations.

5.2.1 Functional Groups in Lignin

The phenylpropane units in the lignin macromolecule have a benzene ring, and an aliphatic side chain, and different functional groups are attached to them. Functional groups on the benzene ring are phenolic hydroxyl, and/or methoxyl. The aliphatic side chain may have aliphatic hydroxyl, aldehyde, alkoxyl, and olephinic functionalities. Further, lignin carbohydrate complexes may also be present.

5.2.1.1 Methoxyl Groups in Lignin with Methods Based on Hydroiodic Acid

In the Zeisel method, lignin is refluxed with concentrated hydroiodic acid to form methyl iodide. The formed volatile methyl iodide is passed into a solution of ethanolic silver nitrate, resulting in the formation of silver iodide precipitate, which is estimated. Sulphur-containing compounds interfere with the method by forming hydrogen sulphide which, in turn, converts into silver sulphide (Zakis 1994). The ASTM D 1166–84 standard wet chemistry method for the determination of methoxyl groups in lignin is based on the modification of Viebock and Schwappach to the Zeisel method. Here, the methyl iodide formed by refluxing the lignin sample with hydroiodic acid through a series of reactions liberates iodine which is titrated against standard sodium thiosulphate. The reactions can be represented as (ASTM D 1166–84 2001).

$$CH_3I + Br_2 \rightarrow CH_3Br + IBr \qquad \{5.1\}$$

$$IBr + 2Br_2 + 3H_2O \rightarrow HIO_3 + 5HBr \qquad \{5.2\}$$

$$HIO_3 + 5HI \rightarrow 3I_2 + 3H_2O \qquad \{5.3\}$$

During the reaction with bromine, any hydrogen sulphide formed is converted to sulphuric acid, thereby eliminating the sulphur interference. Dioxane lignin from alkali-extracted eucalyptus sawdust was found to have 1.64 methoxyl groups per phenylpropane unit (Evtuguin et al. 2001). Methoxyl groups in dioxane lignin from alkali-extracted banana leaf sheath were reported to be 1.09 per phenylpropane unit (Oliveira et al. 2006). In the Adler method, lignin or lignocellulose sample is reacted with sodium metaperiodate at 4 °C in dark. One mole of methanol is produced per methoxyl group which is quantitatively estimated by gas chromatography (Adler and Hernestam 1955; Adler 1977). In a variation of the method, a sample of lignocellulose or lignin is digested at 130 °C for 20 min with 57% aqueous hydroiodic acid. After quenching in an ice bath, ethyl iodide is added as an internal standard. The entire mixture is then extracted with carbon tetrachloride and the organic layer is analysed by gas chromatography. Using this method, the methoxyl content in MWL of Japanese cedar was determined to be 3.5 mmol/g, and that in Japanese birch to be 4.8 mmol/g (Goto et al. 2006). The methoxyl content of ten hardwood species was estimated by this method and found to vary between 1.39 and 1.88 per phenylpropane unit, the highest being for eucalyptus (Santos et al. 2012). In yet another variation, hydroiodic acid digestion time was 30 min, and the unreacted acid was neutralized with sodium hydroxide. The formed methyl iodide was assessed by gas chromatography (Li et al. 2012b). Methoxyl content in softwood sodium lignosulphonate is reported to be 11 mass%, though the method of estimation was not reported (Myrvold 2015). Wheat straw soda lignin was estimated to have 3.59 mmol/g methoxyl groups. This lignin was fractionated by successive solvent extraction with ethyl acetate, 2-butanone, methanol, acetone, and 95 weight% aqueous dioxane. In each solvent extraction step, the solids were contacted with solvent for 2 h followed by filtration of undissolved solids under vacuum, which were used in the next stage of extraction. Lignin fractions were obtained from the filtrate by evaporation of the solvent, resulting in five such fractions in all. Solids insoluble in aqueous dioxane constituted the sixth fraction. The methoxyl groups in the resultant fractions varied between 3.07 and 3.97 mmol/g (Lee et al. 2021).

The abundance and positioning of the methoxyl groups in both native lignin and the isolated technical lignins are important. The benzene ring in the phenylpropane unit can have one methoxyl group at position 3, the guaiacyl-type structure. If the benzene ring has two methoxyl groups at positions 3 and 5, it is a syringyl-type structure. The ratio of these two structural moieties is an important characteristic of a lignin macromolecule. A few benzene rings may have no methoxyl groups and are referred to as the p-hydroxyphenyl type. Studies show that softwood (gymnosperms) lignins are almost exclusively of guaiacyl type whereas lignins from hardwoods (angiosperms) are predominantly mixtures of guaiacyl- and syringyl types. Lignins from the grass family contain all three types of structures: guaiacyl, syringyl, and p-hydroxyphenyl.

5.2.1.2 Methoxyl Groups in Lignin with Alkaline Nitrobenzene Oxidation

One widely used technique for estimating the nature of methoxyl substitution in lignin is oxidation with alkaline nitrobenzene (Creighton et al. 1944; Freudenberg et al. 1940). The method consists of reacting finely ground lignocellulose or isolated lignin at 160–180 °C for 2–4 h with nitrobenzene and a strong alkali such as sodium hydroxide. Sodium hydroxide concentration remains 2N. Different researchers have used the ratio of sodium hydroxide to nitrobenzene between 8.33 and 17.5 (Hepditch and Thring 1997; Katahira and Nakatsubo 2001; Min et al. 2015). Oxidation temperature and time may affect the product yield and distribution to some extent. The result is quantitative cleavage of ether bonds in the lignin macromolecule and formation of the aldehydes vanillin, syringaldehyde, and p-hydroxybenzaldehyde corresponding to guaiacyl-, syringyl-, and p-hydroxyphenyl-type structures. One limitation of the method is the inability of nitrobenzene to break carbon-to-carbon linkages in lignin. Consequently, the guaiacyl structures in condensed entities escape estimation. In technical lignins like soda, or hydrolysis lignins, there may be a greater degree of condensation and corresponding underestimation of guaiacyl moieties. Minor quantities of corresponding carboxylic acids and aceto derivatives may also be formed. A depiction of oxygenated organics produced during alkaline nitrobenzene oxidation is presented in Figure 5.1.

First, the unreacted nitrobenzene is extracted from the reaction mixture with a suitable organic solvent. Chloroform, diethyl ether, and dichloromethane have all been used as the organic solvent for extraction. Then, the aqueous phase is acidified, and the supernatant is again extracted with organic solvent. Products are collected by evaporation of solvent. These are acetylated and estimated by chromatographic separation, identification, and quantification. In a recent modification, the reaction products were derivatized by O-bis(trimethylsilyl)trifluoroacetamide to the trisilynated derivative instead of the usual acetylation (Min et al. 2015).

Wheat straw powder, when oxidized with alkaline nitrobenzene, produced 2.89% guaiacyl-type structures, 2.42% syringyl-type structures, and 0.43% p-hydroxyphenyl structures (Sun et al. 1995). Alkaline nitrobenzene oxidation of spruce, eucalyptus, and bamboo provided the proportions of different methoxyl-substituted moieties in lignin. Spruce, a softwood, had only guaiacyl-type structures. Eucalyptus, a hardwood had a syringyl/guaiacyl ratio of 5.2, and bamboo, a species from the grass family, had a ratio of 0.4:1:1.4 of p-hydroxyphenyl, guaiacyl, and syringyl structures (Katahira and Nakatsubo 2001). The syringyl/guaiacyl ratio in kenaf and beech was determined by nitrobenzene oxidation. In kenaf, the ratio was 0.83–2.91, and in beech, it was 2.4 (Kuroda et al. 2002). The ratio of syringyl/guaiacyl-type structures in lignins from different eucalyptus species was reported between 2.4 and 3.2. For this, extractive-free wood meal was subjected to nitrobenzene oxidation

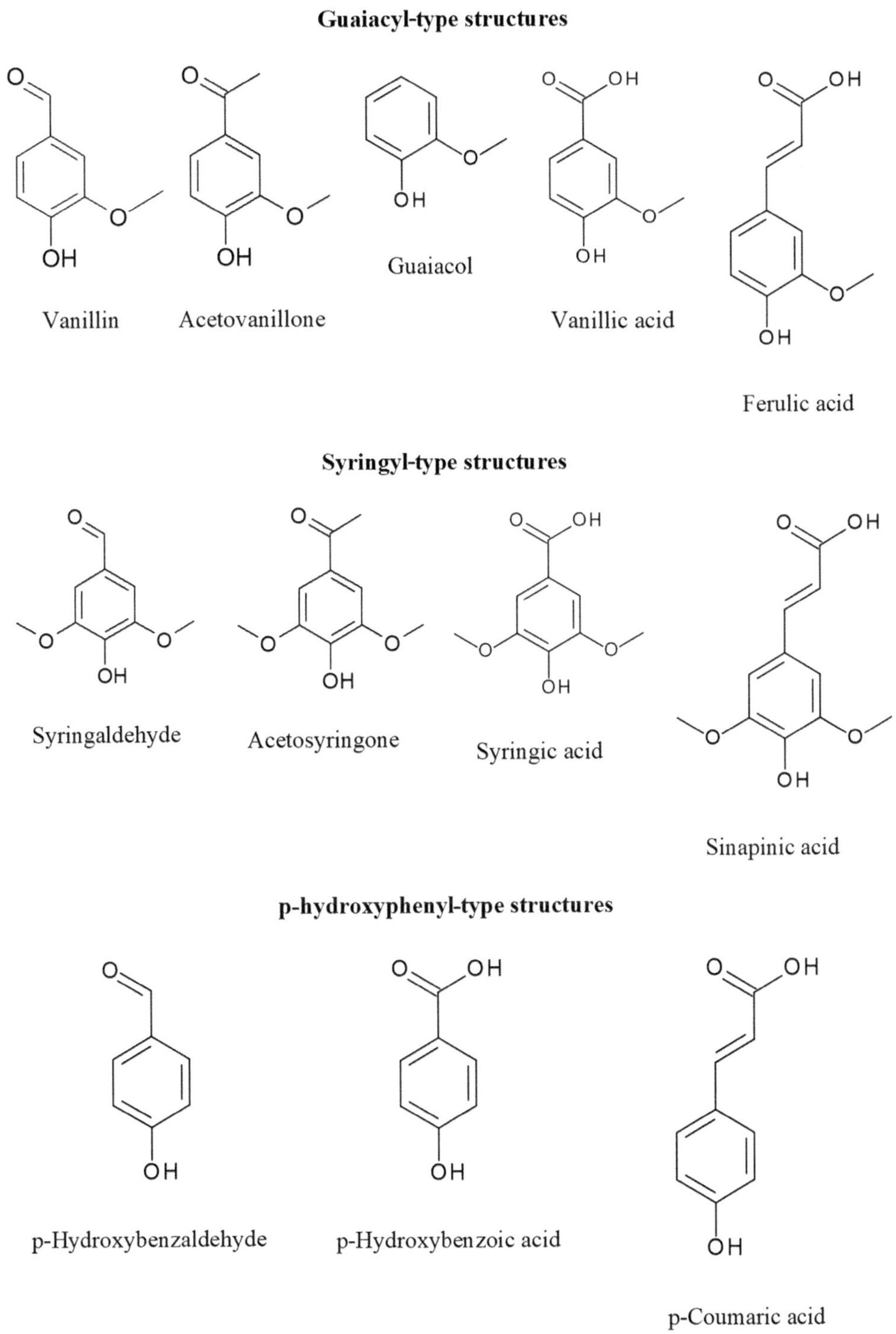

FIGURE 5.1 Oxygenated organics produced during alkaline nitrobenzene oxidation of lignin and lignocelluloses.

(Lima et al. 2008). The syringyl/guaiacyl ratios in different eucalyptus species were reported between 2.7 and 4 by oxidizing ground wood meal in alkaline nitrobenzene (Ohra-aho et al. 2013). The syringyl/guaiacyl ratio in sugarcane bagasse lignin was found to be in the range of 1.02–1.23, as determined by nitrobenzene oxidation of extractive-free bagasse (Lopes et al. 2011). Nitrobenzene oxidation of ground pine, maple, and corn stover was performed to estimate the structural features of their lignins. Pine lignin, besides 41 mol% of guaiacyl structures, had 0.9 mol% of p-hydroxyphenyl entities. Maple lignin had 22.7 mol% guaiacyl, and 31 mol% syringyl amounting to a syringyl/guaiacyl ratio of 1.36. Lignin in corn stover contained 14.1 mol% guaiacyl, 15.8 mol% syringyl, and 11.8 mol% p-hydroxyphenyl structures (Min et al. 2015). Milled wood lignin from green liquor-treated poplar was analysed for guaiacyl and syringyl entities by nitrobenzene oxidation and compared with milled wood lignin from untreated wood. Guaiacyl structures were 0.94 mmol/g in lignin from untreated wood. It decreased to 0.69 mmol/g when the wood was treated with 8% alkali, and further decreased to 0.59 mmol/g when treated with 16% alkali. Syringyl structures in milled wood lignin from untreated wood were 1.33 mmol/g, increased to 1.41 mmol/g after treatment with 8% alkali, and then dropped to 1.28 mmol/g by treatment with 16% alkali. p-hydroxyphenyl structures were 0.07 mmol/g in untreated lignin. It decreased to 0.01 mmol/g when wood was treated with 8% alkali but increased to 0.05 mmol/g after 16% alkali treatment. The authors contended that there was preferential removal of guaiacyl units during green liquor treatment (Tan et al. 2016). Nitrobenzene oxidation of different lignocelluloses was used to estimate the methoxyl moieties in several lignocelluloses. The softwood, Japanese Cedar had a 0.05:1 ratio of p-hydroxyphenyl/guaiacyl and no syringyl. The hardwood species, Japanese beech reported a syringyl/guaiacyl ratio of 2.19 with no p-hydroxyphenyl. Rice straw, lignocellulose from the grass family, was found to have a 0.41:1:0.67 ratio of p-hydroxyphenyl, guaiacyl, and syringyl structures (Takada et al. 2018).

Nitrobenzene oxidation has been extensively employed for elucidation of the molecular structure regarding methoxyl groups in lignin isolated by different methods. Structural elucidation by nitrobenzene oxidation of wheat straw milled-wood lignin revealed 542.3 µmol/g of guaiacyl-type structure, 547.7 µmol/g of syringyl-type structure, and 103 µmol/g of p-hydroxyphenyl-type structure. On the other hand, guaiacyl, syringyl, and p-hydroxyphenyl moieties in enzymatic wheat straw lignin were 453.2, 494, and 108.8 µmol/g, respectively (Billa and Monties 1995). Eucalyptus kraft lignin was oxidized by nitrobenzene to study its structural features. With digestion at 180 °C for 2 h, the syringyl/guaiacyl ratio was reported to be 3.75 on a mass basis (Villar et al. 1997). Nitrobenzene oxidation of Ethanol organosolv lignin from a hardwood mixture yielded 2.5% guaiacyl-type structures, and 6.52% syringyl-type structures amounting to an S/G ratio of 2.6. In kraft lignin, guaiacyl moieties were 5.83%, but syringyl moieties were only 0.03%, which appears to be an unusually low value (Hepditch and Thring 1997). Dioxane organosolv lignin from eucalyptus, when treated with alkaline nitrobenzene, yielded 37.5% oxidized products (vanillin and syringaldehyde) of which 14% was guaiacyl, and 86% was syringyl (Evtuguin et al. 2001). Wheat straw milled-wood lignin yielded 32.49% guaiacyl structures, 38.95% syringyl structures, and 4.95% p-hydroxyphenyl structures upon alkaline nitrobenzene oxidation (Sun et al. 2005). Soda anthraquinone lignin from Japanese cedar yielded 6.6–11.2 moles vanillin – representative of guaiacyl-type structure – per 100 moles phenylpropane as against 10.9 moles vanillin yield per 100 moles phenylpropane from pine kraft lignin (Indulin AT). For comparison, MWL from Japanese cedar yielded 23.4 moles of vanillin per 100 moles of phenylpropane (Ikeda and Magara 2015).

5.2.1.3 Methoxyl Groups in Lignin with Thioacidolysis

During alkaline nitrobenzene treatment, the aliphatic side chain in the lignin phenylpropane backbone is liable to be cleaved. Thioacidolysis is one method that quantitatively estimates the different phenylpropane structures in lignin. Lignin or lignocellulose is reacted in inert atmosphere with a reagent comprising boron trifluoride etherate (2.5% v/v) and ethanethiol (10% v/v) in 1,4-dioxane. The reaction is complete in 4 h at 100 °C followed by cooling to normal temperature and neutralization of the excess acid with a mild base, usually sodium bicarbonate. The reaction products are then extensively extracted with dichloromethane in several stages and the combined extracts are dried in a nitrogen atmosphere. The dried mass is derivatized with N,O-bis(trimethylsilyl)trifluoroacetamide and subjected to gas chromatographic separation, identification, and quantification (Lapierre et al. 1985; Diniz et al. 2019). One shortcoming of the method is the use of foul-smelling ethanethiol. In one modification to the original method, after the neutralization step, the reaction mixture is reacidified to pH 3–4 by the addition of hydrochloric acid followed by extraction with diethyl ether. The organic phase is then analysed by gas chromatography (Yamamura et al. 2012). Alternatively, reacidification to pH 1, and extraction with ethyl acetate, is also practised (Harman-Ware et al. 2016). Recently, a modification to the method was proposed in which the solvent extraction step is eliminated. The entire neutralized reaction products are dried under nitrogen atmosphere and silylated (Chen et al. 2021). The procedure selectively breaks the ether linkages in the lignin macromolecule with the substitution of the SC_2H_5 functional group at α, β, and γ carbons on the aliphatic side chain. The respective product monomers are shown in Figure 5.2.

Thioacidolysis was carried out to determine the syringyl/guaiacyl ratio in kenaf and beech. In kenaf, the ratio varied between 1.35 and 3.72, with the core having a lower ratio than the bast. Syringyl/guaiacyl ratio in beech was 2.02 (Kuroda et al. 2002). The softwood spruce, upon thioacidolysis, yielded 1682 μmol/g of monomers, presumably all guaiacyl (Onnerud and Gellerstedt 2003). Wild poplar and hybrid plantation poplar were analysed by thioacidolysis for their monomer composition. In wild poplar, the abundance of monomers was 613.8 μmol/g guaiacyl, 1322.1 μmol/g syringyl, and 6 μmol/g p-hydroxyphenyl. In comparison, monomer abundance in hybrid plantation poplar was 325.2 μmol/g guaiacyl, 1795.1 μmol/g syringyl, and 9.2 μmol/g p-hydroxyphenyl (Robinson and Mansfield 2009). Pine, spruce, poplar, bamboo, and bagasse were subjected to thioacidolysis to elucidate structural entities. Lignin in softwoods, pine and spruce, turned out to be predominantly guaiacyl type with a minor proportion of p-hydroxyphenyl. Pine had 1020 μmol/g guaiacyl and 45 μmol/g p-hydroxyphenyl, while spruce had 1109 μmol/g guaiacyl and 21 μmol/g p-hydroxyphenyl. Lignin in the hardwood, poplar, was a combination of guaiacyl, 848 μmol/g, and syringyl, 1342 μmol/g, type structures, with no p-hydroxyphenyl. Lignin in the nonwoods bamboo and bagasse had all three structural features with abundance decreasing in the order guaiacyl > syringyl > p-hydroxyphenyl. Bamboo lignin was estimated to have 602 μmol/g guaiacyl, 411 μmol/g syringyl, and 73 μmol/g p-hydroxyphenyl. Bagasse lignin was made up of 303 μmol/g guaiacyl, 265 μmol/g syringyl, and 8 μmol/g p-hydroxyphenyl (Yue et al. 2012). Pine, a softwood, poplar, hardwood, and switchgrass, a nonwood, were analysed for their lignin structures. Pine had 198.6 μmol/g guaiacyl, 5 μmol/g syringyl, and 7.9 μmol/g p-hydroxyphenyl structures. Poplar comprised 195.5 μmol/g guaiacyl, 333.3 μmol/g syringyl, and 3.6 μmol/g p-hydroxyphenyl structures. Guaiacyl, syringyl, and p-hydroxyphenyl entities in switchgrass were 68.2, 57.1, and 5.4 μmol/g (Harman-Ware et al. 2016). Guaiacyl and syringyl units in eucalyptus were determined as 113 and 566.5 μmol/g, respectively, confirming the generally high syringyl/guaiacyl ratio in this hardwood (Saito et al. 2016). Guaiacyl and syringyl units in bagasse were reported to be 130 and 100 μmol/g, respectively.

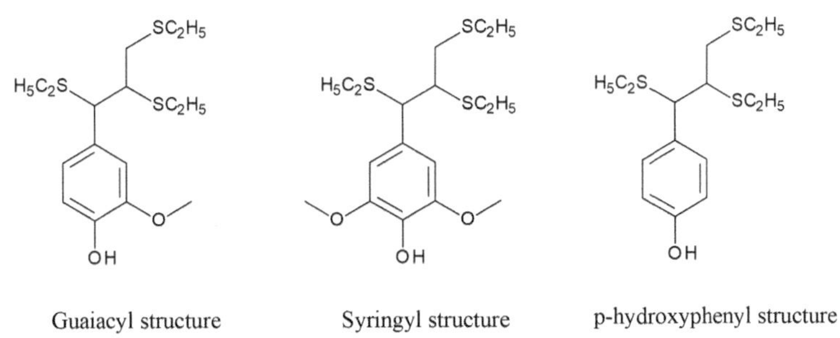

FIGURE 5.2 Lignin monomers produced during thioacidolysis of lignin and lignocelluloses.

For cottonwood, the values were 230 and 290 µmol/g (Chen et al. 2021). Ground pine, corn stover, and several poplar species were analysed for their lignin structural features by thioacidolysis. Pine was estimated to have 205.7 µmol/g of guaiacyl structures and a nominal 6.7 µmol/g of p-hydroxyphenyl structures. Corn stover released 37.9 µmol/g of guaiacyl structures, 47.5 µmol/g of syringyl structures, and 4.8 µmol/g of p-hydroxyphenyl structures. Guaiacyl moieties in poplar varied between 22.3 and 105.6 µmol/g, syringyl moieties between 78.1 and 231.4 µmol/g, and p-hydroxyphenyl entities between 4.5 and 5.5 µmol/g (Happs et al. 2021).

Thioacidolysis analysis of pine lignin revealed 10.71% guaiacyl and 1.8% p-hydroxyphenyl structures. In poplar milled-wood lignin, the guaiacyl structures were 8.36%, and the syringyl structures were 9.37%. In contrast, the enzymatic hydrolysis lignin from poplar was found to contain 9.88% guaiacyl and 19.9% syringyl. This highlights the importance of the isolation method on the lignin characteristics. In this case, the enzymatic hydrolysis lignin had a distinctly higher syringyl/guaiacyl ratio. Wheat straw parietal residue had 10.44% guaiacyl, 16.32% syringyl, and 0.76% p-hydroxyphenyl (Lapierre et al. 1985). Wheat straw milled-wood lignin was determined to have 410.8 µmol/g of guaiacyl-type structure, 369.7 µmol/g of syringyl-type structure, and 53.4 µmol/g of the p-hydroxyphenyl-type structure by thioacidolysis. Enzymatic lignin from wheat straw, by contrast, had 338.1 µmol/g of guaiacyl-type structure, 385.1 µmol/g of syringyl-type structure, and 40.6 µmol/g of p-hydroxyphenyl-type structure (Billa and Monties 1995). Milled wood lignin from spruce exhibited 1119.5 µmol/g of guaiacyl, and 26.7 µmol/g of p-hydroxyphenyl, confirming the predominately guaiacyl nature of softwoods (Marques et al. 1999). Thioacidolysis estimation of eucalyptus dioxane organosolv lignin revealed 300 µmol/g of guaiacyl structures, and 1337 µmol/g of syringyl structures (Evtuguin et al. 2001). Milled wood lignin of spruce had 986 µmol/g of guaiacyl structures, as determined by thioacidolysis (Onnerud 2003; Onnerud and Gellerstedt 2003). Milled wood lignin from coconut coir had a relatively high abundance of guaiacyl moieties at 594 µmol/g, followed by 218 µmol/g syringyl, and 62 µmol/g p-hydroxyphenyl moieties (Rencoret et al. 2013). Enzymatic mild acidolysis lignin from pine showed 1331 µmol/g of guaiacyl structures (van de Pas et al. 2014). Dioxane organosolv lignin from different maize species consisted of 133.7–308.2 µmol/g of guaiacyl, 114.7–482.7 µmol/g of syringyl, and 3.6–8 µmol/g of p-hydroxyphenyl entities (Chazal et al. 2014).

5.2.1.4 Methoxyl Groups in Lignin with Derivatization Followed by Reductive Cleavage

Derivatization Followed by Reductive Cleavage (DFRC) is an advanced technique that combines wet chemistry methods with chromatographic separation and mass spectroscopy (Lu and Ralph 1997; Lu and Ralph 1998). Lignin is reacted for 2 h with acetic bromide dissolved in acetic acid of 8% v/v solution at 50 °C. This causes the formation of the corresponding β-bromo ether structures. After removal of the acid and the bromide, the derivatized material is dissolved in dioxane, and acidified with aqueous acetic acid. Cleavage of ether linkages is carried out by the addition of zinc powder and stirring for 40 min. The reaction mixture is extensively extracted with dichloromethane to transfer all the reaction products to the organic phase. Before gas chromatographic separation and quantitative estimation by mass spectroscopy, the lignin defragmentation products are acetylated by reacting with acetic anhydride.

Milled wood lignin from several lignocellulose materials was characterized by DFRC with conventional acetylation as well as propionylation as a modified protocol. Guaiacyl structures in pine milled-wood lignin were 797 µmol/g with no syringyl structures, the source being a softwood (Holtman et al. 2003). Eucalyptus milled-wood lignin was reported to have 162 µmol/g of guaiacyl-type structures and 278 µmol/g of syringyl-type structures. Pine milled-wood lignin, on the other hand, had only 402 µmol/g of guaiacyl-type structures, and no syringyl moieties. Similarly, spruce milled-wood lignin revealed only 520 µmol/g of guaiacyl-type structures and no syringyl moieties. Milled wood lignin from bamboo had 269 µmol/g of guaiacyl-type structures and 486 µmol/g of syringyl-type structures (del Rio et al. 2007). Milled wood lignin from kenaf stalk was found to have 428 µmol/g of guaiacyl-type structures and 1323 µmol/g of syringyl-type structures. Milled wood lignin from the leaves of sisal had 246 μmol/g of guaiacyl-type structures and 486 μmol/g of syringyl-type structures. In comparison, the milled wood lignin from the leaves of abaca had only 54 μmol/g of guaiacyl-type structures and 276 μmol/g of syringyl-type structures (del Rio et al. 2008). Milled wood lignin from wheat straw had 438 μmol/g of guaiacyl-type structures and 200 μmol/g of syringyl-type structures (del Rio et al. 2012b). DFRC characterization of the milled wood lignin from the elephant grass cortex revealed 164.1 μmol/g of guaiacyl-type structures and 295.6 μmol/g of syringyl-type structures (del Rio et al. 2012a). Japanese cedar, its MWL, and soda anthraquinone lignin were analysed for the type of methoxyl substitution. All these lignins were found to be exclusively of guaiacyl type. The wood meal yielded 12.7 mol% guaiacyl structures based on Klason lignin in wood, the MWL showed 11.9 moles guaiacyl structures per 100 moles phenylpropane, while the soda anthraquinone lignins had 0.3–1.5 moles guaiacyl structures per 100 moles phenylpropane. For comparison, the same study estimated the guaiacyl structures to be 0.9 moles per 100 moles phenylpropane in pine kraft lignin (Ikeda and Magara 2015).

5.2.1.5 Phenolic Hydroxyl Groups in Lignin

The presence of phenolic hydroxyl groups is an important functionality in native and technical lignins. As deliberated in Chapter 2, they influence the course of reaction during lignin isolation through dissolution. Once isolated, the presence and abundance of these groups in the technical lignins affect their utilization through derivatization or other reactive modifications. One effective wet chemistry method for the estimation of phenolic hydroxyl groups in lignin involves its acetylation followed by deacetylation with pyrrolidine. During the reaction, there is preferential deacetylation of acetyl groups on the benzene ring and the acetyl groups at aliphatic carbons are largely unaffected. The reaction may be expressed as

$$RCOOR' + (CH_2)_4NH \rightarrow RCON(CH_2)_4 + R'OH \qquad \{5.4\}$$

Here, $COOR'$ represent the acetyl group on the benzene ring, and $RCON(CH_2)_4$ is the reaction product 1-acetylpyrrolidine. Measuring the reaction product 1-acetylpyrrolidine makes it possible to quantitatively estimate the phenolic hydroxyl groups. Commonly, the sample is acetylated with acetic anhydride in pyridine. The acetylated sample is dissolved in dioxane (20–100 mg/mL concentration) and added to pyrrolidine containing 1-propionylpyrrolidine as an internal standard. The produced 1-acetylpyrrolidine is quantified by gas chromatography. Using this method, the phenolic hydroxyl groups in softwood kraft lignin, hardwood kraft lignin, and softwood milled-wood lignin were estimated to be 3.5, 4.7, and 1.8 mmol/g, respectively (Mansson 1983). The same method was employed for the determination of phenolic hydroxyl groups in several milled wood lignins. The phenolic hydroxyl content varied between 0.95% and 2.89% (Faix et al. 1992). Eucalyptus dioxane organosolv lignin had 2.9 phenolic hydroxyl groups per phenylpropane unit (Evtuguin et al. 2001). Wheat straw soda lignin, pine kraft lignin, wheat straw organosolv lignin, and commercial softwood kraft lignin were analysed for phenolic hydroxyl groups. The observed abundances were 3.4, 4, 2.4, and 3.6 mmol/g, respectively (Serrano et al. 2018). Another method for estimation of the free phenolic hydroxyl groups in lignocellulose, comprising extensive solvent extraction at a liquid-to-solid ratio of 30.8, followed by HPLC separation, identification, and quantification, was proposed. Three different extraction protocols were reported, 5 h with a 1:2 mixture of methanol and toluene, 22 h with a 1:2 mixture of ethanol and toluene, and 5 h with chloroform. 5 h extraction with a 1:2 mixture of methanol and toluene yielded the highest amount of compounds having free phenolic hydroxyls (Lawther et al. 1996).

One corollary of the Adler method for the determination of methoxyl groups is the indirect estimation of the phenolic hydroxyl groups. This is premised on the fact that methanol generated in the process comes from the methoxyl groups ortho to the phenolic hydroxyl. This would be suitable for softwood lignins having only guaiacyl moieties so that one mole of methanol produced is equivalent to one mole of phenolic hydroxyl groups. Using this method, phenolic hydroxyl groups in spruce were estimated as 13 per 100 phenylpropane units. This content was 20 per 100 phenylpropane units for spruce milled-wood lignin (Onnerud and Gellerstedt 2003). The phenolic hydroxyl content of Japanese cedar was reported as 10.4 hydroxyl groups per 100 phenylpropane units. In comparison, MWL from the same species had 16.9 hydroxyl groups per 100 phenylpropane units, and soda anthraquinone lignin had 16.3–20.8 hydroxyl groups per 100 phenylpropane units. Commercial kraft lignin from pine was estimated to have 20 hydroxyl groups per 100 phenylpropane units (Ikeda and Magara 2015). Recently, a rapid titration-based determination of phenolic hydroxyl was proposed. The method prescribes the dissolution of lignin in 0.016 M sodium hydroxide and titrating against standard hydrochloric acid. Conductometric titration is preferred over pH-based titration. Phenolic hydroxyl groups determined by this method were 2.4 mmol/g in wheat straw soda lignin, 1.5 mmol/g in wheat straw organosolv lignin, 2.4 mmol/g in pine kraft lignin, and 2.7 mmol/g in commercial softwood kraft lignin. These values were substantially lower than those measured by aminolysis with pyrrolidine (Serrano et al. 2018).

5.2.1.6 Acetyl Groups in Lignin

While hydroxyl groups are the predominant functionality on the γ carbon of the aliphatic side chain of the phenylpropane units in lignin, some of the γ carbons carry acetyl groups. The wet chemistry method for the determination of acetyl groups involves digestion of the material for 50 min with 0.5 M oxalic acid at 152 °C and chromatographic analysis of the products (Solar et al. 1987). DFRC is an effective characterization tool for estimating the acetyl groups in lignin. Eucalyptus milled-wood lignin was reported to have 8 μmol/g of guaiacyl-type structures having acetylated γ carbons, and 3 μmol/g of syringyl-type structures having acetylated γ carbons. Pine milled-wood lignin, on the other hand, reportedly had no acetylated γ carbon moieties. Milled wood lignin from bamboo had 13 μmol/g of guaiacyl-type structures carrying acetyl groups at γ carbons and 4 μmol/g of syringyl-type structures with acetyl groups at γ carbons (del Rio et al. 2007). Milled wood lignins from kenaf stalk, sisal leaves, and abaca leaves were analysed for acetyl groups on the γ carbons. In kenaf lignin, 38 μmol/g of guaiacyl-type structures, and 780 μmol/g of syringyl-type structures had an acetyl group on the γ carbons. For sisal lignin, 124 μmol/g of guaiacyl-type structures, and 378 μmol/g of syringyl-type structures had an acetyl group on the γ carbons. Abaca lignin had 3 μmol/g of guaiacyl-type structures and 131 μmol/g of syringyl-type structures with an acetyl group on the γ carbons (del Rio et al. 2008). It was observed that milled wood lignin from wheat straw had 52 μmol/g of guaiacyl-type structures with acetylated γ carbons, in comparison to only 2 μmol/g of syringyl-type structures with acetylated γ carbons (del Rio et al. 2012b). Milled wood lignin from the cortex of elephant grass had 4.1 μmol/g of guaiacyl-type structures with acetylated γ carbons. This lignin had 17.7 μmol/g of syringyl-type structures with acetylated γ carbons (del Rio et al. 2012a). Coconut coir milled-wood lignin showed the presence of only 6% of syringyl monomers and 1% of guaiacyl monomers with acetylation at γ carbons (Rencoret et al. 2013).

Another method for the determination of acetyl groups is through quantitative deacetylation and analysis of the product. Zemplen transesterification reaction is one way of deacetylation in which the sample is dissolved in anhydrous methanol containing internal standard – ^{13}C labelled acetyl vanillin – and reacted with sodium methanolate for 20 min at 35 °C. Products are analysed by gas chromatography (Audu et al. 2018).

5.2.2 Intermonomer Linkages in Lignin

Lignin is a biological macromolecule. Like any other polymer, it is an aggregation of a large number of fundamental building blocks, the monomers, which are linked to each other by chemical bonds. The monomeric building blocks in lignin are the phenylpropane units. Different phenylpropane structural entities are joined together by different intermonomer linkages to form the lignin macromolecule. Not only the native lignins in plant cell walls but also the lignins isolated from lignocelluloses by various isolation methods retain the macromolecular structure. Identification and quantification of these linkages is an essential and important part of lignin characterization. Knowledge about the bonds holding together the lignin in plant cell walls is helpful in devising optimal strategies for its isolation. Once isolated, its suitability for different applications or further value addition through physicochemical transformations is also influenced by its macromolecular nature.

5.2.2.1 Alkyl Aryl Ether Linkages in Lignin

The majority of intermonomer linkages in lignin are alkyl aryl ether type where the α or β carbon in one phenyl propane unit is joined through an oxygen bridge to a carbon atom of the benzene ring of an adjacent phenylpropane unit. Principal among the alkyl aryl ether linkages are α-O-4′, and β-O-4′ linkages, of which β-O-4′ are the most abundant. Besides their abundance, ether linkages are relatively easier to break. Therefore, characterizing these linkages is of utmost importance, both for their isolation from lignocellulose material and for putting the isolated lignin to its desired end-use.

5.2.2.2 Alkyl Aryl Ether Linkages by Alkaline Nitrobenzene Oxidation

Oxidation with alkaline nitrobenzene can quantitatively break the alkyl aryl ether bonds in lignin while leaving the carbon-to-carbon bonds intact. The result is the formation of monomeric aldehydes, ketones, and carboxylic acid. Their collective yield tells us about the relative abundance of alkyl aryl ether linkages in lignin macromolecules.

Wheat straw powder, when oxidized with alkaline nitrobenzene, produced 6.89% (based on the mass of wheat straw) monomers arising from the breaking of alkyl aryl ether bonds (Sun et al. 1995). Alkaline nitrobenzene oxidation of spruce, eucalyptus, and bamboo provided the abundance of alkyl aryl ether-linked moieties in lignin. Spruce, a softwood, yielded 29.4 g monomers per 100 g Klason lignin from the breaking of these bonds. Eucalyptus, a hardwood resulted in a high yield of 48.5% monomers, and bamboo, a species from the grass family, provided 31.9% monomers from the cleavage of these bonds (Katahira and Nakatsubo 2001). Nitrobenzene oxidation of ground pine, maple, and corn stover was performed to estimate the alkyl aryl ether linkages of their lignins. Pine lignin had 41.9 mol%, maple lignin 55 mol%, and corn stover lignin 41.7 mol%, of these linkages (Min et al. 2015).

Nitrobenzene oxidation of Ethanol organosolv lignin from a hardwood mixture yielded 9% monomers from the breaking of alkyl aryl ether bonds. In kraft lignin, monomer yield from such entities was 6.03% (Hepditch and Thring 1997). Eucalyptus dioxane organosolv lignin yielded 37.5% monomeric product upon nitrobenzene oxidation indicating that 37.5% of the lignin mass consisted of units linked by alkyl aryl ether bonds (Evtuguin et al. 2001). Monomer yield during nitrobenzene oxidation of pine milled-wood lignin was 31.8 mol% and that from pine cellulolytic enzyme lignin was 32.3 mol% (Holtman et al. 2004). Wheat straw milled-wood lignin yielded 76.4% monomers by cleavage of alkyl aryl ether bonds upon alkaline nitrobenzene oxidation, indicating the predominance of these linkages in lignin macromolecule (Sun et al. 2005). Formic acid organosolv lignin from bamboo exhibited 24.9 mol% alkyl aryl ether-linked entities, while bamboo milled-wood lignin had 39.4 mol% of them (Li et al. 2012a). Soda anthraquinone lignin from Japanese cedar yielded 6.6–11.2 mol% vanillin from the breaking of alkyl aryl ether linkages as against 10.9 mol% from pine kraft lignin. For comparison, MWL from Japanese cedar yielded 23.4 mol% of the corresponding monomer (Ikeda and Magara 2015).

5.2.2.3 Alkyl Aryl Ether Linkages by Thioacidolysis

The fact that thioacidolysis selectively breaks the ether linkages in the lignin macromolecule with the substitution of the SC_2H_5 functional group at α, β, and γ carbons on the aliphatic side chain makes it possible to use it as an analytical tool for quantitative estimation of alkyl aryl ether linkages in lignin. The sum of guaiacyl, syringyl, and p-hydroxyphenyl monomer yield is taken as the measure of alkyl aryl ether linkages. The higher the monomer yield in thioacidolysis, the more the relative proportion of alkyl aryl ether linkages.

The softwood spruce, upon thioacidolysis, yielded 1682 μmol/g of monomers linked through alkyl aryl ether bonds (Onnerud and Gellerstedt 2003a). Thioacidolysis of pine wood meal released 1198 μmol/g of phenyl propane monomer units joined by alkyl aryl ether linkages (Holtman et al. 2003). Thioacidolysis monomer yield from pine milled-wood lignin was 18.1 mol% and that from pine cellulolytic enzyme lignin was 20.3 mol% (Holtman et al. 2004). Wild poplar and hybrid plantation poplar were analysed by thioacidolysis for their monomer composition. In wild poplar, the monomers joined by alkyl aryl ether bonds were 1942 μmol/g. In comparison, monomers joined by alkyl aryl ether bonds in hybrid plantation poplar were 2129 μmol/g (Robinson and Mansfield 2009). Pine, spruce, poplar, bamboo, and bagasse were subjected to thioacidolysis and product desulphurization to examine the abundance of alkyl aryl ether bonds. Lignin in softwoods, pine and spruce, were found to have 1065, and 1130 μmol/g, respectively, of monomers joined by these linkages. Lignin in the hardwood, poplar, had 2190 μmol/g of alkyl aryl ether type structures. Lignin in the nonwoods, bamboo and bagasse, had 1086, and 576 μmol/g, respectively, of monomers joined by alkyl aryl ether linkages (Yue et al. 2012). Recent thioacidolysis studies, however, report a much lower yield of alkyl aryl ether linkages in both softwoods and hardwoods. Pine, poplar, and switchgrass were analysed for their lignin structures. Pine had 211.5 μmol/g, poplar comprised of 532.4 μmol/g, and switchgrass had 130.7 μmol/g monomer units joined by alkyl aryl ether linkages (Harman-Ware et al. 2016). Alkyl aryl ether-linked units in eucalyptus were determined as 679.5 μmol/g (Saito et al. 2016). Ground pine, corn stover, and several poplar species were analysed for their lignin structural features by thioacidolysis. Pine was estimated to have 217.7 μmol/g of phenylpropane units linked by alkyl aryl ether bonds. Corn stover released 97.2 phenylpropane units linked by alkyl aryl ether bonds. Alkyl aryl ether-linked moieties in poplar varied between 104.9 and 329.6 μmol/g (Happs et al. 2021).

Wheat straw milled-wood lignin was determined to have 834 μmol/g of monomers joined by alkyl aryl ether linkages. Enzymatic lignin from wheat straw, by contrast, had 764 μmol/g of such monomers (Billa and Monties 1995). Monomer yields from breaking of alkyl aryl ether linkages in the cellulolytic enzyme lignin of willow and kenaf were 597, and 967 μmol/g, respectively, which were lower than those obtained by DFRC (Lu and Ralph 1998). Milled wood lignin from spruce exhibited 1146.2 μmol/g of alkyl aryl ether-linked structures (Marques et al. 1999). In another study, milled wood lignin of spruce had 986 μmol/g of moieties joined by alkyl aryl ether bond, as determined by thioacidolysis (Onnerud 2003; Onnerud and Gellerstedt 2003). Thioacidolysis estimation of eucalyptus dioxane organosolv lignin revealed 1635 μmol/g of phenylpropane units linked by alkyl aryl ether linkages (Evtuguin et al. 2001). Milled wood lignin from coconut coir had 874 μmol/g moieties in which the lignin monomers had alkyl aryl linkages between them (Rencoret et al. 2013). Phenylpropane units linked by alkyl aryl ether bonds in pine milled-wood lignin were determined to be 995 μmol/g. Dioxane organosolv lignin from different maize species consisted of 299–727 μmol/g of alkyl aryl ether-linked entities, as determined by thioacidolysis (Chazal et al. 2014). Enzymatic mild acidolysis lignin from pine showed 1331 μmol/g of structures where the phenylpropane units were joined by alkyl aryl ether linkage (van de Pas et al. 2014). Alkyl aryl ether-linked units in bagasse were reported to be 230 μmol/g. For cottonwood, the content was 520 μmol/g (Chen et al. 2021).

5.2.2.4 Alkyl Aryl Ether Linkages by Derivatization Followed by Reductive Cleavage

Monomer phenylpropane units linked by alkyl aryl ether bonds can also be estimated by DFRC which breaks the ether bonds to release the corresponding monomers, while the carbon-to-carbon bonds are unaffected. The total monomer yield in DFRC is not always in conformity to that obtained in thioacidolysis, or nitrobenzene oxidation but the method can separately quantify the monomers joined by α-O-4′ bonds from those joined by β-O-4′ linkages.

Cellulolytic enzyme lignin from several lignocellulose materials was subjected to DFRC, and the monomer yields indicated the extent of alkyl aryl ether linkages. Monomer yields were: pine 651 μmol/g, willow 732 μmol/g, aspen 696 μmol/g, kenaf 1258 μmol/g, and bamboo 781 μmol/g

(Lu andRalph 1998). In pine milled-wood, β-O-4′ linked entities were 749.6 μmol/g while α-O-4′ linked entities were only 63.4 μmol/g. Milled wood lignin from the same species showed 735.6 μmol/g of β-O-4′ linked entities and 61.4 μmol/g of α-O-4′ linked entities (Holtman et al. 2003). DFRC monomer yield from pine milled-wood lignin as well as from pine cellulolytic enzyme lignin was 14.5 mol%, which was somewhat lower than those obtained by thioacidolysis, and significantly lower than the results from alkaline nitrobenzene oxidation (Holtman et al. 2004). Upon DFRC, eucalyptus milled-wood lignin was reported to produce 162 μmol monomers per g, pine milled-wood lignin 402 μmol monomers per g, spruce milled-wood lignin 520 μmol monomers per g, and bamboo milled-wood lignin 755 μmol monomers per g, that resulted from breaking of alkyl aryl ether bonds (del Rio et al. 2007). Alkyl aryl ether linkages could be estimated from the quantification of phenylpropane monomers by DFRC. Milled wood lignin from kenaf stalk released 1751 μmol monomers per g. Milled wood lignin from the leaves of sisal produced 732 μmol monomers per g. Monomer yield from milled wood lignin from the leaves of abaca was 330 μmol/g (del Rio et al. 2008). DFRC characterization of the milled wood lignin from the elephant grass cortex produced 459.7 μmol monomers per g (del Rio et al. 2012a) and milled wood lignin from wheat straw released 638 μmol monomers per g upon DFRC (del Rio et al. 2012b), indicating the abundance of alkyl aryl ether linkages in these lignins. Japanese cedar, its MWL, and soda anthraquinone lignin were analysed by DFRC for the extent of alkyl aryl ether linkages. The wood meal yielded 12.7 mol% monomers based on Klason lignin in wood, the MWL resulted in an 11.9 mol% monomer yield, while the soda anthraquinone lignin yielded 0.3–1.5 mol% monomers from the breaking of alkyl aryl ether bonds. (Ikeda and Magara 2015).

5.2.2.5 Aryl Aryl Ether Linkages and Carbon-to-Carbon Linkages in Lignin

Standard alkaline nitrobenzene oxidation, thioacidolysis, or DFRC breaks the alkyl aryl ether linkages in lignin while leaving the other linkages intact. Characterization of linkages other than the alkyl aryl ether types is vital for a complete understanding of the lignin molecular structure. These bonds are much more resistant to chemical degradation than the alkyl aryl types. Therefore, their relative abundance in native lignins has a strong influence on the lignin isolation process. Similarly, if an isolated lignin has more such bonds, it would be comparatively less susceptible to chemical derivatization.

5.2.2.6 Thioacidolysis

Standard thioacidolysis only breaks the alkyl aryl ether bonds and allows for their quantitative estimation. It is also believed that no condensation reaction takes place during thioacidolysis forming new carbon-to-carbon linkages. Thioacidolysis reaction products, therefore, consist of different monomeric compounds having a single phenylpropane skeleton. The reaction products also contain dimeric and oligomeric structures where two or more phenylpropane units are joined by carbon-to-carbon linkage or aryl aryl ether linkage. The identification and quantitative estimation of these dimers and oligomers would make it possible to quantify the linkages other than the alkyl aryl ether type. To accomplish this, the thioacidolysis products extracted into organic solvent (dichloromethane, diethyl ether, or ethyl acetate) are desulphurized. Thioacidolysis extract is mixed with an aqueous slurry of Raney nickel and methanol (1:5:25). Desulphurization proceeds at 80 °C for 4 h followed by dilution with water and acidification to pH 3–4. The reaction products are then extracted with dichloromethane and the organic phase is evaporated to dryness under vacuum. The dried mass is derivatized with N,O-bis(trimethylsilyl)trifluoroacetamide and subjected to gas chromatographic separation, identification, and quantification, as in the original thioacidolysis procedure (Lapierre et al. 1991). Dimeric moieties, carrying carbon-to-carbon and aryl aryl ether intermonomer linkages, produced after thioacidolysis and desulphurization are schematically represented in Figure 5.3.

The desulphurization of spruce thioacidolysis products revealed 318.4 μmol/g of dimers in which the phenylpropane units were linked by different carbon-to-carbon linkages. Of these, 90 μmol/g had 5–5′ linkage, 85.8 μmol/g had β-4′ linkage, 94.6 μmol/g had β-5′ linkage, 27.9 μmol/g had

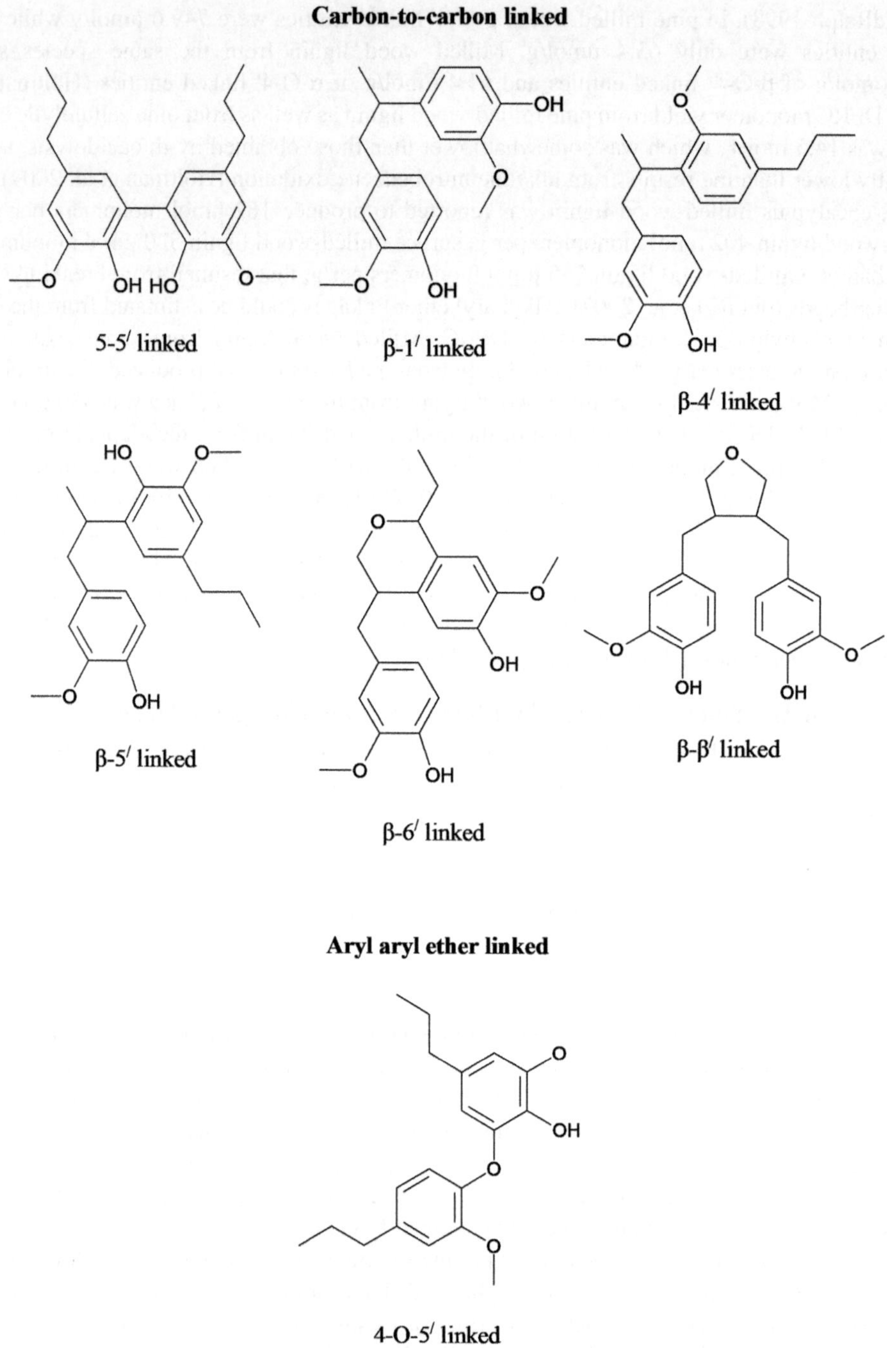

FIGURE 5.3 Lignin dimers produced during thioacidolysis of lignin and lignocelluloses.

β-6′ linkage, and 16.1 μmol/g had β-β′ linkages. In addition, in 20.2 μmol/g of dimers, the phenylpropane units were joined by aryl aryl ether linkage of the type 4-O-5′ (Lapierre et al. 1991). In a later thioacidolysis followed by a desulphurization study on spruce, several carbon-to-carbon linked dimers of phenylpropane units were identified. These were 5-5′, β-4′, β-5′, and β-6′. The aryl aryl ether linkage identified was 4-O-5′ (Onnerud 2003). Milled wood lignin from coconut coir was studied by thioacidolysis and desulphurization of products. As determined by the identification and quantification of dimers, the total entities joined by carbon-to-carbon linkages were 5-5′ 15.9%, β-1′ 22.1%, β-5′ 45.8%, and β-β′ 8.7%. The lignin also had 7.4% of 4-O-5′ aryl aryl ether linkage (Rencoret et al. 2013).

Dimers joined by different carbon-to-carbon, and aryl aryl ether bonds in eucalyptus had the distribution of 50% β-1′, 26% β-β′, 11% β-5′, 4% 5–5′, and 9% 4-O-5′, however, the total yield of dimeric product was not reported (Saito et al. 2016). Pine and spruce, their cellulolytic enzyme lignins, and spruce milled-wood lignin were characterized by thioacidolysis and desulphurization of products. Pine, based on its Klason lignin content, yielded 108.8 μmol/g β-5′ structures, 60 μmol/g β-1′ structures, 55.1 μmol/g 5–5′ structures, 1.2 μmol/g β-β′ structures, and 19.4 μmol/g 4-O-5′ structures. Pine cellulolytic enzyme lignin, on the other hand, yielded 78.6 μmol/g β-5′ structures, 46.1 μmol/g β-1′ structures, 35.1 μmol/g 5-5′ structures, 1 μmol/g β-β′ structures, and 15.2 μmol/g 4-O-5′ structures. Another softwood spruce, based on its Klason lignin content, yielded 103.7 μmol/g β-5′ structures, 58.6 μmol/g β-1′ structures, 49.9 μmol/g 5–5′ structures, 1.6 μmol/g β-β′ structures, and 15.6 μmol/g 4-O-5′ structures. Spruce cellulolytic enzyme lignin, in comparison, yielded 82.3 μmol/g β-5′ structures, 39.9 μmol/g β-1′ structures, 37.5 μmol/g 5–5′ structures, 1.3 μmol/g β-β′ structures, and 16.1 μmol/g 4-O-5′ structures. Spruce milled-wood lignin was also analysed and was found to have 65 μmol/g β-5′ structures, 66 μmol/g β-1′ structures, 62 μmol/g 5–5′ structures, and 16 μmol/g 4-O-5′ structures (Yue et al. 2017).

5.2.3 Lignin Carbohydrate Complexes

Some of the lignin monomers, besides being joined with other lignin monomers, are also linked to carbohydrates present in the lignocellulose or isolated lignin. These lignin carbohydrate complexes, as they are called, need to be characterized to fully elucidate the molecular structure of lignin. They are also important in deciding the purity of a lignin product isolated by any given process. Since complete cleavage of the lignin carbohydrate linkages is not practicable, isolated lignins almost always end up with small amounts of carbohydrates as impurities. The extent of their presence in a lignin formulation may affect its end use. Estimation and characterization of lignin carbohydrate complexes is, therefore, important.

For isolating the lignin carbohydrate complex from milled spruce, it was hydrolysed with endoglucanase. A fraction of lignin carbohydrate complex, mainly containing galactoglucomannan (galactan 12.6%, glucan 5.2%, mannan 33.5%, lignin 39%), was precipitated from the liquid phase of the enzymatic broth by the addition of barium hydroxide. The remaining enzyme-treated material was mixed with 8M urea and allowed to swell for 48 h. The dissolved fraction was separated from the solids and precipitated with barium hydroxide solution to obtain a second lignin carbohydrate fraction with glucomannan, and xylan (glucan 9.3%, mannan 19.2%, xylan 3.1%, lignin 56%). The residue that did not dissolve in urea was mixed with an aqueous solution of 18% sodium hydroxide and 3% boric acid. The undissolved material was the lignin carbohydrate complex containing glucan (glucan 85.2%, lignin 7%). Another fraction of lignin carbohydrate complex containing glucomannan, and xylan (glucan 9.3%, mannan 33.6%, xylan 3.2%, lignin 41%) was precipitated by the addition of barium hydroxide solution (Lawoko et al. 2005). The lignin carbohydrate complex was isolated from pine and birch milled-wood lignin in 30–40% yield. For this, the milled wood lignin was dissolved in 90% aqueous acetic acid and reprecipitated by adding dropwise in water. The unprecipitated solution was evaporated to dryness to obtain the lignin carbohydrate complex. Carbohydrate content in the resulting lignin carbohydrate complex varied between 28% and 40% for pine, with higher milling time resulting in more lignin carbohydrate linkages. Associated carbohydrates were mainly arabinose and galactose. For birch lignin, the carbohydrate complex was 22% with the main carbohydrate entity being xylan (Balakshin et al. 2011). A similar process was used to extract the lignin carbohydrate complex from bamboo. The material so obtained contained 41.7%

xylan, 6.2% arabinan, 4.3% glucan, and 1.4% galactan (Huang et al. 2015). The same method was also adopted to extract lignin carbohydrate complex from Arundo Donax in 1.7% yield based on the lignin content of lignocellulose mass. Its composition was 91.5% lignin, 6.6% xylan, 1.1% glucan, 0.2% galactan, and 0.3% arabinan (You et al. 2015). In one protocol, to fractionate the lignin carbohydrate complexes, extractive-free wood meal was ball-milled and dissolved in a 40% aqueous solution of a mixture of dimethyl sulfoxide and tetra-n-butylammonium hydroxide in equal volume. The resulting solution was added to about five volumes of deionized water. Lignin carbohydrate complexes containing predominantly glucan units were precipitated and collected after thorough washing. The remaining solution was added to about five volumes of saturated barium hydroxide. This time, the lignin carbohydrate complexes containing predominantly glucomannan units were precipitated and collected after thorough washing. The remaining solution was acidified with hydrochloric acid and a precipitate containing lignin carbohydrate complexes with xylan units was collected by dialysis. In spruce lignin, 49.5% of lignin carbohydrate complexes were found to be predominantly glucan type, 30.9% were predominantly glucomannan type, and 12.8% were predominantly xylan type (Du et al. 2013). Another study reports the fractionation of lignin carbohydrate complexes from a ball-milled spruce wood meal. The first fraction was obtained by extracting the ball-milled material with 80 °C deionized water for 4 h at a solvent-to-solid ratio of 10. The extract was treated with Amberlite XAD4 resin and the eluted solution was dried. Spent resin was regenerated with methanol and the second fraction was obtained by evaporating methanol. Residue from water extraction was dissolved in an equivolume mixture of 1-allyl-3-methylimidazolium chloride and dimethyl sulfoxide. To ensure complete dissolution, initially, the residue was stirred in 1-allyl-3-methylimidazolium chloride, at a liquid-to-solid ratio of 2 for 2 h at 80 °C. Next, dimethyl sulfoxide was added with stirring for another 2 h at 80 °C. Another portion of dimethyl sulfoxide was added to the mixture and left overnight at 70 °C. The third fraction of the lignin carbohydrate complex was precipitated by adding water to the formed solution and centrifuged. A fourth fraction was prepared by addition of ethanol to the remaining solution and allowing it to stand overnight at 4 °C for precipitate to form, which was centrifuged. Acetone was added to the solution left after precipitation of the fourth fraction to precipitate the fifth fraction of the lignin carbohydrate complex (Giummarella et al. 2016). The method was also used for the fractionation of lignin carbohydrate complex from birch wood meal with the exception that the fifth fraction was obtained by the addition of water in place of acetone (Giummarella and Lawoko 2016). In a modification to this fractionation method, lignin carbohydrate complexes were isolated from ball-milled pine wood meal and ball-milled pine sulphite pulp. The first fraction was obtained by extracting the ball-milled material with 80 °C deionized water for 4 h at a solvent-to-solid ratio of 10 and freeze-drying the extract. The residue was dissolved in an equivolume mixture of 1-allyl-3-methylimidazolium chloride and dimethyl sulfoxide and precipitated by adding into water to obtain a second fraction of the lignin carbohydrate complex. A third fraction was prepared by precipitation from the remaining solution by the addition of ethanol (Deshpande et al. 2018).

5.3 MOLECULAR WEIGHT AND MOLECULAR WEIGHT DISTRIBUTION IN LIGNIN

Being a polydisperse macromolecular substance, the average molecular weight and the molecular weight distribution among the different macromolecules constituting the sample are characteristic of any lignin formulation. Many physicochemical characteristics of lignin are critically dependent on its molecular weight and molecular weight distribution. Therefore, it also profoundly affects the possible end use and value addition of lignin. The average molecular weight of an isolated lignin tells about the extent of degradation and defragmentation that the lignin has undergone during the isolation process.

5.3.1 Gel Permeation Chromatography

Gel Permeation Chromatography (GPC), also known as Size Exclusion Chromatography, is a widely used technique for the characterization of polymers for their molecular weight distribution. The polymer solution is forced through a chromatographic column consisting of a known and controlled porosity and pore size distribution. Due to steric hindrance, the column differentially adsorbs and desorbs the molecular fractions in the solution according to their hydrodynamic volume that, in turn, relates to the molecular weight. Molecules that are too large to enter the pores in the column are not retained at all and elute quickly. On the other hand, molecules of very small hydrodynamic volume enter the interiors of the pore where they are retained for a long time, therefore, eluting very late. As a result, the polymer fragments with different molecular weights elute from the column at different times and their abundance is tracked by the detector. The obtained chromatogram is a plot, often in a logarithmic scale, between the relative abundance of the polymer fragments against their corresponding molecular weight.

A natural polymer, like lignin, is a polydisperse material, which means that all the macromolecules in the lignin sample are not of the same molecular weight. Valuable information can be extracted from the GPC chromatogram regarding the nature of molecular weight distribution in lignin. Different moments of the molecular weight distribution curve provide different molecular weight averages. The number average molecular weight, M_n, can be defined as (Shrivastava 2018)

$$M_n = \frac{\sum_i N_i M_i}{\sum_i N_i} \tag{5.1}$$

where N_i is the number of molecules in the ith fraction of the sample and M_i is the molecular weight of the molecules in the ith fraction. Similarly, the weight average molecular weight, M_w can be defined as

$$M_w = \frac{\sum_i N_i M_i^2}{\sum_i N_i M_i} \tag{5.2}$$

The ratio of M_w to M_n is the polydispersity – a measure of the spread of the molecular weight distribution curve. The peak of the molecular distribution curve, M_p, represents the molecular weight of the fraction with the highest abundance. A typical unimodal molecular weight distribution for ethanol organosolv lignins is presented in Figure 5.4.

In addition to the average molecular weights, M_w, and M_n a qualitative visual inspection of the chromatogram is also desirable. For polymers like isolated lignins, the molecular weight distribution is far from normal with a lot of skewness. Many lignin samples may have bimodal distributions – molecular weight distribution curves with two peaks. A bimodal distribution can be interpreted as a mixture of two lignin types in terms of peak, and average molecular weight. A typical bimodal molecular weight distribution of lignins is shown in Figure 5.5 for softwood soda anthraquinone lignins delignified after prehydrolysis with different process severities.

5.3.1.1 Chromatographic Procedure

Columns used for GPC are made up of small, nearly spherical beads (gel) from crosslinked polymers or suitable ceramic material. These beads have heterogeneous pores that define their suitability for a particular application. Besides the pore structure of the beads, the size of individual beads is also important. Together, they govern the resolution of separation and the speed of separation. The beads must be chemically inert and mechanically stable under the test conditions. The column

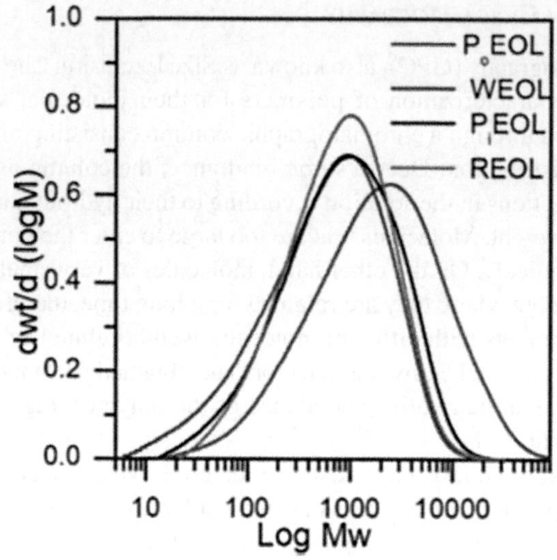

FIGURE 5.4 Unimodal lignin molecular weight distributions for ethanol organosolv lignins. Starting lignocellulose materials are: P_0EOL poplar, WEOL walnut shell, P_iEOL pine, and REOL rice husk. (Reprinted from Sheng, Y, Ma, Z, Wang, X, Han, Y, *Food Chem.*, 376, 131895, 2022, with permission from Elsevier.)

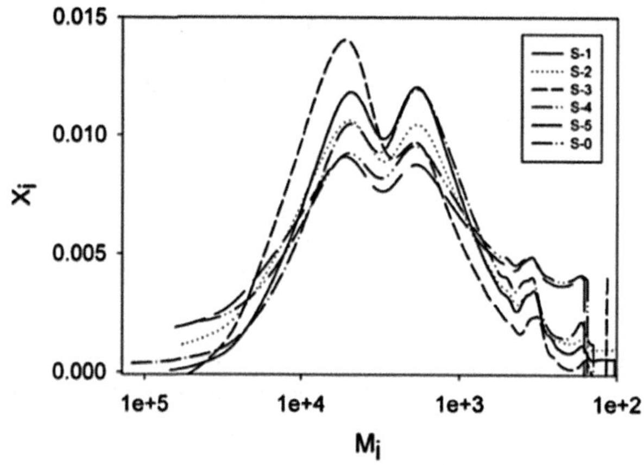

FIGURE 5.5 Bimodal lignin molecular weight distributions for softwood prehydrolysis soda anthraquinone lignins. Prehydrolysis process severities are: S-1 2.64, S-2 3.86, S-3 3.89, S-4 4.26, S-5 4.6, S-0 no prehydrolysis. (Reprinted from Angles, MN, Reguant, J, Garcia-Valls, R, Salvado, J, *Wood Sci. Technol.*, 37, 309–320, 2003, with permission from Springer.)

is integrated into a High-Performance Liquid Chromatography (HPLC) set-up, consisting of a syringe- or reciprocating-type high-pressure accurate discharge pump, a provision to inject a known volume of the lignin solution into the column, and a detector to measure the abundance of lignin fragments in the eluted solution. Additionally, the column may be enclosed in a constant temperature enclosure for better control of separation. There are several reports of using columns packed with crosslinked polystyrene divinylbenzene beads for lignin characterization (Li et al. 2012a; Nitsos et al. 2016; Tan et al. 2016). These columns are available in different configurations with varying column lengths, diameters, and bead sizes (5–10 μm). Beads of different sizes, made up of hydroxylated polymethacrylate, have been used as column packing for GPC characterization (del Rio et al. 2012a; del Rio et al. 2012b; Moniz et al. 2018; Rencoret et al. 2013). Columns packed with dextran–agarose composite beads have been used for GPC characterization of lignin (Chen et al. 2020). Another polymeric bead material used as column packing for lignin GPC characterization is Ethylene-Bridged Hybrid (BEH) particles available in different particle size options (Mun et al. 2021).

At the beginning of the test, the column is equilibrated with the mobile phase, the liquid used to elute the polymer fragments from the column. This equilibration process fills the pores in the beads with the eluent. Next, the equipment is calibrated with a series of polymer solutions of known molecular weights and very narrow molecular weight distribution, noting down the time of elution for each molecular weight. A calibration curve is drawn with molecular weight as the ordinate and elution time as abscissa. Ideally, the calibrating standard should be monodisperse lignin samples of different molecular weights. Practically, however, such standards are rarely available. The most frequently used calibration standard is polystyrene which is available with narrow molecular weight distributions. However, it is important to note that a polystyrene molecule of a given molecular weight would have a hydrodynamic volume different from that of a lignin molecule of the same molecular weight. The range of molecular weight used in the calibration depends on the sample to be analysed. GPC column calibration was done with 162 to 4×10^6 g/mol polystyrene standard for molecular weight estimation of softwood soda anthraquinone and kraft lignins (Angles et al. 2003). For bagasse soda lignin GPC characterization, the column was calibrated with polystyrene standard with molecular weights of 4950–1,188,400 g/mol (Mousavioun and Doherty 2010). GPC calibration with polystyrene standard with a molecular weight range from 890 to 1.86×10^6 g/mol is reported for milled wood lignins from spruce and beech (Crestini et al. 2011). For coconut coir milled-wood lignin elution times were calibrated with monodisperse polystyrene of molecular weight 266 to 3.84×10^6 (Rencoret et al. 2013). Column calibration for GPC characterization of eucalyptus ethanol organosolv lignin was carried out with polystyrene standard with molecular weight range from 580 to 3.6×10^6 g/mol (Yanez-S et al. 2014). The molecular weight of the polystyrene standard ranged from 250 to 70,000 g/mol for GPC calibration used for molecular weight characterization of ethanol organosolv lignins from spruce (Nitsos et al. 2016). Elution times were calibrated with polystyrene standards with molecular weight ranging from 580 to 19,600 g/mol for milled wood lignins from untreated and green liquor-treated poplar (Tan et al. 2016).

A known volume of lignin solution is then injected into the column. The choice of solvent to prepare the lignin solution, and the mobile phase is important. In most cases, both the liquids are the same. Tetrahydrofuran has been frequently used to dissolve lignin and to act as the mobile phase during the GPC run (Crestini et al. 2011; Tan et al. 2016; Yanez-S et al. 2014). Some investigators report the use of aqueous sodium hydroxide solution to dissolve lignin samples and also as the mobile phase for GPC characterization (Moniz et al. 2018; Mousavioun and Doherty 2010; Sameni et al. 2016). There are reports of dimethylformamide with lithium bromide dissolved in it being used as the mobile phase (del Rio et al. 2012a; del Rio et al. 2012b; Rencoret et al. 2013). A recent study reports the use of aqueous acetone with a trace of formic acid as the mobile phase for characterizing organosolv lignins (Ponnudurai et al. 2023). Once dissolved, the lignin macromolecules coil into aggregates. It is the hydrodynamic volume of these aggregates which controls their access into the pores of the beads. Generally, the higher the molecular weight of the lignin macromolecule, the higher the hydrodynamic volume of the formed aggregate. It is important to ensure complete dissolution of lignin in the solvent. Several lignin derivatization protocols are employed prior to GPC characterization to improve the solubility of lignin. Acetylation is the most practised method to improve lignin solubility. Acetylation of ethanol organosolv lignin with acetic anhydride in anhydrous pyridine for 24–72 h is reported. The lignin-to-pyridine ratio varied from 18.75 to 40 mg/mL. A lignin-to-acetic anhydride ratio between 37.5 and 40 mg/mL is reported. At the end of acetylation, the reaction is quenched and acetylated lignin is precipitated by adding ethanol or diethyl ether (Li and Wu 2014; Yanez-S et al. 2014). Another commonly used derivatization practised in recent works is acetobromination (Crestini et al. 2011; Thoresen et al. 2021). In one protocol, lignin was acetobrominated by stirring for 2 h at room temperature in 8% acetyl bromide solution in glacial acetic acid with lignin to solution ratio of 4. The liquid was evaporated after the reaction to obtain the acetobrominated lignin residue which was dissolved in 1 mL of tetrahydrofuran for 2 mg of starting lignin (Crestini et al. 2011).

Detection of the abundance of polymer fragments in the eluted solution is commonly made either by measuring their refractive index or their ultraviolet absorption. In both ways, it is possible to measure the concentration of the polymer. UV detectors, both conventional as well as diode array types, are normally set at 280 nm for estimating lignin concentration. Refractive index detectors typically measure the change in the refractive index of the eluting solution relative to the pure eluent. These are most common in GPC characterization. The advantage of refractive index detectors is their linear response as the refractive index is directly proportional to the solution concentration. A more sophisticated detector is the Multi Angle LASER Light Scattering (MALLS) detector, which works on the principle of scattering of a LASER beam by polymer molecules. Rather than measuring the concentration, these detectors directly measure the molecular weight of the polymer fraction. The use of a MALLS detector in combination with a refractive index detector makes it possible to measure the absolute molecular weights of eluent fractions (Fredheim et al. 2003).

5.3.1.2 Organosolv Lignins

Aqueous formic acid organosolv lignin from bamboo, after acetylation, had weight average molecular weight (M_w) of 10,020, and number average molecular weight (M_n) of 4450 milled wood lignin, resulting in the polydispersity being 2.23. In comparison, M_w and M_n of bamboo milled-wood lignin were found to be 12,090 and 5410, respectively, with a polydispersity of 2.23. It shows that lignin isolation by aqueous formic acid organosolv process causes more degradation to lignin compared to milled wood lignin. For both the lignins, GPC chromatograms showed nearly unimodal molecular weight distribution (Li et al. 2012a). Eucalyptus ethanol organosolv lignins, after acetylation, were subjected to GPC characterization. M_w ranged from 2960 to 4270, M_n ranged from 1170 to 1360, and polydispersity was between 2.53 and 3.16. Ethanol concentration, treatment time, and temperature, all affected the molecular weight, but a clear correlation could not be established (Yanez-S et al. 2014). Ethanol organosolv lignins from spruce, upon GPC assessment, revealed M_w of 1904–11,115, M_n of 603–1656, and polydispersity of 2.6–7.2. For birch ethanol organosolv lignins, these values were M_w 1733–3464, M_n 599–1075, and polydispersity 2.4–4.2. Higher process severity resulted in lower molecular lignin (Nitsos et al. 2016). Ethanol organosolv lignin from rice straw had M_w of 1839–2011 and M_n of 567–606 with polydispersity of 3.2 and 3.3. Organosolv treatment at elevated temperatures caused more degradation of the lignin macromolecule. The study also compared the results with the molecular weight of commercial hardwood ethanol organosolv lignin (Alcell lignin) with M_w, M_n, and polydispersity of 3180, 675, and 4.6, respectively (Moniz et al. 2018). Aqueous ethanol organosolv lignin from spruce, beech, and wheat straw were assessed for their molecular weights and weight distributions by GPC. Spruce organosolv lignin had the highest M_w of 6100 with M_n of 1100 and polydispersity of 5.55. Beech organosolv lignin had M_w of 2500 and M_n of 1000 – polydispersity 2.5. M_w of wheat straw organosolv lignin was 3400 and M_n 1000 with polydispersity of 3.4. For comparison, commercial soda lignin Protobind 1000 was reported to have M_w of 6000 and M_n of 1200 with polydispersity of 5 (Rossberg et al. 2019). Aqueous ethanol organosolv lignins were prepared with different isolation conditions and characterized for their molecular weights. Lignin isolated with 50% aqueous ethanol for 30 min, with no catalyst, had M_w of 7200 and M_n of 1250 with a polydispersity of 5.8, whereas lignin isolated with 70% aqueous ethanol for 30 min, with no catalyst, had M_w of 3550 and M_n of 1300 with a polydispersity of 2.7. Lignin isolated with 60% aqueous ethanol for 15 min, with 0.2% sulphuric acid catalyst, had M_w of 6100 and M_n of 1250 with polydispersity of 4.9, whereas lignin isolated with 60% aqueous ethanol for 15 min, with 1% sulphuric acid catalyst, had M_w of 4400 and M_n of 1600 with polydispersity of 2.7 (Thoresen et al. 2021). Ethanol organosolv lignin from poplar, pine, walnut shells, and rice husk were characterized by GPC. Walnut lignin had the highest M_w, and M_n of 4540, and 590 (polydispersity 7.69). Pine lignin had the next highest M_w, and M_n of 1560, and 320 (polydispersity 4.88). M_w, M_n, and polydispersity of rice husk lignin were 1330, 370, and 3.59, respectively. The lowest molecular weight lignin was that from poplar with M_w, M_n, and polydispersity of 1160, 180, and 6.44, respectively. All four lignins exhibited unimodal molecular weight distribution (Sheng et al. 2022).

5.3.1.3 Lignins from Alkaline Delignification

Acetylated soda anthraquinone lignin from a softwood mixture of pine and spruce was analysed by GPC and exhibited a bimodal molecular weight distribution. A similar distribution was also observed for the commercial softwood kraft lignin Indulin. Soda anthraquinone lignin had M_w of 5490 and M_n of 1260 with a polydispersity of 4.3. Indulin had somewhat lesser M_w of 5403 but M_n of 1379, resulting in polydispersity being 3.9 (Angles et al. 2003). Bagasse soda lignin was estimated to have M_w of 2410 and M_n of 2160 with a polydispersity of 1.12. This lignin was fractionated by solvent extraction into three fractions. The fraction extracted into diethyl ether had M_w of 560 and M_n of 500 – polydispersity 1.12. The second fraction, obtained by extracting the residue from ether extraction with methanol, had M_w of 2670 and M_n of 2380 – polydispersity of 1.12. The lignin residue left after both extractions had M_w of 5990 and M_n of 5350 – polydispersity 1.12 (Mousavioun and Doherty 2010). Soda lignin from *Cunninghamia lanceolata* was fractionated by successive acidification of black liquor to pH 8, 5, and 2. The lignin fractions so isolated were acetylated and analysed by GPC. The fraction isolated at pH 8 was reported to have M_w of 16,115 and M_n of 8780 with a polydispersity of 1.84. The fraction isolated at pH 5 had M_w of 14,218 and M_n of 7679 with a polydispersity of 1.85. Lastly, the fraction isolated at pH 2 consisted of lignin with M_w of 11,001 and M_n of 5934 with a polydispersity of 1.85. It clearly shows that during acid precipitation lignin with higher molecular weight precipitates first (Li and Wu 2014). Acetylated soda lignins from spruce, upon GPC assessment, revealed M_w of 12,007–13,836, M_n of 2887–3301, and polydispersity of 3.64–4.52. For birch soda lignins, these values were M_w 7136–13,902, M_n 1280–3470, and polydispersity 3.98–5.58 (Nitsos et al. 2016). GPC analysis of commercial softwood kraft lignin, Indulin AT, revealed M_w of 6096, M_n of 1191, and polydispersity of 5. Another commercial nonwood soda lignin, Protobind, was reported to have M_w of 5008, M_n of 1084, and polydispersity of 5.12 (Sameni et al. 2016). Commercial soda lignins from wheat straw, and a mixture of wheat straw and sarkanda, were analysed for their molecular weights by GPC. Wheat straw lignin had M_w, M_n, and polydispersity of 5671, 834, and 6.8, respectively. Soda lignin from wheat straw sarkanda mixture had M_w, M_n, and polydispersity of 4495, 743, and 6.1, respectively (Moniz et al. 2018). Eucalyptus kraft lignin was fractionated by dissolution and reprecipitation. Lignin was dissolved in a mixture of methanol and acetone – 70:30 v/v, liquid-to-solid ratio 3.5 mL/g. The first lignin fraction was obtained by reprecipitation into ethyl acetate. The second lignin fraction was obtained by concentrating the remaining solution and reprecipitation into a 1:1 mixture of ethyl acetate and petroleum ether. A third fraction was obtained from concentrating the remaining solution and reprecipitation into petroleum ether. Lignin samples were analysed by GPC after acetylation. Kraft lignin as well as the three fractions returned an unimodal molecular weight distribution. Unfractionated kraft lignin had M_w of 4664 and M_n of 2033 (polydispersity 2.29). The fraction precipitated in ethyl acetate consisted of higher molecular weight, having M_w of 6158 and M_n of 2872 (polydispersity 2.14). The fraction precipitated in a mixture of ethyl acetate and petroleum ether consisted of lower molecular weight having M_w of 2397 and M_n of 1611 (polydispersity 1.49). The last fraction precipitated in petroleum ether consisted of the lowest molecular weight having M_w of 817 and M_n of 620 (polydispersity 1.32) (Chen et al. 2020). Similarly, different fractions were obtained from wheat straw soda lignin by successive dissolution and evaporation of solvent. The solvents used were ethyl acetate, 2-butanone, methanol, acetone, and 95% aqueous dioxane, in that sequence. Unfractionated soda lignin had M_w of 2880 and M_n of 1108 (polydispersity 2.6). Ethyl acetate dissolved fraction had M_w of 1060 and M_n of 610 (polydispersity 1.7). The 2-butanone dissolved fraction was found to have M_w

of 1780 and M_n of 1050 (polydispersity 1.7). M_w and M_n for the fraction dissolved in methanol were 2920 and 1480, respectively (polydispersity 2). Acetone dissolved the lignin fraction with M_w of 4990 and M_n of 2420 (polydispersity 2). Lastly, 95% aqueous dioxane dissolved the lignin fraction with the highest molecular weight, M_w 7790 and M_n 1950 (polydispersity 4). For comparison, the same study reported the M_w, M_n, and polydispersity of LignoBoost kraft lignin to be 4580, 1477, and 3.1, respectively (Lee et al. 2021). Acetylated kraft lignin from mixed hardwoods, upon GPC analysis, was found to have M_w of 3041 and M_n of 1651 with polydispersity of 1.8. For comparison, milled wood lignin from these mixed hardwoods had M_w of 13,314 and M_n of 5657 with a polydispersity of 2.4. The results illustrate that kraft delignification significantly degraded lignin, resulting in molecular fragments of much-reduced size with narrower size distribution (Mun et al. 2021).

5.3.1.4 Lignosulphonates

Ultrafiltered spruce sodium lignosulphonate was estimated to have M_w of 61,000 and M_n of 5000 with polydispersity 12.3. This lignosulphonate, after adjusting the pH to 11–12 and heat treatment, had M_w of 41,000 and M_n of 5100 with polydispersity 11.2. GPC characterization of aspen calcium lignosulphonate determined M_w of 12,000 and M_n of 2200 with 5.3 polydispersity. Eucalyptus calcium lignosulphonate, after heat treatment, had relatively low molecular weight, M_w 5700, M_n 1900, and polydispersity 3 (Fredheim et al. 2003). In another study, M_w, M_n, and polydispersity of spruce calcium lignosulphonate were estimated by GPC to be 43,300, 6500, and 6.7, respectively. This lignosulphonate after ultrafiltration, oxidation, and ion exchange to sodium form, had M_w of 10,500 and M_n of 4700, with a polydispersity of 2.2. Spruce sodium lignosulphonate, without further treatment, had M_w of 53,000 and M_n of 4150, with a polydispersity of 12.8. Eucalyptus calcium lignosulphonate, in comparison, was of a much lower molecular weight with M_w 6900, M_n 2400, and polydispersity 2.9 (Ekeberg et al. 2006). GPC characterization of different sodium lignosulphonates from eucalyptus revealed M_w of 11,600–19,200. Sodium lignosulphonates from acacia had a comparatively lower M_w of 7600–10,700 (Fiskari and Kilpeläinen 2021).

5.3.1.5 Milled Wood Lignins and Hydrolysis Lignins

Milled wood lignin from spruce and beech and enzymatic mild acidolysis lignin from spruce were assessed for their molecular weights by GPC after acetylation. M_n for spruce milled-wood lignin was 14,200 but for enzymatic mild acidolysis lignin, it was only 7300, clearly establishing appreciable lignin degradation during enzymatic and acid treatment. Beech milled wood lignin had an M_n of 9600 (Crestini et al. 2011). Milled wood lignin from the stem of elephant grass was estimated by GPC to have M_w of 6720–6920 and M_n of 2390–2490, with polydispersity varying from 2.7 to 2.9 (del Rio et al. 2012a). Milled wood lignin from wheat straw, another nonwood of the grass family, showed M_w of 4210 and M_n of 1850 with a polydispersity of 2.27 (del Rio et al. 2012b). Coconut coir milled-wood lignin was estimated to have M_w of 7900 and M_n of 2900, with a polydispersity of 2.7 (Rencoret et al. 2013). Eucalyptus milled-wood lignin was reported to have M_w of 6700, M_n of 2600, and polydispersity of 2.6 (Yanez-S et al. 2014). Enzymatic hydrolysis lignin from nonwood after a steam explosion and cellulolytic treatment was found to have a high M_w of 13,488 but a relatively low M_n of only 1093. This lignin, as a result, had a high polydispersity of 12.34. The GPC chromatogram of this lignin exhibited a distinct bimodal character (Sameni et al. 2016). GPC analysis of enzymatic mild hydrolysis lignin from rice straw, a nonwood from the grass family, reported the M_w and M_n to be 8295 and 4732, respectively, with a polydispersity of 1.75. The GPC chromatogram returned a peak molecular weight (M_p) of 10,643 in the polydisperse lignin sample (Lou et al. 2018). Milled wood lignin poplar was analysed by GPC and M_w, M_n, and polydispersity were found to be 9009, 4953, and 1.82, respectively. Milled wood lignin obtained from 8% total alkali green liquor treated poplar had M_w of 6012, M_n of 3222, and polydispersity of 1.87. Green liquor treatment with 16% total alkali before preparation of milled wood lignin further reduced the molecular weight to M_w 5757, M_n 3489, and polydispersity 1.65 (Tan et al. 2016).

5.3.2 Vapour Pressure Osmometry

An older, and presently less prevalent, technique for molecular weight determination of lignin is vapour pressure osmometry. Being a colligative property, the decrease in the vapour pressure of the solution relative to the vapour pressure of the pure solvent is dependent on the number of solute microparticles (molecules, or ions) present in the solution. The same mass of a substance of higher molecular weight will have a smaller number of microparticles than a substance of lower molecular weight. Therefore, at the same mass concentration, a higher molecular weight substance will cause less decrease in the vapour pressure. This allows for the estimation of number average molecular weight of a polymer by measuring the vapour pressure of a polymer solution of known mass concentration and comparing it with the vapour pressure of the pure solvent.

It works on the principle of a decrease in vapour pressure (elevation of boiling point) of a liquid when a solute is dissolved in it. From the principles of thermodynamics, this decrease in vapour pressure can be expressed as (Rudin and Choi 2013)

$$\frac{\Delta p}{p_s^0} = \frac{\mu_s - \mu_s^0}{RT} = V_s^0 C\left[\frac{1}{M_n} + \mathrm{A}_1 C + \mathrm{A}_2 C^2 + \ldots\right] \tag{5.3}$$

where Δp is the decrease in vapour pressure, p_s^0 is the vapour pressure of pure solvent at standard conditions, μ_s is the chemical potential of the solvent in the solution, and μ_s^0 is the chemical potential of pure solvent at standard conditions, T is the absolute temperature, V_s^0 is the molar volume of pure solvent at standard conditions, C is the solute concentration, M_n is the molecular weight of the solute, and A_1, A_2 are constants. Two thermistors at the same temperature are placed in a chamber filled with saturated vapour of the solvent. One thermistor is brought in contact with a drop of pure solvent while the other is contacted with a drop of lignin solution of known concentration. To attain thermodynamic equilibrium, condensation occurs at the thermistor having the solution drop, raising its temperature to the boiling point of the solution at prevailing conditions. The temperature difference between the two thermistors is related to the decrease in vapour pressure of the solution as

$$\Delta p = \frac{\Delta T \lambda_s^V p_s^0}{RT^2} \tag{5.4}$$

where ΔT is the temperature difference between the thermistors and λ_s^V is the latent heat of vaporization of solvent at temperature T. Combining eqs. 5.3 and 5.4

$$\frac{\Delta T}{C} = \frac{V_s^0 RT^2}{\lambda_s^V}\left[\frac{1}{M_n} + \mathrm{A}_1 C + \mathrm{A}_2 C^2 + \ldots\right] \tag{5.5}$$

In practice, the temperature difference is related to the resistance deviation in the electrical bridge containing the thermistors. Similarly, the quantity $\frac{V_s^0 RT^2}{\lambda_s^V}$ is replaced with a lumped instrument constant which is measured experimentally by calibrating against a solution with a solute of known molecular weight. Therefore, for dilute solutions, the working equation after neglecting the concentration terms within the bracket on the right-hand side of eq. 5.5, has the form

$$M_n = \frac{kC}{\Omega} \tag{5.6}$$

Here, k is the instrument constant and Ω is the deviation in the resistance bridge.

Aspen milled-wood lignin and hydrolysis lignins were dissolved in dimethyl formamide and characterized for molecular weight by vapour pressure osmometry, after calibrating the instrument with benzil. M_n of milled wood lignin was found to be 2120 and for different hydrolysis lignins, M_n varied between 970 and 1560 (Chua and Wayman 1979). Soda lignin and ethanol organosolv lignin were isolated from cottonwood and characterized by vapour pressure osmometry. M_n for soda lignin fractions varied between 2250 and 6450. For organosolv lignin fractions, these values were between 835 and 9495 (Pla et al. 1986). Different kraft lignin samples from pine, dissolved in dimethyl formamide, were characterized by vapour pressure osmometry and M_n varied between 1900 and 2400 (Schmidl et al. 1990). Vapour pressure osmometry was used to determine the number average molecular weights of spruce dioxane lignin, spruce oxygen aqueous acetone organosolv lignin, eucalyptus dioxane lignin, and eucalyptus kraft lignin. M_n values were 2400, 1700, 2150, and 650, respectively (Evtuguin et al. 1999). M_n of dioxane lignin from alkali extracted eucalyptus, as determined by vapour pressure osmometry was found to be 2180 (Evtuguin et al. 2001). Several alkali lignins and aqueous ethanol organosolv lignin were characterized by vapour pressure osmometry. Samples were dissolved in dimethyl formamide containing 0.2 M lithium chloride and the instrument was calibrated with glucose pentaacetate. M_n of wheat straw alkali lignin was 1420, for hemp alkali lignin it was 640, and for flax alkali lignin M_n was 600. Mixed hardwood ethanol organosolv lignin had an M_n of 670 (Gosselink et al. 2004). Being an absolute method for molecular weight determination, vapour pressure osmometry was used to determine the number average molecular weights of monodisperse lignin fractions extracted from sugarcane bagasse acetosolv lignin by preparative GPC. The idea was to use these monodisperse lignin fractions as calibration standards in the GPC characterization of lignins. Seven fractions were isolated with M_n of 2270, 2490, 2040, 1160, 770, 630, and 350. Samples were dissolved in tetrahydrofuran before analysis (Botaro and Curvelo 2009). Alcell (ethanol organosolv) lignin from mixed hardwoods was characterized by vapour pressure osmometry to have an M_n of 760. Indulin AT (commercial softwood kraft lignin) had an M_n of 1079 as determined by vapour pressure osmometry (Cateto et al. 2011). Both the values were significantly lower than that obtained by GPC calibrated with polystyrene – 3751 for Alcell lignin and 6165 for Indulin AT (Cateto 2008). For these analyses, lignin samples were dissolved in 2-methoxyethanol and the instrument was characterized with benzil.

5.3.3 LASER Light Scattering

An absolute method of determination of weight average molecular weight of lignin is by LASER light scattering. The method is based on the optical phenomenon of light scattering. When an electromagnetic wave is incident on an isotropic microparticle (a lignin molecule for instance) of dimension smaller than the wavelength of light, it induces polarization in the microparticle, and a fluctuating dipole moment is set up. This causes the microparticle to act as a secondary source of electromagnetic radiation of wavelength equal to that of the incident light. This secondary radiation propagates in all directions, hence the name scattering. The intensity of light scattered from a solution of microparticles in a given direction can be expressed as (Rudin and Choi 2013)

$$I_\theta = \frac{2I_0\pi^2 n_0^2\left(1+\cos^2\theta\right)}{r^2\lambda^4 A}\left(\frac{dn}{dc}\right)^2 M_w C \tag{5.7}$$

Here, I_θ is the intensity of light scattered in a direction making an angle θ from the direction of incident light, I_0 is the intensity of incident light, n_0 is the refractive index of the solvent, n is the refractive index of the solution of microparticles, r is the distance from the microparticle, λ is the wavelength of the incident light, A is the Avogadro number, M_w is the weight average molecular weight, and C is the concentration of microparticles in the solution.

If p is the osmotic pressure of the solution, M_w can be replaced by $\frac{RT}{\left(\frac{dp}{dc}\right)}$. From principles of thermodynamics, using virial equation of state, eq. 5.7 reduces to

$$I_\theta = \frac{2I_0\pi^2 n_0^2\left(1+\cos^2\theta\right)}{r^2\lambda^4 A}\left(\frac{dn}{dc}\right)^2 \frac{C}{\left[\frac{1}{M_w} + A_1C + A_2C^2 + \ldots\right]} \tag{5.8}$$

The equation can be simplified to

$$\frac{1}{M_w} + A_1C + A_2C^2 + \ldots = \frac{KC}{R_\theta} \tag{5.9}$$

The scattering intensity R_θ, an instrument parameter, is defined as

$$R_\theta = \frac{I_\theta r^2}{I_0\left(1+\cos^2\theta\right)} \tag{5.10}$$

It has the dimensions of length^{-1}. The optical constant K is defined as

$$K = \frac{2\pi^2 n_0^2}{\lambda^4 A}\left(\frac{dn}{dc}\right)^2 \tag{5.11}$$

The quantity $\frac{dn}{dc}$ is computed from a separate experiment.

For small solute concentrations, the higher powers of C in eq. 5.9 can be neglected to have

$$\frac{1}{M_w} + A_1C = \frac{KC}{R_\theta} \tag{5.12}$$

Scattering experiments with different solute concentrations yield a linear plot between $\frac{KC}{R_\theta}$ and C. Reciprocal of the intercept on the $\frac{KC}{R_\theta}$ axis gives M_w.

In practice, the light is scattered not from individual solute microparticles but from very small volume elements of a solution consisting of a few microparticles in an envelope of solvent. A correction is required in eq. 5.10 to compensate for scattering by the solvent. The corrected scattering intensity $\overline{R_\theta}$ would be

$$\overline{R_\theta} = R_\theta \text{for solution} - R_\theta \text{for solvent} \tag{5.13}$$

Eq. 5.12 then modifies to

$$\frac{1}{M_w} + A_1C = \frac{KC}{\overline{R_\theta}} \tag{5.14}$$

When the solute microparticles have size comparable to the wavelength of incident light they can no longer be considered as point source. Light scattered from different parts of the molecule differ in phase and interfere as they travel in different directions. Eq. 5.9 is modified to include this effect and we have the effective equation (Dong and Fricke 1993).

$$\frac{1}{M_w} + A_1C + A_2C^2 + \ldots = \frac{KCP(\theta)}{R_\theta} \tag{5.15}$$

and

$$P(\theta) = 1 - \frac{16\pi^2 r_g^2}{3\lambda^2}\sin^2\left(\frac{\theta}{2}\right) \tag{5.16}$$

where r_g is the radius of gyration of the solute microparticle. The equation still reduces to eq. 5.14 in the limit $\theta \rightarrow 0$, and $C \rightarrow 0$. Eq. 51.5 in combination with eq. 5.16 is used in a graphical procedure known as the Zimm formalism where the quantity $\frac{KC}{R_\theta}$ is plotted against $\sin^2\left(\frac{\theta}{2}\right) + kC$, k being an arbitrary constant. A series of points are generated, each corresponding to a given scattering angle θ, and lignin concentration C. Extrapolating to $\theta \rightarrow 0$, and $C \rightarrow 0$ provides the intercept on the $\frac{KC}{R_\theta}$ axis and is equal to $1/M_w$. A typical Zimm plot for hardwood lignosulphonate is shown in Figure 5.6.

One important requirement for the estimation of molecular weight by LASER light scattering is accurate estimation of the quantity $\frac{dn}{dc}$ and its variation with concentration, if any. It depends on the type of lignin sample being characterized, the solvent used to dissolve lignin, and the wavelength of the LASER source. These values are measured with an interferometric refractometer using a LASER source of the same wavelength as used for light scattering. If such a refractometer is not available, one can measure $\frac{dn}{dc}$ at the wavelength being used in the refractometer and calculate $\frac{dn}{dc}$ at the wavelength of light scattering using the Cauchy relation (Zinovyev et al. 2018).

$$\frac{dn}{dc} = A + \frac{B}{\lambda^2} \tag{5.17}$$

where A and B are constants. Normally, the refractive index of a dilute lignin solution varies linearly with concentration. The refractive index of lignin is measured at several concentrations and plotted against the corresponding concentration. Next, a straight line is regressed through these data points and the slope of the line provides the value of $\frac{dn}{dc}$. Figure 5.7 presents such a plot for hardwood kraft lignin and its fractions obtained by sequential solvent extraction.

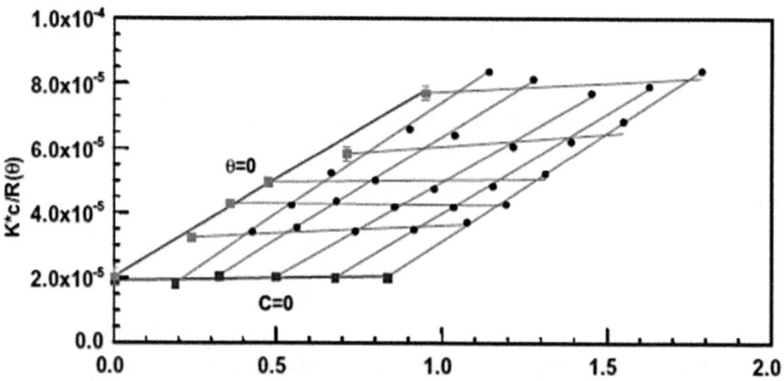

FIGURE 5.6 Zimm plot for hardwood lignosulphonate fraction soluble in 30% aqueous ethanol. (Reprinted from Fredheim, GE, Braaten, SM, Christensen, BE, *J. Wood Chem. Technol.*, 23(2), 197–215, 2003, with permission from Taylor and Francis.)

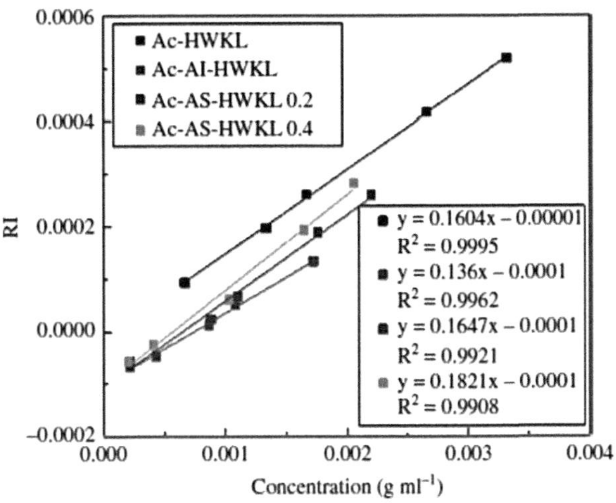

FIGURE 5.7 Graphical determination of $\frac{dn}{dc}$ for eucalyptus kraft lignin and its fractions dissolved in tetrahydrofuran. (Reprinted from Wang, L, Uraki, Y, Koda, K, Gele, A, Zhou, X, Chen, F, *Holzforschung*, 73(4), 363–369, 2019, with permission from De Gruyter.)

Some representative values of $\frac{dn}{dc}$ for different lignins dissolved in various solvents is presented in Table 5.2.

The general procedure described above is rigorously valid only for isotropic substances that do not absorb the incident light or emit light through fluorescence. For lignin solutions, all the three conditions are not valid. Consequently, corrections are to be made. Interference due to fluorescence can be corrected by using a LASER source of wavelength that does not induce fluorescence in lignin. But this is not always practicable. The second alternative is to filter out the fluorescence component from the scattered light before its intensity is measured (Dong and Fricke 1993). For anisotropic solutions and solute microparticles with dimensions comparable to the wavelength of the incident light, the scattering intensity not only depends on the scattering angle θ but also on the azimuthal angle at which the light is scattered. The effect of azimuthal angle on scattering intensity diminishes with a decrease in the scattering angle and in the direction of the incident beam there is no effect of azimuthal angle. Therefore, for accurate measurement of M_w, it is imperative to measure the intensity of scattered light as close to the direction of the incident beam as possible. The Low Angle LASER Light Scattering (LALLS) works on this principle. Yet another approach is to measure the intensities of light scattered in different directions and use the cumulative results to ascertain the molecular weight. MALLS work on this approach. Such a device can also be coupled with GPC to measure the absolute molecular weight of different fractions segregated by the size exclusion chromatography column. It is reported that GPC coupled with MALLS could provide highly resolved molecular weight distribution for kraft lignin compared to refractive index or UV detection. Scattered light intensities were measured at 18 angles around a circular plane (Gidh et al. 2006a; Gidh et al. 2006b).

Different pine kraft lignins, dissolved in dimethyl formamide, had M_w of 13,500, 16,700, and 19,200 as determined by LALLS. For the second and third lignin samples, M_w values were determined as 19,500 and 18,600, respectively, when measurements were done with lignin solution in dimethyl formamide containing 0.1 N lithium bromide (Schmidl et al. 1990). This brings forth the importance of solvent in light scattering-based molecular weight determination as the quantities n_0 and $\frac{dn}{dc}$ are dependent on the solvent used. Using a LASER light source of 632.8 nm wavelength, M_w measured by LALLS, corrected for anisotropy, fluorescence, and absorption, varied between 19,600

TABLE 5.2
$\frac{dn}{dc}$ of Lignins in Different Solvents

Lignin	Solvent	Source Wavelength (nm)	$\frac{dn}{dc}$ (mL/g)	Reference
Pine kraft	0.1 N NaOH	632.8	0.2327	Dong and Fricke 1993
Maple kraft	0.1 N NaOH	632.8	0.2282	Dong and Fricke 1993
Pine kraft	Dimethyl formamide	632.8	0.1911	Dong and Fricke 1993
Pine kraft	Pyridine	632.8	0.1509	Dong and Fricke 1993
Spruce enzyme hydrolysis	Tetrahydrofuran	633	0.1400	Contreras et al. 2008
Spruce enzyme hydrolysis	0.01 N NaOH	633	0.1820	Contreras et al. 2008
Eucalyptus enzyme hydrolysis	Tetrahydrofuran	633	0.1650	Contreras et al. 2008
Softwood kraft (Indulin AT)	Dimethyl sulfoxide with 0.5% LiBr	658	0.1515	Zinovyev et al. 2018
Softwood kraft (LignoBoost)	Dimethyl sulfoxide with 0.5% LiBr	658	0.1550	Zinovyev et al. 2018
Softwood ammonium lignosulphonate	Dimethyl sulfoxide with 0.5% LiBr	658	0.1352	Zinovyev et al. 2018
Sarkanda soda	Dimethyl sulfoxide with 0.5% LiBr	658	0.1503	Zinovyev et al. 2018
Hardwood ethanol organosolv (Alcell)	Dimethyl sulfoxide with 0.5% LiBr	658	0.1666	Zinovyev et al. 2018
Pine milled wood	Dimethyl sulfoxide with 0.5% LiBr	658	0.1050	Zinovyev et al. 2018
Acetylated eucalyptus kraft	Tetrahydrofuran	658	0.1604	Wang et al. 2019
Softwood kraft (Indulin AT)	Dimethyl formamide with 0.05M LiBr	950	0.1340	Gaugler et al. 2021
Hardwood ethanol organosolv (Alcell)	Dimethyl formamide with 0.05M LiBr	950	0.1330	Gaugler et al. 2021
Nonwood soda (Protobind)	Dimethyl formamide with 0.05M LiBr	950	0.1060	Gaugler et al. 2021
Beech acetone organosolv	79.2% aqueous acetone	658	0.1865	Ponnudurai et al. 2023

and 59,528 for pine kraft lignin obtained after delignification to different severities (Dong and Fricke 1993; Dong and Fricke 1995). In comparison, kraft lignin from sugar maple was found to be 12,900. For all these light scattering measurements, lignin samples were dissolved in 0.1 N sodium hydroxide (Dong and Fricke 1993). M_w of different kraft lignin samples were measured using MALLS having a 690-nm LASER source and eighteen detectors at different angles. Lyophilized and unlyophilized lignin samples were dissolved in 0.1 M sodium hydroxide for light scattering. Reported M_w for different lignin samples varied between 1.4×10^4 and 6.19×10^5 (Gidh et al. 2006b). MALLS was combined with GPC to estimate the M_w and molecular weight distribution of several lignin samples. M_w determined for spruce milled-wood lignin was 19,985 and for birch milled-wood lignin was 30,020. Two fractions of larch dioxane organosolv lignin had M_w of 37,775 and 110,400. Steam explosion lignin was extracted from birch with methanol and its M_w was 135,850 (Mikame and Funaoka 2006). Acetylated eucalyptus kraft lignin and its different fractions obtained by sequential solvent extraction were characterized by GPC coupled with MALLS. In one case, a 658-nm LASER source was used while the second set of measurements was made with a 785-nm LASER source. Both the detectors were equipped with filters for correction for fluorescence. Tetrahydrofuran was the solvent used. The authors concluded that the 658 nm source was not suitable for the determination of absolute molecular weights. Using the 785 nm source, M_w of acetylated eucalyptus kraft lignin was determined to be 55,500. This value was significantly higher than the M_w of 8500, determined by a conventional refractive index detector calibrated with polystyrene. For different solvent-extracted fractions, the measured M_w were 135,100, 29,700, and 12,500 (Wang et al. 2019). A similar system of GPC coupled with MALLS having a 785-nm LASER source was used to determine the molecular weights of several lignins. Measured M_w values were: Indulin AT (softwood kraft) 14,000, LignoBoost (softwood kraft) 16,000, Protobind (nonwood soda) 19,000, hardwood organosolv lignin 10,000, pine milled-wood lignin 27,000, and lignosulphonates 33,000–55,000 (Zinovyev et al. 2018). Molecular weight characterization of Indulin AT and Alcell lignin was carried out with GPC coupled with MALLS. One system used a 637-nm LASER source with detection at seven angles with fluorescence correction. Another system worked with a similar LASER source and detection but without fluorescence correction. Dimethyl formamide with 0.05 M lithium bromide was the solvent. For Indulin, measured M_w was 44,600 with fluorescence correction but 319,000 without fluorescence correction. These values were 65,600 and 827,000, respectively, for Alcell lignin. This brings forth the critical importance of fluorescence correction in MALLS estimation of lignin molecular weight (Gaugler et al. 2021). Beech acetone organosolv lignin was characterized by MALLS with a 664-nm LASER source coupled with GPC. M_w, M_n, and polydispersity were measured as 12,600, 8700, and 1.45, respectively. In comparison, the values obtained by relative calibration with polyethylene glycol and polyethylene oxide were 11,000, 4200, and 2.62, respectively (Ponnudurai et al. 2023).

5.4 SOLUTION PROPERTIES OF LIGNIN

5.4.1 Viscosity of Lignin Solutions

Being a macromolecule, one important characteristic of lignin in solution is the viscosity of the formed solution. It depends on the source and isolation method of lignin as well as the solvent used to dissolve lignin. Measurement and analysis of the viscosity of a lignin solution provides important information about the lignin including its molecular weight, interaction among different lignin molecules, and interaction of lignin molecules with the solvent molecules. Coupled with chromatographic separation, viscosity measurement provides another tool for assessing the molecular weight distribution of lignin. The relationship between the intrinsic viscosity of a polymer solution and the molecular weight of the polymer is given by the Kuhn–Mark–Houwink–Sakurada equation (McNaught and Wilkinson 1997)

$$\eta = K\mathrm{M}^{\alpha} \tag{5.18}$$

where η is the intrinsic viscosity of the polymer solution, M is the molecular weight, and K and α are empirical coefficients. The equation is transformed into a linear form by taking logs on both sides, which yields

$$\log\eta = \alpha\log\mathrm{M} + \log K \tag{5.19}$$

A plot of $\log\eta$ against logM provides a linear trace with slope as α, and intercept on $\log\eta$ axis as $\log K$. Besides temperature, the value of α strongly depends on the hydrodynamic characteristics of the polymer microparticles in solution and reveals valuable insight into the nature of the polymer molecules.

The intrinsic viscosity of lignin extracted with dioxane from autohydrolysed aspen and dissolved in dimethyl formamide was measured with a capillary viscometer. At 0.1 g/dL concentration, the viscosities varied between 0.05 and 0.162 dL/g. The parameter α in the Kuhn–Mark–Houwink–Sakurada equation was determined to be 1.6 suggesting linear conformation of lignin microparticles (Chua and Wayman 1979). Several lignin samples were acetylated and formed into a solution in tetrahydrofuran with 3–6 g/L concentration. Acetylated kraft lignins in tetrahydrofuran had intrinsic viscosities between 0.053 and 0.079 dL/g and a Kuhn–Mark–Houwink–Sakurada parameter of α between 0.26 and 0.33.

The intrinsic viscosity of acetylated methanol organosolv lignin solution was measured to be 0.066 dL/g with α of 0.24. For acetylated ethanol organosolv lignin, these values were 0.037 dL/g and 0.21. Acetylated sucrolin – sugarcane bagasse autohydrolysis lignin – had an intrinsic viscosity of 0.066 dL/g and exponent α of 0.18 (Glasser et al. 1993). The intrinsic viscosity of pine kraft lignin was determined with a capillary viscometer by preparing solutions in dimethyl formamide or 0.5 N sodium hydroxide in the concentration range of 1–100 g/L. For both the solvents, the solution viscosity was observed to be strongly dependent on lignin concentration as depicted in Figure 5.8.

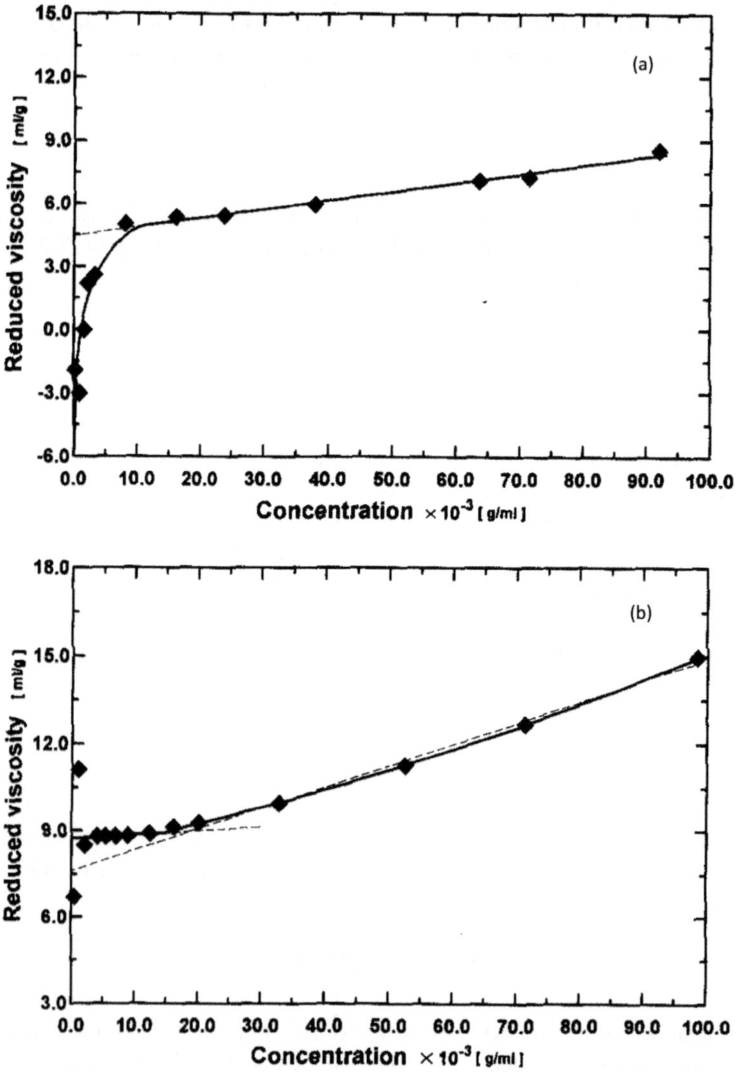

FIGURE 5.8 Concentration dependence of viscosity of lignin solution in (a) 0.5 N sodium hydroxide, (b) dimethyl formamide. (Reprinted from Dong, D, Fricke, AL, *Polymer*, 36(10), 2075–2078, 1995, with permission from Elsevier.)

In 0.5 N sodium hydroxide solution at 303 K temperature, the Kuhn–Mark–Houwink–Sakurada parameters of α and K were found to be 0.23 and 0.4165, respectively. The parameters were 0.11 and 2.5100, respectively, for dimethyl formamide solution at 318 K temperature. Parameters changed to 0.13 and 1.8895 when the temperature of dimethyl formamide solution was increased to 351 K. Low values of α for lignin solutions in dimethyl formamide and 0.5 N sodium hydroxide suggest a compact spherical conformation (Dong and Fricke 1995). Intrinsic viscosities of different molecular weight fractions of acetylated hardwood kraft lignin solutions in tetrahydrofuran or dimethyl sulfoxide were measured with a capillary viscometer. For tetrahydrofuran, the empirical parameters α, and K in the Kuhn–Mark–Houwink–Sakurada equation were determined to be 0.24 and 0.32. These values were 0.26 and 0.142 for lignin solution in dimethyl sulfoxide. Therefore, in both the solvents lignin microparticles acquired the configuration of a compact sphere (Wang et al. 2019). Dioxane lignin from *Serratula coronata* L was fractionated into different molecular weight fractions by dissolution precipitation in a dioxane–benzene mixture. Intrinsic viscosities of the fractions were measured in solution in dimethyl formamide using a capillary viscometer. Intrinsic viscosities decreased from 10.7 to 4.2 mL/g as the molecular weights of the fractions decreased. Exponent α in the Kuhn–Mark–Houwink–Sakurada equation was estimated to be 0.34 (Karmanov et al. 2019). Viscosity measurements on 6 g/L lignin solutions in dimethyl formamide containing 0.05 M lithium bromide were performed with GPC fractionation. Indulin AT (softwood kraft lignin) had the exponent α of 0.09 and Alcell lignin (hardwood ethanol organosolv lignin) had α of 0.13. This indicated solid sphere like structures of lignin microparticles when dissolved in dimethyl formamide containing 0.05 M lithium bromide (Gaugler et al. 2021). Softwood kraft lignin was esterified into different molecular weight fractions and characterized for intrinsic viscosity and molecular configuration as a solution in dimethyl sulfoxide with 0.5% lithium bromide. Fractions with molecular weights between 2200 and 14,700 had very low α in the Kuhn–Mark–Houwink–Sakurada equation suggesting dense solid sphere molecular configuration. For fractions having molecular weights of 82,400 and 740,000, the α values were 0.31 and 0.35, respectively, indicating a branched molecular configuration (Liu et al. 2021).

REFERENCES

Adler, E. *Wood Sci. Technol.* 11:169–218, 1977.

Adler, E, Hernestam, S. *Acta Chem. Scand.* 9:319–334, 1955.

Angles, MN, Reguant, J, Garcia-Valls, R, Salvado, J. *Wood Sci. Technol.* 37:309–320, 2003.

ASTM D 1166-84, 2001.

Audu, IG, Brosse, N, Winter, H, Hoffmann, A, Bremer, M, Fischer, S, Laborie, MP. *Polymers* 10:619, 2018.

Balakshin, M, Capanema, E, Gracz, H, Chang, HM, Jameel, H. *Planta* 233:1097–1110, 2011.

Billa, E, Monties, B. *Res. Chem. Intermed.* 21(3-5):303–311, 1995.

Botaro, VR, Curvelo, AAD. *J. Chromatography A* 1216:3802–3806, 2009.

Cateto, CAB. Lignin-Based Polyurethanes: Characterisation, Synthesis and Applications. Doctoral dissertation. University of Porto. 66, 2008.

Cateto, CA, Barreiro, MF, Rodrigues, AE, Belgacem, MN. *Reactive Functional Polym.* 71:863–869, 2011.

Chazal, R, Robert, P, Durand, S, Devaux, MF, Saulnier, L, Lapierre, C, Guillon, F. *Appl. Spectroscopy* 68: 1342–1347, 2014.

Chen, S, Cheng, H, Wu, S. *Fuel* 270:117520, 2020.

Chen, F, Zhuo, C, Xiao, X, Pendergast, TH, Devos, KM. *Biotechnol. Biofuels* 14:18, 2021.

Chua, MGS, Wayman, M. *Can. J. Chem.* 57:1141–1149, 1979.

Contreras, S, Gaspar, AR, Guerra, A, Lucia, LA, Argyropoulos, DS. *Biomacromolecules* 9:3362–3369, 2008.

Creighton, RHJ, Gibbs, RD, Hibbert, H. *J. Am. Chem. Soc.* 66:32–37, 1944.

Crestini, C, Melone, F, Sette, M, Saladino, R. *Biomacromolecules* 12:3928–3935, 2011.

Cybulska, I, Brudecki, G, Rosentrater, K, Julson, JL, Lei, H. *Bioresource Technol.* 118:30–36, 2012.

del Rio, JC, Marques, G, Rencoret, J, Martinez, AT, Gutierrez, A. *J. Agric. Food Chem.* 55:5461–5468, 2007.

del Rio, JC, Rencoret, J, Marques, G, Gutierrez, A, Ibarra, D, Santos, JI, Jimenez-Barbero, J, Zhang, L, Martinez, AT. *J. Agric. Food Chem.* 56:9525–9534, 2008.

del Rio, JC, Prinsen, P, Rencoret, J, Nieto, L, Jimenez-Barbero, J, Ralph, J, Martinez, AT, Gutierrez, A. *J. Agric. Food Chem.* 60:3619–3634, 2012a.
del Rio, JC, Rencoret, J, Prinsen, P, Martinez, AT, Ralph, J, Gutierrez, A. *J. Agric. Food Chem.* 60: 5922–5935, 2012b.
Deshpande, R, Giummarella, N, Henriksson, G, Germgard, U, Sundvall, L, Grundberg, H, Lawoko, M. *Ind. Crops Prod.* 115:315–322, 2018.
Diniz, CP, Grattapaglia, D, Mansfield, SD, Figueiredo, LFD. *Wood Sci. Technol.* 53:521–533, 2019.
Dong, D, Fricke, AL. *J. Appl. Polymer Sci.* 50:1131–1140, 1993.
Dong, D, Fricke, AL. *Polymer* 36:2075–2078, 1995.
Du, X, Gellerstedt, G, Li, J. *Plant J.* 74:328–338, 2013.
Ekeberg, D, Gretland, KS, Gustafsson, J, Braten, SM, *Fredheim. Anal. Chim. Acta* 565:121–128, 2006.
El Hage, R, Brosse, N, Chrusciel, L, Sanchez, C, Sannigrahi, P, Ragauskas, A. *Polymer Degrad. Stabil.* 94:1632–1638, 2009.
Evtuguin, DV, Domingues, P, Amado, FL, Neto, CP, Correia, AJF. *Holzforschung* 53:525–528, 1999.
Evtuguin, DV, Neto, CP, Silva, AMS, Domingues, PM, Amado, FML, Robert, D, Faix, O. *J. Agric. Food Chem.* 49:4252–4261, 2001.
Faix, O, Grünwald, C, Beinhoff, O. *Holzforschung* 46:425–432, 1992.
Faix, O, Argyropoulos, DS, Robert, D, Neirinck, V. *Holzforschung* 48:387–394, 1994.
Fiskari, J, Kilpeläinen, P. Asia Pacific J. *Chem. Eng.* 16(6):e2707, 2021.
Fredheim, GE, Braaten, SM, Christensen, BE. *J. Wood Chem. Technol.* 23:197–215, 2003.
Freudenberg, K, Lautsch, W, Engler, K. *Chem. Ber.* 73:167–171, 1940.
Gaugler, EC, Radke, W, Vogt, AP, Smith, DA. *J. Anal. Sci. Technol.* 12:30, 2021.
Ghatak, HR, Kundu, PP, Kumar, S. *Thermochim. Acta* 502:85–89, 2010.
Gidh, AV, Decker, SR, Vinzant, TB, Himmel, ME, Williford, C. *J. Chromatography A* 1114:102–110, 2006a.
Gidh, AV, Decker, SR, See, CH, Himmel, ME, Williford, C. *J. Analytica Chim. Acta* 555:250–258, 2006b.
Glasser, WG, Dave, V, Frazier, CE. *J. Wood Chem. Technol.* 13:545–559, 1993.
Gosselink, RJA, Snijder, MHB, Kranenbarg, A, Keijsers, ERP, de Jong, E, Stigsson, LL. *Ind. Crops Prod.* 19:271–281, 2004.
Goto, H, Koda, K, Tong, G, Matsumoto, Y, Meshitsuka, G. *J. Wood Chem. Technol.* 26:81–93, 2006.
Giummarella, N, Zhang, L, Henriksson, G, Lawoko, M. *RSC Adv.* 6:42120–42131, 2016.
Giummarella, N, Lawoko, M. *ACS Sustain. Chem. Eng.* 4:5319–5326, 2016.
Happs, RM, Addison, B, Doeppke, C, Donohoe, BS, Davis, MF, Harman-Ware, AE. *Biotechnol. Biofuels* 14:58, 2021.
Harman-Ware, AE, Foster, C, Happs, RM, Doeppke, C, Meunier, K, Gehan, J, Yue, F, Lu, F, Davis, MF. *Biotechnol. J.* 11:1268–1273, 2016.
Hepditch, MM, Thring, RW. Canadian J. *Chem. Eng.* 75:1108–1114, 1997.
Holtman, KM, Chang, HM, Jameel, H, Kadla, JF. *J. Agric. Food Chem.* 51:3535–3540, 2003.
Holtman, KM, Chang, HM, Kadla, JF. *J. Agric. Food Chem.* 52:720–726, 2004.
Huang, C, He, J, Li, X, Min, D, Yong, Q. *Bioresource Technol.* 192:471–477, 2015.
Ikeda, T, Magara, K. *J. Wood Chem. Technol.* 35:167–177, 2015.
Karmanov, AP, Kocheva, LS, Belyi, VA, Volodin, VV. *Polymer Sci. Series A* 61:53–60, 2019.
Katahira, R, Nakatsubo, F. *J. Wood. Sci.* 47:378–382, 2001.
Kuroda, KI, Nishimura, N, Izumi, A, Dimmel, DR. *J. Agric. Food Chem.* 50:1022–1027, 2002.
Lapierre, C, Monties, B, Rolando, C. *J. Wood Chem. Technol.* 5:277–292, 1985.
Lapierre, C, Pollet, B, Monties, B, Rolando, C. *Holzforschung* 45:61–68, 1991.
Lawoko, M, Henriksson, G, Gellerstedt, G. *Biomacromolecules* 6:3467–3473, 2005.
Lawther, JM, Sun, R, Banks, WB. *J. Agric. Food Chem.* 44:1241–1247, 1996.
Lee, JH, Kim, TM, Choi, IG, Choi, JW. *Nanomaterials* 11:1790, 2021.
Li, H, Chai, XS, Liu, M, Deng, Y. *J. Agric. Food Chem.* 60:5307–5310, 2012b.
Li, MF, Sun, SN, Xu, F, Sun, RC. *Chem. Eng. J.* 179:80–89, 2012a.
Li, X, Wu, S. *BioResources* 9:6277–6289, 2014.
Lima, CF, Barbosa, LCA, Marcelo, CR, Silverio, FO, Colodette, JL. *BioResources* 3(3):701–712, 2008.
Liu, LY, Chen, S, Ji, L, Jang, SK, Renneckar, S. *Green Chem.* 23:4567–4579, 2021.
Lopes, FJF, Silverio, FO, Baffa, DCF, Loureiro, ME, Barbosa, MHP. *J. Wood Chem. Technol.* 31:309–323, 2011.
Lou, R, Lyu, G, Wu, S, Zhang, B, Zhao, H, Lucia, LA. *ACS Sustainable Chem. Eng.* 6:430–437, 2018.
Lu, F, Ralph, J. *J. Agric. Food Chem.* 45:2590–2592, 1997.
Lu, F, Ralph, J. *J. Agric. Food Chem.* 46:547–552, 1998.
Mansson, P. *Holzforschung* 37:143–146, 1983.

Marques, AV, Pereira, H, Meier, D, Faix, O. *Holzforschung* 53:167–174, 1999.
McNaught, AD, Wilkinson, A. IUPAC Compendium of Chemical Terminology, 2nd ed. (the "Gold Book"). Blackwell Scientific Publications, Oxford, UK, 1997.
Mikame, K, Funaoka, M. *Polymer J.* 38:592–596, 2006.
Min, D, Xiang, Z, Liu, J, Jameel, H, Chiang, V, Jin, Y, Chang, HM. *J. Wood Chem. Technol.* 35:52–61, 2015.
Moniz, P, Serralheiro, C, Matos, CT, Boeriu, CG, Frissen, AE, Duarte, LC, Roseiro, LB, Pereira, H, Carvalheiro, F. *Process Biochem.* 65:136–145, 2018.
Mousavioun, P, Doherty, WOS. *Ind. Crops Prod.* 31:52–58, 2010.
Mun, JS, Pe III, JA, Mun, SP. *Molecules* 26:4861, 2021.
Myrvold, BO. *Holzforschung* 69:9–16, 2015.
Nitsos, C, Stoklosa, R, Karnaouri, A, Vörös, D, Lange, H, Hodge, DB, Crestini, C, Rova, USL, Christakopoulos, P. *ACS Sustainable Chem. Eng.* 4:5181–5193, 2016.
Ohra-aho, T, Gomes, FJB, Colodette, JL, Tamminen, T. *J. Anal. Appl. Pyrol.* 101:166–171, 2013.
Oliveira, L, Evtuguin, DV, Cordeiro, N, Silvestre, AJD, Silva, AMS, Torres, IC. *J. Agric. Food Chem.* 54:2598–2605, 2006.
Onnerud, H. *Holzforschung* 57:377–384, 2003.
Onnerud, H, Gellerstedt, G. *Holzforschung* 57:165–170, 2003.
Pla, F, Dolk, M, Yan, JF, McCarthy, JL. *Macromolecules* 19:1471–1477, 1986.
Ponnudurai, A, Schulze, P, Seidel-Morgenstern, A, Lorenz, H. *ACS Sustainable Chem. Eng.* 11:766–776, 2023.
Rencoret, J, Ralph, J, Marques, G, Gutiérrez, A, Martínez, T, del Rio, JC. *J. Agric. Food Chem.* 61: 2434–2445, 2013.
Robinson, AR, Mansfield, SD. *Plant J.* 58:7060–7714, 2009.
Rossberg, C, Janzon, R, Saake, B, Leschinsky, M. *BioResources* 14:4543–4559, 2019.
Rudin, A, Choi, P. *The Elements of Polymer Science and Engineering*. Academic Press, Oxford, UK, 89–148, 2013.
Saito, K, Kaiho, A, Sakai, R, Nishimura, H, Okada, H, Watanabe, T. *J. Agric. Food Chem.* 64:9152–9160, 2016.
Sameni, J, Krigstin, S, Sain, M. *BioResources* 11:8435–8456, 2016.
Santos, RB, Capanema, EA, Balakshin, MY, Chang, HM, Jameel, H. *J. Agric. Food Chem.* 60:4923–4930, 2012.
Schmidl, W, Dong, D, Fricke, AL. *Proc. Mat. Res. Soc. Symp.* 197:21–30, 1990.
Serrano, L, Esakkimuthu, ES, Marlin, M, Brochier-Salon, MC, Mortha, G, Bertaud, F. *Ener. Fuels* 32: 5969–5977, 2018.
Sheng, Y, Ma, Z, Wang, X, Han, Y. *Food Chem.* 376:131895, 2022.
Shrivastava, A. *Introduction to Plastics Engineering*. William Andrew Applied Science Publishers, Oxford, UK, 17–48, 2018.
Solar, R, Kaçik, F, Melcer, I. *Nordic Pulp Paper Res. J.* 2:139–141, 1987.
Sun, R, Lawther, JM, Banks, WB. *Ind. Crops Prod.* 4:241–254, 1995.
Sun, XF, Sun Fowler, P, Baird, MS. *J. Agric. Food Chem.* 53:860–870, 2005.
Takada, M, Rabemanolontsoa, H, Minami, E, Saka, S. *J. Wood. Sci.* 64:802–809, 2018.
Tan, X, Jiang, B, Yang, Y, Min, D, Jin, Y. *Holzforschung* 71:99–108, 2016.
Thoresen, PP, Lange, H, Crestini, C, Rova, U, Matsakas, L, Christakopoulos, P. *ACS Omega* 6:4374–4385, 2021.
van de Pas, DJ, Nanayakkara, B, Suckling, ID, Torr, KM. *Holzforschung* 68:151–155, 2014.
Villar, JC, Caperos, A, Garcia-Ochoa, F. *J. Wood Chem. Technol.* 17:259–285, 1997.
Wang, L, Uraki, Y, Koda, K, Gele, A, Zhou, X, Chen, F. *Holzforschung* 73:363–369, 2019.
Yamamura, M, Hattori, T, Suzuki, S, Shibata, D. *Plant Biotechnol.* 29:419–423, 2012.
Yanez-S, M, Matsuhiro, B, Nunez, C, Pan, S, Hubbell, CA, Sannigrahi, P, Ragauskas, AJ. *Polymer Deg. Stab.* 110:184–194, 2014.
Yang, JM, Goring, DAI. *Canadian J. Chem.* 58:2411–2414, 1980.
You, TT, Zhang, LM, Zhou, SK, Xu, F. *Ind. Crops Prod.* 71:65–74, 2015.
Yue, F, Lu, F, Sun, RC, Ralph, J. *J. Agric. Food Chem.* 60:922–928, 2012.
Yue, F, Lu, F, Regner, M, Sun, R, Ralph, J. *ChemSusChem* 10:830–835, 2017.
Zakis, GF. *Functional analysis of lignins and their derivatives*. Tappi, Atlanta, 31–33, 1994.
Zhao, J, Xiuwen, W, Hu, J, Liu, Q, Shen, D, Xiao, R. *Polymer Deg. Stab.* 108:133–138, 2014.
Zinovyev, G, Sulaeva, I, Podzimek, A, Rössner, D, Kilpeläinen, I, Sumerskii, I, Rosenau, T, Potthast, A. *ChemSusChem* 11:3259–3268, 2018.

6 Spectroscopic Characterization of Lignin

Spectroscopic techniques have immense diagnostic value in material characterization. In principle, any spectroscopic technique is based on the interaction of matter with radiation. These are quantum interactions depending on the nature of radiation and the possible quantum states of the material under study. The entire spectrum of electromagnetic radiation, from radio frequency to gamma rays, has been utilized in different spectroscopic techniques. Similarly, radiation absorbed by matter or emitted by matter can be used as the characterizing tool in spectroscopy. Material can be characterized based on the change in the quantum states of electrons or nucleons. Spectroscopy is a powerful tool for characterizing molecular structures of organic substances. In recent years these methods have not only complemented the wet chemistry methods of analysis of organic substances but also been used as standalone analytical techniques.

The study of lignin chemistry has always posed a challenge to organic chemists owing to its structural heterogeneity and complexity. With the advancement in spectroscopic methods, the scientific community is now in a position to have a much deeper and more thorough understanding of the structural features of lignin macromolecules and the changes occurring during different reactions and chemical modifications. This has opened new avenues for lignin application and its value addition. Techniques like UV spectroscopy, IR spectroscopy, Raman spectroscopy, NMR spectroscopy, and mass spectroscopy are today indispensable analytical tools to characterize lignin.

6.1 LIGNIN CHARACTERIZATION BY UV SPECTROSCOPY

6.1.1 UV Spectroscopy for Determination of Con2centration of Lignin Solution

UV spectroscopy is a technique where material characteristics are elucidated by absorption or emission of electromagnetic radiation of UV frequency. It has wide application in the determination of lignin concentration in solution. The method works on the principle of absorption of UV radiation by lignin solution. When UV light of known intensity is incident on a lignin solution and passes through a fixed path length of the solution, a part of the radiation is absorbed by the solution. Therefore, the UV light emerging from the solution is lower in intensity than the incident light. Difference in intensities of the incident and emergent light measures the extent of radiation absorbed by the solution which is dependent on the solution concentration and path length of light through the solution. Numerically it is expressed by the well-known Beer–Lambert law which states that

$$A = \log\left(\frac{I_0}{I}\right) = \epsilon \mathrm{Cl} \tag{6.1}$$

where A is the absorbance, I_0 is the intensity of incident light, I is the intensity of emerging light, ϵ is the molar absorptivity – also known as extinction coefficient – of the substance being studied, C is the molar concentration, and l is the path length. This possibility of determining the lignin concentration in solution paves the way for UV absorption spectroscopy-based detectors in the chromatographic characterization of lignin and its degradation products, including molecular weight determinations.

 DOI: 10.1201/9781003598664-8

There are quite a few research reports where UV absorbance at 280 nm is used to measure lignosulphonate concentration in spent sulphite liquor and its fractions. Spent sulphite liquor from rice straw delignification was characterized for lignosulphonate concentration with 280 nm UV absorbance (Bhattacharya et al. 2005). A molar absorptivity of 15.3 $M^{-1}cm^{-1}$ was reported for lignosulphonate in ammonium base spent sulphite liquor, but for two different spent liquors from different magnesium base mills, absorptivities of 17.4 and 14 $M^{-1}cm^{-1}$ were determined (Sumerskii et al. 2015). Absorptivities of 6.69 L/(g.cm) for sodium lignosulphonate and 7.64 L/(g.cm) for calcium lignosulphonate were used for concentration measurement with UV absorbance at 280 nm (Demmelmayer et al. 2020). The latter value was also reported by another research group. However, an absorptivity of 13.2 L/(g.cm) was used when measuring lignosulphonate concentration in industrial spent liquor (Kienberger et al. 2020). A recent study reports an absorptivity of 13.01 L/(g.cm) at 280 nm for lignosulphonate concentration measurement in softwood spent sulphite liquor (Bhattacharya et al. 2022). It is also contended that compounds such as furfural and hydroxymethyl furfural also contribute to solution absorbance at this frequency leading to overestimation of lignosulphonate concentration (Schlackl et al. 2020). UV absorbance at 273 nm was used to measure the lignosulphonate content in spent sulphite delignification liquor using an absorptivity of 5.4 L/(g.cm) (Marques et al. 2009). These parameters were also used for lignosulphonate concentration in spent liquor from magnesium base delignification of eucalyptus (Fernandes et al. 2012). One study reports UV absorbance at 232.5 nm to measure the lignosulphonate content of industrial spent liquor from sulphite delignification (Rueda et al. 2014). Spent liquor from sodium base sulphite delignification was fractionated by membrane separation and lignosulphonate concentration in different fractions was determined by UV absorbance at 234 nm and an absorptivity of 31.6 L/(g.cm) (Al-Rudainy et al. 2017). Another study reports maximum absorbance recorded at 283 nm for sodium lignosulphonate (Kumar et al. 2018). It appears that a unanimous consensus is lacking on the choice of UV wavelength and absorptivity when measuring lignosulphonate concentration. Even if a wavelength is selected based on previous research reports, it would be prudent to draw a calibration curve of absorbance against different known concentrations and compute the absorptivity from it.

280 nm is also the wavelength of choice for concentration determination of different alkali lignins in solution by UV absorption spectroscopy. Concentrations of softwood kraft lignin solutions were determined by UV absorption spectroscopy with an absorptivity of 17.662 L/(g.cm) (Gidh et al. 2006). 280 nm absorption with absorptivity of 20.75 L/(g.cm) was used for concentration measurement of Curan 100, a softwood kraft lignin, in an alkaline medium (Norgren and Mackin 2009). For the measurement of lignin concentration in hardwood kraft black liquor, UV absorbance at 280 nm with 20.6 L/(g.cm) absorptivity was used. The absorptivity value for softwood kraft black liquor was 24.6 L/(g.cm) (Ohman et al. 2007). However, in another study for hardwood kraft black liquor, UV absorbance at 280 nm with 21.1 L/(g.cm) absorptivity was reported (Jonsson et al. 2008). Different fractions from softwood kraft black liquor, obtained by membrane filtration, were characterized for their lignin concentration by UV absorbance at 280 nm using absorptivity of 24.6 L/(g.cm) (Arkell et al. 2014). These values were also applicable for measuring the lignin content of mixed softwood kraft black liquor (Zhu and Theliander 2015), liquid lignin formulation from spruce or birch (Li et al. 2021), and in hardwood kraft black liquor or its fractions (Mendes et al. 2021). A marginally higher absorptivity of 24.7 L/(g.cm), used with absorbance at 280 nm, was prescribed for softwood kraft black liquor (Manttari et al. 2015). Another research group measured lignin concentration in softwood kraft black liquor by UV absorbance at 280 nm but with an absorptivity of 23.7 L/(g.cm) (Haddad et al. 2016). Kraft lignin solubilities from eucalyptus, bamboo, and softwood in different solvents were measured with 280 nm absorptivities of 24.23, 24.154, and 26.242 L/(g.cm), respectively (Ma et al. 2021).

Spent liquor from aqueous ethanol organosolv delignification of bagasse was characterized for lignin concentration by 210 nm UV absorption with 78.2 L/(g.cm) absorptivity (Fernando et al. 2010). Other studies suggest 280 nm absorption for concentration measurement (Weinwurm et al.

2016). UV absorption at 280 nm with absorptivity of 13 L/(g.cm) was used for lignin concentration measurement in spent liquid from aqueous dioxane delignification (Fujimoto et al. 2005). The concentration of pine milled-wood lignin and enzyme hydrolysis lignin in aqueous dioxane solution was measured by 280 nm UV absorbance combined with 15.4 L/(g.cm) absorptivity (Hu et al. 2006). 280 nm absorbance and 13 L/(g.cm) absorptivity were used for concentration measurement of sweetgum enzyme hydrolysis lignin extracted with aqueous dioxane (Jiang et al. 2017). Recently, the concentration of carboxylic acid (mixture of aqueous formic and acetic acids) organosolv lignin in different solvents was measured with 280 nm UV absorbance using absorptivity of 25.26 L/(g.cm) (Ma et al. 2021).

Another important application of UV absorption spectroscopy in lignin characterization is the determination of acid-soluble lignin in the standard Klason determination. Acid-soluble lignin in the filtrate from Klason lignin determination of different lignin samples was determined by UV absorption spectroscopy at 205 nm wavelength with absorptivity of 110 L/(g.cm) (Arkell et al. 2014; del Rio et al. 2012; Sameni et al. 2016). One research group suggests using an absorptivity of 113 L/(g.cm) at 205 nm for the measurement of lignin concentration in enzyme hydrolysate (Nitsos et al. 2019) while another group report uses 240 nm absorption for lignin concentration measurement in dilute acid hydrolysate (Chen et al. 2017). A recent study reports the measurement of acid-soluble lignin by UV absorbance at 280 nm but with the same absorptivity of 110 L/(g.cm) (Ela et al. 2020).

The method of UV spectroscopic determination of lignin concentration in solution can also be extended for estimating lignin content in lignin carbohydrate complexes. For this, absorbance at 280 nm is frequently used with different absorptivities reported for different samples. Lignin amount in lignin carbohydrate complex from ginkgo was determined in dimethyl formamide solution with absorptivity of 20.7 L/(g.cm) (Xie et al. 2000). Lignin content in acetobrominated lignin carbohydrate complex from spruce was estimating using absorptivity of 25.7 L/(g.cm) (Giummarella et al. 2016). For a similar lignin carbohydrate complex, another study used an absorptivity of 23.1 L/(g.cm) (Carvalho et al. 2020). Lignin content in the lignin carbohydrate complex from birch was estimated with absorptivity of 13.1 L/(g.cm) (Giummarella and Lawoko 2016). An absorptivity value of 11.9 L/(g.cm) was used for lignin estimation in lignin carbohydrate complexes isolated from lignosulphonate streams (Asikanius et al. 2021).

6.1.2 Lignin Structural Elucidation from UV Spectrum

In equation 6.1, the molar absorptivity ϵ is dependent on the wavelength of the incident light. By recording the absorbance of lignin solution with UV light of different wavelengths, a spectrum is obtained. Analysing the spectrum for absorbance at different wavelengths reveals important information about lignin molecular structure. UV ionization spectroscopy has been used as a quick estimation technique for phenolic hydroxyl content in lignin samples. Phenolic groups are ionized in an alkaline solution while remaining unionized in a neutral solution. Difference in the absorbances in two media provides an ionization difference spectrum which allows for the quantitative estimation of phenolic hydroxyl groups.

Phenolic hydroxyl in milled wood lignins from softwoods, hardwoods, and nonwoods were determined by UV ionization spectroscopy by preparing a solution in a 2:1:1 mixture of 2-methoxyethanol, ethanol, and water. For estimating the percentage of phenolic hydroxyl, UV absorbance at 250 nm was multiplied by a correlation factor of 0.2005. Accordingly, the phenolic hydroxyl content of 42 milled wood lignins was found in the range of 2.2–3.88% (Faix et al. 1992). UV absorption spectra of lignins from different organosolv delignification processes exhibited maxima at 274 and 318 nm except for aqueous ethanol delignification, for which the second maximum was recorded at 316 nm. The first maximum was assigned to non-conjugated phenolics and the second to conjugated structures. 274 nm absorptivity varied between 1.3 and 1.5 L/(g.cm), the highest being for aqueous ethanol and butanol. Absorptivity at the second maximum varied from 1.2 to 1.3 L/(g.cm) (Sun et al. 1998). UV absorption spectrum of dioxane lignin from leaf sheaths of banana plants was obtained in 2-methoxyethanol. π to π* electronic transitions were assigned to 280 nm absorption. Conjugated structures absorbed at 310 nm (Oliveira et al. 2006). In the UV absorption spectra of miscanthus milled-wood lignin and ethanol organosolv lignin, the peak for non-conjugated guaiacyl-type phenolic hydroxyl was observed at 282 nm and the peak for conjugated moieties appeared at 315 nm (El Hage et al. 2009). Corncob milled-wood lignin, dissolved in 0.1 M sodium hydroxide, showed two absorption peaks – 280 nm corresponding to non-conjugated phenolic moieties, and 350 nm from conjugated phenolic entities (Bu et al. 2011). UV spectra of bamboo milled-wood lignin and formic acid organosolv lignin were recorded in ethanol solution. A distinct absorption maximum at 205 nm was attributed to the electronic transition from π to π* molecular orbital in the aromatic rings of phenylpropane units. The low maximum at 286 nm was traced to π to π* transition in guaiacyl units. At this wavelength, the absorptivity of milled wood lignin was determined as 18.1 L/(g.cm). Absorptivities of formic acid lignins ranged from 24.7 to 27.2 L/(g.cm). Another observed low maxima at 312 nm arises from the presence of moieties with carbonyl groups at α carbon which causes n to π* transition. Absorptivities at this wavelength were 15.9 L/(g.cm) for milled wood lignin and 21.2 to 23.8 L/(g.cm) in the case of formic acid lignin (Li et al. 2012). UV absorption spectra of soda anthraquinone lignin from birch showed strong absorption at 220 nm (Kumar et al. 2016). Three prominent shoulders were observed in the UV absorption spectrum of aspen soda lignin dissolved in sodium hydroxide. These were 240 nm (free phenolic), 278 nm (side chain carbonyl conjugated to an aromatic ring), and 348 nm (carboxylic groups on a benzene ring) (Livča et al. 2012). In one protocol for determination of the extent of phenolic hydroxyl in eucalyptus spent sulphite liquor, UV absorbance at 255 nm was utilized along with the adjacent absorbance minimum. The per cent phenolic hydroxyl was reported to be proportional to the difference in absorbance (Alexandri et al. 2016). The choice of solvent is important in lignin characterization by UV spectroscopy. A comparison of UV spectra for lignin dissolved in aqueous dioxane and sodium hydroxide solution is shown in Figure 6.1. Siam weed lignin spectra in alkaline solution had three peaks: 210–220 nm, 240–250 nm, and 280 nm. Of these, the first one was the most prominent. However, the spectra in aqueous dioxane did not have this 210–220 nm peak. Bagasse lignins had an additional peak at 315 nm attributed to conjugated phenolics (Zhao and Liu 2010).

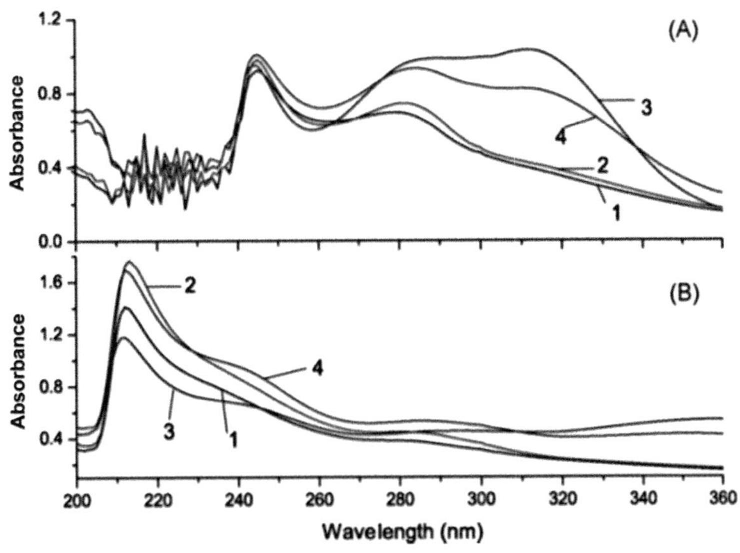

FIGURE 6.1 UV spectra of lignins in different solvents, (A) aqueous dioxane as solvent, (B) sodium hydroxide at 12 pH as solvent. Spectra 1 for siam weed acetic acid organosolv lignin, Spectra 2 for siam weed Milox lignin, Spectra 3 for bagasse acetic acid organosolv lignin, Spectra 4 for bagasse Milox lignin. (Reprinted from Zhao, X, Liu, D, *Ind. Crops. Prod.*, 32, 284–291, 2010, with permission from Elsevier.)

Kraft lignin, dissolved in aqueous dioxane, was characterized by UV absorption spectroscopy. The most prominent peak was observed between 200 and 208 nm arising from conjugated entities. The peak at 285 nm was attributed to syringyl structures and aromatic hydroxyl conjugated with carbonyl at α carbon through the benzene ring (Mohammadpour and Sadeghi 2020). For characterizing aromatic hydroxyl groups in lignin, 0.04 g/L solutions were prepared in aqueous alkaline dioxane, or sodium hydroxide solution buffered to pH 12. In each case, a reference solution was also prepared by buffering at pH 6 for use as the datum for the ionization difference spectrum (Gartner et al. 1999). Ethanol organosolv lignins from grasses and corn stover were analysed for phenolic hydroxyl by UV ionization difference spectra. The authors used the following general equation to determine the abundance of aromatic hydroxyl.

$$\text{Phenolics}\left(\%\right)=\frac{\left(A/\epsilon\right)_{\text{at given wavelength}}}{\text{Cl}}\times 100 \tag{6.2}$$

Absorptivity was determined through the relationship

$$\left(\epsilon\right)_{\text{at given wavelength}}=\frac{1.435\times 10^{5}-4.15\nu}{17} \tag{6.3}$$

Here, ν is the wavenumber of the incident UV light and 17 stands for the molecular mass of the hydroxyl group. Absorbance at 290–300 nm was assigned to guaiacyl and syringyl structures and absorbance at 360–370 nm comes from stilbene moieties (Cybulska et al. 2012). Aromatic hydroxyl in ethanol organosolv lignin from eucalyptus could be estimated by UV ionization difference spectroscopy. Difference spectra exhibited two prominent maxima at 293 and 370 nm. The first of these corresponded to non-conjugated structures and an average molar absorptivity of 4100 (L/mol.cm). The second corresponded to conjugated structures and an average molar absorptivity of 37,250 (L/mol.cm) (Yanez-S et al. 2014). UV ionization difference spectra of soda anthraquinone lignin and kraft lignin from hemp and flax, and various lignosulphonates were recorded by this method. For the determination of nonconjugated phenolics, the absorption peaks at 300 and 350 nm were used along with the following empirical equation (Gartner et al. 1999; Jablonsky et al. 2015).

$$\text{Nonconjugated phenolics}\left(\frac{\text{mmol}}{\text{g}}\right)=\frac{\left[\left\{0.25\times\left(A/\epsilon\right)_{300\,nm}\right\}+\left\{0.0595\times\left(A/\epsilon\right)_{350\,nm}\right\}\right]}{\text{Cl}} \tag{6.4}$$

Similarly, conjugated phenolics were estimated as

$$\text{Conjugated phenolics}\left(\frac{\text{mmol}}{\text{g}}\right)=\frac{0.0476\times\left(A/\epsilon\right)_{350\,nm}}{\text{Cl}} \tag{6.5}$$

The authors contended that 272–282 nm peaks could be attributed to guaiacyl structures while peaks in the 340–350 nm range arise from structures where a double bond between α and β carbons is conjugated with those in the aromatic ring. Their results showed that the abundance of nonconjugated phenolic hydroxyl structures varied between a low of 0.27 to 3.13 mmol/g. This range was 0.05–0.42 mmol/g for conjugated structures (Jablonsky et al. 2015). UV ionization difference spectroscopy was used to characterize softwood kraft lignin (Indulin AT), and Milox organosolv birch lignin for aromatic hydroxyl groups. Solutions with 0.001–0.05 g/L concentration were prepared in aqueous ethylene glycol and buffered to pH 6 and 12. Solutions at each pH were recorded for their

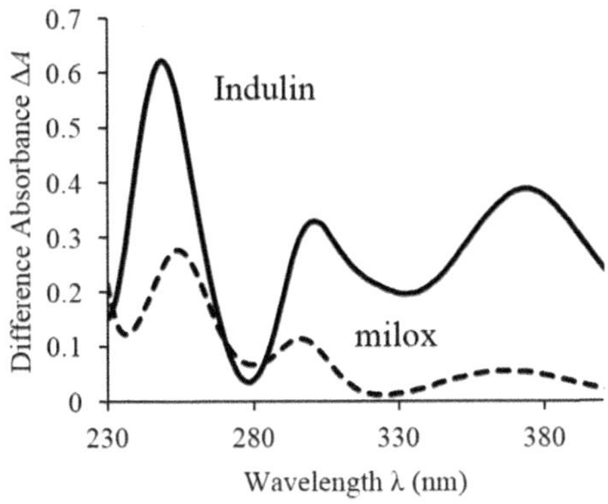

FIGURE 6.2 Ionization difference UV spectra of lignins. (Reprinted from Goldmann, WM, Ahola, J, Mankinen, O, Kantola, AM, Komulainen, S, Telkki, V-V, Tanskanen, J, *Periodica Polytechnica Chem. Eng.*, 61(2), 93–101, 2017, originally published by Budapest University of Technology and Economics under Creative Commons Attribution License.)

UV absorbance at different wavelengths. 0.2 M sodium hydroxide solution was used as a reference. Ionization difference spectra were obtained by taking the difference in absorbance at 12 and 6 pH. These ionization difference UV spectra are shown in Figure 6.2.

From the difference spectrum, the difference absorptivities at each wavelength were calculated as (Goldmann et al. 2017)

$$\Delta\epsilon = \epsilon_{12} - \epsilon_6 \tag{6.6}$$

where $\Delta\epsilon$ is the difference absorptivity and ϵ_{12} and ϵ_6 are absorptivities at pH 12 and 6, respectively. The abundance of different structural moieties in lignin molecules with phenolic hydroxyl functionality was calculated with linear correlations having $\Delta\epsilon$ at 300 and 360 nm. For Indulin AT, $\Delta\epsilon$ at 300 nm was 6.9866 L/(g.cm) and $\Delta\epsilon$ at 360 nm was 7.8966. These values were 1.1253 and 0.7634, respectively, for Milox lignin. UV ionization difference spectrum was used to determine the dissociation constant of phenolic hydroxyl at different pH for softwood kraft lignin Indulin AT. For this purpose, absorbance difference at 305 nm was utilized with the following correlation (Norgren and Lindstrom 2000)

$$\alpha = \frac{A_{\text{Given pH}} - A_{\text{Neutral pH}}}{A_{\text{Strongly alkaline pH}} - A_{\text{Neutral pH}}} \tag{6.7}$$

6.2 LIGNIN CHARACTERIZATION BY FOURIER TRANSFORM INFRARED SPECTROSCOPY

Fourier Transform Infrared (FTIR) spectroscopy is a potent analytical tool in structural characterization of organic molecules. The technique is based on the distortional and/or vibrational and/or rotational periodic movements in organic molecules as they absorb electromagnetic radiation of infrared frequency. An organic molecule is made up of atoms – mainly carbon, hydrogen, and oxygen – joined together primarily by covalent bonds. The electron clouds in these bond regions respond to the incident infrared radiation by absorbing their energy and undergoing distortion and/or vibration as a result. The nature of these alterations in the electron clouds depends on the type of bond, their surrounding environment, and the frequency of incident radiation. Therefore, each alteration in the electron cloud in an organic molecule can be correlated with its characteristic infrared absorption frequency. It turns out that the infrared absorption spectrum of an organic molecule over a range of frequencies is a signature of the molecular structure. In an infrared spectrometer meant for this purpose, the raw absorption data is Fourier-transformed to obtain the spectrum – hence the name FTIR.

Depending on the molecular structure there are a finite number of ways in which a molecule can undergo periodic distortion, vibration, or rotation. These are known as the degrees of freedom. A distinct natural frequency is associated with each degree of freedom. When the frequency of incident radiation matches the natural frequency of a particular degree of freedom, it resonates with strong absorption of photons of the resonant frequency. In quantum parlance, the energy of the molecule goes from the ground to an excited state for that degree of freedom. Various degrees of freedom are possible that can be associated with a large organic molecule with many covalent bonds and functional groups. These may be symmetric or antisymmetric stretching, in-plane bending in the ways of scissoring, wagging, rocking, or twisting, or even out-of-plane bending. All these different degrees of freedom in a molecule absorb infrared photons of different frequencies. The spectrum consists of absorption peaks or bands for each degree of freedom present and reveals the structural details of the organic molecule.

6.2.1 Lignin Structural Elucidation from FTIR Spectrum

Lignin characterization by FTIR spectroscopy is mostly carried out in the solid state by grinding it with spectroscopy-grade potassium bromide and forming it into pellets. The pellet is placed in the path of an infrared beam allowing the lignin particulates to interact. The spectral intensity of the transmitted beam is compared with that of the incident beam to obtain the spectrum. Potassium bromide is used because it does not absorb infrared light. Another way of obtaining the spectrum is to use the attenuated total reflection (ATR) technique. Not only the isolated lignin but the lignin in various spent liquors and solutions can also be analysed. No sample preparation is required, and the lignin material can be characterized as such. The obtained spectrum is independent of the thickness of the sample layer placed on the ATR module as the distance traversed by the infrared light remains the same owing to total internal reflection. In principle, a layer of the sample to be characterized is placed over a detector made up of infrared transparent material with a refractive index higher than the surrounding to make internal reflection possible. Infrared light enters the detector at an angle, travels the thickness of the detector, and rather than emerging out, reflects back into the detector. Depending on the detector geometry, a number of such reflections take place forming nodes of standing wave packets. These wave packets interact with the sample layer causing the signature absorptions. It is common practice to have a prism-shaped detector that results in only one internal reflection node. Diamond, germanium, and zinc selenide are materials suitable for ATR detectors.

Irrespective of the source and isolation method, all lignin molecules have some structural similarities. These are the presence of the benzene ring, aromatic hydroxyl groups, aliphatic hydroxyl groups, methoxyl groups on the benzene ring, and different ether and carbon-to-carbon linkages. Accordingly, there are similarities between FTIR spectra of different lignins in terms of absorption bands. Yet there are subtle differences in absorption bands from similar structural features in different lignins. In addition, there may be absorption bands characteristic of the lignin sample under study based on its source and isolation method. The absorbances for different frequencies are often normalized against the absorbance for hydroxyl stretching to gain insight into the relative abundance of different structural features in the lignin molecule and also to compare the spectra of different lignin samples. The general frequency ranges of characteristic absorption bands in the lignin FTIR spectrum and their assignment to molecular structural features are presented in Table 6.1.

6.2.1.1 Soda and Soda Anthraquinone Lignin

FTIR characterization of soda anthraquinone lignin from steam-exploded spruce and pine revealed characteristic absorption bands for hydroxyl O–H stretching, stretching in C–H bond of methyl and methylene, asymmetric and symmetric stretching in benzene ring skeleton, methoxyl C–H

TABLE 6.1
Characteristic Absorption Bands in the Lignin FTIR Spectrum

Wavenumber cm^{-1}	Assignment	Reference
3425–3460	Stretching in O–H of hydroxyl groups	Reddy and Ghatak 2018
3013	Asymmetric stretching of the double bond system of the benzene ring	Reddy and Ghatak 2018
2843–2941	Stretching in the C–H bond of methyl and methylene groups	Li et al. 2012; Reddy and Ghatak 2018
1702–1731	Stretching in the C=O bond of carbonyls and ester functionalities	Reddy and Ghatak 2018
1593–1605	Symmetric stretching in benzene ring skeleton	Reddy and Ghatak 2018
1505–1515	Asymmetric stretching in benzene ring skeleton	Reddy and Ghatak 2018
1455–1470	Asymmetric bending in the C–H bond of methoxyl groups	Li et al. 2012; Reddy and Ghatak 2018
1420–1430	Stretching in benzene ring skeleton with in-plane bending in the C–H bond	Li et al. 2012; Reddy and Ghatak 2018
1365–1370	Stretching in the C–H bond of methyl groups in the aliphatic side chain	Li et al. 2012; Reddy and Ghatak 2018
1315–1330	Syringyl ring breathing with stretching in the C–O bond of methoxyl	Oliveira et al. 2006; Sameni et al. 2016
1266–1270	Guaiacyl ring breathing with stretching in the C–O bond of methoxyl	Reddy and Ghatak 2018
1210–1221	In-plane bending in the O–H bond in phenolic hydroxyl	Reddy and Ghatak 2018
1124–1126	Stretching in the O–H bond of secondary and tertiary hydroxyl group	Reddy and Ghatak 2018
1028–1035	In-plane bending in the C–H bond of the benzene ring. Stretching in the C–O bond of the primary hydroxyl group	Reddy and Ghatak 2018

asymmetric bending, stretching in benzene ring skeleton with in-plane bending in C–H bond, C–O bond stretching of methoxyl with syringyl or guaiacyl ring breathing, in-plane bending in C–H bond of the benzene ring, and C–O bond stretching of the primary hydroxyl group. However, absorption at 1120 cm^{-1} was also linked to aromatic C–H (Angles et al. 2003). Wheat straw, hemp, and flax soda lignins, in their FTIR spectra, had typical absorption bands for hydroxyl O–H stretching, methyl and methylene C–H stretching, and asymmetric and symmetric skeletal vibrations of the benzene ring. Absorption for C–H stretching in methyl and methylene was very strong and prominent in the flax lignin spectrum. Three bands corresponding to carboxyl groups appeared in the spectra in the 1750–1550 cm^{-1} region. Other characteristic absorption frequencies between 1300 and 1000 cm^{-1} were not individually assigned (Gosselink et al. 2004). FTIR spectra of soda lignins from hemp, flax, jute, sisal, and abaca, reported in the spectral range of 2000–600 cm^{-1}, presented characteristic bands for asymmetric and symmetric aromatic skeletal stretching, methoxyl C–H asymmetric bending, aromatic C–H in-plane bending with ring stretching, and O–H in-plane bending of aromatic hydroxyl. Band for syringyl ring breathing with stretching in C–O bond of methoxyl was observed prominently in all the spectra in the 1324–1328 cm^{-1} region. But similar absorption for guaiacyl moieties, in the region of 1266–1270 cm^{-1} was distinctly present on the spectra of hemp, flax, and jute lignins. In abaca and sisal lignin spectra, this absorption was present only as an indistinguishable shoulder. The study used the ratio of absorbances at 1327 and 1271 cm^{-1} as a measure of the syringyl-to-guaiacyl ratio in these lignins. Finally, the 834 cm^{-1} band was assigned to C–H out-of-plane bending (del Rio et al. 2007). A typical FTIR spectrum of wheat straw soda lignin is shown in Figure 6.3.

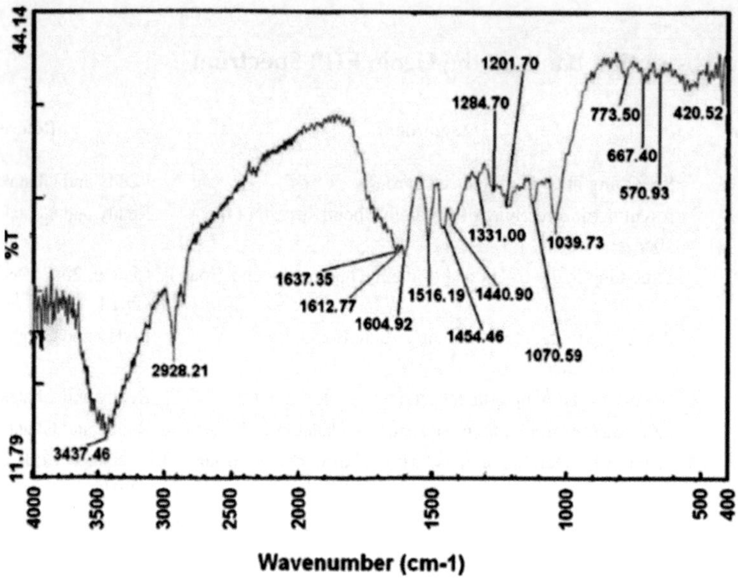

FIGURE 6.3 FTIR spectrum of wheat straw soda lignin. (Reprinted from Ghatak, HR, *Ind. Crops. Prod.*, 28, 206–212, 2008, with permission from Elsevier.)

The characteristic band for hydroxyl appeared at 3437 cm^{-1}. Bands for symmetric and asymmetric aromatic ring stretching and bending in the C–H bond of methoxyl groups were present at their usual frequencies. The absorption band for guaiacyl ring breathing with stretching in the C=O bond appeared at 1284 cm^{-1}. Similar vibration in syringyl structures resulted in a 1331 cm^{-1} band. Bands for vibrations in the benzene ring were present at 1454 and 1440 cm^{-1}. Band for C–O stretching in primary hydroxyl was present at 1039 cm^{-1} (Ghatak 2008). In the FTIR spectrum of soda lignin from acid-hydrolysed aspen, the characteristic hydroxyl stretching, methyl and methylene stretching, symmetric and asymmetric benzene ring stretching, aromatic C–H in-plane bending, and non-conjugated carbonyl C=O bands were present at their usual frequencies of 3434, 2938, 2841, 1607, 1515, 1424, and 1710 cm^{-1}, respectively. The syringyl ring breathing with stretching in the C–O bond of methoxyl was represented by a prominent absorption peak at 1323 cm^{-1}, but the 1266–1270 cm^{-1} band for similar vibrations in guaiacyl moieties was present only as a shoulder. Rather, the very strong absorption at 1216 cm^{-1}, coming from in-plane bending in the O–H bond of phenolic hydroxyl, marked the relative abundance of p-hydroxyphenyl entities. 1033 cm^{-1} peak was attributed to in-plane bending in the C–H bond of the benzene ring as well as stretching in the C–O bond of the primary hydroxyl group (Livča et al. 2012). Soda lignin from *Cunninghamia lanceolata* resulted in usual absorption bands for hydroxyl stretching, methyl and methylene C–H stretching, symmetric and asymmetric benzene ring stretching, aromatic C–H bending, C–H stretching in aliphatic methyl and guaiacyl ring breathing with carbonyl stretching in the FTIR spectrum. Additionally, bands were observed for non-conjugated (1716–1720 cm^{-1}) and conjugated (1649–1654 cm^{-1}) carbonyl stretching, guaiacyl C–H in-plane bending (1147–1152 cm^{-1}), and out-of-plane bending of aromatic C–H (856–859 cm^{-1}) (Li and Wu 2014). Standard absorption peaks for hydroxyl, methyl and methylene, non-conjugated carbonyl, asymmetric benzene ring stretch, methoxyl C–O stretch, benzene ring stretch with C–H bending, breathing in syringyl ring with methoxyl C–O stretch, and O–H in-plane bending of phenolic hydroxyl were observed in the FTIR spectrum of birch soda anthraquinone lignin. Absorptions corresponding to in-plane bending in the C–H bond of the benzene ring and/or stretching in the C–O bond of the primary hydroxyl group were also noticed in the 1038–1060 cm^{-1} window. However, the 1266–1270 cm^{-1} band arising from guaiacyl ring breathing along with methoxyl vibration was missing in the spectrum (Kumar et al. 2016). FTIR spectra of Protobind, a commercial nonwood soda lignin, had characteristic bands for O–H stretching in hydroxyl, methyl and methylene C–H stretching, ketone or ester C=O stretching, aromatic ring skeletal stretching, methoxyl group C–H bending, benzene ring skeletal stretching with in-plane C–H bending, guaiacyl and syringyl ring breathings with methoxyl C–O stretching, and aromatic C–H in-plane bending with C–O stretching of primary hydroxyl. The study assigned the 1365 cm^{-1} band to the in-plane bending of phenolic hydroxyl (Sameni et al. 2016). FTIR spectra of soda lignins, isolated from agricultural residues of wheat straw, barley straw, and fast-growing tree species *Leucaena leucocephala* and *Chamaecytisus proliferus*, were reported to have largely similar absorption bands. These were O–H stretching in hydroxyl, methyl and methylene C–H stretching, asymmetric and symmetric skeletal stretching in benzene ring as well as ring stretching with in-plane C–H bending. Also present were absorption frequencies for asymmetric bending of the C–H bond in methoxyl. C=O stretching frequency in carbonyl at 1714 cm^{-1}, though present in all spectra, was only feeble in the wheat straw spectrum. This was interpreted as low hemicellulose impurities in this lignin. A band at 1034 cm^{-1}, present in all spectra except that of wheat straw lignin, was assigned by the authors to bending vibrations of the C–OH bond in polysaccharides in addition to in-plane bending of aromatic C–H. The presence of conjugated carbonyls, marked by an absorption band at 1649 cm^{-1}, was only detected in soda lignins of agricultural residues. In contrast, the 1211 cm^{-1} band was only observed in wood lignins (Domínguez-Robles et al. 2017). Soda lignins from bamboo, pearl millet, and giant cane were characterized by FTIR spectroscopy. All spectra showed characteristic absorption frequencies (Table 6.1) for hydroxyl, methyl and methylene C–H stretch, C–H bending in methoxyl, benzene ring asymmetric as well as symmetric stretching, and stretching and bending of C–H in aliphatic methyl. 1707 cm^{-1} band was assigned to dimeric carboxylic acid. A band attributed to a guaiacyl-type aromatic ring was present in the spectra at 1329 cm^{-1}. Absorption due to C=O bond stretching of carbonyl appeared in bamboo and giant cane FTIR spectra but was absent in the spectrum of pearl millet. An absorption band at 1166 cm^{-1} in all the spectra was typical of p-hydroxyphenyl entities. 835 cm^{-1} absorption was due to the out-of-plane bending of the aromatic C–H bond (Reddy and Ghatak 2018). A recent study reports FTIR characterization of wheat straw soda black liquor as well as liquor left after lignin precipitation with sulphuric acid. The spectra contained bands for hydroxyl, methyl and methylene, benzene ring skeletal vibration with in-plane C–H bending, syringyl and guaiacyl ring breathings with methoxyl C–O stretch at their characteristic frequencies, though the guaiacyl signals at 1260–1270 cm^{-1} were feeble in intensity. Notably, the peak corresponding to benzene ring symmetrical stretching shifted to 1580 cm^{-1} and only insignificant absorption was noticed in 1505–1515 cm^{-1} region corresponding to asymmetric ring stretching. The study assigned the 1350–1365 cm^{-1} band to O–H bending, and most notably, the 1120 cm^{-1} band was assigned to sodium sulphate produced during acidification (Singh et al. 2023).

6.2.1.2 Kraft Lignin

One study reported lower absorption at 3400 cm^{-1} for hydroxyl O–H stretching but relatively strong bands at 2843–2941 cm^{-1} region for methyl and methylene C–H stretching in the FTIR spectrum of softwood kraft lignin and soda anthraquinone lignin. Bands for symmetric and asymmetric benzene ring stretch were present at their usual positions but the C=O stretching band in carbonyl appeared at 1720 cm^{-1} (El Mansouri and Salvado 2007). Kraft black liquors from the delignification of eucalyptus (hardwood), and pine (softwood) were acidified to different pH to isolate different lignin fractions. FTIR spectra of these lignin fractions, shown in Figure 6.4, are largely similar with subtle differences across wood species and precipitation pH. Peaks for hydroxyl O–H stretching, 2932 cm^{-1} methyl C–H stretching, asymmetric stretching in the aromatic ring, asymmetric bending in the C–H bond of methoxyl groups, and stretching in the benzene ring skeleton with in-plane bending in the C–H bond are similar in all spectra. Methylene C–H stretching, assigned to the 2842 cm^{-1} peak by the authors, is prominently present only in eucalyptus kraft lignin. For pine kraft lignin, a shoulder

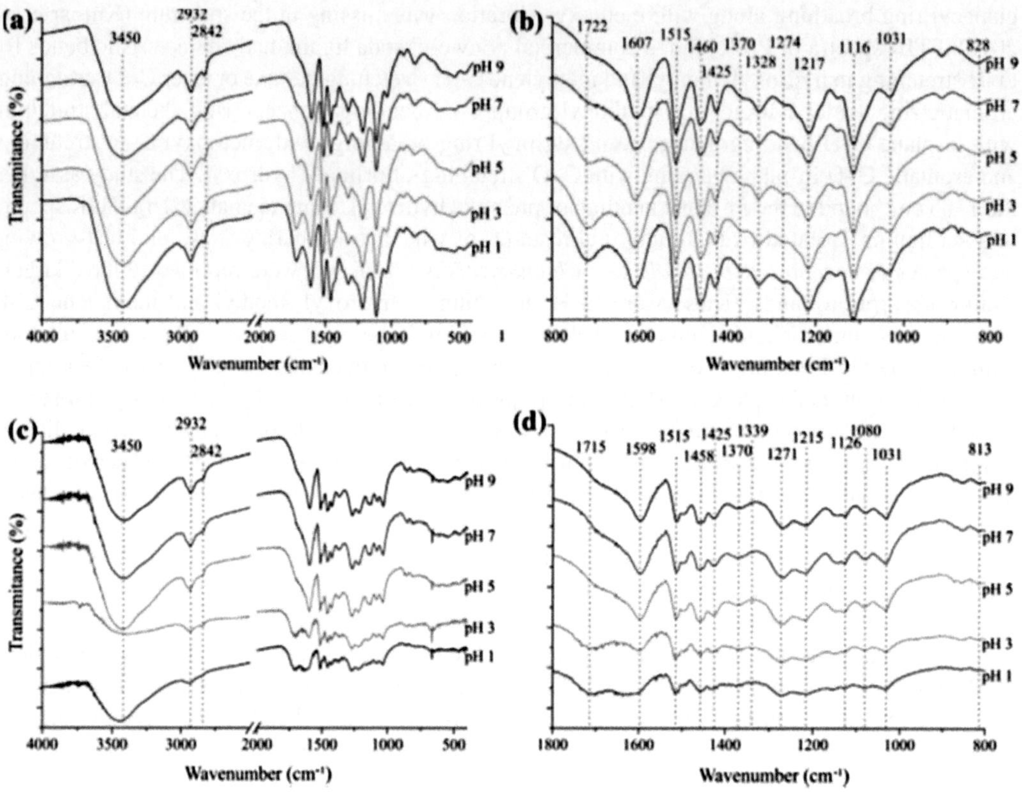

FIGURE 6.4 FTIR spectra of eucalyptus and pine kraft lignins. (a) and (b) eucalyptus, (c) and (d) pine. (Reprinted from Lourençon, TV, Hansel, FA, da Silva, TA, Ramos LP, de Muniz, GIB, Magalhaes, WLE, *Sep. Purific. Technol.*, 154, 82–88, 2015, with permission from Elsevier.)

is observed in the spectra at low precipitation pH which disappears at pH 3 and 1. Carbonyl C=O stretching accounts for a 1722 cm^{-1} band in eucalyptus but shifts to 1715 cm^{-1} in pine. Also, this band grows in intensity as the precipitation pH is lowered. These bands could also originate from hemicellulose inclusion in the samples. Bands corresponding to symmetric stretching in aromatic rings are distinctly observable at 1607 cm^{-1} in eucalyptus but shift to 1598 cm^{-1} for pine. For pine spectra, this band also reduces in strength as the precipitation pH is reduced. 1370 cm^{-1} absorption, assigned to aliphatic methyl C–H vibrations is present only as a shoulder in eucalyptus spectra. As expected, distinct absorption is noticed at 1328 cm^{-1} for syringyl ring breathing with methoxyl C–O stretch for eucalyptus lignin. Corresponding signals for guaiacyl entities are somewhat suppressed at 1274 cm^{-1} in eucalyptus lignin but are prominently present at 1271 cm^{-1} in pine lignin (Lourençon et al. 2015).

Eucalyptus kraft lignin and Indulin AT, a softwood kraft lignin, were subjected to FTIR characterization. Both the spectra had characteristic bands for O–H stretching in hydroxyl, methyl and methylene C–H stretching, ketone or ester C=O stretching, aromatic ring skeletal stretching, methoxyl group C–H bending, benzene ring skeletal stretching with in-plane C–H bending, and aromatic C–H in-plane bending with C–O stretching of primary hydroxyl. This study attributed the 1365 cm^{-1} band to the in-plane bending of phenolic hydroxyl. Spectrum for eucalyptus lignin had bands for both guaiacyl and syringyl ring breathings with methoxyl C–O stretching whereas the 1326 cm^{-1} band for syringyl ring was absent in the Indulin spectrum. The absorption band due to the out-of-plane bending of aromatic C–H in guaiacyl moieties appeared at 851 cm^{-1} in the Indulin spectrum (Sameni et al. 2016). Another recent study on FTIR characterization of eucalyptus kraft lignin and its fractions reported absorption bands for O–H stretching in hydroxyl, C=O stretching of carbonyl, and ring breathing in syringyl moieties with stretching in methoxyl C–O. The study, however, attributed the 1217 cm^{-1} band to methoxyl (Chen et al. 2020). In the FTIR spectrum of mixed hardwood kraft lignin, the absorption bands for hydroxyl O–H stretching, methyl and methylene C–H stretching, C=O bond stretching of ketone, asymmetric and symmetric benzene ring stretching, asymmetric bending in methoxyl groups C–H, benzene ring skeletal stretching with aromatic C–H in-plane bending, guaiacyl ring breathing with methoxyl C–O stretching, O–H in-plane bending in phenolic hydroxyl, and in-plane bending in C–H bond of benzene ring with stretching in C–O bond of primary hydroxyl group were observed at usual frequencies. The spectrum had additional bands at 1320 cm^{-1} (condensed guaiacyl ring), 914 cm^{-1} (aromatic C–H out-of-plane bending), and 827 cm^{-1} (aromatic C–H out-of-plane bending in guaiacyl ring) (Mun et al. 2021). Two eucalyptus kraft lignins were isolated from black liquor acidification to pH 8.1 and 3.5, respectively. In both lignin spectra, characteristic absorptions were reported for hydroxyl O–H stretching, methyl and methylene C–H stretching, and C=O bond stretching of carbonyl (Antonino et al. 2021).

FTIR characterization was performed on lignins isolated from alkaline softwood black liquors by extracting with ionic liquids. The spectra resembled those from lignin isolated by conventional acidification. Characteristic absorption bands corresponded to stretching of O–H in hydroxyl groups, stretching of C–H in methyl and methylene groups, asymmetric and symmetric stretching of the aromatic ring, asymmetric bending in methoxyl groups, and stretching of aromatic ring with in-plane bending of aromatic C–H appeared at their usual positions. Similarly, absorption frequencies. Similarly, syringyl and guaiacyl ring breathings with stretching in the C–O bond of methoxyl were observed at 1331 and 1263 cm^{-1}, respectively. Additionally, the study attributed the bands at 1212 and 1219 cm^{-1} to guaiacyl C–O bond stretch, and 1024 cm^{-1} to C–O bond stretching of the primary hydroxyl group. The presence of hydrocarbons was assigned to the absorption band at 679 cm^{-1} (Bagh and Ray 2022).

6.2.1.3 Lignosulphonate

Softwood lignosulphonates, characterized by FTIR spectroscopy, showed typical absorption bands in the spectra corresponding to O–H stretching in hydroxyl, C–H stretching in methyl and methylene, aromatic ring stretchings, ring vibrations coupled to C–H bending, and guaiacyl ring breathing with C–O stretching in methoxyl. Bands at 1140 cm^{-1} for C–H in-plane bending, and 854 and 817 cm^{-1} for C–H out-of-plane bending were also present. Absorption frequency for non-conjugated carbonyl was found to shift to 1717 cm^{-1} and an additional band at 1771 cm^{-1} was assigned to acetoxyl substituents on the benzene ring. Carbohydrate impurities in the samples showed their presence through absorption peak at 1035 cm^{-1} and other frequencies between 1000 and 1300 cm^{-1} (Boeriu et al. 2004). FTIR spectrum of bagasse sodium lignosulphonate consisted of bands for hydroxyl O–H stretching, methyl and methylene C–H stretching, symmetric and asymmetric aromatic skeletal stretching, asymmetric bending in methoxyl groups C–H, benzene ring skeletal stretching with aromatic C–H in-plane bending, guaiacyl ring breathing with methoxyl C–O stretching, in-plane bending in O–H bond in phenolic hydroxyl, and in-plane bending in C–H bond of benzene ring and stretching in C–O bond of primary hydroxyl group. Additional absorption bands were present at 1118 cm^{-1} for in-plane bending of aromatic C–H in syringyl moieties, 835 cm^{-1} for out-of-plane bending of aromatic C–H bond, and 651 cm^{-1} for vibrations in C–S bond of sulphonic acid group. Notably, the spectrum did not have the absorption band for C–H stretching of methyl groups in the aliphatic side chain at 1365–1370 cm^{-1}, as well as bands for carbonyl and ester entities in the 1640–1750 cm^{-1} region (Singh et al. 2005). Similar absorption bands were observed in softwood lignosulphonate for hydroxyl, methyl, and methylene stretching, aromatic skeletal stretching, methoxyl C–O stretching with guaiacyl ring vibration, and aromatic C–H bending. The sulphonic acid group showed its presence in the spectrum by the S–O stretching band at 620 cm^{-1} but no absorption band related to carbonyls was reported (El Mansouri and Salvado 2007). One study reports FTIR characterization of calcium and magnesium base spent liquor from acid sulphite delignification of eucalyptus. Spectra had usual bands at 3100–3550 cm^{-1} for hydroxyl O–H stretching, 2849–2942 cm^{-1} for methyl and methylene C–H stretching, 1517 cm^{-1} for asymmetric aromatic skeletal stretching, 1464 cm^{-1} for methoxyl C–H asymmetric bending, and 1429 cm^{-1} for aromatic C–H in-plane bending with benzene skeletal vibration. C=O stretch of conjugated carbonyl resulted in an absorption band at 1636 cm^{-1} (Rueda et al. 2014). In the FTIR spectrum of sodium lignosulphonate from corn stover, the major absorption bands present were 3400 cm^{-1} (hydroxyl), 2924 cm^{-1} (methyl C–H), 1315 cm^{-1} (syringyl breathing with methoxyl C–O stretch), 1268 cm^{-1} (guaiacyl breathing with methoxyl C–O stretch), and 1218 cm^{-1} (methoxyl C–O stretch in guaiacyl entities). The authors assigned the distinct band in the 1155–1245 cm^{-1} region to S=O vibrations in sulphonate (Xu et al. 2015). FTIR characterization of the mixture of sodium and ammonium lignosulphonates produced spectra with usual absorption bands for hydroxyl, methyl and methylene, aromatic ring stretching, and primary alcohol. The sulphonate group was associated with the band in the 1190–1210 cm^{-1} window (Karimov et al. 2021). According to one study, the peaks at 1417, 1192, and 1045 cm^{-1} were assigned to S=O stretching in sulphonate groups in the FTIR spectrum of jute lignosulphonate (Bose et al. 2023).

6.2.1.4 Milled Wood Lignin

In an early study, the FTIR spectrum from birch milled-wood lignin was compared with the spectrum of its acetylated derivative. The distinct band for hydroxyl O–H stretching in milled wood lignin was absent for acetylated lignin. Instead, the acetylated lignin resulted in prominent bands in the regions 1710–1790 cm^{-1} (ester C=O stretch), 1330 cm^{-1} (C–H bending in ester methyl group), and 1225 cm^{-1} (C–O and C–C stretching in esters). Moreover, the band for asymmetric stretching of the aromatic ring was weakened due to phenolic hydroxyls converting to acetyls (Faix et al. 1992).

Miscanthus – a nonwood lignocellulose – milled wood lignin generated a FTIR spectrum with typical lignin absorption bands at 3435 cm^{-1} (hydroxyl O–H stretching), 2918 cm^{-1} (methyl C–H stretching), 1709 and 1660 cm^{-1} (non-conjugated and conjugated carbonyl C=O stretch), 1603 and 1508 cm^{-1} (aromatic skeletal stretches), 1463 cm^{-1} (ring stretch with methoxyl C–O stretch), 1422 cm^{-1} (aromatic ring stretch with C–H bending), 1328 and 1267 cm^{-1} (syringyl and guaiacyl breathings, respectively, with C–O stretch), and 1033 cm^{-1} (aromatic C–H in-plane bending). Additional bands appeared at 1768, 1731, and 1650 cm^{-1} for acetate groups and 833 cm^{-1} for out-of-plane bending in aromatic C–H (El Hage et al. 2009). FTIR spectrum of bamboo milled-wood lignin had characteristic absorption bands for hydroxyl groups, C–H stretching of methyl and methylene groups, C–H bending in methoxyl, and symmetric and asymmetric aromatic ring stretching. In addition, for guaiacyl entities, a 1267 cm^{-1} band was assigned to benzene ring breathing, 1034 cm^{-1} to in-plane bending of the C–H bond of the benzene ring, and 920 cm^{-1} to out-of-plane bending in C–H bond of the benzene ring. For syringyl entities, these bands appeared at 1329 cm^{-1}, 1234 cm^{-1}, and 1126 cm^{-1}, respectively. 1168 cm^{-1} absorption was attributed to the C=O bond in the carbonyl group conjugated to the benzene ring double bond system (Li et al. 2012). Similar spectra were also reported in another FTIR characterization of bamboo milled-wood lignin with typical absorptions for hydroxyl O–H, methyl and methylene C–H, aromatic ring stretchings, methoxyl C–H bending, benzene ring stretching coupled to C–H bending, and ring breathings coupled to methoxyl C–O stretch. 1030 cm^{-1} band was assigned to the C–O stretch in glycosidic linkages in addition to the usual C–O stretch in alcohols. Carbonyl C=O stretching frequencies are seen in the spectrum at 1710 and 1654 cm^{-1} (Shi et al. 2019). The spectrum is shown in Figure 6.5.

Luan – a hardwood – milled wood lignin resulted in FTIR spectrum in the 2000–700 cm^{-1} wavenumber span having absorption peaks for benzene ring stretchings, methoxyl groups C–H bending, aromatic C–H in-plane bending coupled to ring stretching, syringyl and guaiacyl ring breathings with C–O stretch of methoxyl, stretching of phenolic hydroxyl, and vibrations of primary, secondary and tertiary alcohols. Further, a shoulder at 1153 cm^{-1} was associated with in-plane C–H bending in guaiacyl moieties (Wang et al. 2014). In the FTIR spectrum of acacia milled-wood lignin, the absorption bands for hydroxyl O–H stretching, methyl and methylene C–H stretching, C=O bond

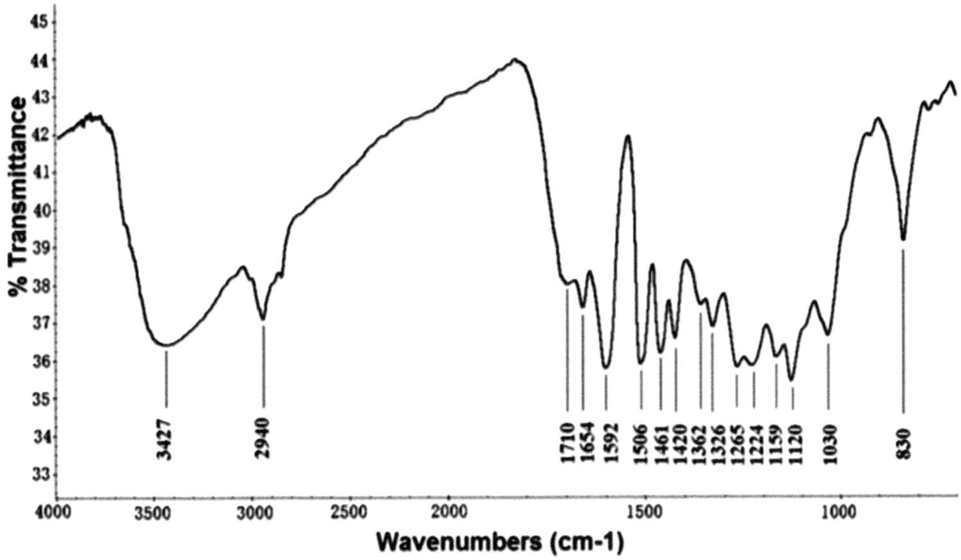

FIGURE 6.5 FTIR spectrum of bamboo milled-wood lignin. (Reprinted from Shi, Z, Shu, G, Deng, J, Dong, M, Murugadoss, V, Liu, C, Shao, Q, Wu, S, Guo, Z, *Green Chem. Lett. Rev.*, 12(3), 235–243, 2019, originally published by Taylor and Francis under Creative Commons Attribution License.)

stretching of esters, asymmetric and symmetric benzene ring stretching, asymmetric bending in methoxyl groups C–H, benzene ring skeletal stretching with aromatic C–H in-plane bending, guaiacyl ring breathing with methoxyl C–O stretching, O–H in-plane bending in phenolic hydroxyl, and in-plane bending in C–H bond of benzene ring with stretching in C–O bond of primary hydroxyl group were observed at usual frequencies. The spectrum had additional bands at 1663 cm^{-1} (conjugated C=O bond stretching in carbonyl, 1328 cm^{-1} (condensed guaiacyl ring), 1089 cm^{-1} (C–O stretching in ether linkages and secondary aliphatic hydroxyl), and 834 cm^{-1} (aromatic C–H out-of-plane bending in syringyl ring) (Mun et al. 2021).

Pine milled-wood lignin FTIR spectrum presented major bands for hydroxyl O–H (3400 cm^{-1}), methyl and methylene C–H (2800–3000 cm^{-1}), carbonyl C=O conjugated to the benzene ring (1663 cm^{-1}), asymmetric and symmetric benzene ring stretching (1606 and 1502 cm^{-1}), guaiacyl breathing with methoxyl C–O (1268 cm^{-1}), guaiacyl C–H bending (1214 and 1140 cm^{-1}), as well as C–OH in primary and secondary alcohols (1082, 1030 cm^{-1}) (Huang et al. 2011).

6.2.1.5 Organosolv Lignin

Mixed hardwood ethanol organosolv lignin (Alcell lignin) showed characteristic absorption frequencies in its FTIR spectrum for hydroxyl O–H stretching, methyl and methylene C–H stretching, and asymmetric and symmetric skeletal vibrations of the benzene ring. Two distinct carboxyl absorptions were also observed in the 1750–1550 cm^{-1} region (Gosselink et al. 2004). Another research group studied Alcell lignin by FTIR spectroscopy and observed a similar absorption pattern regarding characteristic lignin molecular features. 1705 cm^{-1} peak for unconjugated carbonyl C=O vibrations was relatively intense. A weak band at 1650 cm^{-1} was attributed to protein impurities. Bands at 1140 cm^{-1} for C–H in-plane bending, and 854 and 817 cm^{-1} for C–H out-of-plane bending were also present (Boeriu et al. 2004). Miscanthus ethanol organosolv lignin generated an FTIR spectrum with typical lignin absorption bands at 3435 cm^{-1} (hydroxyl O–H stretching), 2918 cm^{-1} (methyl C–H stretching), 1709 and 1660 cm^{-1} (non-conjugated and conjugated carbonyl C=O stretch), 1603 and 1508 cm^{-1} (aromatic skeletal stretches), 1463 cm^{-1} (ring stretch with methoxyl C–O stretch), 1422 cm^{-1} (aromatic ring stretch with C–H bending), 1328 and 1267 cm^{-1} (syringyl and guaiacyl breathings, respectively, with C–O stretch), and 1033 cm^{-1} (aromatic C–H in-plane bending). Additional bands appeared at 833 cm^{-1} for out-of-plane bending in aromatic C–H (El Hage et al. 2009). Aqueous ethanol organosolv lignins were isolated from hot water pretreated rice straw and characterized by FTIR spectroscopy. Characteristic absorption bands for methyl and methylene C–H stretching, C=O stretching in conjugated carbonyls, symmetric and asymmetric aromatic skeletal stretch, guaiacyl and syringyl ring breathings with methoxyl C–O stretch, and aromatic C–H in-plane bending combined with C–O stretch of primary hydroxyl group were reported in the spectra. The band centred at 1115 cm^{-1} was attributed to syringyl moieties C–O Stretch in methoxyl, and aromatic C–H in-plane bending. These authors also used the absorbance ratio of 1326 and 1265 cm^{-1} to ascertain the syringyl-to-guaiacyl ratio in the lignin sample (Moniz et al. 2018). Characteristic absorption bands in the FTIR spectra of birch ethanol organosolv lignin included hydroxyl O–H stretching, C–H bond stretching in methyl and methylene, symmetric and asymmetric stretching in benzene ring skeleton, ring skeletal stretching with in-plane bending of aromatic C–H, and ring breathings in syringyl and guaiacyl structures with stretching in C–O bond of methoxyl. The absorption band at 1218 cm^{-1} was assigned to methoxyl C–O stretching of guaiacyl. Additional peaks present in the spectra came from aromatic C–H stretch at 1113 cm^{-1} (Thoresen et al. 2021). A recent study reported FTIR characterization of aqueous ethanol organosolv lignins from poplar, pine, walnut shell, and rice husk. The spectra are shown in Figure 6.6. Hydroxyl O–H stretching band is centred at 3400 cm^{-1}. C–H bond stretching in methyl and methylene was represented by absorption at 2930 cm^{-1}. Bands for asymmetric and symmetric stretching in benzene ring skeleton appeared at 1600 and 1510 cm^{-1}, respectively. This study assigned the 1659 cm^{-1} peak to vibrations in conjugated entities, 1460 cm^{-1} peak to aromatic ring skeletal vibrations, 1235 cm^{-1} band to C=O stretch in guaiacyl moieties, and 1120 cm^{-1} to C=O stretch in syringyl moieties (Sheng et al. 2022).

In the FTIR spectrum of miscanthus formic acid organosolv lignin strong absorption at 3400 cm^{-1} for hydroxyl O–H stretching and 2843–2941 cm^{-1} region for methyl and methylene C–H stretching was reported. Bands for symmetric and asymmetric benzene ring stretch were present at their usual positions but the C=O stretching band in carbonyl appeared at 1720 cm^{-1} (El Mansouri and Salvado 2007). Bamboo formic acid organosolv lignin showed characteristic absorption bands for hydroxyl

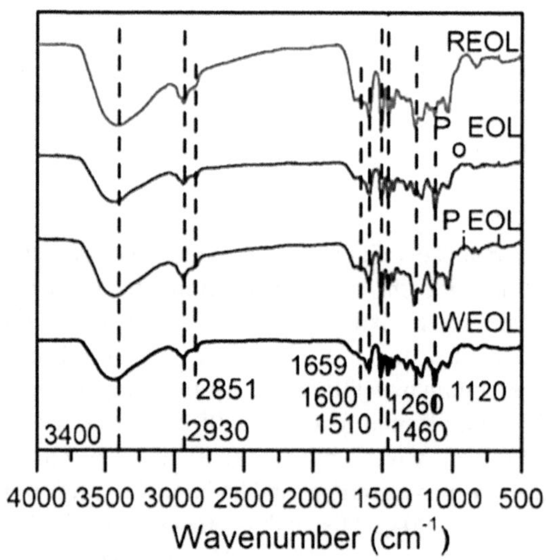

FIGURE 6.6 FTIR spectra of aqueous ethanol organosolv lignins. REOL for rice husk, PoEOL for poplar, PiEOL for pine, and WEOL for walnut shell. (Reprinted from Sheng, Y, Ma, Z, Wang, X, Han, Y, *Food Chem.*, 376, 131895, 2022, with permission from Elsevier.)

groups, C–H stretching of methyl and methylene groups, C–H bending in methoxyl, and symmetric and asymmetric aromatic ring stretching. Benzene ring breathing corresponded to a 1267 cm^{-1} band in guaiacyl moieties and a 1329 cm^{-1} band in syringyl moieties. 1034 and 1234 cm^{-1} absorption arose from in-plane bending of the C–H bond of the benzene ring in guaiacyl and syringyl structures, respectively. Similarly, 920 and 1126 cm^{-1} absorption came from out-of-plane bending of the C–H bond of the benzene ring in guaiacyl and syringyl structures, respectively. The presence of a characteristic formate functional group was recorded in the absorption band at 1709 cm^{-1} (Li et al. 2012).

FTIR characterization of aqueous dioxane organosolv lignin of banana leaf sheath, before and after chloroform extraction, revealed usual absorption bands for hydroxyl O–H stretching, methyl and methylene C–H stretching, C=O stretching in carbonyl and esters, asymmetric and symmetric skeletal vibrations of the benzene ring, methoxy C–H asymmetric bending, stretching in benzene ring skeleton with in-plane bending in the C–H bond, syringyl ring breathing with stretching in C–O bond of methoxyl, phenolic O–H in-plane bending, O–H stretching in tertiary hydroxyl, in-plane bending in aromatic C–H, and stretching in C–O bond of primary hydroxyl group. The spectra are shown in Figure 6.7.

Band for guaiacyl ring breathing with stretching in the C–O bond of methoxyl, normally appearing at 1266–1270 cm^{-1} region is merely noticeable as a feeble shoulder. One characteristic spectral feature of non-purified lignin was rather strong signals at 2938 and 2850 cm^{-1} arising from aliphatic C–H stretching in carbohydrate portion chemically linked to lignin. These bands reduced in intensity when the isolated lignin was purified by chloroform extraction (Oliveira et al. 2006). FTIR spectra of spruce and eucalyptus dioxane lignins in the 1900–700 cm^{-1} range contained typical absorption bands for symmetric and asymmetric breathing of aromatic skeleton, and in-plane C–H bending of methoxyl and aromatic hydrogen in common. The spruce spectrum showed strong absorption for guaiacyl ring breathing with methoxyl C–O stretching (1270 cm^{-1}), secondary and tertiary hydroxyl O–H stretching (1140 cm^{-1}), and in-plane aromatic C–H bending and methoxyl C–O stretching (1032 cm^{-1}). The spectrum originating from eucalyptus, on the other hand, had strong absorption signals for syringyl ring breathing with methoxyl C–O stretching (1328 cm^{-1}), and in-plane bending in O–H bond in phenolic hydroxyl, and secondary and tertiary hydroxyl O–H stretching (1124 cm^{-1}). It was contended that the 1140 cm^{-1} absorption in spruce dioxane lignin was shifted to 1124 cm^{-1} in eucalyptus dioxane lignin. Common absorption bands in the FTIR spectrum of maize dioxane lignin were symmetric and asymmetric breathing of aromatic skeleton (1604 and 1512 cm^{-1}), and in-plane C–H bending of methoxyl and aromatic hydrogen (1460 and 1420 cm^{-1}). Peaks corresponding to both syringyl and guaiacyl ring breathing with methoxyl C–O stretching (1328 and 1270 cm^{-1}) were also present (Chazal et al. 2014).

6.2.1.6 Hydrolysis Lignin

Klason lignin was extracted from Chinese fir (softwood) and maple (hardwood) and characterized by infrared spectroscopy. Characteristic bands at normal frequencies were present for hydroxyl stretching, methyl and methylene C–H stretching, asymmetric stretching in the aromatic skeleton, C–O stretching in guaiacyl methoxyl, and C–O stretching in primary hydroxyl. An additional band for C=O stretching of conjugated carbonyl appeared at the 1600–1615 cm^{-1} region. Authors attributed the 1220 cm^{-1} band to C–O stretching in syringyl methoxyl, present prominently in only maple lignin. They also assigned an additional absorbance frequency at 2358 cm^{-1} in the maple lignin spectrum to C=O stretching of ketone functionality (Zhao et al. 2014).

The lignin-rich residue left after the steam explosion and enzymatic hydrolysis of populus was characterized by FTIR spectroscopy. The study reported lower absorption at 3400 cm^{-1} for hydroxyl O–H stretching but relatively strong bands at 2843–2941 cm^{-1} region for methyl and methylene C–H stretching. Bands for symmetric and asymmetric benzene ring stretch were present at their usual positions but no C=O stretching band in carbonyl was observed (El Mansouri and Salvado 2007). The solid residue from hydrothermal pretreatment of eucalyptus had significant inclusion of

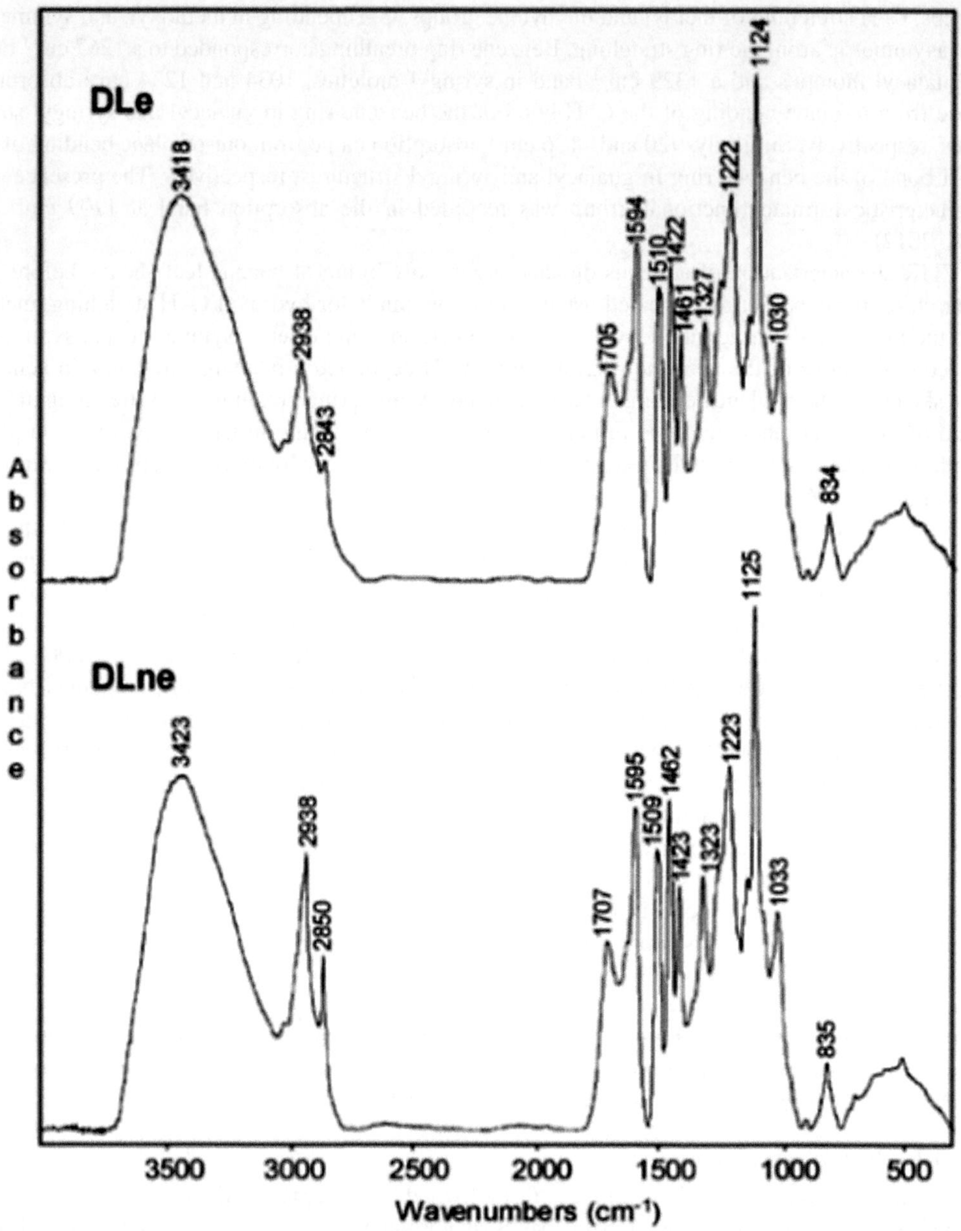

FIGURE 6.7 FTIR spectra of banana leaf sheath aqueous dioxane organosolv lignin. DLe for chloroform extracted sample, DLne for non-extracted sample (Reprinted from Oliveira, L, Evtuguin, DV, Cordeiro, N, Silvestre, AJD, Silva, AMS, Torres IC, *J. Agric. Food Chem.*, 54, 2598–2605, 2006, with permission from American Chemical Society.)

carbohydrate components with lignin, as revealed by their FTIR spectra. Typical absorption frequencies reported for lignin corresponded to symmetric and asymmetric stretching in the benzene ring skeleton, ring skeletal stretching with in-plane bending of aromatic C–H, and syringyl ring breathing with stretching in the C–O bond of methoxyl. The peak at 829 cm^{-1} was attributed to aromatic C–H stretch in p-hydroxyphenyl entities while the 1454 cm^{-1} frequency was assigned to aromatic C–H stretch with ring breathing. The hydroxyl O–H stretching band at 3348 cm^{-1} arose from both lignin and carbohydrates. The bands at 2903, 1107, and 899 cm^{-1} were exclusively assigned to carbohydrates (Shen et al. 2016). FTIR characterization of enzymatic mild hydrolysis lignin from wheat straw resulted in a spectrum with typical absorptions for hydroxyl O–H stretch, methyl and methylene C–H stretch, carbonyls and esters C=O stretch, benzene ring asymmetric and symmetric stretch, syringyl and guaiacyl ring breathing with C–O stretching of methoxyl, and aromatic C–H in-plane bending. Moreover, additional peaks in the spectrum were assigned as 1651 cm^{-1} to C=O stretch in conjugated carbonyls, 1454 cm^{-1} to asymmetric C–H deformations in methyl and methylene, 1156 cm^{-1} to C=O stretch in esters, 1085 cm^{-1} to ether and secondary hydroxyl C–O stretching, and 833 cm^{-1} to C–H out-of-plane bending (Yang et al. 2011). The spectrum is reproduced in Figure 6.8.

In the FTIR spectrum of enzymatic mild hydrolysis lignin from rice straw, characteristic bands were present for stretching in hydroxyl O–H, stretching vibrations in C–H bond of methyl and methoxyl, carbonyl C=O stretching, asymmetric and symmetric benzene ring stretching, C–H asymmetric bending of methoxyl, in-plane aromatic C–H bending coupled with ring stretching, and C–O stretching of methoxyl with syringyl and guaiacyl ring breathing. Additional bands, at 1162 cm^{-1} for conjugated ester C=O and 834 cm^{-1} for out-of-plane bending of C–H in the benzene ring were also present in the spectrum. The carbohydrate inclusion in the sample was confirmed by a 1088 cm^{-1} band that came from glycosidic linkages (Lou et al. 2018).

6.2.1.7 Lignin Carbohydrate Complexes

Lignin carbohydrate complexes were isolated from steam-exploded soda delignified bagasse and steam-exploded modified kraft delignified bagasse. Their FTIR spectra had characteristic absorption bands for hydroxyl O–H stretching, methyl and methylene C–H stretching, symmetric and asymmetric aromatic ring stretching, asymmetric bending in methoxyl C–H, benzene ring skeletal stretching with aromatic C–H in-plane bending, C–H stretching in aliphatic methyl, guaiacyl ring breathing with stretching in methoxyl C–O bond, in-plane bending in phenolic hydroxyl O–H, and C–H in-plane bending of benzene ring with stretching in C–O bond of primary hydroxyl group. Carbonyl stretching in the 1696-1705 cm^{-1} region was assigned to carbohydrate lignin bond. In the spectrum of the modified-kraft derived sample, an additional band at 619 cm^{-1}, though feeble, was due to stretching in the C–S bond of the thioalcohol group. The spectra also had an absorption band at 835 cm^{-1} due to out-of-plane C–H bending in syringyl structures (Singh et al. 2005). Lignin

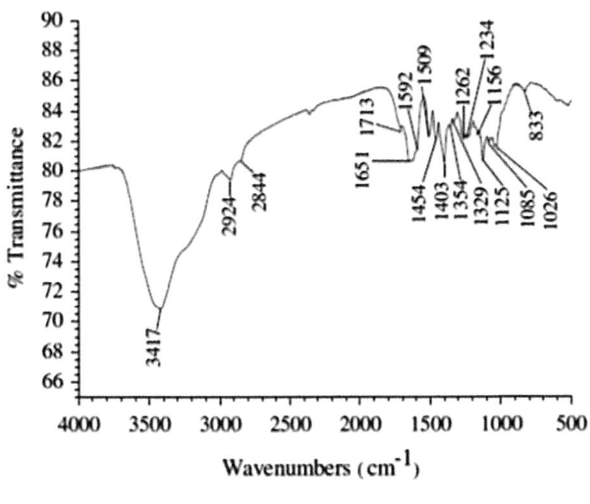

FIGURE 6.8 FTIR spectrum of wheat straw enzymatic mild acidolysis lignin. (Reprinted from Yang, Q, Wu, S, Lou, R, LV, G, *Wood Sci. Technol.*, 45, 419–431, 2011, with permission from Springer.)

carbohydrate complexes were isolated from milled wood lignin and enzyme hydrolysis lignin of Arundo donax, a nonwood species, and subjected to FTIR characterization. These spectra had only feeble characteristic absorptions for symmetric and asymmetric skeletal stretching of benzene ring, and asymmetric bending of methoxyl C–H. More prominent absorptions were observed at 1735 cm^{-1} assigned to C=O stretching in esters linked to xylan or uronic acid of the carbohydrate fraction, 1630 cm^{-1} arising from antisymmetric stretching in ester functionalities on xylan and uronic acid fragments, and 1043 cm^{-1} attributed to hemicelluloses (You et al. 2015).

6.2.1.8 Derivatized Lignin

In an early study on acetylated lignins of several types including alkali lignins, the absorption band at 1766 cm^{-1} was assigned to aromatic acetyl groups while the band at 1743 cm^{-1} was assigned to aliphatic acetyl. A comparison of their normalized intensities could be one of the methods for estimating the ratio of aromatic to aliphatic hydroxyls in lignin. The ratio was 1.56 for Sucrolin bagasse and 1.19 for Indulin AT (Faix et al. 1994). This was based on the presumption that the acetyl groups in acetylated lignins all resulted from hydroxyls present in the original lignin, which may not be rigorously true. A similar analysis was applied to acetylated soda lignins from wheat straw, hemp, and flax by assigning the 1765 cm^{-1} absorption to aromatic esters and 1745 cm^{-1} to aliphatic esters (Gosselink et al. 2004). In FTIR spectra of the acetylated, fluorobenzylated, and fluorobenzoylated lignins of several types, the band intensities for O–H stretching of hydroxyl groups in 3100–3600 cm^{-1} frequencies were significantly weaker than underivatized lignin indicating the conversion of hydroxyl groups into the corresponding derivative. At the same time, the methyl and methylene bands between 2800 and 3000 cm^{-1} increased in strength showing additional presence of these groups after derivatization. The formation of corresponding esters also enhanced the C=O stretching signal at 1700–1900 cm^{-1} for acetylated samples. Spectra of fluorobenzoylated lignins also had additional peaks between 1550 and 1700 cm^{-1} corresponding to carbonyl C=O conjugated to aromatic C=C and between 1080 and 1180 accounting for the formation of new C–O–C bonds (Esakkimuthu et al. 2022).

6.3 LIGNIN CHARACTERIZATION BY RAMAN SPECTROSCOPY

Scattering is an optical phenomenon in which the electromagnetic radiation after interaction with the molecules of the matter is made to propagate not only in the original direction of the incident beam but also in all other directions. It is believed that the energy of the incident photon is momentarily absorbed by the electron cloud of the molecule, driving the vibrational state of the molecule to a metastable higher energy state. In the majority of instances, in the Rayleigh scattering, a photon of equal energy is reemitted as scattered light and the molecule reverts to its ground state of vibration. Though rare, there are, however, instances when the excited molecule returns to a vibrational energy state lower or higher than its original vibrational energy state. If this happens, the reemitted photon has energy – and resulting wavelength or frequency – lower or higher than the incident photon. This phenomenon is known as Raman scattering. If the emitted photon has lower energy – higher wavelength – than the incident photon, it is Stokes–Raman scattering. This leaves the molecule in a higher vibrational energy state. The opposite phenomenon is the anti-Stokes–Raman scattering where the reemitted photon has higher energy – lower wavelength – than the incident photon. This happens when the molecule is initially not at the ground state of vibrational energy but at a higher vibrational energy state, and after the reemission of the photon returns to the ground state. For a single molecule, quantum mechanics postulates equal probability for the two types of Raman scattering. However, in a real sample of material containing a large population of molecules Stokes–Raman scattering is more probable. As in infrared spectroscopy, the energy of the incident and scattered photon is expressed as the reciprocal of their wavelengths, measured as wavenumber in the units of cm^{-1}. The difference between the wavenumbers of the incident photon and the scattered photon is termed the Raman shift.

An organic molecule can have several vibrational degrees of freedom. Each vibrational degree of freedom can have a number of quantum energy states for that particular mode of vibration. In a statistically large population of molecules, the distribution of the number of molecules in these different energy states follows the Boltzmann energy distribution. Depending on the energy of the photons in the incident beam of light and the initial and final vibrational energy state of the molecule, the scattered light consists of photons of different wavenumbers. This leads to a spectrum, known as the Raman spectrum, which is a signature of the molecular structure of the scattering molecules. This is the basis of Raman spectroscopy. The most common in Raman spectroscopy is to use a LASER beam as the source of incident photons. These can be near-infrared, visible, or near-UV. Since Rayleigh scattering is the predominant form of scattering, it is necessary to filter these photons from the scattered light using elaborate optics and electronics. Another possible source of interference is fluorescence occurring due to the absorption of light by the sample. Based on the technique to enhance the intensity of Raman spectra, there are different Raman spectroscopic methods available. Some of these are Surface Enhanced Raman Spectroscopy and Tip Enhanced Raman Spectroscopy where the material to be characterized is placed very close to a metal surface. In resonance Raman Spectroscopy, the energy of the incident photons is equal to or very close to one of the possible energy differences between different energy states of the electron cloud in the molecule.

In Raman spectroscopy of lignin, near-infrared illumination is particularly desirable as the problem of fluorescence is avoided at this incident frequency. As a result, the quality of the Raman spectrum is improved manifold. The use of UV is not desirable due to the strong absorption of UV light by the aromatic backbone in the lignin macromolecule and the resulting fluorescence. In this context, the use of 1064 nm LASER as the source of incident radiation is reported (Agarwal et al. 2011; Agarwal et al. 2019). It is reported that the problem of fluorescence in Raman spectra is much more acute for isolated lignins than in untreated lignocelluloses. Carbon–carbon linkages of 5–5′ type are believed to be an important reason for fluorescence in isolated lignins like milled wood lignin and enzymatic mild acidolysis lignin (Lahdetie et al. 2013).

Raman spectroscopic characterization of softwood (spruce and pine) and hardwood (sweetgum and aspen) milled-wood lignins was performed with 1064 nm near-infrared incident radiation. The spectra are shown in Figure 6.9 for different Raman shift spectral regions. Prominent bands and their assignments were 3068–3071 cm^{-1} (C–H stretching in aromatic carbon), 2939–2941 cm^{-1} (asymmetric C–H stretching in methoxyl), 2844–2847 cm^{-1} (symmetric C–H stretching in methoxyl), 1661–1664 cm^{-1} (conjugated carbonyl C=O stretching), 1595–1598 cm^{-1} (symmetric benzene ring stretching), 1453–1456 cm^{-1} (C–H bending in methoxyl), 1331–1334 cm^{-1} (O–H bending in aliphatic hydroxyl), and 1271–1273 cm^{-1} (C–O stretching in aromatic hydroxyl or methoxyl). Medium intensity peaks at 368–369 cm^{-1} for skeletal vibrations in the benzene ring of syringyl moieties were observed for sweetgum and aspen. But bands in this region were very weak for spruce and pine. In addition, weak to very weak bands appeared at 1501–1508 cm^{-1} ((asymmetric benzene ring stretching), and 1426–1434 cm^{-1} (benzene ring stretching coupled to C–H bending in methyl) (Agarwal et al. 2011).

Surface-enhanced Raman spectrum of spruce milled-wood lignin was obtained by mixing the sample with silver particles in ethanol. A near-infrared source was used for irradiation. Compared to the normal Raman spectrum the peaks in the 250–1800 cm^{-1} region were increased in intensity while those in 2700–3200 cm^{-1} – mostly from C–H stretching – were reduced in intensity. Some shifts in band frequencies were also noticed. Major peaks in the surface-enhanced Raman spectrum were 1581–1589 cm^{-1} for symmetric ring stretching in benzene, 1491 cm^{-1} for asymmetric ring stretching in benzene, 1267–1272 cm^{-1} for C–O stretching in hydroxyl and methoxyl, 1159 cm^{-1} for C–H in-plane bending in the benzene ring, 1125 cm^{-1} for C–O stretching in aliphatic hydroxyl, 925–929 cm^{-1} and 816–820 cm^{-1} for C–H twisting in benzene ring, 641 cm^{-1}, and 416–420 cm^{-1} for deformation in benzene ring. Only a shoulder at 1645 cm^{-1} was present for conjugated carbonyl C=O. The study also highlighted the complementary nature of the Raman spectrum and FTIR spectrum. For example, the 1589 cm^{-1} peak for symmetric ring stretching in benzene in the Raman

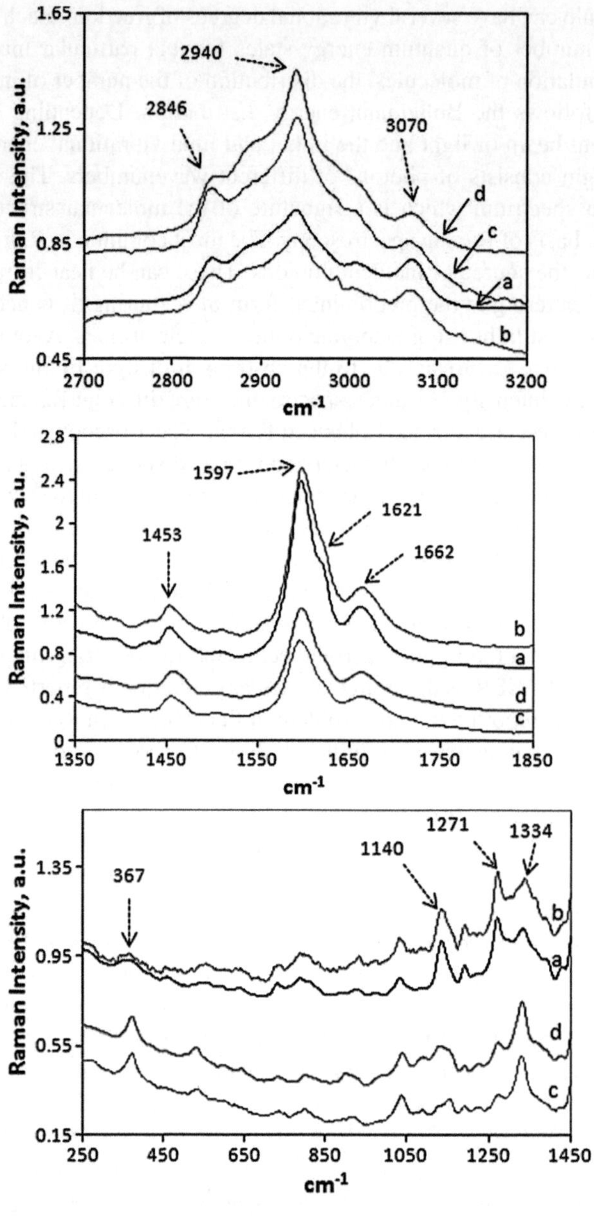

FIGURE 6.9 Raman spectra of milled wood lignins; (a) spruce, (b) pine, (c) aspen (d) sweetgum. (Reprinted from Agarwal, UP, McSweeny, JD, Ralph, SA, J. *Wood Chem. Technol.*, 31, 324–344, 2011, with permission from Taylor and Francis.)

spectrum also appeared in the FTIR spectrum at 1600 cm^{-1}. On the other hand, the peaks at 1159 cm^{-1} for C–H in-plane bending in the benzene ring, and 1125 cm^{-1} for C–O stretching in aliphatic hydroxyl, were observed in Raman spectrum but not in the infrared spectrum (Agarwal and Reiner 2009).

Raman spectroscopy is a very useful tool for the characterization of native lignin in their unmodified state in lignocellulose prior to isolation. Bands in Raman spectra of hardwood and nonwood lignocelluloses were assigned to different structural moieties as 1333 cm^{-1} for syringyl, 1272 cm^{-1} for guaiacyl, 1216 cm^{-1} for hydroxyphenyl, and 1150 cm^{-1} for syringyl and guaiacyl combined. The strong signal at 1600 cm^{-1} accounts for benzene ring stretching. The guaiacyl peak was narrower and sharper in grass lignin with an additional sharp peak at 1177 cm^{-1}. Also, the hydroxyphenyl peak was more prominent in grass lignin compared to hardwood lignin. While hardwood spectra showed a Raman shift at 1665 cm^{-1} for conjugated carbonyl, the grass spectrum had a 1702 cm^{-1} band for acyl groups. The study used the ratio of peak areas of 1333 cm^{-1} and 1270 cm^{-1} as a measure of the syringyl-to-guaiacyl ratio in native lignin of these lignocellulose (Perera et al. 2012). In the Raman spectrum of the bamboo cell wall, lignin structures contributed to bands at 1656 cm^{-1} for olefinic structures conjugated to the benzene ring, 1598 cm^{-1} for benzene ring stretching, 1152 cm^{-1} for guaiacyl, and 1197 cm^{-1} for p-hydroxyphenyl (Zhang et al. 2018). Different parts of the poplar tree stem were characterized by micro Raman spectroscopy to study the relative abundance of cellulose, hemicellulose, and lignin. Lignin exclusively contributed to bands at 1270 cm^{-1} arising from guaiacyl moieties, 1595 cm^{-1} attributed to benzene ring stretching, and 1654 cm^{-1} for conjugated carbonyl and olefinic entities (Zhang 2021).

6.4 LIGNIN CHARACTERIZATION BY NUCLEAR MAGNETIC RESONANCE SPECTROSCOPY

Atomic nuclei are quantum entities made up of fermions. An individual fermion has a non-integer spin quantum number with possible values of +1/2 and –1/2. Non-integer spin quantum number imparts degenerate quantum states to the fermions in the presence of an externally applied magnetic field. The energy difference between these degenerate quantum states depends on the strength of the magnetic field applied. The transition of a fermion between these degenerate quantum energy states produces a spectrum that is characteristic of the fermion and its electromagnetic magnetic environment. Bosons, on the other hand, have integer spins and do not have degenerate quantum states. An ensemble of fermions, like a proton or neutron, or an atomic nucleus, has an integer or non-integer spin quantum number depending on the number of fermions constituting the system. Systems with an odd number of fermions have non-integer spins and act as fermions as a whole. Systems with an even number of fermions have integer spins and act as bosons. Protons and neutrons are made up of three quarks each and, therefore, fermions. Atomic nuclei with an even number of nucleons (protons and neutrons) have integer spin quantum numbers and behave as bosons. Their quantum energy states do not degenerate in external magnetic fields and no spectrum is obtained. On the other hand, atomic nuclei with an odd number of nucleons (protons and neutrons) have non-integer spin quantum numbers and behave as fermions. When such an atomic nucleus is placed in an external magnetic field and irradiated with photons of energy equal to the energy difference between the degenerate energy states it transits between them resulting in a spectrum. This is the basis of Nuclear Magnetic Resonance (NMR) spectroscopy.

The frequency corresponding to the difference in energy of the degenerate quantum states may be calculated as

$$\nu = \frac{\mu \mu_N B_0}{hS} \tag{6.8}$$

where ν is the frequency corresponding to the difference in energy of the degenerate quantum states, μ is the magnetic moment of the nucleus, μ_N is the nuclear magneton, B_0 is the external magnetic field, h is the Plank's constant, and S is the spin quantum number.

The quantum state degeneracy depends on the strength of the applied magnetic field, the type of nucleus, and its electromagnetic surroundings. The latter plays a crucial role in the characterization of organic molecules by NMR spectroscopy. The electronic environment around a nucleus is affected by its electron cloud which depends on the chemical bond the atom forms with other atoms. The energy difference between degenerate quantum states of an isolated atom will differ from the energy difference if the atom is forming a bond with other atoms. The frequency of the absorbed photon also changes accordingly. This shift in absorption frequency is known as a chemical shift and is given by

$$\delta = \frac{\nu_{ref} - \nu_{mol}}{\nu_{irr}} \times 10^6 \tag{6.9}$$

where ν_{ref} is the frequency of absorption by the atom in a reference electromagnetic environment, ν_{mol} is the frequency of absorption by the atom in the molecule under study, and ν_{irr} is the frequency of irradiating photons. The multiplying factor 10^6 in equation (6.6) is introduced to compensate for the very low magnitude – several Hz – of δ in comparison to the different absorption and irradiation frequencies ν_{ref}, ν_{mol}, and ν_{irr} – several MHz. Therefore, in NMR spectroscopy it is common to express δ in ppm. A common reference substance is tetramethylsilane (TMS).

The most abundant atoms in an organic molecule like lignin are carbon and hydrogen. Proton, the nucleus of the common isotope of the hydrogen atom, is a fermion and susceptible to NMR spectroscopy. Proton or ^{1}H NMR spectroscopy is the most common and the oldest of NMR spectroscopy for the characterization of organic molecules. Being the common isotope of hydrogen, proton is naturally abundant and ^{1}H NMR spectrum of an organic molecule of reasonable intensity can be obtained with relatively modest external magnetic fields. The nucleus of the common isotope of carbon has six protons and six neutrons. The nucleus ^{12}C is, therefore, a boson and does not exhibit quantum state degeneracy when placed in an external magnetic field. As a result, no NMR spectrum of ^{12}C can be obtained. However, the isotope of carbon ^{13}C has a nucleus with six protons and seven neutrons. Therefore, it is a fermion and produces an NMR spectrum with an applied external magnetic field. However, due to the low natural abundance of ^{13}C nuclei compared to ^{12}C nuclei, much stronger magnetic fields must be applied to obtain a spectrum of reasonable intensity. ^{1}H and ^{13}C NMR spectroscopy are the two widely used NMR spectroscopy methods to characterize organic molecules including lignin. Another atomic nucleus that is a fermion is the isotope of phosphorous ^{31}P. Phosphorous is present in many natural organic substances and these molecules can be characterized with ^{31}P NMR spectroscopy in addition to the usual ^{1}H and ^{13}C. Organic molecules without phosphorous atoms can also be characterized by ^{31}P NMR spectroscopy by derivatization and introduction of a phosphorous-containing substituent group. There are also a few instances of ^{19}F NMR spectroscopy.

6.4.1 Lignin Characterization by ^{1}H NMR Spectroscopy

^{1}H NMR or proton NMR spectroscopy is the most widely used NMR spectroscopy technique used in the characterization of organic molecules including lignin. The primary reason is the natural abundance of the ^{1}H isotope of hydrogen which allows for sharp and well-resolved spectra even at relatively modest applied magnetic fields. Different protons in the molecule have different electron clouds around them depending on the type of bonds they are participating in. Thus, different protons in the molecule experience different extents of shielding or deshielding of their magnetic moments. The presence of electronegative nuclei in the surroundings shifts the electron cloud away from the proton. This causes deshielding and an increase in chemical shifts. The opposite happens with electropositive centres. This results in their chemical shifts varying according to their position and nature in the molecule. The characteristic spectrum of the molecule contains a number of peaks corresponding to these chemical shifts, each assigned to a type of proton in the molecule. The correct interpretation of these shifts reveals the molecular structure. Besides qualitative information about the molecule, the spectrum has quantitative diagnostic value. The area of each peak in the ^{1}H NMR spectrum is proportional to the number of protons of that type. Suppose there are N_1 protons of type 1, N_2 protons of type 2, and so on in a molecule. The peak corresponding to the protons of type 1 in the NMR spectrum has an area of A_1 units, the peak corresponding to the protons of type 2 in the NMR spectrum has an area of A_2 units, and so on. Then,

$$A_1 = \kappa N_1$$
$$=> N_1 = \frac{A_1}{\kappa} \tag{6.10a}$$

Similarly,

$$N_2 = \frac{A_2}{\kappa} \tag{6.10b}$$

and so on. The proportion of protons of ith type in the molecule can then be computed as

$$\frac{N_i}{\Sigma N_i} = \frac{A_i}{\Sigma A_i} \tag{6.11}$$

This ratio is often used along with the atomic percentage of hydrogen in the molecule obtained from elemental analysis.

Another important aspect of ^{1}H NMR spectroscopy is the spins – and associated magnetic moments – of adjacent protons of different types interacting with each other across chemical bonds. This phenomenon, referred to as a spin-spin coupling, causes the respective peaks in the spectrum to split so that a group of closely spaced peaks is obtained rather than a single peak. A doublet peak indicates that one or more additional protons of a particular type are present in the vicinity of the proton generating the peak. Similarly, triplets and quartets are also possible, and their analyses provide very important information about the molecular structure. It is to be noted that only non-equivalent protons in the neighbourhood would cause peak splitting and not the equivalent protons. The classic case is the protons in the benzene molecule which are all equivalent and produce only one peak. The coupling becomes weaker and weaker for more distant protons and is normally negligible beyond three chemical bonds. Lignin has a complex molecular structure with a lot of heterogeneity across source and isolation methods. Therefore, assigning peaks to a particular proton type in lignin ^{1}H NMR spectra is more complicated compared to simple organic molecules. The general chemical shifts of characteristic peaks in the lignin ^{1}H NMR spectrum and their assignment to molecular structural features are presented in Table 6.2.

6.4.1.1 Soda and Soda Anthraquinone Lignin

In the ^{1}H NMR spectra of acetylated soda anthraquinone lignins from steam-exploded mixed softwoods, chemical shifts were assigned for aliphatic acetoxyl substituents, aromatic acetoxyl substituents, and methoxyl groups at 1.7–2.17, 2.17–2.5, and 3.1–4.1 ppm, respectively. From the relative abundance of aromatic and aliphatic acetoxyl groups, the ratio of phenolic and aliphatic hydroxyls was estimated to be in the range of 0.7–0.89. Methoxyl groups varied between 0.4% and 0.5% (Angles et al. 2003). Aliphatic and phenolic hydroxyl content in soda lignins from hemp, flax, and wheat straw was determined using ^{1}H NMR spectroscopy. Solution state spectra were recorded in deuterated chloroform with acetylated samples and hydrogen signals for aliphatic acetyl (2 ppm) and aromatic acetyl (2.3 ppm) were integrated. The ratio of phenolic to aliphatic hydroxyl in hemp, flax, and wheat straw were 1, 0.78, and 0.5, respectively. However, the values did not correspond well with other characterization methods (Gosselink et al. 2004). Aliphatic and phenolic hydroxyl content in softwood soda anthraquinone lignin was reported as 3.1% and 4.5%, respectively, by ^{1}H NMR spectroscopy. Spectrum was recorded by dissolving acetylated lignin in deuterated chloroform. Signal integration for hydrogen in aliphatic and aromatic acetyl groups was used for the

TABLE 6.2
Characteristic Chemical Shifts in Lignin ¹H NMR Spectrum

Chemical Shift (ppm)	Assignment	Reference
0–1.6	Hydrogen linked to aliphatic carbon with no oxygen	Ghatak et al. 2010; Lou et al. 2018
1.2	Hydrogen in aliphatic hydroxyl	Lou et al. 2018
1.9–2.02	Hydrogen in acetyl group on aliphatic carbon	Lou et al. 2018
2.14–2.32	Hydrogen in acetyl substituent on benzene ring	Lou et al. 2018
3.55–3.95	Hydrogen in methoxyl	Ghatak et al. 2010; Lou et al. 2018
4.18	Hydrogen linked to γ carbon	Lou et al. 2018
4.5–5.2	Hydrogen linked to β carbon	Ghatak et al. 2010
5.2–6.25	Hydrogen linked to α carbon	Ghatak et al. 2010
6.25–7.9. Syringyl 6.81, guaiacyl 6.98, p-hydroxyphenyl 7.36–7.62	Aromatic hydrogen	Ghatak et al. 2010; Lou et al. 2018

¹H NMR spectroscopy is frequently practised in solution states where the substance to be analysed is dissolved in a suitable organic solvent. It is necessary to prevent the hydrogen nuclei in the solvent molecules from interfering. To ensure this, deuterated solvents are to be used in which the common isotope of hydrogen ¹H is substituted with the isotope ²H, i.e., deuterium. A deuterium nucleus is a boson having 2 nucleons in it. Deuterated variants of dimethyl sulfoxide, tetrahydrofuran, chloroform, or water are frequently used in solution state ¹H spectroscopy of lignin. Lignin samples are often acetylated to improve solubility and signal strength.

purpose (El Mansouri and Salvado 2007). Wheat straw and bagasse soda lignins, isolated by conventional acid precipitation, were characterized by ¹H NMR spectroscopy by dissolving in deuterated dimethyl sulfoxide. Combined with elemental analysis and signal integrations in different chemical shift regions, wheat straw lignin conformed to a molecular formula of $C_9H_{16.83}O_{4.71}(OCH_3)_{0.38}$. The molecular formula for bagasse soda lignin was found to be $C_9H_{16.7}O_{4.81}(OCH_3)_{0.36}$. The study also compared these lignins with soda lignins isolated by electrodeposition. Electrodeposited soda lignin from wheat straw had the molecular formula of $C_9H_{11.59}O_{3.37}(OCH_3)_{0.12}$, and that from bagasse had the molecular formula of $C_9H_{10.86}O_{3.36}(OCH_3)_{0.11}$ (Ghatak et al. 2010). The representative spectrum for wheat straw soda lignin is presented in Figure 6.10.

Acetylated bagasse soda lignin was dissolved in deuterated chloroform and characterized by ¹H NMR to determine the methoxyl groups in lignin. For this purpose, NMR signal integrations were acquired for aromatic hydrogen in syringyl (6.28–7 ppm) and guaiacyl (7–8 ppm) structures. Similarly, signal integration for methoxyl hydrogen was performed. The ratio of aromatic hydrogen to methoxyl hydrogen was used to determine the methoxyl content through a statistical linear regression given as

$$\text{Methoxyl}(\%) = 28.28436 - 19.75004\left(\frac{\text{Aromatic hydrogen}}{\text{Methoxyl hydrogen}}\right) \quad (6.12)$$

The reported methoxyl content was 10.9% (Mousavioun and Doherty 2010). Methoxyl content in commercial nonwood soda lignin (Protobind) was estimated as 1.41 per phenylpropane unit. For this, signals were integrated into the ¹H NMR spectrum for methoxyl hydrogen (3.4–3.95 ppm) and used in combination with elemental analysis. Similarly, signal integration for hydrogen in aliphatic acetoxyl (1.6–2.1 ppm) and aromatic acetoxyl (2.1–2.45 ppm) groups estimated the aliphatic and phenolic hydroxyl content, respectively. 0.71 aliphatic hydroxyl groups and 0.77 phenolic hydroxyl groups per phenylpropane unit were reported. The calculation was based on the presence of three hydrogen atoms per methoxyl or acetoxyl group (Sameni et al. 2016). Another study reports

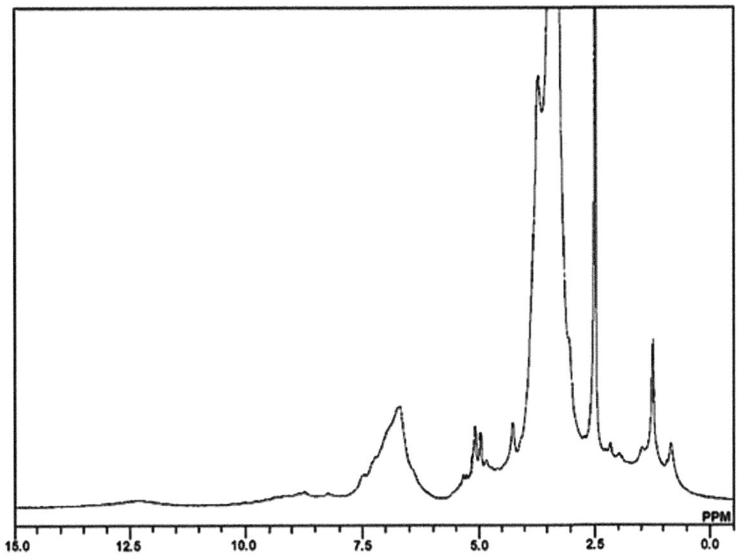

FIGURE 6.10 ^{1}H NMR spectrum of wheat straw soda lignin. (Reprinted from Ghatak, HR, Kundu, PP, Kumar, S, *Thermochimica Acta*, 502, 85–89, 2010, with permission from Elsevier.)

1.8 mmol phenolic hydroxyl groups per g Protobind lignin by ^{1}H NMR spectroscopy. The solution state spectrum was recorded first by dissolving lignin in deuterated dimethyl sulfoxide with a trace of deuterated dimethyl formamide and then again by dilution of the sample with D_2O (Serrano et al. 2018). Soda lignins from spruce (softwood) and birch (hardwood) were acetylated before dissolving in deuterated chloroform and recording ^{1}H NMR spectra. Phenolic and aliphatic hydroxyl groups were estimated based on signal integrations for hydrogen in acetyl groups of aromatic and aliphatic origin. In spruce soda lignin, depending on delignification severity, aliphatic hydroxyl groups varied between 2.53 and 3.11 mmol/g, and phenolic hydroxyl groups between 1.2 and 2 mmol/g. Birch soda lignin had 0.67–6.11 mmol/g aliphatic hydroxyl and 0.17–0.9 mmol/g phenolic hydroxyl (Nitsos et al. 2016).

6.4.1.2 Kraft Lignin

In an early work total and phenolic hydroxyl content in mixed softwood kraft lignin (Indulin AT) was estimated by acetylating the sample, dissolving it in deuterated chloroform, and recording ^{1}H NMR spectrum. Hydrogen signals from acetyl groups in aliphatic (about 1.7–2.2 ppm) and aromatic (about 2.2–2.5 ppm) regions were integrated. These signal integrations relative to methoxyl hydrogen signal integration (about 3.45–4 ppm) were used to estimate 52 phenolic hydroxyl groups and 119 total hydroxyl groups per 100 phenylpropane units. The number of methoxyl groups per 100 phenylpropane units was independently estimated by wet chemistry methods and elemental analysis (Faix et al. 1994). In a modified ^{1}H NMR spectroscopy protocol, Indulin AT was dissolved in deuterated dimethyl sulfoxide and the spectrum was recorded. Thereafter, 20% D_2O was added to the solution and the ^{1}H NMR spectrum was recorded again. Signals were integrated between 11 and 14 ppm for the two spectra and the difference in integration was taken as a measure for hydrogen in the carboxyl group. Similarly, the difference in signal integrations between 8 and 11 ppm provided a measure of hydrogen in phenolic hydroxyl. It was argued that the addition of D_2O leads to all hydrogen in hydroxyl groups (carboxyl or phenolic) being replaced with deuterium. With this method, Indulin AT was found to have 0.11 carboxyl groups and 0.64 phenolic hydroxyl groups per phenylpropane unit (Tiainen et al. 1999). Aliphatic and phenolic hydroxyl content in softwood kraft lignin was reported as 10.09% and 4.1%, respectively, by ^{1}H NMR spectroscopy. Spectrum was recorded by dissolving acetylated lignin in deuterated chloroform. Signal integration for hydrogen in aliphatic and aromatic acetyl groups was used for the purpose (El Mansouri and Salvado 2007). ^{1}H NMR spectra of acetylated kraft lignins from eucalyptus and Indulin AT had assigned chemical shift peaks at usual positions for hydrogen linked to aliphatic carbon with no oxygen, methoxyl hydrogen, hydrogen linked to α, β, and γ carbons, and aromatic hydrogen. Acetoxyl signals were observed at 1.6–2.1 ppm (aliphatic) and 2.1–2.45 ppm (aromatic). Hydrogen in carboxyl and carbonyl entities caused a chemical shift in the 9–12 ppm region. From signal integration in the methoxyl hydrogen region, the methoxyl group per phenylpropane unit was worked out in combination with elemental analysis. For eucalyptus kraft lignin it was 1.45 and for Indulin AT it was 1.02. Similarly, signal integration for hydrogen in aliphatic acetoxyl and aromatic acetoxyl groups estimated the aliphatic and phenolic hydroxyl content, respectively. 0.59 aliphatic hydroxyl groups and 0.86 phenolic hydroxyl groups per phenylpropane unit were reported in eucalyptus kraft lignin. These values were 0.92 and 0.67, respectively, for Indulin AT. The calculation was based on the presence of three hydrogen atoms per methoxyl or acetoxyl group (Sameni et al. 2016). Another study reports 3.1 mmol phenolic hydroxyl groups per g Indulin AT by ^{1}H NMR spectroscopy. The solution state spectrum was recorded first by dissolving lignin in deuterated dimethyl sulfoxide with a trace of deuterated dimethyl formamide and then again after dilution of the solution with D_2O (Serrano et al. 2018). Acetylated mixed hardwood kraft lignin was dissolved in deuterated chloroform to record its ^{1}H NMR spectrum. Peaks for hydrogen in aliphatic side chain carbons without oxygen, hydrogen in methoxyl, and hydrogen in acetate groups were present at their characteristic chemical shifts. Aromatic acetate groups were more abundant than the aliphatic ones as noticed from the peak areas in the NMR spectrum. Peaks arising from aromatic hydrogen in syringyl or guaiacyl entities were very insignificant. So were the peaks for hydrogen linked to β–O–4′ and β–1′ (4.3–4.9 and 5.75–6.25 ppm) (Mun et al. 2021).

One important application of ^{1}H NMR spectroscopy in lignin characterization is the study of its aggregation behaviour in solution. The technique relies on the controlled introduction of spatial inhomogeneity in the applied magnetic field in a pulsed manner. Alkaline solution of softwood kraft lignin Indulin AT was studied for analysing the distribution of self-diffusion coefficients using Pulsed Field Gradient ^{1}H NMR with a field gradient up to 3 T/m (Norgren et al. 2002).

6.4.1.3 Hydrolysis Lignin

Total and phenolic hydroxyl content in cottonwood cellulolytic hydrolysis lignin was estimated by acetylating the sample, dissolving it in deuterated chloroform, and recording ^{1}H NMR spectrum. Hydrogen signals from acetyl groups in aliphatic (about 1.7–2.2 ppm) and aromatic (about 2.2–2.5 ppm) regions were integrated. These signal integrations relative to methoxyl hydrogen signal integration (about 3.45–4 ppm) were used to estimate 19 phenolic hydroxyl groups and 168 total hydroxyl groups per 100 phenylpropane units. The number of methoxyl groups per 100 phenylpropane units was independently estimated by wet chemistry methods and elemental analysis (Faix et al. 1994). Aliphatic and phenolic hydroxyl content in lignin-rich residue, obtained after steam explosion of poplar and hydrolysis of polysaccharides, was reported as 4.73% and 2.65%, respectively, by ^{1}H NMR spectroscopy. Spectrum was recorded by dissolving acetylated lignin in deuterated chloroform. Signal integration for hydrogen in aliphatic and aromatic acetyl groups was used for the purpose (El Mansouri and Salvado 2007). Enzymatic mild acidolysis lignin from wheat straw was dissolved in deuterated dimethyl sulfoxide and characterized by ^{1}H NMR. 2.49 ppm peak was attributed to the solvent. The study assigned a 0.61–1.12 ppm peak to aliphatic hydrogen that is highly shielded and a 1.72–2.03 ppm region to hydrogen in aliphatic acetyl groups and aromatic acetoxyl groups. Hydrogen in acetyl substituent on the benzene ring caused a chemical shift of 2.2 ppm. Syringyl, guaiacyl, p-hydroxyphenyl, and methoxyl peaks appeared at their usual positions. These peaks were, however, weak except for the methoxyl peak. Usual peaks for hydrogen linked to α, β, and γ carbons were also present in the spectrum. A weak signal at 8.27 ppm was assigned to hydrogen in the carboxyl group most likely due to carbohydrate impurities (Yang et al.

2011). Enzymatic hydrolysis lignin after bioethanol production from hardwood and nonwood was subjected to ^{1}H NMR characterization. Signal integration for methoxyl hydrogen (3.4–3.95 ppm) provided an estimation of methoxyl content to be 1.37 per phenylpropane unit. Similarly, signal integration for hydrogen in aliphatic acetoxyl (1.6–2.1 ppm) and aromatic acetoxyl (2.1–2.45 ppm) groups estimated the aliphatic and phenolic hydroxyl content, respectively. 0.85 aliphatic hydroxyl groups and 0.46 phenolic hydroxyl groups per phenylpropane unit were reported. The calculation was based on the presence of three hydrogen atoms per methoxyl or acetoxyl group (Sameni et al. 2016). A typical ^{1}H NMR spectrum of enzymatic mild hydrolysis lignin from rice straw is shown in Figure 6.11. Signal integrations suggested that of the total hydrogen atoms in the lignin macromolecule, 29% came from aliphatic acetyl groups, 10.8% from aromatic acetyl groups, 21% from methoxyl groups, 11.7% from guaiacyl moieties, and 4.1% from p-hydroxyphenyl units. The rest of the hydrogen atoms were attributed to less abundant structural features in the lignin like hydrogen joined to side chain carbons with different intermonomer linkages (Lou et al. 2018).

6.4.1.4 Organosolv Lignin

Total and phenolic hydroxyl content in commercial mixed hardwood aqueous ethanol organosolv lignin (Alcell lignin) was estimated by acetylating the sample, dissolving it in deuterated chloroform, and recording ^{1}H NMR spectrum. Hydrogen signals from acetyl groups in aliphatic (about 1.7–2.2 ppm) and aromatic (about 2.2–2.5 ppm) regions were integrated. These signal integrations relative to methoxyl hydrogen signal integration (about 3.45–4 ppm) were used to estimate 72 phenolic hydroxyl groups and 129 total hydroxyl groups per 100 phenylpropane units. The number of methoxyl groups per 100 phenylpropane units was independently estimated by wet chemistry methods and elemental analysis (Faix et al. 1994). Aliphatic and phenolic hydroxyl content in Alcell lignin was determined using ^{1}H NMR spectroscopy. Solution state spectra were recorded in deuterated chloroform with acetylated samples and hydrogen signals for aliphatic acetyl (2 ppm) and aromatic acetyl (2.3 ppm) were integrated. The ratio of phenolic to aliphatic hydroxyl was close to 1 (Gosselink et al. 2004). However, both studies reported that the values for aliphatic and phenolic hydroxyl determined by ^{1}H NMR spectroscopy did not correspond well with other characterization methods. Organosolv lignins from switchgrass, cordgrass, and corn stover were extracted with an aqueous mixture of ethyl acetate and ethanol, acetylated, dissolved in deuterated dimethyl sulfoxide, and characterized by ^{1}H NMR spectroscopy. An aromatic proton peak at 6.2 ppm was assigned to

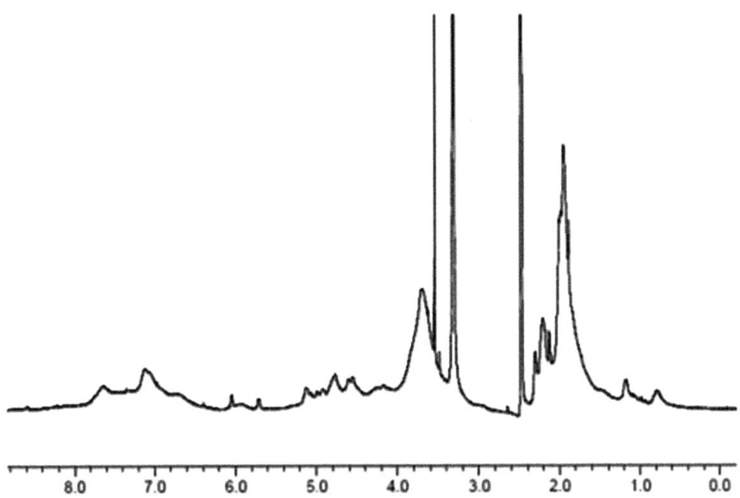

FIGURE 6.11 ^{1}H NMR spectrum of rice straw enzymatic mild hydrolysis lignin. (Reprinted from Lou, R, Lyu, G, Wu, S, Zhang, B, Zhao, H, Lucia LA, *ACS Sustainable Chem. Eng.*, 6, 430–437, 2018, with permission from American Chemical Society.)

syringyl moieties and 6.8 ppm to guaiacyl moieties. These structural entities were observed in all lignins, but cordgrass lignin had lower intensities of these peaks. 7.3–7.4 ppm peak, assigned to aromatic hydrogen in p-hydroxyphenyl moieties, was absent in the cordgrass lignin spectrum. Methoxyl hydrogen recorded its presence with a chemical shift at the usual position, but with different intensities in different lignins (Cybulska et al. 2012). Acetylated aqueous ethanol organosolv lignins from spruce and birch were dissolved in deuterated chloroform and ^{1}H NMR spectra were recorded. Phenolic and aliphatic hydroxyl groups were estimated based on signal integrations for hydrogen in acetyl groups of aromatic and aliphatic origin. In spruce organosolv lignin, depending on delignification severity, aliphatic hydroxyl groups varied between 1.12 and 3.06 mmol/g, and phenolic hydroxyl groups between 2.44 and 3.99 mmol/g. Birch organosolv lignin had 0.62–3.92 mmol/g aliphatic hydroxyl and 1.68–3.84 mmol/g phenolic hydroxyl (Nitsos et al. 2016). Aqueous ethanol organosolv lignin from *Typha capensis* was characterized by solution state ^{1}H NMR spectroscopy by dissolving the sample in deuterated dimethyl sulfoxide with maleic acid as an internal standard. Major assignments in the spectrum were for aliphatic acetate at 1.96 ppm, aromatic acetate at 2.24 ppm, and hydrogen in methoxyl groups at 3.69 ppm. Acetate peaks were rather weak. Peaks that were not assigned in the study but prominently visible in the spectrum were 0.7–0.9 ppm (shielded aliphatic hydrogen), and 1.2–1.3 ppm (hydrogen in hydroxyl group). Peaks for aromatic hydrogen associated with syringyl, guaiacyl, or p-hydroxyphenyl moieties were very weak (Audu et al. 2018).

Aliphatic and phenolic hydroxyl content in miscanthus formic acid organosolv lignin was reported as 3.5% and 3.33%, respectively, by ^{1}H NMR spectroscopy. Spectrum was recorded by dissolving acetylated lignin in deuterated chloroform. Signal integration for hydrogen in aliphatic and aromatic acetyl groups was used for the purpose (El Mansouri and Salvado 2007). ^{1}H NMR characterization of acetylated bamboo formic acid organosolv lignin was performed by dissolving the sample in deuterated dimethyl sulfoxide. Characteristic proton chemical shifts for shielded aliphatic hydrogen, acetate hydrogen in aliphatic side chain and benzene ring, hydrogen in methoxyl group, hydrogen linked to β-O-4′ structures, and hydrogen linked to benzene rings in syringyl, guaiacyl, and p-hydroxyphenyl moieties were observed at normal positions. Hydrogen linked to carboxyl and carbonyl functionalities caused a chemical shift in the 8–11.5 ppm region of the spectrum. Small amounts of hemicellulose inclusion in the sample resulted in proton chemical shifts at 4–4.3 and 4.9–5.2 ppm (Li et al. 2012). One study reports 0.9 mmol phenolic hydroxyl groups per g aqueous formic acid acetic acid organosolv lignin by ^{1}H NMR spectroscopy. The solution state spectrum was recorded first by dissolving lignin in deuterated dimethyl sulfoxide with a trace of deuterated dimethyl formamide and then again after dilution of the solution with D_2O (Serrano et al. 2018).

In the ^{1}H NMR spectrum of acetylated eucalyptus dioxane lignin, hydrogen signals were attributed to aldehyde hydrogen at β carbon (9.3–9.55 ppm), aldehyde hydrogen at γ carbon (9.55–9.75 ppm), and aldehyde hydrogen at α carbon (9.75–10 ppm). Their abundances were reported as 2, 3, and 4 aldehyde groups, respectively, per 100 phenylpropane units. Integration of methoxyl hydrogen signal between 3.55 and 3.95 ppm resulted in the estimation of 164 methoxyl groups per 100 phenylpropane units. Acetyl hydrogen signals were also distinctly visible in the spectrum between 1.9 and 2.3 ppm (Evtuguin et al. 2001). Banana leaf sheath dioxane lignin, acetylated and dissolved in deuterated chloroform, was recorded for ^{1}H NMR spectrum, which is shown in Figure 6.12. Signal integration between 5.8 and 6.2 ppm was attributed to hydrogen at α carbon with β-O-4′ linkage. These linkages were estimated to be 0.26 per phenylpropane unit. Similarly, signal integration between 5.4 and 5.6 ppm was attributed to hydrogen at α carbon with α-O-4′ linkage and these linkages were estimated to be 0.08 per phenylpropane unit. Signal integration between 3 and 3.15 ppm accounted for hydrogen at β carbon with β-β′ linkage. Its abundance was 0.04 per phenylpropane unit. In accordance with other studies, a chemical shift between 0.8 and 1.6 ppm was assigned to hydrogen in aliphatic carbon chains. The authors suggested the presence of suberin in a covalent bond with lignin (Oliveira et al. 2006).

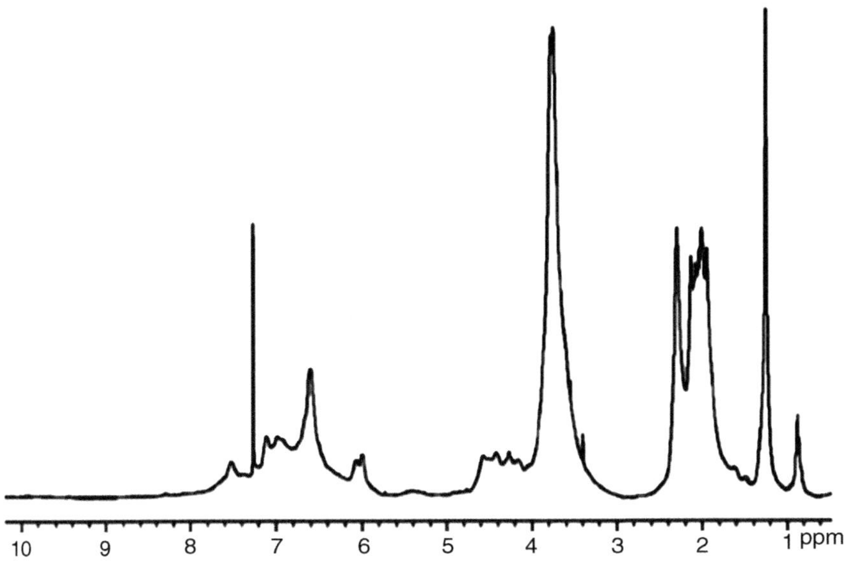

FIGURE 6.12 ^{1}H NMR spectrum of acetylated banana leaf sheath dioxane lignin. (Reprinted from Oliveira, L, Evtuguin, DV, Cordeiro, N, Silvestre, AJD, Silva, AMS, *J. Agric. Food Chem.*, 54, 2598–2605, 2006, with permission from American Chemical Society.)

6.4.1.5 Milled Wood Lignin

In an important early work, acetylated milled wood lignins from 42 lignocellulose species were characterized by ^{1}H NMR spectroscopy. Spectra were recorded after dissolving the samples in deuterated chloroform. The chemical shift peak centred at 2.3 ppm was assigned to aromatic acetoxyl hydrogen, the acetyl groups originating from acetylation of the phenolic hydroxyl. Based on signal integration of these peaks in different spectra the content of phenolic hydroxyl groups was estimated and found to vary between 1.94% and 4.99% (Faix et al. 1992). In a modified ^{1}H NMR spectroscopy protocol, spruce milled-wood lignin was dissolved in deuterated dimethyl sulfoxide and the spectrum was recorded. Thereafter, 20% D_2O was added to the solution and the ^{1}H NMR spectrum was recorded again. Signals were integrated between 11 and 14 ppm for the two spectra and the difference in integration was taken as a measure for hydrogen in the carboxyl group. Similarly, the difference in signal integrations between 8 and 11 ppm provided a measure of hydrogen in phenolic hydroxyl. It was argued that the addition of D_2O leads to all hydrogen in hydroxyl groups (carboxyl or phenolic) being replaced with deuterium. With this method, spruce milled-wood lignin was found to have 0.04 carboxyl groups and 0.23 phenolic hydroxyl groups per phenylpropane unit (Tiainen et al. 1999). Solution state ^{1}H NMR of acetylated lauan milled-wood lignin in deuterated chloroform had normal hydrogen peaks for acetate group hydrogen on aliphatic carbons and benzene rings, hydrogen on γ carbon with hydroxyl group, hydrogen in methoxyl groups, hydrogen in ether bonded α, and β carbons, and aromatic hydrogen in syringyl and guaiacyl structures. Though not reported in the study, the signal integrations for methoxyl, syringyl, and guaiacyl hydrogen can be used to determine the relative abundance of these structural features in the lignin sample (Wang et al. 2014). *Typha capensis* milled-wood lignin was characterized by solution state ^{1}H NMR spectroscopy by dissolving the sample in deuterated dimethyl sulfoxide with maleic acid as the internal standard. Major assignments in the spectrum were for aliphatic acetate at 1.96 ppm, aromatic acetate at 2.24 ppm, and hydrogen in methoxyl groups at 3.69 ppm. All these peaks were prominent and strong, showing a good abundance of acetate groups in the lignin sample. Peaks that were not assigned in the study but prominently visible in the spectrum were 0.8 ppm (shielded aliphatic hydrogen), and 1.2 ppm (hydrogen in hydroxyl group). Peaks for aromatic hydrogen associated with syringyl, guaiacyl, or p-hydroxyphenyl moieties were weak (Audu et al. 2018). ^{1}H NMR spectrum of acetylated acacia milled-wood lignin, dissolved in deuterated chloroform, had characteristic peaks for hydrogen in aliphatic side chain carbons without oxygen, hydrogen in methoxyl, and hydrogen in acetate groups. Aromatic acetate groups were less abundant than the aliphatic ones as noticed from the peak areas in the NMR spectrum. Peaks arising from aromatic hydrogen in syringyl and guaiacyl entities were sufficiently strong and present at their usual chemical shifts. So were the peaks for hydrogen linked to β-O-4′ and β-1′ structures (4.3–4.9 and 5.75–6.25 ppm) (Mun et al. 2021). ^{1}H NMR spectrum of acetylated bamboo milled-wood lignin, dissolved in deuterated dimethyl sulfoxide, recorded typical proton chemical shifts for shielded aliphatic hydrogen, acetate hydrogen in aliphatic side chain and benzene ring, hydrogen in methoxyl group, hydrogen linked to β-O-4′ structures, and hydrogen linked to benzene rings in syringyl, guaiacyl, and p-hydroxyphenyl moieties. As expected, there was a large abundance of acetate groups measured from respective peak integrations. Hydrogen linked to carboxyl and carbonyl functionalities caused a chemical shift in the 8–11.5 ppm region of the spectrum and had reasonable peak areas. Small amounts of hemicellulose inclusion in the sample resulted in proton chemical shifts at 4–4.3 and 4.9–5.2 ppm (Li et al. 2012).

6.4.1.6 Lignosulphonate

In a modified ^{1}H NMR spectroscopy protocol, 60% sulphonated softwood kraft lignin was dissolved in deuterated dimethyl sulfoxide and the spectrum was recorded. Thereafter, 20% D_2O was added to the solution and the ^{1}H NMR spectrum was recorded again. Signals were integrated between 8 and 11 ppm for the two spectra and the difference in integration was taken as a measure for hydrogen in phenolic hydroxyl. It was argued that the addition of D_2O leads to all hydrogen in hydroxyl groups being replaced with deuterium. With this method, sulphonated lignin was found to have 0.35 phenolic hydroxyl groups per phenylpropane unit (Tiainen et al. 1999). Commercial lignosulphonate was dissolved in D_2O for its ^{1}H NMR characterization. The authors assigned the spectral region 0.5–3 ppm to aliphatic hydrogen, 3–5 ppm to hydrogen linked to oxygen, and 7–9 ppm to aromatic hydrogen. The proportion of aromatic hydrogen was estimated to be about 50% (Chen et al. 2018).

6.4.2 Lignin Characterization by ^{13}C NMR Spectroscopy

The backbone of an organic molecule is made up of carbon atoms. The hydrogen atoms and other functional groups are joined to these carbon atoms. If the nature of each carbon atom – or more precisely the carbon nucleus – is known in a molecule, it will provide a very elaborate and unambiguous picture of the molecular structure. This is the motivation behind ^{13}C NMR spectroscopy. ^{13}C NMR spectra are inherently more resolved than ^{1}H NMR spectra but is low in intensity. This is because the isotope ^{13}C is much less abundant naturally – only about 1% of all carbon nuclei present. Normally, a larger sample quantity is used for recording ^{13}C NMR spectra. Another reason for the lower intensity is the smaller magnetic moment of the ^{13}C nucleus compared to ^{1}H which results in a lesser energy difference between degenerate energy states in a magnetic field of comparable strength, according to equation 6.6. The availability of stronger magnets in modern NMR instruments has solved this problem to some extent. A large number of scans, running into several thousand, is normally performed to record a spectrum of acceptable intensity and low noise. The basic principle of ^{13}C NMR spectroscopy remains the same as that for ^{1}H NMR spectroscopy but the chemical shifts are manifold larger, resulting in higher resolution of ^{13}C NMR spectra. Here also, TMS is the most common reference molecule whose chemical shift is taken to be zero.

The spin–spin coupling between adjacent ^{13}C nuclei, though theoretically possible, is practically insignificant due to very little abundance of this isotope. Therefore, peak splitting due to C–C spin coupling is non-existent in ^{13}C NMR spectra. On the other hand, ^{1}H–^{13}C spin coupling is strong enough and ubiquitous. To maintain the sensitivity of ^{13}C NMR spectra it is necessary to break this coupling. One common method of decoupling is to continuously bathe the sample with electromagnetic radiation covering the range of excitation frequencies of protons. In such an environment the protons continuously keep on changing their spin states and their spins do not couple to those of adjacent ^{13}C nuclei. Signal intensity is enhanced due to Nuclear Overhauser Enhancement. Another way of decoupling is the use of inverse gated pulse. One variation of ^{13}C NMR spectroscopy is Distortionless Enhancement by Polarization Transfer (DEPT). In this scheme, a double resonance pulse is used so that in the excited state polarization is transferred from ^{1}H to ^{13}C. Enhancement in signal intensity is much larger than the Nuclear Overhauser Enhancement obtained otherwise. The other advantage of DEPT is the possibility to distinguish between primary, secondary, and tertiary hydrogen attached to carbon by using different proton pulse phase angles.

Normal ^{1}H NMR spectra have quantitative diagnostic value since the area of a peak is always proportional to the number of protons generating the peak with the same proportionality constant in equation 6.8. This happens when the time difference between two radio frequency pulses – known as the pulse delay – exceeds the average time spent by the nuclei in the higher energy spin state before returning to the lower energy spin state – relaxation time. For ^{1}H nuclei, relaxation time is short – typically less than 5 s – allowing for the nuclei to quickly return to their ground states of spin orientation. For this, ^{1}H NMR spectra with quantitative significance can be recorded with relatively short pulse delays of 25–30 s. However, this is not true for a normal ^{13}C NMR spectrum. ^{13}C nuclei have longer relaxation times. Therefore, a normal ^{13}C NMR spectrum – spectrum recorded with normal pulse delay – only has qualitative diagnostic value. The most apparent remedy would be to use a long pulse delay allowing for the nuclei to relax between pulses. However, it has the practical limitation of the exorbitantly long time required to record the spectrum with an acceptable signal-to-noise ratio. Using a relaxant is another option that reduces the relaxation time and hence the requirement for long pulse delays. One frequently used relaxant is chromium (III) acetylacetonate. Inequal Nuclear Overhauser Enhancement for different ^{13}C nuclei during ^{1}H–^{13}C spin decoupling is another problem. This causes different nuclei to have different relaxation times. This problem is overcome with special pulse and decoupling sequences. In the Extended Acquisition Time (EXACT) sequence, short data acquisition periods are separated by cool-off periods when the decoupler and data acquisition are switched off (Al-Aasmi et al. 2022).

One important variation of ^{13}C NMR spectroscopy is cross-polarization magic angle spinning (CPMAS). Normal NMR spectra are recorded in a solution state that requires deuterated solvents and good solubility of the substance to be characterized. Solid-state NMR is free from such limitations. However, the solid-state NMR technique is often plagued by peak broadening and resultant loss in resolution. One remedy for this is to spin the sample in the applied magnetic field so that the spinning axis makes a specific angle – the magic angle – to the direction of the magnetic field. This significantly reduces the peak broadening and gives a sharp and well-resolved spectrum.

Owing to its higher resolution with a wide window of chemical shifts, ^{13}C NMR spectroscopy is more informative for complex organic molecules like lignin. The possibility of peak overlapping is much less in the ^{13}C spectrum than in the ^{1}H spectrum. Different types of carbon atoms in the molecule are clearly distinguished. Typical assignment of ^{13}C NMR peaks to structural features of lignin macromolecule is presented in Table 6.3. One common approach in quantitative ^{13}C NMR spectroscopy of lignin is to integrate the signals for aromatic and vinylic carbon (sp^2 hybridized) in the 103–160 ppm range and consider it equivalent to 6.12 carbon atoms; 6 benzene ring carbons and 0.12 other sp^2 hybridized. Other functional groups are then referenced to this integration and reported on a per-benzene ring basis (Huang et al. 2011).

6.4.2.1 Soda and Soda Anthraquinone Lignin

Soda lignins from wheat straw, hemp, and flax were characterized for their ratio of aromatic and aliphatic hydroxyl groups by quantitative ^{13}C NMR spectroscopy. After the acetylation of samples, the solution state spectrum was acquired in deuterated dimethyl sulfoxide. In the spectrum, signal integration for 166–169 ppm was quantitatively assigned to acetyl carbon from phenolic hydroxyl. Similarly, peak integration for 169–171 ppm provided a quantitative estimation of acetyl carbon from aliphatic hydroxyl. The ratio of phenolic to aliphatic hydroxyl was reported to be 0.5 for wheat

TABLE 6.3
Characteristic chemical shifts in lignin ^{13}C NMR spectrum

Chemical Shift (ppm)	Assignment	Reference
13–42	Methyl and methylene carbons	Lou et al. 2018
52–54	β carbon in carbon–carbon linked units	Lou et al. 2018
54–57	Methoxyl carbon	Holtman et al. 2004; Lou et al. 2018
57–90	α, β, and γ carbons in β–O–4′, β–5′, β–β′ linked units	Lou et al. 2018
90–100	Carbon in lignin carbohydrate linkages	Lou et al. 2018
103–125	Aromatic tertiary carbon	Holtman et al. 2004; Lou et al. 2018
125–141	Aromatic quaternary carbon	Holtman et al. 2004; Lou et al. 2018
141–160	Aromatic carbon attached to oxygen	Holtman et al. 2004; Lou et al. 2018
166	Aromatic carbon attached to oxygen in p-hydroxyphenyl	Lou et al. 2018
162.5–178; 190–195	Carbon in the carbonyl group	Holtman et al. 2004; Lou et al. 2018

straw, and 0.7 for hemp and flax (Gosselink et al. 2004). Solution state ^{13}C NMR spectrum of wheat straw soda lignin, dissolved in deuterated dimethyl sulfoxide, had peaks for aromatic quaternary carbons – having ether linkages or C–C ring condensation in guaiacyl and syringyl moieties – at 152 and 147 ppm, and tertiary aromatic carbons – positions 2, 5, and 6 in guaiacyl and positions 2, and 6 in syringyl – at 115 and 101 ppm. The methoxyl carbon peak was observed at its normal position of 56 ppm. Carbon atoms in the aliphatic side chain were represented by a peak at 72–75 ppm (Ghatak 2008). Aliphatic and phenolic hydroxyl content in softwood soda anthraquinone lignin was reported as 2.45 and 5.31%, respectively, by ^{13}C NMR spectroscopy. Spectrum was recorded by dissolving acetylated lignin in deuterated acetone. Signal integration for carbon in acetyl groups, that originated from primary aliphatic (171 ppm), secondary aliphatic (170 ppm), and aromatic hydroxyl (169 ppm), were used for the purpose. Aliphatic and phenolic hydroxyl content in softwood kraft lignin was reported as 9.8 and 4.99%, respectively (El Mansouri and Salvado 2007). Wheat straw soda lignin was characterized by solid-state quantitative CPMAS ^{13}C NMR spectroscopy. Peaks were assigned for broad chemical shift ranges at usual positions and the abundance of different types of carbon atoms was estimated as 22.7% in methyl and methylene groups, 23.8% aromatic carbon, and 4.3% in carbonyl entities. Methoxyl carbon and carbon in carbohydrates were collectively estimated to be 49.2% (Dominguez-Robles et al. 2017).

6.4.2.2 Kraft Lignin

Softwood kraft lignins, isolated by the LignoBoost process, were analysed by solution state quantitative ^{13}C NMR in deuterated dimethyl sulfoxide. Broad assignment of spectral regions was made as proposed in Table 6.3 without quantitative estimations (Nagy et al. 2010). Acetylated mixed softwood kraft lignins were characterized by ^{13}C NMR. Spectrum was recorded in deuterated dimethyl sulfoxide solution. Methoxyl groups were estimated to be 0.7–0.78 per phenylpropane unit based on signal integration between 54 and 57 ppm. The content of phenolic hydroxyl was determined between 0.8 and 0.89 by integrating aromatic acetoxyl carbon peak between 166 and 169 ppm (Zhu and Theliander 2015). One study reports quantitative ^{13}C NMR spectroscopy of acetylated softwood kraft lignin Indulin AT primarily for estimation of hydroxyl functionalities. Spectrum was acquired in solution state in deuterated dimethyl sulfoxide with trioxane as internal standard (peak at 94 ppm). Peaks for carbon atoms in acetyl groups were resolved according to their originating hydroxyl groups. 168–169 ppm peak was assigned to acetyl group derived from phenolic hydroxyl, 169–170 ppm peak to acetyl group derived from secondary hydroxyl, and 170–171 ppm peak to acetyl group derived from primary hydroxyl. From these signal integrations, primary, secondary, and phenolic hydroxyl groups were determined to be 1.66, 0.98, and 3.84 mmol/g, respectively (Goldmann et al. 2017). A similar ^{13}C NMR characterization of acetylated Indulin AT and another pine kraft lignin was performed for the determination of phenolic hydroxyl content. The spectrum for acetylated pine kraft lignin is presented in Figure 6.13.

The strong peak for methoxyl carbon is immediately noticed at its characteristic position. So are the well-resolved peaks for acetyl carbons originating from primary, secondary, and phenolic hydroxyl. From signal integrations over the aromatic acetyl peak, the phenolic hydroxyl groups in pine kraft lignin were measured as 4.2 mmol/g. They were 3.6 mmol/g in Indulin AT. The values corresponded very well with those obtained by aminolysis (Serrano et al. 2018). Acetylated mixed hardwood kraft lignin was characterized by dissolving in a 9:1 mixture of deuterated acetone and D_2O and recording the ^{13}C NMR spectrum. Spectrum was only analysed qualitatively. Peaks for methoxyl carbons, aromatic tertiary carbons, and aromatic quaternary carbons were sufficiently intense and located at their usual positions. However, the peak for aromatic carbons attached to oxygen (153 ppm) was weak. This was interpreted for extensive demethoxylation of the lignin sample. Similarly, the peaks for α, β, and γ carbons in β-O-4′, β-5′, and β-β′ linked units were feeble in intensity. The carboxyl signal was also very weak (Mun et al. 2021).

6.4.2.3 Organosolv Lignin

Alcell lignin (mixed hardwood aqueous ethanol organosolv) was characterized for its ratio of aromatic and aliphatic hydroxyl groups by quantitative ^{13}C NMR spectroscopy. After the acetylation of lignin, the solution state spectrum was acquired in deuterated dimethyl sulfoxide. In the spectrum, signal integration for 166–169 ppm was quantitatively assigned to acetyl carbon from phenolic hydroxyl. Similarly, peak integration for 169–171 ppm provides a quantitative estimation of acetyl carbon from aliphatic hydroxyl. The ratio of phenolic to aliphatic hydroxyl was reported to be 1.69 which closely corresponded to that determined by wet chemistry methods (Gosselink et al. 2004).

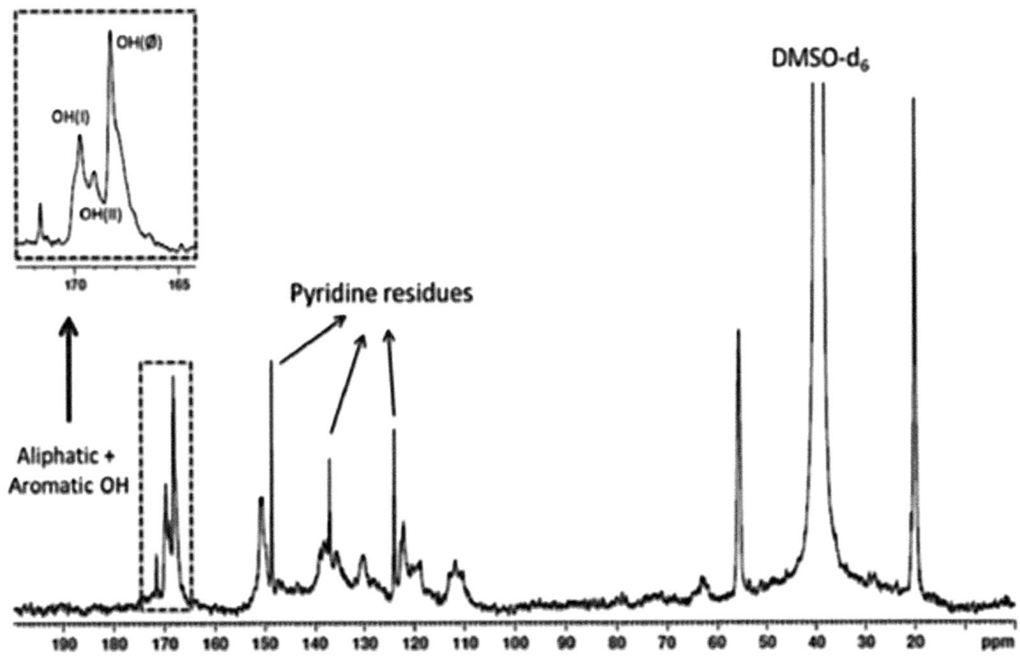

FIGURE 6.13 ^{13}C NMR spectrum of acetylated pine kraft lignin with an enlarged view of 165–173 ppm region. (Reprinted from Serrano, L, Esakkimuthu, ES, Marlin, N, Brochier-Salon, M, Mortha, G, Bertaud, F, *Energy Fuels*, 32, 5969–5977, 2018, with permission from American Chemical Society.)

Spent liquor from aqueous ethanol organosolv delignification of olive tree pruning was sonicated before precipitation of lignin. Lignin was dissolved in deuterated dimethyl sulfoxide and ^{13}C NMR spectrum was recorded for qualitative analysis. A strong methoxyl peak appeared at its normal position. Tertiary and quaternary aromatic carbons were identified for syringyl moieties at positions 2 and 6, and positions 1, 3, 4 and 5, respectively. Associated peaks appeared at 105 ppm for tertiary carbons, and at 134, 137 and 153 ppm for quaternary carbons. In guaiacyl units, a 120 ppm peak was assigned to tertiary carbon at position 6, and a 147 ppm peak to quaternary carbon at position 4. Hemicellulose impurity was observed by the presence of a 63 ppm signal from carbon at position 5 in xylose (Garcia et al. 2012). Quantitative ^{13}C NMR spectra of birch aqueous ethanol organosolv lignins, delignified to different severities, were recorded in solution state in deuterated dimethyl sulfoxide with chromium (III) acetylacetonate as a spin relaxant and trioxane as internal standard. The major peak assignments were: 174–175 ppm for carboxyl carbon, 147–153 ppm for aromatic carbon in the 3 and 5 positions of guaiacyl and syringyl entities, 134–139 ppm for carbon in 1 and 4 positions of guaiacyl and syringyl entities, 109–116 ppm for carbon in 2 and 5 positions of guaiacyl entities, 103–105 ppm for carbon in 2 and 6 positions of syringyl entities, 59–86 ppm α, β, and γ carbons in β–O–4′ linked units, and 55–57 ppm for methoxyl carbons. Their relative abundance in lignin samples was estimated by signal integrations. Methoxyl content varied between 4.56 and 7.02 mmol/g, β–O–4′ linked units between 0.92 and 2.38 mmol/g, and the ratio of tertiary aromatic carbon between 0.64 and 0.91. The presence of the carboxyl group was 0.04–0.1 mmol/g (Thoresen et al. 2021).

Eucalyptus dioxane lignin, dissolved in deuterated dimethyl sulfoxide, was studied by quantitative ^{13}C NMR spectroscopy. The spectrum is shown in Figure 6.14. The methoxyl peak is distinctly observed. The chemical shift region 57–90 ppm, which accounts for α, β, and γ carbons in β–O–4′, β–5′, β–β′ linked units, was categorized into different structural features in lignin macromolecule as 59–61 and 62–64 ppm for γ carbons in β–O–4′ linked entities (0.56 per aromatic ring), 51–54 ppm for β carbons in β–β′ and β–5′ linked entities (0.13 per aromatic ring), and 61.8 ppm peak for cinnamyl alcohol entities (0.02 per aromatic ring). The chemical shift region 103–125 ppm, which

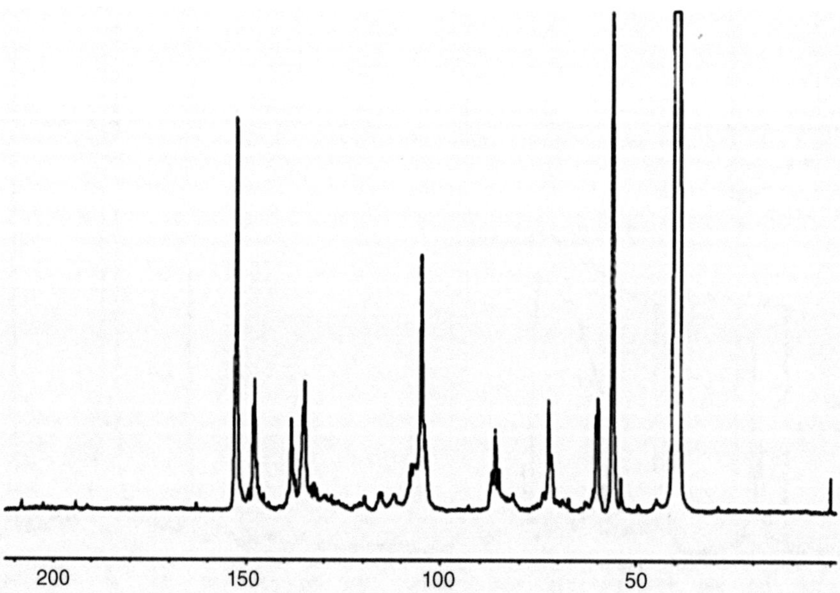

FIGURE 6.14 ^{13}C NMR spectrum of eucalyptus dioxane lignin. (Reprinted from Evtuguin, DV, Neto, CP, Silva, AMS, Domingues, PM, Amado, FML, Robert, D, Faix, O, *J. Agric. Food Chem.*, 49, 4252–4261, 2001, with permission from American Chemical Society.)

accounts for aromatic tertiary carbon, was categorized into syringyl and guaiacyl moieties as 103–110 ppm for syringyl moieties, and 110–123 ppm for guaiacyl moieties. Signal integration between 160 and 162 ppm accounted for p-hydroxyphenyl structures. Accordingly, the ratio of syringyl:guaiacyl:p-hydroxyphenyl was determined as 84:14:2. Carbonyl carbons were subdivided into three types, benzaldehyde type at 191–192 ppm (0.02 per aromatic ring), cinnamaldehyde type at 193.5–194.5 ppm (0.03 per aromatic ring), and ketone type at 195–210 ppm (0.15 per aromatic ring) (Evtuguin et al. 2001).

Aliphatic and phenolic hydroxyl content in miscanthus formic acid organosolv lignin was reported as 3.2 and 3.23%, respectively, by ^{13}C NMR spectroscopy. Spectrum was recorded by dissolving acetylated lignin in deuterated acetone. Signal integration for carbon in acetyl groups, that originated from primary aliphatic (171 ppm), secondary aliphatic (170 ppm), and aromatic hydroxyl (169 ppm), were used for the purpose (El Mansouri and Salvado 2007). Signal integration of peaks in CPMAS ^{13}C NMR spectra of banana leaf sheath dioxane lignin and its chloroform extracted product was analysed to ascertain the effect of chloroform extraction. The extracted product had a somewhat lower methoxyl content of 17.4% compared to 18.3% in parent lignin. Signal at 152 ppm from aromatic tertiary carbons of syringyl moieties (position 2 and 6 in benzene ring) decreased in intensity after chloroform extraction but peak area at 110–120 ppm from aromatic tertiary carbons of guaiacyl moieties (position 2, 5 and 6 in benzene ring) increased. p-hydroxyphenyl entities, identified by aromatic carbon at position 4 attached to oxygen, were not detected after chloroform extraction (Oliveira et al. 2006).

An early work reports ^{13}C NMR characterization of ethylene glycol organosolv lignin from poplar. Acetylated lignin was dissolved in deuterated dimethyl sulfoxide to record the spectrum. Prominent signals that appeared at their usual positions were methoxyl carbon at 56 ppm (peak 22), α, β, and γ carbons in β–O–4′ linked units at 63, 71, 81, and 85 ppm (peaks 15, 16, 18 and 20), tertiary aromatic carbon in guaiacyl entities at 112, 119, and 122 ppm (peaks 11, 12 and 13), tertiary aromatic carbon in syringyl entities at 104 ppm (peak 14), aromatic carbon attached to oxygen in guaiacyl entities at 149, and 152 ppm (peak 5 and 6), aromatic carbon attached to oxygen in syringyl entities at 152 ppm (peak 4), tertiary aromatic carbon in p-hydroxyphenyl entities at 128 and 131 ppm (peak 9 and 10), and aromatic quaternary carbon at 132 and 135 ppm (peak 7 and 8). Carbon in acetyl groups resulted in peaks 168, 169, and 170 ppm (peaks 3, 2, and 1) corresponding to phenolic, secondary, and primary hydroxyl, respectively (Thring et al. 1990). The spectrum is shown in Figure 6.15.

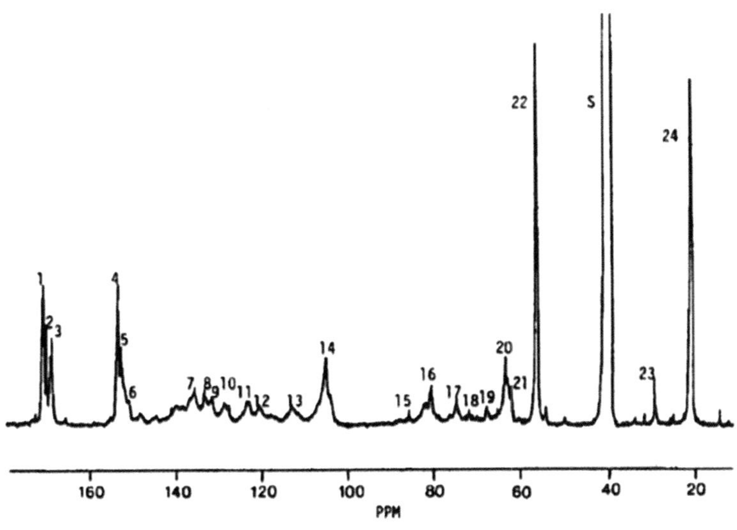

FIGURE 6.15 ^{13}C NMR spectrum of poplar ethylene glycol organosolv lignin. (Reprinted from Thring, RW, Chornet, E, Overend, RP, *Biomass*, 23, 289–305, 1990, with permission from Elsevier.)

6.4.2.4 Milled Wood Lignin

Pine milled-wood lignin was dissolved in deuterated chloroform and characterized by quantitative ^{13}C NMR spectroscopy. Chromium (III) acetylacetonate was the relaxant. The spectrum is shown in Figure 6.16. Of the 30 peaks identified in the spectrum, the important ones are peaks 2, and 3 (191, 194 ppm) for carbonyl at α carbon, peak 4 (181 ppm) for quinone form of p-hydroxyphenyl, peaks 7, 8, 9, and 10 (145–151ppm) for etherified and non-etherified carbons at positions 3 and 4 of guaiacyl moieties, peaks 12, and 13 (130–137 ppm) for carbons at position 1 of guaiacyl moieties, peaks 17, 18, and 19 (108–122 ppm) for carbons with aromatic hydrogen in guaiacyl structures, and peaks 20 to 26 (59–87 ppm) for α, β, and γ carbons in the aliphatic side chain. Peak 27 at 56 ppm as usual represented methoxyl carbon. Methoxyl content was estimated as 0.96 per phenylpropane unit by the integration of this peak. The abundance of aryl aryl carbon–carbon linkages was estimated to be 1.53 per phenylpropane unit. Condensed guaiacyl entities were 0.43 per phenylpropane unit. Of the condensed benzene rings β–5′ linkage (86–90 ppm) accounted for 0.09, and 5–5′ linkage (141–145 ppm) accounted for 0.16 per benzene ring. This lignin had 0.41 β-O-4′ linkages per benzene ring. The lignin macromolecule also had 0.12 carbonyl groups per benzene ring including an aldehyde group at α carbon (0.04 per benzene ring). This indicated degradation of the aliphatic side chain (Holtman et al. 2004).

Another study on pine milled-wood lignin characterized it by quantitative ^{13}C NMR spectroscopy in solution state in deuterated dimethyl sulfoxide. Of all the aromatic carbons, 30.7% were at positions 3 and 4 with methoxyl or hydroxyl groups, respectively. 28.4% of aromatic carbon atoms, at positions 1 and 5, were involved in C–C intermonomer linkages. The rest of the aromatic carbons, at positions 2, 5 and 6, were attached to hydrogen. Methoxyl groups were estimated to be nearly 1 per benzene ring. α, β, and γ carbons in β-O-4′, β-5′, and β-β′ linked units were 1.53 per benzene ring.

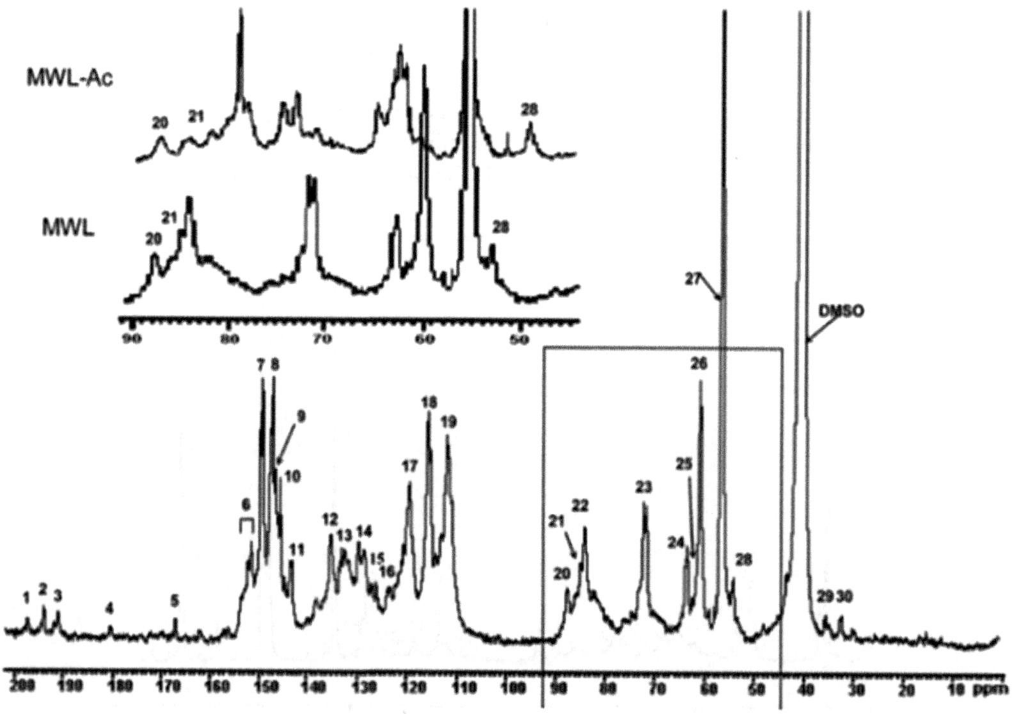

FIGURE 6.16 ^{13}C NMR spectrum of pine milled-wood lignin with an enlarged view of 45–90 ppm region in comparison with acetylated milled wood lignin. (Reprinted from Holtman, KM, Chang, H, Kadla, JF, *J. Agric. Food Chem.*, 52, 720–726, 2004, with permission from American Chemical Society.)

Lignin was free from carbohydrate impurities as suggested by the absence of peaks in 90–100 ppm. Carboxyl groups were also absent (Huang et al. 2011). Qualitative analysis of acetylated lauan milled-wood lignin ^{13}C NMR spectrum, acquired in deuterated dimethyl sulfoxide solution, revealed characteristic methoxyl carbon peak. However, the authors assigned the entire 61–90 ppm region to α, β, and γ carbons in β–O–4′ and α–O–4′ type linkages, which seems to be an oversimplification. Similarly, the identification of syringyl (105–108 ppm) and guaiacyl (112–115 ppm) moieties was limited to only aromatic tertiary carbons at positions 2 and 6. Methyl and methylene carbons had their peaks at normal positions (Wang et al. 2014). ^{13}C NMR characterization of poplar milled-wood lignin was performed by dissolving in deuterated dimethyl sulfoxide with the spin relaxant chromium (III) acetylacetonate. Three distinct peaks were identified for aliphatic carbons related to lignin carbohydrate complexes. These were 96–103 ppm for phenyl glycoside, 78–90 ppm for α carbon in lignin linked to hemicellulose, and 58–65 ppm for γ carbon in phenylpropane forming esters with carboxyl in hemicellulose (Yuan et al. 2011). Similar peaks were identified and quantified, besides usual peak assignments for methoxyl and aromatic carbons, for lignin carbohydrate complexes in the ^{13}C NMR spectrum of acetylated milled wood lignins from poplar (Min et al. 2014). Poplar, untreated as well as green liquor treated, milled wood lignin was characterized by quantitative ^{13}C NMR. Methoxyl signal was present at normal position and accounted for 1.71–1.73 methoxyl groups per phenylpropane unit in untreated as well treated milled wood lignin. Tertiary aromatic carbons were estimated by signal integration between 103 and 123 ppm to be 2.33 per phenylpropane unit in untreated milled wood lignin. This decreased to 2.09–2.17 per phenylpropane unit when the wood was green liquor treated. On the other hand, the abundance of quaternary aromatic carbons, representing carbon–carbon intermonomer linkages, was 1.64 per phenylpropane unit in lignin obtained from untreated wood. These aromatic carbons increased to 1.8–1.88 per phenylpropane unit for milled wood lignin from green liquor-treated wood. For these estimations, spectra were integrated in the region 123–140 ppm. Aromatic carbon attached to oxygen, determined by 140–160 ppm signal integration, remained relatively unchanged at 1.97–1.99 per phenylpropane unit. This study used 58–61 ppm signal integration for quantitative determination of β-O-4′ linked entities. Their abundances per phenylpropane unit were 0.53 for untreated samples and 0.44–0.5 for green liquor samples. Similarly, the 52–54 ppm region was attributed to β-5′ and β-β′ linked units with abundance per phenylpropane unit of 0.06 in untreated milled wood lignin and 0.11 in green liquor treated milled wood lignin (Tan et al. 2017). In qualitative ^{13}C NMR spectrum of bamboo milled-wood lignin, obtained in deuterated dimethyl sulfoxide solution, aromatic carbons in syringyl moieties were identified as peaks at 153 ppm for positions 3 and 5, 138 ppm for position 4 linked with ether bond, 135 and 133 ppm for position 1 linked with ether bond, 132 ppm for position 1 with aliphatic side chain, and 107 and 105 ppm for position 2 and 6. Similarly, aromatic carbons in guaiacyl moieties were identified as peaks at 149–150 ppm for positions 3 linked with ether bond, 145 and 134 ppm for position 1 linked with ether bond, 133 ppm for position 1 with aliphatic side chain, and 111–119 ppm for position 2, 5 and 6. p-hydroxyphenyl entities were identified with their aromatic carbons at position 2 and 6 represented by 128–130 ppm peak. Signals for α, β, and γ carbons in β-O-4′, β-5′, and β-β′ linked units were recorded at their usual positions (Shi et al. 2019).

6.4.2.5 Hydrolysis Lignin

Pine enzymatic hydrolysis lignin was characterized by quantitative ^{13}C NMR as a solution in deuterated chloroform with chromium (III) acetylacetonate as the relaxant. Signal integrations revealed that the lignin macromolecule consisted of 2.64 tertiary aromatic carbon, 1.42 quaternary aromatic carbon, and 2.06 aromatic carbon attached to oxygen, per benzene ring. Methoxyl group abundance was 1.17 per benzene ring. Condensed guaiacyl entities were 0.36 per phenylpropane unit. Of the condensed benzene rings, β-5′ linkage (86–90 ppm) accounted for 0.07 and 5–5′ linkage (141–145 ppm) accounted for 0.12 per benzene ring. This lignin had 0.47 β-O-4′ linkages per benzene ring. The lignin macromolecule also had 0.08 carbonyl groups per benzene ring including the aldehyde group at α carbon (0.02 per benzene ring). This indicated degradation of the aliphatic side chain

(Holtman et al. 2004). Aliphatic and phenolic hydroxyl content in lignin-rich residue, obtained after steam explosion of poplar and hydrolysis of polysaccharides, was reported as 4.55% and 2.7%, respectively, by ^{13}C NMR spectroscopy. Spectrum was recorded by dissolving acetylated lignin in deuterated acetone. Signal integration for carbon in acetyl groups, that originated from primary aliphatic (171 ppm), secondary aliphatic (170 ppm), and aromatic hydroxyl (169 ppm), were used for the purpose (El Mansouri and Salvado 2007). DEPT ^{13}C NMR spectrum of wheat straw enzymatic mild acidolysis lignin was recorded in solution state in deuterated dimethyl sulfoxide with the addition of spin relaxant chromium (III) acetylacetonate. The study reported an elaborate peak assignment, though the spectrum was not well resolved. In the aliphatic region, besides usual signals for methoxyl, peaks were assigned for: γ carbon in β–O–4′ structures (61 and 63 ppm), γ carbon in β–5′ structures (63 ppm), γ carbon in β–β′ structures (73 ppm), β carbon in coniferyl alcohol structures (30 ppm), β carbon in β–1′ structures (80 ppm), α carbon in structures having a carboxyl group at γ position (39 ppm), α carbon in β–β′ structures (70 ppm), α carbon in β–1′ structures (75 and 87 ppm), α carbon in 5–5′ structures (84 ppm), and α carbon in β–5′ structures (88 ppm). In the aromatic region, important carbon peaks were assigned to: tertiary carbons in syringyl units (105 ppm), tertiary carbons in guaiacyl units (111, 120, and 146 ppm), and tertiary carbons in p-hydroxyphenyl units (116, 128, and 131 ppm). Several carbon peaks attributed to lignin carbohydrate linkages were also identified in the spectrum owing to the pyranose ring in xylan and glucuronic acid substituents (Yang et al. 2011). ^{13}C NMR characterization of poplar enzymatic mild acidolysis lignin was performed by dissolving in deuterated dimethyl sulfoxide with the spin relaxant chromium (III) acetylacetonate. Three distinct peaks were identified for aliphatic carbons related to lignin carbohydrate complexes. These were 96–103 ppm for phenyl glycoside, 78–90 ppm for α carbon in lignin linked to hemicellulose, and 58–65 ppm for γ carbon in phenylpropane forming esters with carboxyl in hemicellulose (Yuan et al. 2011). ^{13}C NMR characterization of rice straw enzymatic mild acidolysis lignin revealed the major structural features in lignin macromolecule. These were 13.9% methoxyl, 26.5% β–O–4′ linked entities, 2.5% carbonyl moieties, and 6% lignin carbohydrate linkages. Total aromatic carbon was 33.2% and aliphatic carbon (excluding methoxyl) was 43.9% (Lou et al. 2018).

6.4.3 Lignin Characterization by ^{31}P NMR Spectroscopy

If an organic molecule has phosphorous atoms in its structure, these can be analysed by NMR spectroscopy to study the molecular structure. In ^{31}P NMR spectroscopy, the isotope of phosphorous with atomic mass 31, the natural and stable isotope, is the target nucleus which has a magnetic moment being a fermion. Its magnetic moment is reasonably high, intermediate between ^{1}H and ^{13}C. Thus, ^{31}P NMR spectroscopy combines the advantages of ^{1}H and ^{13}C NMR spectroscopy. The spectrum has a similar high resolution as ^{13}C NMR with a wider range of chemical shifts, and the two spectra often complement each other. At the same time, it has a sufficiently strong intensity due to the natural abundance of ^{31}P. However, the coupling of ^{31}P with ^{1}H nucleus is much stronger than the coupling of ^{13}C with ^{1}H, especially those over one bond distance. Strong decoupling protocols are, therefore required. The inverse gated decoupling is recommended for quantitative signal integrations. Lignin molecule has no phosphorous atom in it and requires derivatization to introduce phosphorous-containing groups. The compound 2-chloro-4,4,5,5-tetramethyl-1,3,2-dioxaphospholane is often used as the phosphitylation agent which is added during the dissolution of lignin in the chosen deuterated solvent. The electronegative chlorine atom in the reagent targets the hydroxyl functional groups of lignin by abstracting the proton from the hydroxyl group and expulsion of a hydrogen chloride molecule. In the resulting product, the lignin molecule is attached to the phosphorous atom of the phosphitylating reagent through an oxygen bridge (Granata and Argyropoulos 1995). Phosphitylation is normally quick and adequate by the time the lignin is dissolved. Spectra is recorded immediately after dissolution. Quantitative ^{31}P NMR spectroscopy has been very effectively used in the structural elucidation of lignin macromolecules with respect to their phenolic hydroxyl content.

6.4.3.1 Kraft Lignin

An early study reports the use of quantitative ^{31}P NMR spectroscopy to estimate the abundance of hydroxylated entities in lignin macromolecules. The solution state spectrum was recorded by dissolving kraft lignin Indulin in a 1.6:1 v/v mixture of pyridine and deuterated chloroform. The solution contained chromium (III) acetylacetonate as the spin relaxant and cyclohexanol as the internal standard. 2-chloro-4,4,5,5-tetramethyl-1,3,2-dioxaphospholane was used as the phosphitylation agent. Condensed structures through carbon–carbon linkages were detected with signals in 140.3–144.4 ppm. Of these, 143.5 ppm peak was assigned to diphenylmethane entities with free aromatic hydroxyl, and 137.8 ppm to p-hydroxyphenyl entities (Granata and Argyropoulos 1995). A quantitative ^{31}P NMR spectrum of softwood kraft lignin (Indulin AT) was obtained with a similar procedure but using cholesterol as an internal standard. It was analysed to estimate the abundance of hydroxyl groups in different structural subunits. It had 4 mmol/g aromatic hydroxyl, 2.59 mmol/g aliphatic hydroxyl, and 0.2 mmol/g carboxyl. Besides, guaiacyl and p-hydroxyphenyl moieties were 95% and 5%, respectively (Sameni et al. 2016). Another ^{31}P NMR study of the same lignin estimated the aromatic hydroxyl as 3.2 mmol/g – 0.2 syringyl, 1.54 guaiacyl, 0.21 p-hydroxyphenyl and 1.25 condensed entities. In comparison, the measured aliphatic hydroxyl content was reported to be 2.06 mmol/g (Serrano et al. 2018). In contrast to other reports, this study claimed this lignin to be SGH type, which is rather unusual. ^{31}P NMR spectrum of eucalyptus kraft lignin was recorded in solution state in a mixture by using the protocol of Granata and Argyropoulos but endo-N-hydroxy-5-norbornene-2,3-dicarboximide as internal standard. The different structural features in lignin macromolecule were quantitatively estimated as aliphatic hydroxyl 1–1.11 mmol/g, syringyl moieties 2.05–2.28 mmol/g, condensed guaiacyl moieties 1.07–1.11 mmol/g, uncondensed guaiacyl moieties 0.9–0.99 mmol/g, and p-hydroxyphenyl moieties 0.13 mmol/g (Antonino et al. 2021).

One study analysed the hydroxyl group profiles of softwood kraft lignins as they were isolated by carbonation to different pH. In lignin present in the black liquor, the distribution of aliphatic hydroxyl groups, hydroxyl groups in condensed phenolic structures, hydroxyl groups in guaiacyl structures, and hydroxyl groups in carboxyl was 1.49, 1.73, 2.46, and 0.69 mmol/g, respectively. In the lignin precipitated at 9.5 pH, this distribution changed to 1.11, 1.14, 1.47, and 0.59 mmol/g, respectively, indicating a decrease in hydroxyl content across all types. The distribution further decreased to 0.91, 0.85, 1.03, and 0.39, respectively, when the precipitation pH was 10.5 (Nagy et al. 2010). In another study, the hydroxyl group profile of softwood kraft lignin was analysed by ^{31}P NMR spectroscopy, when it was precipitated from heat-treated black liquor. Lignin precipitated from non-heat-treated black liquor had 1.9 mmol/g hydroxyl in condensed moieties, 2.2 mmol/g hydroxyl in guaiacyl and catechol moieties, and 2 mmol/g hydroxyl in the aliphatic region. The abundance of hydroxyl groups increased in the aromatic region with the extent of heat treatment. When lignin was precipitated after heat treating the black liquor for 24 h at 200 °C, its hydroxyl content was 2.6 mmol/g hydroxyl in condensed moieties, 4.9 mmol/g hydroxyl in guaiacyl and catechol moieties, and 0.6 mmol/g hydroxyl in the aliphatic region (Pichler et al. 2019). Bamboo kraft lignin was fractionated into ethanol soluble and ethanol insoluble fractions and analysed by ^{31}P NMR spectroscopy for hydroxyl group profile. Hydroxyl groups in unfractionated lignin were distributed between syringyl, guaiacyl, and p-hydroxyphenyl substructures as 0.67, 0.39, and 0.41 mmol/g, respectively. This distribution was 0.89, 0.51, and 0.54 mmol/g in syringyl, guaiacyl, and p-hydroxyphenyl substructures, respectively, in the ethanol-soluble fraction, an increase from the unfractionated sample. In contrast, the ethanol insoluble fraction had a lower abundance of hydroxyl groups with 0.43, 0.27, and 0.29 mmol/g in syringyl, guaiacyl, and p-hydroxyphenyl substructures, respectively (Wang et al. 2019).

6.4.3.2 Soda Lignin

Quantitative ^{31}P NMR spectra of *C. lanceolata* soda lignin fractions revealed different types of hydroxylated entities in the lignin macromolecules as aliphatic hydroxyl 1.43–1.67 mmol/g (146–149 ppm signals), guaiacyl hydroxyl 1.21–1.35 mmol/g (139–140.6 ppm signals), p-hydroxyphenyl 0.01–0.03 mmol/g (137–138 ppm signals), hydroxyl in condensed moieties

0.2–0.27 mmol/g (140.6–145 ppm signals), and hydroxyl group in carboxyl 0.47–0.97 (134–136 ppm signals) (Li and Wu 2014). The quantitative ^{31}P NMR spectrum of nonwood soda lignin (Protobind), with cholesterol as the internal standard, was analysed to estimate the abundance of hydroxyl groups in different structural subunits. It had 2.28 mmol/g aromatic hydroxyl, 2.47 mmol/g aliphatic hydroxyl, and 0.22 mmol/g carboxyl. Besides, syringyl, guaiacyl, and p-hydroxyphenyl moieties were 51%, 40%, and 9%, respectively (Sameni et al. 2016). Another ^{31}P NMR study of the same lignin estimated the aromatic hydroxyl as 2.7 mmol/g – 0.67 syringyl, 0.82 guaiacyl, 0.35 p-hydroxyphenyl and 0.86 condensed entities. In comparison, the measured aliphatic hydroxyl content was reported to be 1.5 mmol/g (Serrano et al. 2018). The spectrum, with its peak assignments, is shown in Figure 6.17.

Miscanthus soda lignin from pilot plant operation was characterized for its hydroxyl group contents, including those in the carboxyl group, by ^{31}P NMR spectroscopy. Spectrum was recorded following the method of Granata and Argyropoulos with cholesterol as the internal standard and chromium (III) acetylacetonate to relax nuclear spin. It had 2.95 mmol/g aliphatic hydroxyl groups, 0.41 mmol/g hydroxyl groups in guaiacyl units, 0.09 mmol/g hydroxyl groups in syringyl and condensed entities, and 0.09 mmol/g hydroxyl groups in p-hydroxyphenyl entities. The lignin also had 0.6 mmol/g of carboxylic groups, a rather high value (Lauwaert et al. 2019).

6.4.3.3 Organosolv Lignin

Miscanthus aqueous ethanol organosolv lignin was analysed for its hydroxyl group content by recording ^{31}P NMR spectrum and determined to have 0.19 aliphatic hydroxyl and 0.5 aromatic hydroxyl per phenylpropane unit. Lignin also had 0.04 carboxylic acid groups per phenylpropane unit (El Hage et al. 2009). ^{31}P NMR characterization of aqueous ethanol organosolv lignin from eucalyptus was performed with endo-N-hydroxy-5-norbornene-2,3-dicarboximide as an internal standard. Different hydroxylated structural features in lignin macromolecule were quantitatively determined by signal integrations as 145.5–150 ppm aliphatic hydroxyl, 140–145 ppm hydroxyl groups in syringyl and condensed moieties, 139–140 ppm guaiacyl hydroxyl, 137–138 ppm

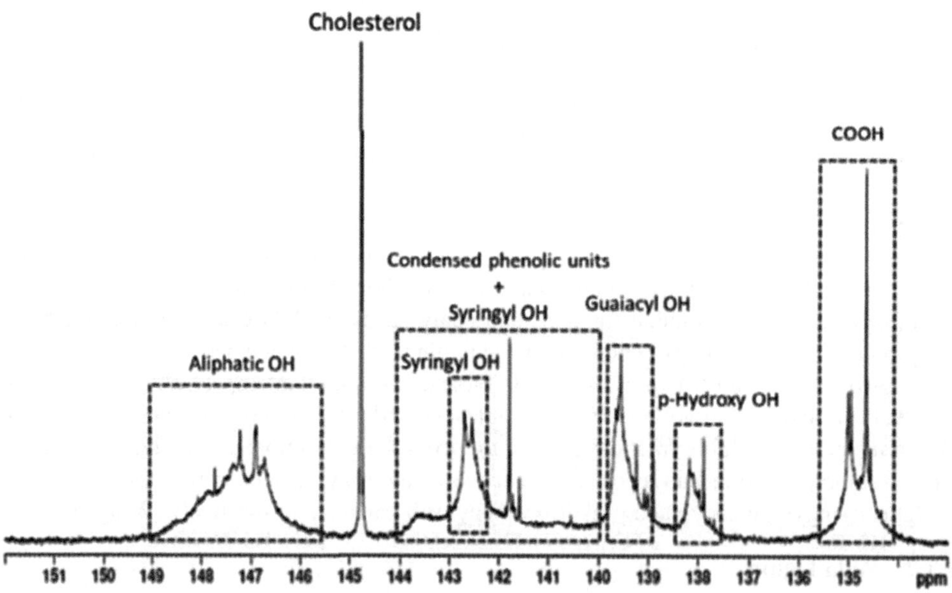

FIGURE 6.17 ^{31}P NMR spectrum of wheat straw soda lignin with hydroxyl moieties associated with chemical shifts. (Reprinted from Serrano, L, Esakkimuthu, ES, Marlin, N, Brochier-Salon, M, Mortha, G, Bertaud, F, *Energy Fuels*, 32, 5969–5977, 2018, with permission from American Chemical Society.)

p-hydroxyphenyl, and 133.6–136.6 ppm hydroxyls in carboxyl groups. Their abundances in lignin samples, depending on the severity of delignification, were 1.92–2.8, 2.54–3, 0.55–0.6, 0.04–0.07, and 0.15–0.26 mmol/g, respectively, Interestingly, catechol was also detected in the lignin samples with signal integration between 138 and 139 ppm giving abundance of 0.09–0.14 mmol/g (Yanez-S et al. 2014). Rice straw aqueous ethanol organosolv lignin was characterized by ^{31}P NMR spectroscopy following the method of Granata and Argyropoulos. The spectrum with peak assignment is shown in Figure 6.18. Quantitative estimation as mmol/g of hydroxyl groups in different types of structural subunits was: syringyl 0.3–0.34, guaiacyl 0.39–0.44, p-hydroxyphenyl 0.45–0.51, condensed substructures 0.26–0.31, aliphatic 2.29–2.55, and carboxylic 0.63–0.73. The study also determined the tricin content of the lignin to be 0.16–0.19 mmol/g (Moniz et al. 2018).

Aqueous ethanol organosolv lignin was obtained from beech and analysed by ^{31}P NMR spectroscopy for molecular structure following the protocol of Granata and Argyropoulos. Its abundance of hydroxyl groups was estimated to be 1.3 mmol/g of aliphatic, 3.1 mmol/g of phenolic, and 0.1 mmol/g of carboxylic (Rossberg et al. 2019). ^{31}P NMR spectrum of spruce aqueous ethanol organosolv lignin revealed 3.7 mmol/g hydroxyl groups in guaiacyl substructures, and 2.73 mmol/g aliphatic hydroxyl groups (Parot et al. 2022).

^{31}P NMR study of aqueous formic acid acetic acid organosolv lignin from wheat straw estimated the aromatic hydroxyl as 1.3 mmol/g – 0.25 syringyl, 0.45 guaiacyl, 0.21 p-hydroxyphenyl and 0.39 condensed entities. In comparison, the measured aliphatic hydroxyl content was reported to be 1.23 mmol/g (Serrano et al. 2018).

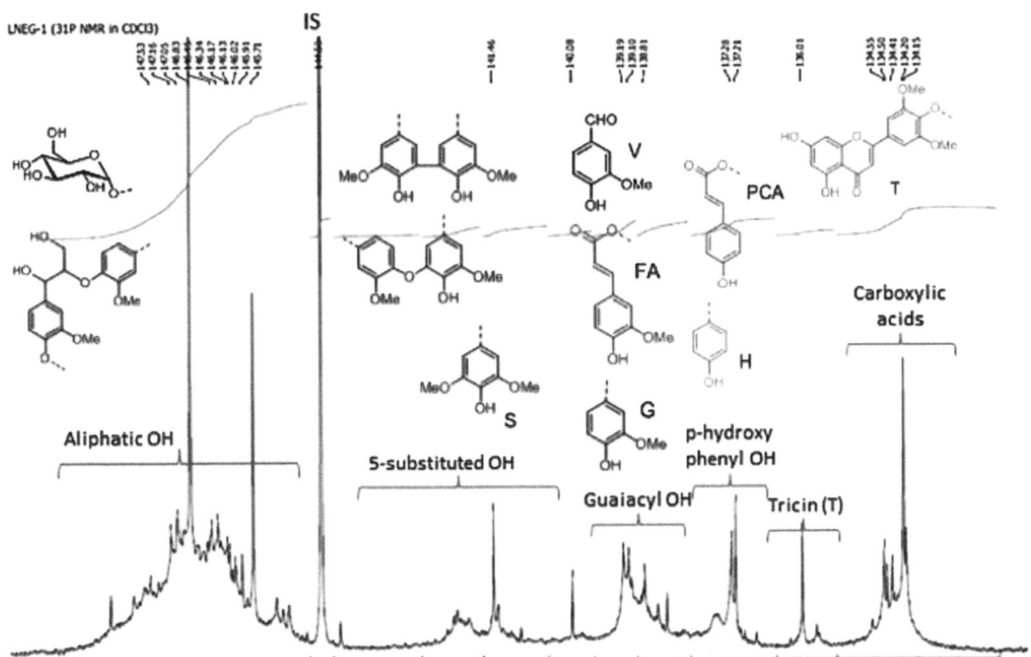

FIGURE 6.18 ^{31}P NMR spectrum of rice straw ethanol organosolv lignin with structural moieties associated with chemical shifts. (Reprinted from Moniz, P, Serralheiro, C, Matos, CT, Boeriu, CG, Frissen, AE, Duarted, LC, Roseiro, LB, Pereira, H, Carvalheiro, F, *Process Biochem.*, 65, 136–145, 2018, with permission from Elsevier.)

6.4.3.4 Milled Wood Lignin

^{31}P NMR spectra of spruce milled-wood lignin were acquired with N-Hydroxynaphthalimide as internal standard and chromium (III) acetylacetonate as relaxant. The study used two phosphitylating agents, namely, 2-chloro-1,3,2-dioxaphospholane and 2-chloro-4,4,5,5-tetramethyl-1,3,2-dioxaphospholane. The former selectively phosphitylated the hydroxyls at α carbons, allowing for the estimation of β–O–4′ substructures. The latter was used for comprehensive phosphitylation of all hydroxyl groups. The spectra revealed a drop of 0.2 mmol/g of β-O-4′ substructures and an increase of 0.6 mmol/g of aromatic hydroxyl groups in uncondensed substructures when the wood was vibratory ball-milled for 96 h compared to 2h (Guerra et al. 2006). Miscanthus milled-wood lignin was analysed for its hydroxyl group content by recording ^{31}P NMR spectrum and determined to have 0.67 aliphatic hydroxyl, 0.11 guaiacyl, 0.11 p-hydroxyphenyl, and 0.04 syringyl per phenylpropane unit (El Hage et al. 2009). Acetylated spruce, and beech milled-wood lignins were characterized by ^{31}P NMR spectroscopy using the method proposed by Granata and Argyropoulos as described above. Phenolic hydroxyls were identified with peaks at 141 ppm – 5–5′ linked units with free hydroxyls at position 4 – and 142 ppm – 4–O–5′ and 5–5′ linked units with free hydroxyl at one benzene ring at position 4. Spruce milled-wood lignin had 18.6 phenolic hydroxyl groups without 5–5′ linkage and 3.4 phenolic hydroxyl groups with 5–5′ linkage per 100 phenylpropane units. These values were 20.3 and 50, respectively, for beech milled-wood lignin (Crestini et al. 2011).

6.4.3.5 Hydrolysis Lignin

Acetylated spruce enzymatic mild acidolysis lignin was characterized by ^{31}P NMR spectroscopy using the method proposed by Granata and Argyropoulos and analysed for aromatic hydroxyl groups as described above. It had 19.8 phenolic hydroxyl groups without 5–5′ linkage and 1.6 phenolic hydroxyl groups with 5–5′ linkage per 100 phenylpropane units (Crestini et al. 2011). A similar method was adopted to record the quantitative ^{31}P NMR spectrum of wheat straw enzymatic mild acidolysis lignin, with the exception that cholesterol was used as the internal standard. Peak assignments for phenolic hydroxyl in different structural subunits in lignin were made as 137–138 ppm for p-hydroxyphenyl moieties, 139–140 ppm for phenolic hydroxyl in guaiacyl moieties, 142–143 ppm for phenolic hydroxyl in syringyl moieties, 143–144.5 ppm for phenolic hydroxyl in condensed structures. Their abundances were 0.301, 0.574, 0.214, and 0.176 mmol/g, respectively (Yang et al. 2011). Quantitative ^{31}P NMR spectrum of lignin-rich residue from hydrolysis of steam-exploded mixed hardwood and nonwood was recorded with cholesterol as internal standard. The spectrum revealed that the lignin sample had 1.51 mmol/g aromatic hydroxyl, 2.22 mmol/g aliphatic hydroxyl, and 0.0.54 mmol/g carboxyl. Besides, syringyl, guaiacyl, and p-hydroxyphenyl moieties were 26%, 31%, and 44%, respectively (Sameni et al. 2016). The authors did not explain the possible reasons for the unusually high p-hydroxyphenyl content. Poplar cellulolytic enzyme lignins were estimated to have 3.84–4.33 mmol/g aliphatic hydroxyl, 0.33–0.53 mmol/g hydroxyl attached to syringyl and condensed entities, 0.2–0.32 mmol/g hydroxyl attached to guaiacyl entities, and 0.28–0.4 mmol/g hydroxyl in p-hydroxyphenyl entities, as revealed by ^{31}P NMR spectroscopy. These lignins also had 0.33–1.5 mmol/g carboxyl groups (Chen et al. 2017). Quantitative ^{31}P NMR spectrum of rice straw enzymatic mild acidolysis lignin, with cholesterol as the internal standard, was analysed to estimate the abundance of hydroxyl groups in different structural subunits. Besides aromatic hydroxyl groups associated with guaiacyl, syringyl, and p-hydroxyphenyl entities and condensed structures, aliphatic hydroxyls were assigned to 146–147 ppm peak and hydroxyl in carboxyl groups were assigned to 134.5–135.5 ppm peak. Accordingly, hydroxyl groups of different types were determined to be 1.04 mmol/g in guaiacyl, 0.28 mmol/g in syringyl, 1.19 mmol/g in p-hydroxyphenyl, 0.58 mmol/g in condensed substructures, 5.35 mmol/g in aliphatic moieties, and 0.09 mmol/g in carboxyl groups (Lou et al. 2018).

6.4.4 Lignin Characterization by Heteronuclear 2D NMR Spectroscopy

We have already discussed the coupling of nuclear spins of the proton and ^{13}C nucleus which affect the energy state degeneracy of both the nuclei. In normal 1D NMR spectroscopy, the spin quantum state of either of the two nuclei is to be observed at a time and this coupling is undesirable. However, heteronuclear 2D NMR spectroscopy takes advantage of this magnetic spin correlation between the two types of nuclei to record a spectrum with two chemical shift axes, one for the proton and one for the heteronucleus. Most commonly, the heteronucleus is ^{13}C. Such a spectrum is much more informative and has quantitative value than the 1D spectrum of either nuclei. Therefore, heteronuclear 2D NMR spectroscopy is a very effective characterization method in molecular structure elucidation of complex organic molecules like lignin.

In ^{1}H–^{13}C heteronuclear single quantum correlation spectroscopy (HSQC), coupling correlations are detected between ^{1}H and ^{13}C nuclei in distinct pairs that are one bond apart. A cross peak is obtained in the spectrum for every unique ^{1}H nuclei that is magnetically coupled to ^{13}C nuclei. The coordinates of the peak correspond to the chemical shifts of ^{1}H and ^{13}C nuclei in that particular pair. The spectrum with all its cross peaks sheds light on the manner different types of hydrogen nuclei are joined to different types of carbon nuclei. Therefore, each cross peak represents a unique substructure in a complex organic molecule like lignin. The technique is based on transferring the ^{1}H magnetization to ^{13}C and transferring back the magnetization to ^{1}H after a time-lapse. The ^{1}H chemical shift axis is termed the direct dimension and the ^{13}C axis is termed the indirect dimension. ^{1}H–^{13}C heteronuclear multiple quantum correlation spectroscopy (HMQC) is similar to HSQC in detecting coupling correlations over one bond distance but differs from HSQC in simultaneously allowing magnetization transfer from one proton to several carbon nuclei in the vicinity. This may lead to lower resolution in the indirect dimension. Also, there is a higher probability of multiplets in the direct dimension. HSQC is, therefore, the method of choice in heteronuclear 2D NMR spectroscopy.

Typical cross-peak assignments in ^{1}H–^{13}C 2D NMR spectra of molecular substructures in lignin macromolecules are presented in Table 6.4. The 2D space can be conveniently subdivided into two regions. One of these corresponds to $C_{aliphatic}$–H correlations in the aliphatic side chain and spans 50–90/2.5–6 ppm $^{13}C/^{1}H$ chemical shift. The second region corresponds to $C_{aromatic}$–H correlations in the benzene ring and spans 100–150/5.5–8.5 ppm $^{13}C/^{1}H$ chemical shift (Shen et al. 2016).

6.4.4.1 Alkali Lignin

^{1}H–^{13}C 2D HSQC NMR characterization of the molecular structure of miscanthus soda lignin, separated in pilot scale, is reported. Spectrum was recorded in deuterated dimethyl sulfoxide. It was estimated to have 44 β–O–4′ linked entities, 5 α–O–4′ and β–5′ linked entities, and 5 α–O–γ′, α′–O–γ, and β–β′ linked moieties per 100 phenylpropane units. The abundance of syringyl, guaiacyl, and p-hydroxyphenyl subunits was 41%, 54%, and 5%, respectively (Lauwaert et al. 2019). One study reports ^{1}H–^{13}C 2D HSQC NMR characterization of LignoBoost softwood kraft lignins and Indulin AT. Besides other molecular structural features, the spectra assigned 81.1/4.78 and 86.9/4.35 ppm $^{13}C/^{1}H$ chemical shift cross peaks to a diastereomer of α–O–γ′, α′–O–γ, and β–β′ linked moieties (Ronnols et al. 2017). Bamboo kraft lignin, dissolved in deuterated dimethyl sulfoxide, was recorded for its ^{1}H–^{13}C 2D HSQC NMR spectrum and qualitatively analysed. Besides the major β–O–4′ intermonomer linkages, this lignin had minor quantities of α–O–γ′, α′–O–γ, and β–β′ linked moieties. However, no cross peaks for α–O–4′ and β–5′ linked entities were detected (Wang et al. 2019). Semi-quantitative solution state ^{1}H–^{13}C 2D HSQC NMR characterization of eucalyptus kraft lignin was performed in deuterated dimethyl sulfoxide. Cross peaks for lignin molecular substructures were referenced against solvent cross peak at 39.5/2.49 ppm $^{13}C/^{1}H$ chemical shift. It had a syringyl/guaiacyl ratio of 2.44 with 3.7 β–O–4′ linkages, 1.2 β–β′ linkages, and 5.4 β–5′ linkages per 100 phenylpropane units. Methoxyl content of this lignin was determined to be 6.86 mmol/g. Hydroxyl group distribution in mmol/g was 1.06 aliphatic, 2.33 attached to syringyl, 0.48 attached to condensed aromatic substructures, 0.96 attached to guaiacyl, 0.04 attached to p-hydroxyphenyl, and 0.54 attached to carboxyl (Chen et al. 2020).

6.4.4.2 Organosolv Lignin

Solution state $^{1}H–^{13}C$ 2D HSQC NMR spectrum of eucalyptus aqueous ethanol organosolv lignin in deuterated dimethyl sulfoxide was recorded and analysed for molecular substructures. Qualitative assignment of different cross peaks was made for substructures reported in Table 6.4 (Yanez-S et al. 2014). Aqueous ethanol organosolv lignin from rice straw, dissolved in deuterated dimethyl sulfoxide, was characterized by $^{1}H–^{13}C$ 2D HSQC NMR spectroscopy. Cross peaks were identified for tertiary aromatic carbons at positions 2 and 6 in syringyl moieties, tertiary aromatic carbons at positions 2, 5, and 6 in guaiacyl moieties, and tertiary aromatic carbon at positions 2 and 6 in p-hydroxyphenyl moieties. Other identified substructures were ferulates, p-coumarates, and tricin as assigned in Table 6.4. Usual cross peaks were also identified for α, β, and γ carbons in different intermonomer linkages conforming to assignments in Table 6.4 (Moniz et al. 2018). Different birch aqueous ethanol organosolv lignins were isolated with different delignification severity and analysed for molecular structure. Solution state $^{1}H–^{13}C$ 2D HSQC NMR spectra were acquired in deuterated dimethyl sulfoxide having spin relaxant chromium (III) acetylacetonate. Molecular substructures

TABLE 6.4
Cross Peak Assignments in 2D NMR Spectra of Lignin

Molecular substructures	$^{13}C/^{1}H$ chemical shift (ppm)	Reference
	HMQC	
α carbon in β-5′ linked moieties	88.1/5.5	Holtman et al. 2004
α carbon in β-O-4′ linked moieties	73.9/6	
α carbon in β-O-4′, α-O-4″, 5′–5″ linked moieties (dibenzodioxocin)	84.5/4.84	Holtman et al. 2004
α carbon in α-O-4′ and β-5′ linked moieties	88.2/5.5	Holtman et al. 2004
α carbon in α-O-γ′, α′-O-γ, and β-β′ linked moieties	85.1–85.4/4.67–4.78	Holtman et al. 2004
β carbon in β-O-4′, α-O-4″, 5′–5″ linked moieties (dibenzodioxocin)	82.6/4.13	Holtman et al. 2004
β carbon in α-O-4′ and β-5′ linked moieties	50.4/3.77	Holtman et al. 2004
β carbon in β-1′ linked moieties	49.9/3.4	Holtman et al. 2004
β carbon in β-O-4′ linked moieties	80/4.63	Holtman et al. 2004
γ carbon in β-O-4′ linked moieties	59.8/3–3.85	Holtman et al. 2004
γ carbon in coniferyl alcohol	62/4.1	Holtman et al. 2004
γ carbon in α-O-γ′, α′-O-γ, and β-β′ linked moieties	71.8–4.27 and 71.9/3.93	Holtman et al. 2004
	HSQC	
β carbon in β-β′ linked moieties	53.7/3.12	del Rio et al. 2008
Methyl carbon in methoxyl	55.6/3.73	del Rio et al. 2012
β carbon in β-1′ linked moieties	56.1/3.09	del Rio et al. 2008
β carbon in α-O-α′ and β-1′ linked moieties	59.5/2.75	del Rio et al. 2012a
γ carbon with hydroxyl in β-O-4′ linked moieties	60/3.38–3.71	del Rio et al. 2008
γ carbon with acetyl in β-O-4′ linked moieties	62.7/3.83–4.19	del Rio et al. 2012
γ carbon in β-β′ linked moieties	71.7–3.81 and 71.7/4.17	del Rio et al. 2008
α carbon in β-O-4′ linked moieties	72.3/4.86	del Rio et al. 2008
α carbon in α-O-β′ linked moieties	79.5/5.5	del Rio et al. 2012a

(Continued)

TABLE 6.4 (Continued)
Cross Peak Assignments in 2D NMR Spectra of Lignin

Molecular substructures	$^{13}C/^{1}H$ chemical shift (ppm)	Reference
α carbon in α-O-α′ and β-1′ linked moieties	81.2/5.01	del Rio et al. 2012a
α carbon in β-1′ linked moieties	82.1/5.12	del Rio et al. 2008
α carbon in α-O-4′, β-O-4″ and 5′–5″ linked moieties	83.3/4.81	del Rio et al. 2012a
α carbon in α-O-γ′, α′-O-γ, and β-β′ linked moieties	84.8–85.4/4.64–4.65	del Rio et al. 2008; del Rio et al. 2012a
β carbon in α-O-4′, β-O-4″ and 5′–5″ linked moieties	85.3/3.85	del Rio et al. 2012a
α carbon in β-1′ linked moieties	85.4/4.8	del Rio et al. 2008
β carbon in β-O-4′ linked moieties with hydroxyl at γ carbon	86.5/4.1	del Rio et al. 2008
α carbon in α-O-4′ and β-5′ linked moieties	86.8–87.7/5.45	del Rio et al. 2008; del Rio et al. 2012a
Tertiary aromatic carbon at positions 2 and 6 in syringyl moieties	103.8/6.68 and 106.7/7.36	del Rio et al. 2008
Tertiary aromatic carbon at position 2 in guaiacyl moieties	111.5/6.99	del Rio et al. 2008
Tertiary aromatic carbon at position 2 in β-1′ linked moieties	111.6/6.23	del Rio et al. 2008
Tertiary aromatic carbon at position 5 in guaiacyl moieties	115.2/6.71 and 115.2/6.94	del Rio et al. 2008
Tertiary aromatic carbon at position 6 in β-1′ linked moieties	118.3/6.19	del Rio et al. 2008
Tertiary aromatic carbon at position 6 in guaiacyl moieties	119.5/6.83	del Rio et al. 2008
Tertiary aromatic carbon at positions 2 and 6 in p-hydroxyphenyl moieties	127.9/7.2	Moniz et al. 2018
Tertiary aromatic carbon at position 2 in ferulates	110.8/7.3	Moniz et al. 2018
Tertiary aromatic carbon at position 6 in ferulates	122.7/7.1	Moniz et al. 2018
Tertiary aromatic carbon at positions 2 and 6 in p-coumarate moieties	129.8/7.47	Moniz et al. 2018
Tertiary aromatic carbon at positions 2 and 6 in tricin moieties	103.87/7.31	Moniz et al. 2018
Tertiary aromatic carbon at position 6 in tricin moieties	98.4/6.21	Moniz et al. 2018
Tertiary aromatic carbon at position 8 in tricin moieties	93.8/6.56	Moniz et al. 2018

were quantified from the 2D spectra in association with ^{13}C spectra. β–O–4′ linked moieties varied between 1.71 and 3.2 mmol/g, β–5′ linked moieties between 0.06 and 0.38 mmol/g, β–β′ linked moieties between 0.15 and 0.84 mmol/g, and β–1′ linked moieties between 0.01 and 0.25 mmol/g. These lignins had 4.23–11.26 mmol/g methoxyl groups, 0.89–1.71 mmol/g guaiacyl entities, and 2.95–5.85 mmol/g syringyl entities (Thoresen et al. 2021). ^{1}H–^{13}C 2D HSQC NMR spectroscopy was performed on aqueous ethanol organosolv lignins originating from poplar, pine, walnut shell, and rice husk. Spectra were recorded by dissolving the samples in deuterated dimethyl sulfoxide. The spectra are shown in Figure 6.19. The main intermonomer linkages identified were β–O–4′ – poplar 77.2%, pine 67.4%, rice husk 58.4%, and walnut shell 67.4%. β–β′ linkages were poplar 10.2%, pine 11.9%, rice husk 3.8%, walnut shell 14.2%. β–5′ linkages were poplar 12.6%, pine 20.7%, rice husk 37.8%, walnut shell 18.4%. Syringyl/guaiacyl ratios were 0.99, 0.05, and 1.06, respectively, in poplar, rice husk, and walnut shell lignins. Pine lignin, as expected, did not show syringyl moieties (Sheng et al. 2022).

^{1}H–^{13}C 2D HSQC NMR spectrum of bamboo aqueous formic acid organosolv lignin was quantitatively analysed to find 64% β–O–4′ linked moieties, 10% α–O–γ′, α′–O–γ, and β–β′ linked moieties, 22% α–O–4′ and β–5′ linked moieties, and 4% syringyl moieties with aldehyde group at γ carbon. Signal assignments for cross peaks somewhat differed from Table 6.4. A checmical shift of 71.7/4.88 ppm $^{13}C/^{1}H$ was attributed to α carbon in β–O–4′ linked moieties and 59.9/3.26–3.75 $^{13}C/^{1}H$ chemical shift was attributed to γ carbon in β–O–4′ linked moieties. Three cross peaks at

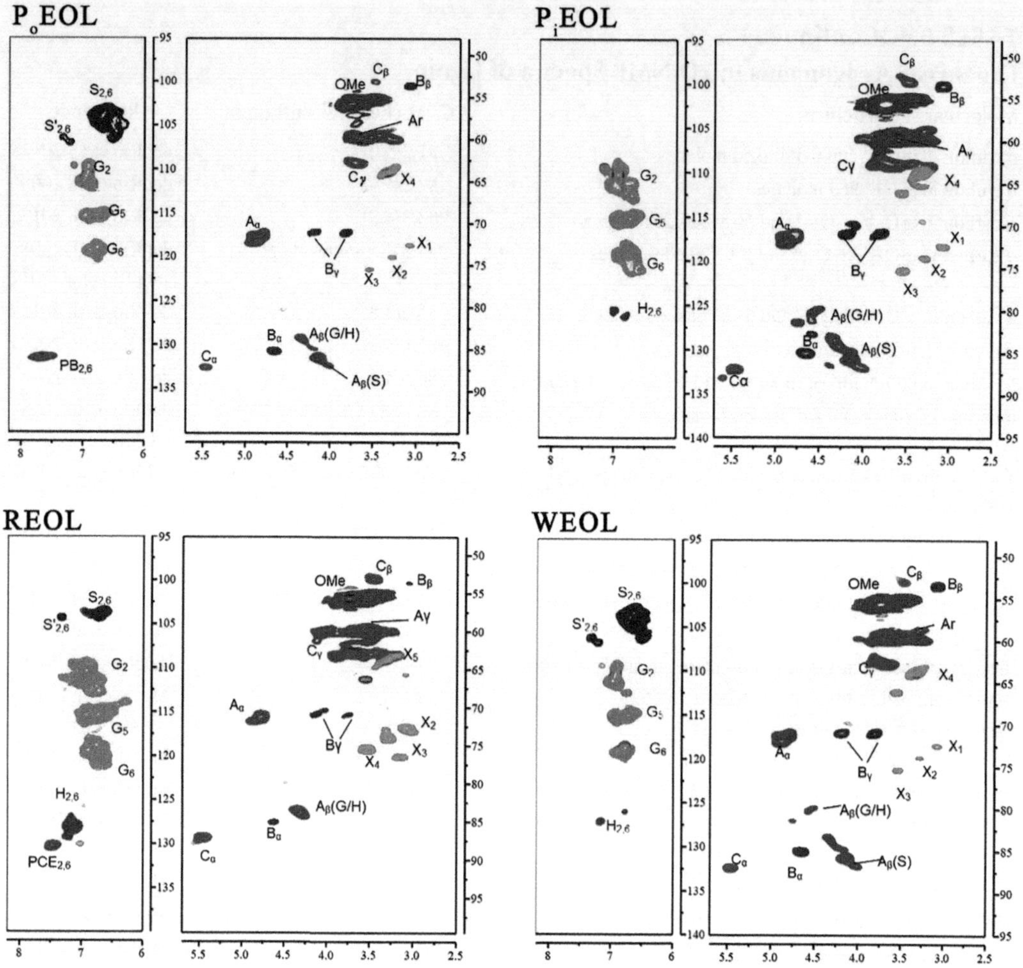

FIGURE 6.19 ^1H-^{13}C 2D HSQC NMR spectra of ethanol organosolv lignins; P_oEOL poplar, P_iEOL pine, REOL rice husk, WEOL walnut shell. (Reprinted from Sheng, Y, Ma, Z, Wang, X, Han, Y, *Food Chem.*, 376, 131895, 2022, with permission from Elsevier.)

83.5/4.3, 86/4.14, and 82.9/5.25 $^{13}C/^1H$ chemical shifts were attributed to β carbon in β–O–4′ linked moieties (Li et al. 2012).

6.4.4.3 Milled Wood Lignin

Pine milled-wood lignin was qualitatively characterized by solution state ^1H–^{13}C 2D HMQC NMR spectroscopy in deuterated chloroform. Cross peaks for lignin molecular substructures were referenced against solvent cross peak at 77.25/7.27 ppm $^{13}C/^1H$ chemical shift. Peak assignments are reported in Table 6.4 (Holtman et al. 2003). In a closely related study, pine milled-wood lignin was qualitatively characterized by solution state ^1H–^{13}C 2D HMQC NMR spectroscopy in deuterated dimethyl sulfoxide. Cross peaks for lignin molecular substructures were referenced against solvent cross peak at 39.5/2.5 ppm $^{13}C/^1H$ chemical shift. Peak assignments are reported in Table 6.4 (Holtman et al. 2004). Milled wood lignins from nonwood species kenaf, sisal, abaca, and curaua were extensively characterized by ^1H–^{13}C 2D HSQC NMR spectroscopy. Spectra were recorded in solution state in deuterated dimethyl sulfoxide which resulted in a solution cross peak at 40.1/2.5 ppm $^{13}C/^1H$ chemical shift. β–O–4′ intermonomer linkage was the most abundant, accounting for 84%,

89%, 94%, and 94%, respectively, for kenaf, sisal, abaca, and curaua. One notable feature of the spectra is that the carbohydrate cross peaks are distinctly identifiable from those of lignin. This is a clear advantage of 2D NMR over 1D NMR. Carbohydrate peaks are the least in abundance and variability in kenaf milled-wood lignin (del Rio et al. 2008). Peak assignments are reported in Table 6.4. In a similar study on wheat straw milled-wood lignin, small quantities (2%) of β–O–4′ and α–O–4″ linked moieties were detected besides other usually found molecular substructures. The lignin was reported to contain non-etherified p-coumarate and ferulate substructures besides syringyl, guaiacyl, and p-hydroxyphenyl. 77% of the intermonomer linkages were β–O–4′ type, 2% of these having carbonyl at α position. Other intermonomer linkages observed were 11% α–O–4′ and β–5′ linked moieties, 4% α–O–γ′ and β–β′ linked moieties, 3% α–O–4′, β–O–4″ and 5′–5″ linked moieties, and 3% α–O– α′ and β–1′ linked moieties. The study also reported for the first time, the presence of tricin in lignin identified by the cross peaks of 94.1/6.56 and 98.8/6.2 ppm $^{13}C/^{1}H$ chemical shift which was also confirmed by HMBC spectrum (del Rio et al. 2012a). In the ^{1}H–^{13}C 2D HSQC NMR spectrum of spruce milled-wood lignin, recorded in deuterated dimethyl sulfoxide, cross peaks were quantitatively integrated to estimate the abundance of different molecular substructures. For every 100 phenylpropane units, there were 3.3 β–O–4′, α–O–4′, 5–5′ linked moieties, 1.5 β–1′ linked moieties, and 3 β–β′ linked moieties. A similar analysis was also performed on beech milled-wood lignin. β–O–4′, α–O–4′, 5–5′ linked moieties were almost non-existent. On the other hand, this lignin had 35 β–1′ linked moieties, and 8 β–β′ linked moieties per 100 phenylpropane units (Crestini et al. 2011). ^{1}H–^{13}C 2D HSQC NMR spectrum of poplar milled-wood lignin, recorded in deuterated dimethyl sulfoxide solution, revealed usual structural moieties in lignin macromolecule. These were quantified per 100 phenylpropane units as 41.5 β–O–4′ linked moieties, 14.6 α–O–γ′, α′–O–γ, and β–β′ linked moieties, 3.7 α–O–4′ and β–5′ linked moieties, and 0.7 α–O–α′ and β–1′ linked moieties. The syringyl-to-guaiacyl ratio was 1.57. It also had the inclusion of a small amount of lignin carbohydrate linkage in the form of 2.1 α–O–C_{carb}, 4.1 4–O–C_{carb}, and 3.4 γ–O–(CO)–C_{carb} per 100 phenylpropane units (Yuan et al. 2011). Pine and birch milled-wood lignins were dissolved in deuterated dimethyl sulfoxide for recording ^{1}H–^{13}C 2D HSQC NMR spectra. Pine lignin was estimated to have 30.2 β–O–4′ linkage, 5.2 β–β′ linkage, and 10.4 β–5′ linkage per 100 phenylpropane units. It also had the inclusion of a small amount of lignin carbohydrate linkage in the form of 2.3 α–O–C_{carb} per 100 phenylpropane units. Birch lignin had 42 β–O–4′ linkage, 10.2 β–β′ linkage, and 2.7 β–5′ linkage per 100 phenylpropane units. Its syringyl-to-guaiacyl ratio was determined to be 2. This milled wood lignin also had lignin carbohydrate complex identified with 1.8 α–O–C_{carb} linkage, 0.6 4–O–C_{carb} linkage, and 5.2 γ–O–(CO)–C_{carb} linkage per 100 phenylpropane units (Balakshin et al. 2011). ^{1}H–^{13}C 2D HSQC NMR spectrum of bamboo milled-wood lignin was quantitatively analysed to find 72% β–O–4′ linked moieties, 15% α–O–γ′, α′–O–γ, and β–β′ linked moieties, 8% α–O–4′ and β–5′ linked moieties, 3% syringyl moieties with hydroxyl group at γ carbon, and 2% syringyl moieties with aldehyde group at γ carbon (Li et al. 2012). Another ^{1}H–^{13}C 2D HSQC NMR study of bamboo milled-wood lignin revealed 79.5% β–O–4′ linked moieties, 7.2% α–O–γ′, α′–O–γ, and β–β′ linked moieties, 5.6% α–O–4′ and β–5′ linked moieties, 4% α–O–α′ and β–1′ linked moieties, and 3.2% syringyl moieties with hydroxyl group at γ carbon. This lignin had 56% syringyl, 35% guaiacyl, and 9% p-hydroxyphenyl substructures. Cross peaks assignments conformed to Table 6.4 (Shi et al. 2019). The spectrum is shown in Figure 6.20.

Milled wood lignin from coconut coir, dissolved in deuterated dimethyl sulfoxide, was recorded for ^{1}H–^{13}C 2D HSQC NMR spectrum and subjected to semi-quantitative analysis. Cross-peak assignments broadly followed Table 6.4. Of all the intermonomer linkages 82% were of β–O–4′ type, 13% α–O–4′ and β–5′ type, 4% α–O–γ′, α′–O–γ, and β–β′ type, and 1% α–O–γ′, α′–O–γ, and β–β′ type. 18% of phenylpropane units conformed to syringyl, 78% to guaiacyl, and 4% to p-hydroxyphenyl (Rencoret et al. 2013). In ^{1}H–^{13}C 2D HSQC NMR spectrum of Arundo donax milled-wood lignin, different intermonomer linked molecular substructures were identified and quantified as 82% β–O–4′ linked, 9% α–O–γ′, α′–O–γ, and β–β′ linked, 7.7% α–O–4′ and β–5′ linked, 1% α–O–α′ and β–1′ linked, and 0.3% α–O–4′ and β–O–4′ linked. Syringyl, guaiacyl, and

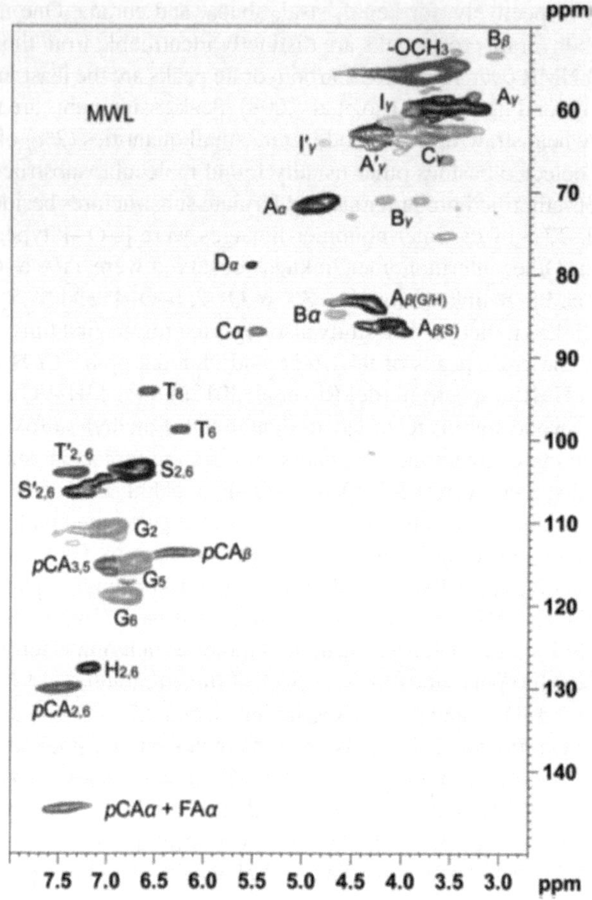

FIGURE 6.20 ^1H–^{13}C 2D HSQC NMR spectra of bamboo milled-wood lignin. (Reprinted from Shi, Z, Xu, G, Deng, J, Dong, M, Murugadoss, V, Liu, C, Shao, Q, Wu, S, Guo, Z, *Green Chem. Lett. Rev.*, 12(3), 235–243, 2019, originally published by Taylor and Francis under Creative Commons Attribution License.)

p-hydroxyphenyl moieties were 39%, 60%, and 1%, respectively. This lignin was also found to have covalent linkages with carbohydrates having 4.4 benzyl ether linkages, and 2.4 γ ester linkages per 100 phenylpropane units (You et al. 2015). Poplar milled-wood lignin was isolated from untreated wood as well as after pretreating the wood with green liquor of different alkali content. The resulting lignins were studied by ^1H–^{13}C 2D HSQC NMR spectroscopy. Untreated milled wood lignin had 0.55 β–O–4′ linkages per phenylpropane unit. This dropped to 0.5 and 0.41 when wood was pretreated with 8% and 16% alkali green liquor, respectively. β–β′ and β–5′ linkages in untreated lignin were reported as 0.09 and 0.03 per phenylpropane unit. β–β′ linkages per phenylpropane unit increased to 0.11 when treated with 8% alkali green liquor and to 0.12 when treated with 16% alkali green liquor. There was little change in β–5′ linkages. Green liquor treatment with 8% alkali increased the syringyl-to-guaiacyl ratio to 2.08 and further to 2.27 when treated with 16% alkali liquor, in comparison to 1.43 in untreated lignin (Tan et al. 2017).

6.4.4.4 Hydrolysis Lignin

In the ^1H–^{13}C 2D HSQC NMR spectrum of spruce enzymatic mild acidolysis lignin, recorded in deuterated dimethyl sulfoxide, cross peaks were quantitatively integrated to estimate the abundance of different molecular substructures. For every 100 phenylpropane units, there were 3.8 β–O–4′, α–O–4′, 5–5′ linked moieties, 2.6 β–1′ linked moieties, and 1.6 β–β′ linked moieties (Crestini et al. 2011). ^{1}H–^{13}C 2D HSQC NMR spectrum of poplar mild acidolysis lignin, recorded in deuterated dimethyl sulfoxide solution, revealed usual structural moieties in lignin macromolecule. These were quantified per 100 phenylpropane units as 43.3 β–O–4′ linked moieties, 12.7 α–O–γ′, α′–O–γ, and β–β′ linked moieties, and 4 α–O–4′ and β–5′ linked moieties. The syringyl-to-guaiacyl ratio was 1.62. It also had the inclusion of a small amount of lignin carbohydrate linkage in the form of 5.8 α–O–C_{carb}, 4.5 4–O–C_{carb}, and 1.3 γ–O–(CO)–C_{carb} per 100 phenylpropane units (Yuan et al. 2011). Pine and birch cellulolytic enzyme lignins were dissolved in deuterated dimethyl sulfoxide for recording ^{1}H–^{13}C 2D HSQC NMR spectra. Pine lignin was estimated to have 35.1 β–O–4′ linkage, 4.8 β–β′ linkage, and 10.7 β–5′ linkage per 100 phenylpropane units. It also had the inclusion of lignin carbohydrate linkage in the form of 5.8 α–O–C_{carb} linkage, 0.6 4–O–C_{carb}, and 2.2 γ–O–(CO)–C_{carb} linkage per 100 phenylpropane units. Birch lignin had 48.1 β–O–4′ linkage, 9.4 β–β′ linkage, and 3 β–5′ linkage per 100 phenylpropane units. Its syringyl-to-guaiacyl ratio was determined to be 3.2. This milled wood lignin also had lignin carbohydrate complex identified with 2.5 α–O–C_{carb} linkage, 0.8 4–O–C_{carb} linkage, and 3.5 γ–O–(CO)–C_{carb} linkage per 100 phenylpropane units (Balakshin et al. 2011). In ^{1}H–^{13}C 2D HSQC NMR spectrum of Arundo donax cellulolytic enzyme lignin, different intermonomer linked molecular substructures were identified and quantified as 86% β–O–4′ linked, 5.6% α–O–γ′, α′–O–γ, and β–β′ linked, 6.4% α–O–4′ and β–5′ linked, and 2% α–O–4′ and β–O–4′ linked. Syringyl, guaiacyl, and p-hydroxyphenyl moieties were 46%, 53%, and 1%, respectively. This lignin was also found to have covalent linkages with carbohydrates having 0.34 benzyl ether linkages per 100 phenylpropane units (You et al. 2015). Deuterated dimethyl sulfoxide solution state ^{1}H–^{13}C 2D HSQC NMR characterization of cellulolytic enzyme lignin from hydrothermal pretreated eucalyptus is reported. Quantitative estimation revealed β–O–4′ linked moieties to be 55.7 per 100 phenylpropane units with no hydrothermal pretreatment. It dropped to 40.4 and 7.1 per 100 phenylpropane units with hydrothermal pretreatment at 140 and 180 °C, respectively. Other molecular structural features were qualitatively assigned to different cross peaks in the usual manner (Shen et al. 2016).

6.4.4.5 Lignin Carbohydrate Complex

Three types are lignin carbohydrate linkages are known to exist in lignin carbohydrate complexes. These are α–O–C_{carb} (benzyl ether), γ–O–(CO)–C_{carb} (γ ester), and 4–O–C_{carb} (phenyl glycoside). ^{1}H–^{13}C 2D HSQC NMR spectroscopy is a reliable characterization method for these linkages and is widely practised. The cross peak at 75–77/6–6.2 ppm $^{13}C/^{1}H$ chemical shift is attributed to benzyl ether linkage. The cross peak at 62–65/4–4.5 ppm $^{13}C/^{1}H$ chemical shift is attributed to γ ester linkage. Similarly, cross peak at 101.4–102.6/4.94–5.17 ppm $^{13}C/^{1}H$ chemical shift is attributed to phenyl glycoside linkage (Balakshin et al. 2007). Lignin carbohydrate complexes were isolated from pine and birch milled-wood lignins. These were dissolved in deuterated dimethyl sulfoxide for recording ^{1}H–^{13}C 2D HSQC NMR spectra. The pine lignin carbohydrate complex was found to have 4.3 benzyl ether linkage, 6.8 phenyl glycoside linkage, and 4.5 γ ester linkage per 100 phenylpropane units. The lignin backbone was estimated to have 22.4 β–O–4′ linkage, 1.7 β–β′ linkage, and 5.1 β–5′ linkage per 100 phenylpropane units. Birch lignin carbohydrate complex had 0.7 benzyl ether linkage, 5.4 phenyl glycoside linkage, and 5.6 γ ester linkage per 100 phenylpropane units. The lignin backbone had 31.7 β–O–4′ linkage, 4.6 β–β′ linkage, and 1.2 β–5′ linkage per 100 phenylpropane units (Balakshin et al. 2011). In milled wood lignins from several hybrid poplar species, lignin carbohydrate linkages were observed and quantified by solution state ^{1}H–^{13}C 2D HSQC NMR spectroscopy in deuterated dimethyl sulfoxide. These linkages varied between 2.22 and 4.24 per 100 phenylpropane units. Of these, 0.18–1.71 were of benzyl ether type, 0.5–1.77 of phenyl glycoside type, and 0.66–1.39 of γ ester type (Min et al. 2014). In another study, lignin carbohydrate complex was isolated from milled bamboo kraft pulp as well as from milled bamboo. They were characterized by solution state ^{1}H–^{13}C 2D HSQC NMR spectroscopy in deuterated dimethyl sulfoxide. The abundance of different lignin carbohydrate linkages in lignin carbohydrate complex from kraft pulp was quantified as 1.84 benzyl ether linkage, 2.78 phenyl glycoside linkage, and 4.01 γ ester linkage per 100 phenylpropane units. Linkages in lignin carbohydrate complex originating from milled bamboo were quantified as 3.11 benzyl ether linkage, 3.95 phenyl glycoside linkage, and 6.42 γ ester linkage per 100 phenylpropane units (Huang et al. 2015). Lignin carbohydrate complex isolated from milled Arundo donax was characterized by ^{1}H–^{13}C 2D HSQC NMR spectroscopy after dissolving the sample in deuterated dimethyl sulfoxide. In this study, cross-peak assignments of 80–81.5/4.3–5.3 ppm $^{13}C/^{1}H$ chemical shift for benzyl ether linkage was different from other reports. The isolated lignin carbohydrate complex had 4 benzyl ether linkage and 0.14 γ ester linkage per 100 phenylpropane units (You et al. 2015). 99.5–105/4.28–5.17 ppm $^{13}C/^{1}H$ chemical shift for phenyl glycoside linkage.

REFERENCES

Agarwal, UP, McSweeny, JD, Ralph, SA. *J. Wood Chem. Technol.* 31:324–344, 2011.

Agarwal, UP, Ralph, SA, Padmakshan, D, Liu, S, Foster, CE. *J. Agric. Food Chem.* 67:4367–4374, 2019.

Agarwal, UP, Reiner, RS. *J. Raman Spectroscopy* 40:1527–1534, 2009.

Al-Aasmi, ZH, Shchukina, A, Butts, CP. *Chem. Commun.* 58:7781–7784, 2022.

Al-Rudainy, B, Galbe, M, Wallberg, O. *Sep. Purific. Technol.* 187:380–388, 2017.

Alexandri, M, Papapostolou, H, Komaitis, M, Stragier, L, Verstraete, W, Danezis, GP, Georgiou, CA, Papanikolaou, S, Koutinas, AA. *Bioresource Technol.* 214:504–513, 2016.

Angles, MN, Reguant, J, Garcia-Valls, R, Salvado, J. *Wood Sci. Technol.* 37:309–320, 2003.

Antonino, LD, Gouveia, JR, de Sousa Jr., RR, Garcia, GES, Gobbo, LC, Tavares, LB, dos Santos, DJ. *Molecules* 26:2131, 2021.

Arkell, A, Olsson, J, Wallberg, O. *Chem. Eng. Res. Des.* 92:1792–1800, 2014.

Asikanius, B, Jääskeläinen, AS, Koivula, H, Oinonen, P, Österberg, M. *Frontiers En. Res.* 9:782545, 2021.

Audu, IG, Brosse, N, Winter, H, Hoffmann, A, Bremer, M, Fischer, S, Laborie, MP. *Polymers* 10:619, 2018.

Bagh, FSG, Ray, S. *J. Ionic Liquids* 2:100028, 2022.

Balakshin, M, Capanema, E, Gracz, H, Chang, HM, Jameel, H. *Planta* 233:1097–1110, 2011.

Balakshin, MY, Capanema, EA, Chang, HM. Holzforschung 61:1–7, 2007.

Bhattacharya, PK, Todi, RK, Tiwari, M, Bhattacharjee, C, Bhattacharjee, S, Datta, S. *Desalination* 174: 287–297, 2005.

Bhattacharya, A, Butler, S, Al-Rudainy, B, Wallberg, O, Stålbrand, H. *Molecules* 27:3207, 2022.

Boeriu, CG, Bravo, D, Gosselink, RJA, van Dam, JEG. *Ind. Crops Prod.* 20:205–218, 2004.

Bose, S, Chakrabarti, SK, Sarma, US. *Proc. Natl. Acad. Sci., India, Sect. A Phys. Sci.* 93:729–735, 2023.

Bu, L, Tang, Y, Gao, Y, Jian, H, Jiang, J. *Chem. Eng. J.* 175:176–184, 2011.

Chazal, R, Robert, P, Durand, S, Devaux, MF, Saulnier, L, Lapierre, C, Guillon, F. *Appl. Spectroscopy* 68: 1342–1347, 2014.

Chen, TY, Wang, B, Shen, XJ, Li, HY, Wu, YY, Wen, JL, Liu, QY, Sun, RC. *RSC Adv.* 7:3376–3387, 2017.

Chen, K, Cao, M, Ding, C, Zheng, X. *RSC Adv.* 8:26782–26792, 2018.

Chen, S, Cheng, H, Wu, S. *Fuel* 270:117520, 2020.

Crestini, C, Melone, F, Sette, M, Saladino, R. *Biomacromolecules* 12:3928–3935, 2011.

Cybulska, I, Brudecki, G, Rosentrater, K, Julson, JL, Lei, H. *Bioresource Technol.* 118:30–36, 2012.

de Carvalho, DM, Lahtinen, MH, Lawoko, M, Mikkonen, KS. *ACS Sustainable Chem. Eng.* 8: 11795–11804, 2020.

del Rio, JC, Gutierrez, A, Rodríguez, IM, Ibarra, D, Martinez, AT. *J. Anal. Appl. Pyrolysis* 79:39–46, 2007.

del Rio, JC, Rencoret, J, Marques, G, Gutierrez, A, Ibarra, D, Santos, JI, Jimenez-Barbero, J, Zhang, L, Martinez, AT. *J. Agric. Food Chem.* 56:9525–9534, 2008.

del Rio, JC, Prinsen, P, Rencoret, J, Nieto, L, Jimenez-Barbero, J, Ralph, J, Martinez, AT, Gutierrez, A. *J. Agric. Food Chem.* 60:3619–3634, 2012.

del Rio, JC, Rencoret, J, Prinsen, P, Martinez, AT, Ralph, J, Gutierrez, A. *J. Agric. Food Chem.* 60: 5922–5935, 2012a.

Demmelmayer, P, Hilgert, J, Wijaya, R, Kienberger, M. *Ind. Eng. Chem. Res.* 59:16420–16426, 2020.

Domínguez-Robles, J, Sánchez, R, Espinosa, E, Savy, D, Mazzei, P, Piccolo, A, Rodríguez, A. *Int. J. Mol. Sci.* 18:327–343, 2017.

Ela, RCA, Spahn, L, Safaie, N, Ferrier Jr., RC, Ong, RG. ACS Sustainable *Chem. Eng.* 8:13997–14005, 2020.

El Hage, R, Brosse, N, Chrusciel, L, Sanchez, C, Sannigrahi, P, Ragauskas, A. *Polymer Deg. Stab.* 94: 1632–1638, 2009.
El Mansouri, NE, Salvado, J. *Ind. Crops Prod.* 26:116–124, 2007.
Esakkimuthu, ES, Marlin N, Brochier-Salon, MC, Mortha, G. *AppliedChem* 2:30–47, 2022.
Evtuguin, DV, Neto, CP, Silva, AMS, Domingues, PM, Amado, FML, Robert, D, Faix, O. *J. Agric. Food Chem.* 49:4252–4261, 2001.
Faix, O, Grünwald, C, Beinhoff, O. *Holzforschung* 46:425–432, 1992.
Faix, O, Argyropoulos, DS, Robert, D, Neirinck, V. *Holzforschung* 48:387–394, 1994.
Fernandes, DLA, Silva, CM, Xavier, AMRB, Evtuguin, DV. *J. Biotechnol.* 162:415–421, 2012.
Fernando, EF, Vallejos, EM, Area, MC. Cellulose *Chem. Technol.* 44:311–318, 2010.
Fujimoto, A, Matsumoto, Y, Chang, HM, Meshitsuka, G. *J. Wood. Sci.* 51:89–91, 2005.
Garcia, A, Alriols, MG, Labidi, J. *Bioresource Technol.* 108:155–161, 2012.
Gärtner, A, Gellerstedt, G, Tamminen, T. *Nordic Pulp Paper Res. J.* 14:163–170, 1999.
Ghatak, HR. *Ind. Crops Prod.* 28:206–212, 2008.
Ghatak, HR, Kundu, PP, Kumar, S. *Thermochim. Acta* 502:85–89, 2010.
Gidh, AV, Decker, SR, See, CH, Himmel, ME, Williford, CW. *Analytica Chimica Acta* 555:250–258, 2006.
Giummarella, N, Zhang, L, Henriksson, G, Lawoko, M. *RSC Adv.* 6:42120–42131, 2016.
Giummarella, N, Lawoko, M. *ACS Sustain. Chem. Eng.* 4:5319–5326, 2016.
Goldmann, WM, Ahola, J, Mankinen, O, Kantola, AM, Komulainen, S, Telkki, VV, Tanskanen, J. Periodica Polytechnica *Chem. Eng.* 61(2):93–101, 2017.
Gosselink, RJA, Snijder, MHB, Kranenbarg, A, Keijsers, ERP, de Jong, E, Stigsson, LL. *Ind. Crops Prod.* 19:271–281, 2004.
Granata, A, Argyropoulos, DS. *J. Agric. Food Chem.* 43:1538–1544, 1995.
Guerra, A, Filpponen, I, Lucia, LA, Saquing, C, Baumberger, S, Argyropoulos, DS. *J. Agric. Food Chem.* 54:5939–5947, 2006.
Haddad, M, Labrecque, R, Bazinet, L, Savadogo, O, Paris, J. *Chem Eng. J.* 304:977–985, 2016.
Holtman, KM, Chang, HM, Jameel, H, Kadla, JF. *J. Agric. Food Chem.* 51:3535–3540, 2003.
Holtman, KM, Chang, HM, Kadla, JF. *J. Agric. Food Chem.* 52:720–726, 2004.
Hu, Z, Yeh, TF, Chang, HM, Matsumoto, Y, Kadla, JF. *Holzforschung* 60:389–397, 2006.
Huang, F, Singh, PM, Ragauskas, AJ. *J. Agric. Food Chem.* 59:12910–12916, 2011.
Huang, C, He, J, Li, X, Min, D, Yong, Q. *Bioresource Technol.* 192:471–477, 2015.
Jablonsky, M, Kocis, J, Haz, A, Sima, J. *Cellulose Chem. Technol.* 49(3-4):267–274, 2015.
Jiang, B, Cao, T, Gu, F, Wu, W, Jin, Y. *ACS Sustainable Chem. Eng.* 5:342–349, 2017.
Jönsson, AS, Nordin, AK, Wallberg, O. *Chem. Eng. Res. Des.* 86:1271–1280, 2008.
Karimov, OK, Teptereva, GA, Chetvertneva, IA, Movsumzade, EM, Karimov, EK. *IOP Conf. Series: Earth Environ. Sci.* 839:022086, 2021.
Kienberger, M, Demmelmayer, P, Weißl, M, Zankl, A, Spirk, S. *Solv. Extract. Ion Exchange* 38(1): 132–141, 2020.
Kumar, H, Alen, R, Sahoo, G. *BioResources* 11:9869–9879, 2016.
Kumar, V, Singh, RK, Chowdhury, P. *J. Ind. Eng. Chem.* 67:109–122, 2018.
Lahdetie, A, Nousiainen, P, Sipilä, J, Tamminen, T, Jääskeläinen, AS. *Holzforschung* 67:531–538, 2013.
Lauwaert, J, Stals, I, Lancefield, CS, Deschaumes, W, Depuydt, D, Vanlerberghe, B, Devlamynck, T, Bruijnincx, PCA, Verberckmoes, A. *Sep. Purific. Technol.* 221:226–235, 2019.
Li, MF, Sun, SN, Xu, F, Sun, RC. *Chem. Eng. J.* 179:80–89, 2012.
Li, K, Kollberg, L, Almqvist, H, Xu, B, Hulteberg, C. *Ind. Crops Prod.* 169:113657, 2021.
Li, X, Wu, S. *BioResources* 9:6277–6289, 2014.
Livča, S, Verovkins, A, Shulga, G, Neiberte, B, Vitolina, S. *Latvian J. Chem.* 51(4):421–427, 2012.
Lou, R, Lyu, G, Wu, S, Zhang, B, Zhao, H, Lucia, LA. *ACS Sustainable Chem. Eng.* 6:430–437, 2018.
Lourençon, TV, Hansel, FA, da Silva, TA, Ramos, LP, de Muniz, GIB, Magalhães, WLE. *Sep. Purific. Technol.* 154:82–88, 2015.
Ma, Q, Wang, L, Zhai, H, Ren, H. *Int. J. Biological Macromolecule* 182:51–58, 2021.
Manttari, M, Lahti, J, Hatakka, H, Louhi-Kultanen, M, Kallioinen, M. *J. Membrane Sci.* 490:84–91, 2015.
Marques, AP, Evtuguin, DV, Magina, S, Amado, FML, Prates, A. *J. Wood Chem. Technol.* 29:322–336, 2009.
Mendes, SF, Rodrigues, JS, de Lima, VH, Botaro, VR, Cardoso, VL, Reis, MHM. *Appl. Biochem. Biotechnol.* 193:3079–3097, 2021.
Min, DY, Yang, C, Chiang, V, Jameel, H, Chang, HM. *Fuel* 116:56–62, 2014.
Mohammadpour, R, Sadeghi, GMM. *J. Polymer Res.* 27:362, 2020.

Moniz, P, Serralheiro, C, Matos, CT, Boeriu, CG, Frissen, AE, Duarte, LC, Roseiro, LB, Pereira, H, Carvalheiro, F. *Process Biochem.* 65:136–145, 2018.
Mousavioun, P, Doherty, WOS. *Ind. Crops Prod.* 31:52–58, 2010.
Mun, JS, Pe III, JA, Mun, SP. *Molecules* 26:4861, 2021.
Nagy, M, Kosa, M, Theliander, H, Ragauskas, AJ. *Green Chem.* 12:31–34, 2010.
Nitsos, C, Stoklosa, R, Karnaouri, A, Vörös, D, Lange, H, Hodge, DB, Crestini, C, Rova, USL, Christakopoulos, P. *ACS Sustainable Chem. Eng.* 4:5181–5193, 2016.
Nitsos, CK, Lazaridis, PA, Mach-Aigner, A, Matis, KA. *ChemSusChem* 12:1179–1195, 2019.
Norgren, M, Lindstrom, B. *Holzforschung* 54:519–527, 2000.
Norgren, M, Mackin , S. *Ind. Eng. Chem. Res.* 48:5098–5104, 2009.
Norgren, M, Edlund, H, Wågberg, L. *Langmuir* 18:2859–2865, 2002.
Öhman, F, Wallmo, H, Theliander, H. *Nord. Pulp Pap. Res. J.* 22:188–193, 2007.
Oliveira, L, Evtuguin, DV, Cordeiro, N, Silvestre, AJD, Silva, AMS, Torres, IC. *J. Agric. Food Chem.* 54: 2598–2605, 2006.
Parot, M, Rodrigue, D, Stevanovic, T. *Bioresource Technol. Reports* 17:100880, 2022.
Perera, PN, Schmidt, M, Chiang, VL, Schuck, PJ, Adams, PD. *Anal. Bioanal. Chem.* 402:983–987, 2012.
Pichler, T, Maitz, S, Kienberger, M. *Holzforschung* 74:286–292, 2019.
Reddy, IAK, Ghatak, HR. *J. Thermal Anal. Calorimetry* 132:407–423, 2018.
Rencoret, J, Ralph, J, Marques, G, Gutiérrez, A, Martínez, T, del Rio, JC. *J. Agric. Food Chem.* 61: 2434–2445, 2013.
Rönnols, J, Jacobs, A, Aldaeus, F. *Holzforschung* 71:563–571, 2017.
Rossberg, C, Janzon, R, Saake, B, Leschinsky, M. *BioResources* 14:4543–4559, 2019.
Rueda, C, Calvo, PA, Moncalián, G, Ruiz, G, Coz, A. *J. Chem. Technol. Biotechnol.* 90:2218–2226, 2014.
Sameni, J, Krigstin, S, Sain, M. *BioResources* 11:8435–8456, 2016.
Schlackl, K, Bischof, RH, Fackler, K, Samhaber, W. *Sep. Purific. Technol.* 250:117177, 2020.
Serrano, L, Esakkimuthu, ES, Marlin, M, Brochier-Salon, MC, Mortha, G, Bertaud, F. *Ener. Fuels* 32: 5969–5977, 2018.
Shen, XJ, Wang, B, Huang, PL, Wen, JL, Sun, RC. *Bioresource Technol.* 206:57–64, 2016.
Sheng, Y, Ma, Z, Wang, X, Han, Y. *Food Chem.* 376:131895, 2022.
Shi, Z, Xu, G, Deng, J, Dong, M, Murugadoss, V, Liu, C, Shao, Q, Wu, S, Guo, Z. *Green Chem. Lett. Rev.* 12:235–243, 2019.
Singh, R, Singh, S, Trimukhe, KD, Pandare, KV, Bastawade, KB, Gokhale, DV, Varma, AJ. *Carbohydrate Polymers* 62:57–66, 2005.
Singh, SP, Jawaid, M, Yadav, B, Sarmin, SN. *Biomass Conv. Biorefinery* 13:10865–10875, 2023.
Sumerskii, I, Korntner, P, Zinovyev, G, Rosenau, T, Potthast, A. *RSC Adv.* 5:92732–92742, 2015.
Sun, RC, Lawther, JM, Banks, WB. *Wood Fiber Sci.* 30(1):56–63, 1998.
Tan, X, Jiang, B, Yang, Y, Min, D, Jin, Y. *Holzforschung* 71:99–108, 2017.
Thoresen, PP, Lange, H, Crestini, C, Rova, U, Matsakas, L, Christakopoulos, P. *ACS Omega* 6:4374–4385, 2021.
Thring, RW, Chornet, E. *Biomass* 23:289–305, 1990.
Tiainen, E, Drakenberg, T, Tamminen, T, Kataja, K, Hase, A. *Holzforschung* 53:529–533, 1999.
Wang, S, Lin, H, Ru, B, Sun, W, Wang, Y, Luo, Z. *J. Anal. Appl. Pyrolysis* 108:78–85, 2014.
Wang, G, Pang, T, Xia, Y, Liu, X, Li, S, Parvez, AM, Kong, F, Si, C. *Int. J. Biological Macromolecules* 133: 156–164, 2019.
Weinwurm, F, Drljo, A, Waldmüller, W, Fiala, B, Niedermayer, J, Friedl, A. *J. Cleaner Prod.* 136:62–71, 2016.
Xie, Y, Yasuda, S, Wu, H, Liu, H. *J. Wood Sci.* 46:130–136, 2000.
Xu, H, Yu, G, Mu, X, Zhang, C, DeRoussel, P, Liu, C, Li, B, Wang, H. *Ind. Crops Prod.* 76:638–646, 2015.
Yanez-S, M, Matsuhiro, B, Nunez, C, Pan, S, Hubbell, CA, Sannigrahi, P, Ragauskas, AJ. *Polymer Deg. Stab.* 110:184–194, 2014.
Yang, Q, Wu, S, Lou, R, Lv, G. *Wood Sci. Technol.* 45:419–431, 2011.
Yuan, TQ, Sun, SN, Xu, F, Sun, RC. *J. Agric. Food Chem.* 59:10604–10614, 2011.
You, TT, Zhang, LM, Zhou, SK, Xu, F. *Ind. Crops Prod.* 71:65–74, 2015.
Zhang, W, Tian, G, Polle, A, Janz, D, Euring, D, Yue, X, Zhao, H, Fei, B, Jiang, Z. *Wood Sci. Technol.* 52: 1331–1341, 2018.
Zhang, X. *RSC Adv.* 11:13124, 2021.
Zhao, X, Liu, D. *Ind. Crops Prod.* 32:284–291, 2010.
Zhao, J, Xiuwen, W, Hu, J, Liu, Q, Shen, D, Xiao, R. *Polymer Deg. Stab.* 108:133–138, 2014.
Zhu, W, Theliander, H. *BioResources* 10:1696–1714, 2015.

7 Thermochemical Characterization of Lignin

In many potential value-added applications of lignin, it is subjected to treatment at elevated temperatures in a controlled environment. As a result, it undergoes thermochemical transformations. For its desirable end use, it is important to examine and understand these thermochemical transformations from a kinetic, mechanistic, and equilibrium standpoint. Analysis of thermochemical degradation products could shed light on the molecular structure of lignin macromolecules. Besides thermochemical transformations, some thermophysical transformations that alter the physical and viscoelastic properties of lignin are also important. Standard analytical techniques based on thermochemical and thermophysical transformations are thermogravimetry (TG), differential thermal analysis (DTA), differential scanning calorimetry (DSC), and differential mechanical analysis (DMA). Pyrolysis-gas chromatography-mass spectroscopy is an advanced characterization technique that combines thermochemical degradation of lignin macromolecule with chromatographic separation of degradation products followed by mass spectroscopic identification and quantification of these products. This is a potent tool in the molecular structure elucidation of lignin and often complements other characterization techniques described in previous chapters.

7.1 LIGNIN CHARACTERIZATION BY THERMOGRAVIMETRY

Thermogravimetry is an analytical technique where a known mass – normally a few mg only – of the sample is placed in a controlled gaseous environment and heated with a preset temperature protocol. The mass of the sample is continuously monitored with a microbalance as it undergoes thermochemical transformation. The gaseous environment is controlled with respect to its composition and flow rate. In the majority of cases, an inert gas like argon or nitrogen is used. However, other gaseous environments like carbon dioxide, oxygen, or other gaseous mixtures are also used in certain procedures. The temperature protocol in most cases is a constant temperature ramp – a constant rate of temperature increase – to the final temperature. Modifications are also sometimes used like one ramp to an intermediate temperature and then a second ramp to the final temperature. Alternatively, an isothermal period may also be included between the two temperature ramps. The result is obtained as the changing sample mass, often expressed as a percentage of the initial mass, plotted against the corresponding temperature. Such a plot is commonly known as the thermogravimetry (TG) plot. Combined with the rate of temperature increase it is possible to simultaneously analyse the effect of reaction temperature and reaction time on thermochemical transformation. It is usual to also plot the first derivative – the rate of change of mass with respect to time – of the thermogravimetry data against temperature to analyse the changes in reaction rates with temperature. Such a plot is referred to as the differential thermogravimetry (DTG) plot. Important lignin characteristics inferred from TG and DTG analyses are onset temperature for thermal degradation, temperature when the rate of mass loss is maximum (T_{max}), the maximum rate of mass loss, and the amount of residual char. Lignin with a low T_{max} and a high maximum rate of mass loss would be easier to degrade and vice versa. The shape of the DTG curve also reveals important information. A curve with more than one peak – often a small but distinct peak or may be a shoulder – indicates the chemical heterogeneity of lignin. Lignin with a high amount of char residue may be desired for certain value-added applications. TG characterization can be correlated with other lignin characteristics like elemental analysis and abundance of different inter monomer linkages. It is also possible

DOI: 10.1201/9781003598664-9

to combine thermogravimetry with spectroscopic techniques like FTIR, or chromatography-mass spectroscopy to analyse the gases and volatiles evolved during the thermochemical treatment.

7.1.1 Soda Lignin

TG analysis of bagasse soda lignin at 10 °C/min in nitrogen is reported. Major thermal degradation started at 289–299 °C with a T_{max} of 390–394 °C. 58–59% material was left as char at 590 °C (Singh et al. 2005). TG thermograms of different fractions of bagasse soda lignins were recorded in 15 mL/min nitrogen flow at 10 °C/min heating rate. This lignin exhibited the onset of thermal degradation at 175 °C and T_{max} of 367 °C, leaving a final char residue of 33% at 900 °C (Mousavioun and Doherty 2010). Wheat straw and bagasse soda lignins were characterized by TG in 100 mL/min nitrogen flow and 10 °C/min heating rate. The onset of thermal degradation occurred at 300 °C. T_{max} for wheat straw lignin was 357 °C with a maximum rate of mass loss recoded as 271.7 μg/min. The values were 361 °C and 286.1 μg/min, respectively. At the end of heating at 800 °C, residual char was 43% from wheat straw lignin and 44% from bagasse lignin (Ghatak et al. 2010). One study reported TG characterization of commercial soda lignin Protobind, obtained from a mixture of wheat straw and sarkanda. Thermograms were obtained in 30 mL/min nitrogen flow at different heating rates between 2 and 200 °C/min. About 30% char was obtained when heated to 900 °C at a heating rate of 200 °C/min. Also, T_{max} shifted to higher temperatures with an increase in heating rate (Jiang et al. 2010). TG characterization of commercial nonwood soda lignin was carried out at 10 °C/min heating rate and 20 mL/min flow of argon. Thermal degradation started at about 200 °C, rate of degradation peaked at 365 °C to moderate thereafter up to about 600 °C. A shoulder was also observed in the DTG plot at about 290 °C, indicating that there might be two stages of degradation. After heating to 1000 °C, 37% char residue was left (Sameni et al. 2014). TG characterization of *Cunninghamia lanceolata* soda lignins, isolated with different precipitation pH of 8, 5, and 2, was performed in nitrogen atmosphere at 20 °C/min. Lignin isolated at the lowest pH exhibited the earliest onset of thermal degradation at 104.3 °C with the rate of mass loss peaking at 379.1 °C. The onset temperature and T_{max} were 109.1 and 381.7 °C for lignin precipitated at pH 5. These were 117.4 and 382.6 °C for lignin precipitated at pH 8. The three lignins resulted in 35.5, 36.1 and 45.1% char residue, respectively, at 700 °C (Li and Wu 2014). TG characterization of commercial alkali lignin at 10 °C/min heating rate and 120 mL/min nitrogen flow is reported. Main thermal degradation occurred between 210 and 570 °C with T_{max} around 380 °C. At 900 °C, 45.7% char residue was obtained (Yang et al. 2007). Another TG study of this alkali lignin at 20 °C/min heating rate and 70 mL/min nitrogen flow is reported. After the initial expulsion of moisture, rapid mass loss was observed between 108 and 466 °C with a T_{max} of 327 °C and the maximum rate of mass loss being 0.24%/°C. Beyond 466 °C the mass loss was gradual leaving 45.29% char residue at 800 °C (Ma et al. 2016). The TG analysis of wheat straw soda lignin was carried out at 10 °C/min in nitrogen flow. Two distinct thermal degradation stages were noticed occurring between 160 and 270 °C and then again between 330 and 460 °C with T_{max} at 232 and 490 °C. 57.6% residue was obtained at 650 °C (Lin et al. 2015). Soda lignins from nonwood species – bamboo, *Arundo donax* (giant cane), and pearl millet – were analysed by TG for their thermochemical degradation behaviour in 200 mL/min nitrogen flow rate up to 800 °C. Experiments were performed at heating rates of 10 and 100 °C/min. After the initial moisture loss, lignins were observed to decompose in two distinct stages as revealed by the bimodal DTG plots. These plots for 10 °C/min are shown in Figure 7.1. For bamboo lignin at 10 °C/min, the first decomposition stage started at 261 °C and the rate of mass loss peaked at 345 °C (T_{max}). The second decomposition stage started at 423 °C with the rate of mass loss peaking at 500 °C. At 100 °C/min, the onset temperatures shifted to 293 and 490 °C, respectively. For pearl millet lignin at 10 °C/min, the first decomposition stage started at 247 °C and the rate of mass loss peaked at 322 °C. The second decomposition stage started at 396 °C with the rate of mass loss peaking at 547 °C. At 100 °C/min, the onset temperatures shifted to 293 and 490 °C, respectively. For giant cane lignin at 10 °C/min, the first decomposition stage started at 219 °C and the rate

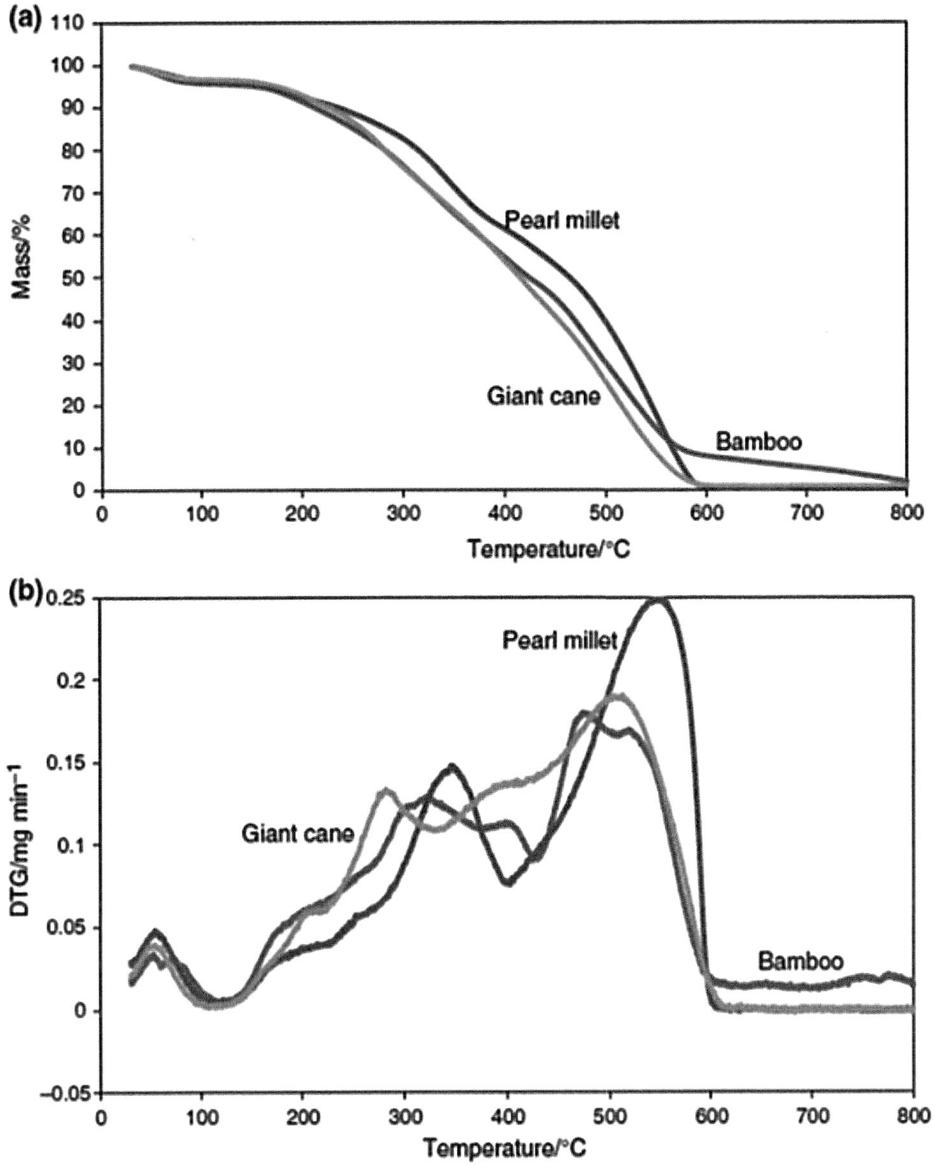

FIGURE 7.1 TG analyses of nonwood soda lignins at 10 °C/min (a) TG (b) DTG. (Reprinted from Reddy, IAK, Ghatak, HR, *J. Thermal. Anal. Calorim.*, 132, 407–423, 2018, with permission from Springer.)

of mass loss peaked at 282 °C. The second decomposition stage started at 417 °C with the rate of mass loss peaking at 504 °C. At 100 °C/min, the onset temperatures shifted to 258 and 490 °C, respectively (Reddy and Ghatak 2018).

TG thermogram of palm kernel shell aqueous soda lignin was recorded in nitrogen flow rate of 45 mL/min and heating rates of 10, 20, 30, and 40 °C/min. This lignin underwent pyrolysis in two different phases – the first one accounting for major mass loss followed by a second phase with lesser but noticeable mass loss. The temperature ranges for these phases changed with the heating rate. At 10 °C/min, the first phase occurred between 193 and 312 °C (T_{max} 270 °C, maximum rate 6.41%/min) and the second phase between 313 and 416 °C (T_{max} 361 °C, maximum rate 1.29%/min). Temperature range, T_{max}, and maximum rate progressively increased with increasing heating rate. At 40 °C/min, the first phase occurred between 221 and 336 °C (T_{max} 286 °C, maximum rate 26.42%/min) and the second phase between 337 and 466 °C (T_{max} 380 °C, maximum rate 4.98%/min) (Ma et al. 2018).

7.1.2 Kraft Lignin

Softwood kraft lignin was characterized by TG at a heating rate of 20 °C/min and nitrogen flow of 200 mL/min. Lignin was pyrolysed in a broad temperature range of 200–450 °C with T_{max} around 370 °C. It left 59% char residue at 500 °C (Hatakeyama et al. 2010). TG characterization of pine kraft lignin was reported in 20 mL/min flowing nitrogen at 10 °C/min. Pyrolysis can be divided into three phases. Initially, two modest volatilization phases occurred with DTG shoulders at 180 and 290 °C. The third and major pyrolytic phase occurred between 300 and 500 °C with T_{max} at 417.5 °C (Chen et al. 2014). Spruce LignoBoost kraft lignin was TG-characterized. In a nitrogen environment, lignin was heated at 10 °C/min. The main mass loss occurred between 200 and 500 °C with T_{max} 387 °C. Char residue at 800 °C was 40.2% (Han et al. 2019). The TG analysis of softwood kraft lignin (Indulin) at 10 °C/min in nitrogen is reported. Major thermal degradation started at 236 °C with a T_{max} of 393 °C. 45.7% material was left as char at 590 °C (Singh et al. 2005). The TG analysis of the pyrolytic behaviour of kraft lignin from softwood and hardwood mixture was carried out in nitrogen flowing at 60 mL/min with heating rates of 20 and 40 °C/min. The observed T_{max} was 376 °C with a decomposition rate of 0.29%/°C at 20 °C/min heating. This increased to 390 °C and 0.31%/°C at 40 °C/min. Char residues were 38.49 and 36.39% at 900 °C, respectively, at the two heating rates (Shen et al. 2010). TG characterization of softwood (commercial) and hardwood kraft lignins was carried out at 10 °C/min heating rate and 20 mL/min flow of argon. In softwood lignin, thermal degradation started at about 200 °C, rate of degradation peaked at 380 °C to moderate thereafter up to about 600 °C. After heating to 1000 °C, 45% char residue was left. Degradation behaviour was markedly different for hardwood lignin. Two very distinct thermal degradation stages were observed. The first stage started at about 200 °C, rate of degradation peaked at 245 °C to moderate thereafter up to about 600 °C. In the first stage of degradation, two shoulders are also noted in the DTG plot one at about 340 °C and another at about 480 °C. The second thermal degradation stage started at 640 °C, and the rate of decomposition peaked at 850 °C, with a shoulder also observed at about 685 °C. After heating to 1000 °C, 24% char residue was left (Sameni et al. 2014). Yet another study reports TG characterization of Indulin AT at 10 °C/min heating rate and 20 mL/min flow of nitrogen. Major thermal degradation was observed between 200 and 700 °C with T_{max} at 395 °C. A feeble shoulder appeared at 347 °C in the DTG plot. Char residue was 40.81% at 900 °C (Amit et al. 2021). TG characterization of commercial kraft lignin, performed in 10 mL/min nitrogen flow at 15 °C/min, revealed major mass loss due to thermal degradation to occur between 250 and 500 °C. T_{max} was observed at 340 °C and a shoulder near 220 °C attributed to the breaking of ether bonds. 50% char was left at 750 °C (El Moustaqim et al. 2018). Eucalyptus kraft lignin was characterized by TG in nitrogen atmosphere at a heating rate of 20 °C/min. T_{max} was observed at 370 °C and it left 40.61% char residue at 700 °C (Chen et al. 2020). A recent study reported TG analysis of mixed hardwood (acacia and Korean hardwoods) kraft lignin. Thermograms were recorded in an inert nitrogen atmosphere with a heating rate of 10 °C/min. The onset of thermal degradation was observed at 288 °C with a maximum mass loss rate occurring at T_{max} of 358 °C. A minor mass loss peak was also noticed near 250 °C which was attributed to the initial breakdown of intermonomer linkages. The amount of char was 34.7% at 800 °C (Pe III et al. 2023).

7.1.3 Organosolv Lignin

TG analysis of bagasse aqueous ethanol organosolv lignin at 10 °C/min in nitrogen is reported. Major thermal degradation started at 277 °C with a T_{max} of 389 °C. 55.6% material was left as char at 590 °C (Singh et al. 2005). Alcell lignin, a commercial aqueous ethanol organosolv lignin, was characterized by TG. The thermogram was recorded in an argon environment at a 15 °C/min heating rate. The onset of major weight loss was observed at 247 °C. Negligible residue was left after 260 °C (Murugan et al. 2008). Another study reported TG characterization of Alcell lignin. Thermograms were obtained in 30 mL/min nitrogen flow at different heating rates between 2 and 200 °C/min. About 30% char was obtained when heated to 900 °C at a heating rate of 200 °C/min. Also, T_{max} shifted to higher temperatures with an increase in heating rate (Jiang et al. 2010). Birch aqueous ethanol organosolv lignins, isolated with different delignification severities, were subjected to TG characterization. Major thermal degradation occurred between 200 and 500 °C with T_{max} lying between 310 and 360 °C. This was correlated to the breaking of ether linkages in the lignin macromolecules. In some of the lignins, a second stage of thermal degradation was also noticed in the 600–800 °C temperature window, attributed to the breaking of carbon–carbon linkages and aromatic ring opening (Thoresen et al. 2021). Oak, pine, and corn stover aqueous ethanol organosolv lignins were analysed by TG. Temperature programme in nitrogen environment was ambient to 105 °C at 10 °C/min, isothermal at this temperature for 40 min, and heating at 10 °C/min to 900 °C. Major mass loss in oak lignin pyrolysis occurred between 160 and 500 °C with a T_{max} of 357 °C. Pine pyrolysis followed a broad range of temperatures between 180 and 520 °C with T_{max} at 376 °C. Corn stover lignin pyrolysed mainly in the 200–500 °C temperature range with T_{max} near 380 °C (Zhou et al. 2016). Bamboo aqueous ethanol organosolv lignin was characterized by TG in 20 mL/min nitrogen flow at 10 °C/min. After the onset of thermal degradation at 230 °C, the material continuously degraded up to 600 °C. T_{max} was observed at around 360 °C (Zhang et al. 2018). TG thermogram of palm kernel shell aqueous ethanol organosolv lignin was recorded in nitrogen flow rate of 45 mL/min and heating rates of 10, 20, 30, and 40 °C/min. This lignin underwent pyrolysis in two different phases, the temperature ranges for which changed with the heating rate. At 10 °C/min, the first phase occurred between 103 and 296 °C (T_{max} 237 °C, maximum rate 3.02%/min) and the second phase between 297 and 415 °C (T_{max} 342 °C, maximum rate 1.69%/min). Temperature range, T_{max}, and maximum rate progressively increased with increasing heating rate. At 40 °C/min, the first phase occurred between 140 and 329 °C (T_{max} 263 °C, maximum rate 13.01%/min) and the second phase between 330 and 473 °C (T_{max} 358 °C, maximum rate 7.12%/min) (Ma et al. 2018). A study reported TG analysis of aqueous ethanol organosolv lignins from poplar, pine, walnut shell, and rice husk in a nitrogen environment at 10 °C/min. Major degradation occurred in a wide temperature window of 200–600 °C. T_{max} was observed at 357.5, 382.8, 365, and 345.6 °C, respectively, for poplar, pine, walnut shell, and rice husk lignins. A second minor thermal degradation was also observed in walnut shell lignin between 625 and 800 °C. These lignins left char residues of 30.9%, 36%, 29.9%, and 37.2% after heating to 800 °C (Sheng et al. 2022).

Pine straw, wheat straw, flax, and alfalfa were delignified in two stages, first with an aqueous mixture of formic acid and acetic acid, followed by delignification with peroxyformic acid and peroxyacetic acid. Organosolv lignins were isolated from mixed spent liquors and subjected to TG characterization in nitrogen at 10 °C/min. The recorded TG and DTG thermograms are shown in Figure 7.2. At 800 °C, the lowest char residue of 29.5% was obtained from pine straw followed by 35% from alfalfa, 39.2% from flax, and 40.4% from wheat straw. The observed T_{max} were 336.1, 331.9, 332.4, and 328.5, respectively (Watkins et al. 2015).

Organosolv lignin was isolated from prairie cordgrass with an aqueous mixture of isobutyl ketone and ethanol. This lignin was characterized by its pyrolytic behaviour by TG in flowing nitrogen at 20 mL/min and a heating rate of 20 °C/min. After an initial devolatilization with a small peak at 216 °C the main pyrolysis occurred in the temperature interval of 300–500 °C with T_{max} 390 °C. 28% residue was left after heating to 900 °C. The same study also analysed aqueous butanol organosolv lignin from aspen. For this also, the main pyrolysis was observed between 300 and 500 °C with T_{max} at 354 °C. Thermogram had a small shoulder peak at 282 °C. Char remain was 34% at 900 °C (Zhang et al. 2012).

TG analyses of larch and Manchurian ash dioxane lignins were carried out at 10 °C/min in a nitrogen atmosphere. In larch lignin, thermal degradation set in at 200 °C. The degradation rate peaked to

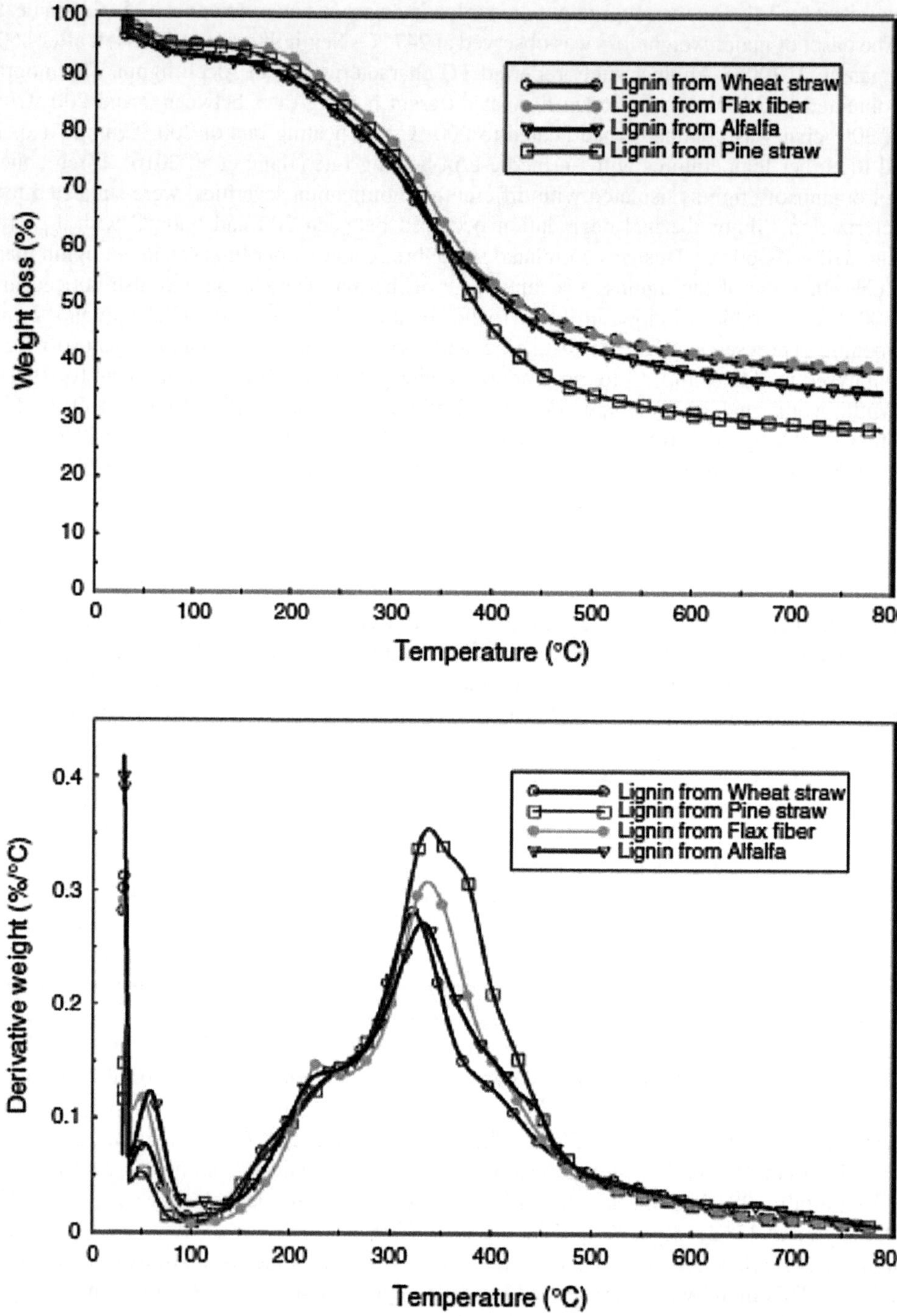

FIGURE 7.2 TG and DTG analyses of aqueous formic acid acetic acid lignins at 10 °C/min. (Reprinted from Watkins, D, Nuruddin, M, Hosur, M, Tcherbi-Narteh, A, Jeelani, S, *J. Mater Res. Technol.*, 4(1), 26–32, 2015, with permission from Elsevier.)

2.06%/min at a T_{max} of 343 °C. Thermal degradation onset for Manchurian ash lignin was at 210 °C. The maximum degradation rate was 4.84%/min at a T_{max} of 317 °C. The two lignins left char residues of 50% and 36%, respectively, at 700 °C (Li et al. 2002). Pine, poplar, and bamboo dioxane organosolv lignins were characterized by TG in nitrogen atmosphere at a 20 °C/min heating rate.

Major mass loss was observed between 150 and 450 °C. Pine lignin had the highest T_{max} of 353.8 °C followed by 335.8 °C for bamboo lignin and 332.8 °C for poplar lignin. Corresponding char residues at 800 °C were 39.33% from pine lignin, 30.19% from poplar lignin, and 20.17% from bamboo lignin (Wang et al. 2021).

7.1.4 Milled Wood Lignin

Bamboo milled-wood lignin was analysed by TG with a 10 °C/min heating rate in nitrogen flowing at 30 mL/min. Reasonable thermal degradation set in at 230 °C and T_{max} was observed at about 297 °C with the rate of mass loss peaking to 3.2%/min. 38% material was converted to char residue at 600 °C (Li et al. 2012). The TG analysis of lauan milled-wood lignin in a 60 mL/min nitrogen atmosphere and 20 °C/min heating rate revealed the onset of thermal degradation at about 160 °C. T_{max} was observed at 365 °C and a shoulder was also observed at 433 °C. 40.7% char residue was left after heating to 800 °C (Wang et al. 2014). The TG analysis of amur linden (a hardwood) milled-wood lignin was carried out at 10 °C/min in nitrogen flow. Thermal degradation mainly occurred between 180 and 500 °C with T_{max} at 342 °C and a DTG shoulder near 240 °C. 41% residue was obtained at 650 °C (Lin et al. 2015). Oak, pine, and corn stover milled-wood lignins were analysed by TG. Temperature programme in the nitrogen environment was ambient to 105 °C at 10 °C/min, isothermal at this temperature for 40 min, and heating at 10 °C/min to 900 °C. DTG thermogram for oak lignin revealed two comparable mass loss events with T_{max} of 268 and 340 °C. Pine pyrolysis followed a broad range of temperatures with T_{max} at 345 °C and a DTG shoulder at around 280 °C. Corn stover lignin pyrolysed in three phases. The first phase peaked at about 159 °C with a small DTG peak. The second and third phases occurred between 200 and 300 °C and 300–450 °C. T_{max} for these two thermal events appeared near 250 and 365 °C (Zhou et al. 2016). TG thermogram of palm kernel shell milled-wood lignin was recorded in nitrogen flow rate of 45 mL/min and heating rates of 10, 20, 30, and 40 °C/min. This lignin underwent pyrolysis in two different phases, the temperature ranges for which changed with the heating rate. At 10 °C/min, the first phase occurred between 201 and 308 °C (T_{max} 246 °C, maximum rate 2.14%/min) and the second phase between 308 and 481 °C (T_{max} 347 °C, maximum rate 1.65%/min). Temperature range, T_{max}, and maximum rate progressively increased with increasing heating rate. At 40 °C/min, the first phase occurred between 199 and 332 °C (T_{max} 284 °C, maximum rate 8.14%/min) and the second phase between 332 and 536 °C (T_{max} 371 °C, maximum rate 6.95%/min) (Ma et al. 2018). A recent study reported TG analysis of milled wood lignins from acacia and Korean mixed hardwoods. Thermograms were recorded in an inert nitrogen atmosphere with a heating rate of 10 °C/min. The onset of thermal degradation was observed at 226 °C for acacia lignin and 212 °C for Korean mixed hardwood lignin. Maximum mass loss rates occurred at T_{max} of 340 and 276 °C, respectively. A DTG peak was also noticed near 360 °C for Korean mixed hardwood lignin. The amount of char was 32.7% and 30.7%, respectively, at 800 °C (Pe III et al. 2023).

7.1.5 Hydrolysis Lignin

The TG characterization of lignin-rich residue left after hydrolysis and ethanol production from mixed hardwood and nonwood was carried out at 10 °C/min heating rate and 20 mL/min flow of argon. Thermal degradation started at about 150 °C. The low onset temperature was believed to be due to the presence of carbohydrates. The rate of degradation peaked at 316 °C to moderate thereafter up to about 600 °C. A shoulder was also observed in the DTG plot at about 190 °C indicating that there might be two stages of degradation. After heating to 1000 °C, 28% char residue was left (Sameni et al. 2014). Lignin was isolated by solvent extraction with sodium hydroxide from the lignin-rich residue left after hydrolysis and ethanol production from corn stover. It was subjected to TG analysis in 60 mL/min nitrogen flow. The temperature programme was ambient to 100 °C at 10 °C/min, held at this temperature for 15 min, and heated to 800 °C at a heating rate of 30 °C/min. Char left at 800 °C from lignin-rich residue from ethanol production was 31.2%. For lignin isolated with sodium hydroxide, the residual char amount was 45.9%. The lignin-rich residue had a sharp DTG peak at T_{max} of 367 °C compared to lignin extracted with sodium hydroxide, which had a T_{max} of 377 °C (Tao et al. 2016). The lignin-rich residue after acid and enzyme hydrolysis of mixed wood was TG-characterized. In a nitrogen environment, lignin was heated at 10 °C/min. The main mass loss occurred between 200 and 500 °C with T_{max} 350 °C. Char residue at 800 °C was 42.2% (Han et al. 2019). TG characterization of lignin, isolated by alkali extraction of lignin-rich residue left after hydrolysis and ethanol production, at 10 °C/min heating rate and 20 mL/min flow of nitrogen is reported. Major thermal degradation was observed between 200 and 700 °C with T_{max} at 355 °C. Char residue was 47.22% at 900 °C (Amit et al. 2021). Enzyme hydrolysis lignin from steam-exploded wheat straw was subjected to TG characterization. The nitrogen flow rate was 50 mL/min and the heating rate was 10 °C/min. Main pyrolytic degradation occurred between 175 and 550 °C with T_{max} at 333 °C. A minor peak was also observed at about 200 °C. Residual char at 800 °C was 30.8% (Wadrzyk et al. 2021). A Lignin-rich residue from acid hydrolysis of rice straw – 47.7% lignin, 36.9% polysaccharides – was characterized by TG for low-temperature thermal degradation. Thermograms were recorded in 80 mL/min nitrogen flow at 20, 30, 40 and 50 °C/min heating rate. Appreciable thermal degradation started at about 250 °C with a T_{max} of 299 °C and a maximum rate of mass loss of 89%/min. With the rate of heating increasing, T_{max} and the corresponding maximum rate of mass loss also increased (Ghatak 2022).

TG characterization was performed on cellulolytic enzyme lignin from spruce and poplar to ascertain their torrefaction behaviour. All experiments were carried out in nitrogen flowing at 50 mL/min. Samples were first heated to 110 °C at 5 °C/min and held at this temperature for 30 min. This ensured the complete drying of the samples. The dried samples were tared and heated at 5 °C/min. After heating to 280 °C, spruce lignin lost 15.2% mass and poplar lignin lost 21.8% mass (Lv et al. 2015). The TG analysis of corn stover enzymatic hydrolysis lignin, obtained from bioethanol production, was carried out at 10 °C/min in nitrogen flow. Thermal degradation mainly occurred between 160 and 480 °C with T_{max} at 305 °C. 57.8% residue was obtained at 650 °C (Lin et al. 2015). Bamboo enzymatic hydrolysis lignin was analysed by TG at heating rates of 10, 20, and 40 °C/min. Thermal degradation mass loss was mainly observed between 225 and 625 °C with two degradation events noticed from the DTG plot. At heating rates of 10 and 20 °C/min, the two DTG peaks were well resolved. At 10 °C/min heating rate, the first thermal event had a T_{max} of 283 and the second thermal event had a T_{max} of 357 °C. At 20 °C/min heating rate, these peaks occurred at 300 and 379 °C. At a higher heating rate of 40 °C/min, the second thermal event had a clearly distinguished peak at 391 °C but the first volatilization only appeared as a shoulder in the DTG plot at 308 °C. Char formation was between 30.2% and 26.7% with increasing heating rate (Yan et al. 2016).

One study reported TG characterization of Klason lignins from beech, willow, and mixed softwoods. Thermograms were obtained in 30 mL/min nitrogen flow at different heating rates between 2 and 200 °C/min. About 40–45% char was obtained when heated to 900 °C at a heating rate of 200 °C/min. Also, T_{max} shifted to higher temperatures with an increase in heating rate (Jiang et al. 2010). The TG analysis of hazelnut shell Klason lignin was performed at 40 °C/min heating rate in nitrogen flowing at 100 mL/min. Two pyrolytic degradation phases were observed. An initial phase with smaller mass loss occurred between about 150 and 230 °C. Major pyrolysis was noticed between 230 and 700 °C with T_{max} 409 °C and a maximum mass loss rate of 6.9%/min. 36.9% char was obtained at 900 °C (Haykiri-Acma et al. 2010). Klason lignins from maple and Chinese fir were characterized by TG in 20 mL/min flowing nitrogen at 20 °C/min heating rate. The material is thermally degraded in two stages. For maple lignin, the first stage started at 180 °C with T_{max} at 230 °C. The onset of the second stage was at 300 °C and T_{max} was 378 °C. For Chinese fir lignin, first-stage T_{max} was observed at 255 °C and second-stage T_{max} at 400 °C. Char residue left after heating to 900 °C was 46.4% for maple lignin and 47.5% for Chinese fir lignin (Zhao et al. 2014). TG thermogram of palm kernel shell Klason lignin was recorded in nitrogen flow rate of 45 mL/min and heating rates of 10, 20, 30, and 40 °C/min. This lignin underwent major pyrolysis in a temperature range that changed with the heating rate. At 10 °C/min, it occurred between 143 and 485 °C with

T_{max} 357 °C and a maximum rate of 2.33%/min. Temperature range, T_{max}, and maximum rate progressively increased with increasing heating rate. At 40 °C/min major pyrolysis weight loss occurred between 172 and 530 °C with T_{max} 373 °C and maximum rate 10.17%/min (Ma et al. 2018).

Strong acid detergent fibres from birch and fir were reported to have more than 99% lignin. These were characterized by TG with heating rates of 20 °C/min in 30 mL/min flowing nitrogen. Thermal degradation exhibited two distinct phases. For birch lignin, these were 158–267 °C and 301–402 °C (T_{max} 227 and 376 °C for the two stages). These degradation temperature ranges were 166–274 °C and 351–481 °C (T_{max} 223 and 425 °C for the two stages) for fir lignin (Liu et al. 2008).

7.1.6 Lignin Characterization by TG-FTIR

In TG-FTIR the gases and volatiles evolved during the TG experiment are passed through a gas cell and characterized in real time for their FTIR absorption spectrum. Important information regarding the chemical nature of the evolved gases and volatiles can be obtained from such spectra. Time-lapse FTIR spectra from different moments of thermal degradation during TG are combined into a 3D composite plot. The time-temperature evolution profiles of different chemical functionalities from thermal degradation can be studied. A representative 3D composite plot from TG-FTIR characterization is shown in Figure 7.3.

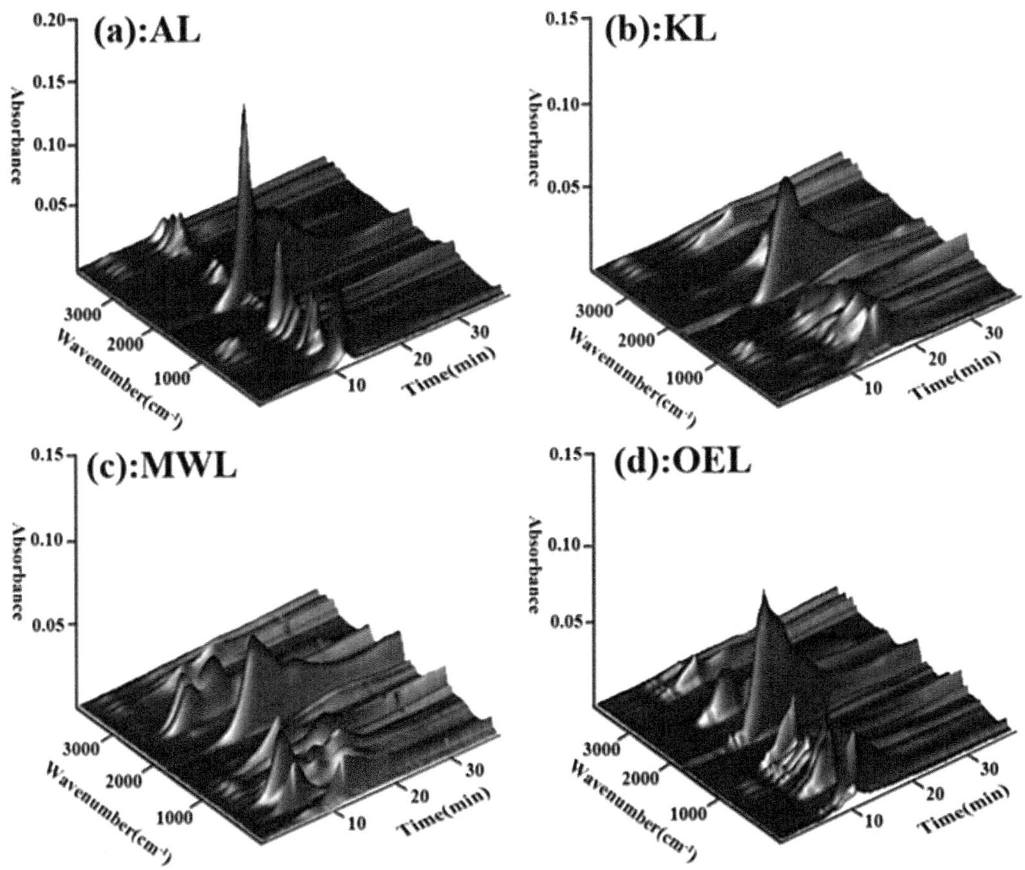

FIGURE 7.3 3D composite TG-FTIR plots of different lignin formulations obtained from palm kernel shell at 20 °C/min; AL – soda lignin, KL – Klason lignin, MWL – milled wood lignin, OEL – ethanol organosolv lignin. (Reprinted from Ma, Z, Wang, J, Zhou, H, Zhang, Y, Yang, Y, Liu, X, Ye, J, Chen, D, Wang, S, *Fuel Process. Technol.*, 181, 142–156, 2018, with permission from Elsevier.)

It is believed that low molecular weight gases like methane, and volatiles evolve from lignin pyrolysis due to the breakdown of intermonomer linkages and removal of substituents from aromatic ring and aliphatic side chain. Demethoxylation and demethylation are mainly responsible for methane formation. Demethoxylation can also lead to methanol formation. Water vapour is mainly formed from hydroxyl functionalities. Decarboxylation gives rise to the evolution of carbon dioxide.

7.1.6.1 Lignin from Alkaline Delignification Processes

TG-FTIR characterization of commercial alkali lignin at 10 °C/min heating rate and 120 mL/min nitrogen flow is reported. Main gaseous products were identified from FTIR traces as methane at 3017 cm^{-1}, carbon dioxide at 2363 cm^{-1}, and carbon monoxide at 2167 cm^{-1}. The peak centred at 1730 cm^{-1} was assigned to C=O stretching in oxygenated organics. Two distinct carbon dioxide evolution phases were observed peaking at around 260 and 700 °C. Carbon monoxide release was significant only above 650 °C. The methane evolution profile was again bimodal with a minor peak near 330 °C and maximum evolution centred near 500 °C (Yang et al. 2007). Another study reported TG-FTIR characterization of this alkali lignin at 20 °C/min heating rate and 70 mL/min nitrogen flow. Water vapour was related to the absorption band centred at 3599 cm^{-1}, methane to 2954 cm^{-1}, carbon dioxide to 2310 cm^{-1}, and carbon monoxide to 2087 and 741 cm^{-1}. Among other volatiles, aromatics in general recorded their presence with 1512 cm^{-1} absorption and phenolics with 1261 cm^{-1} (Ma et al. 2016). Modified soda lignin from palm kernel shell was subjected to TG-FTIR analysis. Main evolved gases and volatiles during pyrolysis were attributed to FTIR absorption bands as: 3651 cm^{-1} water vapour, 2926 cm^{-1} methane, 2360 cm^{-1} carbon dioxide, 2183 cm^{-1} carbon monoxide, 1763 cm^{-1} carbonyl, 1516 cm^{-1} aromatics, 1176 cm^{-1} phenols, and 672 cm^{-1} carbon dioxide. A peak for water vapour evolution occurred at 100 °C. Water vapour evolution again picked up at about 200 °C and reached its maximum evolution at nearly 265 °C. Methane evolution exhibited three evolution peaks nearly at 280, 410, and 540 °C. Maximum carbon dioxide evolution was near 250 °C. Carbon monoxide evolution started at around 240 °C and quickly reached a peak near 265 °C. Its evolution temporarily diminished till about 320 °C to again pick up and gradually go on increasing (Ma et al. 2018). TG-FTIR analysis of the pyrolytic behaviour of kraft lignin from softwood and hardwood mixture was carried out in nitrogen flowing at 60 mL/min with a heating rate of 20 °C/min. Water vapour evolution peaked at 110 °C, identified with absorption bands at 3823–3870 and 1645–1750 cm^{-1}. Two notable evolutions of phenolics and alcohols were observed at 270 and 380 °C. These entities were assigned to absorption bands 3500–3600 (O–H stretching), 1300–1400 (O–H bending), and 974–1058 (C–O stretching). Methane evolution peaks were identified at 420 and 535 °C, mainly attributed to C–H stretching bands at 3020–3190 and 2990–3010 cm^{-1}. Carbon dioxide and carbon monoxide evolutions were maximum at 380 and 480 °C, respectively. Carbon dioxide was identified with C=O stretching vibrations at 2210–2390 and 645–690 cm^{-1}. C–H stretching vibrations from carbon monoxide appeared at 2020–2120 cm^{-1}. Maximum aromatics evolution was observed at 270 °C, attributed to benzene ring skeletal vibrations at 1613 and 1513 cm^{-1} (Shen et al. 2010).

7.1.6.2 Milled Wood Lignin

Milled wood lignin from palm kernel shell was subjected to TG-FTIR analysis. Main evolved gases and volatiles during pyrolysis were attributed similar to modified soda lignin. A peak for water vapour evolution occurred at 100 °C. Water vapour evolution again picked up at about 200 °C and reached a plateau of increased evolution at nearly 450 °C. Methane evolution attained its peak nearly at 440 °C with a minor evolution peak near 600 °C. Maximum carbon dioxide evolution was near 290 °C and maximum carbon monoxide evolution was near 525 °C (Ma et al. 2018).

7.1.6.3 Hydrolysis Lignin

TG-FTIR characterization of Klason lignin from maple and Chinese fir, in 20 mL/min flowing nitrogen at 20 °C/min heating rate, provided spectra for evolved gases and volatiles corresponding to peak rate of mass loss. Bands appeared for methanol at 1057 cm^{-1}, carbon monoxide at 2112 and 2180 cm^{-1}, carbon dioxide at 2360 cm^{-1}, and methane at 3016 cm^{-1}. Methane peak increased in intensity at the cost of methanol peak at higher temperatures. Maple lignin produced more methane and phenolics (absorption in 1330–1370 cm^{-1}) than Chinese fir lignin which was attributed mainly to more methoxyl groups in maple lignin and their easier demethoxylation (Zhao et al. 2014). Klason lignin from palm kernel shell was subjected to TG-FTIR analysis. Main evolved gases and volatiles during pyrolysis were attributed similar to modified soda lignin. A peak for water vapour evolution occurred at 100 °C. Water vapour evolution again picked up at about 200 °C and reached a plateau of increased evolution at nearly 450 °C. Methane evolution attained its peak nearly at 430 °C with a minor evolution peak near 590 °C. Maximum carbon dioxide evolution was near 350 °C and maximum carbon monoxide evolution was near 525 °C (Ma et al. 2018). The lignin-rich residue from acid hydrolysis of rice straw – 47.7% lignin, 36.9% polysaccharides – was characterized by TG-FTIR for low-temperature thermal degradation. Experiments were performed in 80 mL/min nitrogen flow at 50 °C/min heating rate. In the temperature interval of 200–250 °C, absorption bands were not much pronounced. Feeble bands observed accounted for hydroxyl stretching in water at 3550–3900 cm^{-1} and C=O stretching in low molecular weight carboxylic acids at 1790 and 1752 cm^{-1}. In the 250–300 °C temperature range absorption bands were observed at 3550–3900 cm^{-1}, 2929, and 2181 cm^{-1} (C–H stretching in methane and C–O stretching in carbon monoxide), 1514 cm^{-1} (skeletal stretching in benzene ring), and 1378 cm^{-1} (C–H bending in methyl group of acetic acid and methanol). The carbonyl peak appeared at 1752 cm^{-1}. These bands were much more pronounced in the spectra recorded for evolved gases in the 300–350 °C temperature range. Additional bands were also observed at 1363 cm^{-1} for stretching vibrations in alkanes, and 1055 cm^{-1} for C–O stretching in alcohols and phenols. Carbonyl peaks for aldehydes and ketones appeared at 1741 and 1735 cm^{-1} (Ghatak 2022).

Strong acid detergent fibres from birch and fir, with more than 99% lignin, were characterized by TG-FTIR with heating rates of 20 °C/min in 30 mL/min flowing nitrogen. Thermal degradation exhibited two distinct phases. In the first phase the main volatiles and gaseous products were water (3500–3964 cm^{-1}), carbon monoxide (2112 and 2180 cm^{-1}), carbon dioxide (2217–2391 and 586–726 cm^{-1}), formaldehyde (1747 cm^{-1}), formic acid (1786, 1133 and 640 cm^{-1}), and alcohol and phenols (1300–1400 cm^{-1}). In the second and main pyrolysis phase, the main volatiles and gaseous products were carbon monoxide and carbon dioxide, methane and other small hydrocarbons (2850–3200 cm^{-1}), and methanol (960–1085 cm^{-1}). In birch lignin, carbon monoxide evolution was maximum in the temperature range of 477–627 °C while carbon dioxide evolution peaked between 327 and 427 °C. These temperature ranges were 450–577 °C and 377–427 °C for fir lignin. The pyrolysis of birch lignin released a maximum amount of methane between 375 and 475 °C, maximum phenolics between 148 and 260 °C, and maximum methanol between 330 and 410 °C. The pyrolysis of fir lignin released a maximum amount of methane between 375 and 620 °C, maximum phenolics between 180 and 280 °C, and maximum methanol between 220 and 440 °C. (Liu et al. 2008).

7.1.6.4 Organosolv Lignin

Organosolv lignin was isolated from prairie cordgrass with an aqueous mixture of isobutyl ketone and ethanol and from aspen with aqueous butanol. These lignins were characterized for gases and volatiles evolved during pyrolysis by TG-FTIR in flowing nitrogen at 20 mL/min and a heating rate of 20 °C/min. Major product groups and gaseous components were identified from FTIR absorption bands centred at 3018 cm^{-1} for unsaturated organics (aromatics and alkenes), 2969 cm^{-1} for methane and other alkanes, 2384 cm^{-1} for carbon dioxide, 2180 cm^{-1} for carbon monoxide, and 1747 cm^{-1} for carbonyls. In prairie cordgrass lignin, methane and other saturated hydrocarbons mainly evolved between 110 and 200 °C and then again, some evolution was noticed between 350 and 500°C. Unsaturated organics were mainly released between 380 and 680 °C, peaking at around 490 °C. Carbon dioxide started evolving around 150 °C and maximum evolution was observed between 380 and 600 °C, after which it gradually tapered. Appreciable amounts of carbon monoxide were released only after 400 °C. In aspen lignin, methane and other saturated hydrocarbons mainly evolved between 250 and 400 °C with some evolution also noticed between 110 and 200 °C. Unsaturated organics were mainly released between 380 and 600 °C, peaking at around 470°C. Carbon dioxide started evolving around 190 °C and maximum evolution was observed between 300 and 500 °C (peak around 320 °C), after which it gradually tapered. Appreciable amounts of carbon monoxide were released only after 220 °C, peaking around 320 °C, and tapering thereafter (Zhang et al. 2012). Aqueous ethanol organosolv lignin from palm kernel shell was subjected to TG-FTIR analysis. Main evolved gases and volatiles during pyrolysis were attributed similar to modified soda lignin. A peak for water vapour evolution occurred at 100 °C. Water vapour evolution again picked up at about 200 °C and reached a plateau of increased evolution at nearly 270 °C. Methane evolution attained its peak nearly at 430 °C with two minor evolution peaks on either side near 250 °C and 540 °C. Maximum carbon dioxide evolution was near 240 °C and maximum carbon monoxide evolution was near 525 °C (Ma et al. 2018). Pine, poplar, and bamboo dioxane organosolv lignins were characterized by TG-FTIR in nitrogen atmosphere at a 20 °C heating rate. Major compounds and groups of compounds in evolved volatiles were identified from IR absorbance bands as water vapour (3500–3800 cm^{-1}), methane (2990–3010 cm^{-1}), carbon dioxide (2315–2355 cm^{-1}), carbon monoxide (2025–2235 cm^{-1}), carbonyls (1730–1786 cm^{-1}), phenols (1200–1400 cm^{-1}), and methanol and other alcohols (1000–1060 cm^{-1}) (Wang et al. 2021).

7.1.7 Kinetic Interpretation from Thermogravimetry

Lignin, a substance originating from a morphologically complex and chemically heterogeneous biomaterial like lignocellulose, is itself chemically heterogeneous with a morphology of its own. Therefore, its thermochemical transformation is not a single chemical reaction. It is a multiple reaction – most probably a combination of series and parallel reactions – combined with heat and mass transfer steps. In general terms, the rate of such a reaction can be expressed as (Ghatak 2022)

$$-\frac{d(\mathrm{RMF})}{dt} = f_1(T)f_2(\mathrm{RMF}) \tag{7.1}$$

where RMF is the mass fraction of the lignin remaining after time t, and T is the absolute temperature. For a single reaction, the temperature function $f_1(T)$ is the Arrhenius equation where the temperature dependence is characterized by the activation energy which is independent of system composition. Distributed Activation Energy Model (DAEM) extends this concept to lignin thermochemical decomposition, which is a multiple reaction combined with heat and mass transfer steps, with the assumption that the activation energy itself is a function of system composition (Navarro et al. 2009). The function $f_1(T)$ then takes the form (Ghatak 2022)

$$f_1(T) = A\exp\left\{\frac{-E_a(\mathrm{RMF})}{RT}\right\} \tag{7.2}$$

Here, $E_a(\mathrm{RMF})$ represents the RMF-dependent functionality of activation energy with A as the pre-exponential factor.

One of the methods to estimate the activation energies at different RMFs is through isoconversional approximation which assumes that the form of the function $f_2(\mathrm{RMF})$ is independent of RMF (Bardow and Marquardt 2004). Hence, the value of the function $f_2(\mathrm{RMF})$ is invariable at a given RMF and eq. 7.1 and 7.2 can be combined as

$$-\frac{d(\mathrm{RMF})}{dt} = f_2(\mathrm{RMF})A\exp\left\{\frac{-E_a(\mathrm{RMF})}{RT}\right\} = A'\exp\left\{\frac{-E_a(\mathrm{RMF})}{RT}\right\} \tag{7.3}$$

TG experiments are conducted at several heating rates – usually, 4–5 heating rates are sufficient. At each RMF, the rate of thermal degradation $-\frac{d(RMF)}{dt}$ at different temperatures, extracted from TG data, are regressed into eq. 7.3 by nonlinear least square method to estimate activation energy for that RMF. The procedure is repeated for all RMFs to obtain the distribution of activation energy in relation to RMF. A representative plot showing the regression is presented in Figure 7.4.

Alternatively, eq. 7.3 can be linearized as

$$\ln\left[-\frac{d(RMF)}{dt}\right] = \ln A' - \frac{E_a(RMF)}{RT} \tag{7.4}$$

A plot of $\ln\left[-\frac{d(RMF)}{dt}\right]$ against 1/T should result in a straight line.

Activation energies for the pyrolysis of Alcell lignin (commercial hardwood aqueous ethanol organosolv lignin) and softwood kraft lignin were determined according to DAEM. The estimated activation energies for Alcell lignin varied between 129 and 361 kJ/mol. For kraft lignin the range was 80–158 kJ/mol (Ferdous et al. 2002). Another study reported activation energies for the pyrolysis of Alcell lignin for three ranges of conversion according to DAEM. The activation energy was 67.89 kcal/mol for conversion 0.4–0.6, 58.01 kcal/mol for conversion 0.6–0.8, 29.89 kcal/mol for conversion 0.8–0.9, and 8.49 kcal/mol for conversion exceeding 0.9 (Murugan et al. 2008). For lauan milled-wood lignin, the activation energy was assumed to follow a double Gaussian distribution about the mean activation energies E_{a1} and E_{a2}. The two activation energies were estimated as 148.8 and 184.2 kJ/mole with respective standard deviations of 14.66 and 35.42 (Wang et al. 2014). For palm kernel shell milled-wood lignin, the activation energy was assumed to follow a Gaussian distribution about the mean activation energy. This leads to a linear correlation between $\ln\left[\frac{\beta}{T^2}\right]$ and 1/T with the slope of E_a / R for each RMF. Here β is the heating rate in TG experiment. The activation energy was found to attain a minimum of around 0.65 RMF (Ma et al. 2018). Activation energies for pyrolysis of eucalyptus kraft lignin were computed to vary from 127.4 kJ/mol at 20% conversion to 373.9 kJ/mol at 90% conversion (Chen et al. 2020). In one study, the activation energy

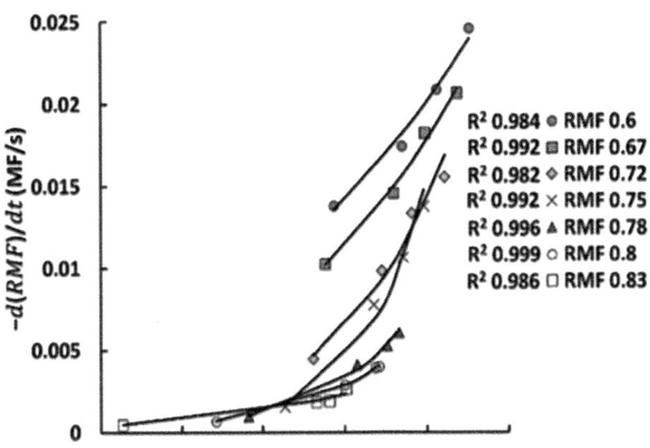

FIGURE 7.4 Computation of activation energy from eq. 7.3 by nonlinear least square regression. (Reprinted from Ghatak, HR, *Period. Polytech. Chem. Eng.*, 66(3), 494–502, 2022, originally published by Budapest University of Technology and Economics under Creative Commons Attribution License.)

for the pyrolysis of lignin-rich residue from acid hydrolysis of rice straw – 47.7% lignin, 36.9% polysaccharides – was found to conform to the following Lorentzian distribution with respect to RMF.

$$E_a(RMF) = \frac{159.8982}{1 + \left(\frac{RMF - 0.7524}{0.0779}\right)^2} \tag{7.5}$$

The estimated activation energy attained a maximum at an intermediate RMF (Ghatak 2022).

In Flynn, Wall, and Ozawa approximation, a linear correlation between $\log \beta$ and 1/T is used to estimate activation energy at each RMF. The distribution of activation energies with RMF for softwood kraft lignin, hardwood kraft lignin, and Alcell lignin, computed from Flynn, Wall, and Ozawa approximation, is shown in Figure 7.5.

In the Coats–Redfern approximation, a linear correlation is obtained between $\ln\left[\frac{-\ln(RMF)}{T^2}\right]$ and 1/T when the reaction is considered first order in RMF. Using this method, hydrolysis lignins obtained by strong acid detergent treatment of fir and birch were analysed for their pyrolysis activation energies. The estimated activation energies not only varied with RMF but also with heating rate. In fir lignin, activation energies varied between 61.4–132.2 and 78.5–143.8 kJ/mol with the heating rate increasing from 10 to 40 °C/min. In birch lignin, activation energies varied between 90.8–144.1 and 82.2–134.8 kJ/mol with the heating rate increasing from 10 to 40 °C/min. The change in activation energy with the heating rate was compensated using the relationship,

$$\ln A = aE_a + b \tag{7.6}$$

and a single activation energy was proposed for each lignin independent of the heating rate. For fir lignin, it was 72.9–136.9 kJ/mol and for birch lignin, it was 87.2–141.7 kJ/mol (Liu et al. 2008). A similar analysis was applied for the estimation of activation energies for the pyrolysis of bamboo enzymatic hydrolysis lignin. Measured activation energy varied with RMF as well as heating rate. For activation energy dependence, the pyrolysis was categorized into two stages based on TG analysis. Accordingly, for low-temperature volatilization activation energy range was 50.6–56.7 kJ/mol with the heating rate changing from 10 to 40 °C/min. The activation energy range was 74.7–97.9 kJ/mol for high-temperature charring (Yan et al. 2016). Activation energies were computed for hydrolysis

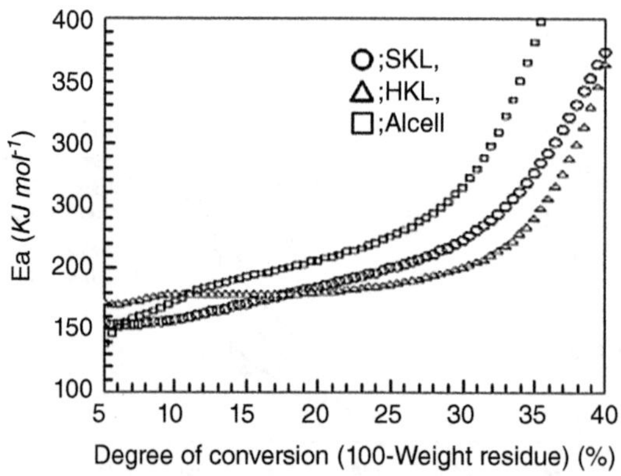

FIGURE 7.5 Distribution of activation energies with RMF from Flynn, Wall, and Ozawa approximation; SKL (softwood kraft lignin), HKL (hardwood kraft lignin), Alcell (hardwood aqueous ethanol organosolv lignin). (Reprinted and adapted from Kubo, S, Kadla, JF, *J. Wood Chem. Technol.*, 28, 106–121, 2008, with permission from Taylor and Francis.)

lignins obtained by strong acid detergent treatment of sunflower stalk, hazelnut shell, and almond shell. For sunflower stalk lignin, the activation energy varied between 34.5 and 60 kJ/mol, for hazelnut shell lignin between 36 and 49.9 kJ/mol, and for almond shell lignin between 41.4 and 59 kJ/mol (Ballice et al. 2020).

In Kissinger's method, a single value of activation energy is computed through a plot of $\ln\left[\frac{\beta}{T_{max}^2}\right]$ against $1/T_{max}$ which gives a linear correlation with the slope E_a / R (Kissinger 1957). Kissinger's method was applied to the pyrolysis of pine aqueous acetic acid organosolv lignin and LignoBoost softwood kraft lignin. The estimated activation energies were 252 and 259 kJ/mol, respectively (Jankovic 2011). Activation energies for pyrolysis of several lignins were estimated by Kissinger's method. Commercial hardwood aqueous ethanol organosolv lignin (Alcell lignin) had an activation energy of 159.5 kJ mole. Nonwood soda lignin had 133.9 kJ/mol activation energy. The activation energy for beech Klason lignin was 151.1 kJ/mol, for willow Klasson lignin 156.5 kJ/mol, and for mixed softwood Klason lignin the activation energy was 145.2 kJ/mol (Jiang et al. 2010). Activation energies for pyrolysis of lignin-rich residue from hydrolysis of corn stover as well as lignin isolated from this residue by alkali extraction were determined by Kissinger's method and found to be 176.6 and 230.1 kJ/mol, respectively (Tao et al. 2016).

7.2 LIGNIN CHARACTERIZATION BY DIFFERENTIAL THERMAL ANALYSIS

DTA is an analytical technique to evaluate the energy dynamics of thermochemical transformations including chemical reactions like pyrolysis. Unlike TG, where the residual mass of the sample is recorded as it is heated at a preset rate, in DTA the temperature of the sample is measured with respect to the temperature of a reference material as the furnace temperature is increased at a preset rate. The reference material, normally α alumina, does not undergo phase change or chemical transformation in the experimental range of temperature. Accordingly, its temperature change is only related to its specific heat. On the other hand, the material to be analysed – lignin in the present case – undergoes both phase change and chemical transformation. Therefore, its temperature change is strongly related to the enthalpy of phase change and chemical reaction. If exothermic transformations take place, the temperature of the material exceeds that of the reference material and vice versa. This way, the temperature difference between the sample and the reference material provides a basis for estimating the enthalpies of different phase changes and chemical transformations. It is customary to subtract the sample temperature from the temperature of the reference material to get ΔT. A negative ΔT on the DTA plot indicates an exothermic process while a positive ΔT indicates an endothermic process. In lignin pyrolysis, it is generally believed that volatilization is endothermic while char formation is exothermic.

7.2.1 Alkali Lignin

Wheat straw and bagasse soda lignins were characterized by DTA in 100 mL/min nitrogen flow and 10 °C/min heating rate. Both the lignins exhibited exothermic degradation. Wheat straw lignin exothermally degraded between 319 and 443 °C with a total of 128 mJ/mg heat evolved, with maximum heat evolution being at 390 °C. Bagasse lignin exothermally degraded between 321 and 447 °C with a total of 196 mJ/mg heat evolved, maximum heat evolution being at 392 °C (Ghatak et al. 2010). Soda lignins from nonwood species – bamboo, *A. donax* (giant cane), and pearl millet – were analysed by DTA for the energetics of their thermochemical degradation behaviour in 200 mL/min nitrogen flow rate up to 800 °C. Experiments were performed at heating rates of 10 and 100 °C/min. After the initial moisture loss, lignins were observed to decompose in two distinct stages as described in Section 7.1. The DTA plots are shown in Figure 7.6. For a heating rate of 10 °C/min, after initial

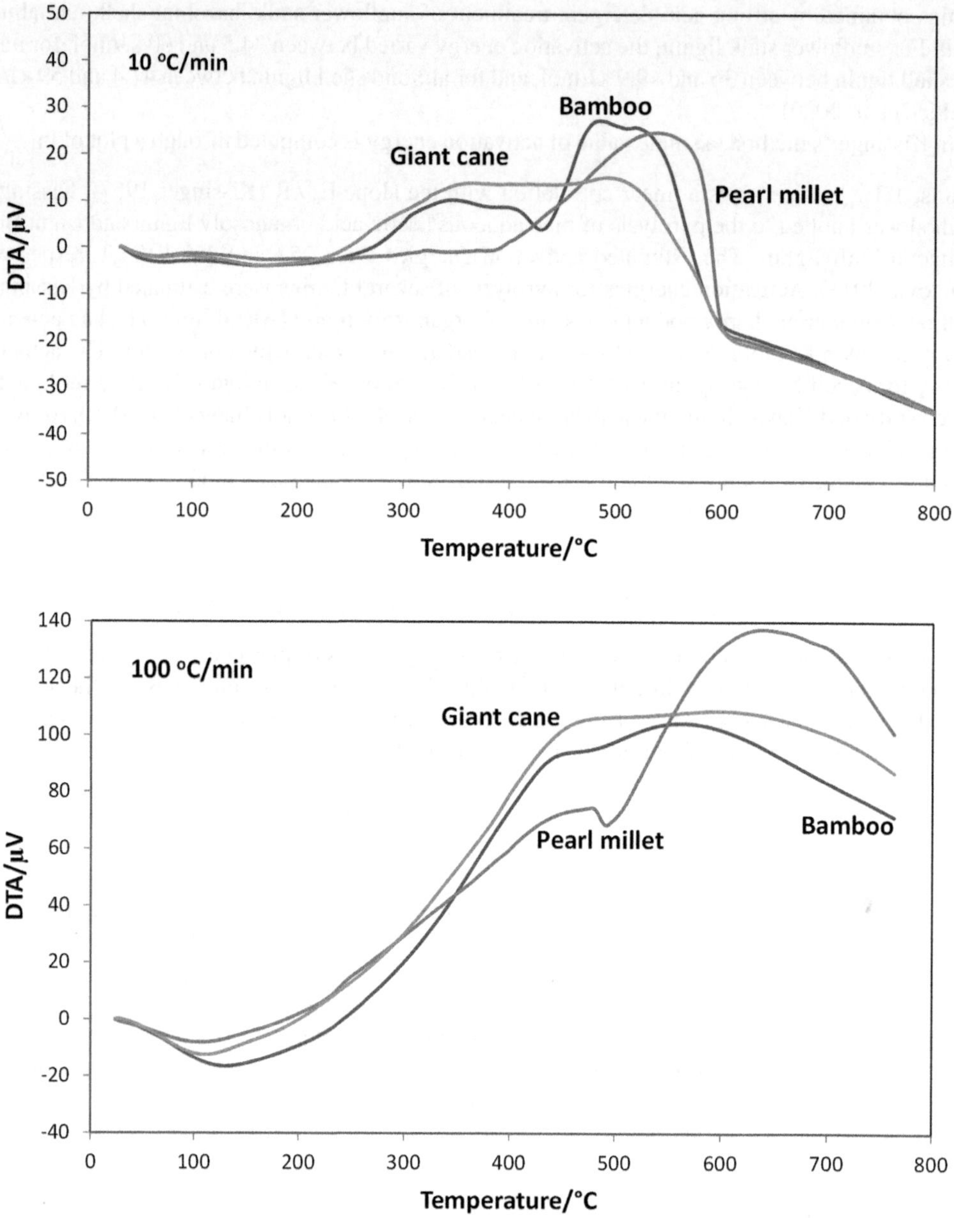

FIGURE 7.6 DTA plots of nonwood soda lignins. (Reprinted from Reddy, IAK, Ghatak, HR, *J. Thermal. Anal. Calorim.*, 132, 407–423, 2018, with permission from Springer.)

moisture loss and evolution of volatiles which is slightly endothermic, the process shifted to exothermic at 276 °C and remained so till about 580 °C. A somewhat similar trend was observed at 100 °C/min heating rate, where the process shifted to exothermic at 250 °C and remained so for the rest of the decomposition. For pearl millet lignin, the process shift from endothermic to exothermic occurred at 400 °C for a heating rate of 10 °C/min and at 205 °C for a heating rate of 100 °C/min. For giant cane lignin, the process shift from endothermic to exothermic occurred at 255 °C for a heating rate of 10 °C/min and at 210 °C for a heating rate of 100 °C/min (Reddy and Ghatak 2018). Energetics of commercial kraft lignin pyrolysis was studied by DTA in 10 mL/min nitrogen flow at 15 °C/min. A major part of thermal degradation was exothermic with an energy evolution peak occurring at 420 °C (El Moustaqim et al. 2018).

7.2.2 Milled Wood Lignin

Bamboo milled-wood lignin was analysed by DTA with a 10 °C/min heating rate in nitrogen flowing at 30 mL/min for the energetics of their thermochemical degradation behaviour. The entire degradation process was found to be endothermic (Li et al. 2012).

7.3 LIGNIN CHARACTERIZATION BY DIFFERENTIAL SCANNING CALORIMETRY

DSC is an analytical technique which, in many ways, is similar in inference to DTA. In DSC, the sample is heated at a preset rate and the instrument simultaneously heats the reference substance such that its temperature matches that of the sample. The differential heat required to do so is measured and plotted against the changing temperature of the sample. If the heat given to the sample exceeds that given to the reference substance, the process is endothermic. As a custom, differential heat is measured by subtracting the heat given to the sample from that given to the reference substance. Therefore, negative differential heat flows on DSC plots indicate endothermic processes.

7.3.1 Lignin from Alkaline Delignification Processes

DSC characterization of commercial alkali lignin at 10 °C/min heating rate and 120 mL/min nitrogen flow is reported. Main pyrolytic degradation was found to be exothermic with a heat evolution peak at 365 °C. The DSC plot contained a minor endothermic peak around 750 °C (Yang et al. 2007). Bagasse soda lignin was characterized for glass transition temperature (T_g) by DSC in a nitrogen flow rate of 15 mL/min. Temperature programme was ambient to 200 °C at 10 °C/min, 200 to –10 °C at 30 °C/min, and –10 °C to 200 °C at 10 °C/min. The measured T_g was 130 °C (Mousavioun and Doherty 2010). Eucalyptus soda lignin was studied for energy transformations during thermal degradation by DSC analysis. Experiments were run in 60 mL/min nitrogen flow at a heating rate of 10 °C/min. The measured T_g was 93 °C (Awal and Sain 2011). DSC characterization of nonwood soda lignin was carried out at a nitrogen flow rate of 15 mL/min. Temperature programme was ambient to 200 °C at 10 °C/min, 200 to –10 °C at 30 °C/min, and –10 °C to 200 °C at 10 °C/min. The onset of polymer structure rearrangements started at 163.75 °C and the measured T_g was 189.34 °C (Sameni et al. 2014).

Softwood kraft lignin was characterized by DSC at a heating rate of 10 °C/min and nitrogen flow of 30 mL/min. The measured T_g was close to 100 °C (Hatakeyama et al. 2010). Softwood kraft lignin was studied for energy transformations during thermal degradation by DSC analysis. Experiments were run in 60 mL/min nitrogen flow at a heating rate of 10 °C/min. The measured T_g was 100 °C (Awal and Sain 2011). One study reported DSC characterization of softwood and hardwood kraft lignins in 50 mL/min nitrogen flow. The sample was heated to 200 °C at 10 °C/min, cooled at the same rate to –60 °C, and reheated at the same rate to 200 °C. The measured T_g for softwood lignin was 148 °C and for hardwood lignin, it was 138 °C (Ropponen et al. 2011). DSC characterization of softwood (commercial) and hardwood kraft lignins was carried out at a nitrogen flow rate of 15 mL/min. Temperature programme was ambient to 200 °C at 10 °C/min, 200 to –10 °C at 30 °C/min, and –10 °C to 200 °C at 10 °C/min. In softwood kraft lignin, the onset of polymer structure rearrangements started at 143.72 °C and the measured T_g was 158.15 °C (Sameni et al. 2014). In hardwood kraft lignin these temperatures were 95.75 and 109.82 °C, respectively (Sameni et al. 2014). A T_g of 148.3 °C was reported from the DSC characterization of pine kraft lignin in 20 mL/min flowing nitrogen. Prior to the DSC run, lignin was desiccated for 24 h at 70 °C. DSC temperature programme was heating to 200 °C at 10 °C/min and cooling to 40 °C at 40 °C/min (Chen et al. 2014). Thermally induced phase change in commercial kraft lignin was analysed by DSC in 10 mL/min flowing argon at 15 °C/min heating rate. T_g was determined to be 180 °C from the thermogram section up to 200 °C. The thermogram is presented in Figure 7.7. The thermogram also showed that a major part of lignin degradation is exothermic with an evolution of 2780 J/g heat (El Moustaqim et al. 2018).

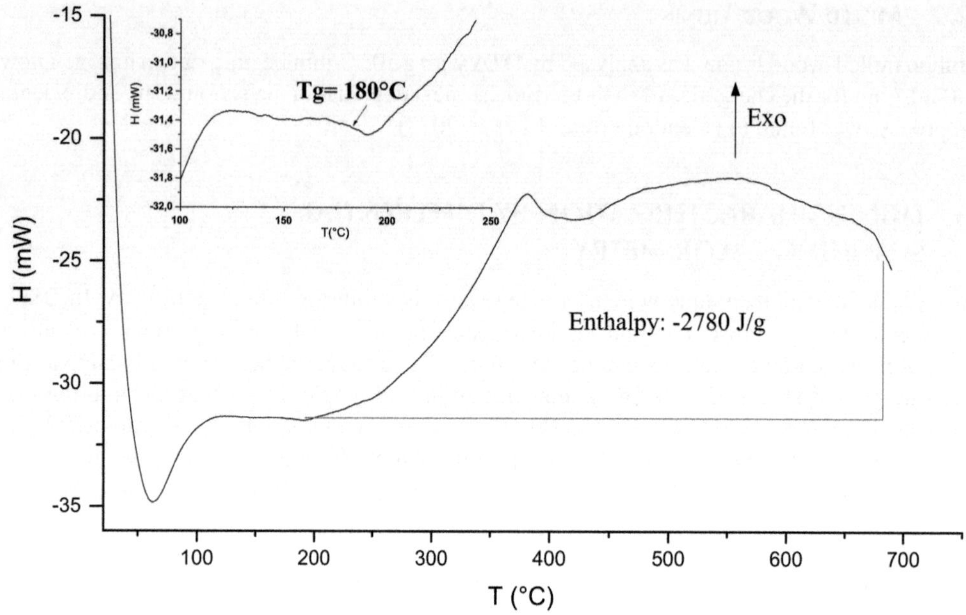

FIGURE 7.7 DSC thermogram of commercial kraft lignin. (Reprinted from El Moustaqim, M, El Kaihal, A, El Marouani, M, Men-La-Yakhaf, S, Taibi, M, Sebbahi, S., El Hajjaji, S, Kifani-Sahban, F, *Sustain. Chem. Pharmacy*, 9, 63–68, 2018, with permission from Elsevier.)

Spruce LignoBoost kraft lignin was DSC characterized. In a nitrogen environment, lignin was heated at 10 °C/min to 105 °C and held for 30 min. It was then cooled at 10 °C/min to –20 °C and reheated at the same rate to 220 °C. The measured T_g was 96 °C. The authors also identified a softening point near 200 °C from the first heating scan (Han et al. 2019). A recent study reported DSC analysis of mixed hardwood kraft lignin. Thermograms were recorded in an inert nitrogen atmosphere with heating to 240 °C at a rate of 10 °C/min, cooling to 40 °C at the same rate, and heating again to 240 °C at 10 °C/min. Two thermal phase transitions were observed, T_g of 63.2 °C and softening temperature of 148.6 °C (Pe III et al. 2023).

One study reports T_g measurement of esterified (acetate) commercial softwood kraft lignin Indulin AT in nitrogen flow with 5 °C/min heating. The sample was annealed overnight before the DSC run. T_g was reported to be 106 °C (Fox and McDonald 2010). DSC characterization of Indulin AT in 20 mL/min flow of nitrogen is reported. Prior to analysis, lignin was annealed for 24 h under vacuum at 60 °C. The temperature programme followed by heating from 20 to 110 °C at 10 °C/min, cooling back to 20 °C at 10 °C/min, and reheating to 400 °C at 15 °C/min. This lignin was characterized by a T_g of 148 °C (Amit et al. 2021). This clearly brings forth the lowering effect on lignin T_g by esterification.

7.3.2 Organosolv Lignin

Energetics of thermal degradation of Alcell lignin, a commercial aqueous ethanol organosolv lignin, was characterized by DSC. The thermogram was recorded in an argon environment at a 15 °C/min heating rate. Endothermic transformation at 113 °C was attributed to melting. Two other minor endothermic peaks were observed at 213 and 252 °C (Murugan et al. 2008). Bagasse aqueous ethanol organosolv lignin was characterized for T_g by DSC in a nitrogen flow rate of 15 mL/min. Temperature programme was ambient to 200 °C at 10 °C/min, 200 to –10 °C at 30 °C/min, and –10 °C to 200 °C at 10 °C/min. The measured T_g was 130 °C (Mousavioun and Doherty 2010). Aqueous ethanol organosolv lignins were obtained from spruce, beech, and wheat straw and their

T_gs were measured by DSC. Temperature programme was 25 to 105 °C at 10 °C/min, 105 to –60 °C at 10 °C/min, –60 °C to 200 °C at 10 °C/min, 200 to –60 °C at 10 °C/min, and –60 to 200 °C at 10 °C/min. The measured T_g were spruce 130 °C, beech 123 °C, and wheat straw 110 °C (Rossberg et al. 2019). A study reported a DSC analysis of aqueous ethanol organosolv lignins from poplar, pine, walnut shell, and rice husk in a nitrogen environment. Temperature protocol was ambient to –40 °C at 20 °C/min, held at this temperature for 1 min, heating to 200 °C at 20 °C/min, held at 200 °C for 1 min, cooling to –40 °C at 20 °C/min, again held at this temperature for 1 min, and final heating to 200 °C at 20 °C/min. Measured T_g were 96.94, 97.06, 94.3, and 95.8 °C, respectively, for poplar, pine, walnut shell, and rice husk lignins (Sheng et al. 2022).

One study reported DSC characterization of phosphinic acid-catalysed aqueous acetic acid organosolv lignins from spruce and birch in 50 mL/min nitrogen flow. The sample was heated to 200 °C at 10 °C/min, cooled at the same rate to –60 °C, and reheated at same rate to 200 °C. The measured T_g for spruce lignin was 192 °C and for birch lignin it was 155 °C (Ropponen et al. 2011).

7.3.3 Hydrolysis Lignin

The DSC characterization of lignin-rich residue, left after hydrolysis and ethanol production from mixed hardwood and nonwood, was carried out at a nitrogen flow rate of 15 mL/min. Temperature programme was ambient to 200 °C at 10 °C/min, 200 to –10 °C at 30 °C/min, and –10 °C to 200 °C at 10 °C/min. The onset of polymer structure rearrangements started at 136.26 °C and the measured T_g was 149.16 °C (Sameni et al. 2014). Lignin-rich residue from wood hydrolysis for ethanol production was DSC characterized. In a nitrogen environment, lignin was heated at 10 °C/min to 105 °C and held for 30 min. It was then cooled at 10 °C/min to –20 °C and reheated at the same rate to 220 °C. The measured T_g was 112 °C (Han et al. 2019). DSC characterization of lignin, isolated by alkali extraction of lignin-rich residue left after hydrolysis and ethanol production, in 20 mL/min flow of nitrogen is reported. Prior to analysis, lignin was annealed for 24 h under vacuum at 60 °C. The temperature programme followed heating from 20 to 110 °C at 10 °C/min, cooling back to 20 °C at 10 °C/min, and reheating to 400 °C at 15 °C/min. The measured T_g was 137 °C (Amit et al. 2021).

One study reports T_g measurement of esterified (acetate) corn stover enzymatic hydrolysis lignin in nitrogen flow with 5 °C/min heating. The sample was annealed overnight before the DSC run. T_g was reported to be 139 °C (Fox and McDonald 2010).

DSC analysis of hazelnut shell Klason lignin was performed at 40 °C/min heating rate in nitrogen flowing at 100 mL/min. The major pyrolytic degradation was exothermic with maximum heat evolution at 554 °C (Haykiri-Acma et al. 2010).

7.3.4 Milled Wood Lignin

A recent study reported a DSC analysis of milled wood lignins from acacia and mixed Korean hardwoods. Thermograms were recorded in an inert nitrogen atmosphere with heating to 240 °C at a rate of 10 °C/min, Cooling to 40 °C at the same rate, and heating again to 240 °C at 10 °C/min. Relatively higher T_g were measured, 167.3 °C for acacia lignin, and 162.3 °C for mixed Korean hardwoods lignin (Pe III et al. 2023).

7.4 LIGNIN CHARACTERIZATION BY PYROLYSIS AND PRODUCT IDENTIFICATION AND QUANTIFICATION

In Chapter 5, we discussed the degradative characterization of lignins like nitrobenzene oxidation, thioacidolysis, and DFRC. These methods break the lignin macromolecule under controlled reaction conditions into low molecular weight products which can be identified and quantified. Understanding the mechanism of degradation pathways and reaction end products helps in the elucidation of the molecular structure of lignin in terms of the abundance of different inter monomer linkages and functional groups. A similar strategy can be adopted for lignin characterization by defragmenting the lignin macromolecule through controlled pyrolysis. Low molecular weight organic molecules evolved as volatiles during pyrolysis are identified and quantified by mass spectroscopy. A common method of analysis is to separate the individual components in the evolved volatiles through gas chromatographic separation before their identification and quantification in the coupled mass spectroscope. The complete characterization protocol is referred to as pyrolysis-gas chromatography-mass spectroscopy (Py-GC-MS). The resultant chromatogram is normally called a pyrogram. During the ionization phase of MS detection, a single molecule may break up into a number of charged fragments. If the peak intensity – height and area – for a compound in the pyrogram is based on cumulative weightage from all the detected ions for that compound, the pyrogram is known as Total Ion Chromatogram. Alternatively, the peak intensity may be based on only a particular ion resulting in a pyrogram known as an Extracted Ion Chromatogram.

Important low molecular weight compounds present in the evolved volatiles are phenol, guaiacol, and syringol along with their derivatives with substitution at position 4. The substituents present could be methyl, or ethyl, or allyl, or propenyl, or vinyl. Carbonyl compounds like vanillin, syringaldehyde, apocynin, acetosyringone, and p-hydroxybenzaldehyde are also present. However, it is believed that guaiacyl units linked by 5–5′ and β–5′ linkages are not released during pyrolysis. This could lead to errors in the estimation of guaiacyl-type moieties (van de Pas et al. 2014). Other compounds that may evolve from lignin pyrolysis are cresols, catechol, methoxycatechol, eugenol, and sinapaldehyde. Monoaromatic hydrocarbons and polyaromatic hydrocarbons may also be produced, especially at higher pyrolysis temperatures, from the deoxygenation of phenols. Carbohydrate impurities can produce furfural and other furan compounds. A mechanistic scheme of the formation of low molecular weight volatile compounds during lignin pyrolysis is presented in Figure 7.8. It is important to note that the product composition in the evolved volatiles is dependent on pyrolysis temperature as well as the heating rate. Different studies report Py-GC-MS at different temperatures and heating rates and a standard protocol in this regard is lacking within the scientific community. A direct correlation with the molecular structure of the parent lignin is to be made with caution.

The compound 4-vinylphenol is of particular importance. It can be produced from the p-hydroxyphenyl moieties in lignin macromolecules as well as from compounds like p-coumaric acid and ferulic acid present in many technical lignins. Therefore, an estimation of 4-vinylphenol in itself is not sufficient to quantify p-hydroxyphenyl moieties in lignin macromolecule as an overestimation can ensue from this compound being released by the pyrolysis of p-coumaric acid and ferulic acid. Pyrolysis in the presence of tetramethylammonium hydroxide (Py/TMAH) provides a solution to this interference. When pyrolysed in the presence of TMAH, decarboxylation is prevented and the phenylpropanoid compounds like p-coumaric acid and ferulic acid are released as their methyl ester derivatives only.

In Curie point pyrolysis, the sample is placed in contact with a ferromagnetic material which is inductively heated rapidly by irradiating with radio frequency electromagnetic pulses. The upper limit of the pyrolysis temperature is controlled by the Curie point of the ferromagnet being 610 °C. Alternatively, the sample is held in a refractory crucible and dropped into a preheated furnace. One variation of Py-GC-MS is double shot Py-GC-MS in which the pyrolysis is carried out in two separated temperature ranges. In the first low-temperature pyrolysis, also known as thermal desorption, the easy-to-degrade part of the lignin macromolecule releases volatiles which are identified. B–O–4′ linkages are mostly cleaved along with demethoxylation. Low molecular weight entities such as p-hydroxybenzoate present in lignin formulations are also released. In the second stage of the high-temperature pyrolysis, the remainder of the lignin macromolecule is defragmented. Carbon–carbon linkages are broken along with extensive deoxygenation leading to the evolution of aromatic hydrocarbons. Main char formation also takes place during this stage.

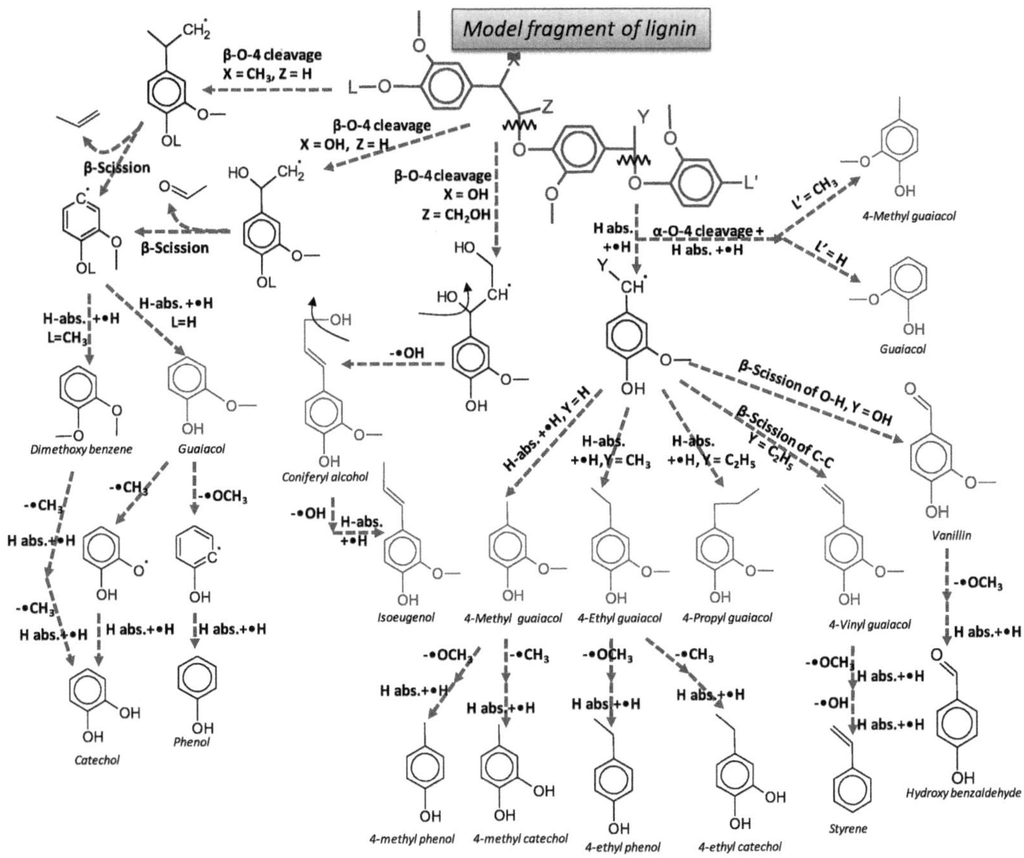

FIGURE 7.8 Mechanistic pathways for the production of low molecular weight volatiles during lignin pyrolysis. (Reprinted from Ojha, DK, Viju, D, Vinu, R, *Bioresource Technol.*, 241, 142–151, 2017, with permission from Elsevier.)

7.4.1 Lignin Characterization by Py-GC-MS

7.4.1.1 Soda Lignin

Soda lignins from jute, flax, hemp, sisal, and abaca were characterized by Py-GC-MS. Guaiacyl entities guaiacol, 4-methylguaiacol, 4-vinylguaiacol, vanillin, apocynin, and trans-isoeugenol were predominant in the evolved volatiles from hemp and flax lignins in comparison to syringyl entities. In contrast, the evolved volatiles from pyrolysis of jute, abaca, and sisal lignins had a predominance of syringyl entities syringol, 4-methylsyringol, 4-vinylsyringol, syringaldehyde, acetosyringone, and trans-4-propenylsyringol. 4-vinylphenol was present in the volatiles from hemp and abaca, more in abaca than in hemp. Accordingly, the syringyl/guaiacyl ratios were estimated as hemp 0.8, flax 0.4, Jute 1.7, sisal 3.4, and abaca 2.9. Py/TMAH was also performed for correct estimation of p-hydroxyphenyl-type moieties in lignin macromolecule as well as the presence of phenylpropanoids independently. p-hydroxyphenyl moieties had the highest 26% abundance in abaca lignin followed by 9% in hemp. Other lignins had less than 5% of these moieties. In abaca lignin, independent phenylpropanoids (p-coumaric acid and ferulic acid) were 39% in abundance. This was 11% in sisal lignin. In other lignins, their abundances were less than 5% (del Rio et al. 2007b).

Py-GC-MS characterization of soda lignin from flax shives at 700 °C revealed guaiacol and 4-vinylguaiacol to be the most abundant compounds in the evolved volatiles. Other compounds present in reasonable relative abundance were 4-methylguaiacol, catechol, homocatechol, vanillin, isoeugenol, and syringol. The estimated syringyl/guaiacyl ratio was 0.08 (Ross and Mazza 2010). *C. lanceolata* soda lignin was subjected to Py-GC-MS characterization by flash pyrolysis at 600 °C with a heating rate of 10,000 °C/s. Three lignin fractions were isolated based on the pH to which the black liquor was acidified. Pyrograms with peaks correlated to individual low molecular wight compounds are shown in Figure 7.9. Guaiacyl and p-hydroxyphenyl-type compounds accounted for about 90% abundance in evolved volatiles. Of these, 4-methylguaiacol, 4-ethylguaiacol, and 4-vinylguaiacol were most abundant. Carbonyl compounds like vanillin, and apocynin were also detected (Li and Wu 2014).

Py-GC-MS was carried out on wheat straw soda lignin by flash pyrolysis at 20,000 °C/min to 450 °C temperature and holding for 30 s. Guaiacyl/syringyl/p-hydroxyphenyl ratio was reported to be 2.3/1/0.9. The main guaiacyl-type compounds in evolved gases were guaiacol, vinylguaiacol, and

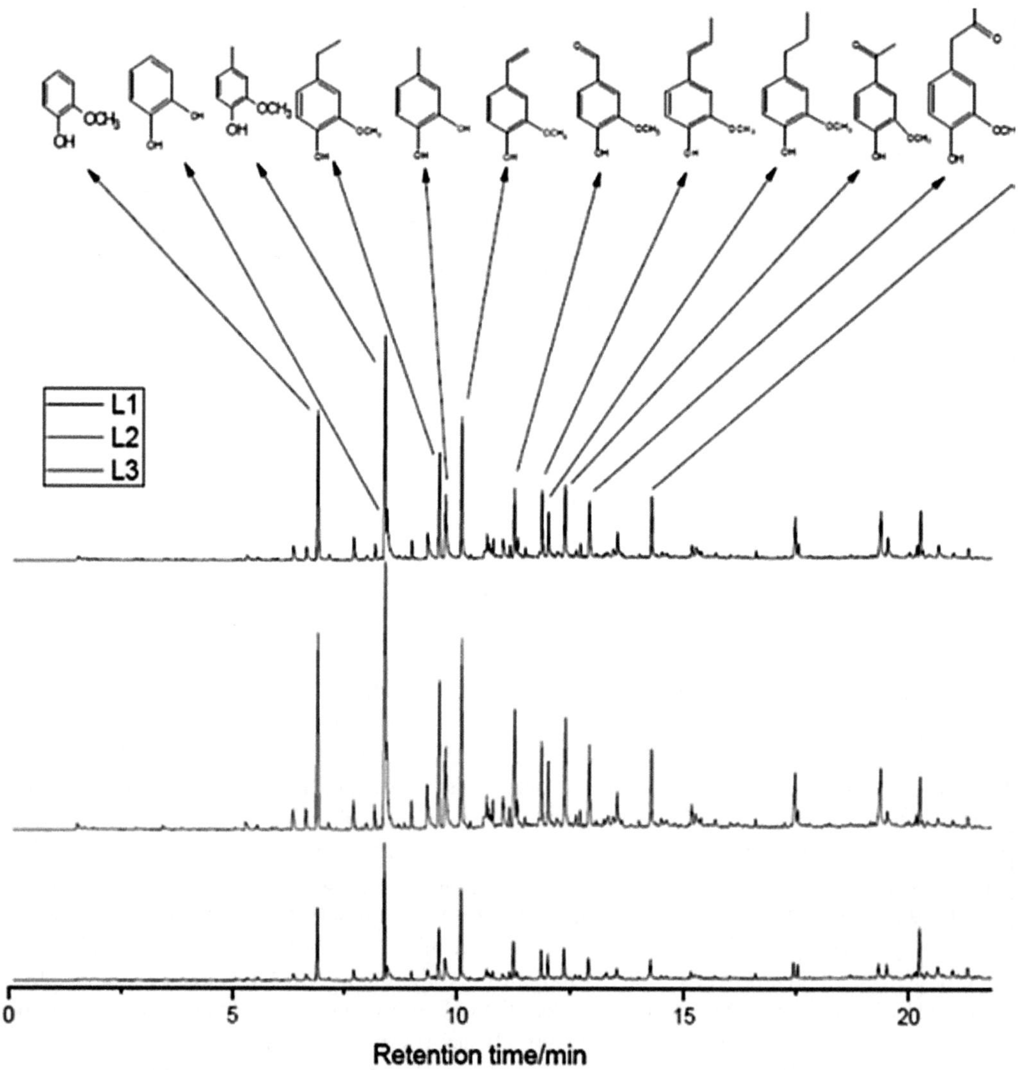

FIGURE 7.9 Total ion chromatogram of evolved volatiles from pyrolysis of *Cunninghamia lanceolata* soda lignins; precipitation pH, L1 8, L2 5, L3 2. (Reprinted from Li, X, Wu, S, *BioResources*, 9(4), 6277–6289, 2014, originally published by BioResources under their open access policy.)

vanillic acid. The main syringyl-type compounds were syringol, propenylsyringol, and syringaldehyde. 4-methylphenol and 4-ethylphenol were the important compounds of the p-hydroxyphenyl-type (Lin et al. 2015). Py-GC-MS analysis of palm kernel shell soda lignin in the temperature range of 450–950 °C is reported. In evolved volatiles, the proportion of guaiacyl and syringyl entities decreased with increasing pyrolysis temperature. The proportion of p-hydroxyphenyl entities was maximum at 750 °C, after which it decreased with further increase in temperature. Aromatic hydrocarbons first appeared in the evolved volatiles at 650 °C pyrolysis temperature and their proportion continuously increased thereafter with increasing temperature. The proportion of catechols was the highest at pyrolysis temperatures of 550 and 650 °C (Ma et al. 2018).

7.4.1.2 Kraft Lignin

Commercial softwood kraft lignin Indulin AT was characterized for molecular substructures by Py-GC-MS at 450 °C pyrolysis temperature and the evolved volatiles were found to consist mainly of guaiacyl-type substructures (Fox and McDonald 2010). Wood kraft lignin was characterized by Py-GC-MS at temperatures of 575, 650, and 700 °C. The main guaiacyl-type compounds in evolved volatiles were guaiacol and its 4-methyl and 4-ethyl derivatives. Vanillin, apocynin and eugenol were also present. The only syringyl-type compound in the volatiles was syringol. Both guaiacyl and syringyl entities decreased in abundance with increasing pyrolysis temperature. Phenol, 4-ethylphenol, cresols, and catechols, on the other hand, increased in abundance with increasing temperature (Shen et al. 2010). Softwood and hardwood kraft lignins were subjected to Py-GC-MS with 2 s pyrolysis at 580 °C. In evolved volatiles from softwood lignin, guaiacol, 4-methylguiacol, vinylguaiacol, trans-isoeugenol, vanillin, acetoguaiacone, homovanillin, dihydroconiferyl alcohol, trans-coniferyl alcohol, and coniferaldehyde were the major low molecular weight compounds. Syringyl moieties were negligible and guaiacyl/p-hydroxyphenyl ratio was estimated to be 0.03:1. In evolved volatiles from hardwood lignin, guaiacol, 4-methylguiacol, vinylguaiacol, trans-isoeugenol, vanillin, acetoguaiacone, homovanillin, and dihydroconiferyl alcohol, were the major guaiacyl compounds. Syringyl compounds present in relatively high abundance were syringol, methylsyringol, 4-vinylsyringol, syringaldehyde, and acetosyringone. p-hydroxyphenyl-type entities were negligible. Syringyl/guaiacyl ratio was estimated to be 1.55:1 (Ropponen et al. 2011). The Py-GC-MS characterization was performed on spruce kraft LignoBoost lignin in the temperature range of 300–700 °C. The highest amount of volatiles evolved at 600 °C with a major contribution from different guaiacols followed by catechols and phenols. Small amounts of hydrocarbons were also present as were furans and pyrans originating from carbohydrate impurities. Interestingly, polycyclic aromatic hydrocarbons were also detected in small amounts (Han et al. 2019). Eucalyptus kraft lignin was pyrolysed for 90 s at 400, 500, and 600 °C temperatures and the evolved volatiles were identified by GC-MS. With increasing temperature, the proportion of guaiacyl compounds increased while syringyl compounds decreased in the evolved volatiles. Phenolic compounds with no methoxyl substituents evolved only at 600 °C (Chen et al. 2020). Mixed hardwood kraft lignin was characterized by Py-GC-MS at 675 °C pyrolysis temperature. The main degradation products of guaiacyl-type were guaiacol, 4-methylguiacol, and 4-vinylguaiacol. The main degradation products of syringyl-type were syringol, 4-methylsyringol, and 4-vinylsyringol along with carbonyl compounds syringaldehyde and acetosyringone. 4-methylphenol and methoxycatechol were also detected. This lignin was estimated to have 47.5% syringyl, 44.5% guaiacyl, and 8% p-hydroxyphenyl moieties (Pe III et al. 2023).

In a variation of the method, several alkali lignins were analysed by TG-thermal desorption-GC-MS. In brief, the evolved volatiles during TG were adsorbed onto a polymer adsorbent matrix contained in a stainless steel tubing. The tubing with loaded adsorbents was then coupled to a gas chromatograph where the adsorbed volatiles were desorbed at high temperatures undergoing chromatographic separation and MS detection. The important variation in this method was the slow rate of heating at 10 °C/min which could influence the volatiles composition. The final temperature was 600 °C. Bagasse soda lignin was estimated to have 18.9% syringyl, 52.5% guaiacyl, and 28.6%

p-hydroxyphenyl entities. Bagasse soda anthraquinone lignin also had similar abundances 19.2% syringyl, 51.8% guaiacyl, and 29% p-hydroxyphenyl entities. Two eucalyptus kraft lignins had 35.1–45.5% syringyl, 52–61.4% guaiacyl, and 2.5–3.5% p-hydroxyphenyl moieties. Pine kraft lignins, as expected, were estimated to have 83.2–84% guaiacyl moieties with 8–8.4% syringyl and 7.7–8.9% p-hydroxyphenyl entities. Besides various phenols and their derivatives, evolved volatiles also contained toluene, anisole, and methyl anisole (Naron et al. 2017).

7.4.1.3 Organosolv Lignin

Organosolv lignins from prairie cordgrass by aqueous methyl isobutyl ketone and ethanol and from aspen by aqueous butanol were characterized by Py-GC-MS. The heating rate was 1000 °C/min and the final temperature was 600 °C. The main low molecular weight compounds in the evolved volatiles from cordgrass lignin, in decreasing order of abundance, were 4-vinyl guaiacol, guaiacol, 4-methylguaiacol, 2,3-dihydrobenzofuran, furfural, acetic acid, phenol, 4-methylphenol, and syringol. The main low molecular weight compounds in the evolved volatiles from aspen lignin, in decreasing order of abundance, were syringol, phenol, guaiacol, 4-methylsyringol, 4-methylguaiacol, 4-vinylguaiacol, acetosyringone, and propenylguaiacol. Besides guaiacyl-, syringyl-, p-hydroxyphenyl-, cresol-, and catechol-type compounds, the evolved volatiles contained aromatic and aliphatic hydrocarbons. Compounds like methoxybenzene, methoxytoluene, and methyl furan were also detected in the evolved volatiles. Presuming that the ratio of guaiacyl-, syringyl-, and p-hydroxyphenyl-type entities in evolved volatiles remain the same as they were in the lignin (not necessarily a true presumption), these ratios were worked out. Cordgrass lignin had 0.15 syringyl and 0.4 p-hydroxyphenyl substructures for every guaiacyl substructure. Aspen lignin had 0.78 syringyl and 0.46 p-hydroxyphenyl substructures for every guaiacyl substructure (Zhang et al. 2012). The molecular structure analysis of poplar alkaline aqueous ethanol organosolv lignin was carried out by Py-GC-MS. Flash pyrolysis with a 10,000 °C/min heating rate was performed at different pyrolysis temperatures between 400 and 700 °C. Important syringyl compounds in evolved volatiles were syringol, 4-methylsyringol, 4-ethylsyringol, 4-vinylsyringol, acetosyringone, sinapaldehyde, and syringaldehyde. Important guaiacyl compounds in evolved volatiles were guaiacol, 4-methylguaiacol, 4-ethylguaiacol, 4-vinylguaiacol, vanillin, and isoeugenol. Important p-hydroxyphenyl compounds in evolved volatiles were phenol and methylphenol. Acetic acid in evolved volatiles originated from acetyl groups in aliphatic side chains of phenylpropane units in lignin. Maximum volatiles evolved at 600 °C (Kim et al. 2013). Beech aqueous ethanol organosolv lignin was characterized by Py-GC-MS at 550 °C temperature. Lignin was qualitatively assessed to have substructures of syringyl, guaiacyl, and p-hydroxyphenyl-types from the identification of low molecular weight volatiles evolved (Hansen et al. 2016). Py-GC-MS analysis of palm kernel shell aqueous ethanol organosolv lignin in the temperature range of 450–950 °C is reported. In evolved volatiles, the proportion of guaiacyl entities decreased with increasing pyrolysis temperature while the proportion of syringyl entities did not follow a fixed pattern. The proportion of p-hydroxyphenyl entities was maximum at 750 °C, after which it decreased with further increase in temperature. Aromatic hydrocarbons first appeared in the evolved volatiles at 550 °C pyrolysis temperature and their proportion continuously increased thereafter with increasing temperature. The proportion of catechols was the highest at a pyrolysis temperature of 650 °C (Ma et al. 2018). Py-GC-MS analysis of bagasse aqueous ethanol organosolv lignin is reported in 400–700 °C temperature range at 100 °C intervals. The ratio of p-hydroxyphenyl/guaiacyl/syringyl compounds in evolved volatiles, 0.8:1:1, did not change appreciably between 400 and 600 °C. However, the abundance of p-hydroxyphenyl-type compounds was much higher at a pyrolysis temperature of 700 °C with p-hydroxyphenyl/guaiacyl/syringyl ratio reported to be 4.5:1:0.7 indicating extensive demethoxylation from aromatic ring at this temperature (Soongprasit et al. 2020). Syringyl, guaiacyl, and p-hydroxyphenyl substructures were estimated in birch aqueous ethanol organosolv lignins isolated at different severities by Py-GC-MS. In the absence of an acid catalyst, the syringyl content varied between 61% and 68%, guaiacyl content between 30% and 36%, and p-hydroxyphenyl content between 2% and 4%.

However, lignins isolated in the presence of an acid catalyst had much lower syringyl content between 37% and 49% with a higher guaiacyl content of 45–60% (Thoresen et al. 2021).

Py-GC-MS characterization of eucalyptus dioxane lignin is reported. The estimated syringyl/guaiacyl/p-hydroxyphenyl ratio was 84.5/15/0.5 (Evtuguin et al. 2001). Pine, poplar, and bamboo dioxane organosolv lignins were characterized by flash pyrolysis at a heating rate of 20,000 °C/min to 400–600 °C temperature. Py-GC-MS result suggested that pine lignin was predominantly guaiacyl-type with guaiacols contributing close to 80% in evolved volatiles. With increasing temperature, the relative content of guaiacols decreased and phenol content increased, but guaiacols were still the predominant product group. Among the guaiacols that evolved from pyrolysis of pine lignin, guaiacol, 4-methylguaiacol, 4-vinylguaiacol, isoeugenol, and vanillin were the most abundant. In poplar lignin, syringyl substructures had the highest abundance but the guaiacyl entities were also present in good amounts. Important syringols were syringol, 4-(2-propenyl) syringol, and 2,5-dimethoxy-4-methyl benzaldehyde. Evolved volatiles contained 19.2% phenol at 400 °C, but this phenol originated from p-hydroxybenzoate as well as from the breakdown of the lignin macromolecule. The phenolic content in evolved volatiles from bamboo lignin was close to 50% at all temperatures, 4-vinylphenol being the most abundant. In this case, the phenolic compounds mostly originated from p-coumaric acid (Wang et al. 2021).

Spruce and birch phosphinic acid-catalysed aqueous acetic acid organosolv lignins (commercial name Lignofibre) were subjected to Py-GC-MS with 2 s pyrolysis at 580 °C. In evolved volatiles from spruce lignin, guaiacol, 4-methylguiacol, vinylguaiacol, eugenol, isoeugenol, vanillin, homovanillin, and guaiacylacetone were the major low molecular weight compounds. Syringyl moieties were negligible and guaiacyl/p-hydroxyphenyl ratio was estimated to be 0.04:1. In evolved volatiles from birch lignin, guaiacol, 4-methylguiacol, vinylguaiacol, eugenol, isoeugenol, and guaiacylacetone were the major guaiacyl compounds. Syringyl compounds present in relatively high abundance were syringol, methylsyringol, ethylsyringol, 4-vinylsyringol, allylsyringol, propenylsyringol, syringaldehyde, acetosyringone and syringylacetone. p-hydroxyphenyl-type entities were negligible. Syringyl/guaiacyl ratio was estimated to be 2.47:1 (Ropponen et al. 2011).

The molecular structure analysis of poplar lignin isolated with the ionic liquid 1-ethyl-3-methylimidazolium acetate was carried out by Py-GC-MS. Flash pyrolysis with a 10,000 °C/min heating rate was performed at different pyrolysis temperatures between 400 and 700 °C. Important syringyl compounds in evolved volatiles were syringol, 4-methylsyringol, 4-vinylsyringol, acetosyringone, sinapaldehyde, and syringaldehyde. Important guaiacyl compounds in evolved volatiles were guaiacol, 4-methylguaiacol, 4-ethylguaiacol, 4-vinylguaiacol, vanillin, and isoeugenol. Important p-hydroxyphenyl compounds in evolved volatiles were phenol and methylphenol. Acetic acid in evolved volatiles originated from acetyl groups in aliphatic side chains of phenylpropane units in lignin. Maximum volatiles evolved at 600 °C (Kim et al. 2013).

7.4.1.4 Hydrolysis Lignin

Corn stover enzymatic hydrolysis lignin and rice straw acid hydrolysis lignin were characterized for molecular substructures by Py-GC-MS. Corn stover lignin was estimated to have 0.9 p-hydroxyphenyl and 0.07 syringyl substructure for every guaiacyl substructure. Rice straw lignin had three syringyl moieties for every guaiacyl moiety but a negligible amount of p-hydroxyphenyl moieties (Fox and McDonald 2010). Py-GC-MS analysis of corn stover enzymatic hydrolysis lignin from bioethanol production was carried out by flash pyrolysis at 20,000 °C/min to 450 °C temperature and holding for 30 s. Guaiacyl/syringyl/p-hydroxyphenyl ratio was reported to be 1.7/1/0.8. The main guaiacyl-type compounds in evolved gases were guaiacol, vinylguaiacol, and vanillic acid. The main syringyl-type compounds were syringol, propenylsyringol, and syringaldehyde. 4-methylphenol and 4-ethylphenol were the important compounds of the p-hydroxyphenyl-type (Lin et al. 2015). Bamboo enzymatic hydrolysis lignin was characterized by Py-GC-MS at 600 °C. Syringyl/guaiacyl/p-hydroxyphenyl ratio in evolved volatiles turned out to be 1.16:1:0.85. Major guaiacyl entities were guaiacol, 4-methylguaiacol, 4-vinylguaiacol, vanillin, and vanillic acid. Syringol, allyl syringol, acetosyringone, and syringaldehyde were the main syringyl entities. The lignin also contained an appreciable quantity of lignin carbohydrate complexes resulting in an abundance of 2,3-dihydrobenzofuran, acetic acid, and furfural (Yan et al. 2016). One study reported Py-GC-MS characterization of pine enzymatic mild hydrolysis lignin. The pyrolysis temperature was 600 °C. The Py-GC-MS characterization of rice straw enzymatic mild acidolysis lignin is reported. Fast pyrolysis was carried out in 400–900 °C temperature range at 100 °C intervals. With an increase in temperature from 400 to 800 °C, guaiacyl compounds decreased from 36.4% to 2.2% in relative abundance. The decrease in syringyl compounds was from 7.6% to 0.9%. An associated increase in p-hydroxyphenyl compounds was observed from 11% to 28%. The amount of 2,3-dihydrobenzofuran, resulting from hemicellulose impurities, steadily decreased from 24.8% at 400 °C to 7% at 800 °C. This study also reported the evolution of naphthalene, methylnaphthalene, vinylnaphthalene, and phenanthrene at high pyrolysis temperatures (Lou et al. 2018). The Py-GC-MS characterization was performed on lignin-rich residue obtained from softwood acid hydrolysis for ethanol production. Pyrolysis was done in the temperature range of 300–700 °C. The highest amount of volatiles evolved at 600 °C with a major contribution from different guaiacols followed by catechols and phenols. Small amounts of furans and pyrans were also present, originating from carbohydrate impurities. Polycyclic aromatic hydrocarbons were also detected in small amounts (Han et al. 2019).

Klason lignins from maple and Chinese fir were characterized by Py-GC-MS by flash pyrolysis at 20,000 °C/min to temperatures of 550, 650, and 800 °C. Product distribution was strongly influenced by pyrolysis temperature. Higher temperatures released more phenolic moieties (phenols, cresols, and catechols) and hydrocarbons compared to syringyl and guaiacyl moieties. Thus, during the pyrolysis of Chinese fir lignin, guaiacyl-type compounds in evolved volatiles decreased from 49.5% at 550 °C to 22% at 800 °C. The corresponding increase in phenolic compounds increased from 0% to 6.2%. The highest evolution of syringyl compounds was observed at 650 °C. In maple lignin, both syringyl and guaiacyl compounds decreased in evolved volatiles with increasing pyrolysis temperature. While catechols were produced at all pyrolysis temperatures, phenols were released only at 800 °C. For both the lignins, aromatic hydrocarbons – benzene, toluene, and xylene – were produced in small quantities at all temperatures but the quantity increased at higher temperatures (Zhao et al. 2014). Py-GC-MS analysis of palm kernel shell Klason lignin in the temperature range of 450–950 °C is reported. In evolved volatiles, the proportion of guaiacyl and syringyl entities did not follow any particular trend with increasing pyrolysis temperature. The proportion of p-hydroxyphenyl entities was maximum at 750 °C, after which it decreased with further increase in temperature. Aromatic hydrocarbons first appeared in the evolved volatiles at 550 °C pyrolysis temperature and their proportion continuously increased thereafter with increasing temperature. The proportion of catechols was the highest at pyrolysis temperatures of 650 and 750 °C (Ma et al. 2018).

7.4.1.5 Milled Wood Lignin

Wheat straw milled-wood lignin structural characterization was done by Py-GC-MS as well as Py/TMAH at 500 °C. In the evolved gases from pyrolysis alone, 4-vinylguaiacol was most abundant, followed by 4-methylguaiacol, 4-vinylphenol, and guaiacol. Other notable compounds present were syringol, vanillin, eugenol, and methyl syringol. From Py/TMAH, the most abundant evolved volatile was trans-3-(4-methoxyphenyl)-3-propenoic acid methyl ester, followed by 3,4-dimethoxybenzaldehyde. Other compounds present in good abundance were 4-methoxystyrene, 3,4-dimethoxytoluene, 3,4-dimethoxystyrene, 3,4-dimethoxybenzoic acid methyl ester, and 3,4,5-trimethoxyacetophenone. Syringyl/guaiacyl ratio was estimated to be 0.5 from Py/TMAH analysis (del Rio et al. 2012). The molecular structure analysis of poplar milled-wood lignin was carried out by Py-GC-MS. Flash pyrolysis with a 10,000 °C/min heating rate was performed at different pyrolysis temperatures between 400 and 700 °C. Important syringyl compounds in evolved volatiles were syringol, 4-methylsyringol, 4-vinylsyringol, acetosyringone, sinapaldehyde, and syringaldehyde. Important guaiacyl compounds in evolved volatiles were guaiacol, 4-methylguaiacol, 4-vinylguaiacol, vanillin, isoeugenol, coniferaldehyde, and coniferyl alcohol. Acetic acid in evolved volatiles originated from acetyl groups in aliphatic side chains of phenylpropane units in lignin. Maximum volatiles evolved at 600 °C (Kim et al. 2013). Coconut coir milled-wood lignin was subjected to Py-GC-MS analysis at 500 °C pyrolysis temperature. The major guaiacyl-type compounds released were guaiacol, 4-methylguaiacol, 4-vinylguaiacol, vanillin, trans-isoeugenol, and coniferaldehyde. Syringyl-type compounds were less abundant with the presence of syringol, 4-methylsyringol, vinylsyringol, and syringaldehyde. Syringyl/guaiacyl ratio was determined as 0.29. Pyrolysis also released a large quantity of phenol with 27.1% abundance in evolved volatiles which could arise from p-hydroxyphenyl substructures in lignin as well as p-hydroxybenzoate present independent of lignin macromolecule. To ascertain their exact source Py/TMAH was conducted as pyrolysis in the presence of TMAH is known to prevent decarboxylation. Thus, p-hydroxyphenyl substructures in lignin would release methoxybenzene while p-hydroxybenzoate would release 4-methoxybenzoic acid methyl ester. The pyrogram is shown in Figure 7.10.

Respective peaks are identified for methoxybenzene and 4-methoxybenzoic acid methyl ester. Integration of these peak areas revealed that most of the phenols in evolved volatiles from normal pyrolysis originated from p-hydroxybenzoate rather than p-hydroxyphenyl substructures in lignin (Rencoret et al. 2013). Py-GC-MS analysis of amur linden (a hardwood) milled-wood lignin was carried out by flash pyrolysis at 20,000 °C/min to 450 °C temperature and holding for 30 s. The guaiacyl/syringyl ratio was reported to be 1.1. The main guaiacyl-type compounds in evolved gases were guaiacol, vinylguaiacol, vanillic acid, and vanillin. The main syringyl-type compounds were syringol, propenylsyringol, and syringaldehyde (Lin et al. 2015). Py-GC-MS analysis of palm kernel shell milled-wood lignin in the temperature range of 450–950 °C is reported. In evolved volatiles, the proportion of guaiacyl and syringyl entities decreased with increasing pyrolysis temperature. The proportion of p-hydroxyphenyl entities was maximum at 750 °C, after which it decreased with further increase in temperature. Aromatic hydrocarbons first appeared in the evolved volatiles at 650 °C pyrolysis temperature and their proportion continuously increased thereafter with increasing temperature (Ma et al. 2018). Korean mixed hardwood milled-wood lignin and acacia milled-wood lignin were characterized by Py-GC-MS at 675 °C pyrolysis temperature. The main degradation products of guaiacyl-type were guaiacol, 4-methylguiacol, 4-vinylguaiacol, and vanillin. The main degradation products of syringyl-type were syringol, 4-methylsyringol, and 4-vinylsyringol along with carbonyl compounds syringaldehyde and acetosyringone. 4-methylphenol and methoxycatechol were also detected. Mixed hardwood lignin was estimated to have 46.3% syringyl, 47.5%

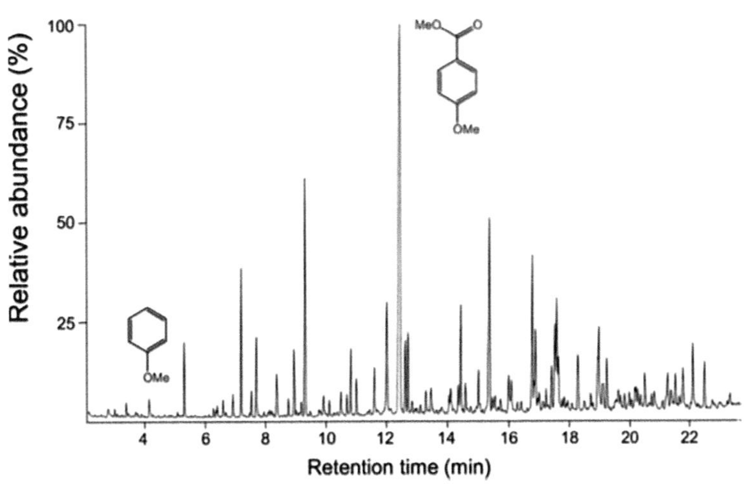

FIGURE 7.10 Chromatogram of evolved volatiles from pyrolysis of coconut coir milled-wood lignin in the presence of TMAH. (Reprinted and adapted from Rencoret, J, Ralph, J, Marques, G, Gutierrez, A, Martinez, AT, del Rio, JC, *J. Agric. Food Chem.*, 61, 2434–2445, 2013, with permission from American Chemical Society.)

guaiacyl, and 6.2% p-hydroxyphenyl moieties and acacia lignin had 31% syringyl, 60.3% guaiacyl, and 8.7% p-hydroxyphenyl moieties (Pe III et al. 2023).

7.5 COMPARISON OF CHARACTERIZATION METHODS, LIGNIN SOURCE, AND ISOLATION METHODS

From the discussion in Chapters 5 and 6 and foregone discussion in this chapter, it is immediately evident that lignins have remarkable variability in terms of functional groups, intermonomer linkages, and molecular substructures. Characterization results also vary according to the method used for the purpose. Representative variability in lignin characteristics based on their source, isolation method, and characterization technique is summarized in Tables 7.1–7.4.

From Table 7.1, we observe that hardwoods generally have a higher abundance of methoxyl groups compared to softwoods. In most instances, softwood lignins are found to have close to 1 or slightly less than 1 methoxyl group per phenylpropane unit which is in accordance with their predominantly guaiacyl nature. One notable reported exception is for pine enzymatic hydrolysis lignin where the estimated methoxyl content is 1.17 per phenylpropane unit compared to 0.96 for milled wood lignin (Holtman et al. 2004). The estimated methoxyl content in eucalyptus dioxane lignin (Evtuguin et al. 2001) is more than that found in eucalyptus kraft lignin (Sameni et al. 2016). However, this difference may also arise from the difference in subspecies of eucalyptus used in the two studies. Among the different characterization methods, heteronuclear 2D NMR reports somewhat more methoxyl content than ^{13}C NMR (Thoresen et al. 2021). One study that compares the wet chemistry method with ^{1}H NMR spectroscopy reports identical results for methoxyl content (Evtuguin et al. 2001).

Not only the total abundance of methoxyl groups but their distribution at different positions on the aromatic ring is also an important characterizing feature of lignin. Accordingly, molecular substructures in lignin may conform to syringyl-type, guaiacyl-type, or p-hydroxyphenyl-type. From Table 7.2, we observe that for most of the softwood lignins, the reported molecular substructures are exclusively guaiacyl-type. In rare cases of some softwood organosolv and milled wood lignins, p-hydroxyphenyl entities are found in less than 5% abundance in addition to the guaiacyl entities. Among the hardwood lignins, eucalyptus lignins are found to have a larger proportion of syringyl substructures irrespective of the isolation method. The remaining molecular structures are guaiacyl-type except for one ^{13}C NMR and Py-GC-MS characterization of dioxane organosolv lignin reporting 0.5–2% abundance of p-hydroxyphenyl moieties. Also, for other hardwood lignins, the methoxyl groups are found distributed between syringyl and guaiacyl moieties with syringyl predominating over guaiacyl in most cases. Nonwood lignins are more heterogeneous in terms of their relative abundance of syringyl, guaiacyl, and p-hydroxyphenyl. Most of these lignins have all three molecular substructures present. Different characterization methods are not always in agreement. For example, lower syringyl content is reported in coconut coir milled-wood lignin when estimated by ^{13}C NMR as compared to thioacidolysis (Rencoret et al. 2013). Similarly, the estimated syringyl, guaiacyl, and p-hydroxyphenyl content in eucalyptus dioxane organosolv lignin differed when characterized by nitrobenzene oxidation, thioacidolysis, ^{1}H NMR, and Py-GC-MS (Evtuguin et al. 2001). Py-GC-MS estimation of syringyl, guaiacyl, and p-hydroxyphenyl content is particularly contentious as it is strongly temperature-dependent. Compared to other characterization methods, estimated p-hydroxymethyl content by Py-GC-MS is appreciably higher. This is not always a reflection of true abundance but is likely to result from the demethoxylation of syringyl and guaiacyl moieties.

The abundance of hydroxyl groups and their distribution between aliphatic and phenolic types is an important lignin characteristic. There is a lot of variability with respect to reported results irrespective of lignocellulose source and isolation method as summarized in Table 7.3. For example, in technical lignin Indulin AT (softwood kraft lignin), some studies report more aliphatic hydroxyl than phenolic hydroxyl (Faix et al. 1994; Sameni et al. 2016) while others report the contrary (Serrano

TABLE 7.1
Methods Comparison for Lignin Characterization for Methoxyl Group Content

Lignin type	Characterization Method			
	Hydroiodic Acid	^{1}H NMR	^{13}C NMR	Heteronuclear 2D NMR
Wheat straw soda lignin	3.59 mmol/g (Lee et al. 2021)	0.38 per phenylpropane unit (Ghatak et al. 2010)		
Nonwood soda lignin (Protobind)		1.41 per phenylpropane unit (Sameni et al. 2016)		
Eucalyptus kraft lignin		1.45 per phenylpropane unit (Sameni et al. 2016)		6.86 mmol/g (Chen et al. 2020)
Softwood kraft lignin (Indulin AT)		1.02 per phenylpropane unit (Sameni et al. 2016)		
Mixed softwood kraft lignins			0.7–0.78 per phenylpropane unit (Zhu and Theliander 2015)	
Birch aqueous ethanol organosolv lignins			4.56–7.02 mmol/g (Thoresen et al. 2021)	4.23–11.26 mmol/g (Thoresen et al. 2021)
Eucalyptus dioxane lignin	1.64 per phenylpropane unit (Evtuguin et al. 2001)	1.64 per phenylpropane unit (Evtuguin et al. 2001)		
Banana leaf sheath dioxane lignin	1.09 per phenylpropane unit (Oliveira et al. 2006)		18.3% (Oliveira et al. 2006)	
Japanese cedar milled-wood lignin	3.5 mmol/g (Goto et al. 2006)			
Japanese birch milled-wood lignin	4.8 mmol/g (Goto et al. 2006)			
Poplar milled-wood lignin			1.72 per phenylpropane unit (Tan et al. 2017)	
Pine milled-wood lignin			0.96 per phenylpropane unit (Holtman et al. 2004)	
			1 per phenylpropane unit (Huang et al. 2011)	
Pine enzymatic hydrolysis lignin			1.17 per phenylpropane unit (Holtman et al. 2004)	
Rice straw enzymatic mild acidolysis lignin			13.9% (Lou et al. 2018)	

TABLE 7.2
Methods Comparison for Lignin Characterization for The Relative Proportion of Syringyl, Guaiacyl, and P-Hydroxyphenyl Moieties

	Characterization Method					
Lignin Type	**Nitrobenzene Oxidation**	**Thioacidolysis**	**DFRC**	**^{13}C NMR**	**Heteronuclear 2D NMR**	**Py-GC-MS**
Abaca soda lignin						Syringyl 55% Guaiacyl 19% p-Hydroxyphenyl 26% (del Rio et al. 2007b)
Jute soda lignin						Syringyl 60% Guaiacyl 36% p-Hydroxyphenyl 4% (del Rio et al. 2007b)
Hemp soda lignin						Syringyl 40% Guaiacyl 51% p-Hydroxyphenyl 9% (del Rio et al. 2007b)
Miscanthus soda lignin					Syringyl 41% Guaiacyl 54% p-Hydroxyphenyl 5% (Lauwaert et al. 2019)	
Wheat straw soda lignin						Syringyl 23.8% Guaiacyl 54.8% p-Hydroxyphenyl 21.4% (Lin et al. 2015)
Bagasse soda lignin						Syringyl 18.9% guaiacyl 52.5% p-Hydroxyphenyl 28.6% (Naron et al. 2017)
Japanese cedar soda anthraquinone lignin	Guaiacyl 100% (Ikeda and Magara 2015)		Guaiacyl 100% (Ikeda and Magara 2015)			
Pine kraft lignin	Guaiacyl 100% (Ikeda and Magara 2015)		Guaiacyl 100% (Ikeda and Magara 2015)			

Eucalyptus kraft lignin	Syringyl 78.9% Guaiacyl 21.1% (Villar et al. 1997)				Syringyl 70.9% Guaiacyl 29.1% (Chen et al. 2020)	
Eucalyptus dioxane organosolv lignin	Syringyl 86% Guaiacyl 14% (Evtuguin et al. 2001)	Syringyl 81.7% Guaiacyl 18.3% (Evtuguin et al. 2001)		Syringyl 84% Guaiacyl 14% p-hydroxyphenyl 2% (Evtuguin et al. 2001)		Syringyl 84.5% Guaiacyl 15% p-hydroxyphenyl 0.5% (Evtuguin et al. 2001)
Pine ethanol organosolv lignin					Guaiacyl 100% (Sheng et al. 2022)	
Poplar ethanol organosolv lignin					Syringyl 49.7% Guaiacyl 50.3% (Sheng et al. 2022)	
Aspen butanol organosolv lignin						Syringyl 34.8% Guaiacyl 44.6% p-hydroxyphenyl 20.6% (Zhang et al. 2012)
Birch acetic acid organosolv lignin						Syringyl 71.2% Guaiacyl 28.8% (Ropponen et al. 2011)
Prairie cordgrass methyl isobutyl ketone and ethanol organosolv lignin						Syringyl 9.7% Guaiacyl 64.5% p-hydroxyphenyl 25.8% (Zhang et al. 2012)
Walnut shell ethanol organosolv lignin					Syringyl 51.5% Guaiacyl 48.5% (Sheng et al. 2022)	
Spruce acetic acid organosolv lignin						Guaiacyl 96.2% p-Hydroxyphenyl 3.8% (Ropponen et al. 2011)

(*Continued*)

TABLE 7.2 (Continued)
Methods Comparison for Lignin Characterization for The Relative Proportion of Syringyl, Guaiacyl, and P-Hydroxyphenyl Moieties

	Characterization Method					
Lignin Type	**Nitrobenzene Oxidation**	**Thioacidolysis**	**DFRC**	**^{13}C NMR**	**Heteronuclear 2D NMR**	**Py-GC-MS**
Spruce milled-wood lignin		Guaiacyl 97.7% p-Hydroxyphenyl 2.3% (Marques et al. 1999) Guaiacyl 100% (Onnerud and Gellerstedt 2003)	Guaiacyl 100% (del Rio et al. 2007a)			
Pine milled-wood lignin			Guaiacyl 100% (Holtman et al. 2003) Guaiacyl 100% (del Rio et al. 2007a)	Guaiacyl 100% (Huang et al. 2011) Guaiacyl 96%p-hydroxyphenyl 4% (Holtman et al. 2004)		
Poplar milled-wood lignin		Syringyl 52.8% Guaiacyl 47.2% (Lapierre et al. 1985)			Syringyl 61.1% Guaiacyl 38.9% (Yuan et al. 2011) Syringyl 58.8% Guaiacyl 41.2% (Tan et al. 2017)	
Eucalyptus milled-wood lignin			Syringyl 63.2% Guaiacyl 36.8% (del Rio et al. 2007a)			
Birch milled-wood lignin					Syringyl 66.7% Guaiacyl 33.3% (Balakshin et al. 2011)	
Coconut coir milled-wood lignin		Syringyl 24.9% Guaiacyl 73.2% p-Hydroxyphenyl 1.9% (Rencoret et al. 2013)			Syringyl 18% Guaiacyl 78% p-Hydroxyphenyl 4% (Rencoret et al. 2013)	Syringyl 16.4% Guaiacyl 56.5% p-Hydroxyphenyl 27.1% (Rencoret et al. 2013)
Wheat straw milled-wood lignin	Syringyl 45.9% Guaiacyl 45.5% p-Hydroxyphenyl 8.6% (Billa and Monties 1995) Syringyl 51% Guaiacyl 42.5% p-Hydroxyphenyl 6.5% (Sun et al. 2005)	Syringyl 44.3% Guaiacyl 49.3% p-Hydroxyphenyl 6.4% (Billa and Monties 1995)	Syringyl 31.3% Guaiacyl 68.7% (del Rio et al. 2012)			Syringyl 33.3% Guaiacyl 66.7% (del Rio et al. 2012)

Bamboo milled-wood lignin		Syringyl 64.4% Guaiacyl 35.6% (del Rio et al. 2007a)	Syringyl 56% Guaiacyl 35% p-Hydroxyphenyl 9% (Shi et al. 2019)	
Kenaf milled-wood lignin		Syringyl 75.6% Guaiacyl 24.4% (del Rio et al. 2008)		
Bamboo enzymatic hydrolysis lignin				Syringyl 38.5% Guaiacyl 33.2% p-Hydroxyphenyl 28.3% (Yan et al. 2016)
Arundo donax milled-wood lignin			Syringyl 39% Guaiacyl 60% p-Hydroxyphenyl 1% (You et al. 2015)	
Poplar enzymatic hydrolysis lignin	Syringyl 66.8% Guaiacyl 33.2% (Lapierre et al. 1985)			
Poplar mild acidolysis lignin			Syringyl 61.8% Guaiacyl 38.2% (Yuan et al. 2011)	
Birch enzymatic hydrolysis lignin			Syringyl 76.2% Guaiacyl 23.8% (Balakshin et al. 2011)	
Wheat straw enzymatic hydrolysis lignin	Syringyl 50.4% Guaiacyl 44.3% p-Hydroxyphenyl 5.3% (Billa and Monties 1995)			
Arundo donax enzymatic hydrolysis lignin			Syringyl 46% Guaiacyl 53% p-Hydroxyphenyl 1% (You et al. 2015)	

TABLE 7.3
Methods Comparison for Lignin Characterization for Aliphatic and Phenolic Hydroxyl Groups

	Characterization Method				
Lignin Type	**UV**	**^{1}H NMR**	**^{13}C NMR**	**^{31}P NMR**	**Heteronuclear 2D NMR**
Miscanthus soda lignin				Aliphatic 2.95 Phenolic 0.59 mmol/g (Lauwaert et al. 2019)	
Protobind		Aliphatic 0.71 Phenolic 0.77 per phenylpropane unit (Sameni et al. 2016)		Aliphatic 2.47 Phenolic 2.28 mmol/g (Sameni et al. 2016) Aliphatic 1.5 Phenolic 2.7 mmol/g (Serrano et al. 2018)	
Spruce soda lignin		Aliphatic 2.53 Phenolic 1.41 mmol/g (Nitsos et al. 2016)			
Birch soda lignin		Aliphatic 1.49 Phenolic 0.9 mmol/g (Nitsos et al. 2016)			
Indulin AT	Phenolic 2.88 mmol/g (Goldmann et al. 2017)	Aliphatic 0.67 Phenolic 0.52 per phenylpropane unit (Faix et al. 1994) Phenolic 0.64 per phenylpropane unit (Tiainen et al. 1999) Aliphatic 0.92 Phenolic 0.67 per phenylpropane unit (Sameni et al. 2016) Phenolic 3.1 mmol/g (Serrano et al. 2018)	Aliphatic 2.64 Phenolic 3.84 mmol/g (Goldmann et al. 2017) Phenolic 3.6 mmol/g (Serrano et al. 2018)	Aliphatic 4 Phenolic 2.59 mmol/g (Sameni et al. 2016) Aliphatic 2.06 Phenolic 3.2 mmol/g (Serrano et al. 2018)	
Pine kraft lignin			Phenolic 4.2 mmol/g (Serrano et al. 2018)		
Eucalyptus kraft		Aliphatic 0.59 Phenolic 0.86 per phenylpropane unit (Sameni et al. 2016)			Aliphatic 1.06 Phenolic 3.81 mmol/g (Chen et al. 2020)

Bamboo kraft lignin				Phenolic 1.47 mmol/g (Wang et al. 2019)
Spruce ethanol organosolv lignin		Aliphatic 1.12 Phenolic 2.78 mmol/g (Nitsos et al. 2016)		
Birch ethanol organosolv lignin		Aliphatic 2.81 Phenolic 2.51 mmol/g (Nitsos et al. 2016)		
Alcell lignin		Aliphatic 0.57 Phenolic 0.72 per phenylpropane unit (Faix et al. 1994)		
Miscanthus formic acid organosolv lignin		Aliphatic 3.5% Phenolic 3.33% (El Mansouri and Salvado 2007)	Aliphatic 3.2% Phenolic 3.23% (El Mansouri and Salvado 2007)	
Miscanthus ethanol organosolv lignin				Aliphatic 0.19 Phenolic 0.5 per phenylpropane unit (El Hage et al. 2009)
Rice straw ethanol organosolv lignin				Aliphatic 0.26 Phenolic 1.77 mmol/g (Moniz et al. 2018)
Beech ethanol organosolv lignin				Aliphatic 1.3 Phenolic 3.1 mmol/g (Rossberg et al. 2019)
Spruce ethanol organosolv lignin				Aliphatic 2.73 Phenolic 3.7 mmol/g (Parot et al. 2022)
Pine milled-wood lignin	Phenolic 3.22% (Faix et al. 1992)	Phenolic 2.71% (Faix et al. 1992)		
Spruce milled-wood lignin		Phenolic 0.23 per phenylpropane unit (Tiainen et al. 1999)		Phenolic 0.22 per phenylpropane unit (Crestini et al. 2011)
Beech milled-wood lignin				Phenolic 0.7 per phenylpropane unit (Crestini et al. 2011)

(*Continued*)

TABLE 7.3 (Continued)
Methods Comparison for Lignin Characterization for Aliphatic and Phenolic Hydroxyl Groups

	Characterization Method				
Lignin Type	**UV**	**^{1}H NMR**	**^{13}C NMR**	**^{31}P NMR**	**Heteronuclear 2D NMR**
Arundo donax milled-wood lignin	Phenolic 2.48% (Faix et al. 1992)	Phenolic 2.65% (Faix et al. 1992)			
Miscanthus milled-wood lignin	Phenolic 2.73% (Faix et al. 1992)	Phenolic 3.41% (Faix et al. 1992)		Aliphatic 0.67 Phenolic 0.26 per phenylpropane unit (El Hage et al. 2009)	
Spruce enzymatic mild acidolysis lignin				Phenolic 0.21 per phenylpropane unit (Crestini et al. 2011)	
Poplar enzymatic hydrolysis lignin		Aliphatic 4.73% Phenolic 2.65% (El Mansouri and Salvado 2007)	Aliphatic 4.55% Phenolic 2.7% (El Mansouri and Salvado 2007)		
Wheat straw enzymatic mild acidolysis lignin				Phenolic 1.27 mmol/g (Yang et al. 2011)	
Rice straw enzymatic mild acidolysis lignin				Aliphatic 5.35 Phenolic 3.09 mmol/g (Lou et al. 2018)	

TABLE 7.4
Methods Comparison for Lignin Characterization for The Relative Proportion of Intermonomer Linkages

	Characterization Method					
Lignin Type	**Nitrobenzene Oxidation**	**Thioacidolysis**	**DFRC**	**^{1}H NMR**	**^{13}C NMR**	**Heteronuclear 2D NMR**
Miscanthus soda lignin						β–O–4′ 0.44 (68.8%) α–O–4′ 0.05 (7.8%) β–5′ 0.05 (7.8%) β–β′ 0.05 (7.8%) α–O–γ′ 0.05 (7.8%) per phenylpropane unit (Lauwaert et al. 2019)
Eucalyptus kraft lignin						β–O–4′ 0.37 (84.9%) β–5′ 0.054 (12.4%) β–β′ 0.012 (2.7%) per phenylpropane unit (Chen et al. 2020)
Pine ethanol organosolv lignin						β–O–4′ 67.4% β–β′ 11.9% β–5′ 20.7% (Sheng et al. 2022)
Birch ethanol organosolv lignin					β-O-4′ 3.2 (68.5%) β–5′ 0.38 (8.1%) β–β′ 0.84 (18%) β–1′ 0.25 (5.4%) mmol/g (Thoresen et al. 2021)	
Poplar ethanol organosolv lignin						β–O–4′ 77.2% β–β′ 10.2% β–5′ 12.6% (Sheng et al. 2022)
Eucalyptus dioxane organosolv lignin	Alkyl aryl ether bonds 37.5% (Evtuguin et al. 2001)	Alkyl aryl ether bonds 1635 μmol/g (Evtuguin et al. 2001)			β–O–4′ 0.56 (81.2%) β–β′, β–5′ 0.13 (18.8%) per phenylpropane unit (Evtuguin et al. 2001)	

(Continued)

TABLE 7.4 (Continued)
Methods Comparison for Lignin Characterization for The Relative Proportion of Intermonomer Linkages

	Characterization Method					
Lignin Type	**Nitrobenzene Oxidation**	**Thioacidolysis**	**DFRC**	**^{1}H NMR**	**^{13}C NMR**	**Heteronuclear 2D NMR**
Banana leaf sheath dioxane organosolv lignin				β-O-4′ 0.26 (68.4%) α–O–4′ 0.08 (21.1%) β–β′ 0.04 (10.5%) per phenylpropane unit (Oliveira et al. 2006)		
Bamboo formic acid organosolv lignin	Alkyl aryl ether bonds 24.9% (Li et al. 2012)					β–O–4′ 64% β–β′ 10% β–5′ 22% (Li et al. 2012)
Pine milled-wood lignin	Alkyl aryl ether bonds 31.8% (Holtman et al. 2004)	Alkyl aryl ether bonds 18.1% (Holtman et al. 2004) Alkyl aryl ether bonds 995 μmol/g (Chazal et al. 2014)	β-O-4′ 735.6 μmol/g (92.3%) α-O-4′ 61.4 μmol/g (7.7%) (Holtman et al. 2003) Alkyl aryl ether bonds 402 μmol/g (del Rio et al. 2007a)		β–O–4′ 0.41 carbon–carbon linkages 1.53 per phenylpropane unit (Holtman et al. 2004)	β–O–4′ 0.3 (66.1%) β–β′ 0.05 (11%) β–5′ 0.104 (22.9%) per phenylpropane unit (Balakshin et al. 2011)
Spruce milled-wood lignin		Alkyl aryl ether bonds 1146.2 μmol/g (Marques et al. 1999) Alkyl aryl ether bonds 986 μmol/g (Onnerud and Gellerstedt 2003) β-5′ 65 μmol/g, β-1′ 66 μmol/g5-5′ 62 μmol/g, 4-O-5′ 16 μmol/g (Yue et al. 2017)	Alkyl aryl ether bonds 520 μmol/g (del Rio et al. 2007a)			β–O–4′ plus α′–O–4 plus 5–5′ 0.033, β–1′ 0.015 β–β′ 0.03 per phenylpropane unit (Crestini et al. 2011)
Japanese cedar milled-wood lignin	Alkyl aryl ether bonds 23.4% (Ikeda and Magara 2015)		Alkyl aryl ether bonds 11.9% (Ikeda and Magara 2015)			

Eucalyptus milled-wood lignin			Alkyl aryl ether bonds 162 μmol/g (del Rio et al. 2007a)			
Poplar milled-wood lignin					β–O–4′ 0.53 Carbon–carbon linkages 1.64 per phenylpropane unit (Tan et al. 2017)	β–O–4′ 0.415 α–O–γ′ plus α′–O–γ plus β–β′ 0.146 α–O–4′ plus β–5′ 0.037 α–O–α′ plus β–1′ 0.007 per phenylpropane unit (Yuan et al. 2011) β–O–4′ 0.55 (82.1%) β–β′ 0.09 (13.4%) β–5′ 0.03 (4.5%) per phenylpropane unit (Tan et al. 2017)
Wheat straw milled-wood lignin	Alkyl aryl ether bonds 76.4% (Sun et al. 2005)	Alkyl aryl ether bonds 834 μmol/g (Billa and Monties 1995)	Alkyl aryl ether bonds 638 μmol/g (del Rio et al. 2012)			β–O–4′ 77% α–O–4′ plus β–5′ 11% α–O–γ′ plus β–β′ 4% α–O–4′ plus β–O–4″ plus 5′–5″ 3% α–O– α′ plus β–1′ 3% (del Rio et al. 2012)
Bamboo milled-wood lignin	Alkyl aryl ether bonds 39.4% (Li et al. 2012)		Alkyl aryl ether bonds 755 μmol/g (del Rio et al. 2007a)			β–O–4′ 72% α–O–γ′ plus α′–O–γ plus β–β′ 15% α–O–4′ plus β–5′ 8% (Li et al. 2012) β–O–4′ 79.5% α–O–γ′ plus α′–O–γ plus β–β′ 7.2% α–O–4′ plus β–5′ 5.6% (Shi et al. 2019)
Coconut coir milled-wood lignin		Alkyl aryl ether bonds 874 μmol/g, 5-5′ 15.9%, β-1′ 22.1%, β-5′ 45.8%, β-β′ 8.7% 4-O-5′ 7.4% (Rencoret et al. 2013)				β–O–4′ 82% α–O–γ′ plus α′–O–γ plus β–β′ 4% α–O–4′ plus β–5′ 13% α–O–4′ plus β–O–4″ plus 5′–5″ 1% (Rencoret et al. 2013)

(Continued)

TABLE 7.4 (Continued)
Methods Comparison for Lignin Characterization for The Relative Proportion of Intermonomer Linkages

Lignin Type	Characterization Method: Nitrobenzene Oxidation	Thioacidolysis	DFRC	^{1}H NMR	^{13}C NMR	Heteronuclear 2D NMR
Pine enzymatic hydrolysis lignin	Alkyl aryl ether bonds 32.3% (Holtman et al. 2004)	Alkyl aryl ether bonds 20.3% (Holtman et al. 2004) β-5′ 78.6 μmol/g, β-1′ 46.1 μmol/g 5-5′ 35.1 μmol/g β-β′ 1 μmol/g4-O-5′ 15.2 μmol/g (Yue et al. 2017)	Alkyl aryl ether bonds 651 μmol/g (Lu and Ralph 1998)		β–O–4′ 0.47 Carbon–carbon linkages 1.42 per phenylpropane unit (Holtman et al. 2004)	β–O–4′ 0.35 (68.6%) β–β′ 0.05 (9.8%) β–5′ 0.11 (21.6%) per phenylpropane unit (Balakshin et al. 2011)
Pine enzymatic mild acidolysis lignin		Alkyl aryl ether bonds 1331 μmol/g (van de Pas et al. 2014)				
Aspen enzymatic hydrolysis lignin			Alkyl aryl ether bonds 696 μmol/g (Lu and Ralph 1998)			
Birch enzymatic hydrolysis lignin						β–O–4′ 0.48 (80%) β–β′ 0.09 (15%) β–5′ 0.03 (5%) per phenylpropane unit (Balakshin et al. 2011)
Wheat straw enzymatic hydrolysis lignin		Alkyl aryl ether bonds 834 μmol/g (Billa and Monties 1995)				
Rice straw enzymatic mild acidolysis lignin					β–O–4′ 26.5% (Lou et al. 2018)	

et al. 2018). For spruce and birch lignins, the isolation method was found to have a strong bearing on the abundance of hydroxyl groups. Spruce soda lignin had higher hydroxyl content than birch soda lignin but spruce ethanol organosolv lignin had less hydroxyl content than birch ethanol organosolv lignin (Nitsos et al. 2016). There is disagreement in results reported by different characterization methods even for the same lignin. The ratio of aliphatic to phenolic hydroxyl in Indulin AT was estimated to be 1.37 by ^{1}H NMR and 1.54 by ^{31}P NMR (Sameni et al. 2016). In poplar enzymatic hydrolysis lignin, the ratio of aliphatic to phenolic hydroxyl was reported as 1.78 by ^{1}H NMR and 1.68 by ^{13}C NMR (El Mansouri and Salvado 2007).

The abundance and relative proportion of different intermonomer linkages in lignin from different sources and isolation methods as characterized by various characterization techniques is summarized in Table 7.4. Barring a few exceptions, there is general unanimity that the alkyl aryl ether bond β-O-4′ is most frequently encountered in a lignin macromolecule. Among the carbon–carbon linkages, β–β′ and β–5′ are the most abundant. Lignins isolated from eucalyptus is found to have a high proportion of β-O-4′ linkages, often exceeding 80%. However, it is difficult to observe a clear trend for the abundance of β-O-4′ intermonomer bonds in lignins sourced from other lignocelluloses. In different studies, it varied between 60% and 75%. Wet chemistry methods mostly estimate a composite value for the occurrence of all alkyl aryl ether bonds including α-O-4′. In one study, DFRC characterization of pine milled-wood lignin estimated the α-O-4′ linkages to be 7.7% of all alkyl aryl ether bonds (Holtman et al. 2003). Among all the characterization methods, 2D heteronuclear NMR gives the most comprehensive account of the abundance and distribution of different intermonomer linkages in lignin macromolecules. Not only is it possible to individually account for the different linkages but the method also characterizes lignin in terms of molecular substructures where phenylpropane units are joined by more than one linkage. Thus, it is possible to estimate the individual abundance of structural subunits joined by a combination of α–O–γ′, α′–O–γ, and β–β′ linkages (resinol), α–O–4′ and β–5′ linkages (phenylcoumaran), α–O–4′, β–O–4″ and 5′–5″ linkages (dibenzodioxocin), and α–O–α′ plus β–1′ linkages (spirodienone). Wet chemistry methods sometimes may underestimate the alkyl aryl ether linkages. For example, there are instances where alkyl aryl ether linkages in different lignin samples are reported between 23.4% and 39.4% by nitrobenzene oxidation (Evtuguin et al. 2001; Holtman et al. 2004; Ikeda and Magara 2015; Li et al. 2012). Similarly, the estimated proportion of alkyl aryl ether linkages in Japanese cedar milled-wood lignin by DFRC characterization was only 11.9% (Ikeda and Magara 2015). There is little correspondence between results estimated by different wet chemistry characterization methods for the same lignin. Alkyl aryl ether linkages in Japanese cedar milled-wood lignin were estimated to be 23.4% by nitrobenzene oxidation compared to only 11.9% by DFRC (Ikeda and Magara 2015). The estimated alkyl aryl ether bonds in pine milled-wood lignin by nitrobenzene oxidation was 31.8% but thioacidolysis of the same lignin estimated 18.1% linkages to be alkyl aryl ether type (Holtman et al. 2004).

REFERENCES

Amit, TA, Roy, R, Raynie, DE. *Current Res. Green Sustain. Chem.* 4:100106, 2021.

Awal, A, Sain, M. *J. Appl. Polym. Sci.* 122:956–963, 2011.

Balakshin, M, Capanema, E, Gracz, H, Chang, HM, Jameel, H. *Planta* 233:1097–1110, 2011.

Ballice, L, Sert, M, Saglam, M, Yüksel, M. *Arabian J. Sci. Eng.* 45:7429–7444, 2020.

Bardow, A, Marquardt, W. *Chem. Eng. Sci.* 59:2673–2684, 2004.

Billa, E, Monties, B. *Res. Chem. Intermed.* 21(3-5):303–311, 1995.

Chazal, R, Robert, P, Durand, S, Devaux, MF, Saulnier, L, Lapierre, C, Guillon, F. *Appl. Spectroscopy* 68: 1342–1347, 2014.

Chen, Y, Stark, NM, Cai, Z, Frihart, CR, Lorenz, LF, Ibach, RE. *BioResources* 9:5488–5500, 2014.

Chen, S, Cheng, H, Wu, S. *Fuel* 270:117520, 2020.

Crestini, C, Melone, F, Sette, M, Saladino, R. *Biomacromolecules* 12:3928–3935, 2011.

del Rio, JC, Marques, G, Rencoret, J, Martinez, AT, Gutierrez, A. *J. Agric. Food Chem.* 55:5461–5468, 2007a.

del Rio, JC, Gutierrez, A, Rodríguez, IM, Ibarra, D, Martinez, AT. *J. Anal. Appl. Pyrolysis* 79:39–46, 2007b.

del Rio, JC, Rencoret, J, Marques, G, Gutierrez, A, Ibarra, D, Santos, JI, Jimenez-Barbero, J, Zhang, L, Martinez, AT. *J. Agric. Food Chem.* 56:9525–9534, 2008.

del Rio, JC, Rencoret, J, Prinsen, P, Martinez, AT, Ralph, J, Gutierrez, A. *J. Agric. Food Chem.* 60: 5922–5935, 2012.

El Hage, R, Brosse, N, Chrusciel, L, Sanchez, C, Sannigrahi, P, Ragauskas, A. *Polymer Deg. Stab.* 94: 1632–1638, 2009.

El Mansouri, NE, Salvado, J. *Ind. Crops Prod.* 26:116–124, 2007.

El Moustaqim, M, El Kaihal, A, El Marouani, M, Men-La-Yakhaf, S, Taibi, M, Sebbahi, S, El Hajjaji, S, Kifani-Sahban, F. *Sustain. Chem. Pharmacy* 9:63–68, 2018.

Evtuguin, DV, Neto, CP, Silva, AMS, Domingues, PM, Amado, FML, Robert, D, Faix, O. *J. Agric. Food Chem.* 49:4252–4261, 2001.

Faix, O, Grünwald, C, Beinhoff, O. *Holzforschung* 46:425–432, 1992.

Faix, O, Argyropoulos, DS, Robert, D, Neirinck, V. *Holzforschung* 48:387–394, 1994.

Ferdous, D, Dalai, AK, Bej, SK, Thring, RW. *Ener. Fuels* 16:1405–1412, 2002.

Fox, SC, McDonald, AG. *BioResources* 5:990–1009, 2010.

Ghatak, HR. Period. Polytech. *Chem. Eng.* 66:494–502, 2022.

Ghatak, HR, Kundu, PP, Kumar, S. *Thermochimica Acta* 502:85–89, 2010.

Goldmann, WM, Ahola, J, Mankinen, O, Kantola, AM, Komulainen, S, Telkki, VV, Tanskanen, J. Periodica Polytechnica *Chem. Eng.* 61:93–101, 2017.

Goto, H, Koda, K, Tong, G, Matsumoto, Y, Meshitsuka, G. *J. Wood Chem. Technol.* 26:81–93, 2006.

Han, T, Sophonrat, N, Tagami, A, Sevastyanova, O, Mellin, P, Yang, W. *Fuel* 235:1061–1069, 2019.

Hansen, B, Kusch, P, Schulze, M, Kamm, B. *J. Polym. Environ.* 24:85–97, 2016.

Hatakeyama, H, Tsujimoto, Y, Zarubin, MJ, Krutov, SM, Hatakeyama, T. *J. Thermal Anal. Calorim.* 101: 289–295, 2010.

Haykiri-Acma, H, Yaman, S, Kucukbayrak, S. *Fuel Process. Technol.* 91:759–764, 2010.

Holtman, KM, Chang, HM, Jameel, H, Kadla, JF. *J. Agric. Food Chem.* 51:3535–3540, 2003.

Holtman, KM, Chang, HM, Kadla, JF. *J. Agric. Food Chem.* 52:720–726, 2004.

Huang, F, Singh, PM, Ragauskas, AJ. *J. Agric. Food Chem.* 59:12910–12916, 2011.

Ikeda, T, Magara, K. *J. Wood Chem. Technol.* 35:167–177, 2015.

Jankovic, B. *Bioresource Technol.* 102:9763–9771, 2011.

Jiang, G, Nowakowski, DJ, Bridgwater, AV. *Thermochimica Acta* 498:61–66, 2010.

Kim, JY, Oh, S, Hwang, H, Kim, UJ, Choi, JW. *Polym. Deg. Stab.* 98:1671–1678, 2013.

Kissinger, HE. *Anal. Chem.* 29:1702–1706, 1957.

Kubo, S, Kadla, JF. *J. Wood Chem. Technol.* 28:106–121, 2008.

Lapierre, C, Monties, B, Rolando, C. *J. Wood Chem. Technol.* 5:277–292, 1985.

Lauwaert, J, Stals, I, Lancefield, CS, Deschaumes, W, Depuydt, D, Vanlerberghe, B, Devlamynck, T, Bruijnincx, PCA, Verberckmoes, A. *Sep. Purific. Technol.* 221:226–235, 2019.

Lee, JH, Kim, TM, Choi, IG, Choi, JW. *Nanomaterials* 11:1790, 2021.

Li, J, Li, B, Zhang, X. *Polym. Deg. Stab.* 78:279–285, 2002.

Li, MF, Sun, SN, Xu, F, Sun, RC. *Chem. Eng. J.* 179:80–89, 2012.

Li, X, Wu, S. *BioResources* 9:6277–6289, 2014.

Lin, X, Sui, S, Tan, , S, Pittman Jr., CU, Sun, J, Zhang, Z. *Energies* 8:5107–5121, 2015.

Liu, Q, Wang, S, Zheng, U, Luo, Z, Cen, K. *J. Anal. Appl. Pyrolysis* 82:170–177, 2008.

Lou, R, Lyu, G, Wu, S, Zhang, B, Zhao, H, Lucia, LA. *ACS Sustainable Chem. Eng.* 6:430–437, 2018.

Lu, F, Ralph, J. *J. Agric. Food Chem.* 46:547–552, 1998.

Lv, P, Almeida, G, Perre, P. *BioResources* 10:4239–4251, 2015.

Ma, Z, Sun, Q, Ye, J, Yao, Q, Zhao, C. *J. Anal. Appl. Pyrolysis* 117:116–124, 2016.

Ma, Z, Wang, J, Zhou, H, Zhang, Y, Yang, Y, Liu, X, Ye, J, Chen, D, Wang, S. *Fuel Process. Technol.* 181: 142–156, 2018.

Marques, AV, Pereira, H, Meier, D, Faix, O. *Holzforschung* 53:167–174, 1999.

Moniz, P, Serralheiro, C, Matos, CT, Boeriu, CG, Frissen, AE, Duarte, LC, Roseiro, LB, Pereira, H, Carvalheiro, F. *Process Biochem.* 65:136–145, 2018.

Mousavioun, P, Doherty, WOS. *Ind. Crops Prod.* 31:52–58, 2010.

Murugan, P, Mahinpey, N, Johnson, KE, Wilson, M. *Energy Fuels* 22:2720–2724, 2008.

Naron, DR, Collard, FX, Tyhoda, L, Görgens, JF. *Ind. Crops Prod.* 101:61–74, 2017.

Navarro, MV, Murillo, R, Mastral, AM. *AIChE J.* 55:2700–2715, 2009.

Nitsos, C, Stoklosa, R, Karnaouri, A, Vörös, D, Lange, H, Hodge, DB, Crestini, C, Rova, USL, Christakopoulos, P. *ACS Sustainable Chem. Eng.* 4:5181–5193, 2016.

Ojha, DK, Viju, D, Vinu, R. *Bioresource Technol.* 241:142–151, 2017.
Oliveira, L, Evtuguin, DV, Cordeiro, N, Silvestre, AJD, Silva, AMS, Torres, IC. *J. Agric. Food Chem.* 54:2598–2605, 2006.
Onnerud, H, Gellerstedt, G. *Holzforschung* 57:165–170, 2003.
Parot, M, Rodrigue, D, Stevanovic, T. *Bioresource Technol. Reports* 17:100880, 2022.
Pe III, JA, Mun, JS, Mun, SP. *BioResources* 18:926–936, 2023.
Reddy, IAK, Ghatak, HR. *J. Thermal. Anal. Calorim.* 132:407–423, 2018.
Rencoret, J, Ralph, J, Marques, G, Gutiérrez, A, Martínez, T, del Rio, JC. *J. Agric. Food Chem.* 61:2434–2445, 2013.
Ropponen, J, Rasanen, L, Rovio, S, Ohra-aho, T, Liitia, T, Mikkonen, H, van de Pas, D, Tamminen, T. *Holzforschung* 65:543–549, 2011.
Ross, K, Mazza, G. *Int. J. Mol. Sci.* 11:4035–4050, 2010.
Rossberg, C, Janzon, R, Saake, B, Leschinsky, M. *BioResources* 14:4543–4559, 2019.
Sameni, J, Krigstin, S, Rosa, DD, Leao, A, Sain, M. *BioResources* 9:725–737, 2014.
Sameni, J, Krigstin, S, Sain, M. *BioResources* 11:8435–8456, 2016.
Serrano, L, Esakkimuthu, ES, Marlin, M, Brochier-Salon, MC, Mortha, G, Bertaud, F. *Ener. Fuels* 32:5969–5977, 2018.
Shen, DK, Gu, S, Luo, KH, Wang, SR, Fang, MX. *Bioresource Technol.* 101:6136–6146, 2010.
Sheng, Y, Ma, Z, Wang, X, Han, Y. *Food Chem.* 376:131895, 2022.
Shi, Z, Xu, G, Deng, J, Dong, M, Murugadoss, V, Liu, C, Shao, Q, Wu, S, Guo, Z. *Green Chem. Lett. Rev.* 12:235–243, 2019.
Singh, R, Singh, S, Trimukhe, KD, Pandare, KV, Bastawade, KB, Gokhale, DV, Varma, AJ. *Carbohydrate Polymers* 62:57–66, 2005.
Soongprasit, K, Sricharoenchaikul, V, Atong, D. *En. Rep.* 6:151–167, 2020.
Sun, XF, Sun, Fowler, P, Baird, MS. *J. Agric. Food Chem.* 53:860–870, 2005.
Tan, X, Jiang, B, Yang, Y, Min, D, Jin, Y. *Holzforschung* 71:99–108, 2017.
Tao, Y, Li, S, Li, P, Wu, Q. *Holzforschung* 70:1175–1182, 2016.
Thoresen, PP, Lange, H, Crestini, C, Rova, U, Matsakas, L, Christakopoulos, P. *ACS Omega* 6:4374–4385, 2021.
Tiainen, E, Drakenberg, T, Tamminen, T, Kataja, K, Hase, A. *Holzforschung* 53:529–533, 1999.
van de Pas, DJ, Nanayakkara, B, Suckling, ID, Torr, KM. *Holzforschung* 68:151–155, 2014.
Villar, JC, Caperos, A, Garcia-Ochoa, F. *J. Wood Chem. Technol.* 17:259–285, 1997.
Wadrzyk, M, Janus, R, Lewandowski, M, Magdziarz, A. *Ren. Energy* 177:942–952, 2021.
Wang, S, Lin, H, Ru, B, Sun, W, Wang, Y, Luo, Z. *J. Anal. Appl. Pyrolysis* 108:78–85, 2014.
Wang, G, Pang, T, Xia, Y, Liu, X, Li, S, Parvez, AM, Kong, F, Si, C. *Int. J. Biological Macromolecules* 133:156–164, 2019.
Wang, TP, Li, H, Yuan, JM, Li, WX, Li, K, Huang, YB, Xiao, LP, Lu, Q. *Ind. Crops Prod.* 171:113912, 2021.
Watkins, D, Nuruddin, M, Hosur, M, Tcherbi-Narteh, A, Jeelani, S. *J. Mater. Res. Technol.* 4(1):26–32, 2015.
Yan, K, Liu, F, Chen, Q, Ke, M, Huang, X, Hu, W, Zhou, B, Zhang, X, Yu, H. *Biotechnol. Biofuels* 9:76, 2016.
Yang, H, Yan, R, Chen, H, Lee, DH, Zheng, C. *Fuel* 86:1781–1788, 2007.
Yang, Q, Wu, S, Lou, R, Lv, G. *Wood Sci. Technol.* 45:419–431, 2011.
You, TT, Zhang, LM, Zhou, SK, Xu, F. Ind. Crops Prod. 71:65–74, 2015.
Yuan, TQ, Sun, SN, Xu, F, Sun, RC. *J. Agric. Food Chem.* 59:10604–10614, 2011.
Yue, F, Lu, F, Regner, M, Sun, R, Ralph, J. *ChemSusChem* 10:830–835, 2017.
Zhang, M, Resende, FLP, Moutsoglou, A, Raynie, DE. *J. Anal. Appl. Pyrolysis*, 98:65–71, 2012.
Zhang, W, Tian, G, Polle, A, Janz, D, Euring, D, Yue, X, Zhao, H, Fei, B, Jiang, Z. *Wood Sci. Technol.* 52:1331–1341, 2018.
Zhao, J, Xiuwen, W, Hu, J, Liu, Q, Shen, D, Xiao, R. *Polymer Deg. Stab.* 108:133–138, 2014.
Zhou, S, Xue, Y, Sharma, AM, Bai, X. *ACS Sust. Chem. Eng.* 4:6608–6617, 2016.
Zhu, W, Theliander, H. *BioResources* 10:1696–1714, 2015.

Section 3

Lignin Value Addition

8 Value-Added Carbons from Lignin

Compared to other lignocellulose constituents, lignin has a higher proportion of carbon. Referring to Table 5.1, carbon accounts for more than 60% of the elemental composition of most lignins. This is also reflected in its thermochemical characterization. TG analyses of different lignins at slow heating rates result in 40–50% char residue for most lignin samples. Therefore, lignin could be a suitable candidate for the production of carbon products substituting the current petroleum-derived carbon products. These include biochar and activated carbons at the low-value end to niche products like carbon fibres, carbon mats, carbon films, carbon nanospheres, carbon nanotubes, and carbon nanorods at the high-value end.

8.1 BIOCHAR AND ACTIVATED CARBON FROM LIGNIN

Biochar is the carbon-enriched residue left after the slow pyrolysis of organic materials of biological origin. The main useful characteristic of biochar is its high specific surface area compared to the starting materials. In addition, the biochar may also have specific chemical functionalities at its surface to make the adsorbent properties more selective. Pyrolysis not only increases the carbon content of the resulting material but also transforms it into graphitic structures with an orderly crystalline framework. Graphitic domains also impart higher thermal and electrical conductivities in biochar compared to starting lignin. Hydrothermal carbonization is another method of producing biochar by treating the biomass in the presence of water at high temperatures and elevated pressure. However, the applied temperature is less than that used in conventional pyrolysis. The high specific surface area of biochar comes from its porous morphology which results from the process of volatilization and carbonization of the precursor material. The specific surface area strongly depends on the morphological characteristics – pore size and pore size distribution. It is possible to pretreat the organic precursor material to obtain a final product having manifold increased specific surface area. Not only the surface area increases but it also makes it possible to have a product with closely controlled pore size and pore size distribution. This process is known as activation and the product is referred to as activated carbon. Moreover, the activation process is not always limited to upgrading the morphology but may also introduce desired chemical modifications on the surface to make it compatible with a particular substrate. Carbonization temperature is usually high for producing activated carbon.

One common method of activation is to impregnate the lignin – the activated carbon precursor – with a liquid activating agent as an aqueous solution. The impregnated lignin is then dried at 150–200 °C to remove the water and incorporate the activating agent within the microstructure of lignin. The impregnated and dried material is then carbonized at elevated temperatures under an inert atmosphere to produce the activated carbon. Once incorporated within the lignin microstructure, it forms a network of incrustation providing a support system and rigidity to the lignin pore structure. This prevents the carbon particles from collapsing on each other and forming a compact microstructure. The result is an activated carbon with much enhanced specific surface area and microporosity. Once the activated carbon is prepared, it can be washed with deionized water to remove the incrusted inorganics to acceptable levels. Orthophosphoric acid is a frequently used activating agent. Another much-used activation agent is zinc chloride. Hydroxides and carbonates of alkali metals like sodium and potassium are also used as activation agents. These compounds are

DOI: 10.1201/9781003598664-11

responsible for activation by inducing partial oxidation of char carbon at high temperatures. Schematic reactions would be

$$C + M_2CO_3 \rightarrow 2M + CO + CO_2 \quad \{1\}$$

$$2C + 2MOH \rightarrow 2M + 2CO + H_2 \quad \{2\}$$

where M represents the alkali metal. In lignins with high ash content, the inorganic impurities may provide some activation effect. However, their efficacy is inferior to that of externally added activation agents. Also, the inorganic elements present are often undesirable in the final end use of the activated carbon.

Another common activation method is to treat the initially formed carbon material in a flow of carbon dioxide at high temperatures. While flowing through the microstructure of carbon at high temperatures, carbon dioxide reacts with carbon according to the Boudouard reaction.

$$C + CO_2 \rightarrow 2CO \quad \{3\}$$

This partial loss of carbon, referred to as burnoff, forms additional micropores within the solid matrix leading to a final product with enhanced specific surface area and microporosity. However, reduced product yield is the price for this activation.

Similar to carbon dioxide activation, activation can be effected with steam. Carbonized material is contacted with steam at high temperatures. It reacts with carbon present in initially formed pores and interstices to produce carbon monoxide and hydrogen.

$$C + H_2O \rightarrow CO + H_2 \quad \{4\}$$

This leads to a partial loss of carbon and an increase in pore volume and specific surface area. Escaping carbon monoxide and hydrogen also induces microporosity.

Slow pyrolysis of lignin can produce a large amount of biochar owing to its high total carbon and fixed carbon content compared to the starting material. Biochar has several important applications, especially in the context of climate change and carbon-neutral or carbon-negative technologies. The application of biochar as a soil conditioner is found to reduce the emission of greenhouse gases from agricultural processes (Ghysels et al. 2019). Though having less specific surface area than activated carbon, biochar is a good low-cost adsorbent for the removal of pollutants from wastewater (Gul et al. 2021). Biochars may transform from electrical insulators to electrical conductors upon application of pressure beyond a threshold (Kane et al. 2021). This is an important biochar characteristic that allows its use as a precursor for preparing conducting carbon that has several high-value applications including in energy storage.

8.1.1 Biochar from Lignin

Biochar was produced from commercial softwood kraft lignin in a nitrogen atmosphere. The material was heated to 400–600 °C at a 10 °C/min heating rate. The produced biochar was ground to 0.15 μm size or less. The biochar yield was 65.2% and 57.5% at 400 and 600 °C, respectively, with 60.2% and 67.1% carbon. Their hydrogen-to-carbon ratios were 0.675 and 0.213 and oxygen-to-carbon ratios were 0.26 and 0.16. There was a marked improvement is specific surface area with increasing temperature; 1.04 m^2/g at 400 °C but 259 at 600 °C. For the latter, a 200 m^2/g area consisted of micropores. Correspondingly, the total pore volume increased from 0.00158 to 0.138 cm^3/g (0.102 cm^3/g micropore volume) and the average pore diameter decreased from 89.1 to 2.68 nm. These biochars could adsorb nitrobenzene from aqueous solution to the extent of 92.2 and 120 mg/g, respectively (Li et al. 2014). In a modified process, pine kraft lignin and pine sodium lignosulphonate were converted to biochar by heating at different temperatures (60–210 °C) in the presence of concentrated sulphuric acid for 6 h. Compared to conventional pyrolysis, these biochars were characterized by increased aromatic content, decreased aliphatic side chain, complete demethoxylation, and increased carboxyl and phenolic hydroxyl content than the starting lignins. Lignin was also extensively depolymerized by breaking of alkyl aryl ether bonds. These biochars were distinctly acidic in character and performed well in adsorbing Pb^{++} ions from aqueous solutions with adsorption capacities ranging from 533 to 679 mg/g. The adsorption capacity was found to increase linearly with an increase in the carboxyl content of biochar (Li et al. 2020). Another study reports the production of biochar from Indulin AT and softwood sodium lignosulphonate by first reacting them with toluene 2,4-diisocyanate to produce the corresponding polyurethane followed by carbonization. For the synthesis of polyurethane, mixtures of lignin and toluene 2,4-diisocyanate – 0.5 to 1.5 g per g lignin – were allowed to react at 60 °C for 24 h. Carbonization was done in two stages. In the first stage, the material was heated to 500 °C at 1 °C/min followed by 1 h holding time. Thereafter, it was heated to 900 °C at 5 °C/min and held at this temperature for 2 h. With Indulin as a starting material, biochar yield varied between 28% and 32%, and with lignosulphonate, it was 29–34%. The ratio of toluene 2,4-diisocyanate to lignin was found to have a profound influence on the specific surface area of the produced biochar. The highest specific surface area of 80.3 m^2/g from Indulin and 70.8 m^2/g from lignosulphonate was obtained when the ratio was 1 g/g. However, at the 1.5 g/g ratio of toluene 2,4-diisocyanate to lignin, there was an abrupt drop in specific surface area to 0.86 and 1.51 m^2/g, respectively (Leitner et al. 2015).

Lignin-rich residue was obtained from poplar after enzymatic hydrolysis of carbohydrates, fermentation of monomer sugars to bioethanol, distillation of ethanol from broth, and anaerobic digestion of residual solids. The lignin-rich residue was converted into biochar through slow pyrolysis at 5–50 °C/min heating rate to 370–450 °C temperature and 5–45 min holding time at the final temperature. The biochar yield varied between 40% and 50% with a hydrogen-to-carbon ratio in the 0.5–0.8 range and an oxygen-to-carbon ratio in the 0.1–0.28 range. Higher pyrolysis temperature yielded less biochar with less hydrogen and oxygen content (Ghysels et al. 2019). Commercial softwood kraft lignin was copyrolysed with red mud – a solid waste from the Bayer process of aluminium extraction from its ore – in a carbon dioxide atmosphere. A 5 g mixture of lignin and red mud with an equal mass ratio was placed in 500 mL/min gas flow and heated to 900 °C at 10 °C/min. Solid residue at the end of heating was 48% of the original mass with a specific surface area of 184.1 m^2/g and an average pore size of 2.76 nm. The most notable transformation during the copyrolysis was the conversion of Fe^{3+} (hematite form of ferric oxide) to Fe^0 (metallic iron) as a result of the reducing action of carbon and carbon monoxide produced. Sodium aluminium silicate was also produced. This solid residue was studied for its effectiveness in reducing p-nitrophenol and Cr^{6+} with reasonable efficacy reported (Yoon et al. 2019). Nano-biochar was prepared from mixed hardwood and wheat straw kraft lignin. Prior to pyrolysis, inorganic impurities were removed by dissolving the lignin in sodium hydroxide and ultrafiltration. Purified lignin was reprecipitated by acidification and spray-dried. Pyrolysis was carried out in 100 mL/min nitrogen flow at a heating rate of 10 °C/min to 800 °C and holding for 2 h. The resulting biochar was pulverized for 4 h in a ball mill to produce the nano-biochar. Compared to the starting lignin, the biochar had markedly reduced hydroxyl groups and other oxygenated functional groups but significantly increased graphitic structures. Biochar had a specific surface are of 83.41 m^2/g compared to only 22.3 for starting lignin. Produced biochar was studied for its effectiveness as a filler in styrene–butadiene rubber as a possible replacement for petroleum-derived carbon black (specific surface area 76.73 m^2/g). At a 40% addition level with styrene–butadiene rubber, the tensile properties and vulcanization rates were comparable to those using carbon black (Jiang et al. 2020). Pyrolysis of softwood lignosulphonate was carried out at 500–900 °C in 150 mL/min nitrogen flow to produce biochar. Char yield was 60.5–38% with increasing temperature. Elemental carbon content remained nearly uniform between 61.6% and 62.6% but the corresponding decrease in hydrogen-to-carbon ratio was from 0.027 to 0.006. Char produced at 900 °C had a specific surface area of 30 m^2/g but increased to 528 m^2/g

upon extensive washing with deionized water. Importantly, the ash content dropped from 31% to 4% upon washing. The major element in the ash was sodium. The obtained biochars were used as catalysts for the dehydrogenation of 2-propanol and methanol. While conversion was close to 100% the selectivity of dehydrogenation was temperature dependent with propylene forming from the dehydration side reaction. It was reasoned that the produced char had both basic and acidic sites on the surface; the former catalysing dehydrogenation and the latter, dehydration (García-Mateos et al. 2017). Modified softwood sodium lignosulphonates – lower sulphonic acid groups and higher carboxylic acid groups – and poplar soda lignin were pulverized by ball milling and pyrolysed by holding for 1 h at 900 or 1100 °C. The heating rate was 10 °C/min to maximum temperature with 30 mL/min nitrogen flow. The biochar yield varied between 45% and 50%. A high peak electrical conductivity of 18.5 S/cm was obtained for compressed biochar obtained at 1100 °C from poplar soda lignin. This biochar had a high carbon content of 83.5%, hydrogen-to-carbon ratio of 0.064, and oxygen-to-carbon ratio of 0.08. The average particle size was about 5.5 μm and the specific surface area was 109 m^2/g. The primary pore size was 7.9 Å. Increased pyrolysis temperature from 900 to 1100 °C increased specific surface area and decreased pore size (Kane et al. 2021). Softwood and hardwood dilute acid hydrolysis lignin were converted to biochar by pyrolysis at 500–700 °C for 1 h with 1 L/min nitrogen flow rate and 24 °C/min heating rate to final temperature. The biochar yield decreased from 42.95% at 500 °C to 37.99% at 700 °C with elemental carbon increasing from 76.7% at 500 °C to 85.39% at 700 °C. The corresponding hydrogen-to-carbon ratio decreased from 0.046 to 0.019 and the oxygen-to-carbon ratio from 0.097 to 0.013. Biochar produced at 600 °C had a specific surface area of 378 m^2/g which increased to 430 m^2/g at 700 °C. The total pore volume was 0.19 and 0.17 cm^3/g but the average pore size was 2 and 1.6 nm, respectively. For comparison, the specific surface area in starting lignin was only 4 m^2/g (Naydenova et al. 2023).

8.1.2 Activated Carbon from Lignin

Commercial kraft lignin was the starting material for producing activated carbons through chemical activation. Lignin was impregnated with orthophosphoric acid at room temperature and air environment for 1–48 h followed by carbonization in air. Impregnated material was first heated to 150 °C and held for 1 h for moisture removal. The dried material was heated to 400–650 °C and held for 2 h to complete the carbonization. Obtained carbon was washed with distilled water to remove residual acid. The yield and specific surface area of produced activated carbons were found to depend on the ratio of orthophosphoric acid to lignin, impregnation time, and carbonization temperature. Higher temperatures reduced yield. With increasing temperature, specific surface area first increased and then decreased after attaining a maximum at 600 °C. At this temperature, the elemental carbon and carbon-to-hydrogen ratio were also the highest. The yield of activated carbon increased with an increase in the ratio of orthophosphoric acid to lignin, reaching 32% at a ratio of 1.4. However, the maximum specific surface area was obtained at a ratio of 1.4 beyond which it reduced with a further increase in acid dose (Fierro et al. 2006). These trends are shown in Figure 8.1.

In a related procedure, softwood kraft lignin was the starting material to produce activated carbon through orthophosphoric acid activation. Pulverized lignin was impregnated with an orthophosphoric acid solution, and the mixture was heated to 170 °C at 10 °C/min. The mixture was kept at this temperature for 1 h in 0.6 l/min air flow to completely remove the moisture. The dry solid mass was, then, heated to 250–500 °C at 2 °C/min in 0.6 l/min air flow. At the end of the heating, the produced activated carbon was removed from the furnace and thoroughly washed with deionized water. Activated carbon yield was close to 70% for an acid-to-lignin ratio of 1.5 even at 500 °C. This high product yield is most likely due to no holding time provided at the final carbonization temperature. The phosphorous content in the activated carbon increased with increased temperature as well as increased acid-to-lignin ratio, varying between 4 and 15 mg/g. This is reflected in the increase in specific surface area with an increase in acid-to-lignin ratio from less than 500 to nearly 1400 m^2/g

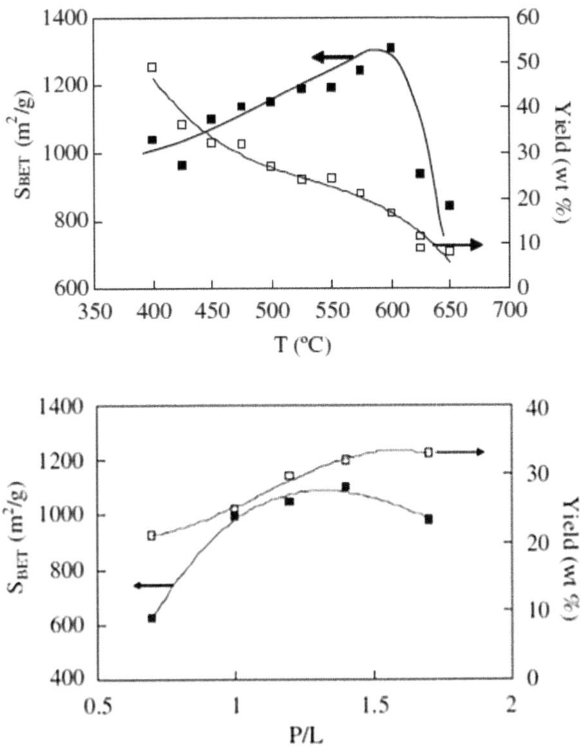

FIGURE 8.1 Change in activated carbon yield and specific surface area with carbonization temperature and the ratio of orthophosphoric acid to kraft lignin. For the upper figure, the acid-to-lignin ratio is 1.4 and the impregnation time is 1 h. For the lower figure, the impregnation time is 1 h and the carbonization temperature is 450 °C. (Reprinted from Fierro, V, Torne-Fernandez, V, Celzard, A, *Microporous Mesoporous Mater.*, 92, 243–250, 2006, with permission from Elsevier.)

at a ratio of 2.5. However, with increasing temperature, the specific surface area reached its maximum of about 1500 m^2/g at 450 °C after which it gradually reduced. A similar trend was also obtained for pore volume, both micropores and mesopores. The produced activated carbons had acidic groups on their surface with the concentration of acidic groups varying with carbonization temperature and acid-to-lignin ratio (Guo and Rockstraw 2006). Recently, a method was proposed to prepare activated carbon from softwood kraft lignin by sodium hydroxide activation. Prior to activation and carbonization, lignin was ultrasonicated for 1 h as an alternative to sulphuric acid leaching to remove inorganic impurities. Ultrasonicated lignin was impregnated with aqueous sodium hydroxide solution and carbonized for 1 h at 600–900 °C under nitrogen or 4% hydrogen mixed Argon. All the activated carbons had a specific surface area close to or exceeding 2000 m^2/g. The highest specific surface area was obtained at 750 °C and argon atmosphere. A specific surface area of 2277 m^2/g with ultrasonication was on the lower side than 2442 m^2/g obtained by acid washing. However, the micropore surface area with ultrasonication was only 69 m^2/g compared to 189 m^2/g with acid washing. The effect of activation was most obvious; the specific surface area of carbon obtained without activation was only 218 m^2/g at 600 °C. The produced activated carbon was used for preparing composite polyurethane foams (Kim et al. 2021). Commercial softwood kraft lignin Indulin AT was the source of activated carbon by slow pyrolysis or hydrothermal carbonization followed by chemical activation of the obtained biochar with potassium hydroxide. Slow pyrolysis was carried out in 10 L/min nitrogen flow and 10 °C/min heating rate to 600 °C and holding for 2 h. The produced biochar was cooled to ambient temperature under nitrogen flow. For hydrothermal carbonization, the lignin and water, in a mass ratio of 2:8, were sealed with in a pressure reactor, heated to 220 °C in 1 h and held for 5 h. The produced hydrochar was filtered and water-washed. For chemical activation, char was mixed with solid potassium hydroxide in a 1:4 mass ratio, heated to 600 °C at 10 °C/min and 10 L/min nitrogen flow. It was then allowed to react for 2 h in flowing nitrogen and subsequently cooled. The biochar yield was about 42% compared to the hydrochar yield of close to 85%. However, the biochar was 91.8% carbon compared to only 69.7% in hydrochar. However, the yield difference was largely eliminated after activation with the final yield of activated carbon being about 33% through the hydrothermal route and 35% from the pyrolysis route. It is evident that hydrothermal treatment in itself was not efficient in carbonization and main carbonization took place during chemical activation of hydrochar. The specific surface area of 2314 m^2/g of the activated carbon from the pyrolysis route was also somewhat higher than the specific surface area of 2183 m^2/g of the activated carbon from the hydrothermal route. Pyrolysis-derived activated carbon had 92% micropores compared to only 71% for hydrothermal carbonization-derived activated carbon (Correa et al. 2017).

Eucalyptus kraft lignin was carbonized to produce activated carbons in the temperature range of 450–900 °C in nitrogen or carbon dioxide atmosphere. Lignin was first carbonized at 350 °C followed by extensive washing with 1% sulphuric acid. This precarbonized and washed char was further carbonized at the desired elevated temperature under a nitrogen atmosphere at 150 mL/min and 10 °C/min heating rate. The material was held for 2 h at the final temperature to complete the carbonization. The yield of activated carbon varied between 47.2% and 39% with increasing temperature. The specific surface area of 496 m^2/g was highest at 550 °C and decreased thereafter. Activated carbons obtained at 550 and 800 °C were further activated at 800 °C in carbon dioxide flowing at 150 mL/min for 4–40 h. The final yield of activated carbons consistently decreased with increased activation time but the specific surface area consistently increased. However, no clear correlation could be observed between carbonization temperature and final yield. The highest yield was 36.2% with 800 °C carbonization temperature and 4 h activation time but with the lowest specific surface area of 641 m^2/g. Conversely, the activated carbon produced at 550 °C carbonization temperature and 40 h activation time had the lowest yield of 10.2% but the highest specific surface area of 1360 m^2/g (Rodriguez-Mirasol et al. 1993a). These authors also studied the kinetics of carbon dioxide activation. From their study, a plot of reaction rate, r, against conversion, X, is presented in Figure 8.2, where the reaction rate is defined on a mass basis, w, as

$$r = -\left(\frac{1}{w}\right)\left(\frac{dw}{dt}\right) \tag{8.1}$$

Using the isoconversional approach and Arrhenius equation for reaction rate dependence on temperature, the activation energies were calculated at different conversions. Activation energy was found to continuously decrease from 239 kJ/mol at 0.2 conversion to 212 kJ/mol at 0.8. It was proposed that the activation reaction was catalysed by sodium present in lignin and the precarbonized intermediate product. Moreover, it was believed that the sodium is very evenly distributed over the carbon mass. As expected, the specific surface area increased, and the yield decreased with increasing burnoff. A maximum specific surface area of 1853 m^2/g was obtained at an activation temperature of 850 °C with a burnoff of 76.8%; corresponding yield being 9.3% (Rodriguez-Mirasol et al. 1993b).

Using the experimental results from this study, the specific surface area could be correlated with activated carbon yield and burnoff percentage with a second-order polynomial as

$$SS = 1.278Y^2 - 96.103Y + 2569.2 \quad R^2\,0.9815 \tag{8.2}$$

$$SS = 0.2037BO^2 - 2.3977BO + 769.16 \quad R^2\,0.9815 \tag{8.3}$$

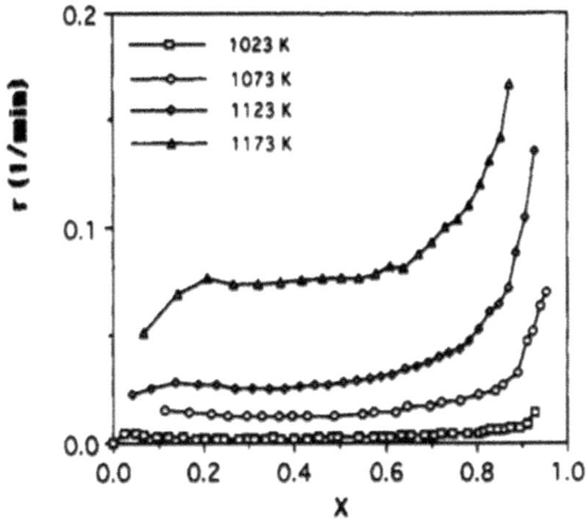

FIGURE 8.2 Reaction rate of carbon dioxide activation of carbon obtained from eucalyptus kraft lignin. (Reprinted from Rodriguez-Mirasol, J, Cordero, T, Rodriguez, JJ, *Energy Fuels*, 7, 133–138, 1993, with permission from American Chemical Society.)

From this study and several other studies, it appears that activation is a direct tradeoff between yield and specific surface area. Activated carbons were prepared from eucalyptus kraft lignin in two stages. In the first stage, lignin was thermally treated for 2 h at 350 °C under nitrogen. The heating rate was 10 °C/min. Residue from the first stage was extensively washed with 1% sulphuric acid to remove inorganic impurities. Thereafter, a second pyrolysis stage was performed in 150 mL/min nitrogen flow for 2 h at 550–800 °C with 10 °C/min heating rate to the final temperature. The obtained carbonized material was activated in 150 mL/min carbon dioxide flow at 750–850 °C with activation time varying from 4 to 50 h. Activated carbon yield varied from 9.4% to 36.2% with 8.1% to 76.8% of the carbonized material lost during activation (burnoff). The specific surface area of the produced activated carbons ranged between 641 and 1853 m^2/g and the pore volume between 0.331 and 2.091 cm^3/g (Cotoruelo et al. 2007). Using the experimental results from this study, the specific surface area could be correlated with activated carbon yield and burnoff percentage with a second-order polynomial as

$$SS = 0.8819Y^2 - 76.125Y + 2295.8 \tag{8.4}$$

$$SS = 0.1796BO^2 - 1.0637BO + 708.33 \tag{8.5}$$

where SS is the specific surface area in m^2/g, Y is the percentage of starting lignin obtained as activated carbon, and BO is the percentage of material from carbonization lost during activation. It is graphically presented in Figure 8.3.

Activated carbon was prepared from corncob soda lignin obtained from alkaline pretreatment of corncob for bioethanol production. Lignin was impregnated with 20 wight% orthophosphoric acid before carbonization in a reduced air environment for 1 h at 400 °C. Post carbonization, the activated carbon was washed with 1 M hydrochloric acid, water washed and dried. The obtained carbon material was further activated by treatment with and incorporation of various manganese compounds – MnO_2, $MnSO_4$, and $KMnO_4$. Obtained activated carbons had 186–547 m^2/g specific surface area; highest in the case of $MnSO_4$ activation, followed by MnO_2 and $KMnO_4$ activation. Pore volume also exhibited a similar trend, varying between 0.342 and 0.52 cm^3/g. The average pore size ranged between 3.129 and 4.434 nm, the smallest for $MnSO_4$ activation. The activated carbons were studied for their effectiveness in adsorbing methylene blue from an aqueous solution. However, adsorption capacity did not follow the specific surface area trend. It was highest at 249.2 mg/g with MnO_2

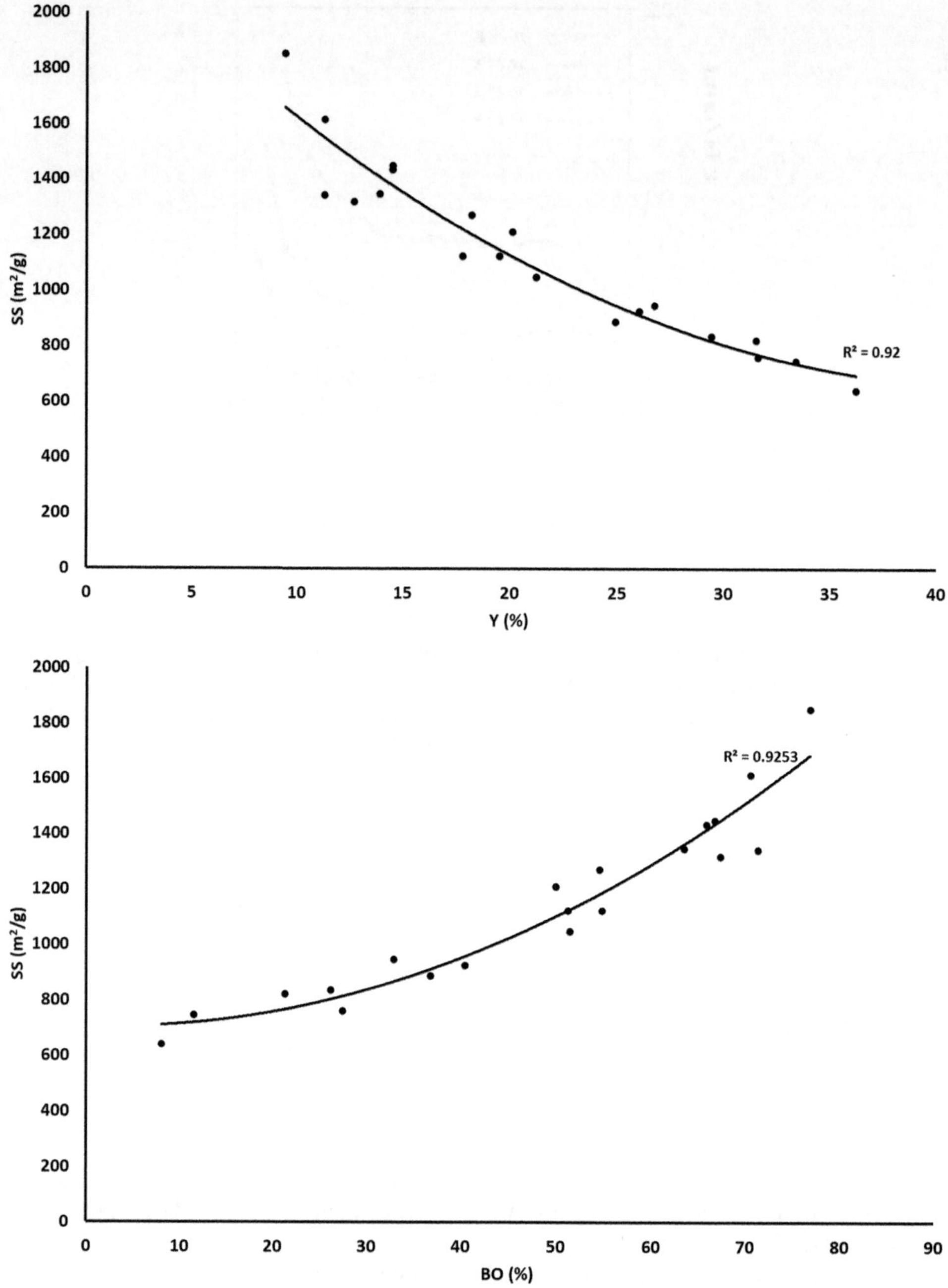

FIGURE 8.3 Variation in specific surface area of activated carbons with yield and burnoff. Source of experimental data Cotoruelo, LM, Marques, MD, Diaz, FJ, Rodriguez-Mirasol, J, Cordero, T, *Sep. Sci. Technol.* 42, 3363–3389, 2007.

activation, followed by 241.35 with $KMnO_4$ activation and 237.82 mg/g with $MnSO_4$ activation. Interestingly, the activated carbon without activation with any manganese compound could still adsorb 234.65 mg/g with a specific surface area of only 48.14 m^2/g (Liu et al. 2021). A recent study reports the preparation of mesoporous carbon from Protobind 1000 – commercial wheat straw soda lignin – by first preparing a crosslinked resin precursor acting as the template for the final carbon product. In the first step, lignin was propargylated. For this, it was dissolved in acetone followed by the addition of potassium carbonate and potassium iodide. The temperature of this mixture was raised to 60 °C and 80% solution of propargyl bromide in toluene was added dropwise under continuous stirring and nitrogen purging. The quantities added were: lignin 200 g, acetone 2 L, potassium carbonate 941 g, potassium iodide 113 g, and propargyl bromide solution 759 mL. The mixture was refluxed under continuous stirring for 24 h to complete propargylation. The product was filtered, washed with acetone and vacuum-dried. The propargylated lignin and a suitable surfactant – 1:3 mass ratio – were dissolved in 2-methythetrahydrofuran followed by evaporation of the solvent. Surfactants used were Pluronic P123, Pluronic F127, Kraton G1652, and Solsperse M387. The mixture was converted into the crosslinked resin by curing at 180 °C for 14 h. The prepared resin was carbonized in two steps. In the first step, it was heated at 1 °C/min under nitrogen to the surfactant decomposition temperature and held for 2 h. In the second step, it was further heated under nitrogen at 2 °C/min to 600–1000 °C and held for 3 h for complete carbonization. Subsequently, the activated carbon was cooled under nitrogen at 5 °C/min to ambient temperature and ground. The specific surface area of the obtained carbons varied between 95 and 371 m^2/g and the pore volume between 0.05 and 0.33 cm^3/g (Jedrzejczyk et al. 2021). It is apparent that the obtained specific surface area and pore volume are not commensurate with the tedious and complicated production method and cannot compete with other commercially available activated carbons.

One study reports the preparation of activated carbons from spruce kraft lignin. Potassium carbonate, potassium hydroxide, sodium carbonate, sodium hydroxide, zinc chloride, and orthophosphoric acid were examined as the activation agents. Lignin was kneaded with an aqueous solution of activation agent and dried at 110 °C. Prepared precursor material had an equal mass ratio of activation agent to lignin. The impregnated precursor material was carbonized for 1 h at 500–900 °C with a 10 °C/min heating rate to the final temperature. When activating with zinc chloride or orthophosphoric acid, the largest specific surface area and pore volume were achieved at 600 °C temperature. The specific surface area was about 1200 m^2/g with zinc chloride and 1000 m^2/g with orthophosphoric acid. The corresponding pore volumes were about 0.6 and 0.5 cm^3/g with most of the pores being micropores. 800 °C was the most effective carbonization temperature for developing specific surface area during activation with sodium and potassium compounds. At this temperature, the highest specific surface area of about 2000 m^2/g was achieved by potassium carbonate activation. The corresponding pore volume was about 0.9 cm^3/g; mostly micropores. A specific surface area of about 700 m^2/g was the least with sodium carbonate activation with a corresponding pore volume of about 0.4 cm^3/g (Hayashi et al. 2000). Spruce kraft lignin with a low degree of sulphonation was impregnated with an aqueous solution of activation agent – sodium hydroxide or potassium hydroxide – followed by heating at 10 °C/min. The heating sequence was to 105 °C, holding for 30 min to moisture removal, then to 850–950 °C. The holding time at the final temperature was 30 min at 850 and 950 °C and 5–60 min at 900 °C. After carbonization, the material was cooled down to room temperature, washed with deionized water, and dried at 105 °C. The entire heating and cooling was in flowing nitrogen at 100 mL/min. At activation agent to lignin ratio of 1, carbon yield with sodium hydroxide varied between 19.8% and 27.2% and with potassium hydroxide between 14.6% and 27.6%; higher temperature lowering yield. All activated carbons produced had more than 95% carbon content. A higher ratio of activation agent to lignin also reduced yield. The specific surface area was highest at 900 °C; 1540 m^2/g with potassium hydroxide activation and 1321 m^2/g with sodium hydroxide activation. Pore volume was the same at 0.43 cm^2/g. Produced carbons were mostly mesoporous in nature (Hu and Hsieh 2017).

Sodium lignosulphonates were converted into activated carbon for electrical capacitive application by heating in an argon environment to 600–800 °C at 5 °C/min heating time and holding at the final temperature for 2 h. It was proposed that activation was achieved during carbonization by the decomposition of inorganic calcium and sulphur compounds into gaseous products which escaped through the reacting mass leaving behind a porous structure. The produced material was leached with a hydrochloric acid solution to remove the inorganic impurities. Final product yields were 29%,

18%, and 15% at 600, 700, and 800 °C, respectively. Corresponding specific surface areas were 628, 1006, and 1362 m^2/g and the pore volumes were 0.37, 0.59, and 0.83 cm^3/g with more mesoporosity with increasing temperature (Chen and Zhou 2012).

Activated carbons were prepared from softwood enzymatic hydrolysis lignin obtained after bioethanol production. Dried and milled lignin was pretreated with a 1 M solution of mineral acid to remove residual inorganic impurities. After acid treatment, the material was thoroughly water-washed to an acid-free state and dried. The pretreated precursor material was then used to prepare the activated carbon using different activation methods. In one approach the precursor was carbonized by heating to 800 °C at a 6.5 °C/min heating rate. At this temperature, steam was fed for 0.5–2 h at feed rates of 0.5–2 g/min. For each run, the ratio of steam to precursor material was maintained at 6 g/g. In the second approach, the precursor material was impregnated with the activation agent – zinc chloride or potassium hydroxide or sodium carbonate – by stirring in an aqueous solution for 3 h at 80 °C. The mass ratio of activation agent to lignin varied between 1 and 4. The material was then dried at 105 °C for 48 h. Dry impregnated material was carbonized in nitrogen flow by heating to 600–800 °C at 6.5 °C/min and holding for 2 h. Produced carbons were refluxed in 1 M hydrochloric acid at 105 °C for 3 h followed by water washing. With steam activation, the produced carbons had a specific surface area of 492–616 m^2/g and a pore volume of 0.21–0.29 cm^3/g. No distinct correlation could be observed between activation time and specific surface area or pore volume. However, in all cases, the micropores were in excess of 93% except when the lignin was extracted with orthophosphoric acid. Produced activated carbons had more than 95% carbon content with yield varying between 25.3% and 44.4%. In comparison, the activated carbons produced by the second approach had a significantly higher specific surface area between 514 and 2054 m^2/g and a pore volume of 0.25–1.36 cm^3/g. The yield, on the other hand, was too low at 2.6–9.2% with carbon content between 58.9% and 96.5%. Activation with zinc chloride resulted in the highest specific surface area and pore volume for the same temperature and ratio of activation agent to lignin. However, the specific surface area and pore volume decreased with the temperature increasing from 500 to 600 °C. These carbons also had low microporosity between 31% and 33% and high mesoporosity. The trend was reversed for potassium hydroxide as the activation agent. Here, both the specific surface area and pore volume increased with increasing temperature and ratio of activation agent to lignin. Microporosity was also higher at 83–94%. Sodium carbonate was the least effective activation agent (Bergna et al. 2022).

8.2 CARBON FIBRES FROM LIGNIN

One of the advanced materials among modern high-performance functional materials is carbon fibre. Known for their excellent mechanical properties, thermal stability, and chemical resilience, these fibres find application in several high-end products. They have a remarkably high strength-to-weight ratio. Being composed almost entirely of carbon, they need a precursor material with a high carbon content. Currently, most of the carbon fibres are produced from petroleum-derived precursor materials like polyacrylonitrile or pitch. However, this is not a sustainable method due to obvious reasons. There is a need to find a sustainable alternative for petroleum-derived precursor material to produce carbon fibres. Lignin uniquely fits into this slot owing to its often large availability as waste and high carbon content. However, the production of carbon fibres is a technologically complex process and adopting it for lignin is a challenging task. A lot of research and development work is currently underway to fine-tune the process of producing carbon fibres starting with lignin.

A carbon fibre is a microscopic strand of pure elemental carbon with a high length-to-diameter ratio with the diameter being a few μm only. The carbon atoms are joined in a hexagonal ring extending into two-dimensional crystalline lamellae. Based on the manner in which the lamella fold together to make the fibrous structure, carbon fibres may be either turbostratic – disorderly folded lamella – or graphitic – orderly folded lamella. Some hybrid carbon fibres may consist of both types. The main steps involved in the production of carbon fibres are spinning and carbonization. In the spinning step, the precursor – often mixed with suitable plasticizing and binding material – is spun into long and thin strands. In the carbonization step, the strands are heated at high temperatures in an inert atmosphere to remove the non-carbon elements. Other intermediate steps may be required depending on the precursor and the intended end use. Carbonization temperature decides the type of carbon fibres produced. Type I fibres are produced at temperatures exceeding 2000 °C and possess a high modulus of elasticity. Carbonization temperature is between 1500 and 2000 °C for Type II fibres having high tensile strength. Type III fibres have lower mechanical properties and are produced at 1000 °C or less carbonization temperature.

Between spinning and carbonization, there is an intermediate step of thermo-stabilization. Initially formed fibrous strands and wound yarn are prone to destabilize and deform during high-temperature carbonization. To prevent this, the precursor yarns are heated in air at 250–300 °C to induce cross-linking that imparts rigidity to the molecular structure. The breaking of alkyl aryl ether bonds and the formation of carbon–carbon bonds between adjacent aromatic rings are mainly responsible for crosslinking. The process is also associated with volatilization and partial oxidation with the evolution of carbon dioxide and water vapour. This leads to a thermoset material that does not deform, melt, or fuse during high-temperature carbonization.

8.2.1 Spinning of Lignin-Based Carbon Fibres

8.2.1.1 Solution Spinning

There are three main spinning methods employed in the production of carbon fibres from lignin. These are solution spinning, melt spinning, and electrospinning. In solution spinning, the polymeric precursor material is dissolved in a solvent at a concentration that gives the right viscosity for spinning. This solution is then extruded through a cluster of very fine nozzles to form the fibre strands. The method is further sub-classified into dry, wet, or dry–wet type. If the solvent is highly volatile, the strands are extruded into a hot stream of air to evaporate the solvent and form the fibres. It is then termed dry spinning. On the other hand, if the solvent is non-volatile, the strands are extruded into a pool of coagulant – a liquid that acts as an antisolvent. This way, the extruded polymer is drawn into a thread that solidifies by coagulation. This is termed as wet spinning. In between there is the dry–wet spinning method where the extruding strands pass through short exposure to hot air before landing into the coagulant pool.

Softwood kraft lignin was blended with bleached softwood kraft pulp in a 7:3 ratio and dissolved in the ionic liquid 1-ethyl-3-methylimidazolium acetate at a concentration of 18%. Precursor fibres were solution spun from this dope at 60 °C. Extruded fibres were initially in air for a path length of 10 cm before entering a refrigerated pool of deionized water at 5 °C. Final precursor fibres were obtained at two thicknesses, 20 and 14 µm, corresponding to draw ratios of 4 and 7. These precursor fibres were finished by extensive washing with deionized water, 45 min drying at 80 °C, followed by a quick dipping in 0.4 M ammonium dihydrogen phosphate (equivalent to 2.5 weight% phosphorous in the fibres). The inclusion of phosphorous controlled the dehydration and dehydrogenation rates during oxidative stabilization without adversely affecting the dimensional stability of the fibres and improved the yield of the carbon fibres. Drying followed for 24 h at 23 °C. Spun, finished, and dried precursor fibres were laid on graphite bridges and oxidatively stabilized in 7 L/min air flow at 250 °C using different heating protocols. One protocol heated the precursor fibres to 200 °C at 0.2 °C/min and then to the final temperature at 1 °C/min. Alternatively, the fibres were heated at a single heating rate of 5 °C/min. The holding time at the final stabilization temperature varied between 1 and 10 h (Bengtsson et al. 2019).

8.2.1.2 Melt Spinning

The melt spinning process for carbon fibre can trace its origin in the textile industry where thermoplastics are heated to convert them into a flowable melt. This melt is extruded through an assembly of fine orifices, known as a spinneret. Melt temperature during extrusion controls the melt viscosity which in turn controls the extrusion velocity. The extruding strands of precursor material are gradually cooled and drawn to solidify. Fibres are then wound into yarn. The distance between the point of emergence of the strands from the spinneret and the point at which the fibres are solidified, known as the spinning line, is an important controlling parameter. Through the spinning line, the fibres are also elongated to some extent and the final diameter of the fibre is less than what emerges from the spinneret. This distance can be computed as (Wang et al. 2022)

$$L = \frac{\rho d v C_p}{4h} \ln \frac{T_0 - T_s}{T_e - T_s} \tag{8.6}$$

where L is the spinning line distance, ρ is the density of the melt, d is the average fibre diameter, v is the average axial velocity of the fibre, C_p is the specific heat capacity of the melt, h is the heat transfer coefficient, T_0 is the melt temperature during extrusion, T_s is the temperature of the surrounding, and T_e is the temperature of the solidified fibre.

An early study reported melt spinning of precursor fibres from different lignins – hardwood kraft lignin, and Alcell lignin – and their blends with polyethylene oxide. Lignins were heat-treated at 145 °C and reduced pressure for 1 h. This removed the volatiles – mainly hemicellulose impurities – and improved spinnability. Hardwood kraft lignin and its blends with polyethylene oxide required spinning temperatures between 150 and 228 °C. The range of spinning temperature varied from 120 to 172 °C for Alcell lignin and its blends. The addition of polyethylene oxide reduced the glass transition temperature of the produced blends over that of lignin and improved spinnability and required spinning temperature. Hardwood kraft lignin had a T_g of 83.3 °C which reduced to 68.1 °C even by the addition of 5% polyethylene oxide. For Alcell lignin this reduction in T_g was from 68.2 to 50.9 °C. Spinnability substantially improved with the polyethylene oxide content increasing to 25%. Melt spinning was carried out by extruding through a 0.79 mm diameter spinneret with a winding speed of 74 m/min. Formed precursor fibres were laid on a steel plate, heated to 250 °C, and stabilized in air for 1 h. The heating rate during stabilization was found to be an important parameter for the dimensional stability and prevention of fibres fusing together during the subsequent carbonization step. Precursor fibres from Alcell lignin required a slow heating rate of 0.2 °C/min and none of its blends resulted in fibres that could be stabilized. Precursor fibres from hardwood kraft lignin blended with up to 12.5% polyethylene oxide could be stabilized but blends with more polyethylene oxide resulted in fibres that fused during stabilization (Kadla et al. 2002). Indulin AT was washed with 1 N hydrochloric acid to remove inorganic impurities and blended with pyrolysed fuel oil. The blend was dissolved in tetrahydrofuran as 8 weight% solution and heated for 4 h at 230 °C. This process evaporated the solvent and some of the volatiles to form a dope ready for melt spinning. Lignin to pyrolysed fuel oil ratio of 7:3 produced the optimal blend with softening temperature of 260–280 °C – glass transition temperature of 90 °C and a melting temperature of 290 °C. Melt spinning under nitrogen was done with conditions: temperature 260–280 °C, extrusion pressure 4.9 kPa, nozzle diameter 0.5 mm, nozzle length 0.5 mm, and winding rate 70–100 m/min. Continuous fibre formation was obtained under these conditions. The yarn was stabilized in an airflow of 5–20 mL/min by heating to 200–280 °C at 1 °C/min. The obtained fibres were about 30 nm in diameter (Kim et al. 2015). In another study, Indulin AT was blended with polyethylene glycol – 80% to 90% lignin – at 180 °C in a twin screw extruder to produce pelletized melt spinnable dope. Blending with polyethylene glycol substantially reduced the glass transition temperature compared to lignin alone – 141 °C for lignin and 79–85 °C for blends – rendering the material amenable to melt spinning. Precursor fibres were melt spun through 0.6 mm spinneret and drawn by winding at 200 m/min. Melt spinning temperature varied between 140 and 180 °C. As expected, higher temperatures were required during melt spinning for blends having a higher percentage of lignin. Authors report successful melt spinning with the starting lignin having 2.69% ash. These fibres were stabilized in air for 1 h at 250 °C by placing them on a graphite plate. The heating rate was 1 °C/min for the blend having 90% lignin.

With an 80% lignin blend, the initial heating to 60 °C occurred at 1 °C/min and thereafter at 0.1 °C/min (Enengl et al. 2021).

8.2.1.3 Electrospinning

In electrospinning, the initial strands extrude through an electrically charged spinneret. A potential difference is applied between the spinneret and a conducting collection electrode of a large surface area. The extruding charged liquid droplet is acted upon by capillary force, viscoelastic force, air drag and electrostatic force. With the potential difference exceeding a threshold, the charge on the electrostatic force overwhelms the other forces, causing the droplet to elongate and take a thread-like shape. This threshold potential difference is (Wang et al. 2022)

$$V^2 = \frac{0.468L^2}{l^2}\left[\ln\left(\frac{2l}{r}\right) - 1.5\right]\pi r\gamma \tag{8.7}$$

Here, V is the threshold potential difference, L is the distance between the spinneret and collector, l and r are the length and radius of the nozzle in the spinneret, and γ is the surface tension of the extruding polymer solution. The ensuing jet follows a helical path to the collector plate, being elongated during its flight, finally forming into a coil of precursor having a diameter of only a few nm.

Electrospinning of precursor fibres from blends of sulphonated spruce kraft lignin and polyethylene oxide was studied. Prior to blending, the lignin was washed with 5 weight% hydrochloric acid solution several times to remove the inorganics. Blends – lignin proportion between 74% and 90% – were dissolved in water with 10–11.7% concentration and extruded through 0.56 mm diameter spinneret at 1 mL/h extrusion rate. Blending with polyethylene oxide was necessary as the lignin solution on its own was not found to be suitable for spinning. The applied potential difference was 10,000–15,000 V across a 12–20 cm spinning length. Electrospun fibre mats, consisting of 0.5–1.5 μm dimeter fibres, were collected over copper mesh (Hu and Hsieh 2013). Aqueous ethanol organosolv lignin was isolated from the hydrolysis residue of corn for use as the carbon source in carbon fibre preparation. An electrospinning solution of 25 weight% was prepared by dissolving equal weights of lignin and polyacrylonitrile in dimethyl formamide. Electrospinning conditions were: applied potential difference 10,000 V, distance between spinneret and collector 20 cm, and feed rate 1 mL/h. The formed fibres were vacuum dried at 60 °C followed by impregnation with iodine vapour for 12 h at 70 °C to form a charge transfer complex with lignin through its aromatic π electron system. Complex formation was confirmed by surface abundance of iodine determined by energy dispersive X-ray as well as FTIR. The effect was an increase in T_g. which ensured fibre integrity during stabilization and carbonization. Oxidative stabilization was performed at heating rates of 0.2 or 2 °C/min to a stabilization temperature of 220 °C and holding for 1 h. Iodine-treated fibres could retain their structural integrity at both heating rates, but the untreated fibres were stable only at a slow heating rate (Dai et al. 2018). In one study, hardwood kraft lignin was heated for 30 min under vacuum at 0.1 kPa pressure and 200 °C. This heat treatment volatilized the hemicellulose impurities and also induced condensation by forming additional alkyl aryl carbon–carbon bonds. Heat-treated lignin was dissolved in sodium hydroxide and fractionated into low and high molecular weight fractions by membrane separation. Low and high molecular-weight lignin were separately acid precipitated from the permeate and the retentate, respectively. The high molecular weight fraction had M_w 1696 g/mol and polydispersity of 2.43 while the low molecular weight fraction had M_w 1183 g/mol and polydispersity of 1.97. These lignins had glass transition temperatures of 199.4, and 150.3 °C, respectively. For comparison, the original lignin had a T_g of 157.9 °C and heat-treated lignin had a T_g of 160.2 °C. For electrospinning, lignin was dissolved in dimethyl formamide at 95 °C to prepare a solution of 500 g/L concentration. Polyethylene oxide was added to this solution in a mass ratio of 1:100 of polyethylene oxide to lignin. Fibres were electrospun from this solution through a needle of 0.17 mm nominal internal diameter at a feed rate of 0.5 mL/h. The applied potential difference was 15,000 V across a 15 cm distance to the collector. The high molecular weight fraction exhibited poor spinnability which was attributed to its high polydispersity. The formed fibres were oxidatively stabilized in air by heating to 270 °C at 0.15 °C/min and holding for 1 h (Bai et al. 2021).

Irrespective of the spinning method, the important characteristics of lignin affecting the spinning process and properties of the resulting fibres are the molecular weight and polydispersity. A polydispersity close to 1 is desirable. For melt spinning the glass transition temperature and the softening behaviour of lignin is important. It is possible to control these parameters by chemical modification of lignin. In most of these modifications, the aliphatic or phenolic hydroxyl groups in the lignin macromolecule are substituted by a suitable functional group like carboxyl, isocyanate, or ester. Alternatively, lignin can be blended with a compatible thermoplastic to tailor its melt flow characteristics. A similar approach could also be adopted for electrospinning by blending lignin with a polymer with superior electrospinning ability. Some examples are polyacrylonitrile, polyethylene oxide, and polylactic acid.

8.2.2 Carbonization of Spun Fibres and Fibre Characteristics

Solution spun, finished, and oxidatively stabilized precursor fibres from softwood kraft lignin and softwood kraft pulp blend were carbonized at 1000 °C in flowing nitrogen at 200 mL/min. Heating rates were: up to 600 °C at 1 °C/min, 600 to 1000 °C at 3 °C/min. Total carbonization time was 11.8 h including cooling time to room temperature. The yield of carbon fibres was close to 50% on a mass basis. With a 14 µm diameter of precursor fibres, the finally prepared carbon fibres were 6–8 µm thick having 76–77 GPa tensile modulus and 1070–1170 MPa tensile strength. In contrast, 13–15 µm diameter fibres – precursor fibre diameter 22 µm – had a tensile modulus of 64–66 GPa and tensile strength of 680–890 MPa. This brings forth the beneficial effect of increased draw during spinning which caused more alignment of polymer chains in the precursor fibres along the fibre axis. One interesting observation during this study was that the tensile properties of carbon fibres prepared without oxidative stabilization were almost comparable to those in thin fibres (Bengtsson et al. 2019).

Melt spun and oxidatively stabilized precursor fibres from hardwood kraft lignin, and Alcell lignin, and their blends with polyethylene oxide were carbonized under nitrogen by heating to 1000 °C at a heating rate of 3 °C/min. Alcell lignin resulted in carbon fibres with a 31 µm average diameter in 40% yield. The average tensile strength of these fibres was 388 MPa and the tensile modulus was 40 GPa. The average carbon fibre diameter from hardwood kraft lignin was 46 µm and the yield was 45.7%. Corresponding tensile strength and tensile modulus were 422 MPa and 40 GPa, respectively. The blending of polyethylene oxide up to 5% did not have much effect on tensile properties (Kadla et al. 2002). Precursor fibres, melt spun from an annealed blend of Indulin AT and pyrolysed fuel oil, and oxidatively stabilized in air, were carbonized under argon by heating to 800 °C or to 2800 °C and holding for 1 h. Fibres obtained at 800 °C had a diameter of about 20 nm and those carbonized at 2800 °C had a 15 nm diameter. The fibres formed at 2800 °C had core shell morphology with the core having disordered but isotropic texture at the nano level comprising 10–15 layers of carbon nanosheets. The outer shell on the other hand had crystallites oriented in fibre direction and the arrangement of carbon nanosheets resembling graphite. It was contended that the fibre core arises from the lignin precursor whereas the outer layer comes from the pyrolysed fuel oil. This was due to phase separation during melt spinning. With an increase in carbonization temperature, the tensile strength increased from 490 to 530 MPa. The increase in Young's Modulus was from 56 GPa at 800 °C to 100 GPa at 2800 °C (Kim et al. 2015). Melt spun and oxidatively stabilized precursor fibres from a blend of Indulin AT and polyethylene glycol – 80% to 90% lignin – were carbonized at temperatures of 900, 1500, 2000, and 2300 °C and held for 30 min at the respective temperatures. The heating protocol was as follows: 200 °C at 10 °C/min, 200 to 600 °C at 1 °C/min, 600 to 900 °C at 10 °C/min, 900 to 1400 °C at 10 °C/min, and 1400 to final carbonization temperature at 5 °C/min. For 900 °C carbonizing atmosphere was nitrogen and for other temperatures it was argon. The yield of carbon fibres varied between 40% and 49.3% and the fibre diameter between 45 and 75 µm. The

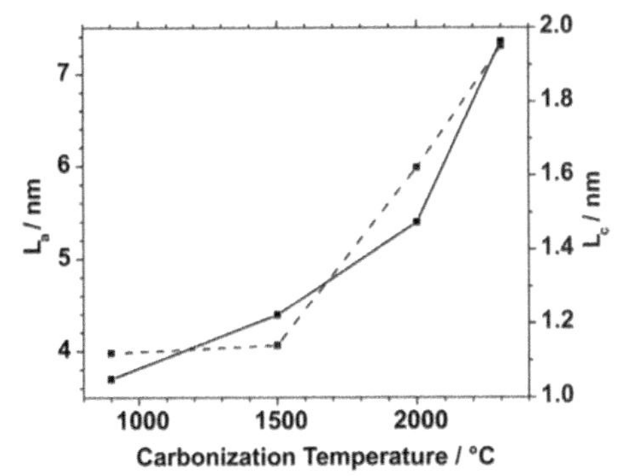

FIGURE 8.4 Dimensions of graphite-like crystallites in carbon fibres obtained from melt spinning of a 90:10 blend of Indulin AT with polyethylene glycol. L_a (solid line) is the lateral dimension and L_c (dashed line) is the thickness. (Reprinted and adapted from Enengl, C, Lumetzberger, A, Duchoslav, J, Mardare, CC, Ploszczanski, L, Rennhofer, H, Unterweger, C, Stifter, D, Fürst, C, *Carbon Trends*, 5, 100094, 2021, with permission from Elsevier.)

higher the lignin content in the blend, the higher the yield, the higher the carbonization temperature, and the lower the yield. The tensile modulus of the fibres lay between 22.3 and 34 GPa and the tensile strength lay between 223 and 311 MPa. An increase in lignin content in the blend decreased the tensile modulus but increased the tensile strength. Increased carbonization temperature increased the tensile properties up to 2000 °C, but beyond this temperature, the tensile properties were reduced. However, the size of the graphite crystallites consistently increased with increasing temperature (Enengl et al. 2021). These trends are shown in Figure 8.4.

Rolled precursor fibre mats, obtained from electrospinning of blends of acid-washed sulphonated spruce kraft lignin and polyethylene oxide, were carbonized at 600–850 °C under flowing nitrogen at 100 mL/min. Initially, the fibres were heated to 105 °C and held for 30 min to remove residual moisture. Subsequently, the fibres were heated at the same rate to the carbonization temperature and held for another 30 min. Carbonized fibres were cooled over a 12 h period under a similar nitrogen flow. Significantly, the study demonstrated that dimensionally stable fibres without fusion could be obtained even in the absence of stabilization of the precursor fibres before carbonization (Hu and Hsieh 2013). Electrospinning of ethanol organosolv lignin mixed with polyacrylonitrile resulted in precursor fibres that were impregnated with iodine vapour. After stabilization, these fibres were carbonized at 1400 °C for 2 h in nitrogen with heating to carbonization temperature at 4 °C/min. Iodine-treated fibres did not fuse during carbonization even when the stabilization heating rate was high, resulting in carbon fibres free from defects. On the contrary, untreated fibres stabilized at 2 °C/min fused during the carbonization step. Even at a stabilization heating rate of 0.2 °C/min, the produced carbon fibres carried structural defects. Also, the carbon fibre yield from iodine treatment was considerably high – 59–61% – compared to that without treatment – 22–38%. Advantages in carbon fibre morphology gained from iodine treatment of precursor fibres reflected in their mechanical properties. These fibres had a tensile strength of 85–89 MPa and a tensile modulus of 5.2–5.3 GPa. Carbon fibres produced without iodine treatment had a tensile strength of 56 MPa and a tensile modulus of 3.2 GPa only (Dai et al. 2018). Oxidatively stabilized electrospun fibres from hardwood kraft lignin, heat-treated hardwood kraft lignin, high molecular weight hardwood kraft lignin, and low molecular weight hardwood kraft lignin, mixed with polyethylene oxide, were carbonized at 1200 °C in argon environment. The heating rate was 3 °C/min. The average diameter of produced yarns were 952, 1197, 1005, and 1080 nm, respectively. The lowest yield of fibres was obtained from low molecular weight lignin. Fibres from low molecular weight lignin had a grain size of 3.39 nm with graphite crystallites and those from high molecular weight lignin had a grain size of 3.7 nm with small graphitization. Carbon fibres from hardwood kraft lignin had a tensile strength of 0.51 MPa which improved to 0.92 MPa when the precursor was heat-treated lignin. Of its two fractions, the lower molecular weight heat-treated lignin precursor resulted in fibres of 0.54 MPa tensile strength. The tensile strength of fibres from high molecular weight fraction was 1.10 MPa (Bai et al. 2021). Scanning electron micrographs of the produced carbon fibre yarns along with size distributions are shown in Figure 8.5.

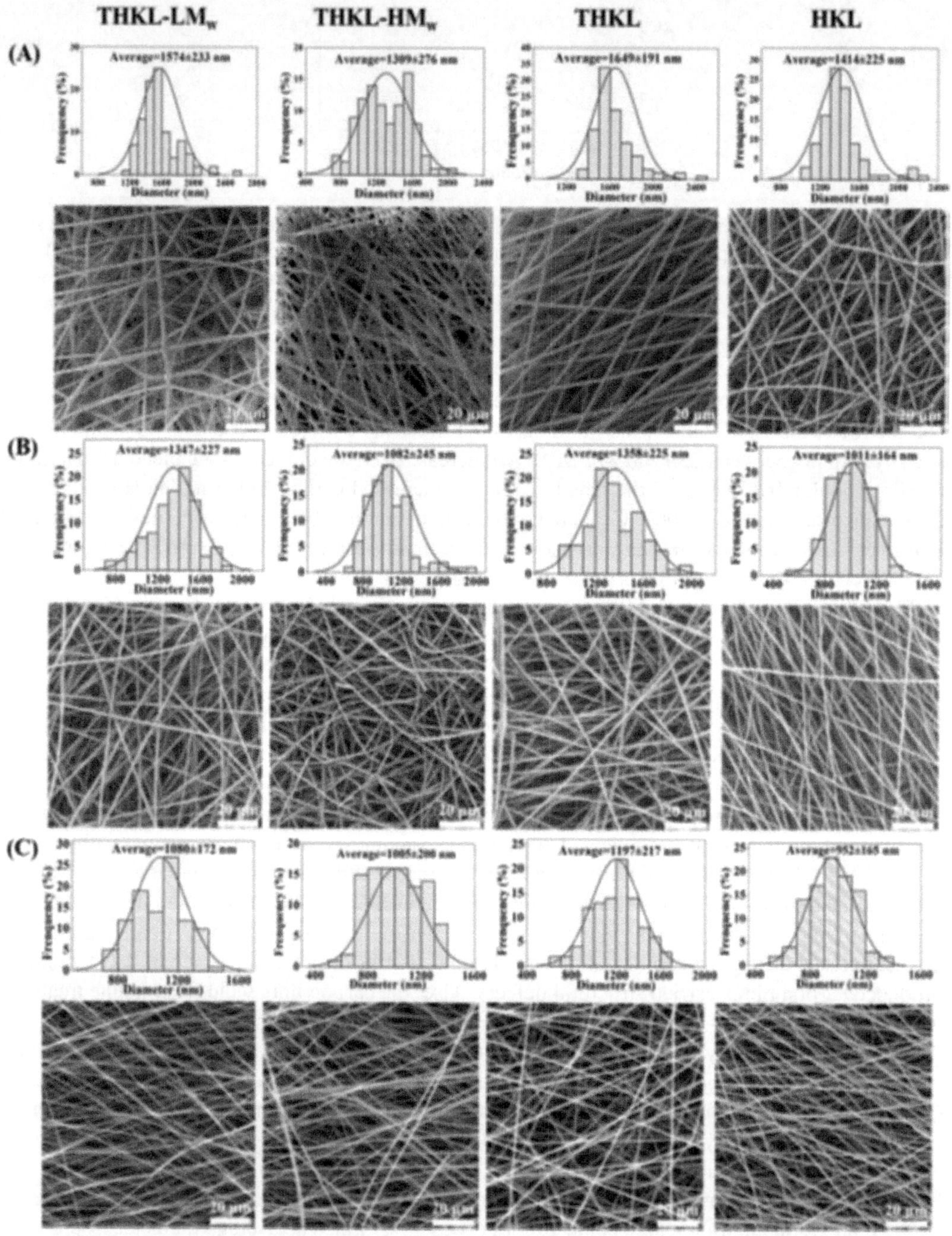

FIGURE 8.5 Scanning electron micrographs of carbon fibre yarns along with size distributions produced from hardwood kraft lignin (HKL), heat-treated HKL (THKL), low molecular weight THKL (THKL-LM$_w$), and high molecular weight THKL (THKL-HM$_w$). (a) electrospun yarns, (b) oxidatively stabilized yarns, and (c) carbonized yarns. (Reprinted from Bai, J, Wang, S, Li, Y, Wang, Z, Tang, J, *Int. J. Biological Macromolec.*, 187, 594–602, 2021, with permission from Elsevier.)

8.3 CARBON NANOTUBES MODIFIED WITH LIGNIN

Another very important microscopic carbon material meant for high-end speciality applications is carbon nanotubes. As discussed in Section 8.2, carbon fibres are microscopic solid entities made of carbon crystallites with a large length-to-diameter ratio. On the other hand, carbon nanotubes are hollow microscopic entities, with a large length-to-diameter ratio, in which a void annulus is surrounded by a wall made of carbon. The wall of a carbon nanotube essentially consists of layers of two-dimensional carbon sheets – graphene. In each graphene sheet, the sp^2 hybridized carbon atoms covalently bond with each other to form honeycomb carbon rings which extend in space in two dimensions. Such a carbon sheet is only one carbon atom in thickness. Mostly, the wall of a carbon nanotube has several layers of graphene sheets packed around a common central axis. These are known as multiwalled carbon nanotubes (MWCNT). Single-walled carbon nanotubes (SWCNT) are those where the tube wall is made of only one graphene sheet. The wall of a SWCNT is, therefore, only one carbon atom thick. The ends of the nanotubes are closed by a half-fullerene dome.

SWCNTs are difficult to synthesize and require sophisticated catalytic chemical vapour deposition techniques. Cobalt or nickel are the catalysts most often used. A catalyst-coated substrate is placed in a chamber at a high temperature (900–1000 °C) under inert atmosphere. A controlled flow of precursor carbon source gases is injected into the chamber where they atomize into carbon atoms. These carbon atoms condense over the catalyst-coated surface of the substrate forming the nanotubes. Methane, acetylene, or carbon monoxide can be used as the carbon source gas. Other synthesis methods are high-energy electrical discharge or LASER ablation. When a sufficiently high direct current is passed across two graphite electrodes with a very small inter-electrode gap (a few mm only), a high-energy electrical arc is produced raising the temperature sufficient to sublime the carbon. An inert atmosphere is maintained in the reaction zone with the help of low-pressure helium or argon to avoid oxidation or other undesirable side reactions. Vaporized carbon rapidly cools to form carbon nanotubes along with fullerenes and amorphous carbon. Catalysts are used to facilitate nanotube formation. In the LASER ablation technique, high-energy LASER partly supplies the energy to vaporize carbon from graphite. Thermal energy is also provided by raising the temperature of graphite to about 1200 °C before irradiating with LASER. The reaction environment consists of low-pressure inert gas along with the requisite catalyst.

Carbon nanotubes have unmatched properties. Since the carbon atoms in the wall are arranged in a rigid ordered lattice, they can only have certain discrete dimeters. The diameter of a SWCNT is fixed by the lattice parameters of the graphene sheet making its wall and the direction of rolling the sheet with respect to the lattice. If two SWCNTs of different lattice parameters, and hence different diameters, are packed into a concentric fashion, it will give rise to a two-layer MWCNT or double-walled carbon nanotube (DWCNT) of 'Russian doll' conformation. The number of layers in the wall can increase if several SWCNTs, each with lattice parameters and diameters different from the others, are packed into concentric arrangements. The conformation is still of a 'Russian doll' type. The gap between the adjacent layers of graphene is a fraction of a nanometre. Alternatively, a large sheet of graphene may roll into a spiral making an MWCNT of 'parchment' conformation.

Owing to the highly ordered arrangement of carbon atoms in their walls, the SWCNTs have excellent tensile properties and thermal conductivity. Extremely high tensile strength and tensile modulus combined with a not-so-high density make these nanotubes material with the highest specific strength. Tensile properties deplete with the number of layers in the wall but are still very high for specific applications. The tensile properties of MWCNTs can be significantly improved by cross-linking between layers. Carbon nanotubes do not have comparable compression strength and are rather soft in radial direction. In addition, SWCNTs may be electrical conductors or semiconductors in the longitudinal direction. Moreover, there is the possibility of grafting specific functional groups onto the carbon backbone to get nanotubes with tailored properties. This allows them to be used as excellent nanocatalyst support. The electrical properties of SWCNTs depend on the lattice parameters. The electrical conductivity of a conducting SWCNT can be several orders of magnitude higher than that of a metal, opening interesting possibilities. Carbon nanotubes are heat conductors in the longitudinal direction – thermal conductivity several orders of magnitude higher than metals – but insulators in the radial direction.

In a graphene sheet, the base hexagonal carbon ring is characterized by the cell vectors $\mathbf{a_1}$ and $\mathbf{a_2}$ as depicted in Figure 8.6. When such a graphene sheet rolls itself into a SWCNT, it can do so in two ways depending on the chiral vector **C**. It can be mathematically represented as

$$\mathbf{C} = n\mathbf{a_1} + m\mathbf{a_2} \tag{8.8}$$

Here, m and n are scalar indices defining the angle of vector **C** with respect to $\mathbf{a_1}$ and $\mathbf{a_2}$. The resulting configuration is called zigzag when m = 0. In such a case, the direction of the chiral vector coincides with the vector $\mathbf{a_1}$. When n = m, the resulting configuration is called armchair configuration. The diameter, d, of a SWCNT, is related to the indices m and n as (Eatemadi et al. 2014)

$$d = \frac{a\sqrt{m^2 + mn + n^2}}{\pi} \tag{8.9}$$

Here, a, the lattice constant of graphene, is 0.246 nm (Yang et al. 2018).

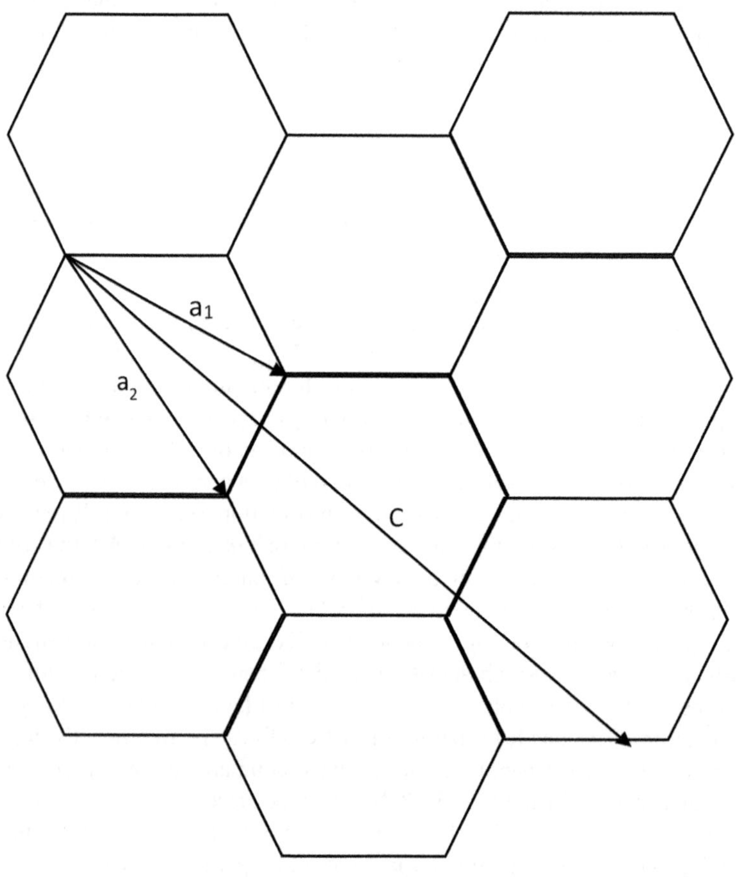

FIGURE 8.6 Schematic representation of graphene with cell vectors a_1 and a_2, and the chiral vector C.

8.3.1 Lignin as a Dispersing Agent for Carbon Nanotubes

The value-added application of lignin in the area of carbon nanotube technology is based on the controlled modification of nanotube characteristics. Individual carbon nanotubes have exceptionally high specific surface area. However, in a population of carbon nanotubes, they have the tendency to cluster together forming larger aggregates. A dispersing agent is required to prevent the aggregation of carbon nanotubes and lignin is one of the potential candidates. Lignin is believed to alter the surface charge distribution of carbon nanotubes by π–π interactions. An early study reported the dispersing ability of kraft lignin for 10–15 nm average diameter MWCNTs by composite formation. Nanotubes were dispersed by 15 min sonication in a 5 g/L solution of kraft lignin in acidified acetonitrile with 0.5 g nanotubes per L of solution. Centrifuged, washed, and dried lignin nanotube composites formed stable dispersion in dimethyl sulfoxide for casting composite film (Milczarek 2010). Another study reported improved dispersibility of MWCNTs in water by modifying them with spruce kraft lignin. Nanotubes were modified with lignin by sonicating them in water with lignin and carbon to lignin ratio between 1 and 6 to form stable dispersions. The dispersion was mixed with polypropylene-grafted-maleic anhydride latex and cast into film. Lignin modification allowed the carbon nanotubes to evenly distribute in the latex matrix without forming agglomerates (Rochez et al. 2013). kraft lignin was used as the dispersing agent for MWCNTs. Lignin, dissolved in tetrahydrofuran, was added to the suspension of MWCNTs and sonicated for 15 min to form a well-dispersed suspension. To this was added palladium nanoparticles, synthesized by LASER ablation. The resulting mixture was further sonicated for 30 min and dried to form composite nanoparticles. These composite nanoparticles demonstrated very good catalytic properties for the reduction of methylene blue by sodium borohydride and the reduction of Cr^{6+} (hexavalent chromium) by formic acid (Arnawtee et al. 2022). Fig 8.7 shows the scanning electron micrographs of carbon nanotubes with and without lignin surface modification with the insets very effectively demonstrating the uniform dispersion of modified nanotubes compared to unmodified ones.

8.3.2 Composites from Lignin and Carbon Nanotubes

Composites from lignin and carbon nanotubes are another application area. Eucalyptus kraft lignin was chemically modified by catalytic oxidation with 500 kPa oxygen pressure, 50 °C temperature, and 2 h reaction time. For this, lignin was dispersed in a 20 mM catalyst solution of sodium decamolybdodivanadophosphate in water. Oxidized lignin was separated by centrifugation and dried. Oxidized lignin was ball-milled with MWCNTs into fine powder. A composite of MWCNTs with lignin-based polyurethane was prepared by reacting the powdered mixture of nanotubes and oxidized lignin with poly(propylene glycol), and toluene 2,4-diisocyanate. Reaction conditions were: ratio of

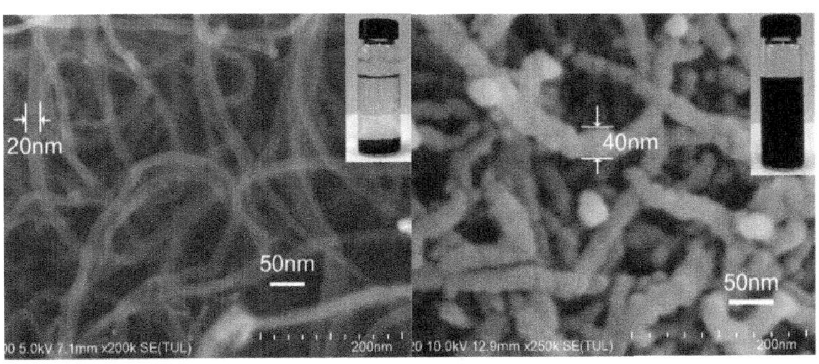

FIGURE 8.7 Scanning electron micrographs of MWCNT with (right) and without (left) surface modification with kraft lignin. Insets show the respective dispersion conditions. (Reprinted and adapted from Li, Z, Chen, J, Ge, U, *Chem. Eng. J.*, 308, 809–817, 2017, with permission from Elsevier.)

isocyanate to hydroxyl 1.5, initial mixing at 40 °C for 40 min to prepare a viscous mixture, heating the mixture to 60 °C with the addition of dibutyltin dilaurate catalyst and stirring for 5–10 min. The resulting composite was spread into film or electrode – platinum or glassy carbon – coating with 4 h curing at 60 °C. Unreacted compounds were removed by extraction with dichloromethane. Compared to lignin-based polyurethane alone, the nanotube-doped composite exhibited significant improvement in electrical conductivity – from 3.13×10^{-9} to about 1×10^{-3} S/m at 0.72 weight% addition of nanotubes (Faria et al. 2012). Kraft lignin solution – 5 mg/mL concentration – was prepared in a mixture of 1 volume of sulphuric acid in 3 volumes of acetonitrile. Carbon nanotubes of 10–15 nm diameter were sonicated in this solution – 0.5 mg/mL concentration – for 15 min to modify their morphology through lignin uptake. Formed composites were centrifuged, thoroughly washed with water, and dried at 40 °C. Dispersion of these composites in dimethyl sulfoxide or ammonium hydroxide solution was applied as a thin film and dried at 60 °C to coat gold electrodes. The electrode was found effective in capacitive charge storage through reversible redox reactions. Compared to a mass-specific capacitance of 75 F/g for unmodified carbon nanotubes, the composite of kraft lignin with nanotubes had a mass-specific capacitance of 143 F/g (Milczarek and Nowicki 2013). There is a report on the doping of carbon nanotube yarns with Alcell lignin to form nanocomposites. Filaments of carbon nanotubes were synthesized by forming MWCNTs – 5 nm average diameter – by chemical vapour deposition and drawing the aerogel of nanotubes through the furnace outlet. Several such filaments entwined to form the carbon nanotube yarn. 2 mm long nanotube yarns were impregnated in lignin solution in tetrahydrofuran by suspending for 5 min. Impregnated yarns were dried for 1 h at 70 °C. The resulting yarns had smoother surface morphology compared to non-doped yarns and a homogeneous coat of lignin on the nanotube surface through π-π interactions. The composites were evaluated for their thermoelectric characteristics. A considerable increase in Seeback coefficient and power factor was noticed with lignin uptake between 10% and 30% with the maximum Seeback coefficient of 98.9 μV/K and highest power factor of 132.2 μW/(m.K^2) obtained at a lignin doping of 23%. Thus, the composite yarn augurs well for thermoelectric energy conversion (Culebras et al. 2020).

Carbon nanotubes have excellent thermal stability. This is useful for imparting fire-retardant properties to otherwise flammable substrates. However, their lack of dispersibility is an impediment to enhancement of properties. As expected, surface modification of carbon nanotubes with lignin is an attractive option to improve the dispersibility and use them in forming fire-retardant composites. Kraft lignin, with a low level of sulphonation to make it water soluble, was formed into an aqueous solution with potassium carbonate – 1 weight% lignin and 1 weight% potassium carbonate. Carbon nanotubes – 9 nm average diameter – were dispersed in the alkaline lignin solution by sonication. This dispersion was used as a coating for making otherwise flammable cotton roving fire retardant. Roving was dipped in the dispersion and dried at 80 °C to evaporate the excess water. Lignin, modified with carbon nanotubes, had improved thermal stability compared to unmodified lignin. This characteristic was further imparted to the cotton roving coated with carbon nanotube-modified lignin. Coating with lignin alone produced only incremental improvement in fire retardant property. Lignin modified with carbon nanotubes improved fire retardance. The best fire retardant properties were observed when the roving was coated with lignin modified with both carbon nanotubes and potassium carbonate (Lukawski et al. 2020).

8.3.3 Lignin-based Carbon Fibres Incorporated with Carbon Nanotubes

One study reported the synthesis of MWCNTs incorporated carbon fibres with different fractions of Indulin AT as the carbon source for fibres. First, lignin and carbon nanotubes were dispersed in dimethyl formamide by 3 h sonication. A 25% concentration solution of lignin, MWCNTs, and poly(ethylene oxide) – 99% lignin, 1% poly(ethylene oxide) – was prepared by heating at 80 °C for 2 h. Precursor fibre mats were electrospun from this solution at an extrusion rate of 0.03 mL/min with a 15,000 V potential difference over an 18 cm distance. The precursor fibre mats were stabilized in air for 1 h at 250 °C with a heating rate of 5 °C/min. Carbonization in nitrogen at 1000 °C followed stabilization. For this, stabilized fibre mats clamped over steel plates were heated at 18.75 °C/min to 250 °C, held for five min, heated at 10 °C/min to carbonization temperature, and held for 1 h. Some important observations from this study were the critical importance of lignin characteristics in adequately dispersing the carbon nanotubes and the ability to form electrospun fibre mats at higher spinning dope concentrations and higher proportions of nanotubes. Indulin AT was fractionated into four fractions with successive solvent extraction with a combination of methylene chloride and methanol – 1st extraction with methylene chloride, 2nd extraction with methanol, and 3rd extraction with 70:30 volume ratio mixture of methanol and methylene chloride. Only the highest molecular weight fraction with M_w of 35,000 and polydispersity of 3.9 – insoluble after the 3rd extraction – was suitable for forming a spinnable dope. Even with this fraction, the proportion of lignin in the dope solution played a crucial role in the uniform dispersion of carbon nanotubes. Up to 4 weight% carbon nanotubes could be dispersed in the spinning solution with the addition of high molecular weight lignin fraction. However, curling of fibres was observed after carbonization when MWCNTs were incorporated compared to the fibres without the nanotubes. This was associated with a decrease in fibre diameter as well as the addition of carbon nanotubes. Interestingly, the fibres without carbon nanotubes had a tensile strength of 45 MPa, but the strength decreased to 37 MPa with the incorporation of 1% nanotubes and further to 28.6 MPa with 4% incorporation. A corresponding decrease in tensile moduli was also observed with values of 6.24, 4.65, and 2.39 GPa, respectively. There was not much improvement in electrical conductivity (Teng et al. 2013). In another study, carbon fibres were first produced using a mixture of kraft lignin and polyacrylonitrile as the carbon source. Carbon nanotubes were subsequently grown on the produced fibres through chemical vapour deposition. For the synthesis of carbon fibres, equal mass of kraft lignin and polyacrylonitrile was dissolved in dimethyl formamide at 15 mass% total concentration by 24 h stirring at 60 °C. 2 mass% catalyst – ferric nitrate or palladium chloride – was also added. Electrospinning conditions were: potential difference 17,000 V, extrusion rate 0.1 mL/h, and electrospinning distance 25 cm. Carbonization conditions were: 10 °C/min to 550 °C and holding for 2 h in a 500:1 mixture of hydrogen and argon, then 10 °C/min to 1000 °C and holding for 1 h in hydrogen. Formed carbon fibres were cooled to 750–850 °C. These catalyst-incorporated carbon fibres acted as the substrate for growing carbon nanotubes through chemical vapour deposition from a 1:3 mixture of acetylene and argon. The choice of catalyst and the temperature for chemical vapour deposition played an important role in nanotube growth and morphology. No carbon nanotubes could grow on the surface of carbon fibres at 850 °C in the absence of a catalyst. However, the heat treatment reduced the average fibre diameter from 1100 nm to 615 nm. At 750 °C, both ferric nitrate and palladium chloride could catalyse the growth of carbon nanotubes on the substrate fibres. Ferric nitrate lost its catalytic ability to grow carbon nanotubes at 850 °C, attributed to rapid catalyst deactivation due to coking. Palladium chloride catalysed nanotube growth at 750 as well as 850 °C with significantly lower nanotube diameter at higher temperatures. All the carbon fibres were thermally stable up to 400 °C. Fibres with carbon nanotubes grown on their surfaces maintained their thermal stability even at 950 °C. Carbon fibre mats and the fibre mats from carbon fibres with nanotube growth exhibited excellent water-repelling capability as revealed by their contact angles (Xu et al. 2014).

There are several research reports highlighting the use of lignin or lignin-based polymers for carbon nanotube surface modification and composite synthesis. However, these reports do not clearly specify the type of lignin in terms of the lignocellulose source and isolation method. Kraft lignin (presumably) was grafted on a carbon nanotube synthesized by chemical vapour deposition with methane as the carbon source in nitrogen flow. Before grafting, the nanotubes were sonicated for 5 h in a 1:3 mixture of nitric acid and sulphuric acid, water washed, and vacuum dried for 24 h at 60 °C. Next, the carbon nanotubes were dispersed in dimethyl formamide – nanotube to medium ratio 8 weight/vol – and reacted with half the volume of thionyl chloride under reflux for 18 h. The excess reactant was distilled under vacuum. This was followed by the dropwise addition of diethylene triamine under stirring at 90 °C. The formed aminated nanotubes were filtered off, water-washed, and vacuum-dried. Finally, lignin was grafted on the aminated carbon nanotubes by mixing an equal mass of nanotubes and lignin in 1 M sodium hydroxide solution. Grafting was facilitated by 6 h continuous stirring at 85 °C with a dropwise addition of formaldehyde. Grafted nanotubes were filtered, washed in water, and vacuum-dried. Grafting resulted in an increase of nanotube diameter from 20 to 40 nm with roughening of surface morphology. These grafted carbon nanotubes were dispersible in water due to negative surface charge and could effectively remove lead ions and diesel from the water matrix (Li et al. 2017). Commercial alkali lignin – presumably kraft lignin – was dispersed in water by sonication. Carbon nanotubes were added to the dispersion and further sonicated to obtain a relatively concentrated dispersion. A coat of this dispersion was applied on both sides of a copier paper. The coating was dried by evaporation of water in air. This coating – 7.5 to 15 g/m^2 – imparted flame retardant properties to the paper with self-extinguishing behaviour. When exposed to a flame for 5 s the coated paper did not show sustained combustion with the burnt area between 11% and 16%. With flame exposure to 15 s, the burnt area increased to 21–36% but the self-extinguishing behaviour was maintained (Song et al. 2017). In one study, MWCNTs were surface-modified by impregnation with an aqueous solution of lignin, but it did not specify the type. However, from the pH, sulphur impurity, and water solubility, it is presumed that the lignin used was most probably sodium lignosulphonate. MWCNTs were suspended in an aqueous solution of lignin having lignin mass 1.5–200 times that of carbon nanotubes and sonicated for several hours at 4 °C. Adequate surface coating of nanotubes and dispersion in water was obtained at a lignin-to-nanotube ratio of 1.5. For producing electrospun carbon nanotube-reinforced fibres, MWCNTs were dispersed in an aqueous solution of poly(vinyl alcohol) and lignin. Electrospinning of nanofibers was carried out at an extrusion rate of 7–10 μL/min with a potential difference of 20,000–30,000 V across 10 cm separation. Spun fibres were dried for 24 h under vacuum at 50 °C. Precursor fibres were cured, first by steaming at 80 °C for 3 h, and then, by heating for 1 h at 200 °C. Carbon nanotube-reinforced fibres had a 190 nm average diameter against 230 nm for non-reinforced fibres. Reinforced fibres also demonstrated a smoother surface. Tensile properties of reinforced fibres were expectedly better than those of non-reinforced fibres; the tensile strength was 3.14 MPa and Young's modulus was 12.3 MPa (Lee et al. 2018). Similarly, lignin of unknown identity was reacted with hexamethylene diisocyanate in the presence of modified iron oxide nanoparticles and carbon nanotubes to produce a polyurethane composite. Initially, a prepolymer was synthesized by adding polyethylene glycol and lignin to the mixture of dimethyl formamide and hexamethylene diisocyanate with 30 min heating at 80 °C. Carbon nanotubes were dispersed in polyvinylpyrrolidone at 5 weight% concentration and added to the prepolymer with iron oxide nanoparticles. Heating was continued for another 30 min and the composite was moulded into the desired shape and characterized for its electromagnetic shielding capability (Hu et al. 2021).

8.3.4 Lignin as Carbon Precursor for Carbon Nanotubes

In rare cases, lignin has also been used as the carbon precursor for the synthesis of carbon nanotube fibres. Initially, nanolignin spheres were synthesized from Indulin AT by dissolving it in dimethyl formamide, atomizing the resulting solution through jet impingement into flowing nitrogen, and finally drying the formed aerosol in a laminar flow heated chamber at 153 °C followed by curing in air flow (Ago et al. 2016). For the synthesis of MWCNTs, the lignin nanoparticles were first stabilized in air at 0.01 °C/min to 250 °C. The stabilized material was carbonized with the heating sequence: 1 °C/min to 500 °C and holding for 1 h, 5 °C/min from 500 to 900 °C and holding for 1 h, and 5 °C/min from 900 to 2000 °C and holding for 1h. The result was carbon nanospheres of 52–83 nm diameter with MWCNTs of 14 nm diameter and about 400 nm length protruding from their surfaces. These MWCNTs were found to have 16 concentric SWCNTs making the tube walls (Gindl-Altmutter et al. 2019). The transmission electron micrograph of a representative MWCNT produced is shown in Figure 8.8.

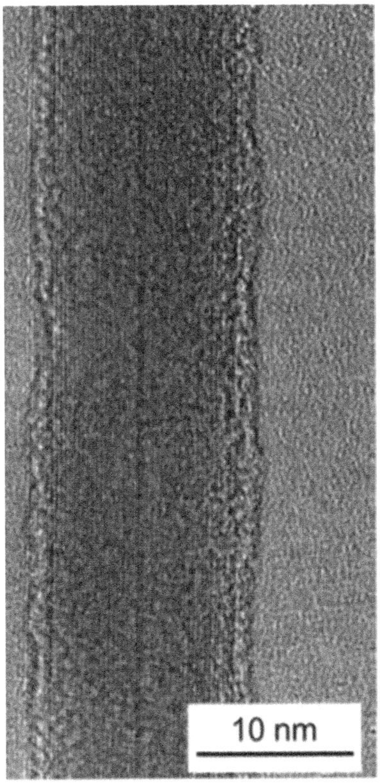

FIGURE 8.8 Transmission electron micrograph of MWCNT produced from carbonization of nanospheres of Indulin AT at 2000 °C. (Reprinted and adapted from Gindl-Altmutter, W, Kohnke, J, Unterweger, C, Gierlinger, N, Keckes, J, Zalesak, J, Rojas, OJ, *Composites Part A*, 121, 175–179, 2019, with permission from Elsevier.)

8.4 OTHER LIGNIN-BASED CARBON PRODUCTS

In a carbon nanotube, the individual graphene layers roll into tubular structures. In the absence of such rolling, the graphene layers may stack together to form multilayer graphene sheets. Catalytic chemical vapour deposition is the process for producing the carbon nanotubes. For nanotube production, a suitable substrate is necessary which promotes the growth of graphene layers in rolled tubular morphology. On the other hand, graphene sheets can be produced from precursor carbon material if a nucleation substrate is not present. There are studies that evaluate the suitability of lignin as a carbon precursor for producing multilayer graphene sheets. Kraft lignin has been used as a carbon precursor for producing multilayer graphene sheets; carbonization is catalysed by iron. A 1 g/mL solution of kraft lignin in tetrahydrofuran was prepared by 2 h stirring. Another solution of ferric nitrate of 2.46 kg/L concentration was prepared in water. The ferric nitrate solution was added dropwise to the lignin solution – 1:3 ratio of the two solutions – for 2 h under stirring. The resulting mixture was kept for 24 h at room temperature and then dried for 24 h at 110 °C. Iron-impregnated lignin was carbonized in different gas environments for 1 h at 1000 °C; the heating rate to the final temperature being 10 °C/min. Product yields were 36.5%, 36%, 31.3%, 28.8%, and 14.7% with the gas environment of methane, natural gas, argon, hydrogen, and carbon dioxide. Graphitization was the highest in natural gas, and methane environments and the lowest in hydrogen and carbon dioxide environments. Graphene nanosheets were found to form in carbonization atmospheres of hydrogen, methane, and natural gas (Yan et al. 2018a). In a related study, the effect of different catalysts on the formation of multilayer graphene sheets from kraft lignin was examined in an argon environment. Copper, nickel, molybdenum, and iron were the catalysts used. Metal ions were incorporated into lignin by coprecipitating solution of lignin in tetrahydrofuran and aqueous solution of the metal ion salt – cupric nitrate, nickel nitrate, ammonium molybdate, and ferric nitrate. Carbonization conditions were: temperature 1000 °C, time 1 h, and heating rate 10 °C/min. Product yield varied between 30% and 34%; highest with copper and lowest with molybdenum. The highest graphitization was achieved with iron followed by nickel, molybdenum, and copper in that order (Yan et al. 2018b).

8.5 LIGNIN-BASED CARBON MATERIALS IN ENERGY STORAGE AND ELECTROCHEMICAL APPLICATIONS

Carbon materials, including those derived from lignin, have unique electrical properties. Porous carbons, with a high specific surface area and pore volume, have dielectric properties. These characteristics make them suitable candidates for use as dielectric material in capacitive energy storage devices. On the other hand, graphitic solid carbons – relatively low specific surface area and mesoporosity – have high electrical conductivities making them useful as electrode materials in storage batteries or electrochemical sensors.

8.5.1 Anode in Rechargeable Cells

8.5.1.1 Lithium-Ion Cells

Lithium-ion cells are ubiquitous in modern electronic and electrical appliances. The anode in a lithium-ion cell stores the lithium in its oxidized form of Li^+. During discharging, these Li^+ ions move towards the cathode where they combine with the electrons. The accumulated electrons move through the external load from the anode to the cathode completing the circuit. During charging, the lithium ions go back to the anode and get stored. Currently, graphite is the anode material in commercial lithium-ion cells storing lithium ions between the individual graphene strata. One of the focus areas of research for lithium-ion cells is the development of renewable graphitic materials with equivalent or more energy storage capacity. Lignin is an obvious choice either as a carbon precursor or as a dispersing agent for other carbon materials like carbon nanotubes.

One study reported the application of carbon nanofibre mat, prepared from the carbonization of electrospun precursor fibre mats of a mixture of lignin (type unknown), polyacrylonitrile, and Pluronic P123 – block polymer of poly(ethylene oxide) and poly(propylene oxide). Ingredients (equal mass of lignin and polyacrylonitrile) were dissolved in dimethyl formamide followed by electrospinning with 20,000 V potential difference over 15 cm collector distance and 1.5 mL/h extrusion rate. Fibre mats were vacuum dried for 12 h at 60 °C, air stabilized for 3 h at 250 °C, and carbonized for 1 h at 1200 °C under nitrogen. The addition of Pluronic P123 resulted in thinner fibres and also promoted the fusion of fibres together resulting in a denser fibre mat. It was contended that a dense fibre mat could enhance electrochemical performance by providing better continuity in electron transfer pathways. These fibre mats were largely mesoporous leading to reduced diffusional resistance to lithium-ion transport through them. As prepared carbon fibre mats, without any additional current collector, were used as anode in the lithium-ion cell. A specific charge capacity of 384 mAh/g was achieved at 20 mA/g current. In comparison, carbon fibre mats without Pluronic P123 achieved a specific charge capacity of only 304 mAh/g. Even at a high current of 5 A/g, the specific charge capacity was a reasonable 128 mAh/g. At high current, the cell could retain 89% of its specific charge capacity after 400 cycles (Shi et al. 2017). The electrochemical performance of lignin-based carbon nanofibre mat anode is depicted in Figure 8.9.

In another study, sodium lignosulphonate was the carbon precursor for producing composites of mesoporous carbon nanospheres and nickel oxide. Precursor lignin nanospheres were prepared by dropping tiny droplets of lignin aqueous solution into isopropyl alcohol. Formed nanospheres were centrifuged and dried at 60 °C. These were carbonized in a nitrogen atmosphere at 800 °C to produce the mesoporous carbon nanospheres – a specific surface area of 868 m^2/g, a pore volume of

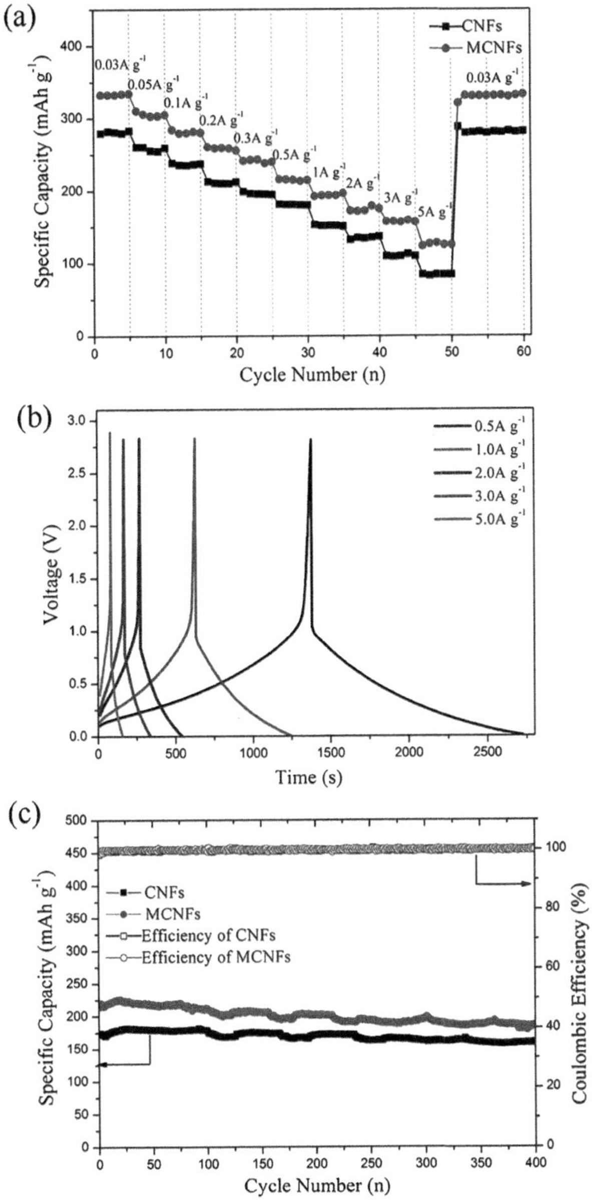

FIGURE 8.9 Electrochemical performance of lignin-based carbon nanofibre mat anode. (Reprinted and adapted from Shi, Z, Jin, G, Wang, J, Zhang, J, *J. Electroanal. Chem.*, 795, 26–31, 2017, with permission from Elsevier.)

0.22 cm^3/g, and an average pore diameter of 2.32 nm. Carbon nanospheres were dispersed in aqueous solution of nickel nitrate and sodium hydroxide – in situ formation of nickel hydroxide – to get nickel hydroxide layered carbon nanospheres. The outer shell of nickel hydroxide was calcined at 450 °C under nitrogen forming composites of mesoporous carbon nanospheres and nickel oxide – 11 weight% nickel oxide, 852 m^2/g specific surface area, 0.16 cm^3/g pore volume, and 2.47 nm average diameter. The produced composite nanospheres were assessed for their suitability as carbon material in the anode of the lithium-ion cell. A mixture of nanospheres, carbon black, poly(vinylidene fluoride), and N-methyl pyrrolidone – 8:1:0.09:0.91 – were stirred into a uniform slurry which was applied as a coat on copper foil. Coated copper foil was vacuum-dried and pressed to form the anode. After 100 cycles of charging and discharging, a storage capacity of 863 mAh/g was obtained at 0.1 A/g, the initial capacity being 1085 mAh/g. To assess the influence of current density, it was increased every 10 cycles. The storage capacities at 0.2, 0.4, 0.6, 0.8, and 1 A/g were 861, 809, 743, 657, and 548 mAh/g, respectively. The storage capacity increased to 839 mAh/g when the current density was reverted to 0.1 A/g (Zhou et al. 2018).

Composites of corn stalk enzymatic hydrolysis lignin and 20–30 nm diameter SWCNTs were prepared by dissolving the lignin in 20% alkaline aqueous ethanol at pH 12 – lignin concentration 2 mg/mL and dispersing carbon nanotubes by 30 min sonication. The complex was precipitated by acidification to pH 2, centrifuged, water-washed, and dried at 80 °C. The composite was activated by impregnation with potassium carbonate and carbonized sequentially under nitrogen and hydrogen atmospheres. The carbonization protocol was 30 min at 250 °C under nitrogen, 2 h at 900 °C under nitrogen, and 1 h at 800 °C under hydrogen. Prepared carbon material was mixed with carbon black and poly(vinylidene fluoride) – carbon material:carbon black:binder 8:1:1 – and fabricated into the anode of the lithium-ion cell. The anode prepared from carbonized lignin had a storage capacity of 1439 mAh/g at 37.2% coulomb efficiency at 200 mA/g. In comparison carbonized composites of lignin and carbon nanotubes (50% nanotube loading) exhibited a lower charge storage capacity of 905 mAh/g but a higher coulomb efficiency of 62.7%. This was attributed to lower specific surface area and higher mesoporosity of the material with carbon nanotube inclusion. The charge storage capacity decreased but coulomb efficiency increased with increasing charging–discharging cycles; 614 mAh/g and 99.5% after 300 cycles (Xi et al. 2019). In another study, microporous carbon was prepared through carbonization of a mixture of lignin – source and characteristics not specified – and a zeolitic imidazolate framework. For preparing the latter, methyl imidazole and cobalt nitrate were separately dissolved in methanol and the two solutions were mixed by stirring to form the precipitate of the zeolitic imidazolate framework. The formed nanoparticles were centrifuged and dried at 60 °C. A mixture of lignin and framework nanoparticles – lignin to nanoparticles mass 10:1 – was dispersed in water and hydrothermally reacted for 12 h at 200 °C. Centrifuged reacted mass was dried at 60 °C and subjected to 3 h carbonization at 750 °C. Following carbonization, embedded cobalt was removed by washing with hydrofluoric acid. The prepared carbon material was applied with the adhesive polyvinylidene fluoride on a copper foil current collector and dried under vacuum to act as the anode in the lithium-ion cell. Compared to carbon particles prepared from lignin alone, the particles prepared from lignin and zeolitic imidazolate framework were uniformly distributed. This resulted in a higher specific surface area with a mesoporous structure which facilitates electrolyte penetration and charge transfer to give improved charging–discharging performance. Specific charge capacities achieved were between 180 and 560 mAh/g with a charging current between 1000 and 100 mA/g. Notably, after charging with 1000 mA/g to full charge and discharging, the battery could again attain a specific charge capacity of 560 mAh/g when recharged at 100 mA/g. The battery retained 99% of its specific charge capacity even after 300 cycles at 100 mA/g charging current which dropped to 85% retention at 1000 mA/g (Li et al. 2023).

Similar to normal lithium-ion batteries, lithium sulphur batteries are currently being developed for their perceived higher specific energy storage capacity. For these applications, a conducting material imbibed with the sulphur is needed. Conducting carbon is an obvious choice. Lignin-based carbons are gaining attention in this context. A study reported kraft lignin as the precursor material to produce the carbon with the desired morphology for sulphur inclusion. Kraft lignin was activated with potassium hydroxide – 2: 3 ratio – by stirring lignin in potassium hydroxide solution and 12 h drying at 80 °C. The resulting material was ground and carbonized at 700 °C in a nitrogen atmosphere for 2 h. Obtained carbon was water-washed to remove excess alkali and dried at 80 °C. The obtained carbon material was partly graphitized and microporous in nature with a specific surface area of 1212 m^2g. Lignin-derived carbon and sulphur were mixed in a 2:3 mass ratio, ground, and heated for 6–10 h under nitrogen at 150 °C for sulphur impregnation. The anode for the lithium sulphur cell was fabricated by applying a 7:2:1 mixture of prepared carbon material, carbon black, and the binder sodium alginate on an aluminium foil. Sulphur loading varied between 0.2 and 1 mg/cm^2. In the cell, the anode was separated by Celgard 2300 from the nickel foam current collector and lithium cathode. The anodes exhibited effective lithiation/delithiation behaviour with a storage capacity of 1370.9 mAh/g at 837.5 mA/g current density. Capacity reduced to 1037.8 mAh/g after 10 cycles. The 1st cycle storage capacity marginally dropped to 1330.7 mAh/g at 1675 mA/g current density (Yu et al. 2017).

8.5.1.2 Sodium Ion Cells

Enzymatic hydrolysis lignin from corn stalk bioethanol production was extracted with acetone to reduce inorganic impurities. This lignin, mixed with 1–15 weight% hexamethylenetetramine, was used to produce carbon microspheres. A mixture of lignin and hexamethylenetetramine was dissolved in ammonium hydroxide by 20 min sonication. Formaldehyde, four times the mass of lignin, was added to the solution, mixed well, and added to olive oil under shear mixing to form water in oil-type emulsion. The emulsion was heated to evaporate water. Microspheres in the aqueous phase of the emulsion contained lignin and formaldehyde which polymerized during evaporation to form precursor microspheres for carbonization. Carbonization was carried out initially at a lower temperature and nitrogen atmosphere followed by 1300 °C heating in a hydrogen atmosphere. The resulting carbon microspheres of 10–20 μm had a very small specific surface area of 8–11 m^2/g confirming their largely solid nature. 80% of these microspheres were mixed with 10% carbon black and 10% poly(vinylidene fluoride) and formed into a slurry in N-methyl-2-pyrrolidone. The slurry was coated on copper foil and cured for 12 h in vacuum at 120 °C to form the anode. Anode without the addition of hexamethylenetetramine exhibited a specific charge capacity of 370 mAh/g which increased to 436 mAh/g with 15% addition of hexamethylenetetramine. The corresponding coulombic efficiencies were 78.2% and 78.7%, respectively (Yu et al. 2018).

For all applications where carbon is to be used as a conducting material, conductivity is an important characteristic. Carbon black, derived from petroleum, is the traditionally used material for conducting applications. However, a search for renewable alternatives is on due to sustainability considerations. In one study, several lignins – kraft or acid hydrolysis – were converted to biochar and compressed beyond their percolation threshold to become electrically conductive. Biochar from 1100 °C pyrolysis of acid hydrolysis lignin attained a conductivity of about 15 S/cm when compressed to a packing fraction of 0.42. In comparison, biochar from kraft lignin at the same pyrolysis temperature attained a maximum conductivity of about 2 S/cm at a packing fraction of 0.6 (Kane et al. 2021).

8.5.2 Capacitive Energy Storage

Sodium lignosulphonate was converted to mesoporous activated carbon by carbonization at 600–800 °C. This material was evaluated for capacitive electrical energy storage performance. A current collector was synthesized by impregnating nickel foam with a mixture of activated carbon, carbon black and polytetrafluoroethylene (80:15:5 weight ratio). Impregnation was carried out with a slurry of the mixture in ethanol at 20 MPa pressure followed by 24 h curing under vacuum at 110 °C. The formed electrodes exhibited the formation of an electrical double layer and were measured to have specific capacitances of 182, 174, and 158 F/g, at 1 mA/cm^2 current density, with activated carbons prepared at 800, 700, and 600 °C respectively. Specific capacitances were 141, 125, and 110 F/g, respectively, at 10 mA/cm^2 current density. Loss of capacitance was less than 5% after 1000 cycles of charging and discharging (Chen and Zhou 2012). The charging–discharging performance and variation of specific capacitance with current density is shown in Figure 8.10.

In a related work, organosolv lignin – an unknown solvent and lignocellulose source – was modified with aniline by dissolving lignin and aniline in ethanol and heating at 80 °C till solvent evaporation. Modified lignin was converted to activated carbon in a two-stage carbonization in a nitrogen atmosphere. First stage carbonization comprised of 10 °C/min heating to 400 °C, holding for 1 h,

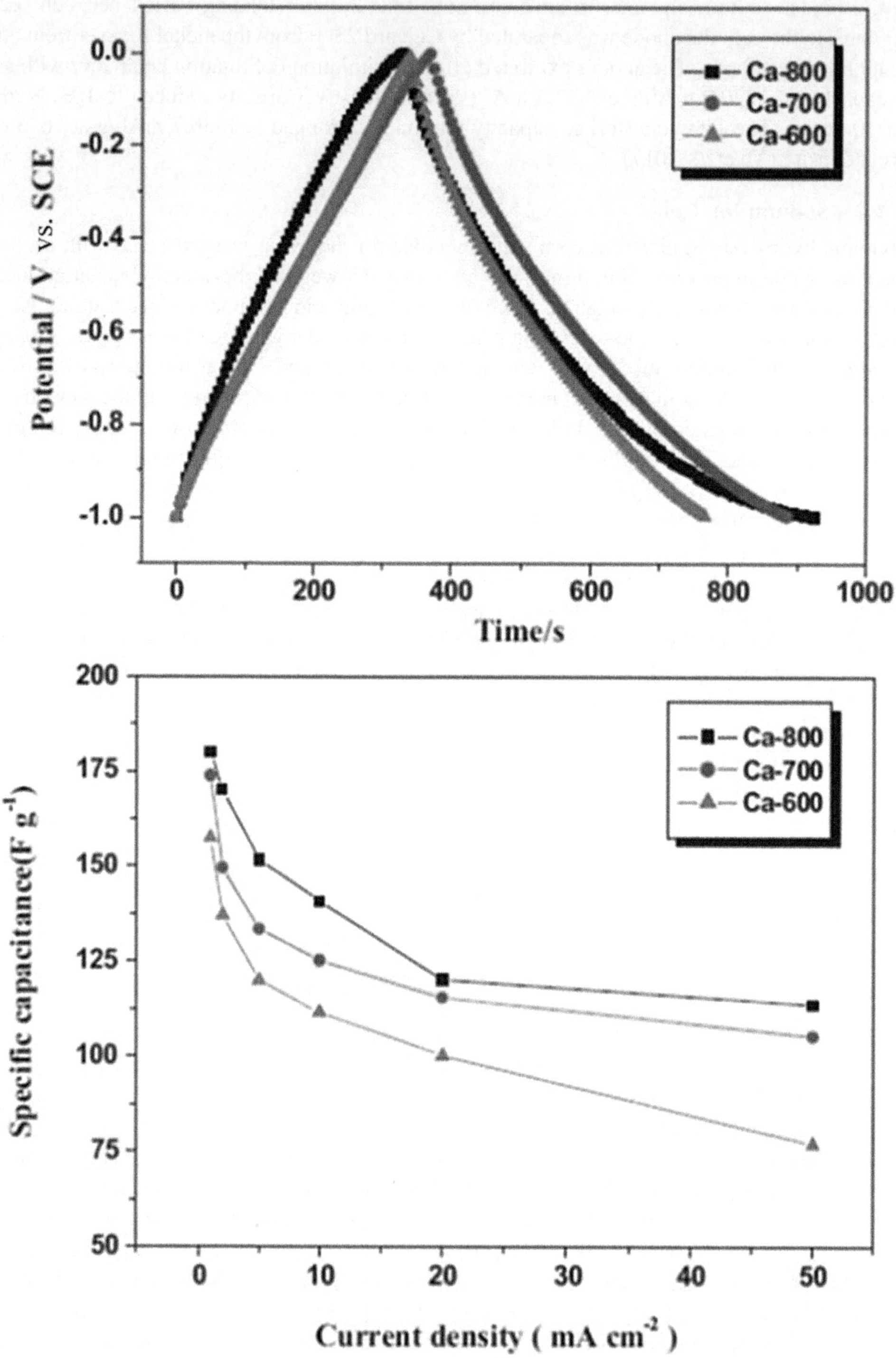

FIGURE 8.10 Electrochemical performance of lignosulphonate-based capacitive energy storage through impregnated nickel foam electrode. (Adapted and reprinted from Chen, XY, Zhou, QQ, *Electrochimica Acta*, 71, 92–99, 2012, with permission from Elsevier.)

and cooling to room temperature. Precarbonized material was impregnated with potassium hydroxide solution and dried for 24 h at 110 °C. Final carbonization was carried out for 1 h at 600–900 °C after heating to the final temperature at a rate of 10 °C/min. Residual potassium hydroxide in activated carbon was removed by water washing. Produced carbon particles had rod-shaped porous morphology. Dry activated carbons were mixed with carbon black and polytetrafluoroethylene – 8:1:1 ratio and imbibed into nickel foam to form the electrodes for capacitive energy storage. The capacitive cell consisted of two electrodes separated by celgard-3501. The highest specific surface area of 2467 m^2/g was obtained at 800 °C carbonization temperature which provided a specific capacitance of 247 F/g. However, the carbon prepared at 700 °C, with a 2265 m^2/g specific surface area, was found to have the highest specific capacitance of 333 F/g. This illustrated the importance of morphological characteristics other than specific surface area in electrical capacitive performance. Almost all of the initial specific capacitance was retained after 1000 cycles of charging and discharging at 1 A/g (Wang et al. 2016).

Similarly, activated carbon in particulate form, from slightly sulphonated kraft lignin, was rolled into a nickel foam current collector in a capacitive cell. Electrical double-layer formation with no redox couple was evident by the cyclic voltammetry characterization. Carbon activated by sodium hydroxide had a specific capacitance of 226 F/g and that activated with potassium hydroxide had a specific capacitance of 132 F/g. The corresponding energy densities were 7.85 and 4.58 Wh/kg, respectively. - Carbon, activated with Sodium hydroxide, could retain 95% of initial specific capacitance after a 5000 charging–discharging cycle. The capacitance was 92% for potassium hydroxide-activated carbon (Hu and Hsieh 2017). Composites of mesoporous carbon nanospheres and nickel oxide, with sodium lignosulphonate as the carbon source (refer to Section 8.5.1), were evaluated for their suitability as electrode material in supercapacitor application. A slurry was prepared in deionized water with nanocomposites, carbon black, and polytetrafluoroethylene – 8:1:0.05. The slurry was coated onto a graphite sheet and dried at ambient temperature, first at atmospheric pressure and then under vacuum. Electrochemical performance was evaluated in 6 M potassium hydroxide with cyclic voltammetry. Carbon in the nanocomposites resulted in electrical double layer formation while the nickel oxide formed a redox couple with its oxidized counterpart NiOOH. The calculated specific capacitance was 508 F/g. 92% of this was retained after 2000 cycles (Zhou et al. 2018).

One particular use of lignosulphonates is in the Negative Active material of lead acid batteries where they prevent electrode passivation. It is common practice to add 0.2–0.3 weight% lignosulphonate in the Negative Active material of lead-acid batteries. Sodium and calcium lignosulphonates from jute sticks were tested in this context. Their performance in stabilizing the charging and discharging cycles of lead acid batteries exceeded that of commercially available wood-based lignosulphonates (Bose et al. 2023).

REFERENCES

Ago, M, Huan, S, Borghei, M, Raula, J, Kauppinen, EI, Rojas, OJ. *ACS Appl. Mater. Interfaces* 8:23302–23310, 2016.

Arnawtee, WH, Jaleh, B, Nasrollahzadeh, M, Bakhshali-Dehkordi, R, Nasri, A, Orooji, Y. *Sep. Purific. Technol.* 290:120793, 2022.

Bai, J, Wang, S, Li, Y, Wang, Z, Tang, J. *Int. J. Biological Macromolec.* 187:594–602, 2021.

Bengtsson, A, Bengtsson, J, Sedin, M, Sjöholm, E. *ACS Sustainable Chem. Eng.* 7:8440–8448, 2019.

Bergna, D, Varila, T, Romar, H, Lassi, U. *Biomass Bioenergy* 159:106387, 2022.

Bose, S, Chakrabarti, SK, Sarma, US. *Proc. Natl. Acad. Sci., India, Sect. A Phys. Sci.* 93:729–735, 2023.

Chen, XY, Zhou, QQ. *Electrochimica Acta* 71:92–99, 2012.

Correa, CR, Stollovsky, M, Hehr, T, Rauscher, Y, Rolli, B, Kruse, A. *ACS Sustainable Chem. Eng.* 5: 8222–8233, 2017.

Cotoruelo, LM, Marque´s, MD, Diaz, FJ, Rodriguez-Mirasol, J, Cordero, T. *Sep. Sci. Technol.* 42: 3363–3389, 2007.

Culebras, M, Ren, G, O'Connell, S, Vilatela, JJ, Collins, MN. *Adv. Sustainable Syst.* 2000147, 2020.

Dai, Z, Shi, X, Liu, H, Li, H, Han, Y, Zhou, J. *RSC Adv.*, 8:1218–1224, 2018.

Eatemadi, A, Daraee, H, Karimkhanloo, H, Kouhi, M, Zarghami, N, Akbarzadeh, A, Abasi, M, Hanifehpour, Y, Joo, SW. *Nanoscale Res. Lett.* 9:393, 2014.

Enengl, C, Lumetzberger, A, Duchoslav, J, Mardare, CC, Ploszczanski, L, Rennhofer, H, Unterweger, C, Stifter, D, Fürst, C. *Carbon Trends* 5:100094, 2021.

Faria, FAC, Evtuguin, DV, Rudnitskaya, A, Gomes, MTSR, Oliveira, JABP, Graca, MPF, Costa, LC. *Poly. Int.* 61:788–794, 2012.

Fierro, V, Torne-Fernandez, V, Celzard, A. *Microporous Mesoporous Mater.*, 92:243–250, 2006.

García-Mateos, FJ, Moulefera, I, Rosas, JM, Benyoucef, A, Rodríguez-Mirasol, J, Cordero, T. *Catalysts* 7:308, 2017.

Ghysels, S, Ronsse, F, Dickinson, D, Prins, W. *Biomass Bioenergy* 122:349–360, 2019.

Gindl-Altmutter, W, Köhnke, J, Unterweger, C, Gierlinger, N, Keckes, J, Zalesak, J, Rojas, OJ. *Composites Part A* 121:175–179, 2019.

Gul, E, Alrawashdeh, AB, Masek, O, Skreiberg, Ø, Corona, A, Zampilli, M, Wang, L, Samaras, P, Yang, Q, Zhou, H, Bartocci, P, Fantozzi, F. *J. Anal. Appl. Pyrol.* 158:105263, 2021.

Guo, Y, Rockstraw, DA. *Carbon* 44:1464–1475, 2006.

Hayashi, J, Kazehaya, A, Muroyama, K, Watkinson, APS. *Carbon* 38:1873–1878, 2000.

Hu, S, Hsieh, YL. *J. Mater. Chem. A*, 1:11279–11288, 2013.

Hu, S, Hsieh, YL. *RSC Adv.* 7:30459, 2017.

Hu, W, Zhang, J, Liu, B, Zhang, C, Zhao, Q, Sun, Z, Cao, H, Zhu, G. *Composites Communic.* 24:100616, 2021.

Jedrzejczyk, MA, Engelhardt, J, Djokic, MR, Bliznuk, V, Geem, KMV, Clercq, AVJD, Bernaerts, KV. *ACS Omega* 6:15222–15235, 2021.

Jiang, C, Bo, J, Xiao, X, Zhang, S, Wang, Z, Yan, G, Wu, Y, Wong, C, He, H. *Waste Manage.* 102:732–742, 2020.

Kadla, JF, Kubo, S, Venditti, RA, Gilbert, RD, Compere, AL, Griffith, W. *Carbon* 40:2913–2920, 2002.

Kane, S, Ulrich, R, Harrington, A, Stadie, NP, Ryan, C. *Carbon Trends* 5:100088, 2021.

Kim, MS, Lee, DH, Kim, CH, Lee, YJ, Hwang, JY, Yang, CM, Kim, YA, Yang, KS. *Carbon* 85:194–200, 2015.

Kim, IT, Sinha, TK, Lee, J, Lee, Y, Oh, JS. *Ind. Eng. Chem. Res.* 60:2439–2446, 2021.

Lee, ES, Kim, YO, Ha, YM, Lim, D, Hwang, JY, Kim, J, Park, M, Cho, JW, Jung, YC. *Euro. Polymer J.* 105: 79–84, 2018.

Leitner, SP, Gratzl, G, Paulik, C, Weber, HK. *J. Carbon Res.* 1(1):43–57, 2015.

Li, J, Li, Y, Wu, Y, Zheng, M. *J. Hazard. Mater.* 280:450–457, 2014.

Li, Z, Chen, J, Ge, Y. *Chem. Eng. J.* 308:809–817, 2017.

Li, Y, Wang, F, Milao, Y, Mai, Y, Li, H, Chen, X, Chen, J. *Bioresource Technol.* 307:123165, 2020.

Li, S, Luo, W, He, Q, Lu, J, Du, J, Tao, Y, Cheng, Y, Wang, H. *Int. J. Mol. Sci.* 24:284, 2023.

Liu, XJ, Li, MF, Singh, SK. *J. Mater. Res. Technol.* 12:1434–1445, 2021.

Lukawski, D, Grzeskowiak, W, Lekawa-Raus, A, Widelicka, M, Lisiecki, F, Dudkowiak, A. *Cellulose* 27: 7271–7281, 2020.

Milczarek, G. *J. Electroanal. Chem.* 638:178–181, 2010.

Milczarek, G, Nowicki, M. *Mater. Res. Bulletin* 48:4032–4038, 2013.

Naydenova, I, Radoykova, T, Petrova, T, Sandov, O, Valchev, I. *Molecules* 28:4842, 2023.

Rochez, O, Zorzini, G, Amadou, J, Claes, M, Richel, A. *J. Mater. Sci.* 48:4962–4964, 2013.

Rodriguez-Mirasol, J, Cordero, T, Rodriguez, JJ. *Carbon* 31:87–95, 1993a.

Rodriguez-Mirasol, J. Cordero, T, Rodriguez, JJ. *Energy Fuels* 7:133–138, 1993b.

Shi, Z, Jin, G, Wang, J, Zhang, J. *J. Electroanal. Chem.* 795:26–31, 2017.

Song, K, Ganguly, I, Eastin, I, Dichiara, AB. *Int. J. Mol. Sci.* 18:2368, 2017.

Teng, NY, Dallmeyer, I, Kadla, JF. *J. Wood Chem. Technol.* 33:299–316, 2013.

Wang, K, Cao, Y, Wang, X, Castro, MA, Luo, B, Gu, Z, Liu, J, Hoefelmeyer, JD, Fan, Q. *J. Power Sources* 307:462–467, 2016.

Wang, S, Bai, J, Innocent, MT, Wang, Q, Xiang, H, Tang, J, Zhu, M. Green *Energy Environ.* 7:578–605, 2022.

Xi, Y, Yang, D, Liu, W, Qin, Y, Qiu, X. *Electrochimica Acta* 303:1–8, 2019.

Xu, X, Zhou, J, Jiang, L, Lubineau, G, Payne, SA, Gutschmidt, D. *Carbon* 80:91–102, 2014.

Yan, Q, Zhang, X, Li, J, Hassan, EB, Wang, C, Zhang, J, Cai, Z. *J. Mater. Sci.* 53:8020–8029, 2018a.

Yan, Q, Li, J, Zhang, X, Hassan, EB, Wang, C, Zhang, J, Cai, Z. *J. Nanopart. Res.* 20:223, 2018b.

Yang, G, Li, L, Lee WB, Ng, MC. *Sci. Technol. Adv. Mater.* 19:613–648, 2018.

Yoon, K, Cho, D, Tsang, YF, Tsang, DCW, Kwon, EE, Song, H. *Chem. Eng. J.* 361:1597–1604, 2019.

Yu, F, Li, Y, Jia, M, Nan, T, Zhang, H, Zhao, S, Shen, Q. *J. Alloys Compounds* 709:677–685, 2017.

Yu, X, Yu, B, Zhang, J, Zhang, Y, Zeng, J, Chen, M, Wang, C. *ChemistrySelect* 3:9518–9525, 2018.

Zhou, Z, Chen, F, Kuang, T, Chang, L, Yang, J, Fan, P, Zhao, Z, Zhong, M. *Electrochimica Acta* 274:288–297, 2018.

9 Value-Added Polymer Applications of Lignin

Synthetic polymers are one of the indispensable functional materials that drive modern human life. With application-specific tailored property sets these unique materials possess unparalleled characteristics and utility. Some have excellent mechanical properties while others have outstanding thermal properties. Yet others have optical characteristics or electrical properties that set them apart. Polymers have made their mark in common day-to-day applications like textiles, paints, footwear, packaging, water treatment, automobiles, and cosmetics, just to name a few. They are equally valuable in niche high-end applications like spacecraft, armaments, biomedicals, and electronics. One of their distinguishing features is remarkable durability compared to substitute natural substances. They are less susceptible to degradation by chemical or photochemical means than their natural counterparts and non-biodegradable. It is the later characteristic that has come to haunt these otherwise benevolent materials. Being non-biodegradable, they are difficult to dispose of and present serious challenges to environmental management and sustainable development. Moreover, the starting materials for these polymers are almost invariably derived from petroleum. This puts another question mark on their use in the context of sustainable development.

In recent years, a lot of research effort has gone into the development of biodegradable polymers using monomers that are obtained from renewable sources. Generally, the presence of one or more distinct functional groups in the monomer makes their polymerization possible. Some of the common functional groups in monomers that participate in the polymerization process are hydroxyl, carboxyl, carbonyl, amine, amide, and isocyanate. Lignin is a renewable material that is abundantly available, often as a waste. It carries the functional groups aliphatic hydroxyl, phenolic hydroxyl, carboxyl, and carbonyl, with their abundance and distribution changing with the lignocellulose source and delignification process. It is, therefore, a suitable candidate for the replacement or partial substitution of petroleum-derived synthetic monomers in polymerization reactions. Another approach is to chemically modify lignin to alter its makeup of functional groups to make it more amenable to specific polymerization reactions. Lignin could also be blended with other compatible polymers to impart certain desirable properties to the parent polymer.

9.1 LIGNIN IN ADHESIVES AND BINDER RESINS

9.1.1 Lignin-Based Phenolic Resins

One very important application of polymers is as adhesives and binders. Phenol formaldehyde is one such resin with widespread use as an adhesive in timber and wood processing industries. Lignin, with its relative abundance of phenolic hydroxyl functionality, has been found to be suitable as a replacement for phenol, fully or partially, in such resins. This is one value-added application of lignin that has seen some commercial success. One drawback with lignin substitution of phenol is its sluggish reactivity towards formaldehyde compared to phenol. This could be surmounted by the chemical modification of lignin. Resins prepared with lignin substitution of phenol have a property mix different from that of the normal phenol formaldehyde resin. The reaction can either be base-catalysed or acid-catalysed, with the former more common. Lignin molecular weight, occurrence of condensed structures through carbon–carbon linkages, and abundance of phenolic hydroxyl groups are important indicators for the suitability of lignin as a phenol substitute. Higher molecular weight

DOI: 10.1201/9781003598664-12

lignin would result in higher viscosity of the formed resin. Carbon–carbon bonds between aryl carbons of adjacent benzene rings render them unavailable for condensation polymerization through reaction with formaldehyde. Lastly, the presence of phenolic hydroxyl makes the ortho carbon susceptible to attack by formaldehyde and making of interlinking methylene bridge between adjacent benzene rings.

Possible phenol substitution with lignin for resin synthesis was noted in early studies. It was observed that lignin derivatization through hydroxymethylation and/or phenolation improves its efficacy in resin synthesis. Hydroxymethylation occurs at unsubstituted positions on the benzene ring adjacent to a phenolated carbon or at the β carbon with a carbonyl entity at α carbon. This derivatization increases the crosslinking ability of lignin during the synthesis of resin. Indulin AT (softwood kraft lignin) and pine acid hydrolysis lignin were hydroxymethylated by adding 37% aqueous formaldehyde solution to a solution of lignin in aqueous sodium hydroxide solution. Reaction temperature was ambient to 70 °C and reaction time 20–72 h; more temperature less reaction time. Derivatized lignin was acid-precipitated, washed, and dried. Indulin AT could be effectively hydroxymethylated due to its predominantly guaiacyl structure but hydroxymethylation was ineffective for acid hydrolysis lignin, ostensible due to condensed structures. The hydroxymethylated lignin was subsequently phenolated by redissolving in an aqueous sodium hydroxide solution, adding phenol and reacting for 2 h at 170 °C. Phenolated lignin was acid precipitated from the solution, washed, and dried. During phenolation, the hydroxymethyl group condenses with the benzene ring of phenol at position ortho to the phenolic hydroxyl. A methylene bridge is formed with the expulsion of a water molecule (Muller and Glasser 1984). Similarly, aqueous acetic acid and an aqueous mixture of formic and acetic acid organosolv lignins were subjected to hydroxymethylation for use as phenol substitutes in phenol formaldehyde resin. An aqueous solution of 15 weight% total concentration of respective lignin, formaldehyde, and sodium hydroxide was prepared. The mole ratio of phenolic hydroxyl to formaldehyde to sodium hydroxide was 1:2.4:1. The mixture was refluxed for 2 h, cooled, and added dropwise into 10 °C acidic water at pH 2 to precipitate the derivatized lignin (Benar et al. 1999). One relatively recent study reported demethylation of the methoxyl groups in kraft lignin to corresponding phenolic hydroxyl. Increased phenolic hydroxyl content of the resulting modified lignin increased its condensation capability with formaldehyde during polymerization. Demethylation was carried out in aqueous alkaline medium – 10% sodium hydroxide on lignin mass – with sodium sulphite acting as catalyst. Optimal demethylation was obtained with a 15% catalyst dose at 80 °C, reaction time being 60 min. After the reaction, the mixture was cooled to 40 °C (Li et al. 2017a). A schematic representation of lignin modification pathways is presented in Figure 9.1.

9.1.1.1 Kraft Lignin

Hydroxymethylated and phenolated Indulin AT was subsequently converted to resin by further reaction with formaldehyde and phenol. Derivatized lignin was dispersed in an aqueous solution of formaldehyde followed by the addition of sodium hydroxide to dissolve the lignin. Phenol was added to the solution and the reaction was carried out under reflux till the desired resin viscosity was achieved. The material ratios were 2.3 for formaldehyde and phenol, 0.61 for sodium hydroxide and phenol, and 2 for derivatized lignin and phenol. This amounted to 36 weight% underivatized kraft lignin in the final resin with 54% phenol substitution. When used as an adhesive for wood blocks, the lignin-substituted resin provided 3240 psi bond strength compared to 3400 for the unsubstituted phenol formaldehyde resin (Muller et al. 1984). Similarly, demethylated kraft lignin – alkaline sulphite catalysed substitution of methoxyl group with phenolic hydroxyl – was used as a 50% phenol substitute in the synthesis of phenol formaldehyde resin. The resin was synthesized with 2.2 moles of formaldehyde per mole of phenolic hydroxyl. Formaldehyde was added in three equal instalments during the reaction. Demethylated lignin, phenol, and the first formaldehyde instalment were mixed in 50% aqueous sodium hydroxide solution and reacted for 50 min. At this point, the second instalment of formaldehyde and sodium hydroxide was added and reacted for 60 min. Thereafter, the third

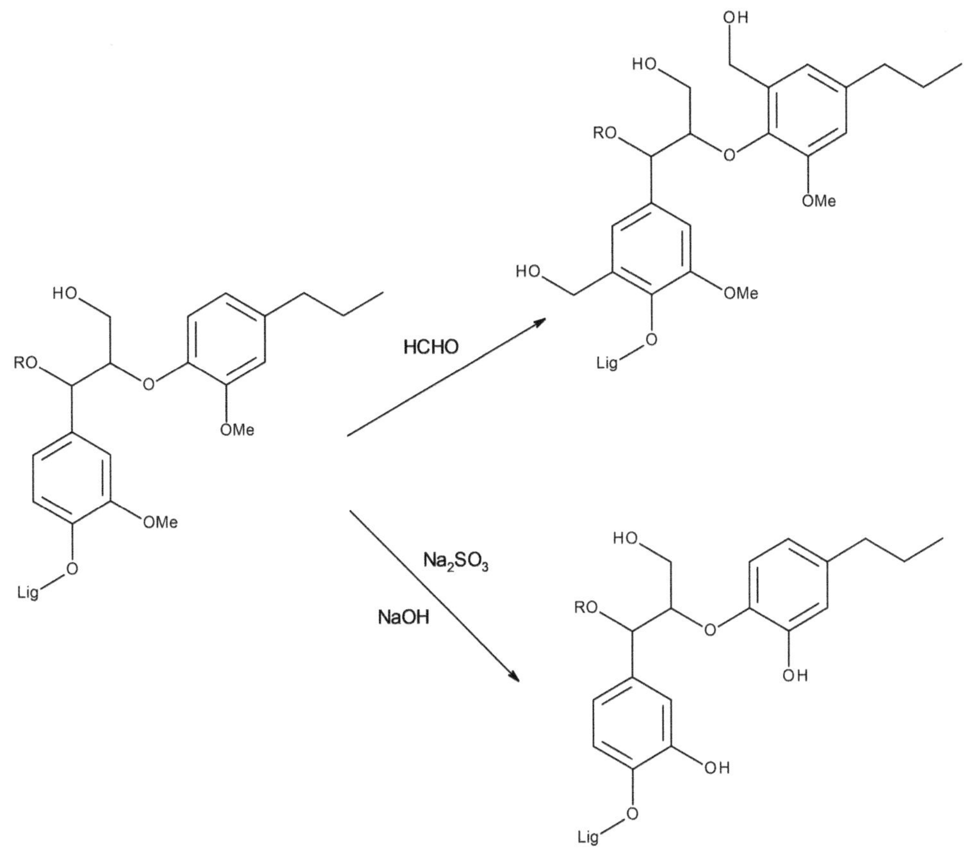

FIGURE 9.1 Lignin hydroxymethylation and demethylation for improving its reactivity during phenol formaldehyde synthesis.

and last formaldehyde portion and sodium hydroxide were added and reacted for 50 min. Finally, the unreacted formaldehyde was consumed by reacting with urea and sodium hydroxide for 30 min. The entire reaction was carried out at 85 °C under stirring. Rapid cooling to 40 °C stopped the reaction. Depending on the extent of demethylation, the tensile strength of the wood panels bonded with the resins varied between 0.62–1.01 MPa. The resin with the highest strength had a gel time of about 8 min (Li et al. 2017a).

Substitution of 20–50 weight% phenol with kraft lignin was reported in the synthesis of phenol formaldehyde resin. Formaldehyde was added in three doses to a mixture of lignin and phenol; the first dose at ambient temperature, the second dose at 65 °C, and the third dose at 80 °C. At this point, sodium hydroxide was added as a polymerization catalyst. The temperature was raised to the final reaction temperature of 95 °C and maintained for 3 h. Prepared resins had a solid content of 31.2–34.2% and gel time between 148 and 173 min – unusually high compared to other reported studies. For comparison, resin prepared without lignin substitution gelled in 140 min. Increased lignin substitution increased the gel time as well as the boiling water resistance of the prepared resins. Wood panels joined with unsubstituted resin exhibited a joint shear strength of 85.6 kg/cm^2. With lignin substitution, joint shear strength decreased, 80.4 kg/cm^2 at 20% substitution and 65.2 kg/cm^2 at 50% substitution (Sarkar and Adhikari 2000). Kraft lignin from kenaf was used for partial substitution – 10 or 50 weight% – of phenol in phenol formaldehyde lignin. For resin preparation, lignin, and phenol were refluxed in an ethanol solution of sodium hydroxide under stirring at 80 °C. Once a homogeneous mixture was obtained, the dropwise addition of formaldehyde was followed – 1.5:1 mole ratio of formaldehyde to phenol. Reaction continued for 2 h under reflux at 80 °C. Finally, the reaction mixture was reduced by 50% volume by evaporating ethanol. At 10% substitution, resin prepared with kraft lignin had solid content almost similar to the resin without lignin substitution. However, the resin with lignin substitution had substantially higher viscosity and lower gel time. Solid content and viscosity further increased, and gel time decreased when the lignin substitution increased to 50%. On the performance front, plywood panels prepared from kraft lignin-substituted resins had higher tensile strength and elastic modules than the panels prepared from resin without lignin substitution (Hussin et al. 2017). Eucalyptus kraft lignin was phenolated by heating at 40 °C a 10–50% mixture of lignin in phenol for 1 h. Modified lignin and 37% aqueous formaldehyde – lignin:phenol:formaldehyde 3–15:50:90 mass ratio – were mixed in 50% aqueous methanol with the addition of sodium hydroxide. The mixture was heated at 80 °C with subsequent additions of sodium hydroxide after 10 and 20 min. The reaction was stopped after 2 h. The resin with no lignin addition had a viscosity of 129 cP and a gel time of 4.52 min. With 10% lignin addition, both viscosity and gel time reportedly dropped to 101 cP and 3.65 min. Thereafter, with successive increases in lignin substitution, the viscosity and gel time of the produced resin did not follow a regular pattern. At 30% lignin substitution, the resin provided the highest shear strength of 3.41 MPa to bonded wood panels as against 2.6 MPa with unsubstituted phenol formaldehyde resin. The shear strength was higher at other substitution levels as well, except for 10% substitution (Galdino et al. 2023).

9.1.1.2 Organosolv Lignin

Eucalyptus aqueous acetic acid organosolv lignin was hydroxymethylated by reacting lignin at 50 °C for 4 h with aqueous alkaline formaldehyde solution – lignin:formaldehyde:sodium hydroxide 1:0.36:0.3. Hydroxymethylated lignin was precipitated from alkaline solution by acidification to pH 4 followed by water washing and drying. Hydroxymethylated lignin was further phenolated under acidic conditions by 1.5 h heating with phenol at 125 °C – lignin:phenol:sulphuric acid ratio of 1:2:0.05. The derivatized lignin was used to substitute phenol to the extent of 20–40% in the synthesis of phenol formaldehyde resin. Polymerization conditions were: reflux at 80 °C, final viscosity 300–1100 mPa.s. The resin was also prepared with unmodified lignin for comparison. The gel time of the resins prepared varied between 11 and 27.4 min depending on lignin content, the ratio of phenol to formaldehyde, and the ratio of phenol to sodium hydroxide. Compared to unmodified lignin, hydroxymethylated or hydroxymethylated and phenolated lignin produced a resin with much lower gel time which was similar to unsubstituted phenol formaldehyde resin (Vazquez et al. 1995; Vazquez et al. 1997). Aqueous acetic acid and an aqueous mixture of formic and acetic acid (formasolv) organosolv lignins, after hydroxymethylation, were used as phenol substitutes in phenol formaldehyde resin. The lignin-to-phenol ratio was between 2:3 and 1:1. Ratio of formaldehyde to phenolic hydroxyl was between 1.5–2.2. For every mole of phenolic hydroxyl, 0.7–1 mole of sodium hydroxide was used. An aqueous mixture of phenol, lignin, and formaldehyde and 58% of total sodium hydroxide was refluxed at 85 °C for 30–150 min for acetic acid lignin, and 60–180 min for formasolv lignin (till a target viscosity was reached). Next, an increment of 25% sodium hydroxide was added, the temperature was adjusted to 75 °C, and the mixture was refluxed for 30–180 min for acetic acid lignin, and 30–150 min for formasolv lignin (till a target viscosity was reached). At this point, the remaining sodium hydroxide was added, and the reaction mixture was cooled to 30 °C. Synthesis concluded after 20 min. Prepared resins had viscosities between 150 and 450 mPa.s. The curing temperature was slightly higher than the unsubstituted resin (Benar et al. 1999). Phenol formaldehyde resins were synthesized with 40% phenol substitution with aqueous acetic acid organosolv lignin from bagasse. First, the lignin was hydroxymethylated by reacting – 70 °C reflux for 1 h – with formaldehyde in the presence of aqueous potassium hydroxide. Phenol and additional formaldehyde were then added to the modified lignin and refluxed for another 90 min at 70 °C. The mass ratio of total formaldehyde, lignin + phenol, and potassium hydroxide was 1.38:1:0.06. The produced resin was measured to have M_w of 10,220 g/mol with polydispersity of 5.87 against M_w of

6590 and polydispersity of 12 for the resin without lignin substitution. But even with higher molecular weight, the lignin-substituted resin was less viscous at 10,775 cP compared to 36,000 cP for unsubstituted resin (Pinheiro et al. 2017).

Aqueous ethanol organosolv lignin-substituted phenol formaldehyde resins were prepared at 10% and 50% substitution levels and compared with resin without lignin substitution. Substituting phenol with 10% organosolv lignin produced a resin with lower solid content but its viscosity was still considerably higher and the gel time lower than the resin without lignin substitution. Increasing lignin substitution to 50% further increased the resin viscosity and lowered the gel time. At a given substitution level, resins produced with organosolv lignin had lower viscosity and higher gel time than the resins produced with kraft lignin substitution. This could be correlated to the higher molecular weight of the kraft lignin compared to that of the organosolv lignin. The mechanical performance of resins prepared with organosolv lignin substitution was not as good (Hussin et al. 2017).

9.1.1.3 Lignosulphonate

Softwood ammonium lignosulphonate was modified by methylolation – introduction of hydroxymethyl group, primarily on the benzene ring. The reaction involved the dissolution of lignosulphonate in hot aqueous sodium hydroxide (1:0.8), cooling the solution to normal temperature, and adding 35% aqueous formaldehyde solution, equivalent to lignosulphonate mass, under reflux at 45 °C. The reaction was continued for 320 min. For resin preparation, a solution of methylolated lignosulphonate, and phenol was prepared in a hot sodium hydroxide (1:0.6) solution. Six equal portions of formaldehyde were added at 15-minute intervals to have 2.6 times the mass of sodium hydroxide. The reaction time duration was 5 h. Excess water was distilled after the reaction. The formed resin cured with similar kinetics as the resin formed without phenol replacement (Alonso et al. 2001; Alonso et al. 2004).

In chapter 8 we have discussed the flame retardant role lignin can play due to its large amount of char formation. With lignin substitution in phenol formaldehyde resin, its thermal stability and ability to resist flame are improved. Whisker silicon-reinforced phenol formaldehyde resin was prepared with 10–20% substitution of phenol with calcium lignosulphonate. Lignin, phenol, water, alkaline catalyst, and whisker silicone were mixed together at 60 °C. Temperature was raised to 75 °C and 1.5 mole of formaldehyde per mole phenolic hydroxyl was added in three 10 min interval doses. The reaction was then carried out for 1.5 h at 85 °C. At the end of polymerization, a small amount of urea was added with 30 min holding at 85 °C followed by cooling to 50 °C. The prepared resin was used to synthesize phenolic foam insulating material by mixing with Tween-80 surfactant (7%), poly(ethylene glycol)-12 polydimethylsiloxane foam homogenizer (3%), and n-pentane (7.5%) as foaming agent. Once a homogeneous mixture was formed, a proprietary acidic curing agent was added and the material was poured into a mould, foamed, and cured. The lowest thermal conductivity of 0.0275 W/(m.K) was obtained for foam without lignin substitution and silicon addition. The addition of silicon and lignin substitution, both increased the thermal conductivity of the foam, but the material was still sufficiently insulating up to 1.8% silicon addition and 20% lignin substitution. But most importantly, the flame retarding characteristic of the foams improved substantially with silicon addition and lignin substitution as measured by the Limit Oxygen Index (LOI) (Zhang et al. 2020). The trend is graphically depicted in Figure 9.2.

9.1.1.4 Hydrolysis Lignin

In one study, enzyme hydrolysis lignin was extracted with aqueous sodium hydroxide from the residue left after ethanol production from corn stalks. This lignin was used as a 5–20 weight% phenol substitute in phenol formaldehyde resin. The resin was prepared by mixing lignin, phenol, formaldehyde, and sodium hydroxide in water and heating for 1 h at 80 °C. The reaction was then stopped by cooling the mixture to 65 °C. The molar ratio of formaldehyde to phenolic hydroxyl was 2. To catalyse the reaction, 2.5–7.5% sodium hydroxide based on the combined mass of lignin and phenol was used. The highest bond shear strength in the wood joint was observed with 10% lignin

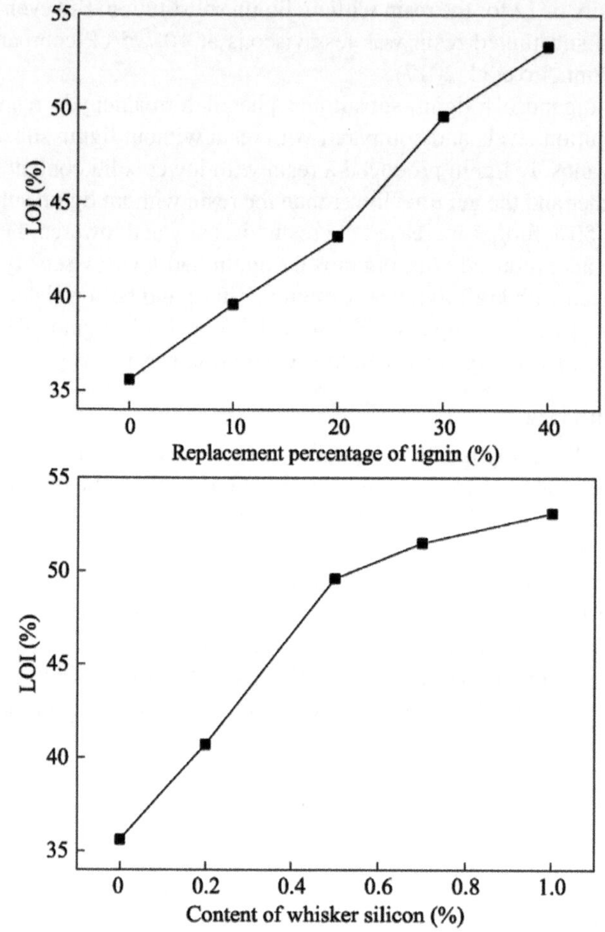

FIGURE 9.2 Increase in Limit Oxygen Index of phenolic resin foam with lignin substitution and whisker silicon addition. (Reprinted from Zhang, N, Li, Z, Xiao, Y, Pan, Z, Jia, P, Feng, G, Bao, C, Zhou, Y, Hua, L, *J. Bioresources Bioprod.*, 5, 67–77, 2020, originally published by Elsevier under Creative Commons Attribution License.)

substitution, beyond which the bond strength decreased (Jin et al. 2010). The use of enzyme hydrolysis lignin-rich residue from agricultural residues for phenol replacement in the synthesis of phenol formaldehyde resin is reported. These lignins were obtained as residue from bioethanol, biobutanol, xylitol, and lactic acid production. Their lignin contents were 39%, 78%, 61%, and 52%, respectively, with phenolic hydroxyl contents of 1.6%, 3.3%, 1.9%, and 1.5%, in that order. These residues had methoxyl group contents of 8.8%, 2.7%, 3.7%, and 3.6%. These differences were due to differences in the isolation process as well as differences in the lignocellulose source. Formaldehyde and sodium hydroxide solution was added under stirring at 80 °C in three equal doses to the mixture of lignin and phenol. After each addition, the reactor content was stirred for 1 h. The reaction was quenched by cooling it to 40 °C. Total formaldehyde addition was 2 moles for every mole of phenolic hydroxyl and total sodium hydroxide addition was 20% based on the mass of phenol and lignin. When used as an adhesive for binding plywood veneers, the resins with lignin substitution returned lower bond shear strength compared to the resin without lignin substitution. Among substitute resins, residue from lactic acid production provided the highest bond shear strength but this resin had a very high viscosity (Zhang et al. 2013). 10% substitution of phenol by straw enzymatic hydrolysis lignin in acid-catalysed synthesis of phenol formaldehyde resin is reported. Oxalic acid in the presence of zinc acetate was the acidic catalyst used. Lignin, phenol, and zinc acetate were mixed for 30 min at 120 °C. The mixture was cooled to room temperature, formaldehyde was added and stirred for 2 h. Oxalic acid was added to this reaction mixture, the temperature was raised to 100 °C, and reacted for another 2 h under reflux. The product was purified by dissolving in ethanol and its subsequent distillation which removed unconverted reactants. Notably, the formaldehyde-to-phenol mole ratio at 0.75–0.85 was much less compared to the usual resin synthesis with an alkali catalyst. Authors claimed the produced lignin to be thermoplastic in nature, but brittle and hard in solid state, in comparison to resins prepared with higher formaldehyde ratio and alkali catalysis. The produced resin was blended with polybutylene succinate to improve its mechanical toughness – lower tensile strength and higher elongation at break (Xie et al. 2023).

Most of the reported studies used lignin as a phenol replacement which was an active ingredient in resin synthesis by reacting with formaldehyde and forming methylene bridges. In a modified approach, micro- and nano-sized lignin particles were prepared from steam explosion enzymatic hydrolysis lignin from Arundo donax. These lignin particles were blended by high shear mixing with a liquid phenol formaldehyde resin with 20% isopropyl alcohol in it. After vacuum degassing, the composite was cured in two stages: 72 h at 80 °C and 1 h at 180 °C. Blends were compared with unmodified phenol formaldehyde resin for their adhesive performance through the shear strength of the lap joint between two wood panels. The shear strength of the joint with unmodified resin was 8.7 MPa. With 5% lignin microparticle blending, the strength improved to 9.6 MPa. It further improved to 10.9 MPa when 5% lignin nanoparticles were blended. Bond strength, however, dropped at a 10% blending level. Increased strength was attributed to copolymerization of lignin and phenol formaldehyde resulting in more crosslinking. This effect was negated due to particle aggregation at higher lignin levels (Yang et al. 2019).

9.1.2 Lignin-Based Epoxy Resins

Another important class of resins with widespread use is epoxy resin which is based on the oxirane structure. It has excellent heat and chemical resistance besides mechanical characteristics of strength and toughness. It finds its use as an adhesive, structural polymer, barrier coating, and as the matrix material in fibre-reinforced composites. All these have high-value applications in the construction, paint, automobile, electronics, and aerospace industries. The resin prepolymer is formed by the condensation of an epoxide functional group with an acidic hydroxyl group. The latter may be provided by a variety of compounds like phenols, glycols, and dicarboxylic acids; phenols being the most used. In its commonest synthesis route, epoxy resins are produced by reacting epichlorohydrin with bisphenol A under alkaline conditions. The properties of the produced resin largely depend on the degree of polymerization which is controlled by the reaction conditions. For its final end use, the resin must be cured by crosslinking. This can be achieved by homopolymerization – resin molecules crosslinking with itself. However, the use of a curing agent is more common. Primary amines, polyphenols, acid anhydrides, and thiols are some of the classes of compounds used as curing agents for epoxy resins. Curing can be done at ambient or elevated temperatures.

Bisphenol A, one of the conventional starting materials for the synthesis of epoxy resins, has associated health and environmental implications. Moreover, it is sourced from petroleum. On both these counts it does not fit into the sustainability paradigm. Search for renewable alternatives is, therefore, on. Lignin, with its phenolic functionality, is one of the options and a lot of research and development work is going on to formulate lignin-based epoxy resins. Almost all major lignin types are capable of epoxidation. Lignin molecular weight and polydispersity are important in these reactions as is the presence of phenolic hydroxyl. Low molecular weight and polydispersity and high phenolic hydroxyl content are preferred. Some studies also suggest the addition of an epoxide group to the γ carbon of the phenylpropane unit carrying a primary hydroxyl if it is activated by the presence of a double bond between α and β carbons (Ferdosian et al. 2016). Others, however, contend that the aliphatic hydroxyls remain unreacted during epoxidation (Gioia et al. 2020). A schematic of the addition of the epoxide group to the lignin structure is presented in Figure 9.3.

Resin is commonly characterized based on its M_w and epoxide content. The latter is expressed either as a percentage of the resin mass or as epoxide equivalent weight. Another way of value-added lignin application in epoxy resin formulations is to use it as a curing agent. Additionally, lignin can also be used as filler in epoxy composites.

9.1.2.1 Soda and Kraft Lignin

There is one mention of softwood kraft lignin, Indulin AT, being crosslinked with glycerol diglycidyl ether having two terminal epoxide groups. The resulting formulation after curing was used as wood adhesive for gluing veneers in plywood production. Lignin and glycerol diglycidyl ether were mixed in a 65:35 mass ratio to match the number of hydroxyl groups on lignin and epoxide groups on the glycerol diglycidyl ether. At 140 °C, the mixture cured into a thermoset in 10 min. Three-ply plywood prepared by using the cured resin as a solid dispersion had bond shear strength comparable to commercial plywood prepared with conventional phenol formaldehyde resin. A liquid resin could also be produced by decreasing the lignin content to 40–50% without compromising on strength (Li et al. 2018).

Soda lignins from wheat straw and sarkanda were reacted with epichlorohydrin to prepare epoxy resins. Lignin was dissolved in 20% concentration aqueous sodium hydroxide; alkali-to-lignin mass ratio varying between 3 and 6. Epichlorohydrin was added to this solution with a mass ratio of epichlorohydrin to lignin as 10. Reaction temperature varied between 50 and 90 °C and time between 3 and 7 h. At the end of the reaction, a 20% concentration solution of sodium dihydrogen phosphate was added, resulting in a two-phase system, a solid water-insoluble resin and a water-soluble resin. Solid epoxy resin was centrifuged, water-washed, and dried. Increasing the reaction temperature and time did not necessarily increase the epoxy content in the resin. With wheat straw lignin, the highest epoxy content of 1.84% was obtained with a 3 h reaction at 50 °C and an alkali-to-lignin ratio of 3. The highest resin epoxy content was 1.65% for sarkanda lignin at 70 °C, other conditions being the same as that for wheat straw lignin. As a modification, lignins were hydroxymethylated prior to epoxidation, but this resulted in lower epoxy content in the formed resins in all cases (Malutan et al.

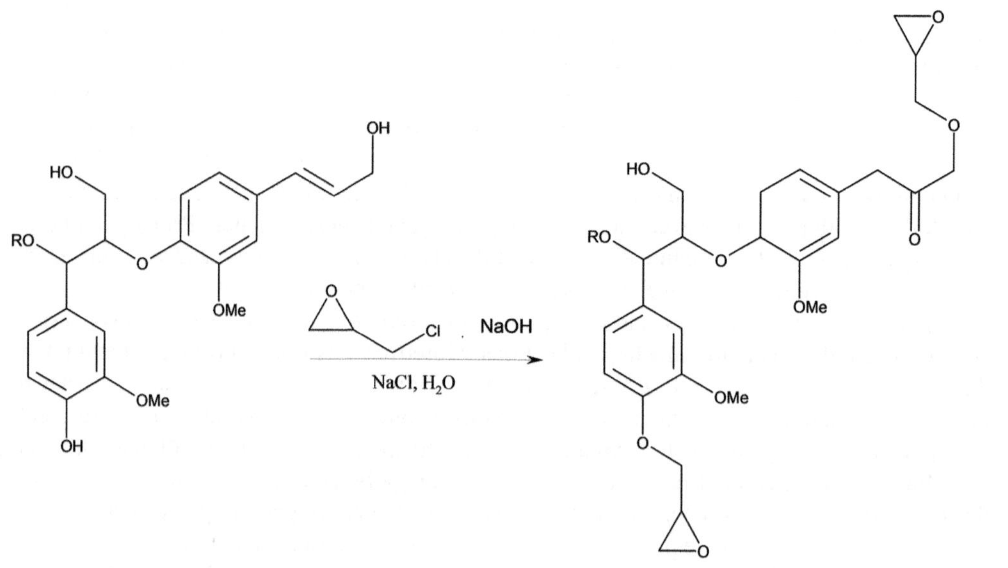

FIGURE 9.3 Epoxidation of phenolic hydroxyl and aliphatic hydroxyl at γ carbon in lignin.

2008). Recently, hardwood soda lignin with M_w 6400 and polydispersity 3.4 was used in a study for epoxy resin synthesis. Its aliphatic hydroxyl content was 1.8 mmol/g, phenolic hydroxyl content was 2.05 mmol/g, and carboxylic content was 1.03 mmol/g. Epoxidation was carried out by dissolving lignin in dimethyl formamide (1:5 mass ratio). Epichlorohydrin, 10 times the mass of lignin, and the catalyst tetrabutylammonium bromide (10% of lignin mass) were added to the lignin solution. The mixture was refluxed under stirring at 60 °C for 3 h. Cooling to ambient temperature was followed by a dropwise addition of 2% aqueous sodium hydroxide solution containing 1.2% tetrabutylammonium bromide. Epoxidation was allowed to complete in 8 h at normal temperature while continuously stirring the mixture. Finally, the resin was precipitated by the addition of a large amount of water. The precipitated resin was filtered, water-washed, and dried. The produced resin had an epoxide equivalent weight of 501 g (8.21%). Authors reported phenolic hydroxyl to be the most reactive to epoxidation followed by carboxyl and aliphatic hydroxyl (Nikafshar et al. 2021).

In one study, kraft lignin as well as its modified forms were evaluated for epoxidation. Hydroxymethylation and glyoxalation were used for chemical modification. Lignin was hydroxymethylated by dissolving it in an aqueous sodium hydroxide solution at pH 13 – 0.18 g alkali per g lignin. Aqueous formaldehyde solution was added to the lignin solution – 0.2 g formaldehyde per g lignin. The reaction was carried out for 8 h under stirring at 50 °C. For glyoxalation, lignin dissolved in sodium hydroxide was charged with an aqueous solution of glyoxal – equal mass of lignin and glyoxal – and reacted at 50 °C for 8 h. Derivatized lignin was acid precipitated and washed with water before drying. For epoxidation, lignin or its derivatized forms were dissolved in a 20% concentration of aqueous sodium hydroxide. Epichlorohydrin was added to this solution with a mass ratio of epichlorohydrin to lignin as 10. Reaction temperature varied between 50 and 90 °C and time between 3 and 7 h. At the end of the reaction, a 20% concentration solution of sodium dihydrogen phosphate was added, resulting in a two-phase system, a solid water-insoluble resin and a water-soluble resin. Solid epoxy resin was centrifuged, water-washed, and dried. The best epoxide content in the formed resins was obtained at 70 °C reaction temperature and 3 h reaction time. This was 2.29% for kraft lignin, 2.95% for glyoxalated lignin, and 3.3% for hydroxymethylated lignin. The increased epoxide group content in resins prepared from glyoxalated or hydroxymethylated lignin could be attributed to their higher aliphatic to phenolic hydroxyl ratio compared to unmodified lignin (El Mansouri et al. 2011). Epoxy resin synthesis from hardwood kraft lignin – 16% KOH equivalent of phenolic hydroxyl and 6% KOH equivalent of aliphatic hydroxyl – was reported. Lignin was dissolved in epichlorohydrin in a mass ratio of 1:3 and added with 2% tetrabutylammonium bromide as the catalyst. The reaction mixture was heated to 60 °C. An aqueous solution of 12% sodium hydroxide concentration was dropwise added – 12.5 mL per g lignin. The reaction was continued for 8 h thermostated at 60 °C. Finally, the prepared resin was filtered, water washed, and vacuum evaporated to remove excess epichlorohydrin. The resin had 0.5 mass% of epoxy content. A mixture of the prepared resin and bisphenol A-based commercially available resin (Epoxide equivalent weight 186 g) in different proportions was cured by the addition of a cycloaliphatic diamine curing agent (amine equivalent weight 186 g). With 15% inclusion of lignin-based resin, the impact strength increased to 2.7 kJ/m^2 compared to 2.2 kJ/m^2 for the commercial resin alone. A corresponding increase was also observed in the shear strength of the bond when two metal sheets were glued with the resin mixture – 8.3 MPa for commercial resin and 12 MPa with 15% lignin-based resin inclusion (Gouveia et al. 2020). A study evaluated different kraft lignins for their suitability for epoxidation and potential replacement of bisphenol A in epoxy resin synthesis. In all, four softwood lignins and three hardwood lignins were studied. Softwood lignins had M_w of 7000–9300 g/mol with a polydispersity of 3.8–4.7. Their aliphatic hydroxyl content was 1.79–2.94 mmol/g, phenolic hydroxyl content was 3.04–3.74 mmol/g, and carboxyl content was 0.42–0.53 mmol/g. Hardwood lignins, on the other hand, had M_w of 3200–12,400 g/mol with a polydispersity of 2.3–4.6. Their aliphatic hydroxyl content was 1.71–2.94 mmol/g, phenolic hydroxyl content was 1.9–3.78 mmol/g, and carboxyl content was 0.12–0.44 mmol/g. Epoxidation of each lignin was carried out by dissolving lignin in dimethyl formamide (1:5 mass ratio). Epichlorohydrin, 10 times the mass of lignin, and the catalyst tetrabutylammonium bromide (10% of lignin mass) were added to the lignin solution. The mixture was refluxed under stirring at 60 °C for 3 h. Cooling to ambient temperature was followed by a dropwise addition of 2% aqueous sodium hydroxide solution containing 1.2% tetrabutylammonium bromide. Epoxidation was allowed to complete in 8 h at normal temperature while continuously stirring the mixture. Finally, the resin was precipitated by the addition of a large amount of water. The precipitated resin was filtered, water-washed, and dried. Resins produced from softwood lignins had epoxide equivalent weight of 430–539 g (7.88–9.81%). Those from hardwood lignins had epoxide equivalent weight of 354–633 g (7–11.98%). The resin produced from hardwood kraft lignin of 3200 M_w and 2.3 polydispersity with 3.78 mmol/g phenolic hydroxyl had the highest epoxide content of 11.98%. This resin was cured with amine-based commercial curing agent GX-3090 and evaluated for mechanical characteristics. It was found to have a storage modulus of 1275 MPa and a loss modulus of 220 MPa at 25 °C (Nikafshar et al. 2021). Softwood kraft lignin – 2.96 mmol/g phenolic hydroxyl – was used as a total replacement of bisphenol A in the synthesis of a biobased epoxy resin in which the epichlorohydrin was also sourced from plants. A mixture of lignin, epichlorohydrin, ethyl lactate (solvent), and tetrabutylammonium bromide (catalyst) – 1:10:5:0.1 mass ratio – was refluxed at 80 °C for 2 h. The reaction was quenched by cooling to 10 °C with dropwise addition of 20 weight% sodium hydroxide solution. The synthesized resin was precipitated by acidification; the resin solution was slowly added to the hydrochloric acid solution at pH 5. The epoxy equivalent weight of the formed resin was 409 g compared to 189 g for the commercial resin EPON 828. The lignin-based resin had a somewhat higher viscosity of 21 Pa.s against 15 Pa.s for EPON 828. Epoxy ring formation was confirmed by 915 cm^{-1} distinct FTIR absorption. The resin was used as matrix material in bamboo fibre-reinforced composite. Cardolite NT-1515 – polyamide type – was the curing agent and Cardolite NC-513 was the diluent. A spreadable mixture of a stoichiometric amount of resin and curing agent was prepared in diluent. Composite was prepared by sandwiching a layer of straightened bamboo fibres between two layers of resin mixture and freeze drying at –54 °C and 4 mbar. Three such sheets were hot-pressed together to form the final composite. These composites exhibited fairly better mechanical properties at all temperatures over their commercial resin counterpart, as measured by their storage modulus and flexural modulus. This was accounted for by higher polydispersity of lignin and better adhesion of the formed resin to the bamboo fibres (Bagheri and Nejad 2023). These behaviours are shown in Figure 9.4.

Other researchers preferred to depolymerize the lignin by catalytic hydrogenation prior to epoxidation. Softwood kraft lignin was hydrogenated to reduce its M_w from 8300 (polydispersity 8.2) to 1400 g/mol with a polydispersity of 2.4. Modified lignin was used to produce epoxy resin through a reaction in an aqueous medium with an epichlorohydrin – mass ratio of epichlorohydrin to lignin 3. Tetrabutylammonium bromide – 0.2% based on lignin mass – was the catalyst used. The reaction temperature was 80 °C, and the time duration was 1 h. The reaction mixture was cooled followed by a dropwise addition of sodium hydroxide. The organic phase was separated and the excess epichlorohydrin evaporated to recover the resin. Purification of the produced resin involved dissolution in acetone, discarding the insoluble impurities, and evaporating the acetone. Epoxide equivalent weight of the resin produced was 768 g and its M_w was 2800 g/mol with a polydispersity of 3.5. The resin was cured with two curing agents, either 4,4′-diaminodiphenyl methane (solid state with amine equivalent weight of 20.6 g) or diethylenetriamine (liquid state with amine equivalent weight of 49.6 g). A stoichiometric amount of curing agent was added to resin heated to 80 °C, mixed for 1 min, and quenched in an ice bath. Diethylenetriamine resulted in curing at a relatively lower temperature – onset 31–33 °C, peak curing rate at 90–102 °C. For 4,4′-diaminodiphenyl methane the temperature was slightly higher – onset 30–39 °C, peak curing rate at 105–128 °C. The former was thermally stable up to 252 °C and the latter up to 290 °C (Ferdosian et al. 2015).

In a variation, kraft lignin from spruce and eucalyptus was fractionated into different molecular weight fractions by sequential solvent extraction. Eucalyptus lignin was first extracted with ethyl acetate – 1 g per 10 mL – for 2 h. Soluble fraction – M_w 1050 g/mol, polydispersity 1.5, and 5.1 mmol/g phenolic hydroxyl – was recovered by vacuum evaporation of the solvent. The insoluble

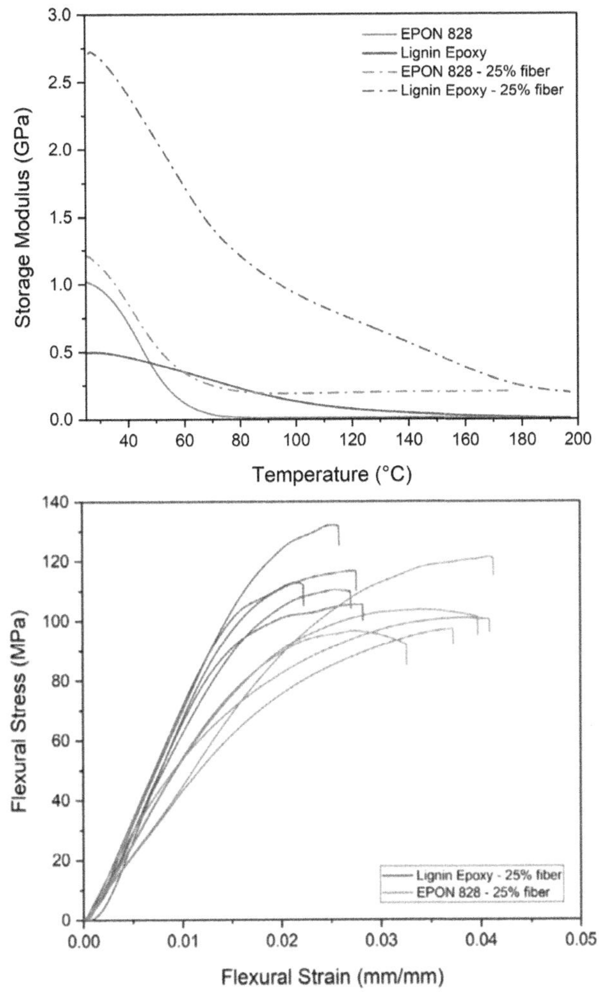

FIGURE 9.4 Comparative mechanical performance of bamboo fibre-reinforced composites from lignin-based epoxy resin and commercial epoxy resin EPON 828. (Adapted and reprinted from Bagheri, S, Nejad, M, *Poly. Composites*, 44, 3926–3938, 2023, with permission from Wiley.)

fraction was further extracted with ethanol and the soluble fraction – M_w 1440 g/mol, polydispersity 1.6, and 4.5 mmol/g phenolic hydroxyl – was recovered by solvent evaporation. The two fractions were separately used for epoxy resin synthesis. For spruce lignin, a single ethyl acetate soluble fraction – M_w 1120 g/mol, polydispersity 1.6, and 4.9 mmol/g phenolic hydroxyl – was used for resin synthesis. The resin was prepared at 50 °C with 50% aqueous acetone as the solvent and reacted for 5 h. The reaction mixture contained the lignin fraction, epichlorohydrin equivalent to 20 moles per mole of phenolic hydroxyl in lignin, and sodium hydroxide equivalent to 3 moles per mole of phenolic hydroxyl in lignin. The reaction was terminated by the addition of water and acidifying to 3.5 pH. The formed resin was filtered, water-washed, redissolved in acetone, and freeze-dried. Resins from ethyl acetate soluble fraction and ethanol soluble fraction of eucalyptus lignin had 5.1 mmol/g epoxide groups. This value was 4.8 mmol/g for the resin prepared with ethyl acetate soluble fraction of spruce lignin. The extent of epoxide groups was comparable to the commercial resin Araldite GY 6010 with 5.3 mmol epoxide groups per g. Resins prepared were cured with poly(propylene oxide) diamine by the tradename Jeffamine D2000 (DP 2000), and Jeffamine D400 (DP 400). The resin and the curing agent, in stoichiometric proportion, were dissolved in acetonitrile, cast in mould, held for 1 h at 50 °C, and dried by solvent evaporation. Curing was completed by 2 h heating at 100 °C and 2 h heating at 150 °C. Thermosets obtained with Jeffamine 2000 had T_g between 49 and 52 °C. However, thermosets obtained with Jeffamine 400 had T_g between 60 and 73 °C, higher than the 53 °C T_g for the thermoset produced from araldite (Gioia et al. 2020).

9.1.2.2 Organosolv Lignin

In a particular study, hardwood organosolv lignin (unknown solvent) was first catalytically hydrogenated in the presence of 4% ruthenium on a carbon catalyst. Lignin was dissolved in acetone in a 1:3 ratio, and the reaction was carried out for 1 h at 100 bar hydrogen pressure and 350 °C. This modification effectively reduced the weight average molecular weight from 2638 to 690 g/mol and polydispersity from 3.58 to 3.01. The hydrogenated lignin was reacted with epichlorohydrin – three times the mass of epichlorohydrin over the mass of lignin – in an aqueous medium in the presence of tetrabutylammonium bromide as the catalyst for the desired reaction time and at a specific temperature. Reaction was terminated by cooling the reaction mixture and dropwise addition of aqueous sodium hydroxide and the organic phase was separated by decantation. Unreacted epichlorohydrin was removed by vacuum evaporation. The obtained crude resin was further purified by redissolving in acetone, filtering, and evaporating the acetone. The highest epoxy content of 6.26% was obtained with a 5 h reaction at 70 °C and a sodium hydroxide-to-lignin ratio of 6.4 (Ferdosian et al. 2014). In a related study, hardwood organosolv lignin was hydrogenated to reduce its M_w to 760 g/mol with a polydispersity of 2. Modified lignin was used to produce epoxy resin through a reaction in an aqueous medium with epichlorohydrin – mass ratio of epichlorohydrin to lignin 3. Tetrabutylammonium bromide – 0.2% based on lignin mass – was the catalyst used. The reaction temperature was 80 °C, and the time duration was 1 h. The reaction mixture was cooled followed by a dropwise addition of sodium hydroxide. Resin recovery and purification were similar to the earlier study. Epoxide equivalent weight of the resin produced was 537 g and its M_w was 1400 g/mol with a polydispersity of 2.8. The resin was cured with two curing agents, either 4,4′-diaminodiphenyl methane or diethylenetriamine. A stoichiometric amount of curing agent was added to resin heated to 80 °C, mixed for 1 min, and quenched in an ice bath. Diethylenetriamine resulted in curing at a relatively lower temperature – onset 28–32 °C, peak curing rate at 87–101 °C. For 4,4′-diaminodiphenyl methane the temperature was slightly higher – onset 45–54 °C, peak curing rate at 128–145 °C. The former was thermally stable up to 228 °C and the latter up to 257 °C (Ferdosian et al. 2015).

One study reported a complete replacement of bisphenol A with organosolv lignins (unknown solvent) from corn stover, bagasse, peanut shell, wheat straw, and hardwood. These lignins had M_w of 5380–15,300 g/mol with a polydispersity of 2.8–5.3. Their aliphatic hydroxyl content was 0.67–2.22 mmol/g, phenolic hydroxyl content was 1.8–3.72 mmol/g, and carboxyl content was 0.23–0.51 mmol/g. Epoxidation of each lignin was carried out by dissolving lignin in dimethyl formamide (1:5 mass ratio). Epichlorohydrin, 10 times the mass of lignin, and the catalyst tetrabutylammonium bromide (10% of lignin mass) were added to the lignin solution. The mixture was refluxed under stirring at 60 °C for 3 h. Cooling to ambient temperature was followed by a dropwise addition of 2% aqueous sodium hydroxide solution containing 1.2% tetrabutylammonium bromide. Epoxidation was allowed to complete in 8 h at normal temperature while continuously stirring the mixture. Finally, the resin was precipitated by the addition of a large amount of water. The precipitated resin was filtered, water-washed, and dried. The resins produced had epoxide equivalent weight of 347–988 g (3.81–12.53%). The resin produced from corn stover organosolv lignin of 5380 M_w and 2.8 polydispersity with 2.2 mmol/g phenolic hydroxyl had the highest epoxide content of 12.53%. This resin was cured with amine-based commercial curing agent GX-3090 and evaluated for mechanical characteristics. It was found to have a storage modulus of 1396 MPa and a loss modulus of 253 MPa at 25 °C (Nikafshar et al. 2021). The variations in mechanical properties with temperature for this resin along with resins produced from hardwood kraft lignin and commercial epoxy resin are shown in Figure 9.5.

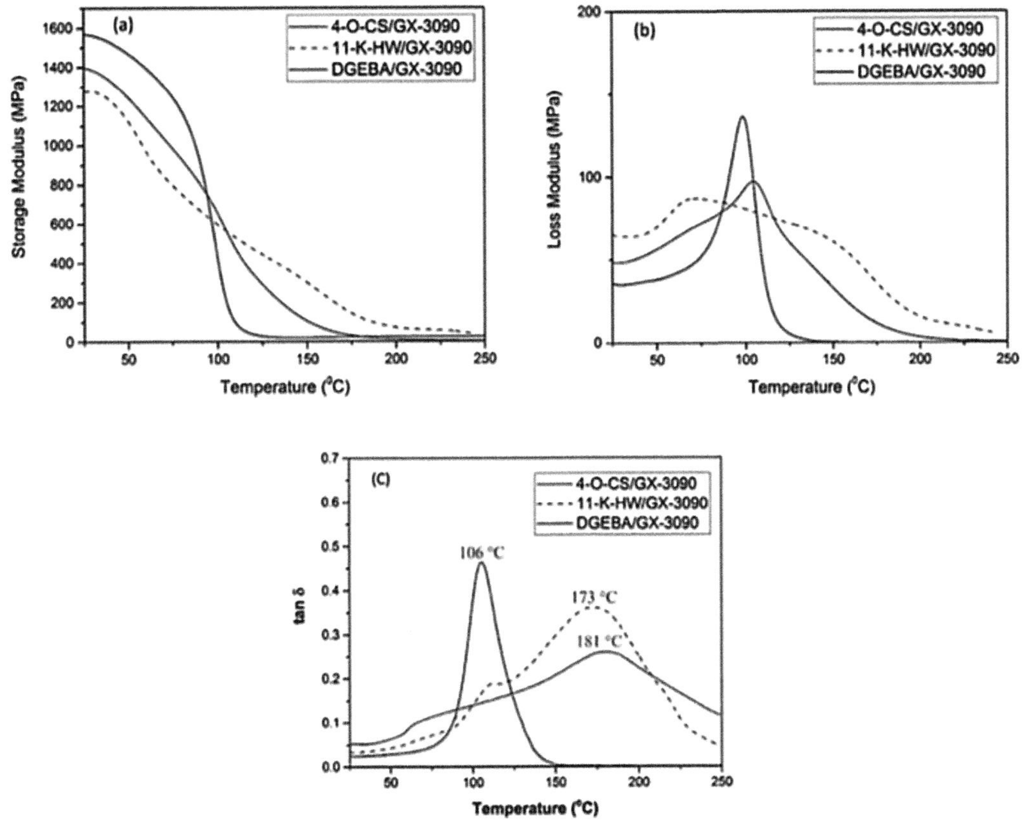

FIGURE 9.5 Comparative mechanical performance of epoxy resins produced from corn stover organosolv lignin, hardwood kraft lignin, and commercial epoxy lignin EPON 828. (Reprinted from Nikafshar, S, Wang, J, Dunne, K, Sangthonganotai, P, Nejad, M, *ChemSusChem*, 14, 1184–1195, 2021, with permission from Wiley.)

9.1.2.3 Hydrolysis Lignin

Lignin was extracted from steam-exploded bamboo by extraction with water and methanol. The lignin so obtained was characterized to have 8.3 mmol/g of total hydroxyl groups. It was mixed with tetrabutylammonium bromide (70% of lignin mass), and epichlorohydrin (16 times the lignin mass) and reacted for 4 h at 80 °C. Contents were cooled to 10 °C and 50% concentration sodium hydroxide (alkali 1.65 times the lignin mass) with dimethyl sulfoxide was added dropwise. The reaction was allowed to continue for 12 h under stirring and then stopped by neutralizing with acetic acid. The synthesized resin was separated into the organic phase, which was water-washed and dried to remove residual epichlorohydrin. This resin had an epoxide equivalent weight of 333 g. In one procedure, the resin was cured, with the bamboo lignin (1:1 ratio of hydroxyl to epoxide groups) itself acting as the curing agent. For comparison, it was also cured with 1-(2-cyanoethyl)-2-ethyl-4-methylimidazole (10% of resin mass). The resin and the curing agent were dissolved in tetrahydrofuran, spread over a PTFE sheet, and vacuum dried at 50 °C. Three-stage curing was performed: 30 min 110 °C, 3 h 150 °C, and 2 h 180 °C. When cured with bamboo lignin, the thermoset was thermally stable up to 260 °C. With 1-(2-cyanoethyl)-2-ethyl-4-methylimidazole as the curing agent, thermal stability increased to 299 °C. However, the flexural strength of the former was 61 MPa against 38 MPa for the latter (Sasaki et al. 2013). Aspen hydrolysis lignin from bioethanol production – 56.7% lignin – was hydrogenated by a proprietary process to reduce its M_w to 2121 g/mol with a polydispersity of 3. Modified lignin was used to produce epoxy resin through a reaction in an aqueous medium with epichlorohydrin – mass ratio of epichlorohydrin to lignin 3. Tetrabutylammonium bromide – 0.2% based on lignin mass – was the catalyst used. The reaction temperature was 80 °C, and the time duration was 1 h. The reaction mixture was cooled followed by a dropwise addition of sodium hydroxide. The organic phase was separated and the excess epichlorohydrin evaporated to recover the resin. Purification of the produced resin involved dissolution in acetone, discarding the insoluble impurities, and evaporating the acetone. The epoxide content of the resin produced was 5.13% of the mass of the resin and its M_w was 5530 g/mol with a polydispersity of 4.6. The resin was cured with 4,4′-diaminodiphenyl methane as the curing agent. A stoichiometric amount of curing agent was added to the resin, mixed, and quenched at 5 °C. The onset temperature for curing was 57–80 °C, and the peak curing rate was attained at 102–122 °C. The cured resin was thermally stable up to 213 °C. The mechanical performance of the prepared resin was evaluated by preparing a glass-fibre-reinforced composite with cured resin as the matrix. The composite attained tensile strength of 138 MPa, Young's modulus of 12.3 GPa, flexural strength of 47 MPa, and flexural modulus of 5 GPa. For comparison, these values were 214 MPa, 17.5 GPa, 266 MPa, and 13 GPa, respectively, when the composite was prepared with commercial epoxy resin DGEBA from Araldite. The composite with commercial resin was also thermally stable up to 360 °C. The primary reason for the significantly better performance of the commercial resin was its higher epoxide content of 14.6% (Ferdosian et al. 2016).

9.1.2.4 Lignosulphonate

Epoxy resin was prepared with phenolated calcium lignosulphonate. First, the calcium lignosulphonate was dissolved in 50% sulphuric acid to convert it into the corresponding lignosulphonic acid. Excess phenol was added to this solution – 40% of lignosulphonate mass and reacted for 3 h at 95 °C. The organic layer was separated, neutralized, and evaporated to remove residual phenol. The resulting gelatinous mass was extracted with water to recover the water-soluble phenolated lignosulphonate and the water-insoluble phenolated lignosulphonate, respectively. The water-insoluble fraction was used for resin synthesis by reacting with epichlorohydrin in the presence of 20% concentration aqueous sodium hydroxide – detailed reaction conditions were not reported. However, the reaction product was a mixture of solid and liquid epoxy resins. The produced liquid resin contained 4.7 mmol/g epoxide groups while the solid resin contained 0.17 mmol/g (Zhao et al. 2001). In a related study, the water-soluble fraction of phenolated lignosulphonate was used for resin synthesis by reacting with epichlorohydrin in the presence of 20% concentration aqueous sodium hydroxide. Liquid epoxy resin with 4.7 mmol/g epoxide groups was reportedly produced. It was cured with maleic anhydride and complete curing could be achieved with a temperature exceeding 110 °C (Sun et al. 2007). There is a report stating the use of ammonium lignosulphonate to prepare epoxy resin by reacting with epichlorohydrin in the presence of 20% sodium hydroxide as a catalyst. The reactant ratios and reaction conditions were not reported but the prepared resin had 0.165% epoxy equivalent. The resin was further modified by phosphorous incorporation by triphenylphosphine catalysed reaction with [2-(6 oxido-6Hdibenz(c,e)oxaphosphorin-6-yl)]-1,4-naphthalendiol at 80–160 °C for 5 h. Phosphorous incorporation was between 4.6% and 5.8%. The prepared resins, with or without phosphorous incorporation, exhibited a Newtonian flow behaviour (Petreus et al. 2008).

9.1.2.5 Lignin-Based Curing Agent for Epoxy Resin

Amination of lignin followed by homo crosslinking is another modification option to increase its propensity for epoxy resin application. Softwood kraft lignin (commercial name BioPiva 100) with M_w 4593 g/mol and polydispersity 2.36 was aminated with aliphatic amines – diethylene triamine, propylene diamine, and ethylene diamine – or ammonia in the presence of excess formaldehyde. Lignin was first dissolved in 0.2 M aqueous sodium hydroxide – lignin to alkali ratio 12.5 – and insoluble impurities to the extent of 16% were removed by centrifugation. Reactant amine was added, and the solution pH was adjusted to 11.2. An excess of 35% aqueous solution of formaldehyde was added to the mixture and reacted for 20 h at 80 °C. At the end of the reaction, the temperature was brought down to ambient, and the aminated homo crosslinked product was precipitated in excess of isopropanol. The effectiveness of the aminating agent, in terms of the amino group introduced, was found to be in the order ammonia (1.2 mmol/g) < diethylene triamine (1.7 mmol/g) < propylene diamine (1.8 mmol/g) < ethylene diamine (2 mmol/g). These aminated products were evaluated for their performance as a curing agent for bisphenol A-based epoxy resin. The aminated lignins were formed into a microsuspension in isopropanol and mixed with the resin and a conventional polyether amine curing agent under vacuum and filled into moulds – 23:38.5:38.5 mass ratio. Curing was completed in two stages: at ambient temperature for 24 h, followed by 70 °C for 4 h. Curing was also done with unmodified kraft lignin to act as the control. Flexural strengths of the cured thermosets varied between 69 and 75 MPa; the highest for ammonia as the aminating agent. Curing with ammonia-modified lignin also provided the highest elastic modulus of 3560 MPa. In comparison resin cured with unmodified lignin only had 57 MPa flexural strength and 3050 MPa elastic modulus (Ott et al. 2021).

Another protocol reported carboxylation of lignin prior to epoxy resin synthesis. Poplar ethylene glycol dioxane organosolv lignin was carboxylated with succinic anhydride. Lignin and anhydride, in a mass ratio of 1:6.3, were dissolved in acetone with the addition of 12% 1-methylimidazole based on lignin mass as the catalyst. Carboxylation was completed in 6 h at 60 °C. The reaction system was cooled to ambient temperature and dropwise added to excess water. Precipitated modified lignin was filtered, water-washed, and dried. Commercial epoxy resin DER 332 – epoxide equivalent weight 171–175 g – was cured by two methods using carboxylated lignin as the curing agent. In the first method, the carboxylated lignin was not precipitated. Rather, the epoxy resin was directly added to the reaction system after carboxylation was complete and the solution was cooled to room temperature. The ratio of resin mass to lignin varied between 2.7 and 12.8. As a catalyst, 1% zinc acetate based on the total mass of the reacting system was added. Curing was carried out at 80 °C for 0.5 h. The reacting system was cooled to ambient temperature and acetone evaporated under vacuum. The pre-cured mixture was filled into mould and curing was completed in three stages: 1 h 120 °C, 3 h 160 °C, 0.5 h 180 °C. In the second procedure, separated carboxylated lignin was redissolved in acetone for its use as the curing agent. Cured thermosets were evaluated for their thermomechanical characteristics. Up to 40 °C – glassy region – the storage modulus improved with the proportion of lignin. Beyond 40 °C – rubbery region – the thermosets with more lignin content had lesser storage modulus. In contrast, the maximum loss modulus increased with increasing lignin content. The behaviour was correlated to the rigid structure of carboxylated lignin compared to the epoxy resin which restricted crosslinking. This was also manifested in the T_g of the thermosets which continuously decreased with increasing lignin content (Zhang et al. 2021).

9.1.2.6 Lignin Epoxy Composites

In one study, enzymatic hydrolysis lignin from corn straw was used to prepare epoxy-based composite material. The epoxy resin used had an epoxy value of 0.51. Composites were prepared by mixing the resin and lignin in equal mass ratio with different amounts of polyamine curing agent at ambient temperature for 10–30 min. The blend was moulded into sheets by hot pressing followed by cold pressing. It was observed that lignin by itself could act as a curing agent for epoxy resin owing to its phenolic hydroxyl groups. A blend of an equiproportion mixture of lignin and epoxy resin cured at 174 °C compared to 110 °C when a polyamine curing agent was used at a 50% level of resin mass. Composite material prepared with the addition of polyamine curing agent had the highest bending strength of 44.5 MPa and impact strength of 4.6 kJ/m^2 when moulded at 130 °C. Increased moulding pressure also improved strength up to 12 MPa (Yin et al. 2012). Another report mentions the preparation of lignin (unknown type) epoxy composite. Bisphenol A-based epoxy resin with 0.51 epoxy value was mixed with 35–70% lignin and 60–120% polyamide-based curing agent. Components were mixed with two roll mills, formed into mould, cold pressed followed by hot pressing. It was observed that the composites deteriorated in terms of their bending strength with increasing lignin content. The highest impact strength was obtained at 45% lignin content (Yin and Di 2012).

9.2 LIGNIN-BASED POLYURETHANES

One of the most versatile polymers, polyurethanes find numerous applications in almost all walks of modern life. With the right choice of starting materials and synthesis conditions, these polymers can be tailored to have a whole range of characteristics to suit a particular application. Thus, we can have polyurethane foams with various porosity and rigidity, polyurethane-based elastomers, polyurethanes for surface applications, and polyurethane-based adhesives, to name a few. In addition, polyurethanes have important biomedical applications.

A urethane linkage – $R_1NHCOOR_2$ – is produced when an isocyanate functional group reacts with a hydroxyl functional group. If one of the reactants carries more than one isocyanate functional group and the other reactant carries more than one hydroxyl group, multiple urethane linkages can form giving rise to the polymer – the polyurethane. The molecular fragments R_1 and R_2 come from the isocyanate-carrying and the hydroxyl-carrying moieties, respectively. These can be aliphatic or aromatic with different structural features which provides flexibility and versatility to the synthesis route. Hydroxyl-carrying moieties – the polyols – used in polyurethane synthesis can have the molecular fragment R_2 consisting of linear alkyls, polyalkylene polyethers, polyester, and acrylics. Frequently used isocyanates for polyurethane synthesis are toluene diisocyanate, 1,5-naphthalene diisocyanate, methylene diphenyl diisocyanate, isophorone diisocyanate, and hexamethylene diisocyanate. If the molecular fragment R_1 is aromatic, it enhances the reactivity of the isocyanate functional group through its electron-withdrawing effect. Reaction is frequently carried out in the presence of catalysts to facilitate synthesis at milder conditions. These may be organometallic compounds or tertiary amines. The ratio of isocyanate and polyol is an important parameter in polyurethane synthesis and governs many of its characteristics. It is also important to ensure homogeneity of the polyol and isocyanate and avoid phase separation. Surfactants may be used to address this. However, like most synthetic polymers, polyurethanes are commonly synthesized using isocyanates and alcohols/phenols sourced from petroleum. This casts doubt on their sustainability. It is, therefore, imperative to search for monomers derived from renewable resources that can replace petroleum-derived ones in polyurethane synthesis. Lignin, with the presence of multiple phenolic and aliphatic hydroxyl groups, is an obvious choice. In addition, the lignin molecule can also have more than one hydroxyl group in the form of carboxyl functionality. Lignin substitution affects the polyurethane characteristics. The difference in characteristics mainly arises from the difference and heterogeneity in reactivity of hydroxyl groups in lignin compared to the hydroxyl groups in petroleum-derived polyols. In a kinetic study, two eucalyptus kraft lignins were evaluated for their reactivity for 4,4′-diphenylmethane diisocyanate. One kraft lignin, precipitated at pH 3.5, had 1.11 mmol/g aliphatic hydroxyl, 4.5 mmol/g phenolic hydroxyl, and 2080 M_w. The other one, precipitated at pH 8.1, had 1 mmol/g aliphatic hydroxyl, 4.14 mmol/g phenolic hydroxyl, and 2830 M_w. Each lignin was dissolved in tetrahydrofuran to prepare a 250 g/L solution. 4,4′-diphenylmethane diisocyanate was added to the lignin solution to maintain a molar ratio of isocyanate to hydroxyl of 8. The reaction was carried out under reflux and conversion of hydroxyl groups was estimated for different reaction times. Conversion of aliphatic hydroxyl was practically complete in 30 min reaction time for both the lignins. In contrast, phenolic hydroxyl conversion was only about 75% after 90 min for the lignin precipitated at pH 8.1. The corresponding conversion was about 95% for the other lignin but unreacted phenolic hydroxyl groups were still present. This clearly brings forth the sluggish reactivity of phenolic hydroxyl towards isocyanate in forming urethane linkages compared to aliphatic hydroxyl. Besides the acidic nature of hydrogen in the phenolic hydroxyl, steric hindrances played a major role in suppressing the reactivity of phenolic hydroxyl. The observed reactivity was the highest for phenolic hydroxyl in p-hydroxyphenyl moieties, followed by non-condensed guaiacyl, condensed guaiacyl, and syringyl moieties (Antonino et al. 2021). The reaction kinetics are presented in Figure 9.6. The aromatic polymeric backbone of lignin also results in differences in crosslinking and resulting characteristics. Besides the difference in physical, morphological, and mechanical characteristics, the lignin-substituted polyurethanes are distinctly darker in colour.

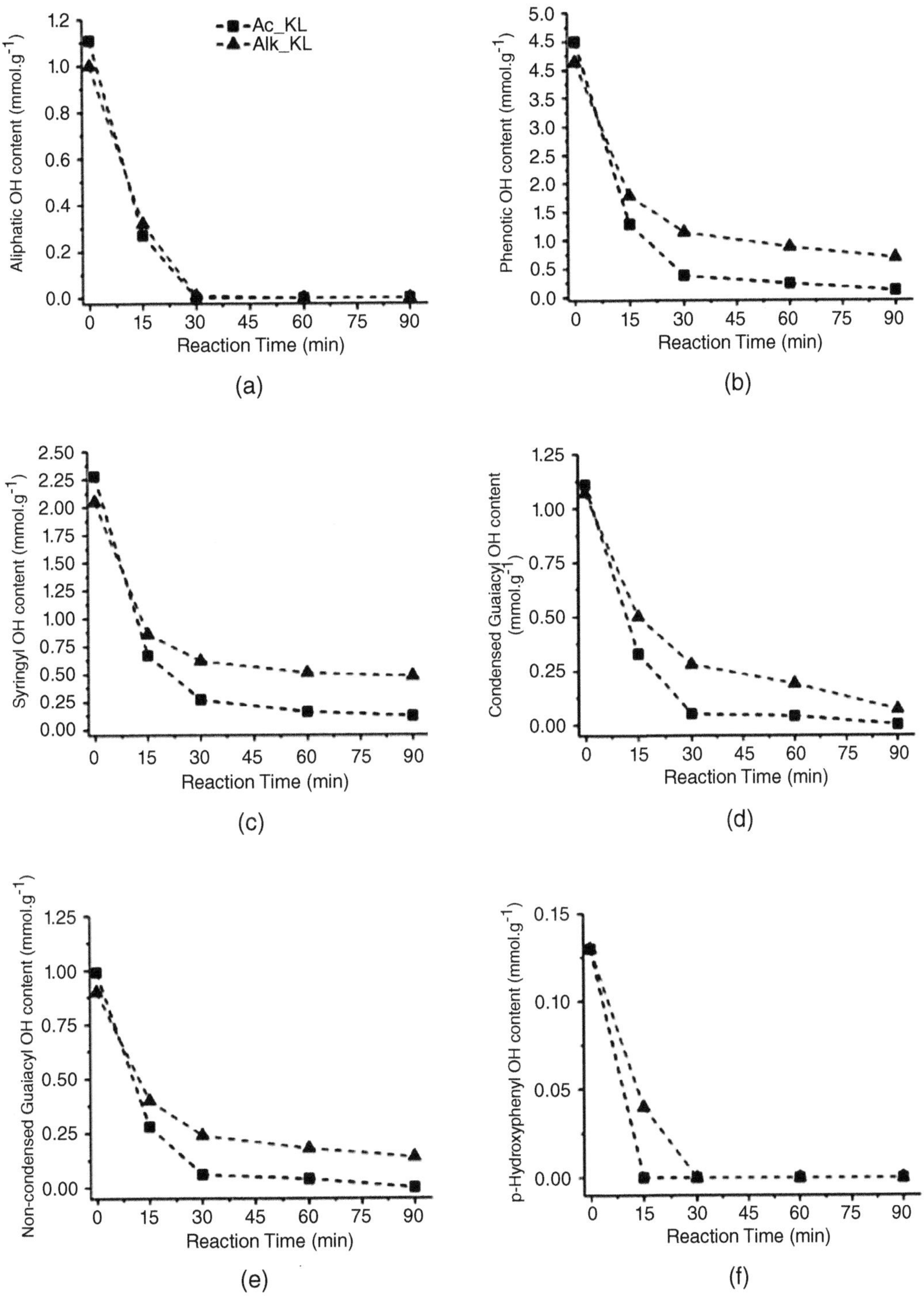

FIGURE 9.6 Reaction kinetics of aliphatic and phenolic hydroxyl groups in lignin towards isocyanate group. Ac_KL kraft lignin precipitated at pH 3.5; Alk_KL kraft lignin precipitated at pH 8.1. (Reprinted from Antonino, LD, Gouveia, JR, de Sousa Junior, RR, Garcia, GES, Gobbo, LC, Tavares, LB, dos Santos, DJ, *Molecules*, 26, 2131, 2021, originally published by MDPI under Creative Commons Attribution License.)

One of the important applications of polyurethanes is in the synthesis of foams. Cream time, gel time, rise time, and tack-free time are frequently measured for a quick subjective estimation of the kinetics of foam formation. Starting from the moment when isocyanate is added to the polyol, cream time refers to the time when the mixture first turns from a clear liquid to a cloudy state. Gel time is the moment when the first visible evidence of the formation of polymeric threads is observed. Rise time is the time till the foam freely expands in volume. Finally, tack-free time denotes the moment when the foam loses its tacky character and can be handled. All these time durations are normally measured in seconds.

9.2.1 Soda and Kraft Lignin

Pine needle soda lignin was the polyol used in polyurethane foam synthesis using 4,4′-diphenylmethane diisocyanate. Lignin was dissolved in dimethyl formamide with the addition of 5% triethylamine, 5% triethanolamine, 6% silicone oil, and 100% water as catalyst, co-catalyst, surfactant, and blowing agent, respectively. 4,4′-diphenylmethane diisocyanate, 1.5 times the lignin mass, was added to the mixture with intense stirring till foam formation. The produced foam had 75% porosity. The foam was used as an adsorbent for malachite green (Kumari et al. 2016). There is one report of oxypropylation of commercial sarkanda soda lignin – 2.41 mmol/g phenolic hydroxyl, and 2.85 mmol/g aliphatic hydroxyl – for its possible subsequent use as a substitute for polyol in polyurethane synthesis. Lignin was reacted with propylene oxide in the presence of a potassium hydroxide catalyst. The lignin-to-propylene oxide mass ratio was varied between 2:8 and 3.5:6.5 and the catalyst amount was 2–5%. The oxypropylation reaction commenced at 160 °C and the exothermic heat of the reaction raised the temperature to 176–254 °C. Grafted polyols had hydroxyl values of 319–346 mg KOH/g (Cateto et al. 2009). As another alternative derivatization approach, nonwood soda lignin Protobind 1075 – 2.38% total hydroxyl – was modified by reaction with 4,4′-diphenylmethane diisocyanate under nitrogen atmosphere. Isocyanate was added dropwise into a 0.25 g/mL solution of lignin in tetrahydrofuran to have a final lignin-to-isocyanate ratio of 2. Initially, the contents were mixed at 40 °C and then the temperature was brought to 70–100 °C. Reaction time varied between 15 and 90 min. Separately, an isocyanate prepolymer was prepared by reacting a mixture of poly(ethylene glycol) and 1,4-butanediol with 4,4′-diphenylmethane diisocyanate under a nitrogen atmosphere. The reaction temperature was 80 °C and the reaction time was 2 h with dibutyltin dilaurate as the catalyst. The mass ratio of poly(ethylene glycol), 1,4-butanediol, and 4,4′-diphenylmethane diisocyanate was 33.5:2.76:100. The modified lignin and the isocyanate prepolymer were further reacted in different proportion – 1–15% lignin in final product – for 10–15 min under intense mixing in a nitrogen atmosphere and presence of dibutyltin dilaurate catalyst. The resulting polyurethane was cast into films on glass, dried for 24 h at ambient conditions followed by vacuum drying, and cured for 2 h at 80 °C. The films increased in their hydrophobic nature with increasing lignin content as illustrated by continuous increase in contact angle. Similarly, the mechanical characteristics of the films improved with increasing lignin content. The tensile strength of the film with 15% lignin was 16.59 MPa compared to 8.55 MPa with no lignin. Tensile modulus increased from 15.94 MPa at no lignin incorporation to 706.39 MPa at 15% lignin inclusion (Chauhan et al. 2014).

Indulin AT – 2.413 mmol/g phenolic hydroxyl and 1.957 mmol/g aliphatic hydroxyl – was used as a partial substitute – 10–25% – of the polyol, polycaprolactonediol, in polyurethane synthesis. Desired proportions of lignin and polycaprolactonediol – molecular weight 400, 750, and 1000 – were mixed. The isocyanate moiety, 4,4′-diphenylmethane diisocyanate, was molten and added to the mixture so as to have an equimolar ratio of isocyanate to hydroxyl groups. Reaction was carried out at 80 °C for 30 min. Isocyanate conversion was more than 95% for all lignin substitution levels with polycaprolactonediol of molecular weight 400. When polycaprolactonediol molecular weight increased to 750, conversion dropped to 85–90%. Conversion further decreased to 72–83% when polycaprolactonediol of molecular weight 1000 was used. Conversion always decreased with increasing lignin proportion (Cateto et al. 2008). Hardwood kraft lignin – phenolic hydroxyl

4.29 mmol/g, aliphatic hydroxyl 4.12 mmol/g, M_w 2400, polydispersity 1.8 – was used to substitute polyether polyol – 4.3 mmol/g hydroxyl – in rigid polyurethane foam synthesis. A mixture of 25–70% lignin and polyol was added with tin-(II)-isooctate catalyst, polyether-modified polysiloxane surfactant and water (blowing agent) and homogenized by stirring. 4,4′-diphenylmethane diisocyanate, with 7.5 mmol/g isocyanate content, was added and quickly mixed by vigorous stirring. The foam was allowed to rise and cure at room temperature for 7 days. The isocyanate-to-hydroxyl mole ratio was 1.1 or 1.3. At a higher ratio of isocyanate to hydroxyl, the prepared foam had lower apparent density and compressive strength. Increasing lignin substitution decreased the apparent density up to 50% substitution level but increased marginally when the lignin substitution further increased to 70%. However, compressive strength continuously decreased with increasing lignin content in foam. Increasing lignin content in the foam resulted in larger and more non-uniform cells in the foam structure which resulted in decreased properties (Pan and Saddler 2013). Similarly, kraft lignin – 356 mg KOH/g hydroxyl value – and polyether polyol – 345–375 mg KOH/g hydroxyl value – blends were prepared by stirring the ingredients with 4% glycerol in the blend. The lignin proportion in the blend varied between 2.5% and 10%. To this blend was added 2.25% of a proprietary surfactant, 1% dimethylcyclohexylamine catalyst, 1.5% water, and 30% of a hydrofluorocarbon-based proprietary blowing agent and mixed thoroughly. Polymeric diphenylmethane diisocyanate was then added with intense mixing for a few seconds and the mixture was transferred to moulds for foam rise. The amount of isocyanate added was equivalent to 1.1-mole isocyanate for every mole of hydroxyl. Kinetic parameters for polyurethane foam formation – cream time, gel time, rise time, and tack-free time – all decreased with increasing lignin content in the foam, indicating increased reactivity. However, the storage modulus and apparent density did not show a regular trend (Hayati et al. 2018). Wheat straw kraft lignin was used as a partial substitution – up to 2.5% – of poly(ethylene glycol) in polyurethane foam synthesis. Lignin in the desired amount was mixed in poly(ethylene glycol). 1.75% water (blowing agent), 1% triethanolamine (co-catalyst), 0.75% dibutyltin dilaurate (catalyst), and 2% silicone oil (surfactant) were added to the mixture with stirring at 80 °C. Next, 40% 4,4′-diphenylmethane diisocyanate was added under thorough stirring and the reacting mass was immediately poured into a mould. Foam was allowed to synthesize for 10 h at 80 °C. The study evaluated the effectiveness of lignin as a crosslinking agent in the polyurethane foam by measuring its crosslinking density in mol/cm³ as (Li et al. 2017b)

$$V_e = -\frac{1}{v_s}\left[\frac{\ln(1-v_p)+v_p+\chi v_p^2}{v_p^{\frac{1}{3}}-0.5v_p}\right] \tag{9.1}$$

where V_e is crosslinking density, v_p is the volume fraction of the polyurethane in the foam, χ is the polyurethane solvent interaction parameter, and v_s is the per cent molar volume of the solvent. Up to 1% addition of lignin, the crosslinking increased beyond which it decreased. This is also reflected in the apparent densities of the foams produced. Apparent density decreased up to 1% lignin addition but increased with further lignin addition. In one protocol, softwood kraft lignin – M_w 13,176, polydispersity 6.3 – was used as a partial substitute for crude glycerol – obtained from transesterification of vegetable oils – in polyurethane synthesis. Glycerol was adjusted to pH 8 and heated to 160 °C. 1 part kraft lignin was added to 9 parts of heated glycerol and reacted for 1.5 h with stirring. Liquefied mass – hydroxyl value 412 mg KOH/g – was then cooled to ambient temperature and added with proprietary gelling catalyst, blowing catalyst, surfactant, and water (blowing agent). All the ingredients were mixed thoroughly. 4,4′-diphenylmethane diisocyanate – isocyanate-to-hydroxyl mole ratio 1.05 – was added to the mixture with stirring. The foam was allowed to rise and cure (Muller et al. 2019). Similarly, kraft lignin (unknown lignocellulose source) was converted to a lignin-based polyol for use in polyurethane synthesis. It was reacted with poly(ethylene glycol) and glycerol – 1:5.6:1.25 mass ratio – in the presence of sulphuric acid catalyst through 5 min 130 °C microwave digestion. Produced lignin polyol – 708 mg KOH/g hydroxyl value – was mixed with 0.4% polypropylene glycol triol, 4.3% water (blowing agent), 4.3% dibutyltin dilaurate catalyst, and 4.3% proprietary catalyst DABCO 33lV. Once a homogeneous mixture was formed, diphenylmethane diisocyanate was added – 1–2 isocyanate-to-hydroxyl molar ratio – and intensely mixed. The mixture was immediately poured into a foaming container and the foam was allowed to rise and subsequently cure for 2 days. The foam rising time duration was 6.24 s with 2-mole isocyanate per mole hydroxyl but reduced to 4.08 s with an equimolar ratio of isocyanate to hydroxyl. Corresponding foam densities were 187 and 220 kg/m^3, respectively. Compressive modulus also increased with increasing isocyanate ratio: 0.67–2.89 MPa (Mohammadpour and Sadeghi 2020).

Several softwood kraft lignins – 3.49–6.7 mmol/g hydroxyl, M_w 4290–12,100, and polydispersity 2.79–5.38 – were used for substituting the polyether-based polyol – 0.5 mmol/g hydroxyl – in polyurethane foam synthesis. Similarly, bagasse kraft lignin with 4.63 mmol/g hydroxyl, 6550 M_w, and 3.28 polydispersity, and hardwood kraft lignin with 6.79 mmol/g hydroxyl, 1740 M_w, and 1.69 polydispersity were also evaluated as polyol substitutes. A mixed diisocyanate consisting of diphenylmethane diisocyanate and polymeric diphenylmethane diisocyanate, having 28% isocyanate, was used for polyurethane foam synthesis. Polyether-based polyol was mixed with water followed by the addition and mixing of a proprietary gelation catalyst (0.5% of polyol), amine-based blowing catalyst (0.3% of polyol), surfactant (0.8% of polyol), and lignin (20% of polyol). Mixed diisocyanate was then added with intense agitation and the mixture was filled into moulds. Free expansion was allowed for an hour at 60 °C. Formed foam was cooled to ambient temperature and curing was completed in 24 h. Apparent densities of the foams, with softwood kraft lignin substitution, varied between 70 and 95 kg/m^3. Foam with bagasse kraft lignin substitution had 72 kg/m^3 apparent density and that with hardwood kraft lignin substitution had 86 kg/m^3 apparent density. All these were higher than the 59 kg/m^3 apparent density for foam without lignin substitution. Polyurethane foams from softwood kraft lignins had compression force deflection between 7 and 14 kPa, that from bagasse kraft lignin 8 kPa, and the foam from hardwood kraft lignin 13 kPa. These are all higher than the 4 kPa compression force deflection measured for polyurethane foam without lignin substitution. Similarly, the polyurethane foams from softwood kraft lignins had tear strengths of 208–299 N/m, that from bagasse kraft lignin 260 N/m, and the foam from hardwood kraft lignin 309 N/m. The tear strengths of lignin-based polyurethane foams also outperformed the foam without lignin substitution as its tear strength was only 190 N/m. Other mechanical characteristics –compression set, and support factor – of polyurethane foams prepared with 20% polyol substitution with kraft lignins were better compared to foam without lignin substitution. Generic improvement in mechanical characteristics of polyurethane foams with kraft lignin substitution was attributed to its higher hydroxyl content and rigid aromatic backbone. All the polyurethane foams with kraft lignin substitution had better thermal stability than the foam without lignin substitution (Gondaliya and Nejad 2021). One study reported modification of Indulin AT – M_w 7600, polydispersity 7 – by oxypropylation prior to its use as a polyol substitute in polyurethane synthesis. Oxypropylation was carried out at 150–285 °C by reacting lignin with propylene oxide in the presence of potassium hydroxide – 1g:4 mL:0.05 g. The reaction was complete in less than 10 min. Unreacted propylene oxide was extracted with hexane. Oxypropylation reduced M_w to 4200 and polydispersity to 2.4 with about 39% polypropylene oxide grafting. There was a commensurate increase in the hydroxyl index from 5.62 for Indulin AT to 6.91 for the oxypropylated lignin. The modified lignin was used as a substitute for sucrose polyol and glycerol polyol. The extent of substitution was 0–100% on the mass of sucrose polyol. A mixture of modified lignin, sucrose polyol, glycerol polyol (60–150% of sucrose polyol mass), dimethylcyclohexylamine catalyst (7.2–18% of sucrose polyol mass), proprietary Mannich base catalyst (3.6–9% of sucrose polyol mass), and silicone surfactant (6–15% of sucrose polyol mass) was prepared. To this mixture was added polymeric diphenylmethane diisocyanate (145–360% of sucrose polyol mass) and the blowing agent pentane (40–100% of sucrose polyol mass) with vigorous mixing. Foams were cured for 48 h at ambient temperature. Similarly, polyurethane foam was prepared with a complete substitution of sucrose polyol with lignin as well as a complete substitution of both sucrose and glycerol polyol with lignin. All the prepared foams had an apparent density of about 30 kg/m^3 but their cell morphologies varied with varying lignin content. These morphological differences are depicted in Figure 9.7 through scanning electron micrographs. The foam prepared with complete substitution of both sucrose polyol and glycerol polyol had the best mechanical properties of yield strength of 0.14 MPa and a compressive modulus of 3.41 MPa (Li and Ragauskas 2012).

There is another report of oxypropylation of two commercial softwood kraft lignins, Indulin AT – 3.95 mmol/g phenolic hydroxyl, and 3.04 mmol/g aliphatic hydroxyl – and Curran 27–11P – 3.63 mmol/g phenolic hydroxyl, and 2.58 mmol/g aliphatic hydroxyl – for their possible subsequent use as a substitute for polyol in polyurethane synthesis. Each lignin was reacted with propylene oxide in the presence of a potassium hydroxide catalyst. The lignin-to-propylene oxide mass ratio was varied between 2:8 and 3.5:6.5 and the catalyst amount was 2–5%. The oxypropylation reaction

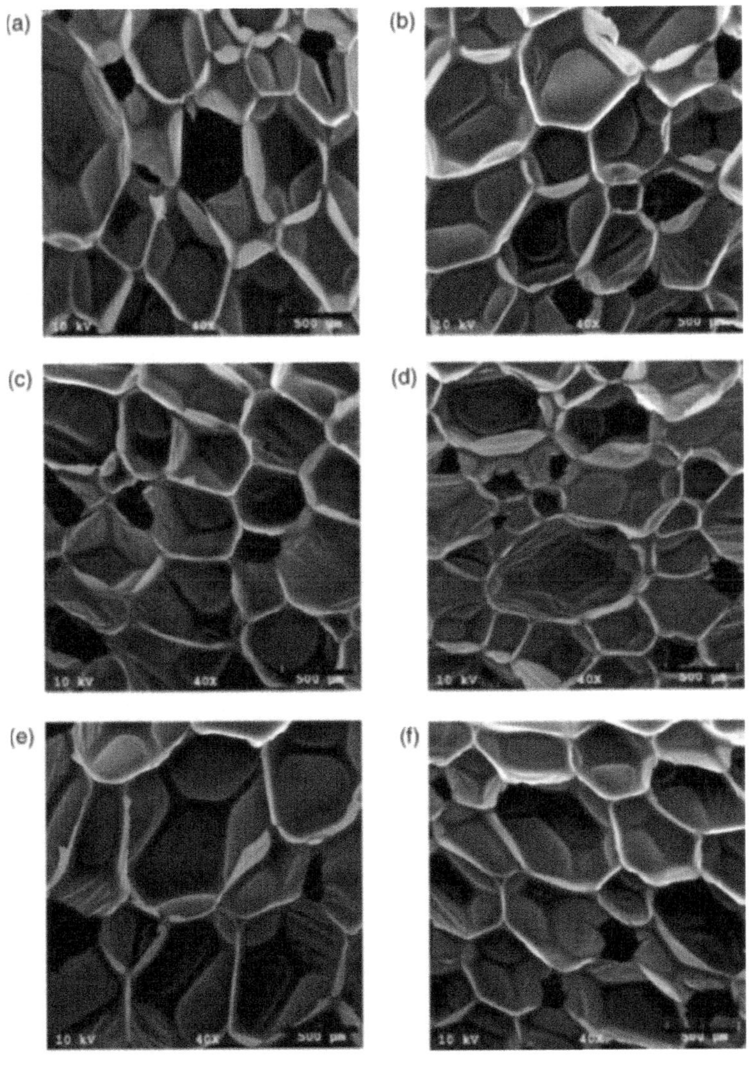

FIGURE 9.7 Scanning electron micrographs of rigid polyurethane foams prepared from the substitution of sucrose polyol with oxypropylated softwood kraft lignin. The other polyol is glycerol polyol. (a) no lignin (b) 10% lignin (c) 30% lignin (d) 60% lignin (e) 100% lignin (f) only lignin, no other polyol. (Reprinted from Li, Y, Ragauskas, AJ, *J. Wood Chem. Technol.*, 32, 210–224, 2012, with permission from Taylor and Francis.)

commenced at 160 °C exothermic heat of the reaction raised the temperature to 248–298 °C. With Indulin AT the grafted polyols had hydroxyl values of 326–355 mg KOH/g. Grafted Curran polyols had hydroxyl values of 366–433 mg KOH/g (Cateto et al. 2009). Another study examined the substitution of diethylene glycol with Indulin AT for the synthesis of polyurethane foam. As a control, a mixture of a 10:1 mass ratio of diethylene glycol and polyethylene glycol was used as the polyol for polyurethane synthesis. In lignin-based foams, diethylene glycol was substituted with 5–20 mass% of Indulin. Lignin was dispersed in diethylene glycol at 120 °C. Based on diethylene glycol plus lignin mass, 2% dibutyltin dilaurate catalyst, 2% polydimethyl siloxane surfactant, and water as a blowing agent was used. All these were mixed into a homogeneous system. Polymeric diphenylmethane diisocyanate – 1.1-mole isocyanate per mole hydroxyl – was then added with intense agitation and quickly transferred to the mould to allow the foam to rise. Curing was continued for 48 h. With lignin substitution of diethylene glycol, the prepared foams increased in their apparent density; 62 kg/m^3 without lignin substitution and 115 kg/m^3 with 20% lignin substitution. However, mechanical properties followed a different pattern with the highest compressive strength of 728 kPa for 5% lignin substitution and the lowest compressive strength of 642 kPa for 20% lignin substitution. A reversed trend was reported for compressive modulus; 10.07 MPa for 5% lignin substitution and 14.72 MPa for 20% lignin substitution. In comparison, the foam with no lignin substitution had a compressive strength of 330 kPa and a compressive modulus of 8.15 MPa (Luo et al. 2020).

In a modified procedure, pine alkali lignin was grafted with polyethylene glycol. Molten polyethylene glycol at 55 °C was mixed with 0.8% boron trifluoride diethyl etherate followed by dropwise addition of 4.6% epichlorohydrin. The reaction was continued with stirring at 55 °C for 2 h to complete epoxidation. A solution of 25% lignin in 1 M aqueous sodium hydroxide was added to the epoxidated glycol and further reacted at 80 °C for 3 h. Unreacted lignin was precipitated by acidification to pH 7 and centrifuged. A second stage of precipitation was carried out by ethanol addition. Finally, the clear liquid was vacuum dried at 50 °C to obtain the glycol grafted lignin to be used for 20–50% substitution of polyethylene glycol for polyurethane foam synthesis. Polyethylene glycol and glycol grafted lignin were mixed in desired proportion at 70 °C followed by the addition of 1% silicone surfactant, 1.2% dibutyltin dilaurate catalyst, and 1.7% blowing catalyst (33:67 mixture of triethylene diamine and dipropylene glycol) and thorough mixing by stirring. 20.7% hexamethylene diisocyanate was quickly added with intense mixing with foam rise at 70 °C allowed for 2 h. The foam was cured at ambient temperature for 24 h. With an increased proportion of grafted lignin, the apparent densities of the prepared foams increased – 119.9 kg/m^3 at 20% substitution and 200.9 kg/m^3 at 50% substitution. Corresponding increases in compression modulus – 88.3 kPa at 20% substitution and 273 kPa at 50% substitution – and compression force deflection – 87.8 kPa at 20% substitution and 201 kPa at 50% substitution – were observed. For comparison, polyurethane foams were prepared with no lignin substitution and 20% substitution of unmodified lignin. Foam with no lignin substitution had apparent density, compression modulus, and compression force deflection of 206.7 kg/m^3, 21.2 kPa, and 35.2 kPa, respectively. These characteristics were 343.9 kg/m^3, 16.1 kPa, and 54.6 kPa, respectively, for foam with 20% substitution of unmodified lignin. However, the thermal stability of the foams was compromised with increased substitution of glycol-grafted lignin (Wang et al. 2019).

9.2.1.1 Organosolv Lignin

An early study reported the use of spruce oxygen aqueous acetic acid organosolv lignin – 0.4 aliphatic hydroxyl and 0.4 phenolic hydroxyl per phenylpropane unit – as polyol substitute in polyurethane synthesis. Lignin was mixed with oligoethyleneoxide diisocyanate at 40 °C. Temperature was gradually increased to 60 °C and 2% dibutyltin dilaurate catalyst was added with continuous stirring. The reaction was allowed to continue for another 8 h. Reactor contents were then extracted with dichloromethane to remove the soluble unreacted ingredients. Films were cast in a similar manner by spreading the reaction mixture into a thin mould just after the catalyst addition and allowed to react and cure in the mould. The authors recommended 50% excess of isocyanate over the stoichiometric requirement for optimal reaction and crosslinking, yielding 68.4% polyurethane (Evtuguin et al. 1998). In one protocol, bagasse aqueous acetic and formic acid organosolv lignin – M_w 3566, polydispersity 4.5 – was used as a partial substitute for crude glycerol – obtained from transesterification of vegetable oils – in polyurethane synthesis. Glycerol was adjusted to pH 8 and heated to 160 °C. 1 part of organosolv lignin was added to 9 parts of heated glycerol and reacted for 1.5 h with stirring. Liquefied mass – hydroxyl value 224 mg KOH/g – was then cooled to ambient temperature and added with proprietary gelling catalyst, blowing catalyst, surfactant, and water (blowing agent). All the ingredients were mixed thoroughly. 4,4′-diphenylmethane diisocyanate – isocyanate-to-hydroxyl mole ratio 1.05 – was added to the mixture with stirring. The foam was allowed to rise and cure (Muller et al. 2019).

Bagasse acetic acid organosolv lignin was hydroxymethylated prior to the synthesis of polyurethane with toluene 2,4-diisocyanate. Hydroxymethylation was carried out at 60–90 °C by the addition of aqueous formaldehyde to a solution of lignin in aqueous sodium hydroxide – molar ratio of lignin:sodium hydroxide:formaldehyde 1:1:2.4. The modified lignin substituted the polyol, poly(ethylene glycol), to the extent of 15–30 mass% during polyurethane synthesis. Hydroxymethylated lignin was dissolved in dimethyl formamide and prepolymerized with toluene 2,4-diisocyanate for 10 min at ambient temperature. The desired quantity of poly(ethylene glycol) was then added to have an equal mass of lignin plus polyol and toluene 2,4-diisocyanate. Reaction completed in 2.5 h. Prepared polyurethanes were used as an adhesive to join brass metal strips. Substitution of 15–25% poly(ethylene glycol) by lignin or hydroxymethylated lignin improved the shear strength of the lap joints. Polyurethane adhesive without any lignin substitution exhibited a bond shear strength of 32.9 MPa. With lignin substitution, strength improved to 35.2–42.3 MPa. With hydroxymethylated lignin substitution, strength improved to 39.4–74.08 MPa. The highest shear strength was for hydroxymethylation temperature of 85 °C and 20% lignin substitution (Chen et al. 2020).

1,4-Butanediol organosolv lignin from spruce – 5.33 mmol/g hydroxyl, 0.25% ash, 3778 M_w, 1.52 polydispersity – was used as the polyol in polyurethane synthesis with 4,4′-diphenylmethane diisocyanate. Lignin and 4,4′-diphenylmethane diisocyanate – 0.67–1.47 mass ratio of isocyanate to lignin – were dissolved in 1,4-dioxane with the addition of triethylamine – 40 μL/g lignin – as the catalyst. Synthesis was carried out at ambient temperature for 1 h. Prepared polyurethane was formed into film and cured for 3 h at 85 °C. The film was thermally stable up to 253.5 °C which was lower than that reported for commercial polyurethane film. The films did not swell much in water, but appreciable swelling was observed in dimethyl sulfoxide which increased with the increase in the ratio of lignin to isocyanate. The film with a maximum tensile strength of 41.6 MPa was produced with an isocyanate-to-lignin mass ratio of 1.134 (Jia et al. 2015).

Use of spruce oxygen aqueous ethanol organosolv lignin – 0.84 aliphatic hydroxyl and 0.34 phenolic hydroxyl per phenylpropane unit – as polyol substitute in polyurethane synthesis is also reported. Lignin was mixed with oligoethyleneoxide diisocyanate at 40 °C. Temperature was gradually increased to 60 °C and 2% dibutyltin dilaurate catalyst was added with continuous stirring. The reaction was allowed to continue for another 8 h. Reactor contents were then extracted with dichloromethane to remove the soluble unreacted ingredients. Films were cast in a similar manner by spreading the reaction mixture into a thin mould just after the catalyst addition and allowed to react and cure in the mould. The authors recommended 50% excess of isocyanate over the stoichiometric requirement for optimal reaction and crosslinking, yielding 75.5% polyurethane. This yield was higher than the yield obtained with acetic acid organosolv lignin (Evtuguin et al. 1998). Hardwood aqueous ethanol organosolv lignin – phenolic hydroxyl 2.76 mmol/g, aliphatic hydroxyl 2.88 mmol/g, M_w 2600, polydispersity 1.62 – was used to substitute polyether polyol – 4.3 mmol/g hydroxyl – in rigid polyurethane foam synthesis. A mixture of 25–70% lignin and polyol was added with tin-(II)-isooctate catalyst, polyether-modified polysiloxane surfactant and water (blowing agent) and homogenized by stirring. 4,4′-diphenylmethane diisocyanate, with 7.5 mmol/g isocyanate content, was added and quickly mixed by vigorous stirring. The foam was allowed to rise and cure at room temperature for 7 days. The isocyanate-to-hydroxyl mole ratio was 1.1 or 1.3. At a higher ratio of isocyanate to hydroxyl, the prepared foam had lower apparent density and compressive strength. Increasing lignin substitution decreased the apparent density up to 50–60% substitution level but increased marginally when the lignin substitution further increased to 70%. However, compressive strength continuously decreased with increasing lignin content in foam. Increasing lignin content in the foam resulted in larger and more non-uniform cells in the foam structure which resulted in decreased properties (Pan and Saddler 2013). In a modified protocol, hardwood aqueous ethanol organosolv lignin – 5.6 mmol/g hydroxyl – was first hydroxymethylated by dissolving in tetrahydrofuran and adding 1.6 mol per kg lignin of 4% aqueous solution of formaldehyde and 18 g per kg lignin of potassium hydroxide. Hydroxymethylation was completed in two stages: 7 h at 55 °C and 17 h at room temperature. This modification reduced the hydroxyl content to 3.99 mmol/g. The modified lignin – 65-80% – was dissolved in 1,4-dioxane and reacted with polybutadiene diisocyanate – 20-35% – in the presence of dibutyltin dilaurate catalyst. The reaction proceeded in two stages; 80 °C 20 h followed by 100 °C 4 h. Prepared thermoplastic polyurethanes were precipitated in isopropanol and recovered by vacuum evaporation of the alcohol. These polyurethanes exhibited two-phase morphology, a rigid phase with a crosslinked lignin backbone and a soft rubbery phase with the polybutadiene moiety. Films prepared from the polyurethanes confirmed this with two distinct loss modulus peaks, the low-temperature peak corresponding to the rubbery phase and the high-temperature peak to the rigid crosslinked lignin phase. A representative plot for 65% lignin-containing polyurethane film is shown in Figure 9.8. Young's modulus of the films consistently increased with increasing lignin content; 45.1 MPa at 65% to 140 MPa at 75% (Saito et al. 2013).

There is one report of oxypropylation of commercial mixed hardwood aqueous ethanol organosolv lignin, Alcell – 3.81 mmol/g phenolic hydroxyl, and 1.45 mmol/g aliphatic hydroxyl – for its possible subsequent use as a substitute for polyol in polyurethane synthesis. Lignin was reacted with propylene oxide in the presence of a potassium hydroxide catalyst. The lignin-to-propylene oxide mass ratio was varied between 2:8 and 3.5:6.5 and the catalyst amount was 2–5%. The oxypropylation reaction commenced at 160 °C and the exothermic heat of the reaction raised the temperature to 234–289 °C. Grafted polyols had hydroxyl values of 292–301 mg KOH/g (Cateto et al. 2009).

The use of spruce oxygen aqueous acetone organosolv lignin – 0.85 aliphatic hydroxyl and 0.38 phenolic hydroxyl per phenylpropane unit – as polyol substitute in polyurethane synthesis is also reported. Lignin was mixed with oligoethyleneoxide diisocyanate at 40 °C. Temperature was gradually increased to 60 °C and 2% dibutyltin dilaurate catalyst was added with continuous stirring. The reaction was allowed to continue for another 8 h. Reactor contents were then extracted with dichloromethane to remove the soluble unreacted ingredients. Films were cast in a similar manner by spreading the reaction mixture into a thin mould just after the catalyst addition and allowed to react

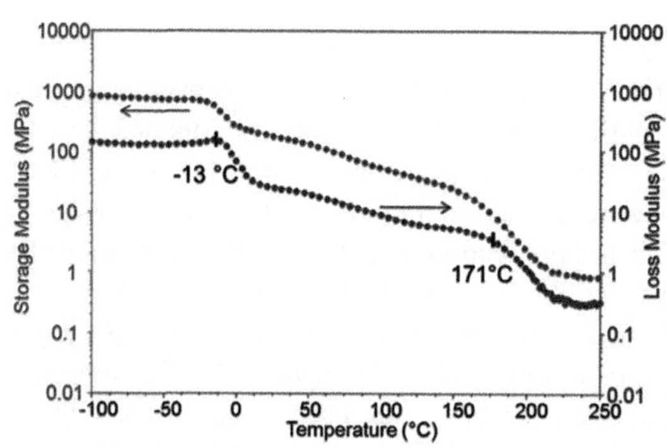

FIGURE 9.8 Mechanical characteristics of polyurethane film prepared with 65% hydroxymethylated ethanol organosolv lignin. (Reprinted from Saito, T, Perkins, JH, Jackson, DC, Trammel, NE, Hunt, MA, Naskar, AK, *RSC Adv.*, 3, 21832–21840, 2013, with permission from Royal Society of Chemistry.)

and cure in the mould. The authors recommended 50% excess of isocyanate over the stoichiometric requirement for optimal reaction and crosslinking, yielding 75.6% polyurethane. This yield was higher than the yield obtained with acetic acid organosolv lignin and comparable to that with ethanol organosolv lignin (Evtuguin et al. 1998).

Several hardwood organosolv (unknown solvent) lignins – 4.08-5.19 mmol/g hydroxyl, M_w 3440–4250, and polydispersity 2.48–2.85 – were used for substituting the polyether-based polyol – 0.5 mmol/g hydroxyl – in polyurethane foam synthesis. Similarly, corn stover organosolv lignin with 3.9 mmol/g hydroxyl, 6240 M_w, and 3.57 polydispersity, and peanut shell organosolv lignin with 3.48 mmol/g hydroxyl, 9080 M_w, and 5.54 polydispersity were also evaluated as polyol substitutes. A mixed diisocyanate consisting of diphenylmethane diisocyanate and polymeric diphenylmethane diisocyanate, having 28% isocyanate, was used for polyurethane foam synthesis. Polyether-based polyol was mixed with water followed by the addition and mixing of a proprietary gelation catalyst (0.5% of polyol), amine-based blowing catalyst (0.3% of polyol), surfactant (0.8% of polyol), and lignin (20% of polyol). Mixed diisocyanate was then added with intense agitation and the mixture was filled into moulds. Free expansion was allowed for an hour at 60 °C. Formed foam was cooled to ambient temperature and curing was completed in 24 h. Apparent densities of the foams, with hardwood organosolv lignin substitution, varied between 58 and 89 kg/m^3. Foam with corn stover organosolv lignin substitution had 67 kg/m^3 apparent density and that with peanut shell organosolv lignin substitution had 63 kg/m^3 apparent density. All these, except for one hardwood organosolv lignin-substituted foam, were higher than the 59 kg/m^3 apparent density for foam without lignin substitution. Polyurethane foams from hardwood organosolv lignins had compression force deflection between 4 and 11 kPa, that from corn stover organosolv lignin 7 kPa, and the foam from peanut shell organosolv lignin 6 kPa. These are all higher than the 4 kPa compression force deflection measured for polyurethane foam without lignin substitution. Similarly, the polyurethane foams from hardwood organosolv lignins had tear strengths of 233–266 N/m, that from corn stover organosolv lignin 207 N/m, and the foam from peanut shell organosolv lignin 215 N/m. The tear strengths of lignin-based polyurethane foams also outperformed the foam without lignin substitution as its tear strength was only 190 N/m. Other mechanical characteristics –compression set, and support factor – of polyurethane foams prepared with 20% polyol substitution with organosolv lignins were better compared to foam without lignin substitution. Generic improvement in mechanical characteristics of polyurethane foams with kraft lignin substitution was attributed to its higher hydroxyl content and rigid aromatic backbone. All the polyurethane foams with organosolv lignin substitution had better thermal stability than the foam without lignin substitution (Gondaliya and Nejad 2021). Similarly, beech organosolv lignin (unknown solvent) was used as a polyol substitute to synthesize polyurethane meant for coating application. A 20–60% suspension of lignin was prepared in 2-methyltetrahydrofuran. A proprietary polyisocyanate formulation by the commercial name DesmodurL75 – containing toluene diisocyanate, diethylene glycol, and trimethylolpropane – was added to the suspension to have isocyanate-to-hydroxyl molar ratios of 0.21–1.96. The reaction was continued for 1 h at room temperature with continuous stirring to prepare the polyurethanes. These were applied as wood coatings. Lignin-based polyurethane coatings had thickness comparable to commercial polyurethane coating but were opaque and dark brown in colour – the higher the lignin content darker the colour. Water repellence was also lower for lignin-based coatings (Bergamasco et al. 2022).

In yet another value-added application of acetic acid organosolv lignin (unknown raw material source), it was cross-linked with polyurethane ionomer to prepare hydrogel. Polyurethane ionomer was synthesized by first mixing poly(ethylene glycol), dimethylol propionic acid, and 1,4-dioxane (solvent) at 40 °C. A mixture of 2,4-toluene diisocyanate – equivalent to 1.5 mole isocyanate per mole hydroxyl – and dibutyltin dilaurate was separately prepared. The second mixture was added dropwise into the first and refluxed under stirring at 80 °C for 2 h to synthesize the ionomer. At this point, the organosolv lignin – 15–55% of ionomer mass – was added to the reacting system. Reaction further continued; 1 h at 60 °C followed by 2 h at 80 °C. Solvent dioxane was evaporated to obtain the hydrogel which was further extracted with acetone to remove the uncrosslinked material. Gel content in the hydrogel continuously decreased with increasing lignin content. But, at all pH, the swelling ratio of hydrogels in water increased with increasing lignin content up to 35% and decreased thereafter. The prepared hydrogels were effective as ammonium sulphate coatings to prepare controlled-release fertilizer formulations (Peng and Chen 2011).

9.2.1.2 Lignosulphonate

Hardwood calcium lignosulphonate – M_w 3688, polydispersity 5.4 – was used as a partial substitute for crude glycerol – obtained from transesterification of vegetable oils – in polyurethane synthesis. Glycerol was adjusted to pH 8 and heated to 160 °C. 1 part lignosulphonate was added to 9 parts of heated glycerol and reacted for 1.5 h with stirring. Liquefied mass – hydroxyl value 592 mg KOH/g – was then cooled to ambient temperature and added with proprietary gelling catalyst, blowing catalyst, surfactant, and water (blowing agent). All the ingredients were mixed thoroughly. 4,4′-diphenylmethane diisocyanate – isocyanate-to-hydroxyl mole ratio 1.05 – was added to the mixture with stirring. The foam was allowed to rise and cure (Muller et al. 2019).

A protocol for synthesizing polyurethane from beech lignosulphonic acid without using isocyanate is reported. Lignosulphonic acid was produced from the acidification of spent sulphite liquor followed by purification through ion exchange and extraction with ethanol. Lignosulphonic acid was mixed with glycerol carbonate (lignosulphonic acid:glycerol carbonate 1g/3.3 mL) and the catalyst 1,8-diazabicyclo[5.4.0]undec-7-en was added. The reaction was carried out overnight under an argon environment at 170 °C. The reaction was stopped by the addition of 15 mL ethanol per g lignosulphonic acid with 1 h stirring at 40 °C. This derivatization produced a mixture of cyclic carbonated lignosulphonate-glycerol and lignosulphonate 1,2-diol which was filtered from the reaction mixture, ethanol washed, and vacuum dried at 40 °C. This mixture was further reacted with dimethyl carbonate in the presence of the catalyst 1,8-diazabicyclo[5.4.0]undec-7-en. 12 h stirring under argon was provided at 75 °C to complete the reaction. Produced cyclic carbonated lignosulphonate-glycerol was centrifuged, ethanol washed, and vacuum dried at 40 °C. To synthesize the polyurethane, a suspension of cyclic carbonated lignosulphonate-glycerol was prepared in 1,6-hexamethylenediamine (1 g per 10 mL). Reaction continued overnight under stirring at 70 °C to yield the polyurethane which precipitated out. Unreacted lignosulphonate and amine were washed with ethanol to get the purified polyurethane (Mimini et al. 2019).

9.3 LIGNIN BLENDED WITH OTHER POLYMERS

A polymer blend refers to a physical intermix of two or more polymers with no or minimal chemical interaction between them. The blend has characteristics inherited from all the constituent polymers. Therefore, in specific end-use performance, they often fare better than individual polymers. In polymer blends, the interaction between the constituent polymers is essentially physical but the extent of this interaction may vary. There are some polymer blends where the interaction between the constituent polymers is feeble. The blend exists as a multiphase heterogeneous system. Such polymer blends are referred to as immiscible polymer blends. The physical properties are non-uniform at the macroscopic level. At a higher level of interaction between the constituent polymers, we have compatible polymer blends, also known as miscible blends. These are blends that still exist as multiphase systems at the microscopic level, but the properties are uniform at the macroscopic level. The compatibility of different polymers is often evaluated in terms of their respective Hansen solubility parameter. The more the difference between solubility parameters, the lesser the compatibility and more chances of forming an immiscible blend. Alternatively, T_g is a good indicative parameter for the nature of the blend. Immiscible blends usually return more than one T_g, each corresponding to a constituent polymer. Compatible blends, on the other hand, show only one T_g (Bozsodi et al. 2016). Sometimes two polymers, which would otherwise form an immiscible blend, can be blended into a compatible blend by the addition of a third component. This third component, known as the compatibilizer, lowers the interfacial tension at the phase boundary between the polymers and facilitates their compatibility. Miscible polymer blends are those having interactions between constituent polymers strong enough to turn them into a single-phase system. The properties are uniform not only at the macroscopic level but also at the microscopic level, often manifested as a single glass transition temperature. The homogeneity or inhomogeneity of polymer blends is strongly dependent on thermodynamic conditions. There are different ways of blending polymers together. They may be blended in a molten state, the process referred to as melt blending. Blends may be prepared by first dissolving one or more of the constituent polymers in one or more solvents, intimately mixing the solutions together, and removing of solvent after mixing. This is known as solution blending. An interpenetrating polymer network (IPN) can be formed if one of the constituent polymers is synthesized in situ in the presence of the other polymer.

Blending lignin with other polymers can help in modifying their properties, like thermal stability, resistance to photodegradation, and flame retardance, while minimizing the cost. This can also mitigate some of the sustainability and environmental concerns, especially when blended with synthetic polymers. Self-interaction between lignin macromolecules is often stronger than the possible interaction with the molecules of the polymer being blended. This may produce immiscible blends. One important application of blending lignin with other polymers is to enhance its spinnability for producing precursor fibres for producing carbon fibres. Lignin is frequently blended with polyethylene oxide or polyacrylonitrile for this purpose. A reference to this can be found in Chapter 8. Specific attention has been paid to lignin blends with thermoplastic polymers like polyolefins. However, lignin compatibility is often a challenge when blending with polyolefins like polypropylene. Lignin molecular weight and abundance of polar hydroxyl groups are major factors in this regard. Polyolefins being non-polar in nature, lack the affinity to polar substances. Blending lignin with other biobased polymers like poly(lactic acid) is another important valorization route.

9.3.1 Soda and Kraft Lignin

Commercial Nylon 6 by the name Durethan B40 F was blended with 0-30% Protobind 1000, a commercial agroresidue soda lignin., Blends were prepared by extrusion through a twin screw extruder running at 300 rpm and 240 °C. Blended pellets were injection moulded at 200–240 °C. Lignin acted as a plasticizer as lesser torque was required for extrusion as the lignin content increased in the blend. Blends exhibited efficient lignin dispersion without aggregation. A single T_g was observed for all the blends which increased with increasing lignin content. This established the formation of miscible blends. A slight increase in Young's modulus was observed with increasing lignin content in the blend (Sallem-Idrissi et al. 2016). 10–30% commercial agroresidue soda lignin, Protobind 2000, was blended with commercial high-density polyethylene by the name DMDA 8965. Blending was done at 150 °C causing the polyethylene to melt and the lignin to soften. Polyethylene was also grafted with maleic anhydride and 5% of the modified polymer was added to some of the blends. Tensile strength, Young's modulus, flexural strength, and flexural modulus all increased with increasing lignin content in the blend while impact strength decreased. The addition of maleic anhydride grafted polyethylene did not significantly influence the mechanical characteristics (Sameni et al. 2018).

An early study reported blending of 5–20% mixed temperate hardwood kraft lignin with polyurethane sealant. The initial set time of the sealant decreased with increasing lignin content; 20 h with 20% lignin in the blend compared to 25 h without lignin addition. The blend hardened faster compared to the pure polyurethane sealant. Similar behaviour was also observed for 10% lignin in the blend. Mechanical properties – toughness and yield stress – of the blends as sealant were evaluated on aluminium, and wood as a substrate. These properties attained their highest values at 5% lignin incorporation but continued to decrease thereafter. Toughness and yield stress for the blend with 20% lignin were lower than pure polyurethane. In contrast, the tangent modulus always increased with increasing lignin content in the blend (Feldman et al. 1988). Another study evaluated the blending of hardwood kraft lignin – 4.1 mmol/g aliphatic hydroxyl, 4.3 mmol/g phenolic hydroxyl, 2400 M_w, and 1.9 polydispersity – with polyvinyl alcohol, polypropylene, polyethylene oxide, or polyethylene terephthalate. Lignin content in the blend varied from 0% to 95%. Lignin, vacuum dried at 160 °C, was mechanically mixed with the chosen polymer in a single screw extruder at 200–250 rpm and drawn at 100 m/min. Mixing temperature varied between 140 and 220 °C depending on the choice of the polymer and the blend composition. Miscible blends with a single T_g were produced with polyethylene oxide or polyethylene terephthalate. T_g of the blend was dependent on its lignin content; higher lignin content resulted in higher T_g. These behaviours are shown in Figure 9.9. T_g of the blend was modelled according to the Gordon Taylor equation.

$$T_g = \frac{w_L T_{gL} + K' w_P T_{gP}}{w_L + K' w_P} \tag{9.2}$$

where T_g is the glass transition temperature of the blend. T_{gL}, and T_{gP} are the glass transition temperatures of lignin and the synthetic polymer, respectively. w_L, and w_P are the weight fraction of lignin and synthetic polymer, respectively, in the blend. K' is the model parameter, 0.331 for polyethylene oxide and 0.123 for polyethylene terephthalate. On the other hand, lignin was found to be incompatible with polyvinyl alcohol, or polypropylene, resulting in immiscible blends with two T_g values corresponding to lignin and the synthetic polymer (Kadla and Kubo 2004).

The same research group reported the mechanical properties of fibres spun from the lignin-polyethylene oxide blend and found the tensile strength and Young's modulus to increase with increasing lignin content (Kadla and Kubo 2003). Kraft lignin was blended with a variety of synthetic polymers, namely, polyethylene, polypropylene, polymethylene oxide, and polyethylene oxide. Polystyrene, polyacrylonitrile, polyvinyl chloride, polyvinyl acetate, polybutylene succinate-adipate and polyester amide. The polymers were selected to have a solubility parameter close to that of lignin – 10–11 $(cal.cm^{-3})^{1/2}$. Lignin was first dissolved in 10% aqueous dioxane. The required amount of the polymer was then mixed – 1% lignin in the blend – with this solution and extruded into 0.5 mm thick films through a slit die. Extruder zones were maintained at 180, 190, and 200 °C and the die temperature was 200 °C. Blend morphologies were found to correlate closely with the solubility parameter of the polymer except for polystyrene. A polymer – for example, polybutylene

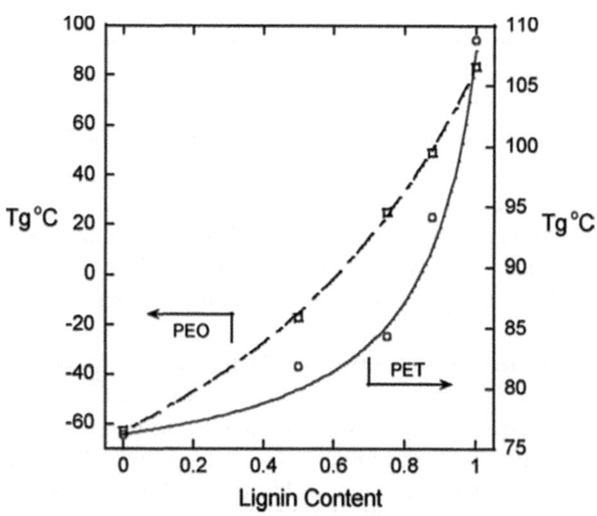

FIGURE 9.9 Variation of T_g with increasing lignin content in miscible blends with polyethylene oxide (PEO), and polyethylene terephthalate (PET). (Reprinted from Kadla, JF, Kubo, S, *Composites: Part A*, 35, 395–400, 2004, with permission from Elsevier.)

succinate-adipate – with a solubility parameter close to that of lignin resulted in a homogeneous blend. Similarly, polyester amide with a solubility parameter of 9.8 $(cal.cm^{-3})^{1/2}$ resulted in a fairly homogeneous blend with lignin. On the other hand, lignin did not blend well with polyethylene having a solubility parameter of 7.7–8.3 $(cal.cm^{-3})^{1/2}$. Polystyrene, though having a solubility parameter of 9.1 $(cal.cm^{-3})^{1/2}$, yielded a homogeneous blend probably due to limited grafting. Lignin was compatible with polyvinyl chloride for blending but not with its plasticized variant. All the blends, however, decreased in homogeneity with increasing lignin content barring the blend with polystyrene. In the latter case, 10% lignin incorporation produced the most compatible blend (Pouteau et al. 2004). A recent study also attempted to correlate the compatibility of pine kraft lignin (Indulin AT) for blending with polylactic acid and polybutylene adipate-co-terephthalate based on their solubility parameters. Maleic anhydride grafted polylactic acid was used – 2% of blend mass – as a coupling agent in the blend. Dry ingredients were added to the mixer in the sequence polylactic acid, coupling agent, polybutylene adipate-co-terephthalate, and lignin and mechanically homogenized for 10 min at 190 °C. Blends were compression moulded into 1 mm thick sheets at 190 °C. The reported solubility parameter for Indulin was 24 $(MPa)^{1/2}$, for polylactic acid 19.5 $(MPa)^{1/2}$, and for polybutylene adipate-co-terephthalate 21.5 $(MPa)^{1/2}$. The resulting blends also presented an immiscible character depicted by distinct T_g for polylactic acid and polybutylene adipate-co-terephthalate. A distinct T_g for lignin was not visible because of its rigid aromatic structure which prevented the detection of phase transformation (Pregi et al. 2023). A blending of 7% softwood kraft lignin with polylactic acid was carried out by compounding the components for 10 min at 60 rpm and 175 °C. The blend had higher Young's modulus but lower tensile strength than pure polylactic acid, indicating increased stiffness due to the presence of lignin. Blending of lignin significantly improved the resistance to photodegradation. Simulated weathering under a mercury lamp for 600 h sharply reduced – almost by 70% – the tensile strength of pure polylactic acid while also recording a drop in Young's modulus. In comparison, the blend could retain about 80% of the tensile strength while recording a small increase in Young's modulus. Improvement in thermal stability was also observed for the blend over pure polylactic acid both before and after weathering (Spiridon et al. 2015). There is an interesting study that reported the effect of softwood kraft lignin particle size and particle morphology on its blending performance with polypropylene. Lignin was subjected to wet ball milling followed by drying. In one instance, the material was dried in a hot air oven. Alternatively, ball-milled lignin was spray-dried. Blending was carried out by premixing the constituents – 10% lignin in the blend – and charging in a twin screw extruder running at 300 rpm and graded temperature zones of 190, 200, 190, 190, 190, and 180 °C. The final blend was obtained as about 0.3 mm thick ribbons after three passes through the extruder. Irrespective of the drying method, and the resulting particle size and morphology, the lignin was found to be incompatible with polypropylene. Blend with oven-dried lignin had higher tensile strength and Young's modulus compared to pure polypropylene. In contrast, tensile strength decreased but Young's modulus increased when polypropylene was blended with spray-dried lignin (Miao and Hamad 2017).

There are several studies reporting chemical modification or derivatization of lignin and blending the modified lignin with a synthetic polymer. One research group derivatized Indulin with either γ-butylolactone or tetrahydrofuran. Reactants were mixed in the ratio of 600 g lignin per L γ-butylolactone or tetrahydrofuran with the addition of 1% sulphuric acid as the catalyst. With γ-butylolactone, the reaction was carried out under stirring in two stages: 1 h at 200 °C and 1 h at 250 °C. With tetrahydrofuran, derivatization under stirring occurred in three stages: 30 min at 50 °C, 30 min at 100 °C, and 1 h at 150 °C. Modified lignin was blended up to 50% with polyethylene terephthalate or polypropylene by kneading for 20 min at 170 °C. The blend was subsequently compression moulded. A blending of tetrahydrofuran-modified lignin with polypropylene reduced the tensile strength but increased the Young's modulus. With the blending of γ-butylolactone derivatized lignin with polyethylene terephthalate, both tensile strength and Young's modulus substantially reduced with increasing lignin content in the blend (Kim et al. 2015). Hydroxypropylated hardwood kraft lignin – 50–90% – was blended with commercial polyurethane by the trade name of Pearlthane

ECO 12T95 in a twin screw microcompounder. The screw speed was 100 rpm with threads extruded through a 1.5 mm opening. The produced blend was again processed through the same compounder at 50 rpm screw speed and extruded through 0.5 mm spinneret to produce carbon fibre precursors. Compounding temperature varied between 175 and 200 °C. Blends were found to be spinnable up to 70% lignin content (Culebras et al. 2018). Eucalyptus kraft lignin as well as its oxypropylated and acetylated derivatives were blended with isotactic polypropylene. For derivatization, lignin was dissolved in 2 M potassium hydroxide. Propylene oxide was added to this solution dropwise to have a final ratio of 1 g lignin per mL propylene oxide. The mixture was stirred for 1 h at 40 °C, then cooled to ambient temperature. The reaction was further allowed to complete in 24 h following which the oxypropylated lignin was precipitated by acidification of the reaction mixture to pH 2.5. For acetylation with maleic anhydride, a solution of 10% maleic anhydride was prepared in acetone. Oxypropylated lignin was mixed with this solution at 5% w/v concentration and refluxed for 7 h at 60 °C. Derivatized lignin was recovered by evaporating the acetone, filtering the solids, water washing, and drying. For acetylation with acetic anhydride, oxypropylated lignin was dissolved in acetic anhydride at 0.5 mg/mL and refluxed for 24 h at 60 °C. The reaction mixture was added to five volumes of water to precipitate the derivatized product. It was filtered, water-washed, and dried. Polypropylene was blended with unmodified kraft lignin or derivatized kraft lignins at 1–5% lignin incorporation. There was negligible change in T_g with blending of lignin or derivatized lignin but a minor reduction in crystallization temperature was observed with increasing lignin content. However, the blend with unmodified lignin had poor lignin dispersion revealed in scanning electron micrographs. All the blends were more hydrophilic than pure polypropylene with the contact angle of water progressively decreasing with increasing content of lignin or its derivative in the blend. The largest drop was noticed with lignin derivative acetylated with maleic anhydride (Bisneto et al. 2022). In another modification scheme, kraft lignin was esterified with oleic acid and the derivatized product was blended with several polymers – polypropylene, polyethylene, or polylactic acid. Oleic acid was mixed with 4-methylpyridine (catalyst) to have 600 g/L concentration and heated to 120 °C. Lignin was added to the solution – 1 kg lignin for 600 g fatty acid – with 10 min stirring followed by the addition of 0.75 L acetic anhydride (activator) per kg lignin. Contents were allowed to react for 1 h at 190 °C. The catalyst and the reaction product acetic acid were evaporated under vacuum and derivatized lignin solvent was extracted with kerosene to remove unreacted oleic acid. Modified lignin was washed with petroleum ether for the final purification step. 20% derivatized lignin was blended with the chosen polymer at 180 °C in a twin screw compounder and, finally, injection moulded into test specimens. Miscible blends were produced with both polypropylene and polyethylene, but phase separation was noticed in the blend with polylactic acid (Orebom et al. 2021).

9.3.2 Organosolv Lignin

There is an early report of blending of Alcell lignin – hardwood aqueous ethanol organosolv lignin having 2200 M_w, 1.7 polydispersity, 3.6 mmol/g aliphatic hydroxyl, 4.3 mmol/g phenolic hydroxyl, and 2.3 syringyl-to-guaiacyl ratio – with polyethylene oxide. Lignin was vacuum dried for 30 min at 160 °C, mechanically mixed with polyethylene oxide, and extruded into fibres through 0.8 mm spinneret at 145–220 °C. Produced fibres did not present a regular variation in mechanical characteristics with increasing lignin content in the blend. However, the produced blends were miscible in nature as manifested by a single T_g that increased with increasing lignin content. Variation of T_g with lignin content followed Gordon Taylor equation (Eq. 9.2) with a K' of 0.37. Most notably, it was possible to spin fibres from pure Alcell lignin without the need to blend with polyethylene oxide (Kubo and Kadla 2004). In another study, Alcell lignin – 50–90% – was blended with commercial polyurethane by the trade name of Pearlthane ECO 12T95 in a twin screw microcompounder. The screw speed was 100 rpm with threads extruded through a 1.5 mm opening. The produced blend was again processed through the same compounder at 50 rpm screw speed and extruded through 0.5 mm spinneret to produce carbon fibre precursors. Compounding temperature varied between 155 and 180 °C. Blends were found to be spinnable up to 70% lignin content (Culebras et al. 2018). Aqueous ethanol organosolv lignin from steam-pretreated corn stover was blended with polymethyl methacrylate either alone or in combination with zinc oxide nanoparticles. The purpose of adding zinc oxide is to act as a photocatalyst that could facilitate the photodegradation of the polymer blend after its intended use. For blending, polymethyl methacrylate was dissolved in dimethylformamide at 0.45 g/mL concentration for 1 h stirring at 60 °C. Lignin – 10% of total blend mass – was added to the solution with mixing for another 1 h at 60 °C. The prepared mix was spread on an aluminium plate and vacuum-dried for 24 h. For blends containing zinc oxide, a suspension of nanoparticles was added, besides the other ingredients, amounting to 5% of the total blend mass. All the blends exhibited very good compatibility of the components resulting in uniform dispersion. Blending of lignin reduced the mechanical characteristics over that of virgin polymethyl methacrylate – Elastic modulus dropping from 2.96 GPa to 2.74 GPa and tensile strength decreasing from 66.9 MPa to 60.5 MPa. The addition of zinc oxide further reduced the elastic modulus to 2.56 GPa and tensile strength to 43.6 MPa (Hu et al. 2021).

A blending of 7% hardwood aqueous acetic acid phosphinic acid organosolv lignin with polylactic acid was carried out by compounding the components for 10 min at 60 rpm and 175 °C. The blend had somewhat higher Young's modulus but lower tensile strength than pure polylactic acid indicating increased stiffness due to the presence of lignin. Blending of lignin significantly improved the resistance to photodegradation. Simulated weathering under a mercury lamp for 600 h sharply reduced – almost by 70% – the tensile strength of pure polylactic acid while also recording a drop in Young's modulus. In comparison, the blend could retain about 80% of the tensile strength while recording a small increase in Young's modulus. Improvement in thermal stability was also observed for the blend over pure polylactic acid both before and after weathering (Spiridon et al. 2015).

There is a report on the esterification of pine aqueous ethanol organosolv lignin and the blending of the derivatized product with polylactic acid. Lignin was dissolved in butyric anhydride or isobutyric anhydride and reacted for 3 h at 50 °C under stirring and in the presence of N-methylimidazole catalyst. Alternatively, lignin was dissolved in 1,4-dioxane and esterified with crotonic anhydride under similar reaction conditions. Modified lignin was precipitated in excess of water, filtered, washed, and dried. 30% of lignin or its esterified derivative was blended with polylactic acid in a twin screw extruder. Compared to pure polylactic acid, blending of organosolv lignin appreciably reduced the tensile strength without a noticeable change in Young's modulus. Lignin esterified with butyric anhydride or isobutyric anhydride produced blends with a similar reduction in tensile strength but significantly higher Young's modulus. However, these blends exhibited much higher strain at break compared to pure polylactic acid. These observations could be attributed to significantly reduced hydroxyl groups in the two esterified lignins – 0.05 mmol/g aliphatic hydroxyl and 0.1–0.11 mmol/g aromatic hydroxyl – compared to non-derivatized lignin – 2.09 mmol/g aliphatic hydroxyl and 4.59 mmol/g aromatic hydroxyl. Blending of lignin esterified with crotonic anhydride did not appreciably alter the mechanical characteristics (Vila et al. 2016).

9.3.3 Hydrolysis Lignin

Enzymatic hydrolysis lignin (unknown lignocellulose source) from bioethanol production, segregated with 55–300 mesh screens, was blended with granular polylactic acid of M_w 1.66×10^5 and polydispersity 1.95. The process employed was melt blending. Initially, manually mixed materials were blended in a twin screw extruder with seven temperature zones at 165–180 °C, forming pellets which were dried before subsequent injection moulding. Specimens for mechanical testing were injection moulded. The temperature profile of the four moulding zones was set at 185–170 °C decreasing at 5 °C intervals. The moulding nozzle was maintained at 185 °C. In the immiscible blends formed, uniform dispersion of lignin in the polylactic acid matrix was observed for all particle sizes, but the morphology of the resulting blends differed. Blends prepared with lignin, sieved through 55 mesh screen, exhibited decreasing tensile strength with increasing lignin content; 63.8 MPa for no lignin addition to 40.1 MPa at 40% lignin in the blend. However, the Young's modulus increased up to 30% lignin addition before decreasing. Impact strength did not vary much with varying lignin content. The shore hardness of the blend was the highest at 20–30% lignin in the blend (Ouyang et al. 2012). These trends are shown in Figure 9.10.

Enzymatic hydrolysis lignin (unknown lignocellulose source) from bioethanol production, M_w 3259 and polydispersity 2.35, was extracted with sodium hydroxide to remove the residual

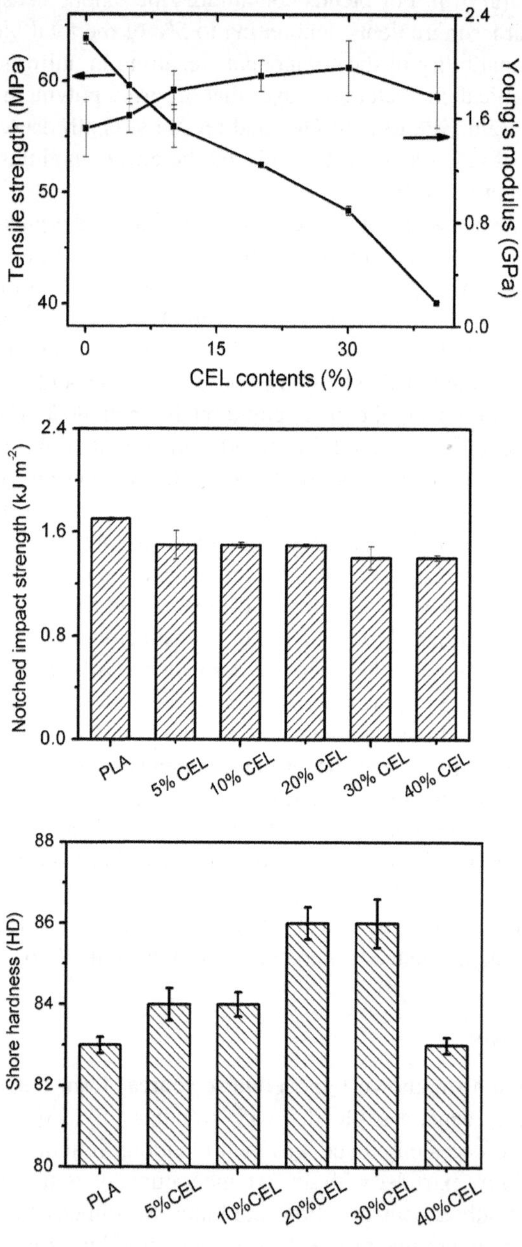

FIGURE 9.10 Mechanical characteristics of blends of poly(lactic acid) with different amounts of enzymatic hydrolysis lignin. (Adapted and reprinted from Ouyang, W, Huang, Y, Luo, H, Wang, D, *J. Poly. Environ.*, 20, 1–9, 2012, with permission from Springer.)

polysaccharides. It was blended with commercial epoxy resin araldite. Desired amounts of lignin and epoxy resin were mixed by vigorous 1 h mixing to prepare a viscous liquid. To this was added 4,4′-diaminodiphenylmethane as the curing agent and stirred for another 1 h. The final blend composition was 75% epoxy, 10% lignin, and 15% curing agent. The mix was spread into preheated mould and cured in two stages: 2 h at 100 °C and 2 h at 150 °C. Cured material had a limiting oxygen index – a measure of flammability – of 24.7 against 23.2 for cured epoxy resin without lignin addition. The burning behaviour of the blended material was compared with the pure resin. Time to ignition decreased from 64 to 43 s with lignin blending. Similarly, total heat released decreased from 108.8 to 57.4 MJ/m^2, the peak for heat release rate decreased from 703.6 to 541.2 kW/m^2, and total smoke produced from 43.2 to 16.3 per m^2. The average effective heat of combustion dropped from 15.07 to 14.66 MJ/kg, and the peak smoke production rate dropped from 0.33 to 0.16 m^2/s. Thus, the blending of lignin appreciably improved the flame retardance characteristics of the material (Lu et al. 2022).

9.3.4 Lignosulphonate

One study reported up to 40% blending of partially desulphonated sodium lignosulphonate (commercial name Vanillex) with poly(ε-caprolactone) thereby producing a blend of complete natural origin. Lignosulphonate and poly(ε-caprolactone) were manually mixed and then compression moulded into 0.2–0.3 mm thick sheets at 150 °C and 5 MPa. Sheets were allowed to season at ambient conditions for 15 days. These sheets had steadily increased Young's modulus and decreased tensile strength with increasing lignosulphonate content in the blend. The authors concluded the lack of compatibility between the blend constituents by analysing the glass transition temperature and storage moduli. However, films cast by dissolving the ingredients in pyridine and evaporating the solvent exhibited compatibility between lignosulphonate and poly(ε-caprolactone) that was attributed to hydrogen bonding through aromatic hydroxyl in lignin and carbonyl group in poly(ε-caprolactone) (Li et al. 2001). Ammonium lignosulphonate was epoxidated to 0.2% equivalent epoxy content for blending with a 30:70 mixture of polyethylene and polypropylene. 2.5–40 weight% modified lignin was blended with the polyolefin mixture along with 5 weight% of a proprietary compatibilizer Exxelor 805 through 10 min open rolling at a temperature of 125 °C and pressure of 120 atm. The highest tensile strength of the blend was obtained at 5% lignin incorporation. The blending of epoxidated lignin facilitated degradation upon burial in soil compared to the blend without lignin addition (Cazacu et al. 2004). Sodium lignosulphonates (unknown lignocellulose source) were blended with polypropylene in 10–70 vol% with the addition of 20% of the mass of lignin maleic anhydride-based proprietary coupling agent to improve lignin interaction with polypropylene. The use of coupling agent was necessitated by the large difference between the Hansen solubility parameters of lignosulphonate (33 (MPa)$^{1/2}$) and polypropylene (16 (MPa)$^{1/2}$). In the absence of a coupling agent, the formed blends were distinctly immiscible with relatively large lignin globules – 10 μm mean size – scattered in a continuous polypropylene matrix. The size of the lignin globules, though reduced considerably from the 80 μm mean size for original lignin, was still large enough to form a sufficiently uniform system. In contrast, when the coupling agent Orevac CA100 was added to the blend, the mean globule size of lignin was reduced to 1 μm. This produced a uniform system with lignin well dispersed in the polypropylene matrix. These behaviours are shown in Figure 9.11 as scanning electron micrographs. As a result of increased lignin polypropylene interactions, the Young's modulus and tensile strengths were both higher with the addition of the coupling agent. The Young's modulus continuously increased with increasing proportion of lignin in the blend. For the blends without the coupling agent, tensile strength progressively decreased with increasing lignin content. When the coupling agent was added, the maximum tensile strength of around 30 MPa was achieved with 30–50% lignin content (Bozsodi et al. 2016).

There is a study on up to 70% blending of softwood calcium lignosulphonate with several commercial synthetic thermoplastics; polypropylene, polystyrene, polymethyl methacrylate, polylactic

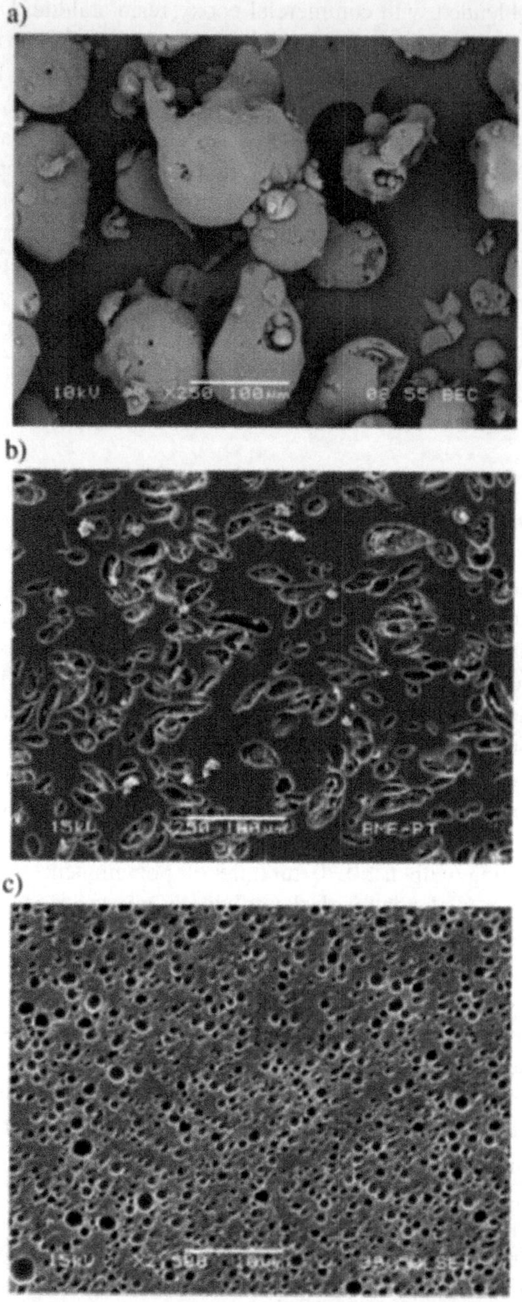

FIGURE 9.11 Scanning electron micrographs of sodium lignosulphonate and its blends with polypropylene with or without coupling agent. (Reprinted from Bozsodi, B, Romhanyi, V, Pataki, P, Kun, D, Renner, K, Pukanszky, B, *Mater. Design*, 103, 32–39, 2016, with permission from Elsevier.)

acid, polycarbonate, glycol-modified polyethylene terephthalate, and an ionomer (copolymer of ethylene and methacrylic acid). In each case, blend components were homogenized for 10 min at 40 rpm. 1 mm thick sheets were compression moulded from the homogenized blend. Homogenization and moulding temperature was 190 °C for all polymers except polylactic acid and polycarbonate. Lignin and polylactic acid were blended and moulded at 180 °C while the temperature was 220 °C for lignin polycarbonate blends. For all blends, Young's modulus increased with increasing lignin content. Variation in tensile strength with increasing lignin content was less regular. With polyethylene terephthalate and polycarbonate, maximum tensile strength was obtained with 10 vol% lignin in the blend while the lignin polystyrene blend had the highest tensile strength at 20 vol% lignin. The tensile strength of the lignin ionomer blend decreased marginally at 10 vol% lignin but gradually increased thereafter. For other blends, increasing lignin content decreased the tensile strength (Romhanyi et al. 2018). These behaviours are presented in Figure 9.12.

The authors also observed poor interaction of lignin with polypropylene and polylactic acid mainly arising from weak dispersion forces. Dispersion forces and hydrogen bonding resulted in partial miscibility in the lignin polymethyl methacrylate blend and dispersion and aromatic interactions were believed to contribute to the interaction between lignin and polystyrene in the blend. Dispersion forces, hydrogen bonding, and aromatic interaction all contributed to component interaction in lignin blends with polycarbonate and polyethylene terephthalate. In the lignin ionomer blend hydrogen bonding and ionic attractions both contributed to the component interaction. These qualitative observations were also correlated with Flory Huggins interaction parameters for the blends, calculated as

$$\chi = \frac{100\left|\left(\delta_L - \delta_P\right)\right|}{RT} \tag{9.3}$$

Here, χ is the interaction parameter, δ_L and δ_P are the solubility parameters of lignin and polymer, respectively, R is the universal gas constant, and T is the absolute temperature. If the two solubility parameters are the same, the interaction parameter is zero indicating complete miscibility. A high value of the interaction parameter denotes poor interaction between the blend constituents and immiscibility. Lignosulphonate ionomer blend had the lowest interaction parameter of 1.2 indicating strong interaction and a near miscible blend. Interaction parameters for the blend with polypropylene and polylactic acid were 11.8 and 7.2, respectively. These blends were, therefore, practically immiscible.

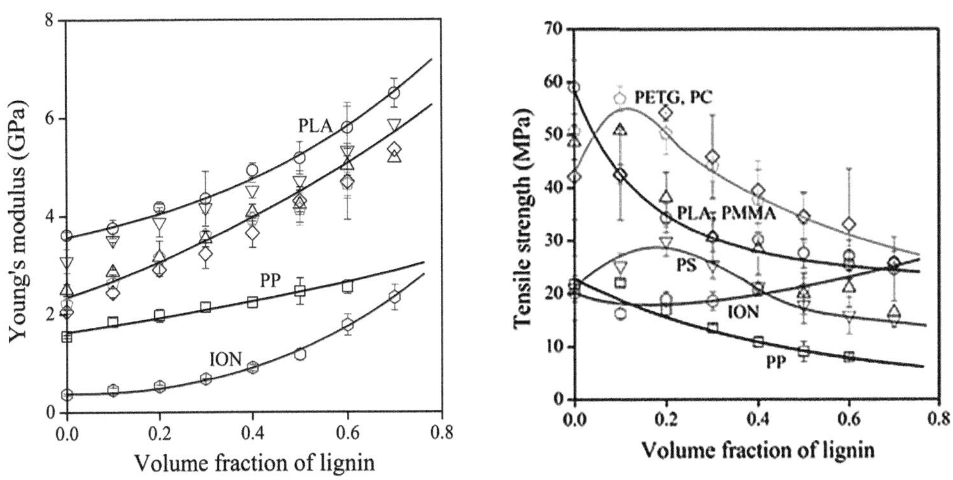

FIGURE 9.12 Variation in mechanical characteristics of blends of softwood calcium lignosulphonate with different thermoplastics with varying lignin content. Polypropylene (PP), polystyrene (PS), poly(methyl methacrylate) (PMMA), poly(lactic acid) (PLA), polycarbonate (PC), glycol-modified poly(ethylene terephthalate) (PETG), and ionomer (ION). (Adapted and reprinted from Romhanyi, V, Kun, D, Pukanszky, B, *ACS Sustainable Chem. Eng.* 6, 14323–14331, 2018, with permission from American Chemical Society.)

REFERENCES

Alonso, MV, Rodriguez, JJ, Oliet, M, Rodriguez, F, Garcia, J, Gilarranz, MA. *J. Appl. Poly. Sci.* 82: 2661–2668, 2001.

Alonso, MV, Oliet, M, Pérez, JM, Rodriguez, F, Echeverria, J. *Thermochimica Acta* 419:161–167, 2004.

Antonino, LD, Gouveia, JR, de Sousa Jr., RR, Garcia, GES, Gobbo, LC, Tavares, LB, dos Santos, DJ. *Molecules* 26:2131, 2021.

Bagheri, S, Nejad, M. *Poly. Composites* 44:3926–3938, 2023.

Benar, P, Goncalves, AR, Mandelli, D, Schuchardt, U. *Bioresource Technol.* 68:11–16, 1999.

Bergamasco, S, Tamantini, S, Zikeli, F, Vinciguerra, V, Mugnozza, GS, Romagnoli, M. *Polymers* 14:416, 2022.

Bisneto, MPD, Gouveia, JR, Antonino, LD, Tavares, LB, Ito, NM, dos Santos, DJ. *Polymers* 14:999, 2022.

Bozsódi, B, Romhányi, V, Pataki, P, Kun, D, Renner, K, Pukánszky, B. *Mater. Design* 103:32–39, 2016.

Cateto, CA, Barreiro, MF, Rodrigues, AE. *Ind. Crops Prod.* 27:168–174, 2008.

Cateto, CA, Barreiro, MF, Rodrigues, AE, Belgacem, MN. *Ind. Eng. Chem. Res.* 48:2583–2589, 2009.

Cazacu, G, Pascu, MC, Profire, L, Kowarski, AI, Mihaes, M, Vasile, C. *Ind. Crops Prod.* 20:261–273, 2004.

Chauhan, M, Gupta, M, Singh, B, Singh, AK, Gupta, VK. *Eur. Poly. J.* 52:32–43, 2014.

Chen, Y, Zhang, H, Zhu, Z, Fu, S. *Int. J. Biologic. Macromolecules* 152:775–785, 2020.

Culebras, M, Beaucamp, A, Wang, Y, Clauss, M, Frank, E, Collins, MN. *ACS Sustainable Chem. Eng.* 6: 8816–8825, 2018.

El Mansouri, NE, Yuan, Q, Huang, F. *BioResources* 6:2492–2503, 2011.

Evtuguin, Andreolety, JP, Gandini, *A. Eur. Poly. J.* 34(8):1163–1169, 1998.

Feldman, D, Lacasse, M, Manley, RSJ. *J. Appl. Poly. Sci.* 35:247–257, 1988.

Ferdosian, F, Yuan, Z, Anderson, M, Xu, C. *RSC Adv.* 4:31745, 2014.

Ferdosian, F, Yuan, Z, Anderson, M, Xu, C. *Thermochimica Acta* 618:48–55, 2015.

Ferdosian, F, Yuan, Z, Anderson, M, Xu, C. *Ind. Crops Prod.* 91:295–301, 2016.

Galdino, DS, Kondo, MY, Araujo, VAD, Ferrufino, GLAA, Faustino, E, dos Santos, HF, Christoforo, AL, Luna, CMR, de Campos, CI. *Polymers* 15:357, 2023.

Gioia, C, Colonna, M, Tagami, A, Medina, L, Sevastyanova, O, Berglund, LA, Lawoko, M. *Biomacromolecules* 21:1920–1928, 2020.

Gondaliya, A, Nejad, M. *Molecules* 26:2302, 2021.

Gouveia, JR, Garcia, GES, Antonino, LD, Tavares, LB, dos Santos, DJ. *Molecules* 25:2513, 2020.

Hayati, AN, Evans, DAC, Laycock, B, Martin, DJ, Annamalai, PK. *Ind. Crops Prod.* 117:149–158, 2018.

Hu, C, Zhao, M, Li, Q, Liu, Z, Hao, N, Meng, X, Li, J, Lin, F, Li, C, Fang, L, Dai, SY, Ragauskas, AJ, Sue, HJ, Yuan, JS. *ChemSusChem* 14:4260–4269, 2021.

Hussin, MH, Zhang, HH, Aziz, NA, Samad, NA, Faris, AH, Abrahim, MNM, Iqbal, A, Latip, AFA, Haafiz, MKM. *Beni-Suef Univ. J. Basic Appl. Sci.* 6:409–418, 2017.

Jia, Z, Lu, C, Zhou, P, Wang, L. *RSC Adv.* 5:53949–53955, 2015.

Jin, Y, Cheng, X, Zheng, Z. *Bioresource Technol.* 101:2046–2048, 2010.

Kadla, JF, Kubo, S. *Macromolecules* 36:7803–7811, 2003.

Kadla, JF, Kubo, S. *Composites: Part A* 35:395–400, 2004.

Kim, S, Park, J, Lee, J, Roh, HG, Jeong, D, Choi, S, Oh, S. *Fibres Polymers* 16:744–751, 2015.

Kubo, S, Kadla, JF. *Macromolecules* 37:6904–6911, 2004.

Kumari, S, Chauhan, GS, Monga, S, Kaushik, A, Ahn, JH. *RSC Adv.* 6:77768–77776, 2016.

Li, Y, Ragauskas, AJ. *J. Wood. Chem. Technol.* 32:210–224, 2012.

Li, J, He, Y, Inoue, Y. *Polymer J.* 33:336–343, 2001.

Li, J, Zhang, J, Zhang, S, Gao, Q, Li, J, Zhang, W. *Polymers* 9:428, 2017a.

Li, J, Wang, B, Chen, K, Tian, X, Zeng, J, Xu, J, Gao, W. *BioResources* 12:8653–8671, 2017b.

Li, RJ, Gutierrez, J, Chung, YL, Frank, CW, Billington, SL, Sattely, ES. *Green Chem.* 20:1459–1466, 2018.

Lu, X, Zhu, X, Dai, P, Robin, HM, Guo, H, Que, H, Wang, D, Liang, D, He, T, Xu, C, Luo, Z, Gu, X. *J. Therm. Anal. Calorim.* 147:5237–5253, 2022.

Luo, S, Gao, L, Guo, W. *J. Wood Sci.* 66:23, 2020.

Malutan, T, Nicu, R, Popa, VI. *BioResources* 3:1371–1376, 2008.

Miao, C, Hamad, WY. *J. Appl. Polym. Sci.* 134:44669, 2017.

Mimini, V, Amer, H, Hettegger, H, Bacher, M, Gebauer, I, Bischof, R, Fackler, K, Potthast, A, Rosenau, T. *Holzforschung* 74:203–211, 2019.

Mohammadpour, R, Sadeghi, GMM. *J. Polymer Res.* 27:362, 2020.

Muller, PC, Glasser, WG. *J. Adhesion* 17:157–174, 1984.

Muller, PC, Kelley, SS, Glasser, WG. *J. Adhesion*, 17:185–206, 1984.

Muller, LC, Marx, S, Vosloo, HCM, Chiyanzu, I. *Polymers. Renew. Resources* 10(1-3):3–18, 2019.
Nikafshar, S, Wang, J, Dunne, K, Sangthonganotai, P, Nejad, M. *ChemSusChem* 14:1184–1195, 2021.
Orebom, A, Francesco, DD, Shakari, P, Samec, JSM, Pierrou, C. *Molecules* 26:3219, 2021.
Ott, MW, Dietz, C, Trosien, S, Mehlhase, S, Bitsch, MJ, Nau, M, Meckel, T, Geissler, A, Siegert, G, Huong, J, Hertel, B, Stark, RW, Biesalski, M. *Holzforschung* 75:390–398, 2021.
Ouyang, W, Huang, Y, Luo, H, Wang, D. J. Poly. Environ. 20:1–9, 2012.
Pan, X, Saddler, JN. *Biotechnol Biofuels* 6:12, 2013.
Peng, Z,Chen, F. *Int. J. Poly. Mater.* 60:674–683, 2011.
Petreus, O, Cazacu, G, Necula, AM, Ciolacu, D. *Cellulose Chem. Technol.* 42:569–576, 2008.
Pinheiro, FGC, Soares, AKL, Santaella, ST, e Silva, LMA, Canuto, KM, Cáceres, CA, Rosa, MD, Feitosa, JPD, Leitão, RC. *Ind. Crops Prod.* 96:80–90, 2017.
Pouteau, C, Baumberger, S, Cathala, B, Dole, P. *Comptes Rendus Biologies* 327:935–943, 2004.
Pregi, E, Romsics, I, Várdai, R, Pukánszky, B. *Polymers* 15:3237, 2023.
Romhanyi, V, Kun, D, Pukánszky, B. *ACS Sustainable Chem. Eng.* 6:14323–14331, 2018.
Saito, T, Perkins, JH, Jackson, DC, Trammel, NE, Hunt, MA, Naskar, AK. *RSC Adv.* 3:21832–21840, 2013.
Sallem-Idrissi, N, Sclavons, M, Debecker, DP, Devaux, J. *J. Appl. Polym. Sci.* 133:42963, 2016.
Sameni, J, Jaffer, SA, Sain, M. *Composites: Part A* 115:104–111, 2018.
Sarkar, S, Adhikari, B. *J. Adhesion Sci. Technol.*, 14:1179–1193, 2000.
Sasaki, C, Wanaka, M, Takagi, H, Tamura, S, Asada, C, Nakamura, Y. *Ind. Crops Prod.* 43:757–761, 2013.
Spiridon, I, Leluk, K, Resmerita, AM, Darie, RN. *Composites: Part B* 69:342–349, 2015.
Sun, G, Sun, H, Liu, Y, Zhao, B, Zhu, N, Hu, K. *Polymer* 48:330–337, 2007.
Vazquez, G, Antorrena, G, Gonzalez, J, Mayor, J. *Bioresource Technol.* 51:187–192, 1995.
Vazquez, G, Gonzalez, J, Freire, S, Antorrena, G. *Bioresource Technol.* 60:191–198, 1997.
Vila, C, Santos, V, Saake, B, Parajo, JC. *BioResources* 11:5322–5332, 2016.
Wang, S, Liu, W, Yang, D, Qiu, X. *Ind. Eng. Chem. Res.* 58:496–504, 2019.
Xie, J, Sun, H, Yang, Y, Liang, J, Li, Y, Hou, D, Lin, X, Zhang, J, Shi, Z, Liu, C. *Int. J. Mol. Sci.* 24:6418, 2023.
Yang, W, Rallini, M, Natali, M, Kenney, J, Ma, P, Dong, W, Torre, L, Puglia, D. *Mater. Design* 161:55–63, 2019.
Yin, Q, Di, M. *Adv. Mater. Res.* 482-484:1959–1962, 2012.
Yin, Q, Yang, W, Sun, C, Di, M. *BioResources* 7:5737–5748, 2012.
Zhang, W, Ma, Y, Wang, C, Li, S, Zhang, M, Chu, F. *Ind. Crops Prod.* 43:326–333, 2013.
Zhang, N, Li, Z, Xiao, Y, Pan, Z, Jia, P, Feng, G, Bao, C, Zhou, Y, Hua, L. *J. Bioresources Bioprod.* 5: 67–77, 2020.
Zhang, Y, Wang, H, Eberhardt, TL, Gu, Q, Pan, H. *Eur. Poly. J.* 150:110389, 2021.
Zhao, B, Chen, G, Liu, Y, Hu, K, Wu, R. *J. Mater. Sci. Lett.* 20:859–862, 2001.

10 Value-Added Low Molecular Weight Organics from Lignin

Native lignin macromolecule is partially depolymerized and chemically modified during its isolation. However, the isolated lignin retains much of the original macromolecular character and the functional groups on the aromatic ring and the aliphatic side chain. It is possible to further depolymerize the lignin with extensive breaking of intermonomer linkages and releasing the constituent monomers. Derivatization often accompanies such extensive depolymerization. The result is the production of several low molecular weight compounds. Considering the phenylpropane-based molecular structure of lignin, many of these compounds are aromatic in nature carrying the functional groups originally present in the lignin macromolecule. New functional groups may also arise along with some ring opening, producing aliphatic compounds. Therefore, the product mix would consist of phenols, aromatic carbonyl compounds, aromatic carboxylic acids, aliphatic carboxylic acids, and aliphatic and aromatic hydrocarbons. Important among these are vanillin, syringaldehyde, apocynin, acetosyringone, p-hydroxybenzaldehyde, vanillic acid, syringic acid, benzene, toluene, xylene, acetic acid, formaldehyde, and methanol. Compounds like vanillin, syringaldehyde, apocynin, and acetosyringone are speciality chemicals with high economic value. They have important uses in food, pharmaceutical, and personal care formulations. Similarly, benzene, toluene, xylene, acetic acid, formaldehyde, and methanol are industrially important chemicals. The current major source of all these chemicals is petroleum. Producing them from lignin opens up a renewable and sustainable alternative.

Oxidative degradation of lignin has been the principal route to produce low molecular weight organic compounds from lignin. Oxidative degradation of lignin is possible with alkaline nitrobenzene, producing the corresponding aromatic carbonyl compounds in a reasonably high yield. However, the use of nitrobenzene is discouraged due to environmental implications. There have been several attempts to oxidize lignin with molecular oxygen in the presence of a suitable catalyst. Transition metal oxides are mostly used to catalyse these reactions and the product composition is similar to that obtained with alkaline nitrobenzene. In recent times, research has focussed on the use of advanced oxidation processes like electrochemical oxidation or photocatalytic oxidation of lignin to produce low molecular organic chemicals. Yet another viable route to produce low molecular weight organic chemicals from lignin is fast pyrolysis. Lignin can be pyrolysed, in the absence or presence of catalysts, to produce a pyrolysis oil that consists of a variety of low molecular weight organic compounds including phenols, carbonyls, acids, alcohols, and hydrocarbons.

All oxidative lignin degradation routes to its valorization essentially involve multiple reactions involving a combination of series and parallel reaction pathways. The target low molecular weight compounds are themselves susceptible to oxidation. It is, therefore, imperative to have close control over reaction conditions. Another key challenge is the separation and purification of the desired compounds from the reaction mixture.

10.1 CATALYTIC OXIDATION OF LIGNIN WITH MOLECULAR OXYGEN

The process is essentially an extension of wet air oxidation practised for wastewater treatment for the oxidative degradation of recalcitrant organic pollutants. An aqueous solution of lignin, mostly under alkaline conditions, is subjected to reaction with molecular oxygen in the presence of a suitable catalyst. Molecular oxygen can be supplied in the form of air or as pure or enriched oxygen.

DOI: 10.1201/9781003598664-13

The product mix after the reaction mainly consists of aromatic aldehydes, ketones, and carboxylic acids. When used with softwood lignin, which is almost exclusively of guaiacyl-type, the major reaction product is vanillin. Some apocynin and vanillic acid are also produced.

10.1.1 Reaction Mechanism and Kinetics

B-O-4′ linkages, the main intermonomer linkage in lignin fragments, are cleaved in the presence of hydroxyl ions at an elevated temperature according to Scheme 1 of Figure 2.2 releasing monomeric lignin fragments. The presence of oxygen, especially in the presence of catalysts, further assists the breaking of intermonomer linkages in lignin. Complete breaking of B-O-4′ linkages, resinol entities, and p-hydroxycinnamyl entities was reported for beech ethanol organosolv lignin when oxidized with molecular oxygen at 10 bar pressure and 135 °C for 40 h. The reaction was carried out in a dissolved state in pyridine at a lignin concentration of 22 g/L. Either 20% copper and vanadium-impregnated hydrotalcites or a mixture of 5% cupric nitrate and 5% vanadium acetate was used as the catalyst (Mottweiler et al. 2015). Several radical and ionic species are produced in an aqueous alkaline system containing dissolved oxygen and lignin ultimately leading to the reduction of molecular oxygen to water. These reactions can be schematically expressed as (Asgarai and Argyropoulos 1998)

$$O_2 + H\bullet \rightarrow HO_2\bullet \qquad \{10.1a\}$$

$$HO_2\bullet + H\bullet \rightarrow H_2O_2 \qquad \{10.1b\}$$

$$H_2O_2 + H\bullet \rightarrow HO\bullet + H_2O \qquad \{10.1c\}$$

$$HO\bullet + H\bullet \rightarrow H_2O \qquad \{10.1d\}$$

Atomic hydrogen driving these reduction reactions may be abstracted from the monomeric lignin fragments from their phenolic hydroxyl (present as phenolate anion in an alkaline system). The resulting radicals from the monomeric lignin fragments can further react and rearrange to produce aromatic aldehydes and other products (Tarabanko et al. 2004). This is represented in Figure 10.1. Further, the oxidant species generated in reaction 10.1 can also undergo secondary reactions producing other radical and ionic species according to the following reactions:

$$O_2 + OH^- \rightarrow O_2^-\bullet + HO\bullet \qquad \{10.2a\}$$

$$O_2^-\bullet + HO_2\bullet \rightarrow O_2 + HO_2^- \qquad \{10.2b\}$$

$$H_2O_2 + HO_2^- \rightarrow O_2^-\bullet + HO\bullet + H_2O \qquad \{10.2c\}$$

In the scheme shown, the quinone methide radical, initially formed by the reaction of molecular oxygen with the phenolate anion, rearranges with the unpaired electron shifting to the β carbon on the side chain. It is then attacked by the superoxide anion radical $O_2^-\bullet$ which ultimately leads to the formation of the corresponding benzaldehyde. Alternatively, the quinone methide radical can also rearrange with the shift of unpaired electrons to carbon atoms at position 1, 3, or 5 on the aromatic ring. These resonant structures could result in the formation of reaction products other than the benzaldehyde as shown in Figure 10.1.

It is believed that the vanillin precursor in the alkaline lignin solution converts to vanillin which in turn is further converted to other end products. For pine kraft lignin, the net rate of vanillin production could be expressed as (Fargues et al. 1996)

$$r_V = k_1 C_{O_2}^{1.75} C_L - k_2 C_{O_2} C_V \quad pH > 11.5 \qquad (10.1)$$

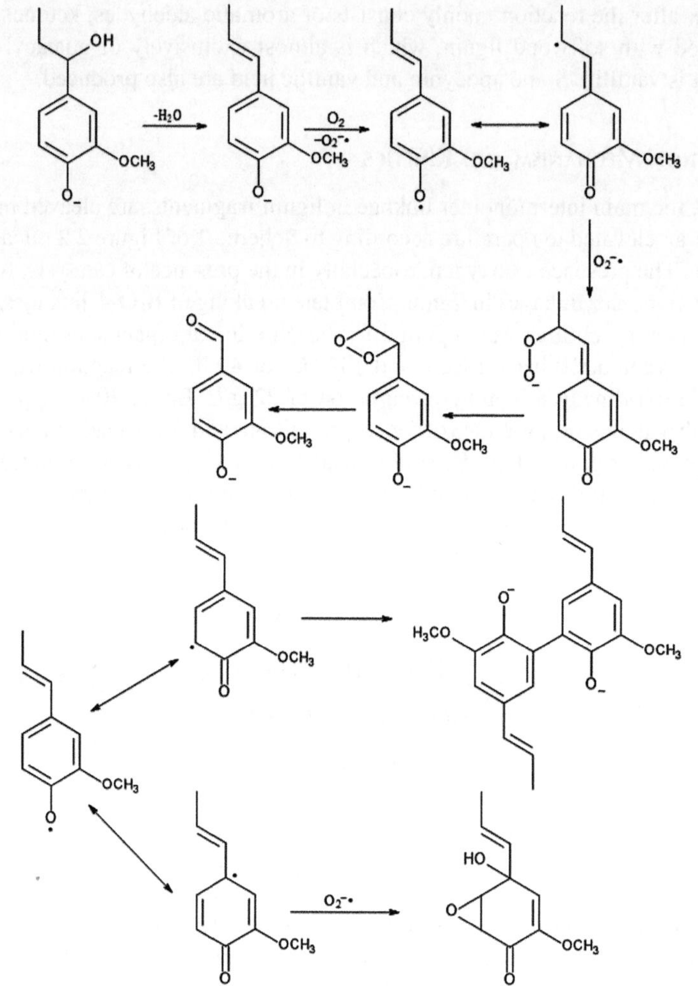

FIGURE 10.1 Reaction pathways in the conversion of lignin to low molecular weight organic compounds with molecular oxygen.

$$r_V = k_1 C_{O_2}^{1.75} C_L - A\left[\frac{BC_{H^+}}{1+BC_{H^+}}\right]^2 C_V^2 \quad \text{pH}<11.5 \tag{10.2}$$

Here, r_V is the net rate of vanillin production, C_{O_2}, C_L, C_V, and C_{H^+} the respective concentrations of dissolved oxygen, lignin, vanillin, and hydrogen ions, k_1 and k_2 are rate constants, and A and B are empirical constants. The temperature-dependent rate constants could be expressed with Arrhenius law with 29.1 kJ/mol activation energy for lignin conversion and 46 kJ/mol activation energy for vanillin oxidation. The respective equations can be written as

$$k_1 = 1.376\times 10^7 \exp\left[-\frac{3502}{T}\right] \tag{10.3}$$

$$k_2 = 4.356\times 10^6 \exp\left[-\frac{5530}{T}\right] \tag{10.4}$$

In another study, the rate of vanillin production from spruce sodium lignosulphonate was found to be first-order in oxygen partial pressure (Pacek et al. 2013). Experiments on bagasse hydrolysis lignin, assuming pseudo first-order reaction with respect to lignin concentration, estimated the activation energies for the formation of vanillin and syringaldehyde to be 63.64 and 66.04 kJ/mol, respectively. Subsequently, these aldehydes oxidized to simpler end products with activation energies of 10.18 and 35.97 kJ/mol (Sales et al. 2006).

10.1.2 Yield and Product Profile

10.1.2.1 Kraft Lignin

An early study reported the production of vanillin from the oxidative degradation of pine kraft lignin with molecular oxygen. An alkaline lignin solution of 30–100 g/L concentration in 40–160 g/L aqueous sodium hydroxide was used. The reaction system was heated to 100–140 °C at 6 bar pressure maintained with nitrogen. Thereafter, oxygen was introduced, increasing the pressure to 9.7 bar. Vanillin concentration reached its maximum in the reaction mixture between 40 and 70 min after oxygen addition. A maximum vanillin concentration of 6 g/L was achieved at 140 °C temperature after about 70 min oxidation with 60 g/L lignin concentration and 80 g/L sodium hydroxide concentration. Increasing lignin concentration from 30 to 60 g/L increased vanillin production but further increase in lignin concentration had a negative effect on vanillin yield. Higher alkali concentration in the 40–160 g/L range consistently improved the vanillin yield. Minor quantities of apocynin as a byproduct were also obtained (Mathias and Rodrigues 1995). Similarly, oxidative degradation of pine kraft lignin, without any added catalyst, was performed with lignin dissolved in 2 M sodium hydroxide. The oxygen partial pressure varied between 1.2 and 5 bar in a total pressure of 9.7 bar maintained with nitrogen. The minimum vanillin yield was 6.3% at 120 g/L lignin concentration, 50 min reaction time, 120 °C temperature, and 2.5 bar oxygen partial pressure. Vanillin yield increased to a maximum of 10.8% when the reaction conditions were 60 g/L lignin concentration, 35 min reaction time, 133 °C temperature, and 2.8 bar oxygen partial pressure. Authors noted that increasing oxygen partial pressure, reaction temperature, and reaction time beyond the optimal level decreased the vanillin yield indicating its subsequent oxidation to undesirable end products (Fargues et al. 1996). Oxidation of pine kraft lignin – 2.3% carbohydrate, 2.6% ash, and 97:3 guaiacyl to p-hydroxyphenyl ratio – with molecular oxygen is reported. Lignin was dissolved in aqueous sodium hydroxide solution to have 60 g/L lignin and 2 M sodium hydroxide concentration. The reaction temperature was 120 °C with 3 bar oxygen partial pressure and 6.7 bar nitrogen partial pressure. A maximum vanillin concentration of nearly 5 g/L was obtained after a 25–30 min reaction time after which the vanillin concentration decreased. The maximum vanillic acid concentration was about 1 g/L, obtained between 30 and 35 min reaction time. Apocynin concentration remained less than 0.5 g/L. pH of the reaction system, close to 14 at the beginning, consistently dropped due to the production of organic acids and settled at 10 after a 2 h reaction (Pinto et al. 2011). The time profiles of different reaction products are shown in Figure 10.2.

In a modified procedure, 500 mg Indulin AT was dissolved in dimethylformamide at 50 g/L concentration and oxidized with molecular oxygen flowing at 10 mL/min. The reaction was carried out for 24 h at 120 °C and 500 rpm agitation in the presence of 40% supported gold nanoparticles as a heterogeneous catalyst. 83% oxidized lignin was obtained after the separation of the catalyst and vacuum evaporation of the solvent. 50 mg of oxidized lignin was dissolved in 1 M sodium hydroxide at 10 g/L concentration by 30 min stirring. Subsequently, the solution was acidified to pH 2 and the mixture extracted three times with ethyl acetate. The total yield of low molecular weight organics as an extract in ethyl acetate was 16.6% of the starting lignin (20% of oxidized lignin). Of these, 3.4% was vanillin, 3.4% vanillic acid, 1.2% coniferyl alcohol, 1.1% ferulic acid, and 0.4% apocynin (Song et al. 2018).

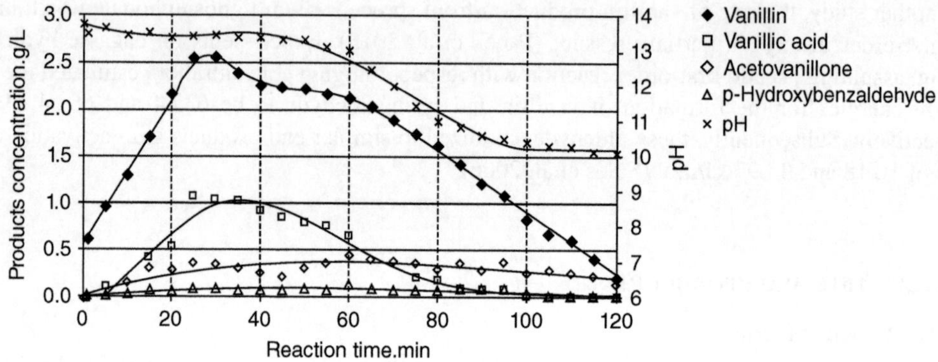

FIGURE 10.2 Time profile of production of low molecular weight oxygenated aromatic compounds from oxidative degradation of pine kraft lignin with molecular oxygen. (Adapted and reprinted from Pinto, PCR, da Silva, EAB, Rodrigues, AE, *Ind. Eng. Chem. Res.*, 50, 741–748, 2011, with permission from American Chemical Society.)

Eucalyptus kraft lignin was dissolved in 0.5–2 M aqueous sodium hydroxide and oxidized at 10–15 atm oxygen partial pressure and 170–190 °C for 0–120 min. Syringaldehyde was the main carbonyl product (around 3% maximum yield) and syringic acid was the main carboxylic acid produced (around 1.8% maximum yield). Vanillin (around 0.8% yield) and vanillic acid (around 0.4% yield) were also produced in reasonable amounts. The study found that increasing alkali concentration increased product yield. Syringaldehyde and vanillin yields improved with increasing oxygen partial pressure, but carboxylic acid yields were not appreciably affected by oxygen pressure. 180 °C was found to be the optimal temperature (Villar et al. 2001). Variation in product yield with oxidation parameters is shown in Figure 10.3.

There is a report of oxidation of Indulin AT with molecular oxygen in a continuous pilot scale operation in a bubble column or a packed column. The column packing was made of a corrugated metal sheet having 750 m^2/m^3 specific surface area and 0.95 voidage. Lignin was dissolved in 2 M aqueous sodium hydroxide to give 60 g/L lignin concentration and pumped upwards through the packed column at 0.99–2.12 L/h and through the bubble column at 2.12–2.5 L/h. Nitrogen was fed into the column bottom at 1000 mL/min. The oxygen feed rate to the bottom of the packed column was 1000 mL/min and to the bottom of the bubble column was 1000–2000 mL/min. Total pressure inside the reactor was maintained at 10 bar. Both the reactors reached their steady state operation in 6 h. The steady-state exit concentration of vanillin from the bubble column was 0.56 g/L at 2.12 L/h liquid flow rate and 1000 mL/min oxygen flow rate with reactor temperature between 167 and 170 °C. Vanillin exit concentration increased to 0.67 g/L at 2.5 L/h liquid flow rate and 2000 mL/min oxygen flow rate with reactor temperature between 159 and 163 °C. The packed-column reactor resulted in a much higher exit concentration of vanillin: 0.73 g/L at 2.12 L/h liquid flow rate with reactor temperature of 170 °C, and 0.89 g/L at 0.99 L/h liquid flow rate with reactor temperature between 155 and 160 °C. It was contended that the transfer of oxygen from the gas to the liquid phase was the rate-limiting step and improved mass transfer in the packed column improved the vanillin yield. Indeed, the vanillin concentration at the exit of the packed-column reactor increased to 1.8 g/L when the nitrogen flow was replaced with oxygen (da Silva et al. 2009).

While lignin oxidation with molecular oxygen in an alkaline medium is the overwhelming norm, there have been attempts to oxidize lignin under acidic conditions. In one such study, Indulin was suspended in aqueous methanol at 10 g/L concentration with the addition of 0.025 M phosphomolybdic acid – oxidation catalyst – to have a pH of 1.13. Oxidation was carried out with 10.8 bar oxygen partial pressure at 170 °C for 20 min. Vanillin and methyl vanillate were the principal low molecular weight aromatic compounds produced, recovered with about 3.5% yield each (Voitl and

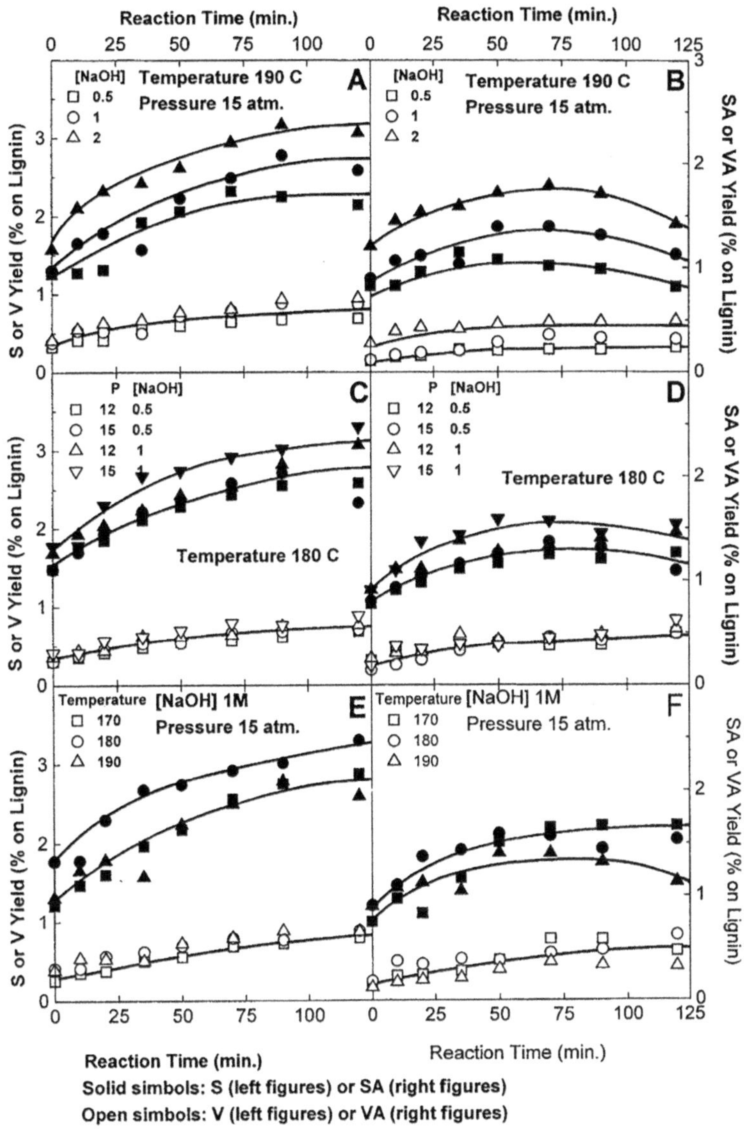

FIGURE 10.3 Effect of oxidation conditions on the production of low molecular weight oxygenated aromatic compounds from eucalyptus kraft lignin with molecular oxygen. (Reprinted from Villar, JC, Caperos, A, Garcia-Ochoa, F, *Wood Sci. Technol.*, 35, 245–255, 2001, with permission from Springer.)

von Rohr 2010). Similarly, Indulin was oxidized as a suspension at 10 g/L concentration in 80% vol/vol aqueous methanol at pH 1 maintained by the addition of sulphuric acid. Oxidation was catalysed by several transition metal salts dissolved in the reaction medium. The reaction temperature was 170 °C, the oxygen partial pressure was 10 bar, and the reaction time was 2 h. Vanillin yield varied between 2.41% and 3.19% and methyl vanillate between 2.33% and 3.15% after 1 h reaction time (Werhan et al. 2011). In continuous experiments, two liquid streams were fed into a capillary microreactor along with an oxygen stream to maintain 32–96 bar oxygen partial pressure. One liquid stream consisted of 80% vol/vol aqueous methanol with Indulin suspended at 5 g/L. The other liquid stream consisted of 80% vol/vol aqueous methanol with 0.2 M sulphuric acid. At equal volumetric flow rates between 25 and 200 μL/min, the liquid streams were mixed in the reactor to have 2.5 g/L lignin concentration and 0.1 M sulphuric acid concentration. Vanillin was the main reaction product; its maximum yield being 1.6% for 48 bar oxygen pressure, 190 °C temperature, and 6 min space time. The other reaction products were methyl vanillate, 5-carboxymethylvanillin, methyl 5-carboxymethyl vanillate, and methyl dehydroabietate. The maximum vanillin yield reached 3% at 210 °C temperature and 1 min space time. The study estimated the activation energy for vanillin formation to be 170.8 kJ/mol and that for vanillin degradation to be 56.3 kJ/mol (Werhan et al. 2013).

10.1.2.2 Organosolv Lignin

Mixed hardwood aqueous acetic acid organosolv lignin was oxidized with molecular oxygen – supplied as air – in the presence of a mixture of metal acetate catalysts at 6.9–30 mM concentration. The reaction was carried out for 1–10 h at 100–220 °C with 69–138 bar air pressure and 3.8–38% lignin concentration dissolved in aqueous acetic acid. The main products of oxidative degradation were vanillin, syringaldehyde, vanillic acid, and syringic acid. The maximum total yield was 10.9% with a 2 h reaction at 180 °C or 8 h reaction at 140 °C using 138 bar air pressure, 8% lignin concentration, and 6.9 mM concentration of mixed metal acetate catalysts. The maximum vanillin yield was 1.36% and the maximum syringaldehyde yield was 3.48% with a 2 h reaction at 140 °C, 69 bar air pressure, 8% lignin concentration, and 6.9 mM concentration of mixed metal acetate catalysts. The maximum vanillic acid yield was 2.91% and the maximum syringic acid yield was 4.51% with 2 h reaction at 180 °C, 138 bar air pressure, 8% lignin concentration, and 6.9 mM concentration of mixed metal acetate catalysts. It is clearly evident that stronger reaction conditions in terms of reaction temperature and oxygen pressure favour the production of corresponding carboxylic acid over the aldehyde (Partenheimer 2009).

In a modified procedure, 500 mg maple γ-valerolactone organosolv lignin was dissolved in dimethylformamide at 50 g/L concentration and oxidized with molecular oxygen flowing at 10 mL/min. The reaction was carried out for 24 h at 120 °C and 500 rpm agitation in the presence of 40% supported gold nanoparticles as a heterogeneous catalyst. 91% oxidized lignin was obtained after the separation of the catalyst and vacuum evaporation of the solvent. 50 mg of oxidized lignin was dissolved in 1 M sodium hydroxide at 10 g/L concentration by 30 min stirring. Subsequently, the solution was acidified to pH 2 and the mixture extracted three times with ethyl acetate. The total yield of low molecular weight organics as an extract in ethyl acetate was 51% of the starting lignin (56% of oxidized lignin). Of these, 2.8% was vanillin, 7.8% vanillic acid, 2.8% coniferyl alcohol, 7.8% ferulic acid, 11.2% syringaldehyde, 5% syringic acid, and 2.8% sinapic acid (Song et al. 2018). In a related study, cotton stalk aqueous ethanol organosolv lignin was dissolved in methanol at 4 g/L concentration and oxidized with 1 MPa molecular oxygen partial pressure at 180 °C for 4 h in the presence of 20% supported gold catalyst. The major low molecular weight organics obtained were vanillin 10.5%, methyl vanillate 6.8%, and methyl syringate 3.4% (Song et al. 2019).

10.1.2.3 Hydrolysis Lignin

Hydrolysis lignin, isolated from sugarcane bagasse, was subjected to catalytic oxidation with molecular oxygen to produce vanillin, syringaldehyde, and p-hydroxybenzaldehyde. Batch experiments were performed with lignin dissolved at 60 g/L concentration in 2 M sodium hydroxide with 2.85% of alumina-supported palladium catalyst. The reaction temperature was 100–140 °C and an oxygen partial pressure of 2–10 bar of a total pressure of 20 bar was maintained with nitrogen. Increasing temperature and oxygen partial pressure allowed for the early attainment of higher product concentration in the reaction mixture. Maximum vanillin, syringaldehyde, and p-hydroxybenzaldehyde concentrations of about 18, 20, and 36 mmol/L, respectively, were achieved after 12, 10, and 45 min reaction time at 140 °C and 5 bar oxygen partial pressure. The authors also performed a continuous oxidation reaction with 30 g/L lignin concentration and 4 bar oxygen partial pressure in a fluidized bed reactor with the catalyst fluidized with concurrent liquid and gas flow from the bottom. Maximum vanillin and syringaldehyde concentrations occurred near 35% downstream along the reactor height from the bottom. The maximum achieved concentration was dependent on reaction temperature, oxygen partial pressure, liquid flow rate, and time of operation. Product concentrations were always higher for lower liquid flow rates and higher temperatures (Sales et al. 2007). Concentration profiles of vanillin and syringaldehyde along the reactor height at different process conditions are presented in Figure 10.4.

Enzymatic hydrolysis lignin was isolated from steam-exploded corn stalk and subjected to oxidative degradation to produce vanillin, syringaldehyde, and p-hydroxybenzaldehyde. Lignin was dissolved in 2 M aqueous sodium hydroxide to 60 g/L concentration and oxidized in the presence of a 5% perovskite-type heterogeneous catalyst on lignin mass. The reacting system was heated to the reaction temperature of 120 °C, 15 bar nitrogen pressure and the reaction was carried out at this temperature by maintaining an oxygen partial pressure of 5 bar (total pressure 20 bar). Maximum vanillin and syringaldehyde yields, 4.3% and 9.3%, respectively, were obtained after 30 min reaction. However, a maximum p-hydroxybenzaldehyde yield of 2% was achieved after a 120-min reaction (Deng et al. 2008). In a related study with another catalyst, the maximum yields of vanillin, syringaldehyde, and p-hydroxybenzaldehyde were 4.55%, 9.99%, and 2.23%, respectively. These yields were achieved after a reaction time of 60 min for vanillin, 50 min for syringaldehyde, and 120 min for p-hydroxybenzaldehyde (Deng et al. 2009). These researchers used a third perovskite-type heterogeneous catalyst and reported vanillin, syringaldehyde, and p-hydroxybenzaldehyde of 4.76–5.3%, 11–12.8%, and 2.71–2.88%, respectively. These yields were achieved after a reaction time of 50 min for vanillin, 40 min for syringaldehyde, and 120 min for p-hydroxybenzaldehyde (Deng et al. 2010). In another study, lignin was isolated from steam-exploded corn stalk by enzymatic hydrolysis. Lignin was dissolved at 60 g/L in 2 M aqueous sodium hydroxide and oxidized at 120 °C at 5 bar oxygen partial pressure in 20 bar total pressure maintained with nitrogen. The maximum vanillin yield was about 3.2% after 60 min reaction and the syringaldehyde yield was about 4.6% after 30 min. 1.4% p-hydroxybenzaldehyde was also obtained after prolonged oxidation for 150 min (Zhang et al. 2009). Bagasse hydrolysis lignin was oxidized with molecular oxygen – supplied as air – in the presence of a mixture of metal acetate catalysts at 6.9–30 mM concentration. The reaction was carried out for 2 h at 100–140 °C with 69–138 bar air pressure and 8% lignin concentration dissolved in aqueous acetic acid. The main products of oxidative degradation were vanillin and syringaldehyde. Some vanillic acid and syringic acid were also produced when the air pressure was less but the catalyst concentration was high.

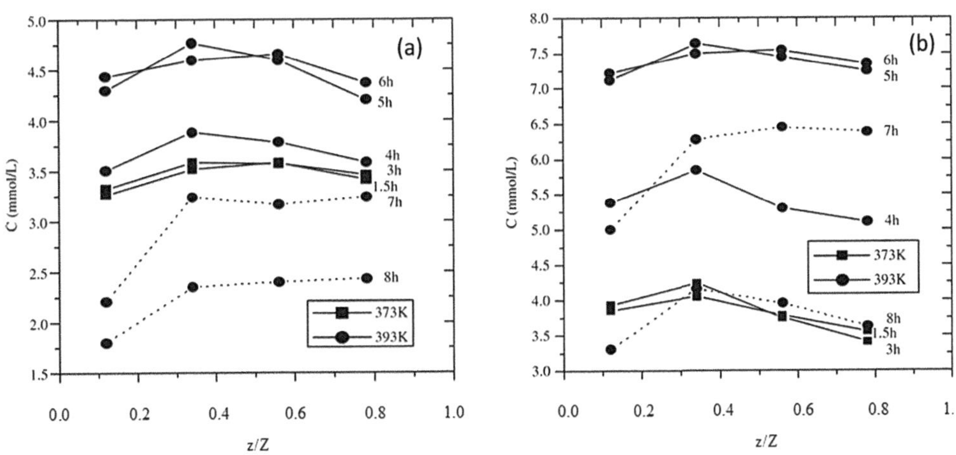

FIGURE 10.4 Concentration profiles of vanillin and syringaldehyde along the reactor height during oxidative degradation of bagasse hydrolysis lignin with molecular oxygen in a fluidized bed reactor. (a) vanillin (b) syringaldehyde. Solid lines represent 5 L/h and dotted lines represent 10 L/h liquid flow rate. (Adapted and reprinted from Sales, FG, Maranhao, LCA, Filho, NML, Abreu, CAM, *Chem. Eng. Sci.*, 62, 5386–5391, 2007, with permission from Elsevier.)

The maximum total yield was 7.53% using 69 bar air pressure, 8% lignin concentration and 30 mM concentration of mixed metal acetate catalysts. A maximum syringaldehyde yield of 2.3% was also attained at these conditions. The maximum vanillin yield was 1.45% using 138 bar air pressure, 8% lignin concentration, and 6.9 mM concentration of mixed metal acetate catalysts (Partenheimer 2009). Oxidative degradation of pine enzymatic hydrolysis lignin – 83.8% Klason lignin and 14.3% carbohydrates – was performed to obtain aromatic aldehyde. The reaction was carried out under stirring with or without Cu(II) catalyst by dissolving lignin at 50 g/L concentration in 2.5 M sodium hydroxide solution. The reaction temperature was 90–140 °C, oxygen pressure 0.1–1.5 MPa, and reaction time 90 min. At 110 °C, the obtained vanillin yield was 5.4% after 20 min reaction in the presence of Cu(II) catalyst but only 3% after 70 min in the absence of catalyst. The reaction was reported to be first order in oxygen pressure up to 2 h reaction, 1.5 MPa oxygen pressure, and 120 °C temperature (Kaygorodov et al. 2018).

10.1.2.4 Lignosulphonate

Early works reported the production of vanillin in 3.6–8.2% yield by the oxidation of softwood sodium lignosulphonate with molecular oxygen in the presence of Cu(II) catalyst. Lignosulphonate was dissolved in 3 M aqueous sodium hydroxide to have 180 g/L lignosulphonate concentration. The reaction was carried out with 0.2 MPa oxygen pressure at 160 °C for 20–60 min. In all these experiments the initial pH was more than 13. Vanillin yield reduced significantly when the initial pH and lignosulphonate concentration were less. At 120 g/L lignosulphonate concentration, it was 2.75% with 10.8 initial pH and 0.75% with 10.3 initial pH (Tarabanko et al. 1995a). Similarly, softwood calcium lignosulphonate was oxidized with molecular oxygen under alkaline conditions using several Cu(II) salts as the catalyst. Lignosulphonate was dissolved in aqueous sodium hydroxide (0.67–0.72 g sodium hydroxide/g lignosulphonate) and copper sulphate solution (0.3–1.5% Cu(II) on lignosulphonate mass) was added. The reaction was carried out at an oxygen pressure of 11 atm and 160–200 °C temperature for 50–110 min under stirring at 850 rpm. The statistical experimental design suggested the amount of sodium hydroxide, Cu(II), reaction time, and the reaction temperature to be significant parameters affecting vanillin yield. The minimum vanillin yield was 2.19% at 0.68 g sodium hydroxide/g lignosulphonate, 0.6% Cu(II) on lignosulphonate mass, 30 min reaction time, and 170 °C temperature. The maximum vanillin yield was 5.41% at 0.69 g sodium hydroxide/g lignosulphonate, 0.9% Cu(II) on lignosulphonate mass, 70 min reaction time, and 180 °C temperature (Bjørsvik and Minisci 1999). Spruce sodium lignosulphonate at 220 g/L concentration in 3M sodium hydroxide was oxidized to vanillin with molecular oxygen in the presence of copper sulphate (equivalent to 0.5% Cu(II) on lignosulphonate mass) as catalyst. Experiments were also carried out without oxygen addition. The study established that a reasonable amount of vanillin is produced from lignosulphonate, even without oxygen addition, when it is heated above 120 °C. At 140 °C and above and 11.5 bar nitrogen pressure, the vanillin concentration reached about 8 g/L after 2 h hydrolysis at the set temperature. Even at 120 °C, the vanillin concentration reached about 5 g/L. The activation energy of hydrolysis of lignosulphonate to vanillin was estimated to be 15.6 kJ/mol. The presence of the catalyst did not appreciably alter the vanillin yield. When the oxidation experiment was carried out with 1.3 bar oxygen partial pressure at 160 °C, the vanillin concentration reached 12 g/L after 50 min and remained constant thereafter till 2 h. It suggested that an additional 4 g/L vanillin could be produced with oxidative degradation over that possible with hydrolysis alone. Still, a higher vanillin concentration of 15 g/L was reached after 40 min at 2.6 bar oxygen partial pressure. This higher vanillin concentration could be achieved even faster by increasing oxygen pressure to 6 bar. The calculated rate of vanillin production was 0.08 g/(L.min) at 1.3 bar oxygen pressure and 0.36 g/(L.min) at 6 bar oxygen pressure (Pacek et al. 2013). Eucalyptus magnesium lignosulphonate yielded 3% vanillin and 7.3% syringaldehyde from its oxidation with molecular oxygen. 1.4% vanillic acid and 0.1% syringic acid were also produced. Lignosulphonate was dissolved at 5 g/L concentration in 0.75 M aqueous sodium hydroxide and reacted for 10 min at

160 °C and 6 bar oxygen partial pressure. The estimated activation energy for the production of vanillin was 70.5 kJ/mol and that for syringaldehyde was 62.6 kJ/mol. The rates of formation of these aldehydes were found to be first order in lignosulphonate concentration and oxygen partial pressure. The reaction order was 1.9 with respect to hydroxyl ion concentration for vanillin production and 1.4 for syringaldehyde production (Santos et al. 2011).

10.1.2.5 Steam Explosion Lignin

In an early work lignin was isolated from steam-exploded poplar by extracting the pulp with aqueous sodium hydroxide at 100 °C and 13 pH for 30 min. Lignin was precipitated from the extract by acidification. This lignin was oxidized at 100 g/L concentration in 125–165 g/L aqueous sodium hydroxide with molecular oxygen partial pressure of 1.38 MPa in the presence of 0.5% ferric chloride catalyst. The reaction time was 2–5 min at 170 °C. A maximum syringaldehyde yield of 6.4% was reported after 3 min and 4% for vanillin after 5 min (Wu et al. 1994). This study calls for special mention as reasonably good aldehyde yield was reported at significantly less reaction time.

10.1.3 Catalysts in Lignin Oxidation with Molecular Oxygen

In the majority of studies, lignin oxidation with molecular oxygen is carried out in the presence of catalysts. Transition metal ions are found to catalyse lignin oxidation with molecular oxygen. These ions can exist in different oxidation states. This allows them to effectively facilitate the rate-limiting electron transfer step from the benzene ring during lignin oxidation to the corresponding carbonyl compound. In the process, the metal ion is reduced to a lower oxidation state. This transformed metal ion subsequently reacts with one of the oxidant species shown in reaction scheme 10.1 or 10.2 to revert back to the original high oxidation state, thereby completing the catalytic cycle. Both homogeneous and heterogeneous catalysts have been used for the purpose. Catalysts not only enhance the rate of oxidation but also influence the product yields and their selectivities.

It is believed that the metal ion in its higher oxidation state initiates the reaction by abstracting an electron from the phenolate anion producing the corresponding radical. The rest of the reaction transforming the phenylpropane entity to the final aldehyde proceeds according to Figure 10.1. The metal ion in the lower oxidation state is re-oxidized to the higher oxidation state either with molecular oxygen or one of the oxidant species produced according to reactions 10.1 and 10.2.

10.1.3.1 Homogeneous Catalysts

Homogeneous catalysts used in lignin oxidation are transition metal salts. Salts of Cu(II), Fe(III), Co(II), and Ce(IV) have all been used to catalyse lignin oxidation with varying results. Catalyst reclamation and recycling are a matter of concern in homogeneous catalysis. An early study on poplar steam explosion lignin was carried out with copper sulphate or ferric chloride as a homogeneous catalyst. Copper sulphate was found to be effective in improving vanillin and syringaldehyde yields in comparison to non-catalytic process. Vanillin yield, at different reaction severities, increased from 3.2–3.4% without a catalyst to 3.3–3.9% in the presence of 5% copper sulphate on lignin mass. Similarly, syringaldehyde yield increased from 3.2–4.7% without catalyst to 5.2–7.9%. p-hydroxybenzaldehyde yield was not affected by the catalyst. Ferric chloride, on the other hand, was reported to be effective only at high reaction severity – high temperature and longer reaction. The best yields of vanillin and syringaldehyde were obtained when a mixture of 5% copper sulphate and 0.5% ferric chloride was used as the catalyst. Under these conditions and 170 °C temperature with a 75% oxygen dose, resulted in the production of 4.7% vanillin and 9.5% syringaldehyde after 10 min oxidation (Wu et al. 1994). However, another contemporary study on pine kraft lignin oxidation in an alkaline medium reported that the addition of 4% copper sulphate based on lignin mass only marginally increased the maximum vanillin yield. Albeit, the presence of a catalyst slowed down the subsequent degradation of vanillin formed at increased reaction times. The reaction was carried out with 60 g/L lignin concentration in 2 M aqueous sodium hydroxide at 120 °C temperature and 4 bar oxygen partial pressure (Mathias and Rodrigues 1995). Homogeneous catalysis of softwood lignosulphonate with a proprietary copper catalyst from Monsanto reportedly increased the vanillin yield from 5–7% to 10% with air as the source of oxygen. This lignosulphonate, at 112 g/L concentration in 3 M aqueous sodium hydroxide, yielded 14.4% vanillin when oxidized with pure oxygen for 15 min. 6.2 g/L copper sulphate was the catalyst used with 0.9 MPa oxygen partial pressure, and 160 °C temperature (Tarabanko et al. 1995b). Catalytic oxidation of softwood lignosulphonate with molecular oxygen reported homogeneous catalysis with Cu(II) or Co(II) to be almost equally effective yielding 5.8–5.9% vanillin. However, the maximum vanillin yield with Cu(II) was achieved after 70 min with 0.21-mole catalyst per kg lignosulphonate against 90 min reaction time and 0.23-mole catalyst per kg lignosulphonate with Co(II). Cu(II) was added as copper sulphate, and Co(II) was added as cobalt acetate (Bjørsvik and Minisci 1999). In one instance, Co(II) added as cobalt acetate in combination with ammonium bromide acted as the homogeneous catalyst for the oxidative degradation of several organosolv lignins to low molecular weight aromatic carbonyl compounds. Bagasse aqueous acetic acid organosolv lignin, eucalyptus aqueous acetic acid organosolv lignin, and bagasse aqueous acetone organosolv lignin were used in the study, dissolved in glacial acetic acid at 100 g/L concentration in the presence of 0.025 M ammonium bromide and 0.005 M Co(II) catalyst. The reaction time was 15 min at 210 °C and 0.5–1.5 MPa oxygen partial pressure. It was contended that Co(II) was oxidized to Co(III) in the presence of molecular oxygen and formed $Co(CH_3COO)_2Br$ by reaction with ammonium bromide. This compound dissociated to regenerate cobalt acetate and bromine atoms. The later initiated the oxidation reaction by forming the organic radical by abstraction of hydrogen. Obtained vanillin yield was 0.2% from bagasse acetic acid organosolv lignin, 1% from eucalyptus acetic acid organosolv lignin, and 0.51% from bagasse acetone organosolv lignin (Goncalves and Schuchardt 1999). Similarly, a mixture of cobalt acetate, manganese acetate, and zirconium acetate in a 1:1:0.1 mole ratio was used as the catalyst in the oxidation of mixed hardwood aqueous acetic acid organosolv lignin dissolved in aqueous acetic acid. At 6.9 mM total catalyst concentration, the yield of aromatic carbonyl compounds varied between 4.4% and 10.9% with varying reaction temperature, acetic acid concentration, lignin concentration, and oxygen partial pressure. With other conditions remaining the same, increasing the catalyst concentration to 30 mM decreased the yield from 7.65% to 4.5% (Partenheimer 2009). This indicates that the presence of a higher amount of catalyst accelerated the degradation of aromatic aldehydes and carboxylic acids more than their formation. In one study cupric hydroxide was the catalyst used for the oxidation of fir lignosulphonate. The reaction was carried out with a 9.75 g/L catalyst concentration at 160 °C with 0.2 MPa oxygen partial pressure. With 180 g/L lignin concentration, vanillin yield was 5.9% after 20 min and 7.7% after 40 min reaction. However, the results were not compared with non-catalytic oxidation (Tarabanko et al. 2004). The addition of 20% copper sulphate to lignin mass as a catalyst substantially increased the aldehyde yield during the oxidation of eucalyptus magnesium lignosulphonate with molecular oxygen. Vanillin yield increased from 2.5% to 4.5% on addition of catalyst and the syringaldehyde yield increased from 11% to 16.1%. The reaction was carried out at 150 °C and 10 bar oxygen partial pressure for 20 min with the lignosulphonate dissolved in 0.9 M aqueous sodium hydroxide at 5 g/L concentration (Santos et al. 2011). Catalytic oxidation of spruce sodium lignosulphonate was carried out with 0.5% copper sulphate on lignin mass added as a catalyst. At 160 °C temperature and 1.3 bar oxygen partial pressure, vanillin yield was 5.5% after 50–120 min reaction time. The initial lignin concentration was 220 g/L in 3 M aqueous sodium hydroxide (Pacek et al. 2013). The addition of 75% copper sulphate – a rather high catalyst dose – on pine enzymatic hydrolysis lignin mass increased the vanillin yield to 5.4% compared to 3% in the absence of a catalyst. Also, the maximum vanillin concentration was attained after 25 min of catalytic oxidation against the 70 min reaction time required without a catalyst. The reaction temperature was 110 °C and oxygen partial pressure was 1.5 MPa in 2.5 M aqueous sodium hydroxide (Kaygorodov et al. 2018).

Most of the studies established the benefit of using homogeneous catalysis in improving the yield of aromatic aldehydes by lignin oxidation with molecular oxygen. One study, however, reports negligible beneficial effects or negative effects of homogeneous catalysis on product yield. Copper sulphate – 0.83 to 3.22 mmol per g lignin – or cobalt chloride – 2.16 to 3.21 mmol per g lignin – was the homogeneous catalyst used for the oxidation of eucalyptus kraft lignin with oxygen. With copper sulphate as the catalyst, the yields of vanillin and syringaldehyde were about 1% and 3.5%, respectively, similar to that obtained without a catalyst. The use of cobalt chloride decreased the respective yields of vanillin and syringaldehyde to about 0.9% and 2.4% (Villar et al. 2001).

For lignin oxidation under acidic conditions, phosphomolybdic acid was found to be effective as a homogeneous catalyst. Indulin AT was dissolved in 80 vol% aqueous methanol in 10 g/L concentration and oxidized in the presence of phosphomolybdic acid at pH 1.13. Vanillin yield was 3.5% after 20 min oxidation at 170 °C and 10.8 bar oxygen pressure. The same amount of vanillic acid was also produced (Voitl and von Rohr 2010). Similarly, Indulin AT was oxidized with molecular oxygen at pH 1 with several homogeneous catalysts. Lignin was dissolved in 80% aqueous methanol at 10 g/L concentration with the addition of catalyst at 0.01 M concentration. The reaction was carried out for 2 h at 170 °C and 10 bar oxygen pressure. A maximum vanillin yield of 3% was achieved with copper sulphate or cobalt chloride as catalyst. However, the maximum vanillin concentration was attained after 1 h with cobalt chloride. Cupric chloride or ferric chloride yielded 2.6% vanillin after only 20 min. Phosphomolybdic acid was the least effective as a catalyst providing only 2.4% vanillin yield after 20 min. Methyl vanillate was the other reaction product with 2–3% yield with different catalysts. For this compound the order of effectiveness of the different catalysts was copper sulphate > cobalt chloride or phosphomolybdic acid > cupric chloride > ferric chloride (Werhan et al. 2011). Oxidation under acidic conditions is challenging because lignin is insoluble in an acid medium and the reaction is essentially heterogeneous.

10.1.3.2 Heterogeneous Catalysts

The use of heterogeneous catalysts is an option to address the issue of catalyst reclamation and recycling during lignin oxidation. Various transition metal oxides have been examined as possible heterogeneous catalysts for lignin oxidation with molecular oxygen. Oxides of copper, manganese, cobalt, and iron have shown catalytic ability for lignin oxidation. Noble metals supported on metal oxide support are also effective in catalysing lignin oxidation to low molecular weight aromatic aldehydes. In one study palladium, supported 2.85% on γ alumina, was used as an oxidation catalyst for bagasse hydrolysis lignin. At a catalyst dose of 4% on lignin mass, the maximum vanillin yield was 4.56%, the maximum syringaldehyde yield was 6.06%, and the maximum p-hydroxybenzaldehyde yield was 7.1%. Oxidation was carried out at 60 g/L lignin concentration dissolved in 2 M aqueous sodium hydroxide at 5 bar oxygen partial pressure, 20 bar total pressure, and 140 °C temperature. These yields were significantly higher than those obtained without the catalyst (Sales et al. 2006; Sales et al. 2007). A study reported catalytic activity of palladium nanoparticles supported on CeO_2 for oxidation of organosolv lignin (type unknown) dissolved in methanol at 2 g/L lignin concentration and 200% catalyst dose on lignin mass. The reaction time was 24 h at 185 °C temperature and 0.1 MPa oxygen partial pressure. Vanillin yield was 5.2% and p-hydroxybenzaldehyde yield was 2.4% (Deng et al. 2015). There is a recent report of gold nanoparticles supported on double layer hydroxide of lithium and aluminium being used as heterogeneous catalyst for lignin oxidation. Indulin AT or maple γ-valerolactone organosolv lignin was dissolved in dimethylformamide and oxidized with molecular oxygen in the presence of a catalyst to yield 83% and 91% oxidized lignin, respectively. The reaction conditions were: 50 g/L lignin concentration, 10 mL/min oxygen flow rate, 24 h reaction time, 120 °C temperature, and 500 rpm agitation in the presence of 40% – rather high catalyst dose – supported gold nanoparticles as heterogeneous catalyst. Oxidized lignin was subsequently dissolved in 1 M sodium hydroxide at 10 g/L concentration and then acidified to pH 2. This allowed the precipitation of unconverted lignin, leaving the low molecular weight products in solution. The respective yields from Indulin AT were: 3.4% vanillin, 3.4% vanillic acid, 1.2% coniferyl alcohol, 1.1% ferulic acid, and 0.4% apocynin. These values from maple γ-valerolactone organosolv lignin were: 2.8% vanillin, 7.8% vanillic acid, 2.8% coniferyl alcohol, 7.8% ferulic acid, 11.2% syringaldehyde, 5% syringic acid, and 2.8% sinapic acid. The study did not compare the results with non-catalytic oxidation (Song et al. 2018). Similarly, gold supported on cerium oxide nanorods at a 20% dose acted as the catalyst for the oxidation of cotton stalk aqueous ethanol organosolv lignin dissolved in methanol. The reaction conditions were: 4 g/L lignin concentration, 180 °C temperature, 4 h reaction time, and 1 MPa oxygen partial pressure. Product yields were: vanillin 10.5%, methyl vanillate 6.8%, and methyl syringate 3.4%. The authors did not compare the results with non-catalytic oxidation (Song et al. 2019).

Recent studies have shown that perovskites of the type ABO_3 possess catalytic activity for lignin oxidation. $LaMnO_3$ was found to be effective in catalysing lignin oxidation. In these, crystalline solid manganese is present in both Mn(III) and Mn(IV) oxidation states while oxygen is bound to the metal as O^{2-}. In addition, several adsorbed oxygen species are also present. When used at 5% of lignin mass (enzyme hydrolysis lignin from steam-exploded cornstalk), catalysed oxidation with 60 g/L lignin concentration, 120 °C temperature, and 5 bar oxygen partial pressure yielded 1.37 times more vanillin, 1.96 times more syringaldehyde, and 1.31 times more p-hydroxybenzaldehyde compared to the non-catalytic process. These yields were also appreciably higher than those obtained with manganese sulphate used as a homogeneous catalyst. Most importantly, the catalyst could be effectively reclaimed after the reaction and used in the subsequent cycle. It was observed that the catalytic activity was preserved even after five cycles of repeated use of 30 min oxidation (Deng et al. 2008). Similarly, $LaCoO_3$ could catalyse the oxidation of the same lignin with molecular oxygen to aromatic aldehydes under similar reaction conditions. These solids had Co(II) and Co(III) in the crystal lattice. Vanillin yield was 1.41 times more, syringaldehyde yield 2.11 times more, and p-hydroxybenzaldehyde yield 1.44 times more compared to the non-catalytic process. These yields were also appreciably higher than those obtained with cobalt sulphate used as a homogeneous catalyst. The catalyst could be effectively reclaimed after the reaction and used in the subsequent cycle. It was observed that the catalytic activity was preserved even after five cycles of repeated use of 50 min oxidation (Deng et al. 2009). The results are shown in Figure 10.5.

It is interesting to compare the catalytic function of the two catalysts from this figure. While the yields of aromatic aldehydes are comparable using the two catalysts, the lignin conversion is about twice with $LaCoO_3$ as a longer oxidation time is provided. In a related work with the same lignin, the perovskite $LaCoO_3$ was further modified with Cu(II) doping to produce crystals of the type $LaCo_{(1-x)}Cu_xO_3$. These were found to have even better catalytic activity than that of the unmodified $LaCoO_3$. Under identical reaction conditions, catalysis with $LaCo_{0.8}Cu_{0.2}O_3$ yielded 17% more vanillin, 28% more syringaldehyde, and 30% more p-hydroxybenzaldehyde over that obtained with undoped $LaCoO_3$. Lignin conversion was also correspondingly higher (Deng et al. 2010). In all these perovskite materials, the enhanced catalytic activity was attributed to adsorbed oxygen species on the active sites of the catalyst which facilitated the reoxidation of Mn, Co, or Cu from the lower oxidation state to the higher oxidation state which completed the catalytic cycle. Another study reported the use of perovskite of the type $LaFe_{1-x}Cu_xO_3$ as a heterogeneous catalyst for the oxidation of the same lignin under similar reaction conditions. Here again, the material was found to be effective in catalysing the production of aromatic aldehydes by lignin oxidation. The composition $LaFe_{0.8}Cu_{0.2}O_3$ was the most effective as a catalyst increasing the vanillin yield 1.42 times, syringaldehyde yield 2.51 times, and p-hydroxybenzaldehyde yield 1.66 times over the non-catalytic process. The catalyst activity was retained after five reuse cycles (Zhang et al. 2009).

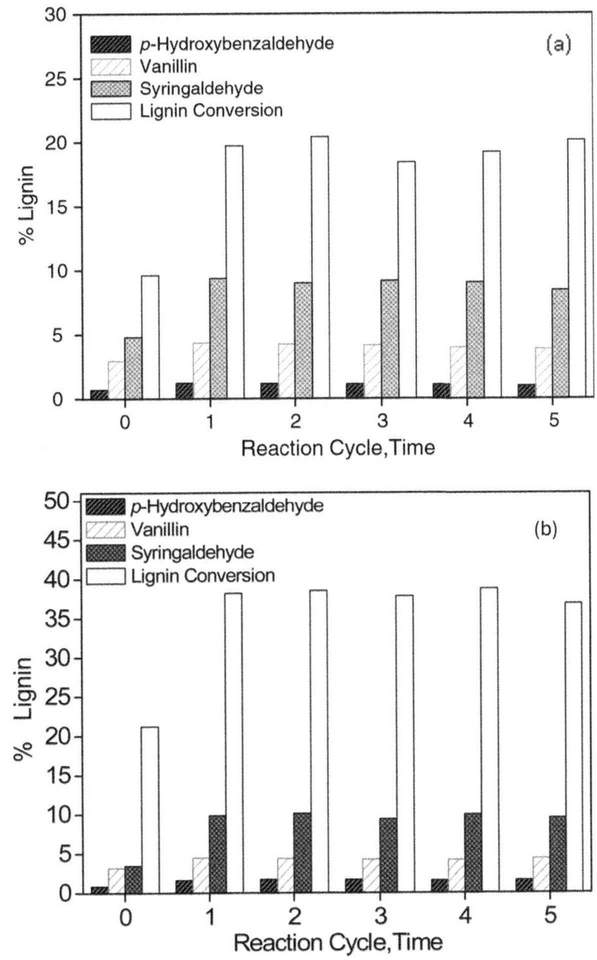

FIGURE 10.5 Reusability of perovskite-type catalyst (a) $LaMnO_3$ and (b) $LaCoO_3$ in lignin oxidation with molecular oxygen. 0 represents the non-catalytic process. (Adapted and reprinted from Deng, H, Lin, L, Sun, Y, Pang, C, Zhuang, J, Ouyang, P, Li, Z, Liu, S, *Catal. Lett.*, 126, 106–111, 2008, and *Energy Fuels*, 23, 19–24, 2009, with permission from Springer and American Chemical Society, respectively.)

10.2 LIGNIN PYROLYSIS TO OBTAIN LOW MOLECULAR WEIGHT ORGANICS

Thermal degradation of lignin in the absence of oxygen produces a liquid fraction – pyrolysis oil – that is a rich combination of organic compounds. Besides a significant amount of water, the pyrolysis oil contains different low molecular weight oxygenated organics – aldehydes, ketones, carboxylic acids, phenols, and alcohols. Many of these are of aromatic nature and could provide a viable substitute for petroleum-derived aromatic chemicals. Moreover, the pyrolysis oil contains a mixture of hydrocarbons, mostly aromatic. Important among these are monoaromatic hydrocarbons, benzene, toluene, xylene, and styrene, as well as polyaromatic hydrocarbons, naphthalene, methyl naphthalene, and dimethyl naphthalene. Due to its higher carbon content, lignin is prone to generating more char during pyrolysis compared to the carbohydrate fractions of lignocellulose. Nevertheless, pyrolysis remains a viable alternative for lignin value addition and has been extensively studied in recent years. Fast or flash pyrolysis is mostly preferred to maximize the yield of pyrolysis oil where the lignin is subjected to a heat treatment of a few milliseconds to a few seconds only. Both fixed bed and fluidized bed operations have been studied with varying success. For many lignins, agglomeration could be an operational problem owing to the relatively low melting temperature. Mixing lignin with different thermally stable inert materials is often practised to prevent agglomeration and improve bed hydrodynamics. Clays, quartz, and silica have been used for this purpose. Pyrolysis in the presence of catalysts can further improve oil yield and also control its composition.

Pyrolysis of lignin is believed to proceed by the homolytic cleavage of intermonomer linkages in the macromolecule forming radical species. These radical species then undergo further degradations and rearrangements to yield the final pyrolysis products. Most important among these end products are the different phenolic entities. Ether bonds are predominantly broken to varying extents based on their bond dissociation enthalpies and the prevailing temperature. A recent study has computed these bond enthalpies for the breaking of β-O-4′ (predominant ether linkage) and 4-O-5′ bonds during the pyrolysis of poplar lignin. For the β-O-4′ bonds where the linked units are of syringyl type, the computed bond dissociation enthalpies vary between 65.07 and 76.65 kcal/mol. For the bond, linking guaiacyl-type units, the computed bond dissociation enthalpy was 64.73 kcal/mol. For the bonds linking one syringyl-type and one guaiacyl-type unit, the computed bond dissociation enthalpies vary between 50.46 and 71.04 kcal/mol. For the 4-O-5′ bond, the study computed the bond dissociation enthalpy to be 86.44 kcal/mol (Azad et al. 2023). The breakdown of the macromolecular structure forms both monomeric and oligomeric phenolic entities. Recombination of initially formed radicals to dimers and oligomers is important in lignin pyrolysis. Radicals with unpaired electrons on the phenolic oxygen or one of the ring carbons are particularly prone to such recombinations, especially if there are no methoxyl groups present on the aromatic ring (Custodis et al. 2014). Lignin molecular weight affects pyrolysis oil yield. Experimental studies with lignins obtained from diverse sources and isolation methods reported the pyrolysis oil yield to decrease with increasing molecular weight of lignin. Higher pyrolysis pressure also decreased oil yield in these experiments. Aqueous ethanol organosolv lignin from wheat straw (M_n 2043), spruce (M_n 2515), and pine milled-wood lignin (M_n 3596) were used in the study. At 500 Pa pyrolysis pressure, the respective oil yields from the three lignins were about 80%, 73%, and 70%. At 10^5 Pa pyrolysis pressure, the corresponding oil yields dropped to about 42%, 39%, and 44%. The pyrolysis was performed at 530 °C and a vapour residence time of 20 ms (Marathe et al. 2019).

Arriving at a universal kinetic model for lignin pyrolysis is less than straightforward. One early study on slow pyrolysis of softwood kraft lignin observed the activation energy to vary from 80 to 158 kJ/mol with temperature increasing from 234 to 503 °C. With Alcell lignin the variation in activation energy was from 129 to 361 kJ/mol with temperature increasing from 272 to 532 °C (Ferdous et al. 2002). A study with Indulin AT and pine aqueous acetic acid organosolv lignin found a reasonable fit for experimental data on the evolution of condensables from lignin pyrolysis assuming first-order kinetics and a value of 10^{13} s^{-1} for the pre-exponential factor k_0. Average activation energy of 192 and 193 kJ/mol were computed for the two lignins, respectively (Beis et al. 2010). In another study on softwood kraft lignin, there was an equally good fit with experimental rate data with reaction orders of either 1, 2, or 3 with the corresponding values of computed activation energies being 47.35, 61.49, and 78.08 kJ/mol, respectively. Similar kinetic interpretations were also arrived at for the pyrolysis of this lignin catalysed by activated carbon with computed activation energies of 45.96, 59.75, and 75.92 kJ/mol for the reaction orders 1, 2, and 3 (Bu et al. 2016).

10.2.1 Non-Catalytic Pyrolysis of Lignin

10.2.1.1 Soda and Kraft Lignin

Fluidized bed pyrolysis of wheat straw and sarkanda soda lignin (Protobind) was performed in a nitrogen and argon atmosphere. 50 g lignin granules of 1–3 mm size were screw fed into the fluidized bed reactor along with 1 L/min nitrogen flow. The material was further fluidized with 20 L/min preheated argon at 300 °C. The reactor was maintained at 400 °C in a bubbling fluidization regime with 1 kg sand as the heating medium. However, the authors reported a great deal of agglomeration leading to collapse of bed fluidization. This could be a major challenge in any future attempt for scaleup. The main oxygenated organics groups in the pyrolysis oil were guaiacols 2.6%, syringols 1.3%, phenols 0.88%, and pyrocatechol 0.24%. Other low molecular weight organics produced were methanol, formic acid, acetone, acetaldehyde, and furaldehyde. However, these yields are not very impressive. For continuous pyrolysis, lignin granules were screw fed at 100 g/h into the fluidized bed reactor along with 1 L/min nitrogen flow. The material was further fluidized with 20 L/min preheated argon at 300 °C. The reactor was maintained at 400 °C in a bubbling fluidization regime with 700 g sand as the bed material. Three pyrolysis oil cuts were collected sequentially by ambient temperature condensation, ambient temperature electrostatic precipitation, and condensation at –20 °C. Pyrolysis oil was produced with a 41% yield, half of this being water. The major part of the organics was phenolic oligomers – about 11% total yield. Syringols were the next major product group followed by catechols, phenols, and guaiacols (de Wild et al. 2009). Similarly, pyrolysis of this lignin in a fixed bed microreactor is reported in a nitrogen atmosphere. The pyrolysis temperature was 800 °C with 120 °C/s heating rate to final temperature and 12 s pyrolysis time. The total liquid yield was 43.1% but only 12.5% could be identified for low molecular weight organics. The major product groups were guaiacols, syringols, and phenols. When the reaction atmosphere was changed to hydrogen the total liquid yield was 74.5% but only 8.1% could be identified for low molecular weight organics (Windt et al. 2009). In another study, the same soda lignin was fractionated by sequential solvent extraction and each fraction was pyrolysed in a microreactor at 10,000 °C/s heating rate to 600 °C and holding at final temperature for 20 s. The highest yield of low molecular weight organics was 13.34% from the ethyl acetate soluble lignin fraction – around 3.6% guaiacols and syringols, around 7.4% alkyl derivatives of phenols, guaiacols, syringols, and catechols, and the rest aromatic aldehydes, ketones, and carboxylic acids. In comparison, the total yield was 10% when the unfractionated lignin was pyrolysed with a reduced proportion of alkyl derivatives of phenols, guaiacols, syringols, and catechols (5.1%), and aromatic aldehydes, ketones, and carboxylic acids (1%). However, these yields should be interpreted in combination with the 22% yield of ethyl acetate soluble fraction of the starting lignin. Therefore, the corrected total yield from the pyrolysis of ethyl acetate soluble fraction based on starting lignin mass would be only 2.9–0.8% for guaiacols and syringols, 1.6% for alkyl derivatives of phenols, guaiacols, syringols, and 0.5% for aromatic aldehydes, ketones, and carboxylic acids. Similarly, the product yields from methanol soluble lignin fraction (53% of starting lignin), and aqueous dioxane soluble lignin fraction (19% of starting lignin) were 8.27% (4.38% based on starting lignin) and 6% (1.14% based on starting lignin). Thus, the sum of product yields from all three fractions adds to only 8.42% based on starting lignin which is less than what is obtained from unfractionated lignin. It is evident that fractionation did not improve overall product yield from pyrolysis but did influence the product distribution in favour of alkyl derivatives, carbonyls, and acids (Kim and Choi 2019). Fast pyrolysis of wheat straw soda lignin – 120 mesh particle size – in a helium environment at 450 °C is reported. The material was heated to the final temperature of 20,000 °C/s to ensure fast pyrolysis. Phenols were the main product group with lesser quantities of ketones, aldehydes, carboxylic acids, esters, alcohols, and furans also present. Phenol, guaiacol, vinyl guaiacol, and syringol were the main individual phenols (Lin et al. 2015). The pyrolysis of wheat straw soda lignin was performed in a downflow fixed bed pyrolizer. The reactor was preheated to 450 °C along with a 150 °C preheated nitrogen flow rate of 20 mL/min. 5 g lignin powder was dropped into the preheated reactor from the overhead hopper. Liquid yield was about 34% (19.7% organic phase) which was slightly reduced, rather than improved, by the addition of 10% alkali – sodium hydroxide, potassium hydroxide, sodium carbonate, or potassium carbonate – with lignin (Peng et al. 2014).

Softwood kraft lignin – 2.25% ash, 43–53 μm particle size – was slow pyrolysed in a fixed bed with a downward flow of helium. The microreactor was filled with 0.5 g lignin and heated at 5–15 °C/min to the pyrolysis temperature of 800 °C and held isothermal for 30 min. A helium flow of 6.7–16.5 mL/min was maintained. One part lignin on a mass basis was mixed with eight parts quartz chips of 4 mm size to prevent bed clogging due to lignin agglomeration. Tar yield was low between 2.7% and 4.6% with increasing heating rates. At a heating rate of 15 °C/min, increasing the helium flow rate to 16.5 mL/min increased the liquid yield to 18.9% (Ferdous et al. 2002). This clearly brings forth the importance of vapour residence time in pyrolysis oil yield. Similarly, pyrolysis of softwood kraft lignin in a fixed bed microreactor is reported in a nitrogen atmosphere. The pyrolysis temperature was 800 °C with 120 °C/s heating rate to final temperature and 12 s pyrolysis time. The total liquid yield was 35.8% but only 6% could be identified for low molecular weight organics. The major product groups were guaiacols, phenols, and monoaromatic hydrocarbons, with minute amount amounts of carboxylic acids also present. Decreasing the pyrolysis temperature to 500 °C increased the total oil yield to 41.6% and 8.5% could be identified for low molecular weight organics. The major product groups were guaiacols, and phenols, with minute amounts of carboxylic acids also present. When the reaction atmosphere was changed to hydrogen at 800 °C, the total liquid yield was 35.1% but only 7.4% could be identified for low molecular weight organics. Identified low molecular organics were mostly guaiacols with some phenols. No acids or hydrocarbons were detected in the pyrolysis oil when the reaction was carried out in a hydrogen environment (Windt et al. 2009). Fast pyrolysis of Indulin AT – particle size below 0.4 mm, 2.43% inorganic content – was carried out in a nitrogen atmosphere at 550 °C. Lignin mass was fed into the fluidized bed reactor at 0.4 g/min along with a stream of nitrogen at 10 L/min. Vapour residence time in the reactor was 1–2 s followed by char filtration at 450 °C and condensation of the pyrolysis oil at 4 °C. Additional oil was collected through electrostatic precipitation of the outgoing gases. Liquid yield was modest at 23% with 39% gaseous products and 41% char. The oil was predominantly aromatic as revealed from signal integrations of ^{13}C NMR spectrum. Interestingly, no carbonyl signals were detected with hydroxyl and methoxyl being the main functional groups. The major components of the gas were 34.5% methane, 27.5% carbon monoxide, and 31% carbon dioxide with small amounts of C2 and C3 hydrocarbons (Beis et al. 2010). In another pyrolysis study on kraft lignin from an undisclosed lignocellulose source the effect of various process parameters, especially the vapour residence time, was examined. Experiments were carried out with 100 g lignin feed in a downflow fixed bed reactor with different nitrogen flow rates and pyrolysis temperatures. Though the authors claim the process to be fast pyrolysis, the reactor dimensions and nitrogen flow rates indicate otherwise. Oil yield was about 35% at 600 mL/min nitrogen flow rate – 1.31 min space time – and 500 °C temperature. There was a nominal increase in oil yield to 37.6% by increasing the temperature to 700 °C. However, the composition of the organic phase in pyrolysis oil was sensitive to temperature. Aromatic carbonyls, carboxylic acids, guaiacols, and furans were the dominant product groups at low temperatures. At high temperatures of 600–700 °C, phenols and catechols were dominant product groups with some presence of aromatic hydrocarbons also observed. At 500 °C, increasing the nitrogen flow rate – decreasing vapour residence time – first increased the oil yield and its organic fraction which declined after attaining a maximum at 600 mL/min flow rate (Shafaghat et al. 2017). Slow pyrolysis – 3 h reaction time – of softwood kraft lignin resulted in the production of negligible liquid product at 350 °C. Liquid yield increased to about 15% at 450 °C and further to about 29% at 550 °C. The major products in the pyrolysis oil were 3 methoxy-5-methyl Phenol, guaiacol, and ethylguaiacol (Borella et al. 2022).

10.2.1.2 Organosolv Lignin

Alcell lignin – 0.15% ash, 43–53 μm particle size – was slow pyrolysed in a fixed bed with a downward flow of helium. The microreactor was filled with 0.5 g lignin and heated at 5–15 °C/min to the pyrolysis temperature of 800 °C and held isothermal for 30 min. A helium flow of 6.7 mL/min was maintained. One part lignin on a mass basis was mixed with eight parts quartz chips of 4 mm size to prevent bed clogging due to lignin agglomeration. Tar yield increased from 14.1% to 21.1% with increasing heating rates (Ferdous et al. 2002). Fluidized bed pyrolysis of hardwood Alcell lignin was performed in a nitrogen and argon atmosphere. 50 g lignin granules of 1–3 mm were screw fed into the fluidized bed reactor along with 1 L/min nitrogen flow. The material was further fluidized with 20 L/min preheated argon at 300 °C. The reactor was maintained at 400 °C in a bubbling fluidization regime with 1 kg sand as the bed material. However, the authors reported a great deal of agglomeration leading to collapse of bed fluidization. This could be a major challenge in any future attempt for scaleup. Product yield was not attractive. Yields of the main oxygenated organics groups in the pyrolysis oil were guaiacols 0.94%, syringols 1.25%, phenols 0.13%, and pyrocatechols 0.54%. Other low molecular weight organics produced were methanol, formic acid, acetone, acetaldehyde, and furaldehyde. Very small amounts of toluene and naphthalene were also produced. For continuous pyrolysis, lignin granules were screw fed at 100 g/h into the fluidized bed reactor along with 1 L/min nitrogen flow. The material was further fluidized with 20 L/min preheated argon at 300 °C. The reactor was maintained at 400 °C in a bubbling fluidization regime with 700 g sand as the bed material. Three pyrolysis oil cuts were collected sequentially by ambient temperature condensation, ambient temperature electrostatic precipitation, and condensation at –20 °C. Pyrolysis oil was produced with 37% yield, with 24% being water. A major part of the organics was phenolic oligomers – about 6% total yield. Syringols were the next major product group followed by catechols, phenols, and guaiacols (de Wild et al. 2009). Similarly, the pyrolysis of Alcell lignin in a fixed bed microreactor is reported in a nitrogen atmosphere. The pyrolysis temperature was 800 °C with 120 °C/s heating rate to final temperature and 12 s pyrolysis time. The total liquid yield was 43.6% but only 7.1% could be identified for low molecular weight organics. The major product groups were syringols, guaiacols, phenols, and monoaromatic hydrocarbons. When the reaction atmosphere was changed to hydrogen at 800 °C, the total liquid yield was 48.7% but only 7.8% could be identified for low molecular weight organics (Windt et al. 2009).

Fast pyrolysis of aqueous acetic acid organosolv lignin – particle size below 0.4 mm, 1.84% inorganic content – was carried out in a nitrogen atmosphere at 550 °C. Lignin mass was fed into the fluidized bed reactor at 0.4 g/min along with a stream of nitrogen at 10 L/min. Vapour residence time in the reactor was 1–2 s followed by char filtration at 450 °C and condensation of the pyrolysis oil at 4 °C. Additional oil was collected through electrostatic precipitation of the outgoing gases. Liquid yield was low at 16% with 21% gaseous products and a high 63% char. The oil was predominantly aromatic as revealed from signal integrations of ^{13}C NMR spectrum. 12.8% signals came from carbonyl carbons followed by 8% of methoxyl and hydroxyl type. The major components of the gas were 36.8% methane, 16.1% carbon monoxide, and 44.8% carbon dioxide with small amounts of C2 to C4 hydrocarbons (Beis et al. 2010). There is a report of fast pyrolysis of a proprietary organosolv lignin – unknown source and solvent used – at 600 °C with a heating rate of 50 °C/s. In terms of carbon yield, the major product group was guaiacols (2-methoxyphenol 16.46%, 2-methoxy-4-methylphenol 6.42%), followed by phenols (phenol 2.69%, 3,5-dimethyphenol 2.93%, 2-methylphenol 2.04%, 4-ethylphenol 0.79%), and aromatic hydrocarbons (toluene 0.65%, p-xylene 1.09%, styrene 0.48%) (Adhikari et al. 2014).

10.2.1.3 Lignosulphonate

Hardwood ammonium lignosulphonate was pyrolysed at 10 °C/min to a final temperature of 550 °C. The pyrolysis oil yield was about 35% along with about 2% tar. This slow pyrolysis also produced about 50% char. Important oxygenated aromatics identified in the pyrolysis oil were phenol, 2-methylphenol, 4-methylphenol, guaiacol, methyl guaiacol, vinyl guaiacol, pyrocatechol, methoxy pyrocatechol, syringol, methyl syringol, vanillin, acetovanillone, vanillic acid, isoeugenol, and acetosyringone. Dimethoxybenzene and trimethoxybenzene also made the aromatic product mix. Several sulphur-containing organic compounds like pyrrole, 2-cyanofuran, 3-methylisothiazole, methyl pyrazine, 2-methyl-1H-pyrrole, and methylpyrrole were also present in the pyrolysis oil. Among the oxygenated aliphatic compounds methanol, and acetic acid were produced (Brebu et al. 2011). Fast pyrolysis of wheat straw lignosulphonate – 120 mesh particle size – in a helium environment at 450 °C is reported. The material was heated to the final temperature at a rate of 20,000 °C/s to ensure fast pyrolysis. Phenols and furans were the main product groups present. Benzofuran, phenol, guaiacol, vinyl guaiacol, and syringol were the main individual compounds (Lin et al. 2015).

10.2.1.4 Hydrolysis Lignin

Pyrolysis of acid hydrolysis lignin from straw in a fixed bed microreactor is reported in a nitrogen atmosphere. The pyrolysis temperature was 800 °C with 120 °C/s heating rate to final temperature and 12 s pyrolysis time. The total liquid yield was 65.7% but only 12.9% could be identified for low molecular weight organics. The presence of carbohydrate impurities in the lignin had an effect on the oil composition. The major product groups were aldehydes and sugars but guaiacols and phenols were present in relatively less amount. When the reaction atmosphere was changed to hydrogen at 800 °C, the total liquid yield was 61.2% but only 16.4% could be identified for low molecular weight organics (Windt et al. 2009). Fast pyrolysis of 30 mg dilute acid hydrolysis lignin – unknown lignocellulose – was performed in a microreactor at 10 °C/s heating rate to 200–900 °C with nitrogen flow to maintain 0.2 s residence time. The major product groups obtained were methoxyphenols, phenols, catechols, furans, and monoaromatic hydrocarbons (Kibet et al. 2012). Fast pyrolysis of corn stover enzymatic hydrolysis lignin – 120 mesh particle size – in a helium environment at 350–650 °C is reported. The material was heated to the final temperature at 20,000 °C/s to ensure fast pyrolysis. Product distribution was strongly dependent on pyrolysis temperature as shown in Figure 10.6. Proportion of phenols in the product mix increased considerably with an increase in temperature to 450 °C and beyond while the proportion of carboxylic acids and alcohols decreased. Aldehydes and furans also increased in proportion with increasing pyrolysis temperature. The proportion of ketones did not change appreciably with temperature. Esters were present in the pyrolysis oil in almost the same proportion with temperatures varying from 350 to 550 °C. The proportion of hydrocarbons decreased up to 550 °C but increased again at 650 °C (Lin et al. 2015).

Spruce and pine hydrolysis lignin – 0.3% ash and 18% carbohydrates – was pyrolysed in an agitated fluidized bed, maintained in a bubbling fluidization regime, with pure nitrogen as fluidizing medium or nitrogen mixed with steam or ethanol vapour. The nitrogen-to-steam ratio varied as 1, 2, or 4.9 and the nitrogen-to-ethanol vapour ratio as 34.6. The total flow rate was controlled to have a space time of 1 s. 300 g aluminium hydroxide of 0.56–0.71 mm was used as the bed material. The reaction conditions were: temperature 500–600 °C, lignin feed rate 500 g/h, and operation time 3 h. The yield of the organic fraction in pyrolysis oil decreased from 40% to 35% with temperature increasing from 500 to 600 °C but the corresponding yield of the aqueous phase increased from 5%

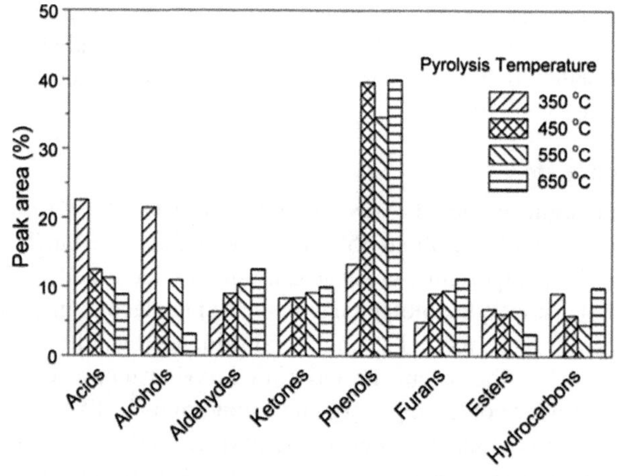

FIGURE 10.6 Temperature dependence of product distribution in pyrolysis of corn stover enzymatic hydrolysis lignin. (Reprinted from Lin, X, Sui, S, Tan, S, Pittman, CU, Sun, J, Zhang, Z, *Energies*, 8, 5107–5121, 2015. Originally published by MDPI under Creative Commons Attribution License)

to 11%. The addition of steam or ethanol vapour during pyrolysis did not appreciably change the liquid yield. The study also reported pilot-scale pyrolysis results in a circulating fluidized bed with 15 kg/h lignin feed rate for a 2-day operation. Bed material was externally heated and circulated 0.1–0.6 mm sand. Superficial gas velocity through the fluidized bed was 7 m/s. Non-condensable syngas fraction from pyrolysis was partly recycled for bed fluidization and partly used as fuel in the combustor to heat the circulating bed material. Char produced in the operation was also burned in the combustor as a heat source. The organic fraction yield in pyrolysis oil was 37% and the aqueous phase yield was 11%. However, the circulating fluidized bed pilot scale operation produced a significantly higher char yield of 43%. The study unequivocally stressed the issue of deposits in gas lines post pyrolysis in both bubbling fluidized beds as well as pilot scale circulating fluidized beds. Pilot scale operation could only run at a stretch for 10 h in the beginning, requiring cleaning shutdown thereafter to remove the deposits formed. The highest deposit formation was observed in the gas line between the char separation cyclone and the condenser. The total available run time was only 64% (Pienihäkkinen et al. 2021). The results are important in drawing inferences about possible operational bottlenecks in large-scale fluidized bed lignin pyrolysis to produce low molecular weight organics.

10.2.1.5 Steam Explosion Lignin

Aspen steam explosion lignin was pyrolysed at 10 °C/min to a final temperature of 550 °C. Pyrolysis oil yield was about 28% along with about 4% tar. This slow pyrolysis also produced about 50% char. Important oxygenated aromatics identified in the pyrolysis oil were phenol, 2-methylphenol, 4-methylphenol, guaiacol, methyl guaiacol, vinyl guaiacol, pyrocatechol, methoxy pyrocatechol, syringol, methyl syringol, vanillin, acetovanillone, vanillic acid, isoeugenol, and acetosyringone. Dimethoxybenzene and trimethoxybenzene also made the aromatic product mix. Among the oxygenated aliphatic compounds methanol, methyl acetate, 2-butanone, acetic acid, methyl propionate, propanoic acid, methyl butanoate, and butanoic acid were produced. Furfural and furfuryl alcohol were produced from the pentosan content of the lignin (Brebu et al. 2011).

10.2.1.6 Milled Wood Lignin

Fast pyrolysis of amur linden milled-wood lignin – 120 mesh particle size – in a helium environment at 450 °C is reported. The material was heated to the final temperature at 20,000 °C/s to ensure fast pyrolysis. Phenols, aromatic aldehydes, and carboxylic acids were the main product group with lesser quantities of ketones, esters, and alcohols also present. 2,6-dimethoxy-4-(2-propenyl)phenol, vanillic acid, guaiacol, vinyl guaiacol, syringol, and syringaldehyde were the main individual compounds in the pyrolysis oil (Lin et al. 2015).

10.2.2 Reactor Configurations and Operational Issues

The two most frequent reactor configurations used in biomass pyrolysis are the fixed bed and the fluidized bed and lignin pyrolysis is no exception. Generally, the fluidized bed operation should be preferred because there is a better possibility to control the temperature uniformity in the bed. The prevalent practice is to maintain the bed in a bubbling fluidization regime. Close control of fluidizing gas flow rate and its distribution across the bed cross section is critical. For design consideration, the vapour residence time is normally taken as equal to the reactor space time which can be written as

$$\text{space time} = \frac{\text{Reactor volume}}{\text{Volumetric feed rate of fluidizing gas}} = \frac{(\text{Reactor cross section area}) \times (\text{Reactor length})}{\text{Volumetric feed rate of fluidizing gas}} \tag{10.5}$$

The equation would be applicable for both fixed bed and fluidized bed reactors. The actual vapour residence time may be different from the space time calculated due to the volume expansion of the gas at pyrolysis temperature, reactor volume occupied by the solids, and nonlinear gas streamlines within the bed. The later effect is more pronounced in a fixed bed. One of the important operational parameters for pyrolizers is the pressure drop across the bed, both fixed and fluidized. Phase transformation behaviour of lignin plays a crucial role in this matter. Lignin starts softening and liquid globules appear in the lignin mass in the 170–200 °C temperature range which is lower than the temperature at which appreciable thermal decomposition starts. This causes the lignin particles to agglomerate and form bigger lumps with reduced porosity. Agglomerated lignin sharply increases the pressure drop across the bed and also induces serious temperature non-uniformities. These temperature non-uniformities may be within the different parts of the bed or within a single lump of lignin. Therefore, the pyrolysis conditions vary leading to reduced yield of pyrolysis oil and unwanted product composition. Over time, lignin agglomeration may take serious proportions requiring reactor shutdown and cleaning of gas passages. The problem of lignin agglomeration and various remedial measures for its redressal have been the focus of study in several research works. The general approach is to mix an inert solid additive with lignin during pyrolysis. The additive functions by dispersing the softened and semi-liquid lignin over its extended surface and prevents the lignin globules from coming together and agglomerating. Moreover, the additive also acts as an effective heat transfer medium that evenly distributes the heat over the biomass bed and helps in maintaining a uniform temperature distribution.

There is mention of another reactor configuration, Pyrolysis Centrifuge Reactor, for fast pyrolysis of wheat straw enzymatic hydrolysis lignin of 78.8% Klason lignin content and less than 0.84 mm particle size. The reactor consisted of a heated steel drum with a rotating tri-winged coaxial agitator at the centre. A total of 204–481 g lignin powder was tangentially fed to the inside of the drum with a screw feeder at 4.1–10.4 g/min. It adhered to the hot inner surface of the drum due to centrifugal forces as it moved forward along the axial direction of the drum. Recycled syngas produced from the pyrolysis was preheated and fed to the reactor which carried the particles along the axial direction. The gas flow rate was adjusted to control the vapour residence time between 0.5 and 5 s. A product mixture of char, vapours, and non-condensable syngas came out of the tangential outlet at the other end of the reactor. Vapours were condensed in isopropyl alcohol maintained at 30–50 °C, after separating the char. No lignin feeding problem was reported but the problem of blockage in the gas passages after the reactor outlet was observed. At vapour residence time of 0.8 s, and lignin feed rate of 10.4 g/min, a reactor temperature of 500–550 °C was the best for getting a high yield of organic fraction – 30–31% – in pyrolysis oil. The yield dropped significantly to less than 20% when the vapour residence time was increased to 4.2 s. Decreasing the lignin feed rate to 5.6 g/min while keeping the residence time at 0.8 s and reaction temperature 550 °C, increased the organic oil yield to 34%. Important individual compounds in the pyrolysis oil were guaiacol and its methyl, ethyl, and vinyl derivatives, syringol, syringaldehyde, acetosyringone, trimethoxybenzene, and acetic acid. Furans and levoglucosan were also present, produced from the pyrolysis of the carbohydrate impurity in the lignin (Trinh et al. 2013).

10.2.2.1 Additives in Soda and Kraft Lignin Pyrolysis

Calcium formate was one such additive studied where it was produced in situ by mixing Indulin AT with calcium hydroxide (4:1 mass ratio) and forming into a slurry with water. Next, formic acid was added into the slurry (with a calcium hydroxide-to-formic acid mass ratio of 2–4) to neutralize the alkali and produce the formate salt. The reaction was completed in 1 h after which the excess acid was neutralized with the addition of calcium hydroxide. Finally, the slurry was dried, ground, and sieved to pass through a 425 μm screen. The formed mixture of lignin and calcium formate was pyrolysed at 500 °C by feeding to the top of the reactor being entrained with nitrogen flowing at 6 L/min to maintain 3 s vapour residence time. A total of 300 g mixture was pyrolysed at a 1–2 g/min feeding rate. A comparison was made by pyrolysing lignin alone or lignin mixed with calcium hydroxide. The pyrolysis of lignin resulted in a 23% oil yield which marginally increased to 23.3% when lignin was mixed with calcium hydroxide at a 4:1 mass ratio. The yield improved to

28.5% when in situ calcium hydroxide was neutralized with twice the amount of formic acid. The yield further increased to 32.5% when the amount of neutralizing formic acid was four times the amount of calcium hydroxide. The most important aspect of mixing lignin with calcium formate was the absence of bed agglomeration and uninterrupted operation without any increase in pressure drop. Another important difference was the organic composition of the pyrolysis oil produced. Guaiacols were the major products in the pyrolysis oil from lignin pyrolysis. When mixed with calcium formate lignin pyrolysis oil mainly contained phenols and alkylphenols. It was suggested that the decomposition of formate produced atomic hydrogen that effectively contributed to the deoxyhydrogenation that removed the methoxyl functionalities (Mukkamala et al. 2012). In one study the lignin was mixed with 30% of magnesium aluminium phyllosilicate type clay and extruded into 2 mm granules. 200 g of these granules were pyrolysed in a continuous fluidized bed pyrolizer operating at 450–550 °C in a bubbling fluidization regime. The total liquid product from pyrolysis of Indulin AT was 34% at 450 °C and 48% at 550 °C. Protobind lignin pyrolysis at 450 °C yielded 42% liquid product. Gas chromatographic characterization of the oils was not entirely successful due to low volatility. The components characterized in pyrolysis oil from Protobind lignin included guaiacols, syringols, phenols, catechols, carboxylic acids, and methanol. Pyrolysis oil from Indulin AT did not have syringols but had a small number of furans when pyrolysed at 550 °C (de Wild et al. 2017). Similarly, the influence of mixing several inert bed materials – silica sand, activated and non-activated biochars, and foamed glass beads – with lignin was examined on the agglomeration behaviour and bed hydrodynamics during fluidized bed pyrolysis. Pulverized kraft lignin from an unknown lignocellulose source was used in the study. To continuously feed the lignin into the reactor it was dropped intermittently through a pinch valve into a horizontal feeding line which opened into the inclined line that led to the fluidized bed reactor. Each dose of lignin into the horizontal line formed a small slug which was swept into the reactor with a nitrogen pulse. A total of 200 g biomass was pyrolysed in each run and fed into the preheated reactor over a period of 20 min. The pyrolysis temperature was 550 °C and vapour residence time was maintained at 1.7 s. The material in the reactor was fluidized with preheated nitrogen and also mechanically agitated to prevent agglomeration. The pyrolysis oil was collected by passing the vapours from the reactor through two stages of condensation with an intermediate electrostatic precipitator. It was observed that pyrolysis oil yield and its organic phase content increased when a bed material was present. Without bed material, the pyrolysis oil yield was 29.5% (14% organic phase). It increased to 36% (22.5% organic phase) with 95 g non-activated lignin biochar as bed material and further increased to 37.3% (24.5% organic phase) with 150 g biochar as bed material. Silica sand as bed material also produced the same amount of oil but with only 21.3% organic phase yield. When 150 g steam-activated lignin biochar was used as bed material, the pyrolysis oil yield was 43% (30.4% organic phase). This organic phase also had the highest proportion of low molecular weight organics, with 50% more phenolics than that obtained with sand as bed material. Foamed glass beads as bed material yielded the least amount of only 33.7% oil (22.9% organic phase). However, it was still higher than the empty bed operation. It is evident that the presence of bed material improved bed hydrodynamics and heat distribution by not allowing the lignin to soften at elevated temperatures to agglomerate. The choice of bed material also had a profound effect on the product composition of the pyrolysis oil organic phase. Even with non-activated lignin char, phenolics yield was 1.19 times that with a sand bed pyrolysis and the carboxylic acid content was only about 0.35 times. The acid content in the organic phase of pyrolysis oil from activated char-supported pyrolysis was half compared to that with sand-supported pyrolysis (Li et al. 2015). These trends are graphically presented in Figure 10.7.

10.2.2.2 Additives in Organosolv Lignin Pyrolysis

The problem of lignin agglomeration was addressed in a work where the lignin was mixed with 30% magnesium aluminium phyllosilicate-type clay and extruded into 2 mm granules. 200 g of these granules were pyrolysed in a continuous fluidized bed pyrolizer operating at 450 °C in a bubbling fluidization regime. Aqueous ethanol organosolv lignin from wheat straw, poplar, or spruce was

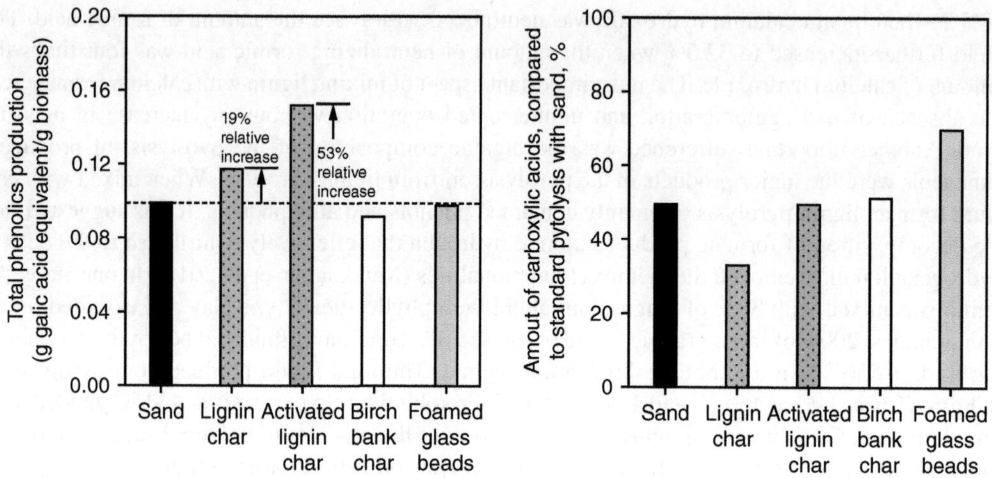

FIGURE 10.7 Effect of bed material on pyrolysis oil quality during fluidized bed pyrolysis of kraft lignin. (Adapted and reprinted from Li, D, Briens, C, Berruti, F, *Bioresource Technol.*, 189, 7–14, 2015 with permission from Elsevier)

used in the study. Total liquid product from pyrolysis of wheat straw lignin was 40%, from poplar lignin 41%, and from spruce lignin 42%. Gas chromatographic characterization of the oils was not entirely successful due to low volatility. The components characterized included guaiacols, phenols, catechols, carboxylic acids, and methanol. Syringols were also present in the oil obtained from wheat straw and poplar lignin. Poplar and spruce pyrolysis oil also had some furans present (de Wild et al. 2017). Similar improvement in bed hydrodynamics was reported for the pyrolysis of beech aqueous ethanol organosolv lignin mixed with different magnesium aluminium phyllosilicate type clay attapulgite, sodium formate, or calcium hydroxide. An equal mass of lignin and the respective clay or sodium formate was formed into a slurry by the addition of 5 L water per kg lignin and mixed for 1 h. For calcium hydroxide, 20% based on lignin mass was mixed to prepare the slurry. The pyrolysis feed mixture was prepared by filtering the slurry, drying the cake, and pulverizing the dry cake. 5 g at a time of pyrolysis feed mixture was fed through a lock hopper into the top of the fluidized bed pyrolizer. The fluidized bed reactor was initially filled with 1.5 kg sand to Assist in fluidization and heat distribution. Mixing of material in the bed was further facilitated with a ribbon mixer. The reactor temperature was 500 °C and the flow rate of fluidizing nitrogen was 283 L/h. Continuous uninterrupted pyrolysis was not possible with unmodified lignin due to severe bed clogging. Calcium hydroxide, though effective in preventing lignin agglomeration, resulted in low pyrolysis oil yield – about 8% aqueous phase and 14% heavy organic phase. With sodium formate as an additive, the liquid yield improved to about 18% aqueous phase and 17% heavy organic phase. The highest liquid yield – about 18% aqueous phase and 30% heavy organic phase – was obtained with the attapulgite additive. In all the heavy organic phase liquids, the main organic product group was alkylphenols followed by catechols and guaiacols (Ghysels et al. 2020).

10.2.3 Catalytic Pyrolysis of Lignin

The total yield of pyrolysis oil and its organic fraction is lower in catalytic lignin pyrolysis compared to non-catalytic pyrolysis. However, catalytic pyrolysis results in a much different product profile in the pyrolysis oil. While oxygenated aromatics are the principal product group from non-catalytic pyrolysis, the predominant product group from catalytic pyrolysis is aromatic hydrocarbon, both monoaromatic and polyaromatic. Catalysts accelerate the deoxyhydrogenation reactions and demethoxylation reactions in comparison to other pyrolytic reactions. This results in fewer methoxyphenol compounds in the pyrolysis oil and more phenol- and catechol-type compounds. Various catalysts have been studied for lignin pyrolysis. Some of them are zeolites, activated carbons, clay, ceramic-supported transition metals, and activated carbon-supported transition metals. There are reports of both in situ and ex situ catalytic pyrolysis of lignin. In situ catalysis refers to the process where lignin and the catalyst are mixed together and then thermally treated. In ex situ catalysis, the vapours from lignin pyrolysis are brought in contact with the catalyst. Alternatively, the pyrolysis oil produced from lignin pyrolysis is catalytically hydrotreated to transform its chemical composition.

10.2.3.1 Soda and Kraft Lignin

An in situ catalytic pyrolysis of nonwood (wheat straw sarkanda mix) soda lignin was studied with HZSM-5, KZSM-5, or Al-MCM-41 as catalysts. 0.45 g of reaction mixture (20% lignin and 80% catalyst) was charged into the microreactor and packed with about 0.08 g quartz wool. The pyrolysis was carried out at 600 °C (heating time 45 s, pyrolysis time 3.5 min) with 15 mL/min helium flow. The highest liquid yield, nearly 49%, was achieved with the HZSM-5 catalyst. Of these 46.7% were monoaromatic hydrocarbons (benzene, toluene, xylene, ethyl benzene, and trimethylbenzene), 46.2% were polyaromatic hydrocarbons (naphthalene, methylnaphthalene, and dimethylnaphthalene), and 7.1% were oxygenated aromatics (methylphenol). Catalysis with KZSM-5 produced about 46% liquid, of which 69.3% were oxygenated aromatics (methylphenol and dimethylphenol), 17.3% monoaromatic hydrocarbons (benzene, toluene, xylene, ethyl benzene, and trimethylbenzene), and 13.5% polyaromatic hydrocarbons (naphthalene and methylnaphthalene). This shows the importance of zeolite acidity in promoting the production of aromatic hydrocarbons in lignin pyrolysis. Al-MCM-41 catalyst yielded about 36% liquid product, the lowest among the three catalysts. Of these, 66.5% were oxygenated aromatics (methylphenol, dimethylphenol, and ethylphenol), 9.3% monoaromatic hydrocarbons (benzene, toluene, xylene, and ethyl benzene), and 24.2% polyaromatic hydrocarbons (naphthalene) (Jackson et al. 2009). Other researchers have also reported the in situ catalytic pyrolysis of this lignin with HZSM-5 catalysts of Si/Al ratio 30. Fast pyrolysis was carried out in a helium atmosphere for 20 s at 600 °C with the heating rate to the final temperature being 10,000 °C/s. The lignin-to-catalyst mass ratio was 1. The total yield of low molecular weight organics was about 11.5–2.94% for aromatic hydrocarbons. The other major product groups were guaiacols, syringols, phenols, aromatic carbonyls, and carboxylic acids. It is evident that the selectivity for aromatic hydrocarbons was not good – only 25.6% (Kim et al. 2015). In another study, the same soda lignin was fractionated by sequential solvent extraction and each fraction was pyrolysed by in situ catalysis in the presence of zeolite catalyst ZeoY. Lignin fraction and catalyst were mixed in equal mass and subjected to fast pyrolysis in a microreactor at 10,000 °C/s heating rate to 600 °C and holding at final temperature for 20 s. The pyrolysis product yield was around 9.9% from unfractionated lignin – 4% guaiacols and syringols, 2.5% alkyl derivatives of phenols, guaiacols, syringols, and catechols, 2.8% aromatic hydrocarbons. In comparison, the total yield of low molecular weight organics was about 2.4% based on starting lignin mass from the ethyl acetate soluble lignin fraction (22% of starting lignin) – around 1% guaiacols and syringols, around 0.33% alkyl derivatives of phenols, guaiacols, syringols, and catechols, and about 1.07% aromatic hydrocarbons. Similarly, the product yields from methanol soluble lignin fraction (53% of starting lignin), and aqueous dioxane soluble lignin fraction (19% of starting lignin) were 3% and 0.6%, respectively, based on starting lignin mass. Thus, the sum of product yields from all three fractions adds to only 6% based on starting lignin which is less than what is obtained from unfractionated lignin. It is evident that fractionation did not improve overall product yield from pyrolysis. Comparing these results with non-catalytic pyrolysis, discussed in Section 10.2.1, it can be concluded that the addition of the zeolite catalyst ZeoY suppressed the overall pyrolysis product yield but considerably increased the proportion of aromatic hydrocarbons in the product mix. (Kim and Choi 2019).

Softwood kraft lignin – < 200 mesh particle size – was subjected to in situ catalytic fast pyrolysis in flowing nitrogen. A 4 g mixture of lignin and catalyst – one to four parts by mass of catalyst per part lignin – along with 3 g glass beads was formed into a fixed bed and fed into the preheated furnace. The catalyst used was acidified bentonite clay with the objective of maximizing the production of aromatic hydrocarbons, benzene, toluene, and xylene. The activation method of the catalyst had a profound effect on product distribution. It was found that bentonite clay acidified with hydrochloric acid maximized the production of desired aromatic hydrocarbons while suppressing the production of phenolic compounds. Optimal operating conditions were 650 °C temperature, 75 mL/min nitrogen flow, and 3 parts catalyst per part lignin. Phenolic compounds were produced in the largest amount with non-catalytic pyrolysis as well as when the clay was acidified with sulphuric acid. Catalytic pyrolysis with unmodified clay or clay activated with orthophosphoric acid resulted in higher production of monoaromatics other than the desired hydrocarbons or phenolic compounds (Elfadly et al. 2017). These results are shown in Figure 10.8.

A pyrolysis study on kraft lignin from an undisclosed lignocellulose source examined the effect of various catalysts. Experiments were carried out with 50 g lignin feed in a downflow fixed bed reactor with 600 mL/min nitrogen flow rate – 1.31 min space time – and pyrolysis temperature of 500 °C. Though the authors claim the process to be fast pyrolysis, the reactor dimensions and nitrogen flow rate indicate otherwise. For all catalysts, oil yield and its organic phase were lower than that for non-catalytic pyrolysis. The catalyst HBeta(25), at a 1:1 mass ratio with lignin, produced the highest amount of oil but it had a lot of aqueous phase in it. However, the organic phase was rich in polyaromatic hydrocarbons and phenols. Some monoaromatic hydrocarbons were also present. Alumina at a 2:1 mass ratio with lignin produced the next higher oil yield with a relatively low amount of aqueous phase. The organic phase predominantly had phenols in it followed by catechols, but the abundance of aromatic hydrocarbons was low. The catalyst HY(30) at a 1:1 mass ratio with lignin produced a somewhat lower amount of pyrolysis oil, and its organic phase was rich in phenols and polyaromatic hydrocarbons. The largest phenol yield was obtained with spent FCC catalyst at a 2:1 mass ratio with lignin. Magnesium oxide and HZSM-5 were less effective in selective production of phenols and aromatic hydrocarbons and the overall yields were also less. In addition, guaiacols were present in all the pyrolysis oils to some extent (Shafaghat et al. 2017). One study compared the in situ and ex situ catalytic pyrolysis of kraft lignin – an unknown lignocellulose source – with an HZSM-5 catalyst. For in situ catalysis, 30% catalyst on lignin mass was mixed with lignin along

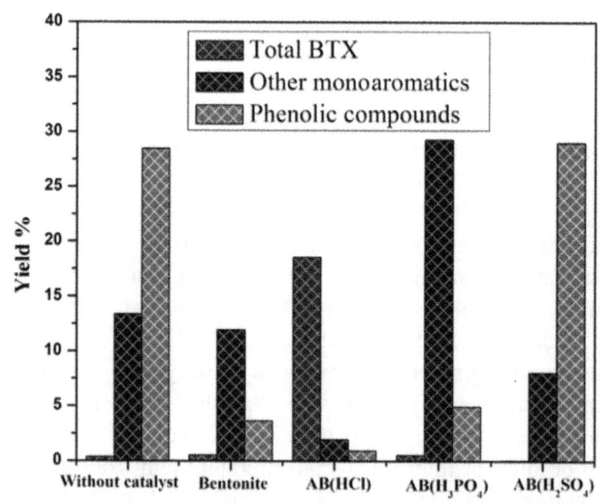

FIGURE 10.8 Product distribution in catalytic pyrolysis of softwood kraft lignin with different bentonite clay-based catalysts and comparison with non-catalytic pyrolysis. AB refers to acidified bentonite. (Reprinted from Elfadly, AM, Zeid, IF, Yehia, FZ, Abouelela, MM, Rabie, AM, *Fuel. Process. Technol.*, 163, 1–7, 2017 with permission from Elsevier)

with silicon carbide particles and pyrolysed at 550 °C in a fixed bed. The pyrolysis oil yield was 28.95% with the major product group being alkyl phenols. The other product groups in the pyrolysis oil, in decreasing order of selectivity, were methoxy phenols, polyaromatic hydrocarbons, phenols, and monoaromatic hydrocarbons. For ex situ catalysis, a mixture of lignin and silicon carbide was pyrolysed at 550 °C and the vapours were passed through a bed of catalyst – 10–40% on lignin mass – maintained at 250–550 °C. The system was vacuumed prior to heating to ensure an oxygen-free environment. At a catalyst dose of 30%, pyrolysis oil yield decreased with increasing temperature from 28.11% to 23.3%. At a catalyst temperature of 450 °C, pyrolysis oil yield decreased with increasing catalyst dose from 29.24% to 21.75%. For comparison, non-catalytic pyrolysis resulted in 34.11% oil yield. Increased catalyst dose decreased the proportion of methoxyphenols in the oil while increasing the proportion of monoaromatic and polyaromatic hydrocarbons. A similar trend was also observed with increasing catalyst temperature except for monoaromatic hydrocarbons, for which the best selectivity was achieved at 450 °C (Fan et al. 2018). Similarly, the in situ catalytic pyrolysis of spruce LignoBoost kraft lignin was studied in the presence of ilmenite (lowest acidity), bentonite (high acidity), or red mud (moderate acidity). 5 g of lignin was mixed with an equal mass of the respective catalyst and pyrolysed at 500–550 °C in 0.5 L/min downward nitrogen flow. The yield of the aqueous fraction of pyrolysis oil was 13.01% with ilmenite, 21.96% with bentonite, and 20.35% with red mud. The corresponding yields of the organic fraction were 18.13%, 7.08%, and 8.83%, respectively. With all the catalysts, the product group with maximum yield was benzenediols (1,2-Benzenediol, 4-methyl-, and catechol the main compounds), followed by phenols, and polyaromatic hydrocarbons, in that order. Small amounts of monoaromatic hydrocarbons and guaiacols were also produced with ilmenite or red mud as the catalyst (Han et al. 2019).

In a modified approach, kraft lignin (unknown lignocellulose source) was catalytically pyrolysed in two stages. In situ catalysis was carried out in the first stage with lignin mixed with natural zeolite (Si/Al ratio 5.4–7.3) in equal mass ratio (3 g each). Vapours from the first stage were catalytically upgraded ex situ by passing over a bed of 1 g HZSM-5 catalyst of Si/Al ratio 30. The entire reactor system was maintained under a nitrogen flow of 50 mL/min. The in situ catalytic pyrolysis was carried out at 500 °C while the temperature was between 500 and 600 °C for ex situ upgradation. In the absence of ex situ upgradation, an in situ catalytic pyrolysis of kraft lignin, with 0.33 g catalyst per g lignin, produced about 34% pyrolysis oil which was hardly any improvement over the 33.6% oil yield in the non-catalytic pyrolysis. Even increasing the catalyst-to-lignin mass ratio to 1 did not improve the oil yield. This exemplified the rather ineffectiveness of natural zeolite in catalysing the pyrolytic degradation of lignin to condensable low molecular weight volatile organics. However, product composition was somewhat altered by in situ catalysis with more abundance of pyrocatechols and alkylphenols and fewer amounts of guaiacols and syringols compared to non-catalytic pyrolysis. With ex situ catalysis – vapours from lignin pyrolysis alone passing over the bed of HZSM-5 at 500 °C – the oil yield was about 28%. Oil yield marginally increased to about 31% with a two-stage combination of in situ (with natural zeolite) and ex situ (with HZSM-5) catalysis. Ex situ catalysis, however, significantly changed the pyrolysis oil composition in favour of alkylphenols, and monoaromatic hydrocarbons – benzene, toluene, xylene, and ethyl benzene. The in situ catalysis with natural zeolite did not produce measurable quantities of aromatic hydrocarbons. In contrast, pyrolysis oil from the single-stage ex situ catalysis contained 5.7% of benzene, toluene, xylene, and ethyl benzene. This quantity increased to 8.7% with two-stage catalysis. On increasing the ex situ catalysis temperature to 600 °C in two-stage catalytic pyrolysis, the oil yield decreased but the produced oil now contained 14.6% of benzene, toluene, xylene, and ethyl benzene (Lee et al. 2016).

10.2.3.2 Organosolv Lignin

In one study, Alcell lignin was pyrolysed at 470 °C or 560 °C and the vapours were catalysed ex situ by passing them over a bed of HZSM-5, FCC, or Olivine maintained at the same temperature. The process was carried out with 40 mg lignin and 40 mg catalyst, in a nitrogen flow of 50 mL/min. The heating time was 40 s and pyrolysis time was 80 s. The product yields – pyrolysis oil, gas, and

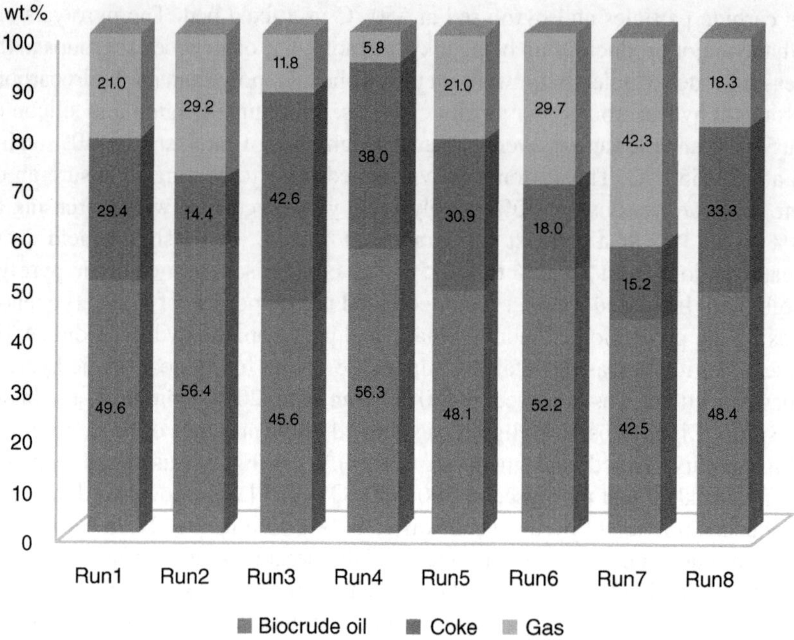

FIGURE 10.9 Product yields in catalytic pyrolysis of Alcell lignin with different catalysts. Run1 and Run5 for non-catalytic pyrolysis at 470 °C and 560 °C. Run2 and Run6 for HZSM-5 catalyst at 470 °C and 560 °C. Run3 and Run7 for FCC catalyst at 470 °C and 560 °C. Run4 and Run8 for Olivine catalyst at 470 °C and 560 °C. (Adapted and reprinted from Choi, HS, Meier, D, Windt, M, *Environ. Progress Sustain. Energy*, 31(2), 240–244, 2012 with permission from Wiley)

char – are shown in Figure 10.9. The highest oil yield was obtained with HZSM-5 and Olivine catalyst at 470 °C. Liquid yield decreased at a higher pyrolysis temperature of 560 °C. With the HZSM-5 catalyst, the char production was minimal while it was quite high with the Olivine catalyst. The lowest oil yield was obtained with the FCC catalyst, even lower than the non-catalytic pyrolysis. Syringols were the largest product group in all the pyrolysis oils followed by guaiacols, with small amounts of phenols, carboxylic acids, and aromatic hydrocarbons also present (Choi et al. 2012).

An in situ catalytic pyrolysis of poplar aqueous ethanol organosolv lignin with HZSM-5 catalyst of Si/Al ratio of 30 is reported. Fast pyrolysis was carried out in a helium atmosphere for 20 s at 600 °C with the heating rate to the final temperature being 10,000 °C/s. The lignin-to-catalyst mass ratio was 1. The total yield of low molecular weight organics was about 13.3–3.17% for aromatic hydrocarbons. The other major product groups were guaiacols, syringols, phenols, aromatic carbonyls, and carboxylic acids. It is evident that the selectivity for aromatic hydrocarbons was not good – only 23.8% of the total yield (Kim et al. 2015). There is a report of in situ catalytic fast pyrolysis of a proprietary organosolv lignin – unknown source and solvent used – with ZSM5 zeolite catalyst. The catalyst was mixed with the lignin in a 1:4 mass ratio and packed into the quartz reactor with quartz wool. The pyrolysis temperature was 500–600 °C with a heating rate of 50 °C/s. Total oil yield was found to increase with pyrolysis temperature from 500 to 550 °C and also with the increasing acidity of the zeolite. Total product yield was actually lower than non-catalytic pyrolysis for zeolites with low acidity. These results are shown in Figure 10.10. In atypical run with highest acidity zeolite as catalyst and 500 °C pyrolysis temperature, the major organic compounds obtained were monoaromatic hydrocarbons (benzene 0.68%, toluene 5.54%, p-xylene 6.14%, ethylbenzene 2.14%, 1-ethyl, 2-methylbenzene 1.46%, 1,2,3-trimethylbemzene 2.63%), polyaromatic hydrocarbons (naphthalene 1.01%, 1-methylnaphthalene 1.75%, 1-ethylnaphthalene 0.59%, 2,6-dimethylnaphthalene 1.6%, 1,6,7-trimethylnaphthalene 0.37%), guaiacol (2-methoxy-4-methylphenol

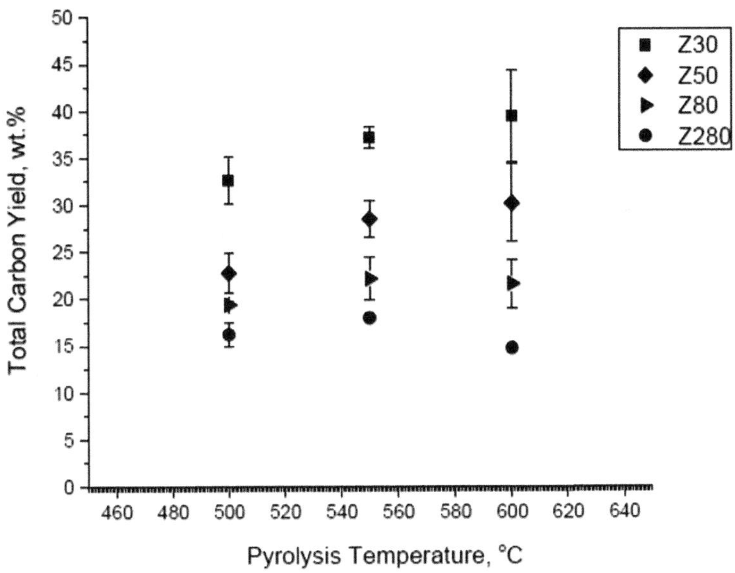

FIGURE 10.10 Product yield in catalytic pyrolysis of proprietary organosolv lignin with ZSM5 zeolites. Z30 highest acidity and Z280 lowest acidity. (Reprinted from Adhikari, S, Srinivasan, V, Fasina, O, *Energy Fuels*, 28, 4532–4538, 2014 with permission from American Chemical Society)

0.29%). Other aromatic hydrocarbon products were indane 1.63% and indene 1.96% (Adhikari et al. 2014).

10.2.3.3 Hydrolysis Lignin

An in situ catalytic pyrolysis of wheat straw hydrolysis lignin with HZSM-5 catalyst of Si/Al ratio of 30 is reported. Fast pyrolysis was carried out in a helium atmosphere for 20 s at 600 °C with the heating rate to the final temperature being 10,000 °C/s. The lignin-to-catalyst mass ratio was 1. The total yield of low molecular weight organics was about 7–2.76% for aromatic hydrocarbons. The other major product groups were guaiacols, syringols, phenols, aromatic carbonyls, and carboxylic acids. The selectivity for aromatic hydrocarbons was reasonable –39.4% of total yield – compared to several other lignin types pyrolysed under similar conditions (Kim et al. 2015). The ex situ catalytic pyrolysis of corn cob Klason lignin was carried out by placing the 0.33–1 g lignin powder – 50 μm particle size – in a wire mesh cup and sliding it into the preheated rection zone of the microreactor. Generated pyrolytic vapours were swept with nitrogen flowing at 100 mL/min and passed through two consecutive beds of catalysts, first activated red mud, and second HZSM-5. The total catalyst amount in the two beds was 2 g with the mass ratio of red mud and HZSM-5 varying between 3 and 0.33. Experiments were performed with a 2 g bed of a single catalyst also, either red mud or HZSM-5. The pyrolysis temperature was studied between 450 and 650 °C. With 1 g lignin feed and a 2:1 ratio of red mud to HZSM-5, the highest pyrolysis oil yield was 20.16% at 550 °C. This oil yield was significantly higher than the yield with red mud as the sole catalyst and marginally better than the yield with HZSM-5 as the sole catalyst. This oil contained 41.27% monoaromatic hydrocarbons – 8.82% benzene and 22.82% toluene – and 22.65% polyaromatic hydrocarbons – naphthalene and its derivatives – along with 11.4% phenols. Increasing the HZSM-5 to red mud ratio to 3 increased the proportion of aromatic hydrocarbons in the oil but slightly decreased the total oil yield (Wang et al. 2019). This research group also studied the ex situ catalytic pyrolysis of the same lignin where the pyrolysis vapours passed through two consecutive beds of catalysts, first Al-SBA-15, and second HZSM-5. Other reaction conditions remained the same. The oil yield and composition did not change appreciably with the change of catalyst in the first bed from red mud to

Al-SBA-15. The highest pyrolysis oil yield was about 24% when Al-SBA-15 was the only catalyst used but it was lower than that obtained in non-catalytic pyrolysis. Oil yield progressively decreased with two consecutive catalyst beds and with increased ratio of HZSM-5 to Al-SBA-15. Yield apart, the pyrolysis oil composition changed with the use of catalysts. When Al-SBA-15 was the sole catalyst, phenols were the most abundant product group followed by guaiacols and syringols. Small amounts of monoaromatic hydrocarbons and aliphatic hydrocarbons were also produced. With two catalyst beds in sequence, monoaromatic and polyaromatic hydrocarbons predominated the product mix while the proportion of aliphatic hydrocarbons and phenols declined. No presence of guaiacols and syringols exhibited the demethoxylation capability of the dual bed catalyst configuration. The highest proportion of monoaromatic hydrocarbons of 42.57% was obtained with the ratio of HZSM-5 to Al-SBA-15 in the two beds as 3 (Wang et al. 2021).

10.2.3.4 Milled Wood Lignin

There is a research report on the in situ catalytic pyrolysis of poplar milled-wood lignin with HZSM-5 catalysts of different acidity. Fast pyrolysis was carried out in a helium atmosphere for 20 s at 500–700 °C with the heating rate to the final temperature being 10,000 °C/s. The catalyst-to-lignin mass ratio varied between 0.6 and 2. The low Si/Al ratio in the catalyst favoured the yield of aromatic hydrocarbons. At a pyrolysis temperature of 600 °C and catalyst-to-lignin ratio of 1, the lowest aromatic hydrocarbon yield of 0.76% was returned with a Si/Al ratio of 280. The highest total yield of low molecular weight organics of about 14.8% was obtained at 600 °C pyrolysis temperature and a catalyst-to-lignin mass ratio of 1 with the catalyst having a Si/Al ratio of 30. However, the yield of aromatic hydrocarbons was only 2.62% representing a selectivity of only 17.7% on total yield. The other major product groups were guaiacols, syringols, phenols, aromatic carbonyls, and carboxylic acids. Increasing the catalysts to lignin ratio to 2 substantially decreased the total yield to about 8.8%. Interestingly, the yield of aromatic hydrocarbons of 3.57% was the highest with this catalyst dose, showing a healthy selectivity of 40.6% (Kim et al. 2015).

10.2.3.5 Catalytic Modification of Pyrolysis Oil from Lignin

Pyrolysis oil obtained from fluidized bed pyrolysis of Indulin AT, wheat straw organosolv lignin, and poplar organosolv lignin (discussed in Section 10.2.1) were hydrotreated in the presence of 5% different heterogeneous catalysts with 100 bar hydrogen partial pressure. Ruthenium supported on carbon was the catalyst used for pyrolysis oils from organosolv lignins. The product organic phase was 80.7% of the wheat straw lignin pyrolysis oil and 69.8% of the poplar lignin pyrolysis oil. The product oil had alkylphenols as the main constituent along with appreciable amounts of cyclic and linear alkanes and aromatic hydrocarbons. Several catalysts were studied for the hydrotreating of Indulin AT pyrolysis oil. For pyrolysis oil obtained at 450 °C, the catalyst was cobalt and molybdenum supported on alumina (4 h reaction at 400 °C) yielding 80.6% organic phase. The reaction selectivity was substantially improved in favour of alkylphenols (much less cyclic and linear alkanes and aromatic hydrocarbons), but the total monomer yield was less. For pyrolysis oil obtained at 550 °C, the catalyst was 20NiMoP/AC (2 h reaction) yielding 78.4% organic phase at 350 °C and 81.2% organic phase at 400 °C. Monomer yield improved – 54.9% at 350 °C and 81.9% at 400 °C – but the selectivities were still low (de Wild et al. 2017).

10.2.4 Catalysts in Lignin Pyrolysis

Catalysts used for lignin pyrolysis are mostly adapted from petroleum refining. The most important among the different categories of catalysts are the zeolites. The catalytic function of zeolites arises from their pore size and shapes which in turn is related to the crystalline structure. Natural zeolite has only a modest specific surface area and acidic active sites. Proprietary zeolites are synthesized to produce materials with a high specific surface area and the right amount of acidic surface sites.

The zeolites can have different cationic forms; ammonium, hydrogen, potassium, and sodium often constitute the cation base of the zeolite catalyst. One of the widely used catalysts is ZSM-5 (Zeolite Socony Mobil–5), the commercial name for a silicon oxide ring-based zeolite catalyst, originally developed and patented by the petroleum refining giant Mobil. A fraction of silicon atoms in the crystal lattice is substituted by aluminium. This generates Bronsted acid sites that are crucial in the selective production of aromatic hydrocarbons and in suppressing the formation of oxygenated aromatics. The pore structure is made of two mutually perpendicular channels with walls made of 10-membered silicon/aluminium rings. This proprietary zeolite can also be doped with transition metals like cobalt and iron to improve selectivity for monoaromatic hydrocarbons. Though having high aromatic hydrocarbon selectivity, Y-type zeolites are less preferred due to less overall pyrolysis oil yield. These zeolites have fauzasite-type crystal structures with perpendicular pores made of 10-membered silicon/aluminium rings. Another catalyst material developed by Mobil is MCM-41 (Mobil Composition of Matter No. 41) whose crystalline structure is entirely made of silicon atoms. Therefore, in its pure form, this material is devoid of any Bronsted acid site on its surface. It is a mesoporous material with unidirectional pores of hexagonal cross section. For catalyst application, it is chemically modified to include aluminium atoms into the lattice and generate acid sites. Such aluminium-modified MCM-41 has been used in lignin pyrolysis with limited success with only lesser production of aromatic hydrocarbons. However, these catalysts are good at demethoxylation and producing alkyl phenols. Similarly, mesoporous silica substrate SBA-15 can be doped with aluminium to induce Bronsted acid sites. The modified material exhibits preferred demethylation and demethoxylation in lignin pyrolysis. Proprietary catalyst materials are often costly and require precisely controlled synthesis protocols. Efforts are ongoing to evaluate naturally occurring materials with relatively low cost for their catalytic effectiveness. One attempted aluminosilicate material for lignin catalytic pyrolysis is bentonite clay. This material has a lamellar structure with alternating sheets of alumina and silica. In its inactivated form it catalyses the lignin demethoxylation during pyrolysis producing catechol and other benzenediols. Its acid-activated form, especially with hydrochloric acid, has shown a propensity for the selective production of benzene toluene, and xylene. Acid activation introduces Bronsted acid sites on the clay surface which is devoid of acid sites in its pristine form. The right amount of these sites is the key to product selectivity. Activation also increases the specific surface area. The mineral ilmenite has been tried as a possible catalyst for lignin pyrolysis. Pure ilmenite has the stoichiometric formula of $FeTiO_3$, with iron providing the acid sites needed for catalytic activity. However, in its natural occurrence, the material is admixed with significant amounts of aluminium, calcium, and magnesium compounds. It is not very selective for hydrocarbons, rather catechol-type compounds are favoured. During the extraction of aluminium from its ore a solid waste rich in alumina, silica, and oxides of iron is produced. This material is commonly known as red mud and is difficult to dispose of. Its use as a catalyst in research studies has shown limited success. It is also chemically modified to reduce its sodium, potassium, and calcium content and increase its specific surface area. Bulk magnesium oxide has very little catalytic activity for lignin pyrolysis owing to its low specific surface area. This shortcoming is overcome to an extent by loading magnesium oxide to a high specific surface area support like alumina, zirconia, or activated carbon. Loading magnesium oxide somewhat reduces the specific surface area of the support material but remarkably increases the acidic sites. This has the overall effect of increasing the proportion of aromatic hydrocarbons in the resulting pyrolysis oil.

Like any heterogeneous catalytic reaction, chemical adsorption of the reactant species on the active sites on the catalyst surface is the key to catalytic pyrolysis of lignin. The primary requirement is a high specific surface area along with the right geometrical disposition of the internal spaces within the catalyst. Therefore, the crystalline structure and the pore structure decide the activity and selectivity of the catalyst for a particular reaction. A descriptive account of lignin pyrolysis catalysts is presented in Table 10.1.

TABLE 10.1
Different Catalysts in Lignin Pyrolysis to Low Molecular weight Organic Compounds

Catalyst	Type/Formation Route	Important Characteristics	Product Profile	Reference
HZSM-5	H-form of ZSM-5 zeolite, obtained by calcining the ammonium form	SiO_2/Al_2O_3 mole ratio: 23–280. Specific surface area: 227–425 m^2/g. Pore structure: 5.5 × 5.1 Å straight pores. and 5.6 × 5.3 Å zig zag pores. Acidity: 0.024–0.129 mmol NH_3/g; 0.153–1.384 mmol isopropylamine/g (increases with decreasing SiO_2 content)	Major product group: mono- and polyaromatic hydrocarbons. Minor product group: phenol	Adhikari et al. 2014; Fan et al. 2018.; Kim et al. 2015; Yu et al. 2012
Metal-doped HZSM-5	H-form of ZSM-5 Zeolite with doping of cobalt or iron	SiO_2/Al_2O_3 mole ratio: 27. Specific surface area: 354.6–363.6 m^2/g. Reduced acid sites compared to unmodified HZSM-5	Major product group: mono- and polyaromatic hydrocarbons. Minor product group: phenol	Jin et al. 2022
KZSM-5	Potassium form Zeolite/ obtained by ion exchange of ammonium form	SiO_2/Al_2O_3 mole ratio: 23–280. Specific surface area: 194–488 m^2/g. Pore structure: 5.5 × 5.1 Å straight pores. and 5.6 × 5.3 Å zig-zag pores	Major product group: alkylphenol Minor product group: mono- and polyaromatic hydrocarbons	Jackson et al. 2009; Lang et al. 2015; Yu et al. 2012
Al-MCM-41	Aluminium-modified MCM-41	Specific surface area: 966 m^2/g. Pore diameter: 2.8 nm	Major product group: alkylphenol Minor product group: mono- and polyaromatic hydrocarbons	Jackson et al. 2009; Wang et al. 2009
Al-SBA-15	Aluminium-modified mesoporous silica	Hexagonal lattice. SiO_2/Al_2O_3 mole ratio: 38. Specific surface area: 696.85 m^2/g	Major product group: phenol Minor product group: methoxyphenol, monoaromatic hydrocarbons	Wang et al. 2021.
ZeoY or HY	Y-type zeolite	SiO_2/Al_2O_3 mole ratio: 5.1–80. Specific surface area: 780–809 m^2/g. Pore structure: 7.4 Å circular pores. Acidity: 1.96 mmol NH_3/g	Overall product yield is low. Major product group: mono- and polyaromatic hydrocarbons. Minor product group: methoxyphenol and catechol	Shafaghat et al. 2017; Park et al. 2019; Yu et al. 2012
H-USY	Steam-activated Y-type zeolite	SiO_2/Al_2O_3 mole ratio: 20. Specific surface area: 670 m^2/g	Major product group: mono- and polyaromatic hydrocarbons	Shen et al. 2016

HBeta	Beta-type zeolite	SiO_2/Al_2O_3 mole ratio: 25 Specific surface area: 680 m^2/g. Pores size: 0.61–0.62 nm Acidity: 1.48 mmol NH_3/g	A large amount of aqueous phase. Major product group: polyaromatic hydrocarbons and phenols	Park et al. 2019; Shafaghat et al. 2017
Acidified bentonite	Clay. Acidification with mineral acid	Specific surface area: 87.55 m^2/g. Pore diameter: 8.53 nm Acidity: 0.491–3.376 mmol NH_3/g	Major product group: monoaromatic hydrocarbons	Elfadly et al. 2017.
Bentonite	Clay	Specific surface area: 12.19 m^2/g	Major product group: benzenediol	Han et al. 2019
Ilmenite	Titanium-iron oxide mineral	Specific surface area: 0.31 m^2/g	Major product group: benzenediol.	Han et al. 2019
Red mud	Solid waste from aluminium extraction from ore	Specific surface area: 8.03 m^2/g	Major product group: benzenediol	Han et al. 2019
Modified red mud	Solid waste from aluminium extraction from ore, modified by acid digestion and reprecipitation	Specific surface area: 212.14 m^2/g	Major product group: phenol and alkylphenols	Wang et al. 2019
Supported MgO	Metal oxide on ceramic or activated carbon support	Specific surface area: 36 (alumina support)-480 (activated carbon support) m^2/g. Acidity: 146 (zirconia support)-343 (alumina support)	Major product group: phenol, methoxyphenol, and alkylphenols. Minor product group: mono- and polyaromatic hydrocarbons	Ryu et al. 2019
Niobium oxide	Mesoporous metal oxide	Specific surface area: 230 m^2/g. Pore diameter: 3.97 nm Acidity: 0.69 mmol NH_3/g	Major product group: monoaromatic hydrocarbons. Minor product group: polyaromatic hydrocarbons, and phenols	Zhang et al. 2022
Niobium oxophosphate	Mesoporous metal oxide	Specific surface area: 440 m^2/g. Pore diameter: 4.93 nm Acidity: 1.25 mmol NH_3/g	Major product group: mono aromatic hydrocarbons. Minor product group: polyaromatic hydrocarbons, and phenols	Zhang et al. 2022

10.3 ADVANCED OXIDATION OF LIGNIN TO OBTAIN LOW MOLECULAR WEIGHT ORGANICS

Advanced oxidation is the process where an organic substance in an aqueous matrix is oxidized with strong oxidizing species. These oxidizing species may be free radicals, or ion radicals, and are generated as secondary reactants in the reacting system through primary reactions involving water, and/or added oxidants, and/or energy stimulus. The main in-situ-generated strong oxidizing species in advanced oxidation processes are hydroxyl radical (•OH), superoxide anion (O_2^-•), hydroperoxide radical (HO_2•), atomic oxygen (•O•), and ozone radical ion (O_3^-•). The primary oxidants used may be hydrogen peroxide, ozone, or a combination of both. These are often used with catalysts to accelerate the production and selectivity of oxidizing species. Energy stimuli in the form of electrical energy, electromagnetic radiation, or ultrasound waves also facilitate the advanced oxidation processes. In the presence of a suitable catalyst, electrical energy or electromagnetic radiation can generate the oxidizing species through water splitting even in the absence of a primary oxidant.

Advanced oxidation processes are mainly attractive for wastewater treatment containing non-biodegradable or biorecalcitrant organic pollutants, which cannot be treated by conventional methods. The strong oxidizing species generated in advanced oxidation processes, especially the hydroxyl radical, are capable of completely oxidizing any organic compound to carbon dioxide, water, and other simple inorganic molecules. However, the reaction is far from a simple, one-step oxidation to carbon dioxide, water, and other simple inorganic molecules. Rather, complete mineralization of a complex organic molecule involves multiple reactions generating several intermediates. Lignin being a biorecalcitrant organic substance is suitable for the application of advanced oxidation processes. Attempts have been made to treat lignin-containing wastewater streams by photocatalytic, electrochemical, and sonochemical oxidation. These processes have shown reasonable success in degrading lignin and mitigating its pollutant potential in wastewater. Subsequently, it was realized that many intermediates formed during the advanced oxidation of lignin have potential industrial and commercial value. Therefore, researchers shifted their attention to the possible production of value-added low molecular weight organics by application of advanced oxidation processes on lignin. Of the various alternatives, electrochemical oxidation of lignin is an attractive proposition for producing low molecular weight organic chemicals from lignin. In some cases, photocatalytic oxidation of lignin is also attempted.

10.3.1 ELECTROCHEMICAL OXIDATION OF LIGNIN

Electrochemical oxidation or electro-oxidation is a variant of advanced oxidation processes where electrical energy is used to provide the energy required to break one of the O–H bonds in the water molecule. This generates hydroxyl radicals that drive the oxidation of organic substances present. The reaction on the anode can be written as (Ghatak 2014a).

$$H_2O \rightarrow \bullet OH + e^- + H^+ \qquad \{10.3\}$$

Other oxidant species are also produced as a result of secondary reactions. One particularly important reaction is the subsequent oxidation of hydroxyl radical to atomic oxygen (Ghatak 2014b).

$$\bullet OH \rightarrow \bullet O \bullet + e^- + H^+ \qquad \{10.4\}$$

Atomic oxygen has lesser reactivity than the hydroxyl radical to react with the organic substances. A significant part of the atomic oxygen would combine to form molecular oxygen.

$$2 \bullet O \bullet \rightarrow O_2 \qquad \{10.5\}$$

In an alkaline reaction environment, reaction {10.2} is also possible as a secondary reaction. The process does not require the addition of any primary oxidant and the only external input is electrical energy. If complete mineralization of the organic substances is the objective, all reactions except reaction {10.3} are undesirable. In such cases one would like to use an anode having a high overpotential for reaction {10.4} to suppress the generation of atomic oxygen, and subsequently molecular oxygen. One anode material that particularly deserves mention for such applications is the Boron-Doped Diamond (BDD) anode. For lignin oxidation to low molecular weight organic compounds, we need an oxidizing environment that is strong enough to break the intermonomer, side chain, and aromatic substitution linkages in lignin, but not so harsh to further oxidize the resulting intermediates. The right combination of anode material and reaction conditions is, therefore, crucial. The important reaction conditions for lignin electro-oxidation are current density, specific charge, and lignin concentration. Current density has a twofold effect on electro-oxidation. One, it is linked to the applied interelectrode potential – higher current density would require higher potential. Interelectrode potential strongly affects the possibility or otherwise of the formation of different oxidant species. Moreover, the electrical energy spent in electrochemical oxidation is directly proportional to the interelectrode potential. Second, it is linked to the rate of formation of the oxidant species and other charge transfer processes. As already mentioned, the oxidant species take part in several reactions: (1) oxidation of lignin to desired low molecular weight compounds, (2) oxidation of desired low molecular weight compounds to undesired organic and inorganic molecular species, and (3) participation in altogether inorganic reactions like reaction {10.5}. With an increase in current density, the rate of formation of oxidant species increases. This enhances the rate of both the desired and undesired reactions. A low current density normally favours the desired reactions over the undesired reactions, provided the associated interelectrode potential is sufficient to support the formation of oxidants. However, one has to use a larger electrode surface area to maintain the same electrolytic current if the current density is low. This would have a bearing on the capital cost of the electrolyser. It must be noted that the low molecular weight oxygenated aromatics – aldehydes, ketones, and phenols – are susceptible to further oxidative degradation/ derivatization. For example, an experimental study reported a second-order degradation of vanillin – an important low molecular weight reaction product in lignin electrochemical oxidation – with a rate constant of 2.29 L/(mol min), when 600 mg/L aqueous solution was electrochemically oxidized at 50 mA/cm^2 current density (Singh and Ghatak 2017). Aldehydes and ketones can oxidize to the respective carboxylic acids and so can the phenols. Demethoxylation is another possible reaction pathway. C_β–C_γ bonds are mostly broken and, in some cases, the C_α–C_β bond may also be broken. Even the C_α–Ar bond may be cleaved leading to the formation of methoxyphenols, and phenols. There are reports of the formation of linear carboxylic acids and even hydrocarbons through aromatic ring openings.

The overall process can be best described as a combination of series and parallel reactions. At any moment, the net rate of formation of a target compound is the difference in its rate of formation and rate of degradation. At the start of the process, lignin concentration is high with negligible or very low concentration of the target compound. Consequently, its rate of formation exceeds its rate of degradation, and the net rate of formation is positive. With the production of the target compound, its concentration in the reacting system increases and so does its rate of degradation. There comes a stage when the two rates are equal, and the net rate of formation is zero. This is the moment when the concentration of the target compound is maximum. Beyond this point, the rate of degradation exceeds the rate of formation, and the net rate of formation is negative. This manifests in a falling concentration of the target compound. These kinetic trends are presented in Figure 10.11 for the electro-oxidation of soda lignins from wheat straw and bagasse on the stainless steel anode.

Some research studies also indicate the possible hydrogenation of lignin during electro-oxidation in undivided cells where the two electrodes are not separated by a diaphragm or membrane. It is postulated that atomic hydrogen (H•) is formed through cathodic charge transfer processes which is responsible for lignin hydrogenation. Formation of methyl anisole was reported during electro-oxidation of aspen lignin – unknown isolation method – in an undivided cell having a lead oxide anode and stainless steel cathode (Wang et al. 2015).

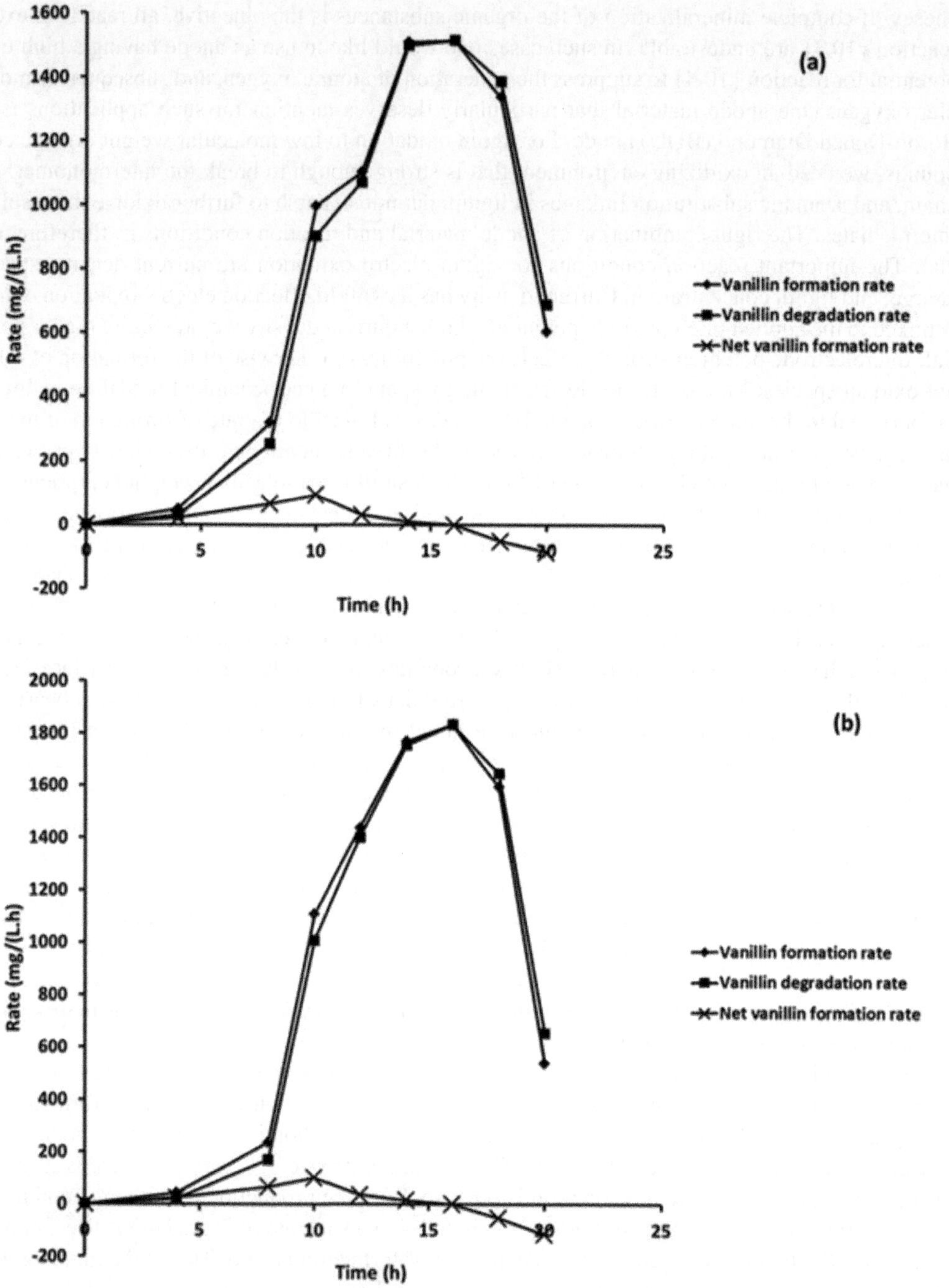

FIGURE 10.11 Kinetics of formation and degradation of vanillin during lignin electrochemical oxidation on stainless steel anode (a) wheat straw lignin (b) bagasse lignin. (Reprinted from Singh, S, Ghatak, HR, *J. Wood Chem. Technol.*, 37, 407–422, 2017 with permission from Taylor and Francis.)

10.3.1.1 Soda and Kraft Lignin

In an early study, kraft lignin from an undisclosed lignocellulose source was electro-oxidized on nickel, platinum, or copper anode. The anode was separated from the cathode by anion exchange membrane. The anode compartment was filled with 172 mL of lignin solution in 1 M sodium hydroxide at 0.5% concentration. Galvanostatic conditions of 70 mA/cm^2 current density were maintained in batch electrochemical cells. The reported vanillin yield was about 2.1% with nickel and platinum anode after 120 min oxidation, after which the yield started to drop. On copper anode, the yield was slightly less but the maximum was attained after only 30 min. The yield improved to about 3.2% on nickel anode on increasing the current density to 85 mA/cm^2. The importance of cell configuration – electrode area for a given solution volume and the interelectrode gap – was evidenced when the electro-oxidation was carried out in a filter press electrochemical cell. In batch electrochemical cells, nickel anode at 30 mA/cm^2 yielded 3.9% vanillin after 3 h oxidation. The yield was substantially increased to 10.9% with lead oxide anode at 60 mA/cm^2 current density. Vanillic acid, guaiacol, apocynin, and hydroxybenzaldehyde were also produced in lesser amounts (Parpot et al. 2000). The electro-oxidation of kraft lignin was carried out at ruthenium-iridium mixed metal oxide anode. Lignin concentration was 500 ppm in 0.5 M aqueous sodium hydroxide. The reaction temperature was 60 °C and the current density was 500 mA/cm^2. The main reaction products were vanillic acid and vanillin during 1 h electro-oxidation (Tolba et al. 2010). Similarly, electro-oxidation of kraft lignin from an undisclosed lignocellulose source – most likely a softwood – was performed at an anode made of lead oxide nanoparticles deposited on titanium dioxide nanotube arrays grown on titanium substrate. Lignin oxidation was observed to be pseudo first order in lignin concentration. Vanillin and vanillic acid were the main reaction products (Pan et al. 2012). Similarly, electro-oxidation of kraft lignin was studied on different anodes at a relatively small current density of 1.9 mA/cm^2 for a specific charge of 2688 C/g. Reported vanillin yields were less than 2% – the highest of 1.8% with cobalt-nickel alloy Stellite 21 – but other byproducts were formed in very small quantities (Schmitt et al. 2015). In one study, softwood kraft lignin with a low degree of sulphonation was electro-oxidized on nickel, cobalt, or nickel–cobalt anodes of 2.25 cm^2 area and 1.55 mg/cm^2 electrocatalyst loading. Lignin was dissolved at 10 g/L concentration in 1 M sodium hydroxide and electro-oxidized for 16 h at 0.598 V against Standard Hydrogen Electrode (SHE). Compared to nickel or cobalt alone, the bimetallic electrocatalyst, with a 20:80 ratio of nickel and cobalt, was reported to give the best electrochemical performance in terms of current density – corresponding specific charge 27,227 mC/g lignin. The main reaction products identified were vanillin and 3-methylbenzaldehyde at 0.51% and 0.36% total yields, respectively, based on lignin mass. The minor reaction products were apocynin, and 1,2-dimethoxy-4-(1-methoxy-1-propenyl) benzene (Ghahremani and Staser 2018). The electro-oxidation of beech kraft lignin, at a current density of 107 mA/cm^2, is reported on nickel foam anode with 5 g/L lignin concentration in 1 M aqueous sodium hydroxide solution. This lignin solution was circulated through the anode chamber of the electrochemical cell at 50 mL/min. The anode and the cathode were separated by an anion exchange membrane. The major low molecular weight organics produced were identified as guaiacol, coniferyl alcohol, vinylguaiacol, vanillin, apocynin, syringaldehyde, and 2,5-di-tert-butyl-p-quinone. Moreover, p-octylacetophenone, dibutyl phthalate, 1,2-benezenedicarboxylic acid, diisooctyl ester, and phenanthrene-carboxylic acid were also identified in the product mix (Stiefel et al. 2016).

Electro-oxidation of wheat straw and bagasse soda lignin was carried out on a stainless steel anode. Lignin was dissolved in 1 M aqueous sodium hydroxide at 10 g/L concentration and electrochemically oxidized at 50 mA/cm^2 current density. The maximum vanillin yields reported were 6% and 6.6%, respectively, from wheat straw, and bagasse lignin. This was obtained at a specific charge of 11,520 C/g lignin. Though vanillin was the major product of electro-oxidation, other low molecular weight organics were also produced in the process. Wheat straw soda lignin yielded acetosyringone in appreciable amounts, besides small quantities of syringaldehyde, apocynin, and phthalic acid. Cyclohexanone was produced in appreciable quantity from bagasse soda lignin along with small quantities of syringaldehyde, acetosyringone, and phthalic acid. Notably, electrochemical oxidation of both the soda lignins resulted in the formation of reasonable quantities of monoaromatic hydrocarbons (alkylbenzenes) and small quantities of C_9 to C_{13} linear alkanes. The main alkyl benzenes resulting from wheat straw lignin was 1,2,3-trimethylbenzene. Small amounts of benzene, styrene, 1-ethyl-2-methylbenzene, and 1-ethyl-3-methylbenzene were also detected. O-xylene was the main monoaromatic hydrocarbon resulting from bagasse lignin (Singh and Ghatak 2017). In a modified procedure, wheat straw soda lignin was thermochemically pretreated in limited air (covered crucible) for 0.5–2 h at 200–300 °C. This thermochemical pretreatment produced aromatic hydrocarbons with a maximum yield of 16–12% o-xylene and 3% toluene – with 2 h treatment at 200 °C. Pretreated lignin was dissolved in 10 g/L concentration aqueous sodium hydroxide – 1 L sodium hydroxide solution for 10 g starting lignin. The soluble part, after filtration, was electrochemically oxidized on a stainless steel anode at 2.24–6.71 mA/cm^2 current density with a specific charge of 3240–19,440 C/g starting lignin. Statistical optimization suggested 2 h thermochemical treatment at 200 °C followed by electro-oxidation at 2.24 mA/cm^2 and 4320 C/g specific charge to be the optimal process condition for getting the maximum yield of aromatic carbonyl compounds of 24.7–29% lignin degradation and 84% of it converting to target compounds. Among the individual compounds, acetosyringone was produced with the highest yield of 11.4–51.5% lignin degradation and 22% of it converting to acetosyringone – at the optimized condition of 2 h thermochemical treatment at 200 °C followed by electro-oxidation at 2.23 mA/cm^2 current density and 3190 C/g specific charge. The maximum vanillin yield was 6.7–21% lignin degradation and 31% of it converting to vanillin – with optimal process conditions of 2 h thermochemical treatment at 200 °C followed by electro-oxidation at 3.05 mA/cm^2 current density and 3774 C/g specific charge. Optimized yields of apocynin and syringaldehyde were 5.4% and 3.2%, respectively (Singh and Ghatak 2018). Without thermochemical pretreatment, the maximum yield of aromatic carbonyl compounds from this lignin was 27.7% after electro-oxidation on stainless steel anode at the optimized condition of 3.3 mA/cm^2 current density and 3505 C/g specific charge. The maximum vanillin yield was 8.1%, obtained under the optimized conditions of 4.77 mA/cm^2 current density and 4529 C/g specific charge. The maximum acetosyringone yield was 12.4%, obtained under the optimized conditions of 3.22 mA/cm^2 current density and 3057 C/g specific charge. Optimized yields of apocynin and syringaldehyde were 2.7%, and 2.6%, respectively. Bagasse soda lignin was also electro-oxidized on a stainless steel anode under similar conditions – dissolved at 10 g/L concentration in 10 g/L aqueous sodium hydroxide solution. The maximum yield of aromatic carbonyl compounds was 30.6% at the optimized condition of 2.24 mA/cm^2 current density and 2897 C/g specific charge. The maximum vanillin yield was 20%, obtained under the optimized conditions of 5.87 mA/cm^2 current density and 17,002 C/g specific charge. The maximum syringaldehyde yield was 11.3%, obtained under the optimized conditions of 6.71 mA/cm^2 current density and 1080 C/g specific charge. Optimized yields of apocynin and acetosyringone were 1.2% and 3.6%, respectively. The two lignins – wheat straw, and bagasse soda lignins – were also electro-oxidized on a tantalum-iridium mixed metal oxide anode for comparison. Optimized yields from wheat straw lignin were much lower; combined aromatic carbonyl compounds 7.2%, vanillin 3.2%, and acetosyringone 4.2%. Bagasse lignin resulted in higher yields; combined aromatic carbonyl compounds 23.1%, vanillin 10.5%, and syringaldehyde 10.5%. However, these yields were still lower than those obtained on stainless steel anode (Singh and Ghatak 2021). It is clearly evident that not only the process conditions during electro-oxidation but also the lignin type, and anode used significantly affected the yield and product composition of aromatic carbonyl compounds.

Kraft lignin (undisclosed lignocellulose source, M_w 7435, polydispersity 3.98) was dissolved at 10 mass% in selected ionic liquids – 1-ethyl-3-methylimidazolium trifluoromethanesulfonate and triethylammonium methanesulfonate – and subjected to electro-oxidation at reticulated vitreous carbon anode. The interelectrode potential was 2.5 V, the interelectrode gap 5 mm and the reaction temperature was 65 °C. Prior to electro-oxidation water was added to the solution – 1 mL water per 11 g solution. The electro-oxidation was continued for 24 h with the liquid level maintained every 1 h by the addition of water. The study confirmed the in situ production of hydrogen peroxide in the system which participated in the lignin oxidation reaction. Vanillin, apocynin, homovanillic acid, and vanillic acid were the main low molecular weight organics produced using 1-ethyl-3-methylimidazolium trifluoromethanesulfonate as the electrolyte. Vanillin, homovanillic acid, vanillic acid, and 3,4-dihydroxybenzaldehyde were the main low molecular weight organics produced using triethylammonium methanesulfonate as the electrolyte. Notably, it was possible to completely recover the ionic liquids after electro-oxidation by the solvent extraction sequence: (1) extracting the reaction mixture with methanol and filtering the solid residue, (2) extracting the filtrate with water and filtering the solid residue, (3) extracting the filtrate with diethyl ether and separating the aqueous phase, and (4) evaporating water from the aqueous phase (Dier et al. 2017).

In an interesting procedure, kraft lignin (undisclosed lignocellulose source) was dissolved in a deep eutectic solvent prepared by mixing ethylene glycol and choline chloride in a 2:1 molar ratio. Lignin concentration in the solution was 5–10 g/L. The solution was then formed into an emulsion by the addition of water or 0.1 M sulphuric acid (3-volume solution for 1-volume aqueous phase) and subjected to electro-oxidation. Used anodes were nickel foam or graphite at an interelectrode potential of 2.5–3.5 V for 1 h treatment. Emulsion stability during the electrochemical run was ensured but the nickel anode was corroded. The study did not report the identities of the low molecular weight compounds formed but their production was confirmed by the molecular weight measurements (Di Marino et al. 2017).

One modification of the lignin electro-oxidation involved the use of pulsed corona discharge. Softwood kraft lignin, dissolved at 80–600 mg/L in aqueous sodium hydroxide (10.5–11.5 pH) was circulated – in the form of droplets – through the corona zone at 15 L/min. The gaseous atmosphere in the corona discharge zone was made of air. Total reaction time was 30 min with each corona pulse lasting for 2–5 min followed by 3 min circulation. Each electrical pulse was equivalent to 0.3 J of energy and the pulse frequency varied between 100 and 840 pulse/s. Increasing pulse frequency increased lignin degradation and aldehyde (mostly vanillin) production. The authors reported a maximum vanillin yield of 0.0122% on lignin mass after 5 min electro-oxidation with 100 mg/L initial lignin concentration and 100 pulse/s discharge frequency. The vanillin yield dropped thereafter due to its subsequent oxidation to simpler compounds or complete mineralization (Panorel et al. 2014).

10.3.1.2 Organosolv Lignin

Organosolv lignin (undisclosed solvent and lignocellulose source, M_w 976, polydispersity 2.52) was dissolved at 10 mass% in selected ionic liquids – 1-ethyl-3-methylimidazolium trifluoromethanesulfonate, and triethylammonium methanesulfonate – and subjected to electro-oxidation at reticulated vitreous carbon anode. The interelectrode potential was 2.5 V, the interelectrode gap was 5 mm, and the reaction temperature was 65 °C. Prior to electro-oxidation, water was added to the solution – 1 mL water per 11 g solution. The electro-oxidation was continued for 24 h with the liquid level maintained every 1 h by the addition of water. The study confirmed the in situ production of hydrogen peroxide in the system which participated in a lignin oxidation reaction. Vanillin, apocynin, homovanillic acid, and vanillic acid were the main low molecular weight organics produced using 1-ethyl-3-methylimidazolium trifluoromethanesulfonate as the electrolyte. Vanillin, homovanillic acid, vanillic acid, and 3,4-dihydroxybenzaldehyde were the main low molecular weight organics produced using triethylammonium methanesulfonate as the electrolyte. Notably, it was possible to completely recover the ionic liquids after the electrochemical oxidation by the solvent extraction sequence: (1) extracting the reaction mixture with methanol and filtering the solid residue, (2) extracting the filtrate with water and filtering the solid residue, (3) extracting the filtrate with diethyl ether and separating the aqueous phase, and (4) evaporating water from the aqueous phase (Dier et al. 2017). In a recent work, beech organosolv lignin (undisclosed solvent), at a concentration of 10 g/L in 0.2 M sodium hydroxide, was electrochemically oxidized at BDD anode. After 3 h reaction at 10 mA/cm^2 current density and 5 mm interelectrode gap, the major products reported in the liquid reaction mixture were vanillin, syringaldehyde, hexadecanoic acid, octadecanoic acid, and 2,4-di-tert-butylphenol. Furthermore, the electrochemically oxidized lignin was acid precipitated from the reaction mixture and analysed to be transformed by conversion of a large part of the methoxyl group to phenolic hydroxyl. The study also reported extensive mineralization at current densities above 40 mA/cm^2 (Conde et al. 2023).

10.3.1.3 Hydrolysis Lignin

One research work reported the electro-oxidation of hydrolysis lignin – lignocellulose source not reported – from the Integrated Biorefinery Research Facility of National Renewable Energy Laboratory, USA. Lignin was dissolved at a concentration of 10 g/L in 1 M sodium hydroxide. Lead oxide (PbO_2) impregnated multiwalled carbon nanotube was the electroactive anode material deposited on a platinum substrate. The authors reported the production of Vanillin, methyl salicylate, 1,3-bis(1,1-dimethylethyl)-benzene, and 2,4-bis(1,1-dimethylethyl)-phenol as the main electro-oxidation products after 24 h and 48 h electro-oxidation with the anode at 0.5 V vs. SHE (Bateni et al. 2021).

There is a report of electro-oxidation of an unknown lignin type on an anode made of cobalt–platinum core–shell nanoparticles deposited on a platinum rotating disc electrode. Vanillin and apocynin were the principal reaction products after constant potential electro-oxidation with the anode at 0.596 V vs. SHE. Acetosyringone, 1,3-bis(1,1-dimethylethyl)-benzene, 2,4-di-tert-butylphenol, guaiacol, heptane, and 2,4-dimethyl-1-heptene were also produced in lesser quantity (Movil-Cabrera et al. 2016).

10.3.1.4 Lignosulphonate

Batch electro-oxidation of spruce lignosulphonate was performed on a nickel electrode at 170 °C temperature and 6.1 mA/cm^2 current density (300 mA current). Lignosulphonate was dissolved in 3 M sodium hydroxide at different concentrations. The highest vanillin yield of 9.6% was observed with 1.04 weight% lignin concentration after 150 min oxidation – corresponding specific charge 3150 C/g. It was followed by electro-oxidation in a continuous flow electrochemical cell with Expamet (a nickel mesh) anode. 1 L lignosulphonate solution, at 61 g/L concentration in 3 M sodium hydroxide, flowed through the cell at 300 L/h with a cell temperature of 145 °C and cell pressure of 4 bar. The best vanillin yield was 5.7% at 1.9 mA/cm^2 current density and 1652 C/g specific charge. A higher specific charge at this current density – more oxidation time – decreased vanillin yield, highlighting its subsequent oxidation to other compounds. The study also highlighted the crucial effect of current density on lignin electro-oxidation. With Retimet (nickel foam) anode, the vanillin yield consistently decreased with increasing current density from 1.9 to 3.8 mA/cm^2. These trends are shown in Figure 10.12. This figure also brings forth the effect of specific charge – time of

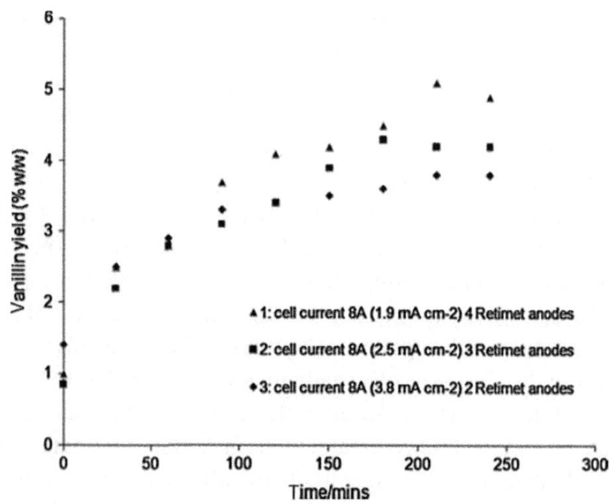

FIGURE 10.12 Effect of current density and specific charge on vanillin yield during electrochemical oxidation of spruce lignosulphonate. (Reprinted from Smith, CZ, Utley, JHP, Hammond, JK, *J. Appl. Electrochem.*, 41, 363–375, 2011 with permission from Springer.)

electro-oxidation – on vanillin yield. For current densities of 1.9 and 2.5 mA/cm^2, the vanillin yield attains a maximum at a particular specific charge – 1652 C/g for 1.9 mA/cm^2 and 1416 C/g for 2.5 mA/cm^2 – and drops thereafter (Smith et al. 2011).

10.3.1.5 Hydrogen Production in the Electrochemical Oxidation of Lignin

One important aspect of the electro-oxidation process is the simultaneous hydrogen evolution at the cathode. The reaction for acidic and alkaline conditions, respectively, can be represented as (Ghatak 2014b)

$$2H^+ + 2e^- \rightarrow H_2 \qquad \{10.6\}$$

$$2H_2O + 2e^- \rightarrow H_2 + 2OH^- \qquad \{10.7\}$$

The electrolytic off-gases from electro-oxidation processes constitute an important energy stream that stores part of the electrical energy spent. Recovering this energy could reduce the net energy consumption in the process (Ghatak 2020).

When the overall reaction (the combination of anodic and cathodic reactions) is water splitting, the minimum interelectrode potential from thermodynamic considerations is 1.23 V. At this condition, the system would be in equilibrium with no net evolution of hydrogen or oxygen. In order to drive the system away from equilibrium in the direction of hydrogen evolution, overpotentials must be surpassed at the cathode and the anode. Therefore, the actual interelectrode potential in practical water electrolysis far exceeds the minimum requirement of 1.23 V. This has implications for the energy requirement of the process as the energy consumption in electrolysis is directly proportional to the applied interelectrode potential. The usual practice to keep the interelectrode potential low involves two strategies: (1) using anodes and cathodes with low oxygen evolution and hydrogen evolution overpotentials, respectively, and (2) reducing the ohmic potential drop in the electrolyser. With lignin electro-oxidation, some of the lignin moieties participate in an anodic charge transfer process having an equilibrium potential lesser than that required for oxygen evolution. Therefore, it should be theoretically possible to work with a lower interelectrode potential compared to conventional water electrolysis (Ghatak 2006; Ghatak 2008). Recently, it was observed that platinum-ruthenium anode was limited in its electrocatalytic activity for lignin oxidation by direct charge transfer. Only the hydroxyl group was oxidized to carboxylic acid (Caravaca et al. 2019). The performance improved slightly by using a nickel carbon composite anode (Beliaeva et al. 2021). Both the anodes, however, deactivated quickly and could not sustain extended operation. One research work reported the electro-oxidation of hydrolysis lignin – lignocellulose source not reported – from the Integrated Biorefinery Research Facility of National Renewable Energy Laboratory, USA. Titanium dioxide nanoparticles were impregnated with a 1:3 ratio mixture of nickel and cobalt and deposited on carbon felt of 10 cm^2 area at 8 mg/cm^2 electrocatalyst loading to form the anode. The cathode was platinum-impregnated carbon. The two electrodes were separated by an anion exchange membrane to form the electrolysis microreactor. Lignin was dissolved at 10–100 g/L concentration in 1 M sodium hydroxide and circulated through the anolyte chamber of the reactor at 1 mL/min – 18 s residence time in contact with the anode. 1 M sodium hydroxide solution flowed through the catholyte chamber at 2 mL/min. Important comparisons were made between lignin electro-oxidation and oxygen evolution at the anode as possible anodic half-reactions to combine with cathodic hydrogen production. At low interelectrode potentials up to 1.5 V and relatively low temperatures up to 60 °C, lignin electro-oxidation at the anode was kinetically favoured over oxygen evolution. A current density of 14 mA/cm^2 was observed at 1.4 V interelectrode potential and 60 °C temperature without any associated oxygen evolution. After 24 h operation, cathodic current efficiencies varied between 95% and 100% and the electrochemical reactor worked with energy efficiencies between 86% and 100% – lower interelectrode potential giving higher energy efficiency (NaderiNasrabadi et al. 2019). Lignin oxidation by direct charge transfer on PbO_2 impregnated

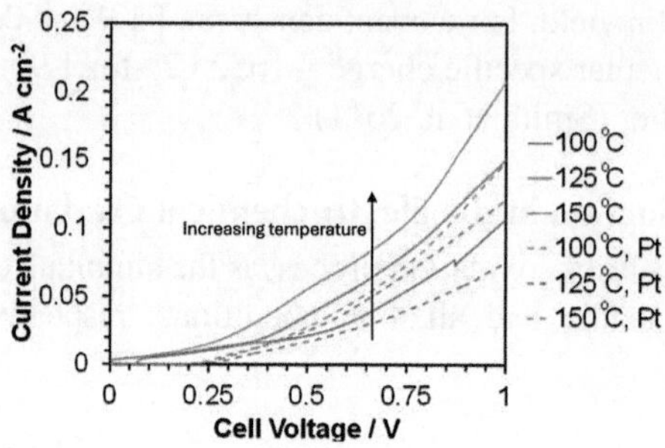

FIGURE 10.13 Comparison of current–voltage relationship during hydrogen evolution from electrochemical oxidation of lignin. Solid traces represent iron-platinum on carbon anode, and the dotted traces the platinated carbon anode. (Reprinted and adapted from Hibino, T, Kobayashi, K, Nagao, M, Teranishi, S, *ChemElectroChem*, 4, 3032–3036, 2017 with permission from Wiley.)

multiwalled carbon nanotube anode occurred at 0.69 V vs. SHE which is a lower potential than oxygen evolution reaction (Bateni et al. 2021). In all these studies, deactivation was attributed to the chemisorption of oxidation intermediates onto the electrocatalyst active sites rendering them inaccessible to subsequent reaction. In a modified approach, lignin electro-oxidation was carried out at elevated temperatures of up to 200 °C to evaluate its effect on hydrogen evolution. A platinated carbon anode was used for the purpose and a mixture of lignin with 85% orthophosphoric acid was applied on it. The cathode was iron-platinum on carbon and the two electrodes were separated by a 0.15 mm electrolyte layer of $Sn_{0.9}In_{0.1}P_2O_7$-polytetrafluoroethylene. Hydrogen evolution could be achieved at about 0.25 V interelectrode potential at 150 °C which is much lower than that required for usual water electrolysis. This onset potential for hydrogen evolution was further reduced when iron-platinum on carbon was used as an anode as well. The process achieved a current density of 208 mA/cm^2 at an interelectrode potential of 1 V (Hibino et al. 2017). The cell performance is graphically shown in Figure 10.13 for iron-platinum on carbon and platinated carbon anodes for comparison.

10.3.1.6 Anodes in Electro-Oxidation of Lignin

A number of anode materials have been used for electro-oxidation of lignin to low molecular weight compounds. The choice of anode depends on several parameters. First and foremost is the overpotential at which the relevant charge transfer processes can occur at an acceptable rate. In other words, the current–voltage relationship at the anode. This will affect the energy consumption in the process and the yield and selectivity of target compounds. An important anode characteristic is its area. From a process engineering point of view, one would like to have as much electrode area as possible for a given volume of electrochemical reactor. More anode area makes it possible to run the electrochemical reactor at a higher current for a specified current density. This would reduce the retention time in batch reactors and space time in flow reactors for a given specific charge. The geometrical arrangement and morphology of the anode are, therefore, important. For planar anodes, the geometrical area and the area available for electrochemical charge transfer are the same. On the other hand, three-dimensional anodes with complex morphologies can provide a large surface area for electrochemical charge transfer for a relatively small geometrical area. Such extended surface areas can be achieved by using metal foams. Another alternative is to deposit high specific surface area particulate electrocatalyst material on conductive substrates. Another consideration for the anode in electro-oxidation is its stability for extended operation. Anode passivation can arise from the irreversible oxidation of the anode material and the formation of an electrochemically inactive oxide layer. An anode can also lose its activity if one or more reaction products or intermediates irreversibly adsorb on its surface. A comparative account of commonly used anode materials for electro-oxidation of lignin is given in Table 10.2.

TABLE 10.2
Different Anodes in Electro-Oxidation of Lignin

Material	Type	Important Characteristics	Reference
Lead oxide (PbO_2) on lead substrate. Pb/PbO_2	Metal oxide on a metal substrate	Planar morphology. Crystal type β-PbO_2	Wang et al. 2015
Lead oxide (PbO_2) impregnated multiwalled carbon nanotubes on a platinum substrate. $Pt/MWCNT/PbO_2$	Metal oxide-carbon nanotube composite on metal substrate	3D morphology Crystal type β-PbO_2, particle size 10 nm. 25–75 weight% PbO_2 loading. Specific surface area 33–52 m^2/g. Average pore diameter 17.4–22.27 nm	Bateni et al. 2021
Ruthenium oxide-iridium oxide on titanium substrate. Ti/RuO_2-IrO_2	Mixed precious metal oxide on a metal substrate	Cracked mud-type microstructure. 30 g/m^2 oxide loading; 30% RuO_2	Tolba et al. 2010
Tantalum oxide-iridium oxide on titanium substrate. Ti/Ta_2O_5-IrO_2	Mixed precious metal oxide on a metal substrate	Planar morphology	Singh and Ghatak 2021
Nickel fleece	Metal fabric	3D morphology. Specific surface 9942 m^2/m^3 Fibre diameter 12 μm	Stiefel et al. 2016
Nickel foam	Metal foam	3D morphology. Specific surface 200,000 m^2/m^3 Porosity 40% Charge transfer area 20–100 times the planar geometric area	Smith et al. 2011; Stiefel et al. 2016
Stainless steel	Alloy SS-304	Planar morphology	Singh and Ghatak 2017; Singh and Ghatak 2018
Copper	Metal	Planar morphology	Parpot et al. 2000
Nickel	Metal	Planar morphology	Parpot et al. 2000; Stiefel et al. 2016
Platinum	Precious metal	Planar morphology	Parpot et al. 2000
Nickel–cobalt on titanium	Mixed metal on a metal substrate	Planar morphology. Mixed metal loading 1.55 mg/cm^2; 20% nickel, 80% cobalt. Grain size is about 100 nm	Ghahremani and Staser 2018
Cobalt–platinum nanoparticles on a platinum substrate	Mixed metal nanoparticles on a metal substrate	Planar morphology. Cobalt core–platinum shell nanoparticles; particle size 4 nm	Movil-Cabrera et al. 2016
Reticulated vitreous carbon foam	Conducting carbon	3D morphology. Porosity 96.5%	Dier et al. 2017

10.4 SEPARATION AND PURIFICATION OF LOW MOLECULAR WEIGHT ORGANIC COMPOUNDS

In all the processes for converting lignin into low molecular weight organic compounds – oxidation with molecular oxygen, pyrolysis, or electro-oxidation – it ends up with a product mix having a number of organic compounds. For a composite process for practicable value-added application of lignin, it is imperative to separate the individual compounds. Moreover, the unreacted lignin must also be separated for possible recycling, other value-added applications, or disposal. With the present state of research and development, it should not be difficult to chemically transform lignin into low molecular weight compounds. Similarly, the separation of unreacted lignin should not be very difficult given the current know-how and process development. The critical aspect and the real bottleneck in developing a composite process for deriving low molecular weight organics from lignin remains the separation and purification of the individual compounds. With their almost exclusive guaiacyl nature, softwood lignins result in a comparatively less diverse product mix. It is more complicated with hardwood and nonwood lignins having both guaiacyl and syringyl moieties. Some research effort has gone into overcoming this challenge, but a realistic solution is still elusive.

10.4.1 Separation of Unreacted Lignin from the Reaction Mixture

Oxidation with molecular oxygen and electro-oxidation are the two lignin conversion processes where the target low molecular weight compounds produced are themselves liable to further react and convert into undesirable or less desirable products. In extreme cases, complete mineralization is also possible. Therefore, these processes are stopped well before the complete lignin is degraded to maximize the yield of target compounds. The undegraded lignin remaining at the end of the reaction must be separated to facilitate the recovery of target products. In an overwhelming majority of cases, lignin oxidation with molecular oxygen, and electro-oxidation is carried out in an alkaline solution. As a result, acidification is the common approach to precipitate the undegraded lignin from the solution. The process fundamentals and steps remain the same as that for lignin separation from the black liquor after the alkaline delignification of lignocelluloses. In brief, the reaction mixture is acidified to the desired pH – 2–4 – with a mineral acid to precipitate the lignin. The precipitated solids are then filtered or centrifuged from the remaining liquid. Overall, it amounts to a recurring input of chemicals to run the entire process – sodium hydroxide to dissolve the lignin before the reaction, and mineral acid to neutralize the alkali after the reaction.

10.4.2 Separation of Low Molecular Weight Compounds from the Reaction Mixture

10.4.2.1 Product Recovery through Bisulphite Complex Formation

The industrial process for vanillin production from softwood lignins – mainly lignosulphonates – involves complex downstream processing of the reaction mixture after the oxidation step with molecular oxygen. The primary reaction product from oxidation of softwood lignins is vanillin with small amounts of apocynin and vanillic acid also present. Vanillin can be separated from the other reaction products by its selective reaction with sodium bisulphite. Vanillin, being an aromatic aldehyde, would form a complex with sodium bisulphite while the ketones (apocynin, for example) and carboxylic acids (vanillic acid, for example) do not form such a complex. The use of sodium bisulphite can be substituted by passing sulphur dioxide into the reaction mixture. Under the prevailing conditions, the aldehyde complex remains in solution while the impurities, including unconverted lignin, precipitate out. After filtering the impurities, the aqueous solution of the vanillin bisulphite complex is acidified to break the complex and recover the vanillin. Sulphur dioxide is released in the process (Sandborn et al. 1936). The process can only be adapted to sulphite delignification processes with relative ease because of the use of sulphur dioxide as a delignification chemical. The filtration step poses serious difficulty in the process owing to the nature of the lignin precipitate. In a modification to the process, it was proposed to extract the low molecular weight reaction products with a suitable organic liquid – butyl alcohol or isopropanol, for example – and recycle the alkaline aqueous phase back to the oxidation step. The organic phase containing the reaction products including vanillin can either be steam-distilled or extracted with water to transfer the organic products to the aqueous phase. Vanillin can be recovered from this aqueous phase by usual complex formation with bisulphite, filtering out the precipitated byproducts, and breaking the complex by acidification (Major et al. 1977; Sandborn et al. 1938). The final purification of vanillin is performed by crystallization.

The above method of product recovery is difficult to use with lignin sourced from hardwoods and nonwoods, and isolated with a process other than the sulphite process. Hardwood and nonwood lignins have syringyl entities in appreciable amounts besides the guaiacyl structures. In addition, some p-hydroxybenzyl entities are also present. As a result, the reaction mixture after the oxidation contains syringaldehyde, and p-hydroxybenzaldehyde, along with vanillin as aldehyde products. All of them would form the bisulphite complex and the separation of individual compounds would not be possible. Moreover, the bisulphite complex-based product recovery process is difficult to integrate with the lignin isolation process when it is other than the sulphite process. Extensive research and development work is currently underway to devise a viable process for product recovery after lignin oxidation that can substitute the bisulphite complex-based process. The majority of these studies are based on synthetic aqueous solutions of a single compound – mostly vanillin or syringaldehyde – or their mixture.

10.4.2.2 Product Recovery through Liquid Extraction

In experiments with 0.05 M synthetic aqueous solutions of vanillin, it was extracted with a solution of monooctylamine in heptane. pH of the vanillin aqueous solution was maintained with acetate buffer. The maximum distribution coefficient was achieved at 8–10 pH of the vanillin solution. The maximum distribution coefficient remarkably increased with increased amine concentration. Even with a small amine concentration of 0.36 M, the distribution coefficient was several orders of magnitude higher than that obtained with pure heptane as an extracting solvent. It was proposed that monooctylamine reacted with the carbonyl functional group in vanillin to form a complex (Tarabanko et al. 2007). Similarly, vanillin was extracted from its synthetic aqueous solution with butane-1-ol, hexane-1-ol, heptane-1-ol, or octane-1-ol. The use of hexane-1-ol resulted in the highest distribution coefficient of 30.5 at an acidic pH of 1–4. At this pH range, the lowest distribution coefficient of 20.3 was for octane-1-ol. Distribution coefficients drastically dropped to less than 1 for all the solvents in the alkaline pH range of 11–14 (Kaygorodov et al. 2010).

Extraction of low molecular weight reaction products from lignin oxidation with ionic liquids is another possibility being explored. In one study, vanillin was extracted from a synthetic aqueous matrix by extraction with several ionic liquids. It was observed that a mixture consisting of a 60 weight% aqueous solution of vanillin, 15 weight% of potassium orthophosphate, and 25 weight% of ionic liquid formed a two-phase system – a potassium orthophosphate rich aqueous phase and an ionic liquid-rich aqueous phase. In this system vanillin preferentially transferred from the potassium orthophosphate-rich phase to the ionic liquid-rich phase. Vanillin concentration in the aqueous solution varied between 0.5 and 7.5 g/L. Extraction was effected by vigorous stirring of the constituents followed by phase separation for 12 h to attain equilibrium. The highest partition coefficient of 49.59 was achieved with 1-hexyl-3-methylimidazolium chloride as the ionic liquid used at 25 °C and 1 g/L initial vanillin concentration. The ionic liquids 1-butyl-3-methylimidazolium chloride, 1-benzyl-3-methylimidazolium chloride, and 1-heptyl-3-methylimidazolium chloride also had partition coefficients above 42. The partition coefficient significantly increased with an increase in the initial concentration of vanillin (Claudio et al. 2010).

In an interesting procedure, electrochemical lignin oxidation was combined with simultaneous product separation by emulsion liquid membrane. Kraft lignin (undisclosed lignocellulose source) was dissolved in a deep eutectic solvent prepared by mixing ethylene glycol and choline chloride in a 2:1 molar ratio. Lignin concentration in the solution was 5–10 g/L. The solution was then diluted by the addition of water (3 volume solution for 1 volume water). Methyl isobutyl ketone was added to this solution as extractant phase maintaining a ratio of 1–1.15 between the amounts of the extractant and the rest of the system. The formed emulsion with 5 g/L lignin concentration was stable for 1 h while that with 10 g/L lignin was stable for 24 h. The composite emulsion was subjected to electro-oxidation over nickel foam or graphite anodes at an interelectrode potential of 2.5–3.5 V for 1 h. After the electro-oxidation, the emulsion was centrifuged to separate the phases; organic methyl isobutyl ketone phase, aqueous eutectic solvent phase, and precipitated unconverted lignin phase.

The study did not report the identities of the low molecular weight compounds formed but their production was confirmed by the molecular weight measurements. Low molecular weight products migrated into the extractant phase (Di Marino et al. 2017). This preliminary study is significant as it can devise a composite process of lignin electro-oxidation and product separation. Since the products are transferred to the extractant phase as they are formed, the chances of their further oxidation to undesired products should be minimized. At the same time, having an emulsion with a large amount of organics may cause a major increase in the ohmic potential drop. This will incur additional energy consumption in the process.

10.4.2.3 Product Recovery through Adsorption and Ion Exchange

Adsorption of vanillin from synthetic aqueous solution – 3.3 mmol/L concentration, 5.3 pH – to non-ionic polymeric resin adsorbent SP206 was carried out. The resin had a specific surface area of 556 m^2/g, particle size of 0.4 mm, and pore radius of 10–15 nm. The resin showed a saturation adsorption capacity of 7.534×10^{-4} mol/g for vanillin. Adsorption continuously decreased when the solution pH was increased up to 9.5 with the addition of sodium hydroxide. The authors contended that the addition of sodium hydroxide converted a part of vanillin into its vanillate form which did not adsorb onto the resin. It was further confirmed by fixed bed adsorption studies where the exhausted bed could be effectively regenerated by elution with 1 M sodium hydroxide solution (Zabkova et al. 2006). Vanillin and syringaldehyde were separated from the spent liquid from alkaline oxygen delignification of wheat straw pulp by adsorption. The spent liquor had a vanillin concentration of 215 mg/L, and a syringaldehyde concentration of 198.5 mg/L. Additionally, apocynin, acetosyringone, and p-Hydroxyacetophenone were also present in reasonable concentrations of 27.7, 134, and 39 mg/L, respectively. The adsorbent used was a non-polar polydivinylbenzene-based resin D101. The spent liquid was adjusted to pH 4.5 and the combined saturation adsorption capacity for vanillin and syringaldehyde was determined to be 2.6×10^{-5} mol/g at 27 °C. Adsorbed aldehydes could be desorbed by elution with ethyl ether. Recoveries of vanillin and syringaldehyde were reported to be 96.2% and 94.7%, respectively (Wang et al. 2010). Batch and fixed bed continuous adsorption experiments with synthetic aqueous solutions of vanillin, and syringaldehyde established the effectiveness of the resin SP700, polystyrene-divinylbenzene-based non-polar resin, as adsorbent. The resin was characterized to have the characteristics of 0.45 mm average particle size, 2.3 mL/g pore volume, 1200 m^2/g specific surface area, and 9 nm average pore size. The maximum adsorption capacity of SP700 for vanillin and syringaldehyde was estimated to be 0.663 and 0.707 g/g dry resin, respectively. Isosteric adsorption enthalpy was determined as –20.5 kJ/mol for vanillin, and –20.1 kJ/mol for syringaldehyde. These values confirm physical adsorption. The exponent in Freundlich adsorption isotherm was computed to be 2.7 for vanillin and 2.8 for syringaldehyde. Both the compounds could be effectively desorbed with deionized water after fixed bed continuous adsorption experiments. The reported extent of desorption was 83–85% (Mota et al. 2016a). Adsorption of these compounds onto SP700 from a solution in 90 vol% aqueous ethanol was less effective. The exponent in Freundlich adsorption isotherm was computed to be 1.1 for both vanillin and syringaldehyde. In column studies, 90 vol% aqueous ethanol was found to be very effective in desorbing the compounds. The extent of desorption was 87–99% for vanillin, and 96–100% for syringaldehyde (Mota et al. 2016b).

10.4.2.4 Product Recovery through Membrane Separation

Synthetic aqueous alkaline solutions of lignin and vanillin were filtered through different ceramic ultrafiltration membranes of 1000, 5000, and 15,000 Da molecular weight cutoff. A solution of 60 g/L lignin and 6 g/L vanillin at pH 12.5 experienced a rapid flux decline when ultrafiltered with 1000 Da molecular weight cutoff membrane at a transmembrane pressure (TMP) of 1.55×10^5 Pa and a feed rate of 130 L/h. Only 10% of the feed permeated through the membrane. Reasonable permeate flux could be maintained for 17 h ultrafiltration of 60 g/L lignin and 5 g/L vanillin solution at pH 12.5 through 5000 Da membrane operated at 1.55 × 10^5 Pa TMP and feed rate of 120 L/h. 27% of the feed permeated through the membrane. Retentate had the lignin concentration increased to 90 g/L with the vanillin concentration remaining almost constant at 5 g/L. Permeate was substantially enriched in vanillin with its concentration remaining at 5 g/L but the lignin concentration decreasing to less than 3 g /L. Good separation was also achieved with a 15,000 Da molecular weight cutoff membrane with solutions of different compositions and pH. 34% of feed volume permeated from a solution of 60 g/L lignin and 6 g/L vanillin at pH 12.5 after 17 h. Vanillin concentration remained essentially the same at 6 g/L in both retentate and permeate. Lignin concentration in retentate increased to about 100 g/L while it dropped to about 8–12 g/L in the permeate. Permeated volume was 31.3% of feed at pH 8.5 after only 9.5 h with retentate and permeate lignin concentrations of about 80 and 4 g/L, respectively (Zabkova et al. 2007). In one study, oxidized eucalyptus kraft black liquor was subjected to sequential membrane separation. To better simulate the reaction mixture from lignin oxidation the oxidized black liquor was enriched with aromatic aldehydes and carboxylic acids commonly produced during lignin oxidation. Enriched black liquor was first filtered through a 50,000 Da molecular weight cutoff ceramic membrane with a zirconia active layer supported on titania–alumina. Next, permeate from the membrane was filtered through a 5000 Da molecular weight cutoff ceramic membrane with titania active layer on patented titania–alumina–zirconia support. Finally, permeate from the second membrane was filtered through a third ceramic membrane of 1000 Da molecular weight cutoff and similar active layer and support as the second membrane. Each membrane separation step was operated at ambient temperature, 1.4 bar TMP, and 120 L/h retentate flow rate. The first membrane retained 38% volume of the respective feed, the second membrane retained 29% volume of the respective feed, and the third membrane retained 33.3% volume of the respective feed. The initial feed had a total solids concentration of 86.5 g/L of which 2.77% were target aromatic aldehydes and carboxylic acids, and 45.4% unconverted lignin and undesirable organic products. In the final permeate after three-stage membrane filtration, the total solids concentration was 58.6 g/L with 3.67% of total solids being target aromatic compounds, and 32.34% unconverted lignin and undesirable organic products. The total recovery of target compounds turned out to be around 80% (Mota et al. 2018). Another study reported 2–8.4% single-pass vanillin recovery from a synthetic aqueous solution of vanillin by membrane filtration with a polypropylene microporous membrane. Membrane characteristics were: 200 μm fibre thickness, 0.6 vol% fibre porosity, 2 fibre tortuosity, and 300 μm fibre inside diameter. Aqueous vanillin solution at 25 mg/L concentration flowed at 1–45 L/h through either the tube side or shell of the membrane module. An organic extraction liquid flowed on the other side of the membrane at a flow rate of 10–45 L/h countercurrent to the aqueous phase flow. The lowest recovery was achieved with 2-ethyl 1-hexanol as the organic phase. The highest recovery was reported with butyl acetate as the organic phase with the aqueous phase flowing through the tube side (Cao et al. 2021). It must be noted that the actual reaction mixture obtained after lignin oxidation would be much different from the synthetic aqueous phase used in this study both in terms of physical characteristics and chemical composition. Needless to say, the extent of separation and purification falls far short of what would be expected for obtaining the individual low molecular weight compounds. Even the unconverted lignin is not separated effectively from the low molecular weight compounds.

One particular study combined electro-oxidation with membrane separation to continuously separate the low molecular weight products of electro-oxidation and prevent their degradation to undesirable products. Beech kraft lignin at 5 g/L lignin concentration in 1 M aqueous sodium hydroxide solution served as the initial feed and was fed to the anode chamber of the electrochemical cell at 50 mL/min. The anode and the cathode were separated by an anion exchange membrane. Post electrochemical oxidation, the solution was filtered through a 1700 Da molecular weight cutoff membrane at a flux of 2.2 L/(m^2.h) across the membrane. Permeate was taken as the final product while the retentate was recirculated to the anode chamber. It was observed that the permeate was largely enriched in low molecular weight products (Stiefel et al. 2016).

10.4.2.5 Composite Process for Product Recovery, Separation, and Purification

In an important study, a reaction mixture from the oxidation of tobacco stalk butanol organosolv lignin with molecular oxygen was first filtered through a 5000 Da molecular weight cutoff membrane at a TMP of 20 bar. Permeate from the membrane – 23.5 g/L organics and 105.8 g/L inorganics – was fractionated by adsorption on polystyrene-divinylbenzene-based non-polar resin SP700. The low molecular weight reaction products in the permeate were p-hydroxybenzaldehyde (11–15 mg/L), vanillin (56–114 mg/L), syringaldehyde (17–57 mg/L), vanillic acid (146–165 mg/L), syringic acid (142–155 mg/L), apocynin (36–40 mg/L), and acetosyringone (3–35 mg/L). Adsorption was carried out for 3 h in resin packed column of 0.37 porosity, 540 s space time, 555 Peclet number, and 10 mL/min liquid flow rate. It was followed by desorption, first with deionized water for 1 h, then with ethanol for another 1 h. An analysis of the liquid from the column outlet revealed that the carboxylic acids did not effectively adsorb on the resin and their outlet concentration quickly reached the feed concentration and remained nearly constant through the adsorption run. Accordingly, the outlet concentration of acids dropped to zero soon after the adsorption run. On the contrary, practically no ketones were detected in the outlet liquid during the adsorption run indicating their strong and near-complete adsorption on the resin. Adsorption of aldehydes was also reasonably strong – but weaker than that of ketones – and their concentrations in the outlet liquid were negligible during the adsorption run. Therefore, it was possible to effectively separate the carboxylic acids with negligible aldehydes and ketones contamination from the adsorption run on the resin. Desorption with deionized water preferentially removed the aldehydes but small amounts of ketones were also present in the outlet liquid from the column. Finally, when desorption was carried out with ethanol, ketones were the major components at the column outlet with a small presence of aldehydes also detected. The concentration profiles of different compounds in the column outlet with the passage of time are presented in Figure 10.14 for a permeate pH of 10. Recovery of individual aldehydes and ketones depended on the pH of the permeate from the membrane filtration step. Vanillin recovery was highest and close to 100% at pH 10. Syringaldehyde recovery was highest and close to 90% at pH 12. Apocynin recovery was highest and close to 85% at pH 9. Acetosyringone recovery was highest and close to 90% at pH 9 (Gomes et al. 2018).

In a related work, the reaction mixture from the oxidation of Indulin AT with molecular oxygen was first filtered through a 1000 Da molecular weight cutoff membrane at 8 bar TMP followed by filtration of the permeate through a 600 Da molecular weight cutoff membrane at 30 bar TMP. Permeate from the second membrane was similarly fractionated by adsorption on resin SP700. The aldehyde – mainly vanillin as Indulin AT – is a softwood kraft lignin-rich liquid phase from desorption with deionized water and was further purified by two-stage extraction with ethyl acetate and crystallization to obtain pure vanillin. The aldehyde-rich fraction had a vanillin concentration of 2.16 g/L after acidification with sulphuric acid to pH 6. One volume of ethyl acetate was used for one volume of acidified aldehyde-rich liquid. Vanillic acid (0.48 g/L), apocynin (0.12 g/L), and p-hydroxybenzaldehyde (0.09 g/L) were present in the liquid along with vanillin. Vanillin extraction was 97% after the first stage extraction and complete after the second stage extraction. Only 13% vanillic acid was extracted resulting in a significant purification during extraction itself. Solvent from the combined extract phase was evaporated under vacuum to obtain concentrated crude product which was redissolved in water to initiate nucleation. A low vanillic acid concentration positively affected the nucleation and growth of vanillin crystals. The best vanillin recovery yield was 80% at 96% purity (Gomes and Rodrigues 2020).

Separation of the two main aldehyde products, vanillin and syringaldehyde, can be achieved through crystallization. There is a report of the separation of vanillin and syringaldehyde from their 214 g/L aqueous solution – vanillin to syringaldehyde mass ratio 29.5:70.5 – by controlled addition of 20% aqueous ammonia and preferential crystallization of syringaldehyde. Initially, the solution of vanillin and syringaldehyde was brought to a pH of 7.8 by the addition of ammonia solution. Thereafter, aqueous ammonia was added dropwise to the aldehyde solution which caused selective

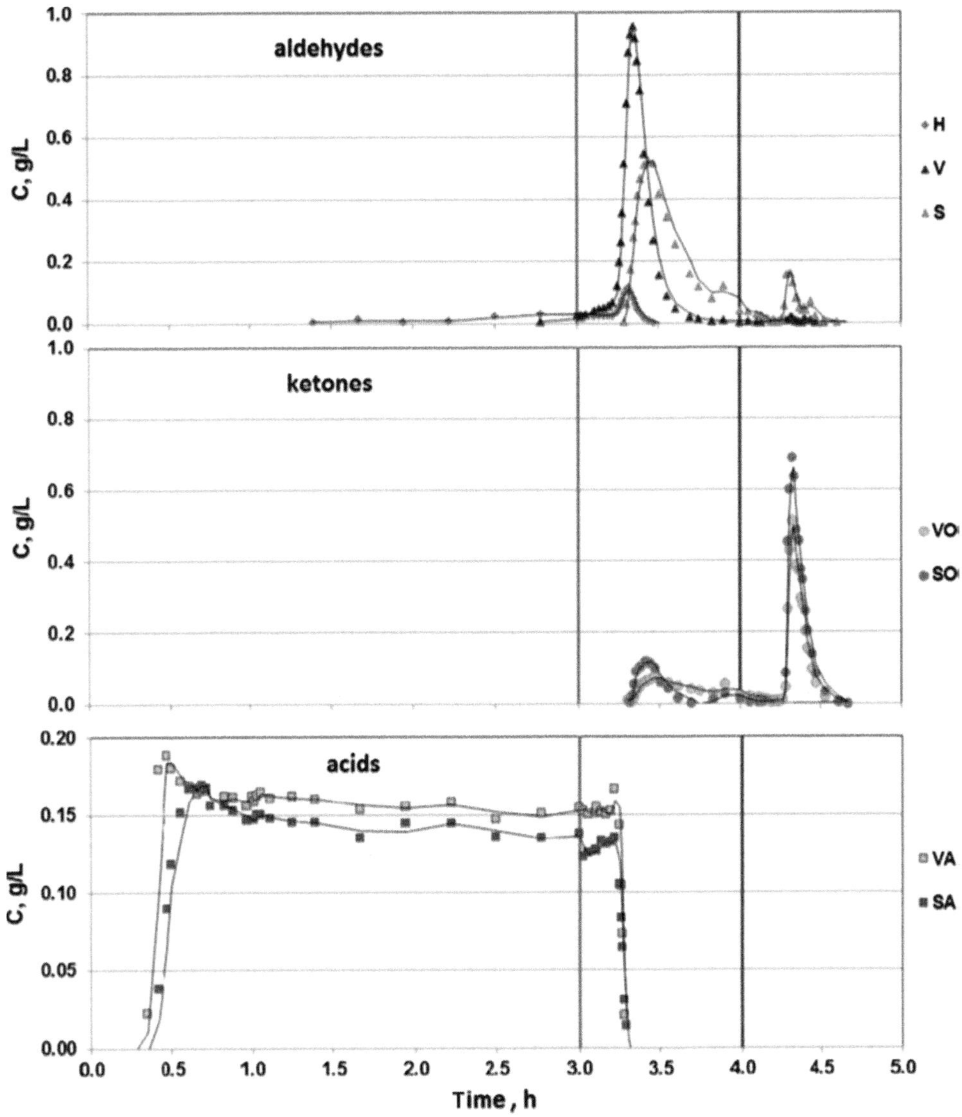

FIGURE 10.14 Time-dependent concentration profiles of aldehydes, ketones, and carboxylic acids at the SP700 resin-filled column outlet during fractionation by adsorption-desorption. The first vertical line represents the start of desorption with deionized water and the second vertical line represents the start of desorption with ethanol. Product legends: H, p-hydroxybenzaldehyde; V, vanillin; S, syringaldehyde; VO, apocynin; SO, acetosyringone; VA, vanillic acid; SA, syringic acid (Reprinted from Gomes, ED, Mota, MI, Rodrigues, AE, *Sep. Purific. Technol.*, 194, 256–264, 2018 with permission from Elsevier.)

crystallization and precipitation of white syringaldehyde-ammonia complex. The precipitate was filtered, water washed, dispersed in water, and acidified to pH 3.5 with sulphuric acid. This caused the expulsion of ammonia and recrystallization of about 99% pure syringaldehyde. Filtrate from the dropwise ammonia addition step and the precipitate washings were acidified to pH 3.5 with sulphuric acid and allowed to stand still for several hours. This resulted in the crystallization of 97–98% pure vanillin. Syringaldehyde recovery was 95.3% and vanillin recovery was 57.4% (Tarabanko et al. 2013).

REFERENCES

Adhikari, S, Srinivasan, V, Fasina, O. *Energy Fuels* 28:4532–4538, 2014.

Asgarai, F, Argyropoulos, DS. *Can. J. Chem.* 76:1606–1615, 1998.

Azad, T, Auad, ML, Elder, T, Adamczyk, AJ. *Energy Fuels* 37:401–423, 2023.

Bateni, F, Ghahremani, R, Staser, JA. *J. Appl. Electrochem.* 51:65–78, 2021.

Beis, SH, Mukkamala, S, Hill, N, Joseph, J, Baker, C, Jensen, B, Stemmler, EA, Wheeler, MC, Frederick, BG, van Heiningen, A, Berg, AG, DeSisto, WJ. *BioResources* 5:1408–1424, 2010.

Beliaeva, K, Grimaldos-Osorio, N, Ruiz-Lopez, E, Burel, L, Vernoux, P, Caravaca, A. *Int. J. Hyd. Energy* 46:35752–35764, 2021.

Bjørsvik, HR, Minisci, F. *Organic Proc. Res. Dev.* 3:330–340, 1999.

Borella, M, Casazza, AA, Garbarino, G, Riani, P, Busca, G. *Energies* 15:991, 2022.

Brebu, M, Cazacu, G, Chirila, O. Cellulose *Chem. Technol.* 45:43–50, 2011.

Bu, Q, Lei, H, Qian, M, Yadavalli, G. *RSC Adv.* 6:100700–100707, 2016.

Cao, Y, Ghadiri, M, Rezakazemi, M, Marjani, A, Pishnamazi, M, Shirazian, S. *J. Molecular Liquids* 323:114606, 2021.

Caravaca, A, Garcia-Lorefice, WE, Gil, S, de Lucas-Consuegra, A, Vernoux, P. *Electrochem. Commun.* 100: 43–47, 2019.

Choi, HS, Meier, D, Windt, M. *Environ. Progress Sustain. Energy* 31:240–244, 2012.

Cláudio, AFM, Freire, MG, Freire, CSR, Silvestre, AJD, Coutinho, JAP. *Sep. Purific. Technol.* 75:39–47, 2010.

Conde, JJ, Gonzalez-Rodríguez, S, Chen, X, Lu-Chau, TA, Eibes, G, Pizzi, A, Moreira, MT. *Biomass Bioenergy* 169:106693, 2023.

Custodis, VBF, Hemberger, P, Ma, Z, van Bokhoven, JA. *J. Phys. Chem. B* 118:8524–8531, 2014.

da Silva, EAB, Zabkova, M, Araujo, JD, Cateto, CA, Barreiro, MF, Belgacem, MN, Rodrigues, AE. *Chem. Eng. Res. Des.* 87:1276–1292, 2009.

Deng, H, Lin, L, Sun, Y, Zhuang, J, Ouyang, P, Li, J, Liu, S. *Catal. Lett.* 126:106–111, 2008.

Deng, H, Lin, L, Sun, Y, Pang, S, Zhuang, J, Ouyang, P, Li, J, Liu, S. *Energy Fuels* 23:19–24, 2009.

Deng, H, Lin, L, Liu, S. *Energy Fuels* 24:4797–4802, 2010.

Deng, W, Zhang, H, Wu, X, Li, R, Zhang, Q, Wang, Y. *Green Chem.* 17:5009–5018, 2015.

de Wild, P, der Laan, RV, Kloekhorst, A, Heeres, E. Environ. Progress Sustain. *Energy* 28:461–469, 2009.

de Wild, PJ, Huijgen, WJJ, Kloekhorst, A, Chowdari, RK, Heeres, HJ. *Bioresource Technol.* 229:160–168, 2017.

Dier, TKF, Rauber, D, Durneata, D, Hempelmann, R, Volmer, DA. *Scientific Rep.* 7:5041, 2017.

Di Marino, D, Aniko, V, Stocco, A, Kriescher, S, Wessling, M. *Green Chem.* 19:4778–4784, 2017.

Elfadly, AM, Zeid, IF, Yehia, FZ, Abouelela, MM, Rabie, AM. *Fuel. Process. Technol.* 163:1–7, 2017.

Fan, L, Chen, P, Zhou, N, Liu, S, Zhang, Y, Liu, Y, Wang, Y, Omar, MM, Peng, P, Addy, M, Cheng, Y, Ruan, R. *Bioresource Technol.* 247:851–858, 2018.

Fargues, C, Mathias, A, Rodrigues, A. *Ind. Eng. Chem. Res.* 35:28–36, 1996.

Ferdous, D, Dalai, AK, Bej, SK, Thring, RW. *Energy Fuels* 16:1405–1412, 2002.

Ghahremani, R, Staser, JA. *Holzforschung* 72:951–960, 2018.

Ghatak, HR. *Int. J. Hyd. Energy* 31:934–938, 2006.

Ghatak, HR, Kumar, S, Kundu, PP. *Int. J. Hyd. Energy* 33:2904–2911, 2008.

Ghatak, HR. *Crit. Rev. Environ. Sci. Technol.* 44:1167–1219, 2014a.

Ghatak, HR. *Environ. Technol.* 35:2483–2492, 2014b.

Ghatak, HR. *Int. J. Hyd. Energy* 45:31466–31480, 2020.

Ghysels, S, Dubuisson, B, Pala, M, Rohrbach, L, den Bulcke, JV, Heeres, HJ, Ronsse, F. *Green Chem.* 22: 6471–6488, 2020.

Gomes, ED, Mota, MI, Rodrigues, AE. *Sep. Purific. Technol.* 194:256–264, 2018.

Gomes, ED, Rodrigues, AE. *Sep. Purific. Technol.* 247:116977, 2020.

Goncalves, AR, Schuchardt, U. *Appl. Biochem. Biotechnol.* 77–79:127–132, 1999.

Han, T, Ding, S, Yang, W, Jönsson, P. *Chem. Eng. J.* 373:846–856, 2019.

Hibino, T, Kobayashi, K, Nagao, M, Teranishi, S. *ChemElectroChem* 4:3032–3036, 2017.

Jackson, MA, Compton, DL, Boateng, AA. *J. Anal. Appl. Pyrol.* 85:226–230, 2009.

Jin, T, Wang, H, Pang, J, Wu, Y, Huang, Z, Tian, X, Ding, M. *Fuel Process. Technol.* 230:107201, 2022.

Kaygorodov, KL, Chelbina, YV, Tarabanko, VE, Tarabanko, NV. *J. Siberian Federal Univ.* 3(3):228–233, 2010.

Kaygorodov, KL, Tarabanko, VE, Chernyak, MY, Chelbina, YV, Tarabanko, N, Smirnova, MA. *Russian J. Bioorganic Chem.* 44:839–844, 2018.

Kibet, J, Khachatryan, L, Dellinger, B. *Environ. Sci. Technol.* 46:12994–13001, 2012.

Kim, JY, Choi, JW. *Fuel* 240:92–100, 2019.

Kim, JY, Lee, JH, Park, J, Kim, JK, An, D, Song, IK, Choi, JW. *J. Anal. Appl. Pyrol.* 114:273–280, 2015.
Lang, L, Zhao, S, Yin, X, Yang, W, Wu, C. *Int. J. Hyd. Energy* 40:9924–9934, 2015.
Lee, HW, Kim, YM, Jae, J, Sung, BH, Jung, SC, Kim, SC, Jeon, JK, Park, YW. *J. Anal. Appl. Pyrol.* 122: 282–288, 2016.
Li, D, Briens, C, Berruti, F. *Bioresource Technol.* 189:7–14, 2015.
Lin, X, Sui, S, Tan, S, Pittman Jr, CU, Sun, J, Zhang, Z. *Energies* 8:5107–5121, 2015.
Major FW, Nicolle, FMA. US Patent 4021493, 1977.
Marathe, PS, Westerhof, RJM, Kersten, SRA. *Appl. Energy* 236:1125–1137, 2019.
Mathias, AL, Rodrigues, AE. *Holzforschung* 49:273–278, 1995.
Mota, MIF, Pinto, PCR, Loureiro, JM, Rodrigues, AE. *Chem. Eng. J.* 288:869–879, 2016a.
Mota, MIF, Pinto, PCR, Loureiro, JM, Rodrigues, AE. *Chem. Eng. J.* 294:73–82, 2016b.
Mota, MIF, Pinto, PCR, Ribeiro, AM, Loureiro, JM, Rodrigues, AE. *Sep. Purific. Technol.* 197:360–371, 2018.
Mottweiler, J, Puche, M, Rauber, C, Schmidt, T, Concepcion, P, Corma, A, Bolm, C. *ChemSusChem* 8: 2106–2113, 2015.
Movil-Cabrera, O, Rodriguez-Silva, A, Arroyo-Torres, C, Staser, JA. *Biomass Bioenergy* 88:89–96, 2016.
Mukkamala, S, Wheeler, MC, van Heiningen, ARP, DeSisto, WJ. *Energy Fuels* 26:1380–1384, 2012.
NaderiNasrabadi, M, Bateni, F, Chen, Z, Harrington, PB, Staser, JA. *J. Electrochem. Soc.* 166:E317–E322, 2019.
Pacek, AW, Ding, P, Garrett, M, Sheldrake, G, Nienow, AW. *Ind. Eng. Chem. Res.* 52:8361–8372, 2013.
Pan, K, Tian, M, Jiang, ZH, Kjartanson, B, Chen, A. *Electrochimica Acta* 60:147–153, 2012.
Panorel, I, Kaijanen, L, Kornev, I, Preis, S, Louhi-Kultanen, M, Sirén, H. *Environ. Technol.* 35:171–176, 2014.
Park, YK, Jung, JS, Jae, J, Hong, SB, Watanabe, A, Kim, YM. *Chem. Eng. J.* 377:119742, 2019.
Parpot, P, Bettencourt, AP, Carvalho, AM, Belgsir, EM. *J. Appl. Electrochem.* 30:727–731, 2000.
Partenheimer, W. *Adv. Synth. Catal.* 351:456–466, 2009.
Peng, C, Zhang, G, Yue, J, Xu, G. *Fuel Process. Technol.* 124:212–221, 2014.
Pienihäkkinen, E, Lindfors, C, Ohra-aho, T, Lehtonen, J, Granström, T, Yamamoto, M, Oasmaa, A. *Energy Fuels* 35:14758–14769, 2021.
Pinto, PCR, da Silva, EAB, Rodrigues, AE. *Ind. Eng. Chem. Res.* 50:741–748, 2011.
Ryu, HW, Lee, HW, Jae, J, Park, YK. *Energy* 179:669–675, 2019.
Sales, FG, Maranhao, LCA, Filho, NML, Abreu, CAM. *Ind. Eng. Chem. Res.* 45:6627–6631, 2006.
Sales, FG, Maranhao, LCA, Filho, NML, Abreu, CAM. *Chem. Eng. Sci.* 62:5386–5391, 2007.
Sandborn LT, Salvesen, J, Howard, GC. US Patent 2057117, 1936.
Sandborn, LT. US Patent 2104701, 1938.
Santos, SG, Marques, AP, Lima, DLD, Evtuguin, DV, Esteves, VI. *Ind. Eng. Chem. Res.* 50:291–298, 2011.
Schmitt, D, Regenbrecht, C, Hartmer, M, Stecker, F, Waldvogel, SR. *Beilstein J. Org. Chem.* 11:473–480, 2015.
Shafaghat, H, Rezaei, PS, Ro, D, Jae, J, Kim, BS, Jung, SC, Sung, BH, Park, YK. *J. Ind. Eng. Chem.* 54: 447–453, 2017.
Shen, D, Zhao, J, Xiao, R. *Energy Conv. Manage.* 124:61–72, 2016.
Singh, S, Ghatak, HR. *J. Wood Chem. Technol.* 37:407–422, 2017.
Singh, S, Ghatak, HR. *Holzforschung* 72:187–199, 2018.
Singh, S, Ghatak, HR. *Iranian J. Chem. Chem. Eng.* 40:1814–1839, 2021.
Smith, CZ, Utley, JHP, Hammond, JK. *J. Appl. Electrochem.* 41:363–375, 2011.
Song, Y, Mobley, JK, Motagamwala, AH, Isaacs, M, Dumesic, JA, Ralph, J, Lee, AF, Wilson, K, Crocker, M. *Chem. Sci.* 9:8127–8133, 2018.
Song, WL, Dong, Q, Hong, L, Tian, ZQ, Tang, LN, Hao, W, Zhang, X. *RSC Adv.* 9:31070–31077, 2019.
Stiefel, S, Schmitz, A, Peters, J, Marino, DD, Wessling, M. *Green Chem.* 18:4999–5007, 2016.
Tarabanko, VE, Fomova, NA, Kuznetsov, BN, Ivanchenko, NM, Kudryashev, AV. *React. Kinet. Catal. Lett.* 55(1):161–170, 1995a.
Tarabanko, VE, Koropatchinskaya, NV, Kudryashev, AV, Kuznetsov, BN. *Russian Chem. Bulletin* 44: 367–370, 1995b.
Tarabanko, VE, Petukhov, DV, Selyutin, GE. *Kinet. Catal.* 45:569–577, 2004.
Tarabanko, VE, Chelbina, YV, Sokolenko, VA, Tarabanko, NV. *Solvent Extract. Ion Exchange* 25:99–107, 2007.
Tarabanko, VE, Chelbina, YV, Kudryashev, AV, Tarabanko, NV. *Sep. Sci. Technol.* 48:127–132, 2013.
Tolba, R, Tian, M, Wen, J, Jiang, ZH, Chen, A. *J. Electroanal. Chem.* 649:9–15, 2010.
Trinh, TN, Jensen, PA, Sárossy, Z, Dam-Johansen, K, Knudsen, NO, Sørensen, HR, Egsgaard, H. *Energy Fuels* 27:3802–3810, 2013.
Villar, JC, Caperos, A, Garcia-Ochoa, F. *Wood Sci. Technol.* 35:245–255, 2001.
Voitl, T, von Rohr, PR. *Ind. Eng. Chem. Res.* 49:520–525, 2010.
Wang, Z, Liu, Y, Shi, P, Liu, C, Liu, Y. *Appl. Catal. B: Environ. Ener.* 90:570–577, 2009.

Wang, Z, Chen, K, Li, J, Wang, Q, Guo, J. *Clean – Soil, Air, Water* 38:1074–1079, 2010.
Wang, Y, Yang, F, Liu, Z, Yuan, L, Li, G. *Catal. Communic.* 67:49–53, 2015.
Wang, S, Li, Z, Bai, X, Yi, W, Fu, P. *Bioresource Technol.* 278:66–72, 2019.
Wang, S, Li, Z, Yi, W, Fu, P, Zhang, A, Bai, X. *Ren. Energy* 163:1673–1681, 2021.
Werhan, H, Mir, JM, Voitl, T, von Rohr, PR. *Holzforschung* 65:703–709, 2011.
Werhan, H, Assmann, N, von Rohr, PR. *Chem. Eng. Process.: Process Intensif.* 73:29–37, 2013.
Windt, M, Meier, D, Marsman, JH, Heeres, HJ, de Koning, S. *J. Anal. Appl. Pyrol.* 85:38–46, 2009.
Wu, G, Heitz, M, Chornet, E. *Ind. Eng. Chem. Res.* 33:718–723, 1994.
Yu, Y, Li, X, Su, L, Zhang, Y, Wang, Y, Zhang, H. *Appl. Catal. A: Gen.* 448:115–123, 2012.
Zabkova, M, Otero, M, Minceva, M, Zabka, M, Rodrigues, AE. *Chem. Eng. Process.* 45:598–607, 2006.
Zabkova, M, da Silva, EAB, Rodrigues, AE. *J. Membrane Sci.* 301:221–237, 2007.
Zhang, J, Deng, H, Lin, L. *Molecules* 14:2747–2757, 2009.
Zhang, Z, Li, Z, Zhang, H, Ma, C, Zhang, Z, Xie, Y, Liu, S, Wang, Q, Pittman, Jr., CU. *Fuel Process. Technol.* 235:107382, 2022.

Index

Pages in *italics* refer to figures and pages in **bold** refer to tables.

A

Abaca, 123–127, 155, 194–195, 221, **230**
Absorption, 31, 77, 115, 134, 141, 148–154, **155**, 156–170, 209–211, 286
Absorptivity, 148–153; *see also* Extinction coefficient
Abundance, 2–4, 9, 13–15, 115–134, 152–156, 169–201, 220–228, 241, 259, 277, 303, 338–339
Acacia, 21–22, 33–34, 136, 160, 178, 204–207, 219, 227–228
Acanthopanax Senticosus, 44
Accellerase, 71, **104**
Accelerate, 23, 46, 336, 346
Acetic anhydride, 122–123, 133, 306
Acetic bromide, 122
Acetobrominate / Acetobromination, 133–134, 150
Acetone, **32**, 41, 44–54, *55*, 56–58, 62, 84, **85**, **92**, 93–96, 117, 130–138, **142**, 143, 180–186, 255, 273, 286–291, 300–306, 324, 329–331
Acetosolv, 8, 138
Acetosyringone, 220–227, 314, 331–334, 349–352, 358–360, *361*
Acetovanillone, *318*, 331, 333; *see also* Apocynin
Acetyl / Acetylated / Acetylation, 118, 122–124, 133–136, **142**, 166, 171, **172**, 173–176, *177*, 178–180, *181*, 183, *184*, 185–186, 190, **192**, 224–227, 306
Acid / Acidic / Acidity, 27–40, 44–48, 64, 67–68, 83, **84**, 88–93, 96, 101, 249–251, 278–283, 292, 318, 325, 337–340, *341*, 342, **344–345**, 353, 357
  Acetic acid, 8, 44–58, 76, **84–85**, **92**, 95–96, 104, 108–110, 122, 129, *151*, 176, 189, 205, *206*, 211, 215, 219, 224–226, **231**, 278–280, 289, 296, 299–301, 306–307, 314, 320–334
  Acid catalysed, 27–28, *29*, 44–47, 51, 56–60, 83, 93–96, 219, 225–227, 283
  Acidification, 7, 69, *70–71*, 72–81, 97, 127, 135, 157–159, 249, 272, 280, 286, 296, 302, 306, 323, **345**, 356, 360
  Acid hydrolysis, 3, 8, 101–106, **116**, 208, 211, 214, 225–226, 250, 273, 278, 332
  Acid precipitation, 68–77, 96, 135, 172
  Acid sulphite, 27–37, 64, 83–86, *87–88*, 89–91, 159
  Carboxylic acid, 95–96, 103, 125, 150, 157, 188, 250, 318–320, 335, *336*, 337, 340–342, 347–349, 353, 356, 359–360, *361*
  Ferulic acid, 220–221, 317, 320, 326
  Formic acid, 8, 44–58, **85**, 96, 125, 133–134, 151, 162, 176, 183, 189, 193, 205–*206*, 211, **235–238**, 265, 299, 329, 331, 334–335
  Hydrochloric acid, 46–47, 57–58, 72, 76, 96, 102–106, 121, 124, 130, 253, 255–259, 286, 338, 343
  Hydroiodic acid, 117, **229**
  Maleic acid, 176–177
  Mineral acid, 47, 75, 81, 102, 256, **345**, 356
  Nitric acid, 33, 47, 267
  Oxalic acid, 124, 283
  p-Coumaric acid, 220–225
  Perfluorosulfonic acid, **79**
  Phosphomolybdic acid, 318, 325
  Phosphoric acid, 47, 102
  Succinic acid, 84, **85**
  Sulphuric acid, 46–47, 57–60, 70–76, 96, 101–103, 107, 117, 134, 157, 249–253, 266–267, 280, 290, 295, 305, 319–320, 338, 351, 360–361
  Sulphurous acid, 27
  Syringic acid, 314, 318–322, 326, 360–360, *361*
  Vanillic acid, 223–227, 314–317, *318*, 326, 331–333, 349–351, 356, 360, *361*
Acidolysis, 27, **116**, 122, 126, 136, *165*, 167, 174, 186, 190, 196–197, 226, **229**, **233**, **236**, **240**
Acrylonitrile butadiene styrene, 9
Activity, 46, 325–326, 343, 353–354
Additive, 5, 14, 16, 334, 336
Adduct, 16
Adhesive, 6, 272, 277–283, 299
Adsorbent / Adsorption, 223, 247–248, 294, 358, *361*
  Saturation adsorption capacity, 358
Agent
  Delignification agent, 28, 43, 45
  Depolymerizing agent, 13–14
  Dispersing agent, 265, 270
  Tanning agent, 5
Agglomerate / Agglomeration, 69–70, *71*, 72–75, 93, 265, 328–336
Aggregate / Aggregating, 68, 93, 133, 265
Agitation, 34, 70–72, 83–88, 95–96, 105, 296, 298, 301, 317, 320, 325
Agroresidue, 13, **15**, **66**, 303
Alcell, 5, 8, 134, 138, **142**, 143–145, 161, 175, 181, 204–205, 213, *214*, 215–218, **235**, 258, 260, 266, 300, 306, 328, 330–331, 339, *340*
Alcohol
  Cinnamyl alcohol, 182
  Coniferyl alcohol, 4, 15, 186, **192**, 223, 226, 317, 320, 326, 349
  Higher alcohol, 43, 57
  Monohydroxy alcohol, 41
  Polyhydroxy alcohol, 43
Aldehyde
  Cinnamaldehyde, 183
  Coniferyl aldehyde, 3–4
Alfalfa, 205
Aliphatic
  Aliphatic hydroxyl, 4, 117, 154, 161, 167–171, **172**, 173–181, 187–190, 228, 277, *284*, 285–288, 292–307
  Aliphatic side chain, 117, 120–125, **155**, 159, 174–180, 184–185, 191, 210, 249, 314

Alkali
- Active alkali, 21, 65, **66**, 67
- Alkalinity, 21, 67
- Effective alkali, 21

Alkoxide, 15
Alkoxyl, 117
1-allyl-3-methylimidazolium chloride, 130
Al-MCM-41, 337, **344**
Almond shell, 41, 45–46, 49–54
Al-SBA-15, 341–342, **344**
Alumina, 215, 320, 325, 338, 342–343, **345**, 359
Aluminium chloride, 48
Amine
- Aminolysis, 124, 181
- Decylamine, 89
- Diethanolamine, 45
- Dioctylamine, 89
- Dodecylamine, 89
- Hexamethylenetetramine, 273
- Monoethanolamine, 45
- Trioctylamine, 89–90
- Tris-2-ethylhexyl amine, 103

Ammonium, 7, 28, **142**, 149, 160, 257, 266, 270, 273, 281, 290, 302, 309, 324, 331, 343, **344**
Amorphous, 18, 83, 263
Amur linden, 207, 227, 333
Analytical, 5, 115, 125, 148, 153, 201, 215, 217
Anion
- Anion exchange membrane, 348–349, 353, 359
- Carbanion, 18
- Mercaptide anion, 15
- Phenolate anion, 15, 315, 323
- Superoxide anion, 315, 346

Anisotropy, 32, 141
Anode, 80–81, 270, *271*, 272–273, 346–347, *348*, 349–353, *354*, 359
Anthrahydroquinone, 16
Anthraquinone, 14, 16, *17*, 20–26, *27*, 64–65, 69–71, **116**, 120–127, 131, *132*, 133–135, 151–157, 171, 179–180, 224, **230**
Antioxidant, 9
Apocynin, 220–223, 314–317, 326, 349–352, 356–360, *361*
Aqueous phase, 88, 93, 97, 103, 118, 273, 332–338, **345**, 351, 356–359
Aqueous solution, 22, 27, 41–45, 56, 75, 81, 89–90, 129–130, 247–248, 253–256, 266, 268–271, 278, 285, 290, 300, 314, 347, 356–360
Aprotic, 101, 108
Arabinose, **67**, **84**, 129
ARBO, 7
ARBOFORM, 9
Area
- Cross section area, 333
- Specific surface area, 33, 247–250, *251*, 252–253, *254*, 255–256, 265, 270–275, 318, 342–343, **344–345**, 354, **355**, 358

A-Recovery+, 76
Argon, 105–106, 201–207, 217–218, 251, 255, 260–263, 267–269, 302, 328–331
Aromatic, 2–9, 115, 151–166, **172**, 173–179, **180**, 181–192, **193**, 205, 210, 220, 223–228, 249, 257–259, 292, 296, 301, 305–311, 314–315, *318–319*, 322–343, **345**, 347, 350, 356, 359
ARPA Bloom, 9
Arrhenius law, 24, 316
Arundo donax, 41, 130, 166, 195, 197–198, 202, **233**, **236**, 283
Ash, **19**, **116**, 205–206, 248, 250, 258, 299, 317, 329–332
Aspen, 34, 126, 136, 138, 143, 151, 156, 167, *168*, 205, 211–212, 224, **231**, **240**, 289, 333, 347
Aspergillus Niger, **104**
Asphalt, 8–9
Assay, 101
Association, 13, 101, 193
Attenuated total reflection (ATR), 154
Atomization, 91
Azeotrope, 103

B

Bacteria, **85**
Bagasse, 2, **19**, 20–25, **32**, 41–58, 94–96, 105–107, 120–121, 126, 133, 135, 138, 144, 149, *151*, 159, 165–166, 172, 202–204, 215–218, 223–224, **230**, 280, 288, 296, 299, 317, 320, *321*, 324–325, 347, *348*, 349–350
Bamboo, **15**, **19**, 21, **32**, 44–56, 96, 110, 118, 121–129, 134, 149, 151, 157, 160, *161*, 162, 169, 176–178, 185–187, 191–195, *196*, 197–198, 202, 205–208, 212–217, 225, **233**, **235**, **238**, **239**, 286, *287*, 289
Banana leaf sheath, 117, 163, *164*, 176, *177*, 183, **229**, **238**
Barium hydroxide, 129–130
Base, 7, 14–15, 27–28, 30, 33–34, 48, 50–51, 56, 59, 83, 89–91, 120, 149, 159, 264, 277, 296, 343
Basswood, 106
Beech, 33–35, 43–44, 47–55, 59, 86, 93–96, 118–121, 134, 136, **142**, 143, 189–190, 195, 206, 215, 218–219, 224, **235**, 301–302, 315, 336, 349, 351, 359
Beneficial, 20, 95, 260, 325
Bentonite, *338*, 339, 343, **345**
Benzene, 15, 30, **32**, 97, 115–118, 123–125, 145, 151–154, **155**, 156–171, **172**, 174, 177–178, 183–185, 190–191, 210–211, 226, 278, 281, 314, 323, 327, 337–343, 349, 352
Benzylic, 38–39
Bimodal, 75, 131, *132*, 135–136, 202, 210
Binder, 6–9, 272–273, 277
Biochemical, 84
Bio-choice, 7
Biodegradable, 1, 277, 346
BioLAN, 9
Biomass, 2, 9, 45, 48, 50–51, 56, 64, 105, 109, 115, 247, 333–335, *336*
BioPiva, 7, 290
Biorefinery, 2–3, 7, 38, 63, 92, 115, 352–353
Birch, 33–34, 44–47, 50–51, 55, 71, 96, 117, 129–130, 134–135, 142, 149–152, 156, 160, 162, 173, 176, 182, 192, 195–197, 205, 209, 211, 214, 219, 224–225, **229**, **231–240**, 241, *336*
Bisulphite, 27–28, 31–36, *37*, 64, 83, 86, 89, 91, 356–357
Bitumen, 8–9
Black liquor, 6, 15, 64–65, **66–67**, 68–69, *70*, 71–72, 75–78, **79**, 80–81, *82*, 135, 149, 157, 159, 222, 356, 359

Blend / Blending, 258–260, *261*, 291, 295, 302–303, *304*, 305–307, *308*, 309, *310–311*
Blowing agent, 294–302
Boiling, 40, 43, 45, 50–51, 92–93, 137, 279
Bond
  Bond breaking, 38, 48, 53
  Covalent bond, 13, 176
  Double bond, 16, 30, 152, **155**, 160, 284
  Ether bond, 13, 32, 38–39, 44, 46, 118, 125–129, 204, **237–240**, 241, 249, 257, 328
  Glycosidic bond, 31
  Hydrogen bond, 309, 311
  Intermonomer bond, 4, 241
BoreGro, 8
Boron Doped Diamond (BDD), 347, 351
Boron trifluoride, 120, 298
Boson, 169–172
Bretax S, 8
Brewers spent grain, 105
Broth, 85, 105–108, 129, 249
Bubble / Bubbling, 72, 77, 318, 328–335
Buddleja Davidii, 43, 46, 52, 54, 57
Buffer, 105–110, 357
Burnoff, 248, 252–253, *254*
Butanediol, 44, 294, 299
Butanol, 43, 51, 55–56, **85**, 150, 205, 211, 224, **231**, 360
Butyl acetate, 97, 359

C

Calcium
  Calcium bisulphite, 27
  Calcium chloride, 48
  Calcium ion, 27
Carbohydrate, 6–8, 13, 16, 20–25, 32–33, 37–38, 46, 53–54, 60–69, 77–80, 92, 101–106, 117, 129–130, 150, 159, 163–166, 174, **180**, 185–186, 195–198, 220–226, 317, 327, 332–334
Carbon
  Activated carbon, 247–250, *251*, 252–253, *254*, 255–256, 273–275, 328, 337, 343, **345**
  α carbon, 16, 30–31, 38, 151–152, **172**, 176, 184–186, **192–193**, 278
  β carbon, 15–16, 30, 125, **172**, 176, **180**, 186, **192–193**, 194, 278, 315
  γ carbon, 38, 121, 124–125, **172**, 176–177, **180**, 181–186, **192–193**, 195, 283, *284*
  Carbon–carbon linkage, 4, 7, 39, 167, 184, 187, 205, 220, **238–240**, 241, 277
  Carbon dioxide, 1, 47, 53, 55, 69–72, 75–77, 85, 107, 115, 201, 210–212, 248–252, *253*, 257, 269, 330–331, 346
  Carbon mat, 247
  Carbon monoxide, 210–212, 248–249, 263, 330
  Carbon tetrachloride, 117
  Quaternary carbon, 180–183
  Reticulated vitreous carbon, 350–351, 355
  Tertiary carbon, **180**, 182–183
Carbonization, 247–250, *251*, 252–261, 267, *269*, 270–275
Carbonyl, 38, 151–152, **155**, 156–169, 174–**180**, 183–186, 195, 210–211, 220–223, 227, 277–278, 309, 314, 318, 323–324, 330–331, 350, 357
Carboxylate, 68, 291
Catalyse / Catalyst, 13, 39, 44–47, 51–53, 57–60, 64, 92, 107, 134, 224–225, 263–267, 278–291, 294–307, 314–326, *327*, 338–339, *340*, 341–343, **344–345**, 346
  Autocatalyzed, 56
  Catalytic, 22, 46–48, 92, 263–269, 286, 314, 320, 323–326, *327*, 336–337, *338*, 339, *340–341*, 342–343
  Electrocatalyst, 349, 353–354
Catechol, 187–189, 220–224, 337–339, 343, **344**
Cathode, 80–81, 270, 273, 347–349, 353–354, 359
Cedar, 44, 117, 120, 123–127, **229–230**, **238**, 241
Cellic® CTec, **104**
Cellophane, 27
Cellulose, 2–3, 6, 9, 13, 20, 23, 27–28, 35–41, 56, 71, 83, 86, 90, 97, 101–105, 169
Cellulysin, **104**, 105
Centrifuged / Centrifugation, 75–76, 96, 105–109, 130, 265–266, 270–272, 284–285, 298, 302, 356–357
Ceramic, 78, **79**, 80–81, *82*, 83, 86, *87–88*, 97, 103, 131, 337, **345**, 358–359
Chamaecytisus Proliferus, 71, 157
Char, 201–208, 215, 220, 247–252, 281, 327, 330–335, *336*, 340
  Biochar, 247–252, 273, 335
Characterize / Characteristics / Characterization, 4–9, **15**, 18, **19**, 31–32, 38–40, 64–65, **66**, 67–73, 77–78, **79**, 83, 88–94, 106, 115–138, 143, 147–154, 157–171, 174–178, 181–188, 191, 197, 201–210, 215–228, **229–240**, 241, 247, 260, 265–277, 283, 286–288, 291–298, *300*, 301–307, *308*, 309, *311*, 335–336, **344–345**, **355**, 358–359
Charge, 15, 21–22, 30, 33–36, 39, 57, 68–69, 77, 81, 93, 259, 265–273, 347–350, *352*, 353, **355**
  Charge dislocation, 15, 30
  Specific charge, 270–273, 347–350, *352*, 353–354
Chemical
  Chemical constituent, 2–3, 9
  Chemical modification, 6, 101, 260, 277, 285, 305
  Chemical recovery, 3, 65
  Chemical shift, 170, **172**, 174–178, **180**, 182, *188–189*, 191, **192–193**, 194–198
  Delignification chemical, 18–28, 31–38, 41, 64–67
  Pulping chemical, 20, 31
  Speciality chemical, 6, 314
Chemistry, 3, 5, 15, 27, 38, 101, 108, 117, 122–124, 148, 173–175, 181, 228, 241
  Lignin chemistry, 3, 100, 108, 148
Chip, 20, 24, 33–34, 43–44, 49–53, 330
Chloroform, 118, 123, 163, *164*, 171–178, 183–187, 194
2-chloro-1,3,2-dioxaphospholane, 190
2-chloro-4,4,5,5-tetramethyl-1,3,2-dioxaphospholane, 186–187, 190
Cholesterol, 187, 190
Cholinium, 45, 50–51, 56, 96
Chromatography
  Chromatogram, 131, 136, 220, *222*, *227*
  Gas chromatography, 117, 121–124, 200, 220
  Gel permeation chromatography (GPC), 5, 131–138, 141–145
  High Performance Liquid Chromatography (HPLC), 123, 132

Chromium (III) acetylacetonate, 179, 182–192
Circulation, 22, 351
Classification, 5
Clay, 335–337, *338*, 343, **345**
Cleavage / Cleaved, 4, 13–18, 28, 30, 33, 38, *39–40*, 44–48, 118–122, 125, 129, 220, 315, 328, 347
Clogging, 75, 330, 336
Coagulate / Coagulating / Coagulation, 68–69, 76, 257
Cobalt, 263, 272, 324–326, 342–343, **344**, 349, 352–353, **355**
Coconut coir, 122–128, 133, 136, 195, *227*, 228, **232**, **239**
Coefficient
  Distribution coefficient, 88–89, 357
  Drag coefficient, 73
  Extinction coefficient, 148
  Mass transfer coefficient, 34, 89
Colloidal, 4, 68–69
Colour, 3, 15, 292, 301
Column
  Bubble column, 77, 318
  Packed column, 318, 360
Combined SO_2, 28, 33–34
Combustion, 3, 115, 268, 309
Compatible / Compatibility, 6, 247, 260, 277, 302–309
Complex, 18, 31, 65, 68, 89–90, 103, 129–130, 150, 171, 179, 191, 195–198, 212, 256, 259, 272, 346, 354–357, 361
Component, 1–5, 13, 38, 41, 45, 48, 64–67, 73, 81, 86, 89–90, 141, 164, 211, 220, 291, 302, 305–311, 330–331, 335–336, 360
Composition, 18, 21, 26, 31–36, *42*, 57, 65–69, 83, 90, **92**, 121, 126, 130, 201, 212, 220, 223, 247, 304, 309, 314, 326–343, 350, 359
Composite, 9, 41, 86, 133, *209*, 241, 251, 265–272, 283, 286, *287*, 290–291, 353, **355**, 360
Compound, 40–41, 47–48, 67, 76, 117, 123, 127, 149, 186, 212, 220–227, 247, 253–255, 266, 283, 292, 314, *316–319*, 323–327, 331–334, 337–340, 343, **344**, 347, 351–360
Compression force deflection, 296, 298, 301
Concrete admixture, 8
Conductivity, 115, 250, 263–267, 273, 281
Configuration, 3, 8, 132, 145, 264, 333–334, 342, 349
Conjugate base, 30
Conjugated, 150–152, 156–169
Contact, 20, 31, 90, 102, 137, 220, 267, 294, 306, 337, 353
Convert / Conversion / Converted, 2–6, 13–16, 20, 34, 83–85, 102–104, 107, 117, 166, 207, 213, *214*, 249–257, 266, 273, 278, 290–295, 315, *316*, 326, 351, 356, 358
Copper, 115, 259, 269–273, 315, 322–325, 348–349, **355**
Cordgrass, 175–176, 205, 211, 224, **231**
Corn
  Corncob, 96, 105, 109, 151, 253
  Corn stover, 45, 105, 120–126, 152, 160, 175, 205–208, 215, 219, 225, 288, *289*, 301, 307, 322, *332*
Cosmetic, 9, 276
Cost, 3, 34, 45, 50, 80, 86, 211, 248, 303, 343, 347
Cottonwood, 122, 126, 138, 174
Countercurrent, 103, 359
Coupling, 171, 178, 186, 191, 305, 309, *310*
Cresol, 220, 223–226
Crosslinked / Crosslinking, 131–132, 254–257, 263, 278, 283–284, 290–295, 299–301
Cross polarization magic angle spinning (CPMAS), 179–180, 183
Crystallite, 260, *261*, 263
Crystallization, 306, 356, 360–361
Cunninghamia Lanceolata, 135, 156, 202, *222*
Curaua, 194–195
Cure / Cured / Curing, 255, 266–268, 273, 280–291, 294–302, 309
Current density, 81, 272–273, 347–351, *352*, 353–354
Cyclohexanol, 187

D

Damage, 20, 56, 81, 110
Date palm, 43, 47, 50, 53, 55, 58
Deactivated / Deactivation, 267, 353–354
Decanol, 43
Degradation, 4–5, 18, 21–24, 32, 35–38, 47–48, 53, 64–67, 108, 127, 130, 134–136, 148, 184–185, 201–219, 223, 227, 277, 309, 314, 317, *318*, 320, *321*, 322–324, 327, 339, 347, 348, 350–351, 359
  Carbohydrate degradation, 21–24, 32, 37, 53, 64, 67
  Oxidative degradation, 314, 317, *318*, 320, *321*, 322–324, 347
Degree
  Degree of delignification, 36
  Degree of dissociation, 69
  Degree of freedom, 154, 167
  Degree of polymerization, 18, 283
Dehydrogenation, 4, 250, 257
Delignify, 13, 20, 25, 32, 43–44, 49, 51, 54
  Acidic delignification, 28, 32–38, 83, **84**
  Alkaline delignification, 4–7, 18–27, 32–33, 64–65, **66**, 67–68, 135, 210, 217, 356
  Delignification agent, 28, 38, 43–45
  Delignification chemical, 18–41, 64–67
  Delignification kinetics, 21, 23
  Delignification rate, 34, 51
  Delignification temperature, 23–25, *26*, 35–36, *37*, 51, 53, 93
  Kraft delignification, 21–25, *26*, 27, 136
  Organosolv delignification, 8, 38, *39–40*, 41, *42*, 43–51, *52*, 53–54, *55*, 57–58, *59–60*, 64, 91, **92**, 93, *94*, 95–97, 149–150, 182
  Soda delignification, 7, 13, 21–25, 70
  Sulphite delignification, 4, 7, 27, 32–37, 85, 149, 356
Demethoxylation, 181, 210–211, 220, 224, 228, 249, 336, 342–343, 347
Demethylation, 210, 278, *279*, 343
Dense / Density, 32, 41, 65, 68–75, 81–83, 145, 258, 263, 270–273, 295–303, 347–351, *352*, 353–354
Deoxygenation, 220
Deoxyhydrogenation, 335–336
Depolymerize / Depolymerization / Depolymerized / Depolymerizing, 13–18, 25, 27–28, 33, 38, 40–41, 53, 64, 67, 249, 286, 314
Derivative / Derivatized / Derivatization, 3, 5, 27–28, 30, 64–65, 67, 83, 118, 121–127, 133, 160, 166, 170, 186, 201, 220, 223–224, 278–280, 285, 294, 302, 305–307, 314, 329, 334, 337, 341, 347

Desirable, 2–3, 24, 88, 115, 131, 167, 201, 272, 277, 356
Desorbed / Desorption, 220, 223, 358–360, *361*
Detected / Detection / Detector, 5, 115, 131–134, 141–143, 154, 157, 183, 187–191, 195, 220–227, 305, 330, 349, 360
Determine / Determined / Determination, 4–5, 20, 23, 41, 47, 57, 68, 70, 91, 101, 117, 121–129, 136–138, *141*, 143–145, 148–153, 171–172, 175, 177, 180–191, 195–197, 213–217, 227, 259, 358
Dextran-agarose composite, 133
Diameter, 74, 248, 256–268, 271–272, **344–345**, 355, 359
Dibenzodioxocin, **192**, 241
Dibutyltin dilaurate, 266, 294–301
Dichloroethane, 90, 108
Dichloromethane, 118, 121–122, 127, 266, 298–300
Differential mechanical analysis (DMA), 201
Differential scanning calorimetry (DSC), 201, 217, *218*, 219
Differential thermal analysis (DTA), 201, 215, *216*, 217
differential thermogravimetry (DTG), 201–202, *203*, 204–205, *206*, 207–208
Diffuse / Diffusion, 20, 32–33, 70, 75, 106, 174
Digested / Digestion / Digester, 3–5, 34, 101, 117, 120, 124, 249, 296, **345**
Dimethyl
  Dimethylamine, 75
  Dimethyl formamide, 45, 52, 56–57, 138, 141, **142**, 143, *144*, 145, 150, 173–176, 259, 266–270, 285, 288, 294, 299
  Dimethyl sulphide, 6
  Dimethyl sulfoxide, 106, 130, **142**, 145, 172–186, 191–198, 265–266, 289, 299
  N,N-Dimethylacetamide, 106
Dimension, 33, 75, 138, 191, *261*
Dimer, 127, *128*, 129, 328
Dioxane, 5, 41, 46, 49, 52–54, 96, 105–110, 117–126, 135–138, 142–145, 150, *151*, 152, 163, *164*, 176, *177*, *182*, 183, 205–206, 212, 225, 228, **229**, **231**, **237–238**, 291, 299–301, 304, 307, 329, 337
4,4′-diphenylmethane diisocyanate, 292–295, 299, 302
Direction
  Fibre direction, 20, 33, 35, 260
  Grain direction, 20
  Tangential direction, 20
Disparity, 20, 38, 93
Dispersed, 75–76, 88, 265–271, 278, 298, 309, 361
  Dispersing agent, 265, 270
  Polydisperse, 68, 77, 130–131, 136
Dissociate / Dissociation, 28, 35, 38, 68–69, 153, 328
Dissolved
  Dissolved lignin, 15, 22, 43, 64–65, 83, 91, 96
  Dissolved solids, 64–65, **66**, 68–69, *70*, 71–72, 75–81, *82*, 83, 86–92, 95–97
Dissolution, 4–5, 13–14, 17–18, 25, 27–28, 31, 38–41, 46–48, 51, 53, 65, 67, 106, 108, 123–124, 130, 133–136, 145, 186, 281, 286, 290
Distance, 71, 73, 138, 154, 186, 191, 258–260, 266–267, 270
Distillation, 43, 102–103, 249, 283
Distortionless Enhancement by Polarization Transfer (DEPT), 179, 186
Distributed Activation Energy Model (DAEM), 212–213
Distribution
  Molecular weight distribution, 130–131, *132*, 133–135, 141–143
  Product distribution, 226, 329, *332*, *338*
  Size distribution, 70, *71*, 73, 75, 91, 136, 247, *262*
Dodecane, 103
Droplet, 77, 91, 259, 270, 351
Dropwise, 72, 105–110, 129, 255, 267–269, 278–279, 285–291, 294, 298, 301, 306, 360–361
Drying, 7, 91–93, 130, 208, 257, 268, 272, 280, 285–286, 294, 305–306, 336

E

Economic / Economical, 1–3, 6–9, 40, 46, 51, 54, 64, **66**, 88, 91–92, 102, 314
Efficiency, 3, 44, 85, 89, 272, 353
Elastomer, 8–9, 292
Electrochemical
  Electrochemical cell, 349, 352, 359
  Electrochemical oxidation, 314, 346–347, *348*, 349–351, *352*, 353, *354*, 359
Electrodialysis, 80–81
Electrolysis, 80, 353–354
Electrolyte, 80, 272, 350–351, 354
Electro-oxidation, 81, 346–354, **355**, 356–359
Electrostatic precipitator / precipitation, 329–331, 335
Element / Elemental, 15, 115, **116**, 139, 171–175, 201, 247–250, 256–257
Elimination, 30, 77
Elute / Eluted / Elution, 90–91, 130–134, 358
Emission, 1, 22, 45, 69, 76, 148, 248
Emulsion, 88–90, 273, 351, 357–358
Endoglucanase, **104**, 129
Endo-N-hydroxy-5-norbornene-2,3-dicarboximide, 187–188
Endothermic, 215–218
Energy
  Activation energy, 23, 26, 35, 38, 47, 53, 212, *213*, 214–215, 252, 316, 320, 322–323, 328
  Dissociation energy, 38
  Energy consumption, 1, 20, 22, 49, 51, 83, 353–354, 358
  Energy density, 41
  Energy efficiency, 3, 353
  Energy state, 166–167, 191
Engineering, 49–50, 65, 76, 354
Enthalpy, 215, 328, 358
Entity, 9, 15, 17, 129, 278, 323
Environment / Environmental, 1, 33, 36, 38, 40, 64, 88, 103, 105, 108, 153, 169–170, 178, 201, 204–208, 218–219, 250, 253, 255, 261, 263, 269, 277, 283, 302–303, 314, 329–333, 339, 346–347
Enzyme, 71, 75, 84, **104**, 105–110, **116**, 125–129, **142**, 150, 166, 190, 197, 208, 281–282, 326
  Cellulase enzyme, 104–108
  Cellulolytic enzyme, **104**, 125–129, 190, 197, 208
  Enzymatic hydrolysis, 8, 101, 104–106, 122, 136, 163, 175, 185, 208, 214, 219, 225, 228, **229**, **233**, **236**, **240**, 241, 249, 256, 272–273, 282–283, 291, 307, *308*, 321–324, *332*, 334
Epichlorohydrin, 75, 283–290, 298

Epoxy, 284–286, *287*, 288, *289*, 290–291, 309; *see also* Epoxy resin
Equilibrium / Equilibria, 28, 31–35, 89, 93, 137, 201, 353, 357
Equation, 35, 73–75, 137–140, 143–145, 150–152, 170, 178, 179, 212, 252, 304–306, 333
Ester, 38, **155**, 157–166, 185–186, 196–198, 220, 226–227, 260, 329, 332–333, 349
  Esterified, 145, 218, 219, 306–307
  Esterification, 218, 307
Estimate / Estimated / Estimation, 1–5, 8, 23, 28, 34–37, 41, 65, 101, 115–118, 120–129, 133–137, 140–145, 150–152, 171–191, 195–197, 212–215, 220–228, 241, 292–294, 317, 320–323, 358
Ethane, 30
Ethanol, 2–5, 8, 20, 41–58, *59–60*, 84, **85**, **92**, 93, *94*, 95–97, 108, **116**, 120, 123–125, 130–131, *132*, 133–134, 138, *140*, **142**, 144–145, 149–152, 161, *162*, 167, 175–176, 181–182, 187–188, 192–193, *194*, 204–208, *209*, 211–213, *214*, 215, 218–219, 224–226, **229–231**, **235–237**, 241, 249, 259–261, 272–273, 279–283, 287, 298–299, *300*, 301–302, 306–307, 315, 320, 326–328, 332–336, 340, 358–360, *361*
  Bioethanol, 3, 6–8, 46, 84, 105, 175, 208, 225, 249, 253, 256, 273, 282, 289, 307–308
  Methoxyethanol, 138, 150–151
Ethanethiol, 120–121
Ether, 4, 13–15, *16–17*, 18, 27–28, *29*, 30–33, 38–39, 44–48, 97, 108, 118, 121–122, 125–129, 133–135, 154, 161, 165, 177, 180, 185, 196–198, 204–205, **237–240**, 241, 249, 257, 284, 306, 328, 351, 358
  Diethyl ether, 97, 118, 121, 127, 133–135, 351
  Enol ether, 15
  Etherified, 15, 184, 195
  Petroleum ether, 108, 135, 306
Ethyl
  Ethyl acetate, 93, 106, 117, 121, 127, 135, 175, 286–287, 317, 320, 329, 337, 360
  Ethyl iodide, 117
1-Ethyl-3-methylimidazolium acetate, 106, 225, 257
1-Ethyl-3-methylimidazolium trifluoromethanesulfonate, 350–351
Ethylene glycol, 43–56, 93, 96, 152, *183*, 281, 291, 294–295, 299–301, 351, 357
Eucalyptus, **19**, 21–22, **32**, 33–36, 41–57, 72–81, 85–86, 89–97, 108, 117–129, 133–138, *141*, **142**, 143, 149–152, 157, *158*, 159–163, 174–176, *182*, 187–192, 197, 204, 213, 217, 223–228, **229–234**, **237–239**, 241, 252, *253*, 265, 280, 286–287, 292, 306, 318, *319*, 322–325, 359
Eugenol, 220, 223–226
Evaporate / Evaporated / Evaporation, 65, 75, *82*, 83, 92–93, *95*, 96–97, 105, 108–110, 118, 127–129, 134–135, 255–258, 266–268, 273, 285–288, 290–291, 300–301, 306, 317, 320, 360
Exoglucanase, **104**
Exothermic, 215–219, 294, 298–300
Experiment / Experimental, 3–4, 20, 25, 33, 48, 51, 57, 60, 70, *71*, 81, 90–91, 95, 107, 139, 202, 208–217, 252–253, *254*, 317–322, 328–330, 338, 341, 347, 357–358
Exponent / Exponential / Exponentially, 24, 68, 144–145, 212, 328, 358
Exposed, 20, 24, 268
Expulsion, 15, 186, 202, 278, 361
Extent
  Extent of carbohydrate degradation, 21
  Extent of delignification, 21, 25–26, 30–32, 36, 44, 48, 52, 65
Extract / Extracted / Extraction, 1, 13, 44–47, 50–56, 85, 88–89, 92, 103–106, *109*, 110, 117–118, 121–123, 127, 130, 135, 140–142, 163, *164*, 175, 183, 207–208, 215, 219, 249, 266–267, 286, 289, 302, 317, 320, 323, 329, 337, 343, **345**, 351, 356–360
Extractive, **32**, 64, 92, 101–110, 118–120, 130
Extrusion, 257–259, 266–270, 303

F

Feedstock, 15
FENIX NTM Bloom, 9
Ferment / Fermentation / Fermented, 5, 84–86, 249
  Fermentable sugar, 6, 27
Fermion, 169–170, 186
Ferric chloride, 48, 323–325
Fibre, 3, 9, 33, 86, *87–88*, 257–261, *262*, 266, 270, 283, 286, *287*, 290, 306–307, **355**, 359
  Carbon fibre, 247, 256–260, *261–262*, 263, 266–267, 270, 303, 306–307
  Cellulose fibre, 20
  Fibre wall, 20
FibreZyme® G4, **104**, 105
Film, 3, 6, 86, 90, 95, 247, 265–266, 294, 298–299, *300*, 304, 309
  Carbon film, 247
Filter / Filtered / Filtering / Filtration, 69–80, 86, *87–88*, 92–94, 96–97, 101, 104–107, 118, 141, 149, 167, 252, 255, 268, 285–288, 291, 302, 306–307, 330–331, 336, 349–351, 356–361
  Filterable / Filterability, 76, 93
  Filter cake, 74–76, 92
  Filtrate, 72–74, 95, 118, 150, 351, 361
Fine, 9, 50, 103, 108, 110, 256–257, 265
Fir, 32, 163, 208–211, 214, 226, 324
Fire retardant, 266
Fixed bed, 327–334, 338–339, 358
Flax, 138, 152, 155, 166, 171, 179–180, 205, 221
Flexible / Flexibility, **19**, 36, 292
Flexural modulus, 286, 290, 303
Float / Floatation, 94, 96
Flocculation / Flocculated, 73, 75
Flow, 68, 73–74, 78, 86, 91, 106–107, 201–211, 215–219, 248–253, 256–257, 260–263, 267–268, 290, 318–320, *321*, 325, 328–339, 352, 354, 359–360
Fluid, 106
  Fluidization / Fluidized / Fluidizing, 320–321, 328–335, *336*, 342
  Newtonian fluid, 68
Fluorescence, 141, 143, 167
Flux, 78, 81, *82*, 83, 86, *87–88*, 358–359
Foam, 273, *274*, 275, 281, *282*, 294–302, 349–352, **355**, 357

Force, 70, 73–74, 77, 259, 296–298, 301, 311, 334
Buoyant force, 73
Centrifugal force, 73
Drag force, 73–74
Gravitational force, 73
Formaldehyde, 6, 9, 15, 30, 38, 211, 268, 273, 277–278, *279*, 285, 290, 299–300, 314
Formasolv, 8, 280
Formation
Complex formation, 89, 259, 356
Crud formation, 89
Deposit formation, 333
Fibre formation, 258
Formulation, 1, 6–9, 91, 129–130, 149, *209*, 284, 301
Asphalt formulation, 8–9
Polymer formulation, 8
Fraction / Fractionated / Fractionation, 6–8, 13, 20, 38, 45, 77–83, 86, *88*, 89–91, 96–97, 101, 106, 117–118, 129–131, 134–136, *140*, *141*, 145, 149, 166, 187, 212, 259–263, 267, 273, 286–287, 290, 295, 304, 327–339, 343, 360, *361*
Fragment, 13–14, 18, 25–27, 33, 38–41, 45, 64, 67–69, 83, 92–93, 131–136, 166, 220, 292, 315
Defragmentation, 15, 33, 46, 122, 130
Free SO_2, 28, 33–34
Fuel, 1–8, 15, 38, 258–260, 333
Fullerene, 263
Function / Functional, 1–9, 26, 30, 41, 64, 73, 115–117, 121, 125, 154, 163, 178–179, 186, 212, 220, 228, 249, 256, 260, 263, 277, 283, 292, 314, 326, 330, 342, 357
Fungal, 33–36
Furfural, 38, 46, 83, **84**, 103, 149, 220, 224–226, 333

G

Galactose, **67**, **84**, 129
Galvanostatic, 348
Gas
Evolved gases, 209–212, 222, 225–227
Flue gas, 69, 76, 115
Gas constant, 40, 311
Gaseous, 1, 31, 34, 115, 201, 210–211, 255, 330–331, 351
Gasification, 115
Generate / Generated / Generation, 1–3, 6–8, 18–20, 64, 67, 81, 95, 102–104, 123, 140, 160–161, 315, 341–343, 346–347
Globule / Globular, 88, 91, 309, 334
Glucomannan, 2, 37, 83, 129–130
Glucose, **67**, **84**, 85, 104, 138
β-glucosidase, **104**, 105, 110
Glycerol, 43–58, *59*, 284, 295–296, *297*, 299, 302
Glycerol diglycidyl ether, 284
Grape, 44, 50–51, 56, 107
Graphene, 263, *264*, 269–270
Graphite, 257–260, *261*, 263, 270, 275, 351, 357
Grass
Congress grass, 20–25
Elephant grass, 123–124, 127, 136
Reed canary grass, 20–25
Switchgrass, 121, 126, 175
Green liquor, 120, 133, 136, 185, 196
Grinding / Ground, 3–5, 20, 49–51, 55, 101–110, 115, 118–122, 125–126, 154, 166, 179, 248, 255, 272, 334
Group
Functional group, 4–7, 30, 64, 115–117, 121, 125, 154, 163, 178–179, 186, 220, 228, 249, 260, 263, 277, 283, 292, 314, 330, 357
Leaving group, 15, 30
Guaiacol, 5, 220–227, 329–334, 340, 349, 352
Guaiacyl, 32, 118–125, 151–152, **155**, 156–165, 168–169, **172**, 174–192, **193**, 195–197, 220–228, **230–233**, 278, 292, *293*, 306, 315–317, 328, 356–357

H

Hazelnut shell, 208, 215, 219
Heat / Heating, 18, 22–26, 31, 34–36, 40, 49, 54–58, 69, 75, 83, 102–103, 136–137, 187, 202–226, 247–261, *262*, 264–270, 273–275, 280–283, 288, 294, 298, 300, 309, 327, 329–342
Heat exchange / Heat exchangers, 22–24
Heat transfer, 22, 34–35, 83, 258, 334
Latent heat, 40, 137
Preheated, 54, 102, 220, 309, 328–331, 334–335, 338, 341
H-factor, 26, *27*
Helium, 115, 263, 329–333, 337, 340–342
Hemi-cellulase, **104**
Hemicellulose, 2–3, 18, **19**, 23, 27–28, 33–35, *37*, 38, 43, 46–47, 51–54, 67, 71, 80, 83, 96–97, 101–105, 157–158, 166, 169, 176–178, 182, 185–186, 226, 258–259
Hemp, 138, 152, 155, 166, 171, 179, 180, 221, **230**
Heptanol, 43
Heterogeneous / Heterogeneity, 20–24, 31, 36, 41, 48, 115, 131, 148, 171, 201, 212, 228, 292, 302, 317, 320–326, 342–343
Hexamethylene diisocyanate, 268, 292, 298
Hexane, 92, 96, 296, 357
Hexose, 83–84, **85**
Homologue, 41
Hydrated, 28, 77
Hydrocarbon, 159, 211–212, 220, 223–227, 314, 327, 330–332, 336–343, **344–345**, 347–350
Hydrodynamic, 69, 72–73, 131–133, 143, 328, 335, 336
Hydrogen / Hydrogenated / Hydrogenetaion, 28–30, 44, 60, 70, 81, 90, 103, 106–108, 115–117, 153, 163, 170–171, **172**, 173–179, 184–186, 191, 248–251, 267–269, 272–273, 286–289, 292, 309, 315–316, 324, 329–332, 335, 342–343, 346–353, *354*
Hydrogen peroxide, 44, 346, 350–251
Hydrolysis
Acid hydrolysis, 3, 8, 101–106, **116**, 208, 211, 214, 225–226, 250, 273, 278, 332
Autohydrolysis, 27, 38, 144
Carbohydrate hydrolysis, 20, 33, 101, 106
Enzymatic hydrolysis, 8, 101, 104–106, 122, 136, 163, 175, 185, 208, 214, 219, 225, 228, **229**, **233**, **236**, **240**, 241, 249, 256, 272–273, 282–283, 291, 307, *308*, 321–324, *332*, 334
Hydrolysate, 102–107, 150
Prehydrolysis / Prehydrolysed, 22, 55, 103, 131, *132*

Hydrophilic / Hydrophilicity, 7, 17, 27, 30–33, 83, 86, 306
Hydrophobic, 7–8, 43, 64, 90, 294
Hydrotreated / Hydrotreating, 337, 342
Hydroxyl
  Aliphatic hydroxyl, 4, 117, 154, 161, 166–168, 171, **172**, 173–176, 179–181, 187–190, 228, 277, *284*, 285, 288, 292–300, 304–307
  Hydroxyl group, 4, 15, 38, 102, 123–124, 150–152, **155**, 156, 159–166, 172–181, 184–191, 195, 228, **234–236**, 241, 249, 260, 277, 283–284, 289–292, *293*, 294, 303, 307, 353
  Hydroxyl ion, 13–15, *16*, 18, 21, 68, 81, 90, 315, 323
  Hydroxyl radical, 346
  Phenolic hydroxyl, 4, 15–17, 38, 68–69, 117, 123–124, 150–153, **155**, 156–165, 171–183, 186, 190, 228, **234–236**, 241, 249, 260, 277–283, *284*, 288, 291–292, *293*, 294, 297–300, 304–306, 315, 351
HZSM-5, 337–**344**

I

Identify / Identification, 5, 115–118, 121–124, 127–129, 185, 201, 219–220, 224
Ilmenite, 339, 343, **345**
Immiscible / Immiscibility, 45, 302–311
Impeller, 71–72
Impregnate / Impregnated / Impregnation, 24–25, 33–36, 54–56, 90, 103, 247, 250, *251*, 256, 259–261, 266, 268–269, 272–273, *274*, 275, 315, 352–353, **355**
Improve / Improved / Improvement, 3–5, 9, 46–53, 56, 71, 76, 85, 103, 106, 133, 167, 172, 248, 257–258, 261–266–267, 272, 281–283, 291, 294–296, 299–301, 305–309, 317–318, 328–329, 334–339, 342–343, 349, 353
Impurity / Impurities, 7–8, 76, 83, 102, 115, 129, 157–161, 174, 182, 185, 220, 223, 226, 248–259, 268, 273, 286, 290, 332–334, 356
Incubation, 104–106
Inductively coupled plasma, 115
Indulin, 5–7, 69, 120, 135, 138, **142**, 143–145, 152–153, 158, 166, 173–174, 180–181, 187, 191, 204, 218, 223, 228, **229**, **234**, 241, 249–251, 258–260, *261*, 266–268, *269*, 278, 284, 294–298, 305, 317–319, 325–330, 334–335, 342, 360
Industrial, 2–8, 27–28, 34–35, 72, 84, 96, 105, 149, 346, 356
Influence / Influenced, 3, 21–27, 36, 46–49, 54, 57–58, 65, 69, 72, 89, 123–127, 223, 226, 249, 272, 303, 323, 329, 325
Infrastructure, 35
Ingredient, 6, 9, 44, 47, 270, 283, 295, 298–302, 305–309
Inject / Injected / Injection, 9, 77, 107, 132–133, 263, 303, 306–307
Inorganic, 2–3, 46, 48, 65, **66**, 67, 76–77, 80, 83, 102, 115, 248–258, 273, 330–331, 346–347
  Inorganic chemical, 3
  Inorganic component, 65, 67
Intensity, 138–141, 148, 154, 157–158, 163, 167, 170, 178–183, 186, 211, 220
Intramolecular, 15, 30
Insoluble, 3, 64, 86, 103, 118, 187, 267, 284–286, 290, 325
Interaction, 5, 25–26, 36, 57–58, *59*, 86, 143, 148, 166, 295, 302–303, 309–311
  π-π interaction, 265–266
Interface, 31, 90
Interlink, 25, 278
Intermediate, 15, 21, 24, 30, 38, 51, 58, 69, 75–77, 186, 201, 214, 252, 257, 335
Interpenetrating polymer network, 303
Inverse gated pulse, 179
Ion
  Bisulphite ion, 27
  Carbocation, 38
  Calcium ion, 27
  Carbonium ion, 30, 38–39
  Counterion, 27, 68, 89
  Hydrogen ion, 28–30, 60, 70, 90, 316
  Hydronium ion, 28
  Hydrosulphide ion, 4, 15, *17*, 18, 21
  Intermediate ion, 15
  Ionic equilibria, 28
  Ion exchange, 90–91, 136, 302, **344**, 358
  Ionic liquid, 38, 45, 50–51, 56, 96, 105–106, 159, 225, 257, 350–351, 357
  Ionic strength, 72, 75, 80
  Oxonium ion, 28
  Phenolate ion, 16, 68
  Sulphite ion, 27–28
  Thiosulphate ion, 15, 83
Iron, 48, 249, 268–270, 325, 343–**345**, *354*
Iridium, 349–350, **355**
Isoconversional approximation, 212
Isolate / Isolation / Isolated, 3–5, 9, 13, 36–38, 41–42, 95–96, 101–108, 115, 118–131, 134–135, 138, 143, 150, 154, 157–159, 162–169, 171–172, 180, 187, 192, 196–198, 202, 205–208, 211, 215, 219, 222–225, 228, 241, 259, 267, 282, 314, 320, 323, 328, 347, 357
Isopropanol, 85, 89, 291, 300, 356
Isotherm / Isothermal, 201, 205–207, 329–330, 358

J

Jute, 41, 155, 160, 221, **230**, 275

K

Kenaf, 118, 121–127, 194–195, **233**, 279
Ketone, 30, 38, 125, 157–159, 163, 183, 205, 211, 224, **231**, 315, 327–329, 332, 333, 347, 356–357, 360, *361*
  Hibbert's ketone, 38
Kinetics, 21–23, 53, 56, 252, 281, 292, *293*, 294, 315, 328, *348*
  Delignification kinetics, 21–23
  Process kinetics, 22
Kraft, 4–9, 14, 20–25, *26*, 27, 64–65, **66**, 67–70, *71*, 72, 75–81, *82*, 83, **116**, 120, 123–125, 133–140, *141*, **142**, 143–145, 149, 152–153, 157, *158*, 159, 165, 173–174, 178–180, *181*, 187, 191, 197, 204, 210, 213, *214*, 215–217, *218*, 223–224, 228, **229–231**, **234–237**, 248–250, *251*, 252, *253*, 255–261, *262*, *265*, 266–269, 272–275, 278–281, 284–288, *289*, 290–292, *293*, 294–296, *297*, 301–306, 315–317, *318–319*, 323–325, 328–330, 334–335, *336*, 337, *338*, 339, 348–351, 357–360

L

Laboratory, 20, 49, 75, 90–91, 352–353
Laminate, 8–9
Larch, 42, 47, 50, 53–55, 60, 94, 110, 142, 205
LASER, 5, 134, 138–143, 167, 263–265
  LASER ablation, 263–265
  LASER light scattering, 134, 138–141
Lattice, 263–264, 326, 343, **344**
Lauan, 177, 185, 207, 213
Leach, 13, 251, 255
Length, **19**, 33, 49, 78, 139, 148, 256–259, 263, 268, 333
Leucaena Leucocephala, 71, 157
Lignin
  Alcell lignin, 134, 138, 143–145, 161, 175, 181, 204–205, 213–215, 218, **235**, 256–260, 266, 306, 328, 330–331, 339, *340*
  Enzyme lignin, 125–129, 197, 208
  Hardwood lignin, 4–6, 169, 204, 207, 217, 223, 227–228, 285–286
  Hydrolysis lignin, 8, 122, 136, 150, 163–166, 174, *175*, 185, 190, 196, 207–210, 214, 219, 225–228, **229**, **233**, **236**, **240**, 241, 250, 256, 272–273, 278, 281–283, 289–291, 307, *308*, 317, 320, *321*, 322–326, *332*, 334, 341, 352–353
  Industrial lignin, 6
  Kraft lignin, 5–9, 65, 69, 120, 123–125, 135–140, *141*, 142–145, 149, 152–153, 157–159, 173–174, 178–180, *181*, 187, 191, 204, 210, 213, *214*, 215–217, *218*, 223, 228, **229–231**, **234–237**, 248–250, *251*, 252, *253*, 255–261, *262*, *265*, 266–269, 272–275, 278–281, 284–288, *289*, 290–292, *293*, 294–296, *297*, 301–306, 315–317, *318–319*, 323–325, 328–330, 334–335, *336*, 337, *338*, 339, 348–351, 357–360
  Lignin carbohydrate complex, 92, 104, 117, 129–130, 150, 165–166, 185–186, 195–198, 226
  Lignosulphonate, 4–8, 27, 32, 83–86, *87–88*, 89–91, 117, 136, *140*, **142**, 143, 149–152, 159–160, 178, 249–250, 255, 268–270, 273, *274*, 275, 281, 290, 302, 309, *310–311*, 317, 322–324, 331, *352*, 356
  Lignin formulation, 129–130, 149, *209*, 220
  Liquid Lignin, 77, 92, 149, 334
  Milled wood lignin, 5, 108–*109*, 120–129, 133–138, 142–143, 150–151, 160, *161*, 166–167, *168*, 177–178, *184*, 185, 190, 194–195, *196*, 197, 207, *209*, 210, 213, 217–219, 226, *227*, 228, **229**, **232–239**, 241, 328, 333, 342
  Native lignin, 7, 41, 64, 101, 115, 118, 168–169, 314
  Organosolv lignin, 6–8, 46, 50, 58–60, 120–126, 131, *132*, 133–134, 138, 142–145, 150, *151*, 152, 161, *162*, 163, *164*, 175–176, 181, *183*, 188, *189*, 192–193, *194*, 204–206, *209*, 211–213, *214*, 215, 218–219, 224–225, 228, **229**, **231**, **235–238**, 241, 259–261, 273, 278–281, 288, *289*, 291, 298–299, *300*, 301, 306–307, 315, 320, 324–331, 335–336, 339–340, *341*, 342, 351, 360
  Steam explosion lignin, 142, 323, 333
  Residual lignin, 36
  Soda lignin, 7, 65, 117, 123–124, 133–135, 138, 151, 155, *156*, 157, 166, 171–172, *173*, 179–180, 187, *188*, 191, 202, *203*, *209*, 210–215, *216*, 217, 221, *222*, 223, **229–230**, **234**, **237**, 241, 250, 253–254, 284–285, 294, 303, 328–329, 337, 347–350
  Softwood lignin, 4–6, 123, 204, 217, 223, 228, 285–286, 315, 356
LIGNIFLEX, 9
LIGNISET, 9
LIGNISOL, 9
LignoBase, 9
LignoBoost, 69–70, *71*, 72, 76, 136, **142**, 143, 180, 191, 204, 215, 218, 223, 339
LignoForce, 76
Lignocellulose, 2–9, 13–15, 18, **19**–27, 31, **32**, 38, 41, 45–56, 64–65, **66**, 67–69, 72, 83, 86, 90, 93, 96, 101–103, **104**, 105–110, 115–118, *119*, 120, *121*, 122–126, *128*, 129–130, *132*, 160, 167–169, 177, 212, 228, 241, 247, 267, 273, 277, 282, 295, 307–309, 327, 330–332, 335, 338–339, 348–353, 356–357
Lignode, 7
LignoLite, 9
LignoPol, 8
Limit Oxygen Index (LOI), 281, *282*
Lineo, 7–9
Linkage
  5–5′ linkage, **15**, 127, 184–185, 190
  α-O-4′ linkage, **15**, 30, 38, 176, 241
  β-O-4′ linkage, 13, **15**, 17, 30, 125–126, 176, 184–185, 191, 195–197, 220, 241, 315
  β-4′ linkage, 127
  β-5′ linkage, **15**, 127, 184–185, 191–197, 220, 241
  β-β′ linkage, **15**
  γ ester linkage, 196–198
  Benzyl ether linkage, 196–198
  Carbon–carbon linkage, 4, 7, 39, 167, 184, 187, 205, 220, **238–240**, 241, 277
  Ether linkage, 4, 13–15, *16–17*, 18, 27–28, *29*, 30, 33, 38, 121–129, 161, 180, 196–198, 205, 241, 328
  Intermonomer linkage, 4, 7, 13, *14*, 115, 124–127, 175, 184–185, 191–195, 204, 210, 228, **237–240**, 241, 314–315, 328
  Inter unit linkage, 13
  Phenyl glycoside linkage, 197–198
Liquid
  Liquid membrane, 89–90, 357
  Liquid reactant, 18
  Liquid to solid ratio, 18, 22–26, 31, 34, 39, 48–50, 57–58, 102–110, 123, 130, 135
  Organic liquid, 8, 38, 57, 356
  Polar liquid, 7
  Pulping liquid, 4–5, 43
  Spent liquid, 3–7, 23, 27, 43, 64–65, 83, 93, 101, 150, 358
Liriodendron Tulipifera, 41
Lithium
  Lithium bromide, 133, 141–145
  Lithium chloride, 106, 138
Log mean, 78
Loss modulus, 286–288, 291, 300
Lupinus rotundiflorus, 103
Lysine, 45

M

Macromolecule, 2–5, 13, *14*, 15–18, 38–41, 44, 115–118, 121, 124–125, 130–134, 143, 148, 167, 175, 179, 182–188, 191, 195–197, 201, 205, 219–221, 225–227, 241, 260, 303, 314, 328
Magnesium, 7–8, 27–28, 31, 48, 89, 149, 159, 322–324, 335–338, 343
  Magnesium chloride, 48
  Magnesium aluminium phyllosilicate, 335–336
Magnitude, 73, 90, 170, 263–264, 357
Maize, 122, 126, 163
Manganese, 48, 253–254, 324–326
Magnetic moment, 169, 178, 186
Maleic anhydride, 265, 290, 303–306, 309
Manchurian ash, 205–206
Mannose, **67**, 83, **84**
Maple, 120, 125, **142**, 163, 208, 211, 226, 320, 325–326
Market, 5–8, 64
Mass
  Mass balance, 77
  Mass transfer, 18–25, 31–36, 49–50, 70, 89, 103, 212, 318
  Molar mass, 5
  Reacting mass, 24–25, 34–36, 255, 295
Material
  Raw material, 2–6, 18–22, 25, 31–36, 44–49, 56, 65, 102, 105, 301
Mathematically, 37, 78, 264
Matrix
  Lignocellulose matrix, 46, 103, 106
  Solid matrix, 13, 22–24, 27, 31–33, 39, 248
Measurement, 4, 115, 141–143, 149–150, 218–219
Mechanism / Mechanistic, 4, 38, 69, 75, 201, 219–220, *221*, 315
Medium / Media, 13, 18, 27–28, *30*, 31–34, 38, 46, 71–74, 149–150, 167, 267, 278, 286–290, 318–319, 323–325, 329, 332–334
Membrane, 1–3, 64, 75–78, **79**, 80–81, *82*, 83, 86, *87–88*, 89–92, 97, 149, 259, 347–349, 353, 357–360
Mesh, 20, 49–50, 105–109, 259, 307–308, 329–333, 337, 341, 352
Metabolic, 13
Metal
  Metal foam, 354, **355**
  Metal oxide, 48, 314, 325, **345**, 349–350, **355**
  Transition metal, 314, 319, 323–325
Methane, 210–212, 263, 267–269, 286–290, 330–331
Methoxyl, 4, 117–123, 154, **155**, 156–167, 171, **172**, 173–178, **180**, 181–186, 191, **192**, 193, 211, 223, 228, **229**, 278, 282, 328–331, 335, 351
Methylene, 154, **155**, 156–166, **180**, 185, 253, 265–267, 278, 283, 292
Methyl iodide, 117
Methyl isobutyl ketone, 224, **231**, 357
METNIN
  METNIN NANOpolyol, 9
  METNIN Resin, 9
Microfiltration, 97
Microorganism, **85**, **104**
Mill
  Ball mill, 105–110, 249–250, 305
  Hammer mill, 20
Mineralization, 346–347, 351, 356
Miscanthus, 42, 45–57, 60, *70*, 71, 94, 97, 105, 109, 151, 160–162, 176, 183, 188–191, **230**, **234–237**
Miscibility, 43–45, 57, 88, 311
Model / Modelling, 47–48, 57, 89–90, 212, 304, 328
Modification, 5–8, 14, 94, 101–102, 117–118, 121–123, 130, 148, 201, 247, 260, *265*, 266–267, 277–278, 284–285, 288–290, 296, 300, 305–306, 342, 351, 356
Module, 77, 86, 154, 280, 359
Moiety / Moieties, 4–5, **15**, 18, 28–34, 38, 41, 69, 77, 118–127, 151–163, 167–169, 175–187, *188–189*, 190–191, **192–193**, 194–197, 220, 225–228, **230–232**, 292–294, 300, 353, 356
Moisture, 202, 215–216, 250, 255, 261
Molecule
  Molecular mass, 5, 97, 152
  Molecular size, 17, 64, 77–78
  Molecular structure, 4–5, 13, 115, 120, 127–129, 150, 154, 167, 170–171, 178, 186, 189–192, 201, 220, 224–226, 257, 314
  Molecular weight, 38, 68, 77–78, 83, 91–92, 115, 130–131, *132*, 133–145, 148, 210–211, 219–220, *221*, 223–225, 259–261, *262*, 267–277, 281–283, 286–288, 294, 303, 314, *316*, 317, *318–319*, 320, 324–342, **345**–351, 354–360
Monodisperse, 133, 138
Monomer / Monomeric, 13, *14*, 39, 46, 64, 67, 83–86, 92, 102–104, *121*, 124–129, 201, 220, 249, 277, 292, 314–315, 328, 342
Monosaccharide, 31–38, 43, 46, 83–86, 89–90, 101–102
Morphological / Morphologically, 18, **19**, 70, 73, 90, 212, 247, 275, 292, 297
Mould / Moulded / Moulding, 9, 268, 281, 288, 291, 295–310
  Compression moulded, 9, 305, 309–310
  Injection moulded / Injection moulding, 303, 306–307
Movement, 20, 31, 153

N

N-Hydroxynaphthalimide, 190
N-methylimidazole, 106, 307
Nano
  Nanofiltration, 77–80, *82*, 86–88, 97
  Nanorod, 247, 326
  Nanosheet, 260, 269
  Nanosphere, 247, 268, *269*, 270–271, 275
  Nanotube, 263–269, 272, 349, 352–354, **355**
Naphthalene, 226, 292, 327, 331, 337, 340–341
NeoLigno, 7–9
Neutralize / Neutralized / Neutralization, 46, 68, 105, 117, 120–121, 290, 334–335, 356
Nickel, 127, 263, 269–273, *274*, 275, 348–353, **355**, 357
Nitrile, 45
Nitrobenzene, 4–5, 118, *119*, 120, 125–127, 219, 228, **230–233**, **237–240**, 241, 248, 314
Nitrogen, 1, 94, 103–110, 115, 121, 201–212, 215–219, 248–261, 267–273, 294, 317–322, 328–341
Nuclear Magnetic Resonance (NMR), 4–5, 148, 169–171, **172**, *173*, 174, *175*, 176, *177*, 178–179, **180**, *181–184*, 185–187, *188–189*, 190–191, **192–193**, *194*, 195, *196*, 197–198, 228, **229–240**, 241, 330–331

Nonanol, 43
Nonconjugated, 152
Nonwood, 7, 45, 50, **116**, 121, 126, 135–136, **142**, 143, 150, 157, 160, 166–168, 172, 175, 188–190, 194, 202, *203*, 207, 215, *216*, 217–219, 228, **229**, 294, 337, 356–357
Novozyme, **104**, 105
Nozzle, 91, 258–259, 307
Nuclear Overhauser Enhancement, 179
Nucleophile / Nucleophilic, 15, 18, 28, 30, 33, 38–39
Number
  Avogadro number, 138
  Kappa number, 36, 46, 52–57
  Reynolds number, 73, 86, *87*
Nutraceutical, 8–9

O

Oak, 20–25, 107, 205–207
Octanol, 43, 89–90
Oil palm, 43–45, 49, 52, 56, 104
Olefinic, 169
Oligomer, 67, 83, 92, 127, 328–331
Oligosaccharide, 31, 90
Olive, 43–44, 47–52, 55–56, 96, 182, 273
Optimal / Optimum, 2–3, 9, 18–20, 47, 57–58, 65, 93, 125, 258, 278, 299–301, 317–318, 338, 350
Organic
  Organic chemical, 15, 38, 314, 346
  Organic component, 3, 64–67
  Organic liquid, 8, 38, 57, 356
  Organic media, 38
  Organic solvent, 5, 8, 40–41, 44–58, 89, 93, 118, 127, 172
Organosolv
  Organosolv delignification, 8, 38, *39–40*, 41, *42*, 43–51, *52*, 53–54, *55*, 57–58, *59–60*, 64, 91, **92**, 93, *94*, 95–97, 149–150, 182
  Organosolv pulping, 5
Order
  First order, 34, 53, 214, 317, 322–323, 328, 349
  Reaction order, 34, 323
  Second order, 34, 252–253, 347
Orientation, 73, 179
Oxirane, 15, 283
Oxygen
  Oxidation, 4–5, 76, 81, 118, *119*, 120, 125–127, 136, 219, 228, **230–233**, **237–240**, 241, 248, 257, 263–265, 314–318, *319*, 320–326, *327*, 346–347, *348*, 349–351, *352*, 353, *354*, **355**, 356–360
  Oxygenated, 118, *119*, 210, 249, *318–319*, 327–333, 336–337, 343, 347
  Oxygen bridge, 13, 38, 125, 186

P

Palladium, 265–267, 320, 325
Panel, 8–9, 279–280, 283
Paper / Paperboard, 3, 6–9, 20, 27, 36, 41, 64, 75, 105, 268
Parameter, 5, 8–9, 18–22, 25–26, 31, 34–36, 39–41, 44–45, 48–50, 53, 57–58, *59*, 60, 65, 68, 70, 78, 139, 144–145, 149, 258–260, 263, 292, 295, 302–305, 309–311, 318, 322, 330, 334, 354
Particle
  Lignocellulose particle, 20, 25, 31, 34
  Microparticle, 137–145, 283
  Particle size, 5, 20, 25, 49–50, 73–75, 91, *94*, 102–108, 133, 250, 305–307, 329–334, 337, 341, **355**, 358
Paulownia Fortune, 108
Peach, 44, 50–51, 56
PEARLLEX, 7
Pentane, 96, 281, 296
Pentose, 83–84, **85**
Permeability / Permeate / Permeated, 75–81, *82*, 83, 86, 97, 259, 358–360
  Semipermeable, 77
Perovskite, 321, 326, *327*
Petrochemical, 1, 5
Petroleum, 1, 5–6, 9, 108, 135, 247–249, 256, 273, 277, 283, 292, 306, 314, 327, 342–343
pH, 28, 31–35, 46–48, 60, 64, **66**, 67–69, *70*, 71–72, 75–78, 81–86, 89, 93–97, 104–110, 121, 124, 127, 135–136, *151*, 152–153, 157–159, 187, 202, *222*, 268, 272, 278–280, 285–287, 290–292, *293*, 295, 298–299, 302, 306, 315–317, *318*, 319–326, 351, 356–361
Phase
  Liquid phase, 18, 21–22, 31, 34, 47–49, 64, 72–73, 77, 88, 93, 129, 318, 360
  Mobile phase, 133
  Phase separation, 45, 89, 260, 292, 306, 357
Phenolate, 278–280, 290
Polyphenol / Polyphenolic, 32, 283
Phenylcoumaran, 241
Phenylpropane, 3–4, 30, 115–120, 123–128, 151, 172–180, 184–191, 195–198, 224–228, **229**, **234–240**, 241, 283, 298–300, 314, 323
Phosphitylation, 186–187, 190
p-hydroxybenzaldehyde, 118, 220, 314, *318*, 320–326, 357, 360, *361*
Physical, 5–8, 13, 18, 31, 65, 68, 83, 88, 91, 101, 106, *107*, 201, 292, 302, 358–359
Physicochemical, 3, 9, 64, 73, 77, 101, 115–117, 125, 130
Pilot plant / Pilot scale, 8, 43, 47, 50, 53, 56, 71, 75, 76, 95, 101, 107, 188, 191, 318, 333
Pine, **19**, 20–26, **32**, 33–36, *37*, 41, 45–49, 52–54, 72, 105–108, 120–130, *132*, 134–135, 138, 141, **142**, 143–144, 150, 154, 157, *158*, 161, *162*, 167, *168*, *181*, *184*–185, 193, *194*, 195–197, 204–207, 212, 215–219, 224–228, **229–240**, 241, 248, 278, 294, 298, 305–307, 315–317, *318*, 322–324, 328, 332
Plant, 2–9, 13, 24–25, 71, 76, 95, 102–103, 115, 124–125, 150, 188, 286
Platinum, 266, 348–349, 352–353, *354*, **355**
Polar / Polarization / Polarized, 7, 30, 38–40, 77, 81, 103, 138, 179, 303, 358–360
Polyacrylonitrile, 256, 259–261, 267, 270, 303–304
Polyamide, 86, 97, 286, 291
poly(ε-caprolactone), 309
Polycaprolactonediol, 294
Polydispersity, 5, 131, 134–136, 143, 259–260, 267, 280–290, 295–296, 299–308, 350–351
Polyethersulphone, **79**
Polyethylene, 6, 9, 143, 258–260, *261*, 268, 298, 303, *304*, 305–306, 309–311

Polyethylene glycol, 143, 258–260, *261*, 268, 298
Polyethylene oxide, 143, 258–261, 303, *304*, 306
Polyethylene terephthalate, *304*, 305, 310–311
Polymer / Polymeric, 1, 4–8, 64, 68, 77–78, **79**, 80–81, *82*, 86, 124, 131–134, 137, 143–144, 217–219, 223, 257–260, 267, 270, 277, 283, 292–307, 310–311, 358
Polymerization, 18, 277–283
Polymethacrylate, 132
Polypropylene, 6, 9, 75, 265, 296, 303–306, 309, *310–311*, 359
Polysaccharide, 2–3, 8, 38, 83, 102–104, 157, 174, 186, 208, 211, 214, 309
Polystyrene, 132–133, 138, 143, 304–305, 309, *311*, 358–360
Polysulphone, **79**, 86
Polytetrafluoroethylene / PTFE, **79**, 273–275, 289, 354
Polyurethane, 6, 9, 249–251, 265–268, 292–296, *297*, 298–299, *300*, 301–306
poly(vinylidene fluoride), 271–273
Poplar, **19**, 21, **32**, 41–50, 53, 56–58, 92–93, 105–106, 110, 120–122, 126, *132*, 133–136, *162*, 169, 174, *183*, 185–186, 190, 193, *194*, 195–197, 205–208, 212, 219, 224–226, **229–233**, **236–239**, 241, 249–250, 291, 323, 328, 335–336, 340–342
Pore / Porous / Porosity, 18–20, 73–78, 81, 90, 131–133, 247–256, 270–271, 275, 292–294, 334, 342–343, **344–345**, **355**, 358–360
Position
4′ position, 17
α position, 15, 195
γ position, 15, 186
Potassium
Potassium bromide, 154
Potassium hydroxide, 48, 251–251, 255–256, 272, 275, 280, 294–297, 300, 306, 329
Potential
Interelectrode potential, 347, 350–354, 357
Overpotential, 347, 353–354
Zeta potential, 4, 69, 93
Powder / Powdered, 50–51, 83, 91, 101, 105–110, 118, 122, 125, 265, 329, 334, 341
Precipitate / Precipitated / Precipitation, 45, 68–69, *70*, 71–72, 75–77, 81–83, 92–93, *94*, 95–97, 105–110, 117, 129–135, 145, 157–158, 172, 182, 187, 202, 222, 259, 272, 278–288, 291–292, *293*, 298, 302–307, 323, 326, 329–331, 351, 356–357, 361
Pre exponential factor, 212, 328
Pressure
Osmotic pressure, 139
Pressure drop, 74–77, 81, 334–335
Partial pressure, 28, 31–37, 317–326, 342
Pressurized, 43, 47, 50, 53–55, 71, 77, 102
Transmembrane pressure (TMP), 77–81, *82*, 86, *87*, 88, 97, 358–360
Vapour pressure, 45, 137–138
Pretreated / Pretreatment, 8, 36, 43, 48–49, 53–54, 58, 85, 104–108, 162–163, 196–197, 253, 256, 307, 350
Printed circuit board, 6
Process
Process kinetics, 22
Process parameter, 8, 18–21, 25, 31, 34–36, 39, 50, 53, 57–60, 65, 330
Product / Production, 1–9, 13, 20, 27, 30–38, 46, 54, 64–69, 75–77, **79**, 81–84, **85**, 86, 91, 95, 102–103, **104**, 105, 109, 115, 118–129, 148, 175, 183, 186, 201, 207–211, 219–227, 247–257, 269–270, 273, 281–284, 289–291, 294, 306–309, 314–331, *332*, 333–337, *338*, 339, *340–341*, 342–343, **344–345**, 347–350, 355–360, *361*
Productivity, 25, 36, 85
Propanol, 43, 250
Property, 4–6, 9, 45, 68, 74–76, 106–107, 137, 143, 201, 247–249, 256–257, 260–270, 277, 283, 286–288, 295–304
Propionylation, 122
Propylene glycol, 44–46, 49–51, 54, 265
Protobind, 7, 134–135, **142**, 143, 157, 172–173, 188, 202, **229**, **234**, 254, 294, 303, 328, 335
Protocol, 24–26, 36, 96, 103, 106, 122–123, 130, 133, 151, 173, 177–178, 186–189, 201, 219, 220, 257, 260, 272, 291, 295, 299, 300–302, 343
Proton, 15, 68, 169–171, 175–179, 186, 191
Pulp / Pulping, 3–7, 13, 27, 36–37, 45, 54–59, 64–65, 76, 130, 197, 257, 260, 323, 358
Kraft pulping, 4, 20
Organosolv pulping, 5
Pulping chemical, 20, 31
Soda pulping, 13–15
Wood pulping, 3
Pulse delay, 179
Pulverized, 45, 249–250, 335
Pure / Purification / Purified, 3, 7–9, 28, 41–46, 53, 72, 76–77, 83, 91–97, 105–109, 134, 137, 163, 249, 256, 283, 286–290, 302–309, 314, 324, 332, 343, 355–361
Purity, 54, 67–72, 75–76, 91, 104–110, 129, 360
Pyridine, 123, 133, 142, 187, 309, 315
Pyrolysis
Fast pyrolysis, 226, 314, 329–334, 337–342
Pyrolysis-gas chromatography-mass spectroscopy (Py-GC-MS), 201, 220–228, **230–233**
Pyrolysis oil, 314, 327–335, *336*, 337–343
Pyrrolidine, 123–124

Q

Quantity / Quantitative, 4–5, 73–74, 78, 101, 106, 115–127, 137–140, 150, 170, 178–197, 226–227, 299, 339, 349, 352
Quantum, 148, 154, 166–170, 191
Quartz, 328–330, 337, 340
Quinone methide, 15–16, 315

R

Radical, 315, 323–324, 328, 346
Radius, 4, 140, 259, 358
Hydraulic radius, 4
Radial, 20, 73, 263–264
Raffinate, 88–89
Raman shift, 166–169
Ramp, 24, 201
Rapeseed, 20–25
Rate
Diffusion rate, 20, 32

Rate constant, 26, 53, 75, 347
Rate determining, 38
Rate limiting, 38, 318, 323
Rate of delignification, 18, 24, 53
Rate of mass loss, 201–203, 207–208, 211
Ratio
Liquid-to-solid ratio, 18, 22–26, 31, 34, 39, 48–50, 57–58, 102–110, 123, 130, 135
Stability ratio, 75
Syringyl to guaiacyl ratio / Syringyl/guaiacyl ratio, 32, 118–122, 155, 162, 169, 191–197, 221–227, 306
Volumetric ratio, 41, 44, 89–90
Rayon, 13, 27
Reactant, 267, 290–292, 343
Reaction
Competing reaction, 39
Condensation reaction, 18, 30–32, 38–39, 46, 53, 56, 127
Multiple reaction, 212, 314, 346
Peeling reaction, 16, 23–24
Reaction rate, 22–23, 26, 31, 34–35, 50–51, 201, 252, *253*
Reaction condition, 8, 38–39, 83, 101, 219, 265, 283, 290, 307, 314, 317, 320, 325–326, 332, 341, 347
Reaction equilibria, 31
Reaction intermediate, 38
Reaction medium, 13, 33–34, 38, 319
Reaction Ordinate, 58, *59*
Reaction product, 4, 13, 20, 31, 67, 118–123, 127, 290, 306, 315–317, 320, 325, 347–349, 352–357, 360
Reaction site, 18–20, 24, 31
Reaction temperature, 18, 22–26, 31, 35–37, 51–60, 76, 104–106, 201, 278–279, 284–290, 294, 317–324, 334, 349–351
Reaction time, 18, 25, *26*, 31, 36, *37*, 54–58, 76, 102–104, 107, 201, 265, 278, 281, 285, 288, 292, *293*, 294, 317, *318*, 319–326, 330, 351
Reactive extraction, 13, 89
Reactive solvent, 13
Secondary reaction, 18, 315, 346
Side reaction, 15, 18, 30, 37–38, 83, 250, 263
Reactivity, 102, 277, *279*, 292, 295, 346
Reactor, 20–25, 34–35, 49, 53–54, 77, 92, 102–107, 252, 282, 298–300, 318–320, *321*, 328–340, 353–354
Centrifuge Reactor, 334
Delignification reactor, 20–22
Rearrangement, 30, 217–219, 328
Recalcitrance / Recalcitrant, 18, 27, 32, 101, 104, 314
Recovery, 3, 23, 41–47, 56–60, 64–65, 76, 90–96, 103–110, 288, 356–361
Chemical recovery, 3, 65
Lignin recovery, 41–42, 47, 56–60, 77, 96, 104, 108–110
Recycle / Recycled, 77, 95, 333–334, 356
Red mud, 249, 339–343, **345**
Reflux / Refluxed / Refluxing, 53, 101, 106, 117, 255–256, 267, 278–288, 292, 301, 306
Refractive index, 134, 138–143, 154
Regenerate / Regenerated / Regeneration, 16, 86, 106, 130, 324, 358
Reinforcement, 9
Relaxant, 179, 182–187, 190–192
Relaxation time, 179
Renewable, 1–3, 6–9, 15, 38, 45, 270, 273, 277, 283, 292, 314, 352–353
RENOL, 9
Residue
Agricultural residue, 7, 56, 157, 282
Solid residue, 3, 8, 52, 65, 101–107, 110, 163, 249, 351
Resin
Anion exchange resin, 90–91
Binder resin, 6–8, 277
Cation exchange resin, 90–91
Epoxy resin, 6, 283–286, *287*, 288–291, 309
METNIN Resin, 9
Phenol formaldehyde resin, 6, 9, 277–284
Resinol, 241, 315
Resistance, 22, 74–75, 103, 137, 270, 279, 283, 303–307
Resource, 1–2, 6–9, 38, 76, 292
Retentate, 77–80, 86–88, 97, 259, 359
Revenue, 64
Reversible, 28, 266
Rheological, 65, **66**, 67–68
Rigid / Rigidity, 2–3, 13, 247, 257, 263, 291–292, 295–296, *297*, 299–301, 305
Rotation / Rotational, 73, 153–154
Ruthenium, 288, 342, 349, 353, **355**

S

Saccharification, 3, 8, 33, 41
Saccharomyces Cerevisiae, 84, **85**
Salix Schwerinii, 94
Salt, 45–48, 270, 319, 322–323, 334
SAN X, 7
Sarkanda, 7, 135, **142**, 202, 284, 294, 328, 337
Saw dust, 22
Scanning Electron Micrograph, *262*, *265*, *297*, 306, 309, *310*
Scattering
Rayleigh scattering, 166–167
Raman scattering, 166
Screen / Screened, 20, 49–50, 307–308, 334
Sediment / Sedimentation, 70–72
Segregation, 77
Select / Selective / Selectivity, 22–23, 27, **32**, 38, 40–41, *42*, 45–48, 51, 58–60, 80–81, 88–89, 247, 250, 337–343, 346, 354–356, 360
Separate / Separated / Separation, 3–9, 43–45, 64–72, 75–78, **79**, 80–83, 86–97, 102, 105–106, 110, 118, 121–123, 127–132, 139, 143, 149, 179, 191, 201, 220, 223, 259–260, 265, 268, 273–275, 286–292, 306, 314, 317, 320, 333, 347–349, 353–360
Sequential / Sequentially, 20, 28, 31, 69, 72, 77, 90–91, 97, 140–142, 272, 286, 329–331, 337, 359
Serratula Coronata L, 145
Severity, 39, 58–60, 93, *94*, 96, 134, 173, 176, 189, 192, 323
Severity factor, 58–60, 94
SHIELD, 9
Shape / Shaped, 70, 73, 77, 91, 154, 201, 259, 268, 275, 342
Si/Al ratio, 337–342
Siam weed, 47–51, 56, *151*
Silica / Silicon, 115, 281, *282*, 328, 335, 339, 343, **344**

Silicone oil, 294–295
Silver, 117, 167
Silylated / Silylation, 121
Simulation, 5, 65
Sisal, 123–124, 127, 155, 194–195, 221
Site
  Reaction site, 18–20, 24, 31
Size
  Molecular size, 17, 64, 77–78
  Particle size, 5, 20, 25, 49–50, 73–75, 91, *94*, 102–108, 133, 250, 305–307, 329, 330–334, 337, 341, **355**, 358
  Size distribution, 70, *71*, 73–75, 78, 91, 131, 136, 247, *262*
  Size fraction, 20, 49
  Size reduction, 20
Slurry, 70, 73–77, 94, 107–108, 127, 271–275, 334–336
Soaked, 44, 55
Soda anthraquinone, 14, 20–26, *27*, 65, 69–71, **116**, 120, 123–127, 131, *132*, 133–135, 151–157, 171, 179–180, 224, **230**
Sodium
  Sodium acetate, 104–105, 108
  Sodium azide, 105–106
  Sodium oxide, 21
  Sodium hydroxide, 13–14, 21–22, 43, 48, 65, 72, 77, 81, 89–90, 117, 124, 129, 133, 142, *144*, 145, *151*, 152–153, 207–208, 249–251, 255, 259, 268, 271, 275, 278–290, 298–299, 308, 317–326, 329, 348–353, 356–359
  Sodium metaperiodate, 117
  Sodium sulphate, 90, 157
  Sodium sulphide, 14, 21–22
Softening, 27, 93, 218, 258–260, 334
Solid
  Dissolved solid, 64–65, **66**, 68–69, *70*, 71–72, 75–81, *82*, 83, 86–92, 95–97
  Porous solid, 18
Solubility
  Solubility parameter, 5, 40–41, 44–45, 302–305, 309–311
  Water solubility, 27, 268
Solution
  Alkaline solution, 14, 18, 32, 72, 150–151, 174, 280, 356–358
  Polymer solution, 131–133, 137, 143, 259
  Stripping solution, 89–90
Solvent
  Antisolvent, 45, 92–97, 257
  Deep eutectic solvent, 351, 357
  Reactive solvent, 13
Solvolysis, 38, 46
Sonicated / Sonication, 93, 182, 265–268, 272–273
Sorghum, 43, 46, 50, 53–55
Source
  Carbon source, 259, 263, 266–267, 275
  Energy source, 2–3
  LASER source, 140–143
  Lignocellulose source, 115, 228, 267, 273, 277, 282, 295, 307–309, 330, 335, 338–339, 348–353, 357
Spectroscopy
  ^{1}H NMR spectroscopy, 4, 170–178, 228
  ^{13}C NMR spectroscopy, 4, 170, 178–186
  ^{19}F NMR spectroscopy, 170
  ^{31}P NMR spectroscopy, 5, 170, 186–190
  Atomic absorption spectroscopy, 115
  Fourier Transform Infrared spectroscopy (FTIR), 153–154, **155**, *156*, 157, *158*, 159–160, *161–162*, 163, *164–165*, 166–168, 202, *209*, 212, 259, 286
  Heteronuclear 2D NMR spectroscopy, 191, 228–**240**
  Heteronuclear multiple quantum correlation spectroscopy (HMQC), 191, **192**, 194
  Heteronuclear single quantum correlation spectroscopy (HSQC), 191, **192**, 193, *194*, 195, *196*, 197–198
  Mass spectroscopy, 122, 148, 201–202, 220
  Raman spectroscopy, 148, 166–169
  UV ionization spectroscopy, 150
  UV spectroscopy, 4, 148, 151
Sphere, 74, 145, 268
  Sphericity, 70, 74
  Spheroidal, 70
Spin quantum number, 169
Spinning
  Electrospinning, 257–261, 267–270
  Melt spinning, 257–260, *261*
  Solution spinning, 257
Spiral, 78, 263
Spirodienone, 241
Spruce, **19**, 20–25, **32**, 33–37, 43–45, 49, 52–54, 88, 94–96, 104–105, 108–110, 118, 121–130, 133–138, **142**, 149–150, 154, 163, 167, *168*, 173, 176–177, 189–190, 195–196, 204, 208, 218–219, 223–225, **231–238**, 241, 255, 259–261, 265, 286–287, 298–300, 317, 322–324, 328, 332, 335–336, 339, *352*
Stalk
  Corn stalk, 272–273, 281, 321
  Cotton stalk, 320, 326
  Kenaf stalk, 122, 124, 127
  Rapeseed stalk, 20–25
  Sunflower stalk, 215
  Tobacco stalk, 360
Standard Hydrogen Electrode (SHE), 349, 352–354
Statistical / Statistically, 57, 70, 167, 172, 322, 350
Steam, 22–25, 35, 44, 49, 76, 102–105, 136, 142, 154, 163–165, 171, 174, 186, 190, 208, 248, 256, 283, 289, 307, 321–323, 326, 332–335, **344**, 356
Stem, 2–3, 51, 110, 136, 169
Storage modulus, 286–288, 291, 295
Straw
  Barley straw, 70–71, 157
  Rice straw, **19**, **32**, 45, 50–51, 56, 120, 134–136, 149, 162, 165, *175*, 186, *189*, 190–192, 208, 211, 214, 225–226, **229**, **235–236**, **240**
  Rye straw, 107
  Wheat straw, 7, **19**, 21–25, **32**, 41–58, 70–71, 94–97, 107, 110, 117–127, 134–138, 155, *156*, 157, *165*, 166, 171–172, *173*, 174, 179–180, 186, *188*, 189–190, 195, 202, 205, 208, 215, 218–219, 222, 226, **229–233**, **236**, **239–240**, 249, 254, 284, 288, 295, 328–331, 334–337, 341–342, 347, *348*, 349–350, 358
Stream, 15, 27, 64, 77, 80, 106–107, 115, 150, 257, 319, 330–331, 346, 353
Stripped / Stripping, 85, 89–90, 103
Strong acid detergent fibre, 209–211
Structure
  Microstructure, 18–20, 31, 247–248, **355**

Molecular structure, 4–5, 13, 115, 120, 127–129, 148–150, 154, 167, 170–171, 178, 186, 189–192, 201, 220, 224–228, 257, 314
Morphological structure, 18
Structural, 1–5, 13, 18, 35, 41, 69, 81–83, 108–110, 118–126, 148–150, 153–154, 168, 171, 175–177, 179, 182, 186–188, *189*, 190–191, 195–197, 226, 241, 259–261, 283, 292
Thiirane structure, 15
Subcritical, 106–107
Suberin, 176
Substitution, 9, 30, 118, 121–125, 220, 277–281, *282*, 292–296, *297*, 298–301, 347
Substrate, 18, **85**, 106, 247, 263, 266–269, 303, 343, 349, 352, **355**
Sugar, 6, 27, 46, 64, 67, 83–86, 89–91, **92**, 93, 97, 102–104, 142, 249, 332
Sugi wood, 106
Sulphate, 48, 70, 83, 90, 157, 302, 322–326
Sulphite, 4–7, 27–28, 31–37, 64, 83–84, **85**, 86, *87–88*, 89–91, 130, 149–151, 159, 278, 302, 356–357
Sulphonate / Sulphonated / Sulphonation, 4, 7, 27–28, *30*, 33, 64, 160, 178, 255, 259–261, 266, 275, 349
Sulphur
Desulphurization, 126–129
Elemental sulphur, 15
Sulphidity, 22, 26
Sulphur bridge, 15
Sulphur dioxide, 27–28, 31–35, 115, 356
Supercritical, 106–107
Supernatant, 75, 105, 118
Superoxide, 315, 346
Superplasticizer, 6
Surface
External surface, 20, 31, 34
Specific surface area, 33, 247–250, *251*, 252–253, *254*, 255–256, 265, 270–275, 318, 342–343, **344–345**, 354, **355**, 358
Surface tension, 65–68, 259
Surfactant, 70, 105, 255, 281, 292–302
Susceptible / Susceptibility, 13, 28, 127, 170, 277–278, 314, 347
Sustainable, 1–6, 9, 15, 38, 92, 103, 256, 277, 314
Sustainable development, 1, 5, 9, 15, 38, 103, 277
Sweetgum, 21–22, 72, 104, 108–109, 150, 167, *168*
Swell / Swelling, 32, 101, 109, 129, 299, 302
Synthesis / Synthesize / Synthesized, 1, 4–6, 9, 45, 249, 263–268, 273, 278, *279*, 280–303, 342–343
Syringaldehyde, 4, 118–120, 220–227, 314, 317–320, *321*, 322–326, 333–334, 349–351, 357–360, *361*
Syringic acid, 314, 318–322, 326, 360, *361*
Syringyl, 118–124, 152, **155**, 156–169, **172**, 174–192, **193**, 195–197, 221–228, **230–233**, 292, *293*, 306, 328, 356–357

T

Tangential, 20, 73, 334
Tantalum, 350, **355**
Technology / Technological, 1–5, 9, 76–77, 84, 265
Techno-economic, 3, 54, 65, 102
Technique, 4–5, 72, 77–**79**, 83, 91, 97, 105, 117–118, 122, 131, 137, 148–150, 153–154, 167, 170, 174, 179, 191, 201, 215–217, 228, 241, 263
Temperature
Delignification temperature, 23–25, *26*, 35–36, *37*, 51–53, 93
Glass transition temperature (T_g), 217, 258–260, 303–304, 309
Onset temperature, 201–202, 207, 290
Reaction temperature, 18, 22–26, 31–37, 51–60, 76, 104–106, 201, 278–279, 284–290, 294, 317–324, 334, 349–351
Tensile modulus, 260–263, 294
Tensile strength, 257, 260–263, 267–268, 279–280, 283, 290, 294, 299, 303–311
Tetrabutylammonium bromide, 285–290
Tetra-n-butylammonium hydroxide, 130
Tetrahydrofuran, 93, 96, 133–134, 138, *141*, **142**, 143–145, 172, 258, 265–266, 269–270, 289, 292–294, 300, 305
Tetramethylammonium hydroxide (TMAH), 220–221, 226, *227*
Tetramethylsilane, 170
TG-FTIR, *209*, 210–212
Thermal conductivity, 115, 263–264, 281
Thermochemical, 201–202, 212, 215–217, 247, 350
Thermogram, 202–208, 217, *218*, 219
Thermogravimetry (TG), 201–202, *203*, 204–205, *206*, 207–298, *209*, 212–215, 223, 247
Thermoplastic, 6–9, 257, 260, 283, 300, 303, 309, *311*
Thermoset / Thermosetting, 6, 9, 257, 284, 288–291
Thickness, 33, 49, 74, 90, 154, 257, *261*, 263, 301, 359
Thioacidolysis, 4–5, 120, *121*, 122, 125–127, *128*, 129, 219, 228, **230–233**, **237–240**, 241
Threshold, 21, 25, 248, 259, 273
Throughput, 24, 64–65, 78, 91
Time
Cream time, 294–295
Gel time, 279–281, 294–295
Hold time, 25
Reaction time, 18, 25, *26*, 31, 36, *37*, 54–58, 76, 102–104, 107, 201, 265, 278, 281, 285, 288, 292, *293*, 294, 317, *318*, 319–326, 330, 351
Rise time, 294–295
Space time, 107, 320, 330–333, 338, 354, 360
Tack free time, 294–295
Treatment time, 54, *55*, 56–58, *59*, 60, 134
Titanium, **345**, 349, 353, **355**
Toluene, 5, 97, 108–109, 123, 224–226, 255, 292, 301, 314, 327, 331, 337–343, 350
Toluene 2,4-diisocyanate, 249, 265, 299
Torrefaction, 208
Tortuosity, 90, 359
Total SO_2, 28, 33–36
Toxicity, 40–41
Transesterification, 124, 295, 299, 302
Transformation, 30, 84, 125, 201, 212, 215–218, 249, 305, 334
Transmission electron micrograph, 268, *269*
Trichoderma Reesei, **104**
Trichoderma Viridae, **104**
Tricin, 189, 192, **193**, 195
Trioxane, 180–182
Tubular, 78, 86, 97, 269
Turbulence, 22, 31, 34, 49
Typha Capensis, 176–177

U

Ultrafiltration, 77–81, *82*, 86–88, 93, 97, 136, 249, 358
Uniformity, 22–25, 34–35, 333
Unimodal, 131, *132*, 134–135
Unit operation, 72, 90
Upgradation, 339
Utilize / Utilization / Utilized, 2–6, 9, 13, 23–24, 38, 64–65, 91, 101, 115, 123, 148, 151–153

V

Vacuum, 75, 92–97, 102–103, 108–110, 118, 127, 218–219, 255, 259, 267–275, 283–291, 294, 298–307, 317, 320, 360
γ-valerolactone, 97, 320, 325–326
Valorization, 6, 303, 314
Value
  δ value, 41–44, 57
  Threshold value, 21, 25
  Value added / Value addition, 2, 6–8, 45, 64, 83, 86, 101, 115, 125, 130, 148, 201, 247, 265, 277, 284, 301, 314, 327, 346, 355
Vanadium, 315
VANILLEX, 7, 309
Vanillic acid, 223–227, 314–317, *318*, 320–322, 325–326, 331–333, 349–351, 356, 360–361
Vanillin, 4–5, 8, 118–120, 124–125, 220–227, 314–326, 331–333, 347–353, 356–360, *361*
Vapour
  Vaporization, 40, 137
  Vapour residence time, 328–335
Vapour pressure osmometry, 137–138
Variable / Variability, 8, 13, 18, 57–58, 195, 228
Vector, 73, *264*
Velocity
  Crossflow velocity, 78–81, *82*, 88, 97
  Superficial velocity, 74
Versatility, 40, 54, 292
Vessel, 71–72
Viability, 3, 9, 46, 65
Viscous / Viscosity, 40, 50, 65, **66**, 67–69, 73–74, 78, 83, 86, 143, *144*, 145, 257, 266, 278–282, 286, 309
Volatility, 40, 92, 335, 336
Volatilization, 204, 208, 214–215, 247, 257
Volume
  Molar volume, 41, 137, 295
  Volume reduction, 81, 97
  Volumetric, 41–46, 74, 78, 89–90, 319, 333

W

Wall, 20, 33, 124–125, 169, 263, 268, 343
Walnut shell, 43, 47, 50, 56, 92, *132*, 134, *162*, 193, *194*, 205, 219, **231**
Wash / Washed / Washing, 65, 75–76, 91–94, 101–106, 109–110, 130, 247, 250–261, 265–268, 272, 275, 278–280, 284–291, 302, 306–307, 361
Waste, 49, 96, 249, 256, 277, 343, **345**
Wastewater, 248, 314, 346
Wavelength, 138–141, **142**, 149–153, 166
Willow, 126, 208, 215
Wood
  Hardwood, 4–8, 13, **15**, 20–25, 32–33, 45, 56, **66**, 69–72, 75–81, *82*, 83, **84**, 85–86, *87–88*, 93, 104, **116**, 117–126, 134–138, *140*, **142**, 143–145, 149–150, 157–163, 167–169, 173–175, 181, 190, 204, 207, 210, 213, *214*, 215–219, 223, 227–228, 249–250, 258–261, *262*, 285–288, *289*, 294–296, 299–307, 320, 324, 330–331, 356–357
  Marabou wood, 95
  Softwood, 4–9, 13, **15**, 21–23, 32, 37, 41, 45, 56, **66**, 69–72, 75–81, *82*, 86, 91, 104, **116**, 117–126, 129–131, *132*, 133, 135, 138, **142**, 143–145, 149–153, 157–159, 163, 167, 171–174, 178–180, 187, 191, 204, 208–210, 213, *214*, 215–218, 223, 226–228, **229**, 248–251, 256–257, 260, 278, 281, 284–286, 290, 295–296, *297*, 305, 309, *311*, 315, 322–324, 328–330, 337, *338*, 349–351, 356–360
  Wood meal, 5, 26, 105–108, 118–120, 123–127, 130

X

Xylan, 2, 23, 37, *71*, 72, 75, 83, 129–130, 166, 186
Xylanase, 110
Xylene, 103, 226, 314, 327, 331, 337–340, 343, 349–350
Xylose, **67**, 83, **84**, 85, 103, 182

Y

Yeast, 84, **85**
Yield
  Biochar yield, 248–252
  Carbon yield, 250, *251*, 252–255, 331
  Lignin yield, 56, 69–72, 75, 94, 105–108, *109*, 110
  Monomer yield, 125–127, 342
  Oil yield, 328–343
  Product yield, 54, 91, 118, 248–250, 255, 269–270, 318, 323–326, 329–331, 337–340, *341*, **344**
  Pulp yield, 57–59
Young's Modulus, 260, 268, 290, 300, 303–311; *see also* Tensile modulus

Z

Zeolite, 337–343, **344–345**
Zimm plot, *140*
Zinc, 48, 122, 154, 247, 255–256, 283, 291, 307

For Product Safety Concerns and Information please contact our EU representative GPSR@taylorandfrancis.com Taylor & Francis Verlag GmbH, Kaufingerstraße 24, 80331 München, Germany

Batch number: 10519288

Printed by Printforce, the Netherlands